Handbook of Climate Change Resilience

Walter Leal Filho
Editor

Handbook of Climate Change Resilience

Volume 2

With 563 Figures and 494 Tables

Springer

Editor
Walter Leal Filho
International Climate Change Information
and Research Programme (ICCIRP)
Faculty of Life Sciences
Hamburg University of Applied Sciences
Hamburg, Germany

Research and Transfer Centre Sustainable Development and
Climate Change Management
Hamburg University of Applied Sciences
Hamburg, Germany

ISBN 978-3-319-93335-1 ISBN 978-3-319-93336-8 (eBook)
ISBN 978-3-319-93337-5 (print and electronic bundle)
https://doi.org/10.1007/978-3-319-93336-8

This Springer imprint is published by the registered company Springer Nature Switzerland AG.
The registered company address is: Gewerbestrasse 11, 6330 Cham, Switzerland

Preface

As the Fifth Assessment Report (AR5) produced by the Intergovernmental Panel on Climate Change (IPCC) has shown, climate change is expected to have widespread impacts on societies.

The fact that the impacts of climate change are already being felt, not only in terms of increases in temperature but also in respect of agriculture (with lower crop yields) and the availability of water resources, among others, suggests that action is urgently needed, in order to increase the resilience of both the physical environment and communities.

In particular, there is a perceived need to identify, test, and deploy appropriate processes, methods, and tools that may help societies and nations to better adapt to the many impacts of a changing climate and become more resilient. There is also a perceived need to showcase successful examples of how this can be achieved and to illustrate how the quest to pursue climate resilience may face the many social, economic, and political problems associated with it.

It is against this background that the *Handbook of Climate Change Resilience* has been prepared by numerous scientists from across all geographical regions, under the auspices of the International Climate Change Information and Research Programme (ICCRIP). This handbook is a truly interdisciplinary publication, mobilizing scholars from over 30 countries undertaking research and/or executing climate change projects around the world.

This book focuses on "fostering resilience and capacity to adapt," meaning that it serves the purpose of showcasing experiences from research, field projects, and best practice in climate change adaptation across many countries, which may be useful and suitable for replication of perhaps even upscaling elsewhere.

Consistent with the need for more cross-sectoral interactions among the various stakeholders working in the field of climate change adaptation, this book aims to

(i) Provide research institutions, universities, NGOs, and enterprises with an opportunity to display and present their works in the field of climate change adaptation and resilience
(ii) Disseminate information, ideas, and experiences acquired in the execution of climate change adaptation projects, especially successful initiatives and good practice

(iii) Introduce methodological approaches and experiences deriving from case studies and projects, which aim to show how climate change adaptation may be implemented in practice

Last but not least, a further aim of this book is to document and disseminate the wealth of experiences on climate change resilience available today.

Enjoy your reading!

August 2019
Germany

Walter Leal Filho

Acknowledgments

The editor would like to thank the valuable support of the Deputy Editor, Professor Desalegn Ayal, as well as the many members of the Editorial Board. Thanks to your hard work, expertise, and constructive criticisms, we were able to produce a publication with the highest standards.

I would also thank the authors for their willingness to share their knowledge, know-how, and experiences, as well as the many peer reviewers, who have helped us to ensure the quality of the manuscripts.

Finally, thanks are due to the International Climate Change Information and Research Programme (ICCIRP) at the Hamburg University of Applied Sciences, Germany, for the coordination of the project and mobilization of a wide range of stakeholders and authors around the world.

Contents

Volume 1

Volume 4

About the Editor

Prof. Walter Leal Filho (B.Sc., Ph.D., D.Sc., D.Phil., D.L., D.Litt., D.Ed.) is Professor and Head of the Research and Transfer Centre "Sustainable Development and Climate Change Management" at the Hamburg University of Applied Sciences in Germany and holds the Chair of Environment and Technology at Manchester Metropolitan University, UK. He is a Lead Author at AR6's Working Group II (Climate Change Adaptation) at the Intergovernmental Panel on Climate Change (IPCC), is Founding Editor of the *International Journal of Climate Change Strategies and Management*, and heads the International Climate Change Information and Research Programme (ICCIRP). He is also Editor-in-Chief of the Climate Change Management series with Springer. Professor Leal serves on the editorial board of various journals. He has in excess of 400 publications to his credit, among which are ground-breaking books such as *Universities and Climate Change*; *The Economic, Social and Political Aspects of Climate Change*; the *Handbook of Climate Change Adaptation*; and the *Handbook of Climate Change Communication*. He has nearly 30 years of field experience on project management and has a particular interest in the connections between sustainability, climate change adaptation, and human behavior.

About the Deputy Editor

Desalegn Yayeh Ayal is Assistant Professor of Geography and Environmental Studies at the Center for Food Security Studies, College of Development Studies, Addis Ababa University. Desalegn holds a Ph.D. in Geography. Desalegn serves as the Deputy Editor of *International Journal of Climate Change Strategies and Management*. He has published more than 30 publications including books, book chapters, and refereed journal articles and is East Africa Vice President for Interconnections for Making Africa Great Empowered and Sustainable Initiative. Desalegn has also presented papers on climate adaptation and related issues at many international and national conferences. His principal areas of research interest include climate change and adaptation, climate resilience, climate change mitigation and related issues, indigenous weather forecasting, integrated natural resources rehabilitation and management, livelihoods, and food security nexus, among others. He thoroughly understands the link between natural and human-induced hazards with sustainable development and works hard to familiarize with current tools of climate change impact assessment on livelihood and the wider environment. Desalegn has been actively involved in climate resilience and integrated natural resources rehabilitation and management research, and development interventions to improve food security.

Editorial Board

Editorial Advisory Board

Contributors

Ghulam Abbas Department of Environmental Science, COMSATS University Islamabad, Vehari Campus, Vehari, Pakistan

Omer A. Abdallah Department of Agricultural Engineering, Faculty of Agriculture, University of Khartoum, Khartoum, Sudan

M. O. Abioja Department of Animal Physiology, College of Animal Science and Livestock Production, Federal University of Agriculture, Abeokuta, Ogun, Nigeria

Praise Abraham Department of Social Sustainability, Centre for Sustainable Development, University of Ibadan, Ibadan, Oyo, Nigeria

Bashir Yusuf Abubakar Department of Botany, Ahmadu Bello University, Zaria, Nigeria

Sarah Achola International Centre of Insect Physiology and Ecology, Nairobi, Kenya

Mariola Acosta International Institute of Tropical Agriculture (IITA), Kampala, Uganda

Myriam Adam Centre de Coopération Internationale en Recherche Agronomique pour le Développement (CIRAD), UMR AGAP/PAM, Montpellier, France

Institut de l'Environnement et de la Recherche Agricole (INERA), Bobo Dioulasso, Burkina Faso

International Crops Research Institute for Semi-Arid Tropics (ICRISAT), Bamako, Mali

Bamidele Adanikin Department of Plant Science and Biotechnology, Ekiti State University, Ado Ekiti, Ekiti State, Nigeria

Oluwabunmi Opeyemi Adejumo Institute for Entrepreneurship and Development Studies, Obafemi Awolowo University, Ile-Ife, Nigeria

M. O. Adekunle Department of Animal Physiology, College of Animal Science and Livestock Production, Federal University of Agriculture, Abeokuta, Ogun, Nigeria

Ibidun O. Adelekan Department of Geography, Faculty of Social Science, University of Ibadan, Ibadan, Oyo State, Nigeria

Mosunmola Lydia Adeleke Department of Fisheries and Aquaculture Technology, FUTA, Akure, Nigeria

Romola Adeola Centre for Human Rights, Faculty of Law, University of Pretoria, Pretoria, South Africa

Peter Adegbenga Adeonipekun Department of Botany, University of Lagos, Akoka-Yaba, Lagos, Nigeria

O. M. Adesope Department of Agricultural Economics and Extension, University of Port Harcourt, Choba/Port Harcourt, Nigeria

William Adzawla Faculty of Agribusiness and Communication Sciences, University for Development Studies, Tamale, Ghana

Oludare Agboola Department of Botany, University of Lagos, Akoka-Yaba, Lagos, Nigeria

Savita Aggarwal Department of Development Communication and Extension, Institute of Home Economics, University of Delhi, New Delhi, India

Yaw Boakye Agyeman Department of Ecotourism, Recreation and Hospitality, University of Energy and Natural Resources, UENR, Sunyani, Brong Ahafo Region, Ghana

Abeer Ahmed Qaed Ahmed Pharmacy Department, Al-Saeed University, Osaifrah, Taiz, Yemen

A. Ajao Amsol Bio Company, Lagos, Nigeria

Oluyede Clifford Ajayi Technical Centre for Agriculture and Rural Co-operation, Wageningen, The Netherlands

I. A. Ajibefun Department of Agriculture and Resource Economics, FUTA, Akure, Nigeria

The Federal University of Technology, (FUTA), Akure, Nigeria

Javaid Akhtar Institute of Soil and Environmental Sciences, University of Agriculture, Faisalabad, Pakistan

J. G. Akinbomi Department of Chemical and Polymer Engineering, Faculty of Engineering, Lagos State University, Lagos, Nigeria

Akinlabi Akintuyi Department of Geography, University of Lagos, Akoka-Yaba, Lagos, Nigeria

Bosede Olufunmilayo Akinwalere Department of Agricultural Extension and Communication, FUTA, The Federal University of Technology, Akure, Nigeria

Tunrayo Alabi Geographic Information Systems Unit, International Institute of Tropical Agriculture, Ibadan, Nigeria

Erwin A. Alampay National College of Public Administration and Governance, University of the Philippines, Diliman, Quezon City, Philippines

Mohammed Al-Azzawi Department of Civil, Geo and Environmental Engineering, Technical University of Munich, Garching, Germany

Eike Albrecht Brandenburg University of Technology Cottbus – Senftenberg, Senftenberg, Germany

Nigatu Alemayehu Consultant, Addis Ababa, Ethiopia

Mohammad Al-Saidi Center for Sustainable Development, College of Arts and Sciences, Qatar University, Doha, Qatar

Zakou Amadou Faculty of Agricultural Sciences, Department of Rural Economics and Sociology, Tahoua University, Tahoua, Niger

T. T. Amos Department of Agriculture and Resource Economics, FUTA, Akure, Nigeria

Peter Adewale Amosun Department of Teacher Education, Faculty of Education, University of Ibadan, Ibadan, Oyo State, Nigeria

Benjamin Anabaraonye Benjy Poetry and Music Global Concepts, Awka, Anambra State, Nigeria

Isaac Gershon Kodwo Ansah Faculty of Agribusiness and Communication Sciences, University for Development Studies, Tamale, Ghana

Toyib Aremu Department of Environmental Sustainability, Centre for Sustainable Development, University of Ibadan, Ibadan, Oyo, Nigeria

Kinfe Asayehegn School of environment, Gender and development Studies, Hawassa University, Hawassa, Ethiopia

Center for International Cooperation in Agricultural Research and Development (CIRAD), Montpellier, France

Department of Agricultural Economics, Universidad Politécnica de Madrid, Madrid, Spain

S. O. Asifat Lopek Engineering and Construction Ltd, Lagos, Nigeria

Joanes Atela African Centre for Technology Studies, Nairobi, Kenya

Musiliyu K. Atoyebi National Centre for Technology Management, Federal Ministry of Science and Technology, Obafemi Awolowo University, Ile-Ife, Nigeria

Amina Nihad Awartani College of Arts and Sciences, Qatar University, Doha, Qatar

David Olufemi Awolala Department of Agricultural and Resource Economics, Federal University of Technology Akure, Akure, Nigeria

Research Fellow, Adaptation Finance Fellowship Programme of the Thailand Research Development Institute Bangkok, Thailand and Frankfurt School of Finance and Management, Frankfurt, Germany

Abiodun Emmanuel Awoyemi Department of Agricultural Economics, Faculty of Agriculture and Forestry, University of Ibadan, Ibadan, Oyo State, Nigeria

Desalegn Yayeh Ayal Center for Food Security Studies, College of Development Studies, Addis Ababa University, Addis Ababa, Ethiopia

Ayansina Ayanlade Department of Geography, Obafemi Awolowo University, Ile-Ife, Nigeria

Neda Azam Department of Water Engineering, College of Agriculture, Isfahan University of Technology, Isfahan, Iran

Olubukola Oluranti Babalola Food Security and Safety Niche Area, Faculty of Natural and Agricultural Sciences, North-West University, Mmabatho, Mafikeng, South Africa

Oladapo Opeyemi Babalola Afe Babalola University, Ado-Ekiti, Nigeria

Mansur Matazu Bako Nigerian Meteorological Agency (NiMet), Abuja, Nigeria

Mohammed Lawal Balarabe Department of Biology, Ahmadu Bello University, Zaria, Nigeria

Mulubrhan Balehegn Department of Animal, Rangeland and Wildlife Sciences, Mekelle University, Mekelle, Ethiopia

Paul Basudde Sector Planning and Policy Analysis Department, Ministry of Energy and Mineral Development (MEMD), Kampala, Uganda

Lin Bautze Research Institute of Organic Agriculture (FiBL), Frankfurt am Main, Germany

Jules Bayala World Agroforestry Centre, West and Central Africa Regional Office – Sahel Node, Bamako, Mali

Paulin Bazié Institut de l'Environnement et de Recherche Agricole (INERA), Ouagadougou, Burkina Faso

Tamrat Bekele Department of Plant Biology and Biodiversity Management, College of Natural and Computational Sciences, Addis Ababa University, Addis Ababa, Ethiopia

Kassa Belay Pan African University Africa Union, Addis Ababa, Ethiopia

Adedoyin Bello Department of Urban and Regional Planning, Crescent University, Abeokuta, Ogun State, Nigeria

Ramkumar Bendapudi Watershed Organisation Trust (WOTR), Pune, India

Mohan Kumar Bera Institute of Economic Growth, New Delhi, India

Hillary K. Bett Department of Agricultural Economics and Agribusiness Management, Egerton University, Njoro, Kenya

Martha Bohm School of Architecture and Planning, State University of New York at Buffalo, Buffalo, NY, USA

Adjoua Nadège Boko-Koiadia Department of Sociology, University of Félix Houphouët-Boigny, Abidjan, Côte d'Ivoire

P. Bosbœuf Lab'Urba – Paris-Est University, Marne-la-Vallée, France

G. D. Bothun Department of Physics, University of Oregon, Eugene, OR, USA

Bennett Brooks Consensus Building Institute, New York, NY, USA

Kathryn Brown Marine Research Centre, School of Environment, Science and Engineering, Southern Cross University, Lismore, NSW, Australia

Claudia Capitani Environment Department, University of York, York, UK

Tales Carvalho Resende University of Strathclyde, Glasgow, UK

Winifred Chepkoech Center for Rural Development (SLE), Humboldt Universität zu Berlin, Berlin, Germany

Dumisani Chirambo Department of Civil and Public Law with references to Law of Europe and the Environment, Brandenburg University of Technology Cottbus-Senftenberg, Cottbus, Germany

Lieven Claessens International Institute of Tropical Agriculture (IITA), Arusha, Tanzania, Kenya

Helene Jacot Des Combes PaCE-SD, Lower Campus, Laucala, USP, Suva, Fiji

Florence Crick Grantham Research Institute on Climate Change and the Environment, London School of Economics and Political Science, London, UK

Tijana Crnčević Institute of Architecture and Urban and Spatial Planning of Serbia, Belgrade, Serbia

Marcella D'Souza The WOTR-Centre for Resilience Studies (W-CReS), Pune, India

Taiwo M. Dada Department of Aquaculture and Fisheries Management, University of Ibadan, Ibadan, Nigeria

Petr Daněk Department of Geography, Masaryk University, Brno, Czech Republic

Kyle Frankel Davis The Earth Institute, Columbia University, New York, NY, USA

The Nature Conservancy, New York, NY, USA

Data Science Institute, Columbia University, New York, NY, USA

Dennis dela Torre Center for Local and Regional Governance, NCPAG, University of the Philippines, Diliman, Quezon City, Philippines

Solomon Desta MARIL (Managing Risk for Improved Livelihood) Consulting Firm, Addis Ababa, Ethiopia

Aliou Diouf Climate Change, Natural Resource Management, Rural and Urban Development Enda Energie-Environnement-Développement, Dakar, Senegal

Samuel A. Donkoh Faculty of Agribusiness and Communication Sciences, University for Development Studies, Tamale, Ghana

Felix Kwabena Donkor College of Agriculture and Environmental Sciences, University of South Africa (UNISA), UNISA Science Campus. Corner of Christiaan de Wet Road and Pioneer Avenue, Florida, South Africa

School of Animal, Plant and Environmental Sciences, University of the Witwatersrand WITS, Johannesburg, South Africa

Jörg E. Drewes Department of Civil, Geo and Environmental Engineering, Technical University of Munich, Garching, Germany

Jack Alban Dyer Australia Maritime College, University of Tasmania, Launceston, TAS, Australia

Abule Ebro International Livestock Research Institute, Addis Ababa, Ethiopia

Glory Ikponmwosa Edwards Climate Change, Development and Policy, School of Global Studies, University of Sussex, Falmer, UK

Jackson Efitre Department of Zoology, Entomology and Fisheries Sciences, Makerere University, Kampala, Uganda

Emmanuel Ejim-Eze National Centre for Technology Management, Obafemi Awolowo University, Ile-Ife, Ile-Ife, Nigeria

M. U. Ekong Department of Forestry and Wildlife Resources Management, University of Calabar, Calabar, Cross River State, Nigeria

Ali W. Elamin Department of Agricultural Engineering, Faculty of Agriculture, University of Khartoum, Khartoum, Sudan

Haitham R. Elramlawi Center of Dryland Farming Research and Studies (CDFRS), Faculty of Agricultural and Environmental Sciences, University of Gadarif, Gadarif, Sudan

David Yisrael Epstein HaLevi Department of Educational Policy and Leadership, University at Albany, Albany, NY, USA

M. O. Esilaba Department of Environmental Science, Egerton University, Egerton, Kenya

Saeid Eslamian Department of Water Engineering, College of Agriculture, Isfahan University of Technology, Isfahan, Iran

Nsikak-Abasi A. Etim Department of Agricultural Economics and Extension, University of Uyo, Uyo, Akwa Ibom State, Nigeria

NseAbasi N. Etim Department of Animal Science, Akwa Ibom State University, Obio Akpa Campus, Abak, Akwa Ibom State, Nigeria

K. N. N. Ezike Department of Agricultural Economics, Management and Extension, Ebonyi State University, Abakaliki, Ebonyi State, Nigeria

O. A. Fagbenro Department of Fisheries and Aquaculture Technology, FUTA, Akure, Nigeria

Augustine E. Falaye Department of Aquaculture and Fisheries Management, University of Ibadan, Ibadan, Nigeria

Belarmain A. Fandohan Université nationale d'agriculture École des sciences et techniques de conservation et de transformation des produits agricoles BP 114, Sakété, Bénin

T. Fane Faculty of Economic Sciences and Management, University of Bamako, Bamako, Mali

M. Farrely Tanzania Organic Movement for Agriculture, Dar es Salaam, Tanzania

Mayowa Fasona Department of Geography, University of Lagos, Akoka-Yaba, Lagos, Nigeria

Aryo Feldman Crops for the Future (CFF), Jalan Broga, Selangor Darul Ehsan, Malaysia

Sabastine Dekaa Francis National Weather Forecasting and Climate Research Centre, Nigerian Meteorological Agency, Abuja, Nigeria

Chao Fu Key Laboratory of Ecosystem Network Observation and Modeling, Institute of Geographic Sciences and Natural Resources Research, Chinese Academy of Sciences, Beijing, China

Hubert Fudjumdjum Research and Transfer Centre Sustainable Development and Climate Change Management, Hamburg University of Applied Sciences, Hamburg, Germany

Charles K. K. Gachene Department of Land Resource Management and Agricultural Technology (LARMAT), University of Nairobi, Nairobi, Kenya

Kenya Agriculture and Livestock Research Organisation, Nairobi, Kenya

Eva N. Gacheru Kenya Agriculture and Livestock Research Organisation, Nairobi, Kenya

Prithviraj Gaikwad Watershed Organisation Trust (WOTR), Pune, India

WOTR Centre for Resilience Studies (W-CReS), Pune, India

Thomas Gaiser Institute of Crop Science and Resource Conservation, University of Bonn, Bonn, Germany

Kate Elizabeth Gannon Grantham Research Institute on Climate Change and the Environment, London School of Economics and Political Science, London, UK

Komla Kyky Ganyo Centre d'Etude Régional pour l'Amélioration de l'Adaptation à la Sécheresse (CERAAS), BP, Thiès, Sénégal

Institut Togolais de Recherche Agronomique (ITRA), Lomé, Togo

Weyessa Garedew Jimma University College of Agriculture and Veterinary Medicine, Jimma, Ethiopia

Andreas Gattinger Faculty 9 – Agricultural Sciences, Nutritional Sciences, and Environmental Management, Justus Liebig University (JLU), Institute of Agronomy and Plant Breeding II, Gießen, Germany

Tosin K. W. Gbadegesin Centre for Sustainable Development (CESDEV), University of Ibadan, Ibadan, Nigeria

Yohannes GebreMichael Department of Geography and Environmental Studies, Addis Ababa University, Addis Ababa, Ethiopia

Getachew Gebru International Livestock Research Institute, Addis Ababa, Ethiopia

Patrick T. Gicheru Kenya Agriculture and Livestock Research Organisation, Embu, Kenya

Victoria Gioto Institute of Climate Change and Adaptation (ICCA), University of Nairobi, Nairobi, Kenya

Laurent Gnonlonfin Doctoral School of Agronomic and Water Sciences/Department of Sustainable Management of Natural Resources, Laboratory of Ecology and Forest Research (LERF), University of Parakou, Parakou, Borgou, Republic of Benin

Daphne Gondhalekar Department of Civil, Geo and Environmental Engineering, Technical University of Munich, Garching, Germany

Degye Goshu Department of Economics, Faculty of Business and Economics, Kotebe Metropolitan University (Asso.), Addis Ababa, Ethiopia

Gérard Gouwakinnou Laboratory of Ecology, Botanic and Plant Biology (LEB), University of Parakou, Parakou, Borgou, Republic of Benin

Aliou Guissé Département de Biologie Végétale, Université Cheikh Anta Diop de Dakar (UCAD), Dakar, Sénégal

Marwa R. Hafez Institute of Graduate Studies and Research (IGSR), Alexandria University, Alexandria, Egypt

Ghrmawit Haile Ministry of Environment Forest and Climate Change and Addis Ababa University, Addis Ababa, Ethiopia

Mitiku Haile Department of Land Resource Management and Environmental Protection, Mekelle University, Mekelle, Ethiopia

Razlin Azman Halimi Crops for the Future (CFF), Jalan Broga, Selangor Darul Ehsan, Malaysia

Southern Cross Plant Science, Southern Cross University, East Lismore, NSW, Australia

Jarso Harou Ministry of Water, Energy and Climate Change, Isiolo County Government, Isiolo, Kenya

Ali Hasantabar-Amiri Department of Civil Engineering, Lenjan Branch, Islamic Azad University, Lenjan, Iran

Peni Hausia Havea PaCE-SD, Lower Campus, Laucala, USP, Suva, Fiji

Sarah L. Hemstock Bishop Grosseteste University, Lincoln, UK

A. Henri-Ukoha Department of Agricultural Economics and Extension, University of Port Harcourt, Choba/Port Harcourt, Nigeria

J. S. (Pat) Heslop-Harrison Department of Genetics and Genome Biology, University of Leicester, Leicester, UK

James Hope Ahmadu Bello University, Zaria, Kaduna State, Nigeria

Md. Shahadat Hossain Institute of Water Modeling, Dhaka, Bangladesh

Towanou Olivier Houetchegnon Doctoral School of Agronomic and Water Sciences/Department of Sustainable Management of Natural Resources, Laboratory of Ecology and Forest Research (LERF), University of Parakou, Parakou, Borgou, Republic of Benin

Mark Howells Unit of Energy Systems Analysis, KTH – Royal Institute of Technology, Stockholm, Sweden

Imad Antoine Ibrahim Institute of Law, Politics, and Development (Dirpolis), Sant'Anna School of Advanced Studies, Pisa, Italy and gLAWcal – Global Law Initiative for Sustainable Development, Essex, UK

Mustapha Adeojo Ibrahim Department of Biology, Ahmadu Bello University, Zaria, Nigeria

Funmilayo C. Idowu National Biotechnology Development Agency, Ogbomoso, Nigeria

N. M. Ifebueme Department of Forestry and Wildlife Resources Management, University of Calabar, Calabar, Cross River State, Nigeria

Samuel Aderemi Igbatayo Department of Economics and Management Studies, Afe Babalola University, Ado-Ekiti, Nigeria

Simeon Igbe Association of Entrepreneurs and Technology Managers of Nigeria (AETMAN), Eco-system Based Adaptation For Food Security Assembly (EBAFOSA), Makurdi, Nigeria

Ana Iglesias Department of Agricultural Economics, Universidad Politécnica de Madrid, Madrid, Spain

Oluwatosin E. Ilevbare National Centre for Technology Management, Federal Ministry of Science and Technology, Obafemi Awolowo University, Ile-Ife, Nigeria

Christopher O. Ilori Crop Protection and Environmental Biology, University of Ibadan, Ibadan, Nigeria

Masengo Francois Ilunga Department of Civil Engineering, University of South Africa, Pretoria, South Africa

Antonina Ivanova Universidad Autonoma de Baja California Sur, La Paz, Mexico

Abdulai Jalloh West and Central African Council for Agricultural Research and Development (CORAF), Dakar, Senegal

Adamu James National Weather Forecasting and Climate Research Centre, Nigerian Meteorological Agency, Abuja, Nigeria

S. Jari Department of Crop Production and Protection, Federal University Dutsin-Ma Katsina State, Katsina, Nigeria

Safieh Javadinejad Water Resource Management, University of Birmingham, Edgbaston, UK

A. Jebiwott Department of Environmental Science, Egerton University, Egerton, Kenya

Obert Jiri Agricultural Practice, Faculty of Agriculture, University of Zimbabwe, Harare, Zimbabwe

School of Agricultural, Earth and Environmental Sciences, University of KwaZulu-Natal, Pietermaritzburg, South Africa

Lara Johannsdottir Environment and Natural Resources, School of Business, University of Iceland, Reykjavík, Iceland

Tino Johansson Department of Geosciences and Geography, University of Helsinki, Helsinki, Finland

Eshwer Kale The WOTR-Centre for Resilience Studies (W-CReS), Pune, India

Antoine Kalinganiré World Agroforestry Centre, West and Central Africa Regional Office – Sahel Node, Bamako, Mali

A. M. Kane Egerton University, Njoro, Kenya

Institut d'Economie Rurale (IER), Bamako, Mali

Rebekah Karimi Enonkishu Conservancy, Narok County, Kenya

Joshua Kayode Department of Plant Science and Biotechnology, Ekiti State University, Ado Ekiti, Ekiti State, Nigeria

Hugh Kelly Permezone, Santa Barbara, CA, USA

Jagriti Kher Department of Development Communication and Extension, Institute of Home Economics, University of Delhi, New Delhi, India

Method Kilasara Department of Soil and Geological Sciences, Sokoine University of Agriculture, Morogoro, Tanzania

Stephen K. Kimani KALRO- Muguga South, Nairobi, Kenya

Caroline King-Okumu The Borders Institute (TBI), Nairobi, Kenya

GeoData Institute, Southampton, UK

Geography and Environment, University of Southampton, Southampton, UK

Isaac V. Kinhonhi Electricity Regulatory Authority (ERA), Kampala, Uganda

Monicah Kinuthia Ministry of Devolution and Planning, Nairobi, Kenya

Rukia A. Kitula Institute of Marine Sciences, University of Dar es Salaam, Zanzibar, Tanzania

L. Kleerekoper Amsterdam University of Applied Sciences, Amsterdam, Netherlands

J. Kluck Amsterdam University of Applied Sciences, Amsterdam, Netherlands

Dominic Kniveton Climate Science and Society (Geography), University of Sussex, Falmer, UK

Aristea Kounani Research Centre of Environmental Education and Communication, Department of Environment, University of the Aegean, Mytılene, Greece

Nitin Kumbhar Watershed Organisation Trust (WOTR), Pune, India

WOTR Centre for Resilience Studies (W-CReS), Pune, India

Catherine Ky-Dembele World Agroforestry Centre, West and Central Africa Regional Office – Sahel Node, Bamako, Mali

Stéphane La Branche PACTE – Grenoble Institute of Political Studies, Grenoble, France

Peter Läderach International Centre for Tropical Agriculture (CIAT), International Center for Tropical Agriculture (CIAT) – Asia Regional Office c/o Agricultural Genetics Institute, Hanoi, Vietnam

J. K. Lagat Egerton University, Njoro, Kenya

Makarius C. S. Lalika Department of Geography and Environmental Studies, Solomon Mahlangu College of Science and Education, Sokoine University of Agriculture, Morogoro, Tanzania

J. K. Langat Egerton University, Njoro, Kenya

Walter Leal Filho International Climate Change Information and Research Programme (ICCIRP), Faculty of Life Sciences, Hamburg University of Applied Sciences, Hamburg, Germany

Research and Transfer Centre Sustainable Development and Climate Change Management, Hamburg University of Applied Sciences, Hamburg, Germany

David K. Lelgut KALRO- Muguga South, Nairobi, Kenya

Chunlan Li School of Geographic Sciences, East China Normal University, Shanghai, China

Wu Liang Research Center for World Geography and Resources, Institute of Geographic Sciences and Natural Resources Research, Chinese Academy of Sciences, Beijing, China

Tran Thi Kim Lien Deutsche Gesellschaft für Internationale Zusammenarbeit (GIZ) GmbH, Ha Noi, Vietnam

Ondari Lilian University of Toledo, Toledo, OH, USA

Clayco Construction, Inc., Chicago, IL, USA

J. F. Lindahl International Livestock Research Institute, Uppsala University and Swedish University of Agricultural Sciences, Nairobi, Kenya

Crispino Lobo Watershed Organisation Trust (WOTR), Pune, India

R. Loeve Amsterdam University of Applied Sciences, Amsterdam, Netherlands

Hermann Lotze-Campen Potsdam Institute for Climate Impact Research (PIK), Potsdam, Germany

Department of Sustainable Land Use and Climate Change, Humboldt-Universität zu Berlin, Berlin, Germany

Kytt MacManus Center for International Earth Science Information Network (CIESIN), The Earth Institute, Columbia University, New York, NY, USA

Malgosia Madajewicz Center for Climate Systems Research, The Earth Institute, Columbia University, New York, NY, USA

Paramu L. Mafongoya School of Agricultural, Earth and Environmental Sciences, University of KwaZulu-Natal, Pietermaritzburg, South Africa

A. Q. M. Mahbub Department of Geography and Environment, Faculty of Earth and Environmental Sciences, University of Dhaka, Dhaka, Bangladesh

Athuman Juma Mahinda Department of Agricultural Engineering, University of Dar es Salaam, Dar es Salaam, Tanzania

Department of Land Resource Management and Agricultural Technology (LARMAT), University of Nairobi, Nairobi, Kenya

Environmental Science and Technology, Kyoto University, Kyoto, Japan

Tanzania Agricultural Research Institute (TARI)-Makutupora Centre, Dodoma, Tanzania

Riffat Mahmood Department of Geography and Environment, Faculty of Life and Earth Sciences, Jagannath University, Dhaka, Bangladesh

Daniel Mailumo Department of Government and Public Administration, College of Advanced and Professional Studies, Makurdi, Nigeria

Association of Entrepreneurs and Technology Managers of Nigeria (AETMAN), Eco-system Based Adaptation For Food Security Assembly (EBAFOSA), Makurdi, Nigeria

Pius Mailumo Association of Entrepreneurs and Technology Managers of Nigeria (AETMAN), Eco-system Based Adaptation For Food Security Assembly (EBAFOSA), Makurdi, Nigeria

University of Calabar, College of Education, Katsina-Ala, Nigeria

Immaculate Maina KALRO Headquarters, Nairobi, Kenya

S. M. Makindi Department of Environmental Science, Egerton University, Egerton, Kenya

Erick Maklago Research and Business Development, National Construction Authority, Nairobi, Kenya

Michael L. Mann Department of Geography, George Washington University, Washington, DC, USA

Kristin Marcell NYS Department of Environmental Conservation and Cornell University, New Paltz, NY, USA

Wilfred Mariki Selian Agriculture Research Institute, Arusha, Tanzania

Mary Masafu University of South Africa, Pretoria, South Africa

Sani Abubakar Mashi Nigerian Meteorological Agency (NiMet), Abuja, Nigeria

Sean Mayes Crops for the Future (CFF), Jalan Broga, Selangor Darul Ehsan, Malaysia

Kevin Mearns College of Agriculture and Environmental Sciences, University of South Africa (UNISA), UNISA Science Campus, Florida, South Africa

Matthias Meier Department of Socio-Economic Sciences, Research Institute of Organic Agriculture (FiBL), Frick, Switzerland

Neil R. Miller Canadian Foodgrains Bank, Arusha, Tanzania

Jane Mills Center for International Earth Science Information Network (CIESIN), The Earth Institute, Columbia University, New York, NY, USA

Greg William Misiaszek Institute of Education Theories, Beijing Normal University, Beijing, People's Republic of China

Sara Abdelhakim Mohammad College of Engineering, Qatar University, Doha, Qatar

Hassan I. Mohammed Department of Agricultural Engineering, College of Agricultural Studies, Sudan University of Science and Technology, Khartoum, Sudan

Mohammad Mousavi Department of Civil Engineering, Najafabad Branch, Islamic Azad University, Najafabad, Iran

Monicah Mucheru-Muna Kenyatta University, Nairobi, Kenya

Paschal Arsein Mugabe College of Engineering and Technology (COET), University of Dar es Salaam, Dar es Salaam, Tanzania

Charles Mugarura Broadfield Enterprises Uganda – Permaculture Group (Resilient African Network- MUK), Kampala, Uganda

Jayne Mugwe Kenyatta University, Nairobi, Kenya

Kingstone Mujeyi International Maize and Wheat Improvement Center (CIMMYT), Southern Africa Regional Office (SARO), Harare, Zimbabwe

Angeline Mujeyi College of Agricultural Engineering and Science, Discipline of Agricultural Economics, University of KwaZulu-Natal, Pietermaritzburg Campus, Scottville, South Africa

Adrian Muller Department of Socio-Economic Sciences, Research Institute of Organic Agriculture (FiBL), Frick, Switzerland

Bertrand Muller Centre d'Etude Régional pour l'Amélioration de l'Adaptation à la Sécheresse (CERAAS), BP, Thiès, Sénégal

Centre de Coopération Internationale en Recherche Agronomique pour le Développement (CIRAD), UMR AGAP/PAM, Montpellier, France

M. Mulumpwa Sengabay Fisheries Research Centre, Salima, Malawi

Nancy W. Mungai Department of Crops, Horticulture and Soils, Njoro, Kenya

M. Muntaka Department of Agricultural Extension and Rural Sociology, Federal University Dutsin-Ma, Katsina State, Nigeria

Fumiaki Murakami Rural and Agriculture Department, Nippon Koei Co., Ltd, Tokyo, Japan

Albanus Mutiso Enonkishu Conservancy, Narok County, Kenya

Fiona Mwaniki Climate Change Communication, Kilimo Media International, Nairobi, Kenya

Majaliwa Mwanjalolo Department of Geography, Geo Informatics and Climatic Sciences, College of Agriculture and Environmental Science, School of Forestry, Environmental and Geographical Science, Makerere University (Asso.), Kampala, Uganda

Caroline Mwongera International Centre for Tropical Agriculture (CIAT), Africa Regional Office, Nairobi, Kenya

Chris M. Mwungu International Centre for Tropical Agriculture (CIAT), Africa Regional Office, Nairobi, Kenya

R. T. Nabinta Department of Agricultural Extension and Rural Sociology, Federal University Dutsin-Ma, Katsina State, Nigeria

Herve Alain Napi Wouapi Department of Agricultural Extension and Rural Sociology, Dschang School of Agronomy and Environmental Sciences, University of Dschang (UDs), Dschang, Cameroon

Sarah Ndonye International Centre of Insect Physiology and Ecology, Nairobi, Kenya

Alice Neht Institute of Urban and Transport Planning, RWTH Aachen University, Aachen, Germany

Mohsen Nekooei Department of Water Engineering, College of Agriculture, Isfahan University of Technology, Isfahan, Iran

Sileshi Nemomissa Department of Plant Biology and Biodiversity Management, College of Natural and Computational Sciences, Addis Ababa University, Addis Ababa, Ethiopia

Humphrey M. Ngibuini Forestry Development Trust, Iringa, Tanzania

Rosemary N. Ngotho-Esilaba Veterinary Sciences Research Institute, Veterinary Research Centre, Kenya Agricultural and Livestock Research Organisation, Nairobi, Kenya

Gian L. Nicolay Department of International Cooperation, Research Institute of Organic Agriculture (FiBL), Frick, Switzerland

Pinghouinde Michel Nikiema Agence Nationale de la Météorologie du Burkina Faso, Ouagadougou, Burkina Faso

F. Njau Institute of Rural Development Planning (IRDP), Dodoma, Tanzania

Peterson N. M. Njeru KALRO- Muguga South, Nairobi, Kenya

Mwangi Peter Njiiri Kenya Wildlife Service, Nairobi, Kenya

Julie Noble City of Kingston, Kingston, NY, USA

Ijaz Rasool Noorka Department of Plant Breeding and Genetics College of Agriculture, University of Sargodha, Sargodha, Pakistan

Alison Nord Department of Plant, Soil and Microbial Sciences, Michigan State University, East Lansing, MI, USA

Chinedum Nwajiuba Alex Ekwueme Federal University Ndufu-Alike Ikwo (AE-FUNAI), Ikwo, Nigeria

Nkem J. Nwosu Department of Agronomy, University of Ibadan, Ibadan, Oyo State, Nigeria

George Nyarko Faculty of Agriculture, University for Development Studies, Tamale, Northern Region, Ghana

Martine Nyeko Faculty of Agriculture and Environment, Gulu University, Gulu, Uganda

Emmanuel C. Nzegbule Nigerian Environmental Study Action Team (NEST), Ibadan, Nigeria

Michael Okpara University of Agriculture Umudike, Umudike, Nigeria

Patrick J. O'Reilly Crops for the Future (CFF), Jalan Broga, Selangor Darul Ehsan, Malaysia

Nelson Ifeanyi Obine Centre for Sustainable Development, University of Ibadan, Ibadan, Nigeria

Kehinde Abraham Odelade Food Security and Safety Niche Area, Faculty of Natural and Agricultural Sciences, North-West University, Mmabatho, Mafikeng, South Africa

Idowu Kolawole Odubote School of Agricultural Sciences, Zambian Open University, Lusaka, Zambia

Samson Samuel Ogallah Solidaridad/Pan African Climate Justice Alliance (PACJA), Nairobi, Kenya

Chika Ogbonna Brandenburg University of Technology Cottbus – Senftenberg, Senftenberg, Germany

G. M. Ogendi Department of Environmental Science, Egerton University, Egerton, Kenya

Dryland Research Training and Ecotourism Centre, Chemeron, Egerton, Kenya

A. U. Ogogo Department of Forestry and Wildlife Resources Management, University of Calabar, Calabar, Cross River State, Nigeria

Olusegun Michael Ogundele Research Group in Health and Environment- Africa, (RGHE-AFRICA), Ibadan, Nigeria

International University Network on Cultural and Biological Diversity, Nigeria International Students' Coordination Unit, (IUNCBD N-ISCU), University of Ibadan, Ibadan, Nigeria

Oluwatoyin Ogundipe Department of Botany, University of Lagos, Akoka-Yaba, Lagos, Nigeria

Damilola Grace Ogunrotimi Department of Plant Science and Biotechnology, Faculty of Science, Ekiti State University, Ado Ekiti, Nigeria

Stephen M. Ojebisi Department of Geography, Obafemi Awolowo University, Ile-Ife, Nigeria

Olumuyiwa Idowu Ojo Department of Civil Engineering, University of South Africa, Pretoria, South Africa

Ladoke Akintola University of Technology, Ogbomoso, Nigeria

Joachim Chukwuma Okafor Department of Political Science, University of Nigeria, Nsukka, Enugu State, Nigeria

David Okali Nigerian Environmental Study Action Team (NEST), Ibadan, Nigeria

Paul B. Okon Department of Soil Science, University of Calabar, Calabar, Nigeria

Yasmein Okour Department of City Planning and Design, Jordan University of Science and Technology, Irbid, Jordan

O. Oladeji Amsol Bio Company, Lagos, Nigeria

O. Adeola Olajide Department of Agricultural Economics, Faculty of Agriculture and Forestry, University of Ibadan, Ibadan, Oyo State, Nigeria

Kehinde Oluwatosin Olajubu Department of Fisheries and Aquaculture Technology, FUTA, Akure, Nigeria

Samuel Adelowo Olakojo Institute of Agricultural Research and Training, Obafemi Awolowo University, Ibadan, Nigeria

Abdulafeez Olalekan Olaniyan Ladoke Akintola University of Technology, Ogbomoso, Nigeria

Olumide Olaniyan National Weather Forecasting and Climate Research Centre, Nigerian Meteorological Agency, Abuja, Nigeria

Olawale Emmanuel Olayide Centre for Sustainable Development, University of Ibadan, Ibadan, Nigeria

Sunday J. Olotu National Centre for Technology Management, Federal Ministry of Science and Technology, Obafemi Awolowo University, Ile-Ife, Nigeria

Christopher Oludhe Department of Meteorology and Institute of Climate Change and Adaptation (ICCA), University of Nairobi, Nairobi, Kenya

O. J. Oluwatosin Department of Fisheries and Aquaculture Technology, FUTA, Akure, Nigeria

Anne Nyatichi Omambia National Environment Management Authority, Nairobi, Kenya

J. N. Ombui Department of Public Health, Pharmacology and Toxicology, College of Agriculture and Veterinary Sciences, University of Nairobi, Nairobi, Kenya

Ademola Omojola Department of Geography, University of Lagos, Akoka-Yaba, Lagos, Nigeria

Gbenga Emmanuel Onibi Department of Animal Production and Health, FUTA, The Federal University of Technology, Akure, Nigeria

J. O. Onono Department of Public Health, Pharmacology and Toxicology, College of Agriculture and Veterinary Sciences, University of Nairobi, Nairobi, Kenya

Robert Onyeneke Alex Ekwueme Federal University Ndufu-Alike Ikwo (AE-FUNAI), Agriculture (Agricultural Economics and Extension Programme), Ikwo, Nigeria

Victor A. Orindi ADA, National Drought Management Authority (NDMA), Nairobi, Kenya

Isimemen Osemwegie International University of Grand Bassam, Grand-Bassam, Côte d'Ivoire

Kaveh Ostad-Ali-Askari Department of Civil Engineering, Isfahan (Khorasgan) Branch, Islamic Azad University, Isfahan, Iran

Christine N. A. Ouinsavi Doctoral School of Agronomic and Water Sciences/ Department of Sustainable Management of Natural Resources, Laboratory of Ecology and Forest Research (LERF), University of Parakou, Parakou, Borgou, Republic of Benin

Emmah Owidi International Centre of Insect Physiology and Ecology, Nairobi, Kenya

Oyediran O. Oyebola Department of Aquaculture and Fisheries Management, University of Ibadan, Ibadan, Nigeria

Obasanjo Joseph Oyedele Department of Mass Communication, Faculty of Social and Management Sciences, Bowen University Iwo, Osun State, Nigeria

Olawale Oyemade Oyekanmi Department of Teacher Education, University of Ibadan, Ibadan, Oyo State, Nigeria

Maureen Papas Faculty of Law, The University of Western Australia, Perth, WA, Australia

Nyeko Pen-Mogi Gulu University, Gulu, Uganda

Anuradha Phadtare The WOTR-Centre for Resilience Studies (W-CReS), Pune, India

Surendra Poonia ICAR-Central Arid Zone Research Institute, Jodhpur, Rajasthan, India

Adithya Pradyumna Watershed Organisation Trust (WOTR), Pune, India

Society for Community Health Awareness Research and Action (SOCHARA), Bangalore, India

Kathrin Prenger-Berninghoff Institute of Urban and Transport Planning, RWTH Aachen University, Aachen, Germany

Geeta Punhani Department of Development Communication and Extension, Institute of Home Economics, University of Delhi, New Delhi, India

Muhammad Abdur Rahaman Climate Change Adaptation, Mitigation, Experiment and Training (CAMET) Park, Noakhali, Bangladesh

Mohammad Mahbubur Rahman Network on Climate Change, Bangladesh (NCC,B) Trust, Dhaka, Bangladesh

Nicholas B. Rajkovich School of Architecture and Planning, State University of New York at Buffalo, Buffalo, NY, USA

Laxmi Ramasubramanian Department of Urban Policy and Planning and The Institute for Sustainable Cities, Hunter College, CUNY, New York, NY, USA

Eunice Pereira Ramos Unit of Energy Systems Analysis, KTH – Royal Institute of Technology, Stockholm, Sweden

A. S. Rao ICAR-Central Research Institute for Dryland Agriculture, Hyderabad, India

Lance Robinson International Livestock Research Institute, Nairobi, Kenya

Sajal Roy Institute for Culture and Society, Western Sydney University, Sydney, NSW, Australia

Department of Women and Gender Studies, Begum Rokeya University, Rangpur (BRUR), Bangladesh

Onkangi N. Ruth Research and Business Development, National Construction Authority, Nairobi, Kenya

Ruchi Sachan School of International Studies, Centre for African Studies, Jawaharlal Nehru University, New Delhi, India

Sakshi Saini Department of Development Communication and Extension, Institute of Home Economics, University of Delhi, New Delhi, India

A. Saleem Khan Center for International Earth Science Information Network (CIESIN), The Earth Institute, Columbia University, New York, NY, USA

Zakari Saley Bana Disaster Risks Reduction Program Officer (DRR/O), Catholic Agency For Overseas Development (CAFOD), Niamey, Niger

David Olabanjo Sanni Department of Geography, Obafemi Awolowo University, Ile Ife, Nigeria

Maruf Sanni National Centre for Technology Management, Federal Ministry of Science and Technology, Obafemi Awolowo University, Ile-Ife, Nigeria

Josias Sanou Institut de l'Environnement et de Recherche Agricole (INERA), Ouagadougou, Burkina Faso

Muhammad Saqib Institute of Soil and Environmental Sciences, University of Agriculture, Faisalabad, Pakistan

Sheena Shah Permaculture Research Institute Kenya, Nairobi, Kenya

Kelvin M. Shikuku International Centre for Tropical Agriculture (CIAT), Africa Regional Office, Nairobi, Kenya

Samson Shimelse Department of Plant Biology and Biodiversity Management, College of Natural and Computational Sciences, Addis Ababa University, Addis Ababa, Ethiopia

College of Dryland Agriculture and Natural Resources, Mekelle University, Mekelle, Ethiopia

Yogesh Shinde The WOTR-Centre for Resilience Studies (W-CReS), Pune, India

Arega Shumetie Department of Economics, College of Business and Economics, Haramaya University, Haramaya, Ethiopia

Constantina Skanavis Research Centre of Environmental Education and Communication, Department of Environment, University of the Aegean, Mytılene, Greece

Sef Slootweg Natural Resources Management Project – Ngorongoro District Council, Deutsche Gesellschaft für Internationale Zusammenarbeit (GIZ) GmbH, Loliondo, Tanzania

Sieglinde Snapp Department of Plant, Soil and Microbial Sciences, Michigan State University, East Lansing, MI, USA

Alabi Soneye Department of Geography, University of Lagos, Akoka-Yaba, Lagos, Nigeria

Vignesh Sridharan Unit of Energy Systems Analysis, KTH – Royal Institute of Technology, Stockholm, Sweden

Silke Stöber Center for Rural Development (SLE), Humboldt Universität zu Berlin, Berlin, Germany

Adeniyi Suleiman Gbadegesin Ladoke Akintola University of Technology, Ogbomoso, Nigeria

E. Y. Swai Agricultural Research Institute (ARI) Hombolo, Dodoma, Tanzania

Nava Tabak Scenic Hudson, Poughkeepsie, NY, USA

Abdel Aziz A. M. Taha Department of Agricultural Engineering, University of Gadarif, Gadarif, Sudan

Hosein Talebmorad Department of Water Engineering, College of Agriculture, Isfahan University of Technology, Isfahan, Iran

Constantinos Taliotis Unit of Energy Systems Analysis, KTH – Royal Institute of Technology, Stockholm, Sweden

Premsagar Tasgaonkar Watershed Organisation Trust (WOTR), Pune, India

Azage Tegegne International Livestock Research Institute, Addis Ababa, Ethiopia

B. Teme Institut d'Economie Rurale (IER), Bamako, Mali

Ludovic Temple Center for International Cooperation in Agricultural Research and Development (CIRAD), Montpellier, France

Molu Tepo Merti Integrated Development Programme (Mid-P), Nairobi, Kenya

Jérôme Ebagnerin Tondoh WASCAL and UFR des Sciences de la Nature, Nandjui Abrogoua University, Abidjan, Côte d'Ivoire

Rosemary Egodi Ubaekwe Research Group in Health and Environment- Africa, (RGHE-AFRICA), Ibadan, Nigeria

International University Network on Cultural and Biological Diversity, Nigeria International Students' Coordination Unit, (IUNCBD N-ISCU), University of Ibadan, Ibadan, Nigeria

Paul Akeh Ugbah National Weather Forecasting and Climate Research Centre, Nigerian Meteorological Agency, Abuja, Nigeria

Collins Ugochukwu Alberta Environmental and Parks, Government of Alberta, Edmonton, AB, Canada

Gloria Ujor Nigerian Environmental Study Action Team (NEST), Ibadan, Nigeria

Aliyu Tambuwal Umar Department of Geography, Usmanu Danfodiyo University, Sokoto, Nigeria

Philippe Vaast Center for International Cooperation in Agricultural Research and Development (CIRAD), Montpellier, France

Swapnil Vyas The WOTR-Centre for Resilience Studies (W-CReS), Pune, India

Hafiz Abdul Wahab Institute of Soil and Environmental Sciences, University of Agriculture, Faisalabad, Pakistan

James R. Wallace Group Communications and Corporate Responsibility, Allianz, Munich, Germany

Liam James Walsh Sydney, NSW, Australia

David Walunya Ong'are National Environment Management Authority, Nairobi, Kenya

Shem Wandiga Department of Chemistry, University of Nairobi, College of Biological and Physical Sciences, Nairobi, Kenya

James M. Warner International Food Policy Research Institute, Addis Ababa, Ethiopia

H. O. Wesonga Veterinary Sciences Research Institute, Veterinary Research Centre, Kenya Agricultural and Livestock Research Organisation, Nairobi, Kenya

Andreas Witte Institute of Urban and Transport Planning, RWTH Aachen University, Aachen, Germany

Muluneh Woldetisadik Addis Ababa University, Addis Ababa, Ethiopia

Lippa Wood Mara Training Centre, Narok County, Kenya

Omonlola Nadine Worou The International Crops Research Institute for the Semi-Arid Tropics (ICRISAT), Bamako, Mali

Dipak Zade The WOTR-Centre for Resilience Studies (W-CReS), Pune, India

Libby Zemaitis NYS Department of Environmental Conservation and Cornell University, New Paltz, NY, USA

Jelena Živanović Miljković Institute of Architecture and Urban and Spatial Planning of Serbia, Belgrade, Serbia

Part II

Climate Resilience and Public Safety, Health, Agriculture, Food Security, and Emergency Management

Resilience of Greek Communities Hosting Climate Migrants: The Perceptions of Environmental Educators

31

Aristea Kounani and Constantina Skanavis

Contents

A. Kounani (✉) · C. Skanavis
Research Centre of Environmental Education and Communication, Department of Environment, University of the Aegean, Mytlene, Greece
e-mail: kounani@env.aegean.gr; akounani@yahoo.gr; cskanav@aegean.gr

W. Leal Filho (ed.), *Handbook of Climate Change Resilience*,
https://doi.org/10.1007/978-3-319-93336-8_1

Abstract
Today's rapid increase of climate change impacts has led, undeniably, to a sharp rise of climate migratory flows, a fact that affects host communities as well. Given the intense pressures that the hosting regions are facing, due to such massive arrivals of refugees, it is of tremendous importance the appropriate training of hosting communities in order to raise their resilience and mitigate their vulnerability. Education for sustainable development is imperative in order to avoid the negative consequences of refugees' inflow and maintain the stability of the region. This research aimed to evaluate the perceptions of the trainees of the second Summer Academy of Environmental Educators toward the refugee crisis that Greece is being confronted with. This Academy, stationed at Skyros Island, is being conducted under the administration of the multi-awarded "Skyros Project," which is the result of the innovative partnership of the Research Unit of Environmental Education and Communication of the University of the Aegean with the Skyros Port Authority. Thirty environmental educator trainees were interviewed, using a semi-structured questionnaire. Their opinions toward refugee crisis, the solutions that could help Greeks to avoid destructive impacts of refugee crisis, and whether the use of educational programs could contribute to building resilience and adaptation to Greek communities hosting refugees were evaluated.

Keywords
Climate change · Climate migrants · Greek communities' resilience · Environmental educators · Syrian refugee crisis · Skyros

Introduction

Climate change (CC) is regarded to be a phenomenon that occurs beyond the globe and influences every single part of the planet. As the Arctic is heating up in a twofold rate than the rest of the areas of the planet, it can influence weather thousands of miles away, causing events such as floods, droughts, heat waves, cold spells, etc., in regions like Europe, Africa, Asia, and the USA (Hirsch 2016). For instance, the defrosting of glaciers in the Himalayas are already causing sea level rise. It is palpable that Bangladesh may lose up to 20% of its land by 2030 as an outcome of flooding. Subsequently, therefore, Bangladesh is one of the most vulnerable countries in the world, as climate change is a tragedy for the 150 million people, most of whom are indigent (Kadian et al. 2013).

Obviously, the least-developed or developing countries, as well as low-lying islands and areas, are being influenced foremost (Hirsch 2016). Developing countries have slightly benefited from the emission of greenhouse gases, though those who have emitted greenhouse gases are developed countries that will presumably undergo much less consequences from CC. Some states, like Canada and Russia, could even benefit from global warming, whereas some northern regions will turn out to be more inhabitable or exploitable (Mayer 2011).

Unquestionably, the alterations in climate and social-ecological systems will disturb people's insights of the risks and benefits related with staying as compared to those associated with migrating (Barnett and Webber 2010). Migration is a way that individuals can respond to climate threats, whereas there may exist people that adapt without migrating or may not adapt at all (Mc Adam 2012). Wherever climate change exacerbates morbidity and mortality, condenses incomes, and decreases access to important forms of natural capital, people may be more likely than otherwise to select to migrate to areas which they perceive to compromise prospects for a better life. Thus, to the extent that there is indication about environmental modification being a factor in migration, there are grounds for concern about mass movements of people in response to CC. However, the number of moving people, their place of origin, and their final terminus are undefined. All approximations of the augmented amount of migrants that are expected to be stimulated by CC are based on very wide scale valuations of exposure to risk, rather than on systematic evidence about the sensitivity of migration patterns to environmental changes (Barnett and Webber 2010). The capacity of a social system to respond and moderate the undesirable impacts of CC is based on financial resources, effectiveness and legitimacy of governance, information to anticipate climate risks, devices of appropriate adaptations, social resources, infrastructure, and technology (Mc Adam 2012).

Migration, in many cases, is thought to be a positive coping and/or adaptation strategy to environmental and economic stresses. Slow-onset environmental crisis compounded by environmental alterations, like climate uncertainty and economic/political instability, often result in longer-term adaptation strategies and more permanent migration (Sudmeier-Rieux et al. 2017). Evidence to this is the islands of Kiribati in the Pacific Ocean, where survival is threatened by rising sea level in such an extent that they expected to be underwater in some decades. Furthermore, an entire village has already been lost to the sea, forcing people to move inland. The environmental degradation in the Pacific is occurring gradually, giving the island nation the opportunity to adaptation plans, in an effort to stave off the necessity of evacuation (Hirsch 2016).

Undeniably migration movements and resettlements not only affect migrants but also the receiving and hosting communities, which are experiencing the impacts of CC and global financial crisis, as well. Communities under stress, which receive large amounts of migrants, are confronting social, economic, and environmental pressures that could trigger conflicts between the inhabitants and migrants. One way to ease such tensions is through the implementation of educational programs that benefit both the local communities and newcomers, since by training them to adapt to the new condition could mitigate their vulnerability and increase their resilience. Environmental education by its definition is "a process aimed at developing a world population that is aware of and concerned about the total environment and its associated problems, and has the attitudes, motivations, knowledge, commitment and skills to work individually and collectively towards solutions of current problems and the prevention of new ones" (Tbilisi Declaration, as summarized in

Stapp 2001). The recent expansion of environmental education to encompass education for sustainable development, which seeks to "encourage changes in behavior that will create a more sustainable future in terms of environmental integrity, economic viability, and a just society for present and future generations" (Papaspiliou et al. 2014), further suggests a need for educational approaches that take place within the context of communities, including cities, so as to better foster learning about social as well as ecological processes (Tidball and Krasny 2007; Papaspiliou et al. 2014).

On the other hand, cautiously managed migration can encourage economic growth and innovation in destination locations and poverty reduction in origin locations. At the same time, migration can offset labor surpluses in sending countries. Receiving countries are taking on related multiple challenges, including managing labor markets (Cohen et al. 2013).

The combination of mitigation of changing climate patterns, adaptation to global warming, and creation of resilience requires cooperation between the top-down and bottom-up approaches (Cohen et al. 2013). Environmental education and education for sustainable development in empowering local societies with knowledge, elasticity, and tolerance to the environmental resilient skills have a vital contribution (Papaspiliou et al. 2014).

This research intends to evaluate the insights of the trainees of the second Summer Academy of Environmental Educators toward global environmental issues, climate change, and refugee crisis that affects Greece. The specific environmental educators are imminent introducers of educational patterns that will attempt to amplify resiliency and adaptation skills to the locals of hosting areas, since the migration pressure has stimulated their environmental concerns and seek ways to restrict potential risks. Moreover, the framework presented in this chapter intends to advocate how environmental education might become linked to practices that foster resilience and sustainability while enabling the participants to experience the value of environmental education at their community (Papaspiliou et al. 2014).

Climate Change and Migration

Environmentally induced migration has the potential to become a phenomenon of unparalleled scale and scope. Its implications on the global economy, international development, and national financial plans could be profound, with significant consequences for almost all extents of human safety, in addition to political and state security (Black 2001). Environmental migration is considered to be a global issue of tremendous importance, since such environmental issues like climate change threaten to enhance human mass movements both within the states and across international borders (Mc Adam 2012).

It has been problematic to distinguish "environmental refugees" from "economic migrants" (Boano et al. 2007). Environmental refugees are defined, initially by the UNEP researcher Essam El-Hinnawi in (El-Hinnawi 1985), as "the persons who compelled to flee from their areas of habitual residence, temporarily or permanently, because of the evolving natural and/or human induced environmental disruption, facts that put in jeopardy their existence and/ or seriously affect their quality of life" (Bates 2002). In other words, environmental refugees are "persons who no longer gain a secure livelihood in their traditional homelands because of what are primarily environmental factors of unusual scope" (Black 2001) and consequently are persons whose basic human rights have been trampled. They are often differentiated into refugees from natural disasters, degradation of land resources, involuntary resettlement, industrial accidents, the aftermath of war, and climatic changes (Sakellari and Skanavis 2012; Lonergan 1998). In case that the displaced people have to leave their habitats, immediately or in the near future, because of sudden or gradual alterations in their natural environment related to at least one of three impacts of climate change, sea level rise, extreme weather events, and drought and water scarcity, they are defined as "climate refugees" (Sakellari and Skanavis 2012; Global Governance Project 2008).

Bearing in mind migrant workers, political refugees, and climate migrants as alternative categories may not adequately replicate the complexity of individual decisions. Moreover, the association among unambiguous bad environmental conditions and climate change may perhaps be challenging to establish: a bad yield and famine often lead to migration, whether or not they result from climate change, war, bad governance, or any other reason (Mayer 2011). The complexity of migration patterns and the absence of straight correlations to endorse the assumption that climate change is encouraging emigration do not permit simple associates among perceived extreme events and mass movements. For instance, a flood event following periods of high rainfall, where in a short period of time vulnerable people lose everything and have to move to a safer place, makes it easier to link migration with natural hazards. The effect of drought is a slow procedure, and emigration only occurs after several dry years, when crop productivity is decreased, harvests have been lost, and the natural fertility of the soil is depleted. Once farmers are confronted with famine, they frequently move to urban areas or to other states to search for a steadier source of livelihood (Cohen et al. 2013).

So, with climate change progressing at such tremendous rate, huge numbers of migrants are relocating within their countries and internationally, while they are rummaging a sustainable future. In the imminent years, prosperous developed countries like the USA, Australia, and many European countries are mainly prospective to confront an influx of migrants from Africa and Asia. Migration will be seen on a massive scale, with more people on the move rather than any other time in human history (Hirsch 2016). People have historically forsake regions with raucous or deteriorating conditions, whether this is in terms of political upheaval, high unemployment, poor precipitation, or some permutation of these or other adverse factors (Black 2001). Given this situation, the interdependence between violent conflicts and environmental alteration has fascinated much attention recently. World history proves that the most dramatic migration movements were, firstly, the one of 12 million Germans migrating from Eastern Europe to Western Europe after the Second World War (Bahcekapili and Cetin 2015) and, secondly, the one caused by the Syrian war. The conflict in Syria continues to produce the gravest displacement crisis in the world today, generating dramatic levels of suffering and shattering the lives of many Syrian people – along with their hopes and dreams.

Even though the particular numbers of climate refugees are extremely difficult to be calculated, the available literature indicates that the climate refugee crisis will surpass all known refugee crises in terms of the number of influenced people (Kraler et al. 2011). However, the exact number of who will actually be displaced or forced to migrate will be contingent on the level of investment, planning, and resources (Stern 2007). This could occur by stepping up social protection, improving access to sustainable livelihoods, providing better infrastructure to people vulnerable to climate change and planning relocation (Kartiki 2011), and creating targeted educational programs that decrease vulnerability and build resiliency, both for inhabitants of states receiving migrants and migrants.

Tackling Climate Migration

Environmental change influences drivers of migration in the coinciding economic, social, political, environmental, and demographic spheres (Foresight 2011; Black et al. 2011). Understanding these drivers will help deal with migration on a policy level more meaningfully. Different factors in these five spheres, as well as disasters, development projects, environmental degradation, shortages and monetary compulsions, and market changes, can frequently have additive actions. They amplify the vulnerabilities of indigenous communities and work as triggers for migration (Piguet 2008; Martin et al. 2013).

As the literature has revealed, humanity can follow two paths in order to respond to climate change, the one of mitigation and the other of adaptation. Mitigation encompasses declining the extent and the rapidity of climate change by targeting the decrease of greenhouse gas emissions, through actions such as using renewable energy resources. On the other hand, adaptation tangles limiting vulnerability to the alterations that are already occurring (Hirsch 2016). Thus, the magnitude to which climate change will displace people is not just defined by the magnitude of climatic changes but also by the implementation and effectiveness of adaptation measures that assist communities and populations to cope with climatic changes (Cohen et al. 2013).

Nevertheless, it is also noteworthy to acknowledge that migration can also act as an operational adaptive response, best if people move voluntary, which means being their own decision-makers (Scheffran et al. 2012; Mc Adam 2012). As a matter of fact, for several civilizations, migration and planned resettlement is an integral part of their livelihood strategies and an adaptive strategy to reduce population stress in certain areas with a fragile environment. Thus, it is being understood as an imperative and inevitable strategy for seriously affected populations (Scheffran et al. 2012; Cohen et al. 2013). A typical example of this is the adaptation strategy of the Maldives including the "Safer Island Strategy" which would resettle populations from smaller and vulnerable islands to better protected ones (Cohen et al. 2013). Consequently, understanding the specifics of the affected regions will define whether

Table 1 Measures to be adopted at a national and international governance structures to address climate migration

Measures at national level	Measures at international level
Implementing relevant government policies to mitigate impacts of climate change and to meet the relevant international targets	Adhering to the existing frameworks and commitments to mitigate impacts of climate change
Reducing the climate vulnerability through adaptation measures	Agreeing upon a convention or legal framework to protect climate migrants
Developing alternative evacuation plan	Broadening the scope of climate refugees and migrants in already existing conventions
Being prepared by emergency response team or disaster management team	Settlement of climate migrants across the globe
Controlling illegal migration	Financial assistance
Improving infrastructure	

Kadian et al. (2013)

the accurate policy measure will be able to support adaptation, to prevent migration, or to assist migration as an adaptive measure.

The role of the international community in addressing climate migration is vital. There are three techniques in which the international community can support addressing climate migration: a) assisting migrant hosting states to cover the costs of adaptation measures, b) supporting the migrants' home states to prevent migration, and c) assisting migrants themselves through a coordinating agency or aid organizations (Docherty and Giannini 2009). Table 1, which was created based on information taken from Kadian et al. (2013), summarizes some of these measures and divides the responsibilities of actions among the international and national governance structures.

Adaptation and Migration

As history has shown, human migration has always been an adaptive response to poverty and social deprivation, as well as to environmental and climatic alterations. Inhabitants and migrants are designated as dynamic social agents who shape their livelihood underneath changing environmental conditions (Scheffran et al. 2012). Adaptation is considered "the adjustment in natural or human systems in response to actual or expected climatic stimuli or their effects, which moderates harm or exploits beneficial opportunities" (IPCC 2007). Resilient communities are active agents that affect their environment, anticipate and withstand to future stresses, and recreate themselves according to their drives and competences. Combining absorption, resistance, and recreation in social resilience transcends the reactive nature of adaptation toward anticipatory learning (Tschakert and Dietrich 2010; Scheffran et al. 2012). On the other hand, migration is an adaptation measure that diminishes stress and offers impending prospects in climate hot spots. With these resources, migrant networks can strengthen the social capital, livelihood, and resilience of their origin communities and develop innovative approaches for climate adaptation. The 2010 Cancun Accord provides a first step toward addressing climate change-induced migration on the international agenda. With this novel framework, the foundations have been laid for accepting migration as a legitimate adaptation possibility which, when taken seriously, would also require financial and institutional tools to facilitate the migration-as-adaptation pathway. As the literature has proved, migration is considered as an essential part of livelihood for several societies for centuries. In vision of this fact, adaptation strategies should be de-linked from measures to halt migration. As an alternative to this fact, developing adaptation can be advantageous to reduce the pressure on youth to migrate by conserving and improving their livelihood opportunities (Scheffran et al. 2012).

The combination of mitigation of changing climate patterns, adaptation to global warming – of which human migration is one form – and the creation of resilience requires cooperation between the top-down and bottom-up approaches. Consequently, a development policy orientated in the direction of poverty alleviation should be considered in any plan for a climate change strategy, bearing in mind an interdisciplinary approach to natural resource management.

Adaptation of Hosting Communities

Undeniably climate change is no longer merely an environmental policy issue. Having changes on climate in such an accelerating manner, massive migration flows are expected to occur, greater than in any other time in history, specifically to Europe and North America (Hirsch 2016; Welzer 2012). Migration movements have always been a cause of pressure and argument among the supporters of the open-door policy and those who are suspicious of the environmental, social, and economic effects of migration. As people settle in new communities, they put pressures on environmental, social, and economic issues (Bahcekapili and Cetin 2015). Consequently, as soon as dealing with climate migration, it is imperative to acknowledge how migration will influence the hosting communities and the migrants, as well (Hirsch 2016).

The international community has turned out to be progressively worried about the consequences of climate change on migration and the capability of current institutions to accomplish augmented influxes so that they do not influence on the stability of hosting areas. They seldom identify the impending for impulsive and planned adaptations to moderate vulnerability to environmental alteration, and they do not sufficiently recognize that migration is itself a strategy to sustain livelihoods in the face of environmental and economic perturbations and change (Barnet and Webber 2010). The perception of migration as a threat to national and international security is a discourse that has gained precedence in migrant hosting communities in Europe and North America during the most recent migration influxes (Brzoska and Fröhlich 2016).

In addition to the security consideration, other complications of mass migrants' arrivals on host communities extend on social, economic, and environmental factors. Regarding the environmental pressures of migration comprise augmented demands on natural resources, land degradation and deforestation for refugee camps, increased waste production, and energy consumption (Skanavis and Kounani 2016). Providing economic prospects to the migrants could be a challenge, since the absence of economic opportunities may increase criminality and consequently put additional pressure on security sector of hosting areas (Hirsch 2016). Therefore, social integration tactics are imperative, both for the migrants to feel welcomed and for the locals to get over the fear and prejudices of these newcomers. Moreover, learning the native language and feeling socially acknowledged in a community can have noteworthy mental health profits for the migrants, who often have built-up anxiety from their ambiguous life.

On the other hand, in a community under stress, tension and conflict can spring up among inhabitants and migrants. A peaceable and well-organized and planned resettlement must take into account prerequisites of the host communities. One way to ease tensions is with inclusive programs that benefit the locals and the migrants as well. Offering such programs to all inhabitants could implement ease resettlement and hostility from the hosting regions (Hirsch 2016).

Building Resilience in Hosting Regions through Environmental Education

Social vulnerability can be distinguished analytically from lack of resilience (Enarson 2012). Within this context, it is clear that reducing the impacts of climate migration and managing the issues of response is one of the significant modern world's immediate concerns. For example, inequalities related to income of certain social groups and the lack of adequate infrastructure, in both local and national level, have contributed to increasing the vulnerability of receiving and hosting regions. In this sense, the resistance of a local society from such stresses reducing the vulnerability of specific groups living within it, it is not only a matter of the construction of infrastructures that host the migrants. It is also related to the resilient mechanisms that make locals more impervious to the environmental, economic, and social pressures and less vulnerable to the adverse consequences resulting there from the large migrants' inflows (Papaspiliou et al. 2014; Kahn 2005).

As resilience is defined "the process of adapting well in the face of adversity, trauma, tragedy, threats or significant sources of stress," it implies "bouncing back" from difficult experiences. Resilience is not a trait; it involves behaviors, thoughts, and actions that can be learned and developed in any individual (APA 2018).

Emphasis should be given on environmental education strategies, since effective environmental awareness can promote active citizen participation in the prevention, preparedness, and mitigation of negative impacts of mass migrant arrivals. Environmental education to the extent that equips citizens with experience handling complex problems, arising directly from the daily contact with its surroundings, is considered one of the main components of a modern system design, where the participation of all interested parties is mandatory (Skanavis 2005, summarized in Papaspiliou et al. 2014). The awareness, improved educational programs, effective cooperation between institutions, mobilization of expertise, and governance at multiple scales are some of the elements that reduce the vulnerability of a society and build resilience and tolerance in case of such risks (Tompkins und Hurlston 2012, summarized in Papaspiliou et al. 2014; Carpenter et al. 2012).

The Case of Refugees' Settlements in Greece

The Syrian Civil War

In early 2006, Syria and the Fertile Crescent experience the worst 4-year drought in the instrumental record. The drought exacerbated existing water and agricultural insecurity and caused massive agricultural failures and livestock mortality. The most noteworthy consequence was the migration of as many as 1.5 million people from rural farming areas to the peripheries of urban center. De Chatel (2014) explicates how the 2006–2010 droughts in Syria directed to the uprising and the ensuing augmentation of emigrants. Predominantly, it was not the drought per se that generated the uprising followed by a conflict and millions of refugees but the government's failure to respond to humanitarian essentials exaggerated by environmental reasons (Kelley et al. 2015).

In 2015, Global Risks Report has identified water as the most imperative challenge worldwide and the biggest societal and economic risk for the next 10 years. While the world, as a whole, has abundant freshwater resources, spatial disparity and seasonal scarcity of freshwater, compounded by climate change, an acute threat for many regions of the planet, are emerging (Zhang 2015). In Syria the government's agricultural policy is considered one of the numerous factors that formed Syria's vulnerability to drought. Notwithstanding the growing water scarcity and the frequent droughts, the government of President Hafez al-Assad (1971–2000) initiated policies, which jeopardized Syria's water security by exploiting the restricted land and the water resources, regardless of sustainability (De Chatel 2014). One grave consequence of these unsustainable policies is the decline of groundwater (Kelley et al. 2015). Overuse of groundwater has been blamed for the recent drying of the Khabur River in Syria's northeast (Rodriguez et al. 1999). The reduced supply of groundwater dramatically augmented Syria's exposure to drought (Kelley et al. 2015).

Characterizing risk as the product of vulnerability and hazard severity, in the case of Syria, its vulnerability to drought and the social effects of the contemporary drought led to the onset of the Syrian civil war (Kelley et al. 2015). As climate change is a causative factor to conflict (Welzer 2012; Collier and Hoeffler 2005), acting in several circumstances as a "threat multiplier" (CNA 2007), negligible changes in climate are linked to proliferations in violence and warfare in human populations (Hsiang et al. 2013; Su et al. 2016). So, there is plenty evidence that the 2007–2010 drought contributed to the conflict in Syria (Kelley et al. 2015).

The Syrian civil war started in spring 2011, well known as the Arab Spring. Pro-democracy protests, all over the country, stood against President Bashar al-Assad, who was characterized by authoritarianism, political corruption, and human rights violation. Protesters were armed with the help of opposition militias who have been fighting against governmental forces. In 2012 the conflict has erupted in full-grown civil war between the pro-government ones against the ones opposing to it (Holliday 2013). At the same time, it is a religious war between different ethnic groups in the Arabian world (Bahcekapili and Cetin 2015). This war is ongoing and causes massive migration of people from Syria and the surrounding countries, as well as to Europe (The Lancet 2014).

The Syria crisis has displaced 5.4 million Syrian refugees into the Republic of Turkey, the Lebanese Republic, the Hashemite Kingdom of Jordan, the Republic of Iraq, and the Arab Republic of Egypt, while there are millions internally displaced people within Syria. Turkey hosts more Syrian refugees than any other countries – some 3.3 million, Lebanon one million, and Jordan 655,000; Iraq hosts nearly 246,000 Syrian refugees, as well as 3.2 million internally displaced Iraqis; and Egypt hosts around 126,000 Syrian refugees along with refugees from many other countries (UNHCR 2017a).

On the authority of the United Nations High Commissioner for Refugees (UNHCR), Syrians, along with migrants and refugees from other war-torn countries, have fled to Turkey and attempted to cross Mediterranean Sea to seek refuge in Europe, overwhelming receiving countries, since they weren't ready or willing to cope with this abrupt outcome (UNHCR 2017b).

Impacts of the Refugees' Settlements in Greece

Syrian crisis is considered as the biggest humanitarian and refugee crisis of recent years, an ongoing causation of suffering for millions of people (UNHCR 2016). This immense immigration is well known as the "Middle East Refugee Crisis," and evidently it has influenced all the neighboring countries to Syria, including Greece. In Fig. 1 are presented the total refugees' influxes for the last 10 years and for 2018 until the 9th of March.

For more than 10 years, refugees and migrants from Syria, Afghanistan, and Iraq have been crossing Mediterranean Sea heading to Greece. However, Greece is not their final European destination since they wish to continue their journey to northern and western EU countries. All these years, Greek islands have been facing an enormous pressure to deal with such large number of refugee arrivals.

The islands which are located closely to Turkey coastlines, such as Lesvos, Kos, Chios, and Samos, happen to have restricted economic resources and a fragile environmental sustainability structure making them confoundedly destructible in emergency circumstances, like the one associated with the Middle East Refugee Crisis.

The refugees' footprint can be summarized as pressure on water and energy demand, soil destruction, air pollution, deforestation, and waste production. However, every refugee crisis is unique and needs to be observed and managed as an individual condition. When it comes to solid waste, there is a large increase because of them staying for days into camps. Therefore, overpopulation of Greek islands is projected that it could cause significant environmental degradation. Greece was not

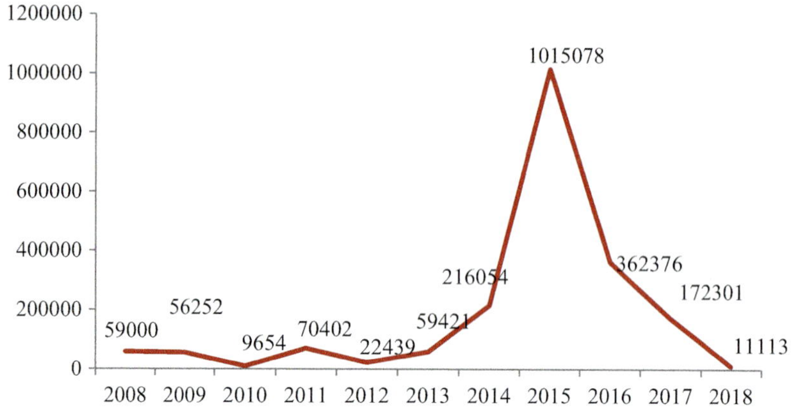

Fig. 1 Total sea arrivals to Aegean Sea per year from 2014 to 2018 (Based on data received in 9th March 2018 by UNHCR)

well prepared to deal with mass refugees arrivals while being also in a harsh economic situation (Skanavis and Kounani 2016).

Greece in the last decade has been seriously affected by the global economic turmoil (Kentikelenis et al. 2011). This crisis has caused a dramatic impact on everyday life, since the reduction or absence of income caused losses in prosperity and pushed large sections of the population in poverty (Knight 2013). The national health system has been in jeopardy, which makes even harder the delivery of services to the latest refugee flow. This entire unfavorable situation that Greece is addressing makes impossible to secure basic human rights, such as food, clean water, housing, etc., for these evicted populations, and often the scenario is too burdensome even for the inhabitants of the hosting areas.

Project Skyros 2017: Environmental Communication Model

The Project "Skyros 2017," a multi-awarded educational program, is based on a successful cooperation between two governmental organizations, the University of the Aegean and Skyros Port Authority. The geographic area that takes place is at the Linaria port of Skyros Island, a boutique port that has been assessed as the best one of its category in Greece.

This year's program, Skyros 2017, similar to the last year's one, consisted of a group of researchers from the Research Center of Environmental Communication and Education of the University of the Aegean that aimed to best implement a comprehensive environmental campaign focusing on disseminating information and raising environmental awareness. It included environmental campaigns targeted to citizens, tourists, boat passengers, and children, providing printed information about the biodiversity of Skyros and several environmental issues, as well. A tourism observatory has been set up as well as an environmental camp. The environmental activities included overseeing that the Linaria Port was clean and well taken care of. Various outdoor activities in collaboration with the Port Authority of Skyros were organized in order to attract residents and visitors' environmental interest. In addition to this, a Summer Academy for Environmental Educators was conducted, following the guidelines of excellence of the North American Association for Environmental Education, in order to give further training to individuals who wish to act as environmental educators.

Methodology

Research Area

The research area was the island of Skyros, and specifically the port of Linaria, where the second Summer Academy of Environmental Educators was conducted. The small tourist port of Linaria is considered a key port that has adopted a new way of sustainable approach to the management of its environmental impact, without spending a lot of its limited financial resources. Skyros Island is the southernmost island of the Sporades group of islands of Aegean Sea. It is located east of Euboea. It is the largest island of the Sporades (210 km^2), with a population of about 2888 residents, according to the 2011 population census by the Hellenic Statistical Authority (2014).

Research Sample

The research sample was the trainees of the second Summer Academy of Environmental Educators that was conducted in Linaria Port, Skyros Island.

Research Instruments

The data were collected through personal interviews with each of the trainee of the second Summer Academy of Environmental Educators. The interview lasted almost an hour with each participant. All interviewees were asked the same sets of semi-structured questions, in the same order. The questions, utilized at the interview, had been used before in other similar research projects (Lasso De La Vega 2004; Simpson 1998; Legault and Pelletier 2000; Botetzagias et al. 2015; De Feo and De Gisi 2010; Mensah 2006; Hiramatsu et al. 2014) and had been adjusted though to the peculiarities of the specific target group. The total number of questions was 33. The first eight questions applied data about the demographics (e.g., gender, age, educational status, marital status, region of residency, and origin). The next five questions were about the trainees' perceptions toward general environmental issues. In the sequel, they were asked five questions toward climate change, while the last 14 questions gave data about the participants' opinions toward refugee crisis, the consequences of refugee inflows in Greece, and the solutions they proposed.

Results

Demographics

The participants, the total of whom were 30, were 40% men and 60% women, aged between 20 and 26 years old, and they all were single. Moreover, on their vast majority (85%) were undergraduate students and former graduates of the Department of Environment, University of the Aegean. Five percent were graduates of the Departments of Education, and only 10% were of postgraduate degrees. Sixty percent were residents of provincial areas (30% of them were residents of the Aegean Islands), and the 40% resided at urban areas (Athens and Thessaloniki). Also, only a minority of them (20%) were descendants of individuals that migrated from another country. All of the trainees of the specific educational program were individuals who had a declared desire to raise their knowledge toward the environmental education, and most of them (65%) wanted to develop appropriate skills in order to be occupied in this field. Figure 2 is presenting this trend, by describing the responses of the participants to the question toward "the reason they selected the 2nd Summer Academy of Environmental Educators to participate in."

Participants' Perceptions toward General Environmental Issues

Initially, the trainees were asked to express their opinions toward the current condition of the environment, and the majority of them (90%) responded that "the environment is in a bad condition, but a lot of effort if invested might reverse it," while the other 10% said that "it's in some trouble and can be reversed with a little effort." In the sequel, they were asked to express their thoughts toward how serious are the following problems for the planet, on a scale of 1 (not serious at all) to 5 (extremely serious) (Table 2).

In Table 3 are depicted their sentiments toward the same problems for Greece.

Moreover, their views toward how much they agreed with the following statements, on a scale of 1 (strongly disagree) to 5 (strongly agree), are presented in Table 4.

Participants' Perceptions Toward Climate Change

At the third part of the questionnaire, participants were asked some questions toward climate change. So, Fig. 3 depicts their views about what would best describe the climate change issue.

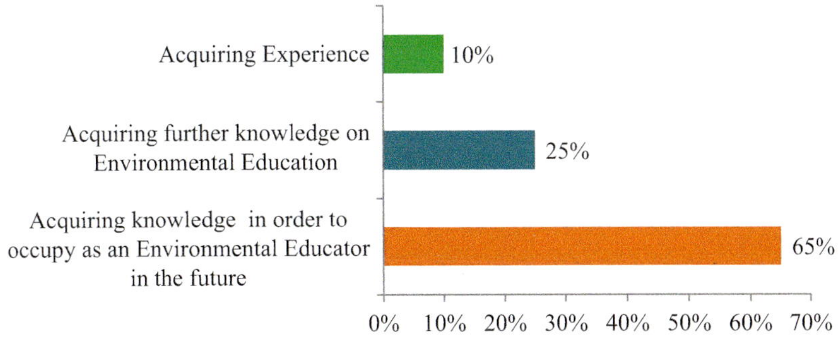

Fig. 2 Trainees' perception on "why they choose the specific educational program"

Table 2 Trainees' beliefs about the significance of the following problems globally

	N	Mean	S. D.	Min	Max
Poverty, lack of food and drinking water	30	4.75	0.55	3	5
Climate change	30	4.7	0.73	2	5
The proliferation of nuclear weapons	30	4.6	0.59	3	5
The spread of infectious diseases	30	4.35	0.67	3	5
The increase in world population	30	4.35	0.87	3	5
Armed conflicts	30	4.05	1.14	1	5
Major economic downturn	30	3.95	0.68	2	5
The illegal movement of refugees and migrants	30	3.65	1.2	2	5

Table 3 Trainees' opinions about the significance of the following problems for Greece

	N	Mean	S. D.	Min	Max
Major economic downturn	30	4.25	0.71	3	5
The illegal movement of refugees and migrants	30	4.25	0.91	3	5
Climate change	30	4.05	0.99	2	5
Poverty, lack of food and drinking water	30	3.9	1.07	2	5
The spread of infectious diseases	30	3.65	1.34	1	5
The increase in world population	30	3.35	1.03	2	5
Armed conflicts	30	3.25	1.40	1	5
The proliferation of nuclear weapons	30	3.1	1.37	1	5

In the sequel, they were asked to express their estimation about climate change. Thus, most of them (60%) responded that climate change is a very important issue that will have negative consequences for the next generation, 25% said that it will already have negative consequences for the present generation, and only 15% answered that it will have undesirable costs for further generations. Additionally, the interviewees' estimations for the most important impact of global climate change are presented in Fig. 4.

Table 4 How much trainees agreed with the following statements

	N	Mean	S. D.	Min	Max
If things continue on their present course, we will soon experience a major ecological catastrophe	30	4.55	0.60	3	5
When humans interfere with nature, it often produces disastrous consequences	30	4.2	0.95	2	5
We are approaching the limit of the number of people the earth can support	30	3.6	1.14	1	5
The earth has plenty of natural resources if we just learn how to treat them	30	3.45	1.35	1	5
The earth has very limited room and resources	30	3	1.07	1	4
Science and technology can overcome any environmental problem	30	2.9	1.44	1	5
Humans have the right to modify the natural environment to suit their needs	30	2.15	1.03	1	4

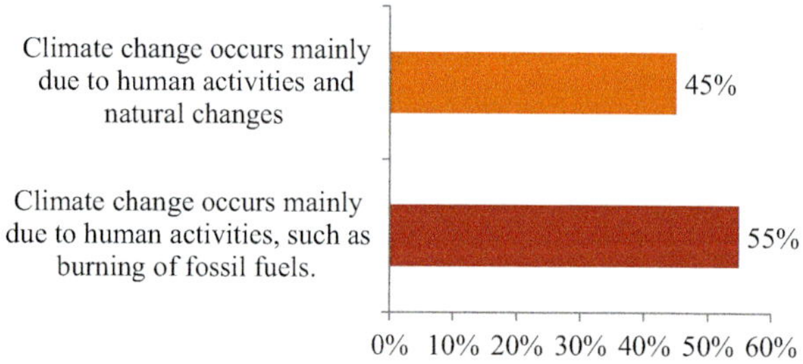

Fig. 3 What best describes participants' view toward climate change

In Table 5 are presented the responses when they were asked to express their thoughts concerning "to what extent climate change will... (on a scale from 1 = not at all, to 5 = very much)."

Participants' Perceptions Toward Refugee Crisis

In order to express their observations about the refugee crisis, the trainees primarily were asked to express their beliefs concerning the extent that Greece is affected by refugees' arrivals. The results of their response showed that they believe on average that Greece is quite affected by refugees' arrivals. Moreover, their perception toward which sectors are affected and to what extent is depicted in Table 6.

The 95% appeared to be aware of the effects of mass refugee inflows that Greece is facing. Furthermore, their opinions toward the most significant consequences of mass refugees' arrivals to Greece are accessible in Fig. 5.

In Table 7 are presented their perceptions toward the issue of "special wastes" which is a problem of great importance in regions that receive mass inflows of refugees.

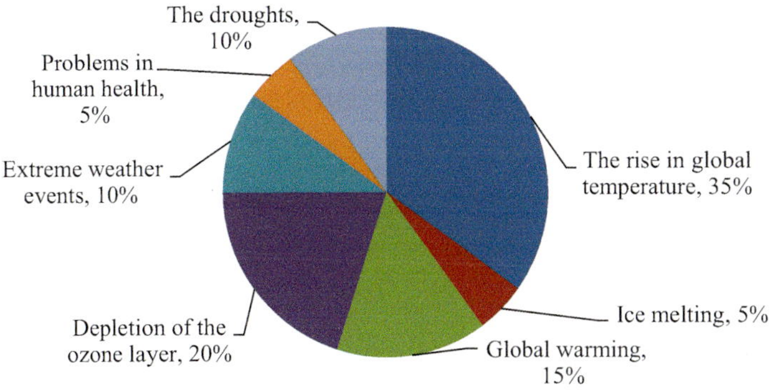

Fig. 4 Trainees opinions' toward the most important impact on global climate change

Table 5 To what extent climate change will...

	N	Mean	S.D.	Min	Max
Intensify the problem of migration globally	30	4.15	0.58	3	5
Cause climate wars globally	30	3.8	0.76	3	5

Table 6 Trainees' views concerning the sectors that are affected by the refugees' arrivals

	N	Mean	S. D.	Min	Max
The environmental pressures experienced by receiving and resettling refugees are due to the inability of state authorities to operate under emergency and pressure (lack of staff, etc.)	30	4.35	0.58	3	5
The advent of large refugee inflows has caused environmental pressures in the receiving areas (pollution from waste, water pollution, deforestation, etc.)	30	4.3	0.73	4	5
The advent of refugees will create polarization and disputes among residents in the host or resettlement areas (those who are in favor of the establishment and those who are against)	30	3.75	0.96	1	5
The arrival of refugees will cause political instability	30	3.6	1.04	1	5
The arrival of refugees causes a deterioration in the economy of the place of establishment	30	2.75	1.20	1	5
The arrival of refugees will provide a solution to many rural areas where many cheap labor hands will emerge	30	2.5	1.10	1	5
The arrival of refugees aggravates unemployment problems because refugees will compete with jobs in reception and resettlement areas	30	2.3	0.97	1	4

In the sequel, they were asked to express their opinion to what extent a problem of lack of food, water, energy, etc., which may often be created in areas receiving refugee inflows, could be solved (on a scale of 1 = not at all, up to 5 = completely), and their responses are presented in Table 8.

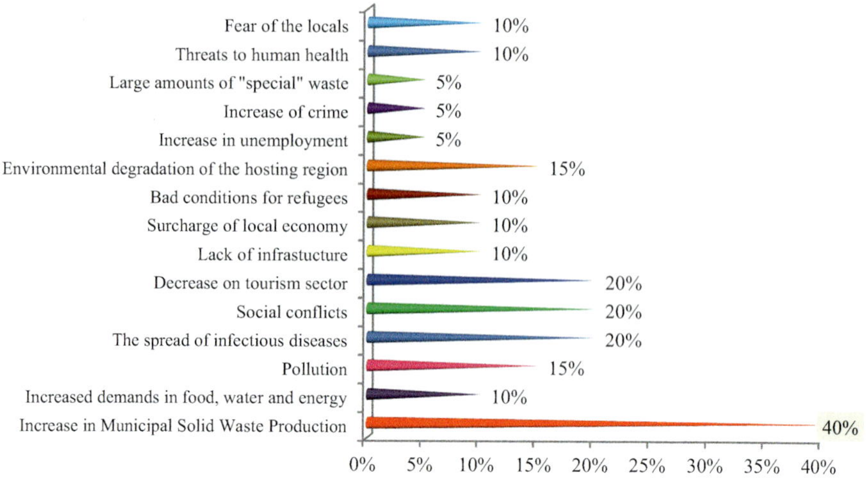

Fig. 5 Trainees' perceptions toward the most important effects of mass refugee flows for Greece

Table 7 Participants' perceptions toward the issue of "special waste"

	N	Mean	S. D.	Min	Max
I am aware of the huge quantities of "special waste" (life jackets, rubber, and fiberglass) in the receiving areas	30	4.1	0.91	2	5
In order to address the problem of "special waste," the refugee crisis needs to be addressed primarily in general	30	3.45	1.27	1	5
To deal with the problem of "special waste," more cleaners are needed to clean the shores and to properly discard waste	30	3.95	0.75	2	5
In order to address the problem of "special waste," appropriate recycling infrastructure for the specific waste is needed	30	4.3	0.73	3	5
To address the problem of "special waste," more landfills are needed	30	2.65	1.13	1	5
To address the problem of "special waste," the "burning" process could be used	30	2.3	1.26	1	5
I am worried about how "special" waste is collected and disposed of	30	4.4	0.68	3	5
Disposal of "special waste" should be addressed urgently to avoid further environmental degradation	30	4.55	0.60	3	5
Insufficient collection and disposal of "special waste" could affect public health	30	4.3	0.65	3	5
Insufficient collection and disposal of "special waste" affects the region's tourism sector	30	4.2	0.83	3	5

In Table 9 are presented the trainees' perceptions about how Greece handled the large refugee influx.

On average they appear to believe that the refugee issue is very worrying for Greece.

Table 8 Participants' opinion to what extent a problem of lack of food, water, energy, etc., which may often be created in areas receiving refugee inflows, could be solved

	N	Mean	S. D.	Min	Max
By reducing energy consumption	30	2.75	0.78	2	5
By reducing water consumption or food to be sufficient for everyone	30	2.65	0.87	1	5
By changing the way of life of local societies to help each one individually in the situation	30	3.7	1.03	2	5
By implementing local education programs to adapt local communities to the new situation	30	3.95	0.88	2	5

Table 9 Trainees' perceptions about how Greece handled the large refugee influx

	N	Mean	S. D.	Min	Max
Greece was ready to receive large refugee flows	30	1.4	0.50	1	2
The way in which the refugee issue has been addressed is satisfactory	30	1.95	0.99	1	4
Greece, due to the economic crisis, did not have the appropriate structures and infrastructure to cope with refugee flows	30	3.95	0.99	2	5

Moreover, their points of view concerning the solution to refugee crisis that Greece is confronting with are depicted in Table 10 and Fig. 6.

The majority of the trainees (95%) have the willingness to go voluntarily to a receiving area to help the refugees that just arrive in Greece, while all of them would volunteer to participate in programs that will help the refugees to integrate into Greek society. Furthermore, 85% believe that educating the local Greek communities that receive large refugee influxes could decrease their vulnerability and raise their resilience to the specific emergent condition. According to trainees, the content of such a program should include the issues that appear in Fig. 7.

Discussion

All of the trainees of the specific educational program were individuals who were dedicated toward the field of environmental education, as it is depicted in Fig. 1. Most of them, 65%, wanted to develop appropriate skills in order to be employed in this profession in the near future.

Perceptions toward General Environmental Issues

As it was aforementioned, the participants were asked some questions on general environmental issues in order to give the interviewers a more comprehensive view of their environmental vision and their environmental profile.

Table 10 Trainees' opinions toward who should give the solution to refugee crisis in Greece

	N	Mean	S. D.	Min	Max
The refugee issue is a global issue for which solutions need to be addressed at the level of states worldwide	30	4.55	0.51	4	5
The refugee issue is an issue that concerns all the member states of the European Union and should therefore be resolved at member state level	30	4.2	0.83	2	5
The refugee issue is an issue that only concerns refugees (e.g., Mytilene) and resettlement areas (e.g., Thessaloniki, Serres), so solutions should be provided at local level	30	1.7	1.03	1	4
The refugee issue is an issue that must concern all of us, and for that reason, we should put pressure on the government to find solutions	30	4.6	0.50	4	5

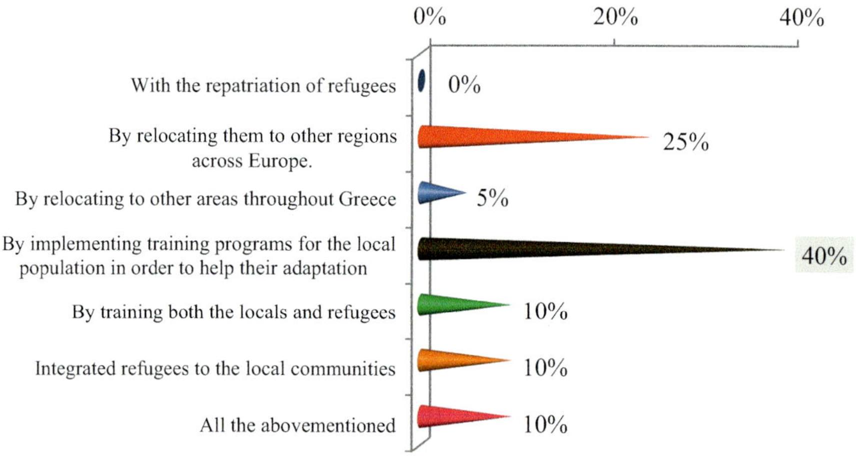

Fig. 6 Trainees' perceptions toward the solution that could solve the refugee crisis in Greece

The analysis of the results revealed that the participants were well informed on general environmental issues, appearing to be environmentally aware young scientists (Tables 2 and 3). Probably their knowledge is derived from attending the pertinent courses at the University of the Aegean, as most of them were undergraduate and graduate students of the Department of Environment at this institution.

Perceptions Toward Climate Change

On the contrary, the interviewees appeared to have moderate knowledge toward the significant issue of climate change, which is concluded from the results of Figs. 2 and 3. Table 4, though, points out that on an average, their views on climate

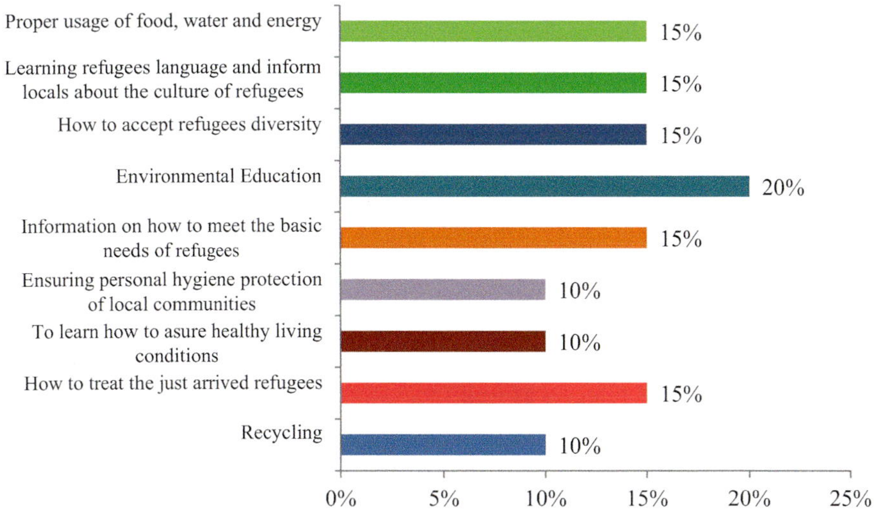

Fig. 7 What the future environmental educators would include in an educational program for the local society

migration are explained by the following beliefs: "climate change is reaching a global spectrum" and "greatly increases the probability of wars due to such change."

Perceptions Toward Refugee Crisis

In order to express their observations about the refugee crisis, the trainees primarily were asked to express their beliefs concerning the extent that Greece is affected by refugees' arrivals, and all of them expressed that Greece is quite affected by the incoming of refugees.

The literature argues that emergency conditions, such as the sudden appearance of a large number of refugees, have a major impact and exert pressure on the social, environmental, and political state of the receiving area. Particularly, in case the hosting nation does not have sustainable and integrated management plans regarding these concerns, the outcome is quite complicated (Dyrholm and Lindholm Mikkelsen 2013). This research showed that generally the receiving regions are confronted with environmental pressures due to the inability of state authorities to operate when there is an emergency and unexpected pressure (Table 5). The responses showed that unemployment rates are not elevated due to availability of cheap labor from refugees. The economy is perceived to be affected, and political instability could possibly be experienced.

Contrary to this, in terms of the pressures that Greece is receiving daily from the mass refugee flows, the most important impact appeared to be (40%) the increase in waste production, while a small portion (15%) believed that the most challenging effect is the environmental degradation in hosting regions. Although they thought that they were aware of the problem, they seemed to be misinformed toward the issues of the wastes that were produced by refugees' arrivals in regions like Lesbos ("special wastes"). Skanavis and Kounani (2016) have proven that the worst environmental challenge that receiving areas are confronting with is the management of the tremendous quantities of "special wastes" like the fiber boats and life jackets.

On the other hand, they appeared to be very worried about how "special wastes" are collected and disposed and felt that this urgent issue should be addressed in order to avoid further environmental degradation and influences on public health and tourism sector.

Furthermore, they said that dealing with imminent problems, like lack of food, water, energy, etc., there is a need for education, in order to promote adaptation measures to locals and refugees as well.

As Skanavis and Kounani (2016) have noticed, Greece, due to the economic crisis, hasn't had the appropriate structures and infrastructures to cope with refugee flows. Table 9 showed that the environmental education trainees, who were interviewed in this study, had similar perceptions. Additionally, they expressed the opinion that the way in which the refugee issue has been addressed is not satisfactory; it should be a global concern so that attention must be directed to this problem at the level of states worldwide. Every citizen around the globe should put pressures for reaching an internationally accepted solution, as well.

According to Kherfan (2016), governments should primarily have a sustainable and integrated management system for the three principal environmental concerns which are natural resources, solid waste, and wastewater. Meeting the essential needs of people displaced by disasters, wars, climatic changes, etc. should be the responsibility of every civilized society. Nongovernmental organizations and governments in the host countries should do their absolute best to accommodate the refugee flows of current days.

Regarding the proposed solution to the refugee crisis, a great portion of interviewees (40%) expressed that through appropriate awareness campaigns and education, Greeks could adapt and be resilient to the outcomes they have been confronting due to refugees' incoming. They stressed the fact that refugees should be relocated in a safe manner to other areas across Europe, as well.

Adaptation and mitigation policies for offsetting the impacts of climate change should focus on promoting integrated and sustainable water management, food and energy consumption, and waste management policies (Cohen et al. 2013).

Conclusion

Syrians are presumed to be climate migrants since the Inter-Agency Standing Committee (IASC) Working Group discriminates between four main causes for climate-related movements: hydrometeorological extreme hazard events; environmental degradation and/or slow-onset extreme hazard events; significant permanent losses in state territory as a result of sea level rise, etc.; and armed conflict/violence over shrinking natural resources (Kraler et al. 2011).

Nowadays, almost more than 850 million people globally are undernourished, fact that is expected to boost due to shrinkage of farmland in the wake of climate change. Such trends give a foretaste of what is going to be happening when the climate change boosts the flow of refugees. Resource conflicts due to global warming will fundamentally alter the shape of Western communities in the next few years. Climate change is an extremely urgent issue for environmental policies' implementation, and it will be the greatest challenge of the modern age, threatening the existence of millions of human beings, forcing them into mass migration movements. Europe and the USA, as they receive a tremendous pressure by the migration flows, will have to do more in order to protect themselves, from the millions of refugees who are expected to follow climate change consequences. Hunger, water problems, wars, and desertification will exert incalculable pressure on the West and North European and North American prosperity (Welzer 2012).

Migration is going to influence the lives of all human beings of the planet. Demonstrably, resettlements affect not only the people moving but also the host communities. Unfortunately, most of the host cities, however, are not well equipped to support such a large influx of migrants. As people crowd into cities, many are forced to live in cramped, unhealthy living conditions in the poorest neighborhoods. Cities are often unable to provide jobs for the newcomers, resulting in high unemployment. Crime can rise as people struggle to afford the costs of their everyday needs. Host communities, in many cases, will be experiencing the impacts of climate change themselves (Hirsch 2016).

Given all these pressures that the receiving and hosting regions are facing, due to such massive refugee movements, it is of tremendous importance for hosting countries to build resilient communities. We are in need of launching awareness campaigns to locals that tackle environmental concerns and ethics, hygiene, and proper applications. Citizens need to socially cope with the presence of refugees in their country by taking part of fundraising projects if possible and allocating resources for providing appropriate help to those in need (Kherfan 2016), while in return they will experience the beauty of assisting those who are in despair. At the same time, locals should also feel protected by the governmental services and policies that would safeguard their well-being as well and keep the host communities safe and clean.

Greek universities implementing projects like "Skyros Project" could train specialized environmental educators in order to communicate mitigating strategies and educate locals and refugees to increase their resilience and decrease their vulnerability.

References

American Psychological Association (APA) (2018) The road to resilience. Available online at http://www.apa.org/helpcenter/road-resilience.aspx

Bahcekapili C, Cetin B (2015) The impacts of forced migration on regional economies: the case of Syrian refugees in Turkey. Int Bus Res 8(9):1–15. https://doi.org/10.5539/ibr.v8n9p1

Barnett J, Webber M (2010) Accommodating migration to promote adaptation to climate change. In: The World Bank Development economics office of the senior vice president and chief

economist, Policy Research Working Paper, pp 1–62, Available online at http://econ.worldbank.org
Bates DC (2002) Environmental refugees? Classifying human migrations caused by environmental change. Popul Environ 23(5):465–477
Black R (2001) Environmental refugees: myth or reality? Working paper no. 34. New issues in refugee research, University of Sussex, Falmer, Brighton. Available via UNHCR. http://www.unhcr.org. Accessed 12 Nov 2017
Black R, Kniveton D, Schmidt-Verkerk K (2011) Migration and climate change: towards an integrated assessment of sensitivity. Environ Plan A 43:431–450
Boano C, Zetter R, Morris T (2007) Environmentally displaced people: understanding linkages between environmental change, livelihoods and forced migration. Refugee Studies Centre, Oxford Department of International Development. Available via University of Oxford. http://conflictrecovery.org/bin/rsc_Dec2007.pdf. Accessed 12 Nov 2017
Botetzagias I, Dima AF, Malesios C (2015) Extending the theory of planned behavior in the context of recycling: the role of moral norms and of demographic predictors. Resour Conserv Recycl 95:58–67
Brzoska M, Fröhlich C (2016) Climate change, migration and violent conflict: vulnerabilities, pathways and adaptation strategies. Migr Dev 5(2):190–210
Carpenter S, Arrow Kenneth R, Barrett J, Biggs S, Reinette B, William A, Crepin AS, Engstrom G, Folke C, Hughes TP, Kautsky N, Chuan-Zhong L, McCarney G, Meng K, Maler KG, Polasky S, Scheffer M, Shogren J, Sterner T, Vincent JR, Walker B, Xepapadeas A, de Zeeuw A (2012) General resilience to cope with extreme events. Sustainability 4(12):3248–3259 https://doi.org/10.3390/su4123248
Center for Naval Analyses (CNA) (2007) National security and the threat of climate change. Available via CAN. http://securityandclimate.cna.org//. Accessed 12 Mar 2018
Cohen IS, Spring UO, Padilla DG, Paredes JC, Inzunza Ibarra MA, Lopez RL, Diaz VG (2013) Forced migration, climate change, mitigation and adaptive policies in Mexico: some functional relationships. Int Migr 51(4):53–72. https://doi.org/10.1111/j.1468-2435.2012.00743.x
Collier P, Hoeffler A (2005) Resource rents, governance and conflicts. J Confl Resolut 49:625–633
de Chatel F (2014) The role of drought and climate change in the Syrian uprising: untangling the triggers of the revolution. Middle East Stud 50(4):521–535
De Feo G, De Gisi S (2010) Public opinion and awareness towards MSW and separate collection programmes: a sociological procedure for selecting areas and citizens with a low level of knowledge. Waste Manag 30(6):958–976
Docherty B, Giannini T (2009) Confronting a rising tide: a proposal for a convention on climate change refugees. Harvard Environ Law Rev 33:349–403
Dyrholm L, Lindholm Mikkelsen LM (2013) Finding global solutions: alternatives to long-term encampment of refugees, 1st master module on global studies. Roskilde Universite, Trekroner
El-Hinnawi E (1985) Environmental refugees. United Nations Environment Programme, Nairobi. Available via UNEP. http://hdl.handle.net/20.500.11822/2651
Enarson E (2012) Women confronting natural disaster: from vulnerability to resilience. Lynne Rienner Publishers Inc, Colorado
Foresight (2011) Migration and global environmental change: future challenges and opportunities. Final project report. The Government Office for Science. Department for Business, London. Available https://www.gov.uk/government/uploads/system/uploads/attachment_data/file/287717/11-1116-migration-and-global-environmental-change.pdf. Accessed 1 Nov 2017
Global Governance Project (2008) Forum on climate refugees. Available via GLOGOV. http://www.glogov.org/?pageid=80. Accessed 15 Nov 2017
Hellenic Statistical Authority (2014) Amendment of decision No 6519/31.07.2012 (Government Gazette 2230/B/31.07.2012). Results of population-housing census 2011 related In the legal population (citizens) of the country
Hiramatsu A, Kurisu K, Nakamura H, Teraki S, Hanaki K (2014) Spillover effect on families derived from environmental education for children. Low Carbon Econ 5:40–50

Hirsch RE (2016) Climate migrants: on the move in a warming world. Twenty-First Century Books, Minneapolis

Holliday J (2013) The Assad regime: from counterinsurgency to civil war. Middle East security report 8. Institute for the Study of War, Washington, DC. Available at http://www.understandingwar.org/sites/default/files/TheAssadRegime-web.pdf. Accessed Nov 2017

Hsiang SM, Burke M, Miguel E (2013) Quantifying the influence of climate on human conflict. Science 341(6151):1235367

Intergovernmental Panel on Climate Change (IPCC) (2007) Climate change 2007: climate change impacts, adaptation and vulnerability. Geneva. Available at https://www.ipcc.ch/pdf/assessment-report/ar4/wg2/ar4-wg2-intro.pdf. Accessed 15 Nov 2017

Kadian A, Kumari P, Sharma A (2013) Navigating climate change: extenuating strategies to combat climate change migration threats. University of Petroleum & Energy Studies, Bidholi, Dehradun, Uttrakhand, India, Ontario International Development Agency (OIDA) International Journal of Sustainable Development. ISSN 1923–6662 (online), pp 53–66. Available http://www.ssrn.com/link/OIDA-Intl-Journal-Sustainable-Dev.html

Kahn M (2005) The death toll from natural disasters: the role of income, geography and institutions. Rev Econ Stat 87:271–284

Kartiki K (2011) Climate change and migration: a case study from rural Bangladesh. Gend Develop 19(1):23–38. https://doi.org/10.1080/13552074.2011.554017

Kelley CP, Mohtadi S, Canec MA, Seager R, Kushnir Y (2015) Climate change in the fertile crescent and implications of the recent Syrian drought, ResearchGate, pp 1–6. Available online at https://www.researchgate.net/publication/273060165_Climate_Change_in_the_Fertile_Crescent_and_Implications_of_the_Recent_Syrian_Drought

Kentikelenis A, Karanikolos M, Papanicolas I, Basu S, McKee M, Stuckler D (2011) Health effects of financial crisis: omens of a Greek tragedy. Lancet 378(9801):1457–1458. https://doi.org/10.1016/S0140-6736(11)61556-0

Kherfan RA (2016) What are the most pressing environmental concerns of refugee camps in conflict-zones? American University of Beirut, Faculty of Health Sciences, Department of Environmental Health, p 14. https://doi.org/10.13140/RG.2.1.3245.4162

Knight DM (2013) The Greek economic crisis as trope. Berghahn J Focaal 65:147–159

Kraler A, Cernei T, Noack M (2011) Climate refugees: legal and policy responses to environmentally induced migration. European Parliament Committee on Civil Liberties, Justice and Home Affairs, pp 1–88. Available online Europarl. http://www.europarl.europa.eu/activities/committees/studies.do?language=EN. Accessed 15 Nov 2017

Lasso De La Vega E (2004) Awareness, knowledge, and attitude, about environmental education: responses from environmental specialists, High School Instructors, Students and Parents, University of Central Florida, Dissertation that submitted in partial fulfillment of the requirements for the degree of Doctor of Education in Curriculum and Instruction in the College of Education

Legault L, Pelletier LG (2000) Impact of an environmental education program on students' and parents' attitudes, motivation and behaviours. Can J Behav Sci 32(4):243–250

Lonergan S (1998) The role of environmental degradation in population displacement. Environ Change Secur Proj Rep 4:5–15

Martin M, Kang YH, Billah M, Siddiqui T, Black R, Kniveton D (2013) Policy analysis: climate change and migration Bangladesh. Working paper 4. Brighton. Available http://migratingoutofpoverty.dfid.gov.uk/files/file.php?name=wp4-ccrm-b-policy.pdf&site=354. Accessed 12 Nov 2017

Mayer B (2011) The international legal challenges of climate-induced migration: proposal for an international legal framework. Colorado J Int Environ Law Policy 22(3):357–416 Available http://www.colorado.edu/law/sites/default/files/Mayer%20%28Corrected%29-S.pdf. Accessed 12 Nov 2017

Mc Adam J (2012) Climate change and displacement: multidisciplinary perspectives. Hart Publishing, Oxford/Portland

Mensah A (2006) People and their waste in an emergency context: the case study of Monrovia, Liberia. Habitat Int 30(4):754–768

Papaspiliou K, Skanavis C, Giannoulis C (2014) Environmental education and resilient societies: a survey of forest fires in Greece. J Educ Train Stud 2(2):48–57
Piguet E (2008) Climate change and forced migration, new issues in refugee research. Research paper no. 153. Geneva. Available via UNHCR. http://www.unhcr.org/research/working/47a316182/climate-change-forced-migration-etienne-piguet.html. Accessed 12 Nov 2017
Rodriguez A, Salahieh H, Badwan R, Khawam H (1999) Groundwater use and supplemental irrigation in Atareb, Northwest Syria (International Center for Agricultural Research in the Dry Areas, Aleppo, Syria). Soc Sci Pap 7. Available via UNEP. http://agris.fao.org/agris-search/search.do?recordID=US201300049566. Accessed 12 Nov 2017
Sakellari M, Skanavis C (2012) An overview of the climate refugees' issues and scenario. In: Leal Filho W, Manolas E (eds) English through climate change. Democritus University of Thrace Publications, Orestiada
Scheffran J, Marmer E, Sow P (2012) Migration as a contribution to resilience and innovation in climate adaptation: social networks and co-development in Northwest Africa. Appl Geogr 33:119–127. https://doi.org/10.1016/j.apgeog.2011.10.002
Simpson B (1998) Environmental awareness survey. Global Issue Language Educ 30:12–14
Skanavis C (2005) Environment and community. Kalidoskopio, Athens, p 246
Skanavis C, Kounani A (2016) The environmental impacts of the refugees' settlements at Lesvos Island. In: 3th International Conference on Protection and Restoration of the Environment, At Mykonos Island
Stapp WB (2001) The concept of environmental education. In: Hungerford HR et al (eds) Essential readings in environmental education. Stipes publishing. LLC, Champaign, pp 33–36
Stern N (2007) The economics of climate change: the stern review. Cambridge University Press, Cambridge
Su Y, Liu L, Fang XQ, Ma YN (2016) The relationship between climate change and wars waged between nomadic and farming groups from the Western Han dynasty to the tang dynasty period. Clim Past 12:137–150. https://doi.org/10.5194/cp-12-137-2016
Sudmeier-Rieux K, Fernández M, Gaillard JC, Guadagno L, Jaboyedoff M (2017) In: Sudmeier-Rieux K, Fernández M, Penna I, Jaboyedoff M, Gaillard J (eds). Identifying emerging issues in disaster risk reduction, migration, climate change and sustainable developmentExploring linkages between disaster risk reduction, climate change adaptation, migration and sustainable development. Springer International Publishing, Cham
The Lancet (2014) The war on Syrian civilians. Lancet 383(9915):383
Tidball KG, Krasny ME (2007) From risk to resilience: what role for community greening and civic ecology in cities? In: Wals AEJ (ed) social learning towards a more sustainable world. Wagengingen Academic Press, Wagengingen
Tompkins EL, Hurlston LA (2012). Public-Private Partnership in the Provision of Environmental Governance: A Case of Disaster Management In: *Adapting Institutions: Governance, Complexity and Social-Ecological Resilience*. Boyd E, Folk C (Eds.). Cambridge: Cambridge University Press.
Tschakert P, Dietrich KA (2010) Anticipatory learning for climate change adaptation and resilience. Ecol Soc 15(2):11 Available at http://www.ecologyandsociety.org/vol15/iss2/art11/
UNHCR (2016) Syria conflict at five years. Available via UNHCR. http://www.unhcr.org/56e6e1b991.html. Accessed 11 Nov 2017
UNHCR (2017a) Regional refugee and resilience plan 2017–2018 (3RP). Available via UNHCR http://www.3rpsyriacrisis.org/. Accessed 11 Nov 2017
UNHCR (2017b) Syria regional refugee response. Available via UNHCR. http://data.unhcr.org/syrianrefugees/regional.php. Accessed 11 Nov 2017
UNHCR (2018) Figures at a glance. Available via UNHCR. http://www.unhcr.org/figures-at-a-glance.html. Accessed 9 Mar 2018
Welzer H (2012) Climate wars: why people will be killed in the twenty-first century. Polity Press, Cambridge
Zhang H (2015) Sino-Indian water disputes: the coming water wars. Wiley Interdiscip Rev Water 3:155–166. https://doi.org/10.1002/wat2.1123

Effect of Climate Variability on Crop Income in Central Highlands of Ethiopia

32

Arega Shumetie, Kassa Belay, Degye Goshu, and Majaliwa Mwanjalolo

Contents

Abstract

Ethiopian agriculture is a vulnerable sector to climate variability, wherein this study identified how strong is the effect on smallholders' crop income in central highland farming system of the country. The study used fixed effect regression to

A. Shumetie (✉)
Department of Economics, College of Business and Economics, Haramaya University, Haramaya, Ethiopia
e-mail: ashueconomist@gmail.com

K. Belay
Pan African University Africa Union, Addis Ababa, Ethiopia

D. Goshu
Department of Economics, Faculty of Business and Economics, Kotebe Metropolitan University (Asso.), Addis Ababa, Ethiopia

M. Mwanjalolo
Department of Geography, Geo Informatics and Climatic Sciences, College of Agriculture and Environmental Science, School of Forestry, Environmental and Geographical Science, Makerere University (Asso.), Kampala, Uganda

W. Leal Filho (ed.), *Handbook of Climate Change Resilience*,
https://doi.org/10.1007/978-3-319-93336-8_2

analyze the unbalanced panel data (1994–2014) collected for eight rounds. The model result shows that individual increment in cropping season rainfall or temperature has negative and significant effect on households' crop income in the farming system. Parallel increment of the two climate variability related inputs had positive and significant effect on crop income. Other variables like flood, frost, storm, and rainfall inconsistency in the onset and cessation time affected households' crop income negatively and significantly. Similarly, draft power and human labor, which are critical inputs in the crop production of Ethiopian smallholders, have positive and significant effect on crop income as to the model result. Thus, this study recommended that there should be supplementation of rainwater through irrigation, water harvesting, and other activities to have consistent water supply for the crop production. Additionally, long-lasting measures like afforestation should be adopted to counterbalance the negative effect of temperature increment.

Keywords

Climate variability · Smallholder · Crop income · Rainfall · Temperature

Introduction

Global climate change is the recurrent and most important challenge that societies are confronted with (Chérif et al. 2017). Scientific findings from different directions of the world argued that the global temperature is increasing because of greenhouse gas emissions (IPCC 2007), and the problem will continue unless there are human interventions to reduce it. Besides the continuous temperature increment, there are also extreme events such as floods and droughts in most parts of the world (Chérif et al. 2017). Global warming promotes floods, drought, hailstorm, and other unexpected shocks that have devastating effect on the world economy (IPCC 2014). There is a steep rise in global temperature especially in the twenty-first century (IPCC 2012), which is an initial manifestation of climate variability. This increment results in significant precipitation variability and severe moisture stress and widespread changes in extreme temperatures, drought, and frequency of heavy precipitation mostly in the tropics.

Developing countries especially African countries are those that seriously suffer from climate variability and change (Leal Filho 2011). Those nations experienced serious food shortage and starvation due to climate variability, while developed ones face relatively simple problems (FAO 2008). The nature-dependent agriculture of developing nations shows frequent variation and inconsistency as a result of rainfall and temperature variability. Majority of the chronically starved people of the world reside in rural areas of developing countries, whose livelihood purely depends on agriculture (FAO 2009). The close connection of developing countries' agriculture with nature results in huge economic performance reduction when there is weather variability (Sara 2012). Fischer et al. (2005) and Cline (2007), respectively, forecasted that agricultural yields of developing countries would decline by 15% and

20% in 2080 due to climate variability. In 2020 African cereal production will decline by 2–3% (Rosenberg et al. 2003), and this would result in significant crop income variation. The UNDP (2007) reported that climate variability could exacerbate income inequalities among countries. In connection to this, IPCC (2007) reported that revenue from crops could fall by 90% in 2100 for some African countries.

Dell et al. (2008) showed that a 1 °C temperature rise would result in 1.1% annual economic growth reduction for poor countries because of widespread effects of higher temperatures on agricultural productivity. According to Schlenker and Lobell (2010), significant yield decline in 2050 would cause huge crop income reduction nearly in all countries of Sub-Saharan Africa. If the current trend of temperature increment continues, vulnerability among the rural livelihood would be exacerbated (Arjan et al. 2011). Mathilde and Zaneta (2014) identified that climate variability should be perceived as an important source of risk for rural households in developing countries. In connection to this, those suffering most are subsistence farmers, whose income is very low and already in starvation and absolute poverty.

Many areas of Ethiopia are facing lower rainfall and higher temperature indicating that climate variability is mainly manifested in those directions (Ethiopian National Meteorology Agency (NMA) 2007)). The agency notified that major climate hazards observed in the country include drought, flood, and disease outbreak. Unfavorable rainfall and temperature conditions can have devastating effects on farm households if agriculture is purely dependent on rain (Paul et al. 2011), which means successive variability of them would have an overwhelming effect on smallholders' productivity. Ethiopian smallholders' vulnerability from the problem is severe due to their low adaptive capacity and high dependency on the nature-based agriculture. Exposure to drought risk is high, and coping capacities are limited due to lower income diversification (Arjan et al. 2011). These all would result in agriculture productivity and income variability.

Though there are few studies done before in connection about the effect of climate variability on agricultural productivity on Ethiopian households, those studies took the national economy rather than smallholders who suffer most from the problem. Thus, this study adopted panel data to examine the effect of climate variability on smallholders' crop income in central highland farming system of Ethiopia.

Methodology

Description of the Data Source and Sampling

Ethiopian Rural Household Surveys (ERHS) were conducted for seven rounds by the International Food Policy Research Institute (IFPRI), Centre for the Study of African Economies (CSAE) at Oxford University, Economics Department of Addis Ababa University, and Ethiopia Development Research Institute (EDRI) on smallholders from the four dominant regional states (Tigray, Amhara, Oromia, and SNNP)

Table 1 Agroecological zone and average rainfall of sampled villages

Village name	Location	Main crops grown	Average rainfall	Agroecology	Sample household
Yetmen	Gojjam	*Teff*, maize, wheat	1241	Moderate	51
D/Berhan	North Shoa	Wheat, barley, beans	919	Cold	168
Dinki	North Shoa	*Teff*, maize, sorghum	1664	Warm	79
Sirbana	Shoa	*Teff*, wheat, barley	672	Moderate	82

Source: Dercon and Hodinott 2011

of the country. The survey was started in 1989 when IFPRI team visited 450 households from 7 villages in Central and Southern Ethiopia (Dercon and Hoddinot 2004). The survey covered 15 villages and became well organized in 1994 so as to consider the main agroclimatic zones and farming systems of the country. About 1477 households were considered in 1994, and they have been reinterviewed in the late 1994 as well as in 1995, 1997, 1999, 2004, and 2009 (Dercon and Hoddinot 2004). Central highland of the country accounts about 40% of the total population based on Central Statistics Agency (CSA) in 1994 and 56.4% sampling share of ERHS in 1994 survey (Dercon and Hoddinott 2011). ERHS considered crop production systems as an important stratification base than administrative boundaries in selecting sample villages. Sampled smallholders were randomly selected from each village in order to have representative sample size. This research considered the 4 villages (Table 1) and 380 smallholders for one-round survey in 2014 that enabled to have eight-round unbalanced panel data for central highland farming systems of Ethiopia.

Type and Methods of Data Collection

Both primary and secondary data were considered in this study, wherein the former were collected from randomly selected sampled households. The data collected cover almost all features of smallholders like demographic, consumption, production, asset holding, purchases and sales, landholdings, livestock ownership, and others. The seven-round surveys from 1994 to 2009 were from high- and midland rural part of the country where mixed farming is practiced (crop production and animal husbandry). Each survey followed a similar format and contents so it is relatively easy to quickly get a sense of what can and cannot be done with these data. The eighth round survey in 2014 was to take into account the recent circumstances of smallholders through the questionnaires used by the previous institutions in 2009 survey. The study revisited all sampled smallholders (380) in 2014 from the four villages of central highland farming system of Ethiopia to have eight-round unbalanced panel data for the analysis.

Methods of Data Analysis

Previous studies adopted Ricardian approach to assess the effect of climate variability on rural household income. The approach argued that environmental factors affect production and costs and farmer's choice that would result in net revenue variation from each plot of land. The approach argued that long-term net revenue accumulation determines land value. A key assumption of this approach is that land used for a given purpose is best used for it. Land values in this model represent the present value of rents when it is used for its optimal purpose. Additionally, it assumes that both product and input markets function perfectly, and climate changes are instantaneous. Based on the approach, each farmer wishes to maximize income subject to exogenous conditions. The effect of climate variability on farm households' income in economic terms can be measured in many ways such as the gross value of production, cost, production net value, and/or farm income (John 2009). Di Falco et al. (2011) implemented Ricardian analysis and used net revenues as a dependent variable, wherein yields of different crops are aggregated to generate income. Nhemachena et al. (2010) considered net farm revenues to measure economic impact of climate change on African agricultural production systems. Thus, rural income from crop production can be formulated as:

$$Y_{hrt} = \sum P_{irt} Q_{iht}(X_{iht}, L_{iht}, H_{iht}, K_{iht}, C_{iht}) - \sum I_{it} X_{it} \quad (1)$$

where:

Y_{hrt} = Crop income of household (h) in a specific village (r) at time t. It was the product of quantity sold and price of crop i.

P_{irt} = Market price of crop i in each village (r) at time t. The research assumed that households in each village have a common market where prices are the same for households within the village on the specific survey year t.

Q_{iht} = Marketed quantities of crop i for household (h) at time t

X_{iht} = A vector of inputs like seeds, fertilizer, and pesticides used by household (h) for crop i.

L_{iht} = Human labor input budgeted by household (h) for producing crop i

H_{iht} = Demographic variables that affect crop income at time t

K_{iht} = Capital input budgeted by household (h) for the producing crop i at time t

C_{it} = Climate variability related explanatory that affect crop i at time t

I_{it} = Inputs quantity for producing each crop at time t

Based on the above equation, marketed quantity is an endogenous variable that is directly affected by vector of inputs, household-related variables, and climatic factors.

$$Q_{iht} = F(X_{iht}, L_{iht}, H_{iht}, K_{iht}, C_{it}) \quad (2)$$

Equation 2 implies that climate variations have a direct effect on marketed quantity and consequently have an indirect effect on income. Variability in temperature and rainfall patterns affects crop yield directly and ultimately smallholders' revenue (IPCC 2001). If a farmer chooses the crop that provides the highest net income and manages the endogenous inputs to maximize income, then income will be just a function of exogenous variables only (Mendelsohn et al. 1994). Therefore, the above model could be:

$$Y_{hrt} = (P_{irt}, X_{iht}, L_{iht}, H_{iht}, K_{iht}, C_{it}, I_{iht}) \quad (3)$$

Smallholder's crop income variability is the result of change in one of those variables or simultaneous change in two or more of them at once. Price volatility may increase in the future, since effects of climate variability are likely to increase uncertainty and food crop production instability in the tropics (Blein and Longo 2009). Additionally, households' income may be affected by the available inputs and other demographic shocks, thus Eq. 3 would be rewritten in the following form:

$$Y_{hrt} = (P_{irt}, X_{iht}, L_{iht}, H_{iht}, K_{iht}, C_{it}, I_{iht}, D_{hrt}) \quad (4)$$

D_{hrt} = Demographic related shocks for household (h) at a specific time t

Deressa (2007) and Jain (2007) considered net farm revenue to assess economic impact of climate change. Similarly, this research approximated smallholder's income by gross margin from crop production and reformulated the above equation as:

$$\begin{aligned} \ln Y_{hrt} = {} & \theta_0 + \theta_1 \ln P_{irt} + \theta_2 \ln L_{iht} + \theta_3 \ln H_{iht} + \theta_4 \ln K_{iht} + \theta_5 \ln C_{it} \\ & + \theta_6 \ln I_{iht} + \theta_7 \ln D_{hrt} + \mu_i + \varepsilon_{it} \end{aligned} \quad (5)$$

The study considered crop season rainfall and temperature variation as an approximation to climate variability (C_{it}), which may take the right-hand share in affecting the crop income. Extreme shocks or weather events like droughts, flooding, and frost severely affect smallholders' crop production, and they were included in the regression. Considering all these the above equation can be transformed to the following form:

$$\begin{aligned} \ln Y_{hrt} = {} & \theta_0 + \theta_1 \ln P_{irt} + \theta_2 \ln L_{iht} + \theta_3 \ln H_{iht} + \theta_4 \ln K_{iht} + \theta_5 \ln R_{rt} \\ & + \theta_6 \ln T^0_{rt} + \theta_7 \ln I_{iht} + \theta_8 \ln S_{ht} + \theta_9 \ln D_{hrt} + \mu_i + \varepsilon_{it} \end{aligned} \quad (6)$$

R_{rt} and T^0_{rt} connote crop season rainfall and temperature level in each village r at time t.

S_{ht} is a dummy variable which refers shocks related to climate variability that affect crop production of household h at time t.

Results and Discussion

Crop Income and Rainfall Trend in Central Highland Ethiopia

Crop income was very low before 2004 production period, which may be due to either production or price level reduction in which both of them were lower as compared to the recent level. Villages that are near to the central market have better and continuously increasing crop income than those located distantly. There was crop price increment since 2008, which may be one reason for the sharp income increment for sample households especially for those near to the central market. Households' average income was 1894 Birr with a standard error of 1068 Birr.

Crop season rainfall of each station showed high variation in each survey year. All stations had relatively higher rainfall in 1999, and there was continuous reduction after this year in some villages. Though there was subsequent reduction in crop season rainfall starting from 1999, but the crop income exhibited continuous increment. At the medium rainfall level in 1995 and 2004, crop income was relatively better wherein there was positive interaction between them for those years though it is difficult to generalize consistent relationship. However, rainfall was lower in 1997 production year as compared to years both before and after, but the crop income collected was better. Sample smallholders collected lower crop income in 1995 wherein the cropping season rainfall was also the lowest. Accordingly, farmers received a better income when rainfall level was lower in 1995, 1997, and 2004. The higher rainfall in 1994b and 1999 production years correlated with lower crop income as compared to others like 1997 and 2004. Cropping season rainfall level of 1995 and 2009 was relatively equal, but there was sharp increment in crop income in the latter production year that may be due to higher crop price. This logic of price and climate variability can be strengthened by considering 1999 production year on which crop income was better than what it was in 2004. According to Fig. 1, there is inverse relationship between cropping season rainfall and crop income for larger proportion of the survey years. Thus, it may be possible to generalize that crop income increment may be sourced from either quantity or crop price rise.

Model Diagnostic Test

Crop production variability due to weather variation would finally result in significant net income variation from the sub-sector (Susanne 2012). This implies that climate variability has a strong effect on smallholders' income if crop production is the main source. This research assumed that unlike demographic and idiosyncratic

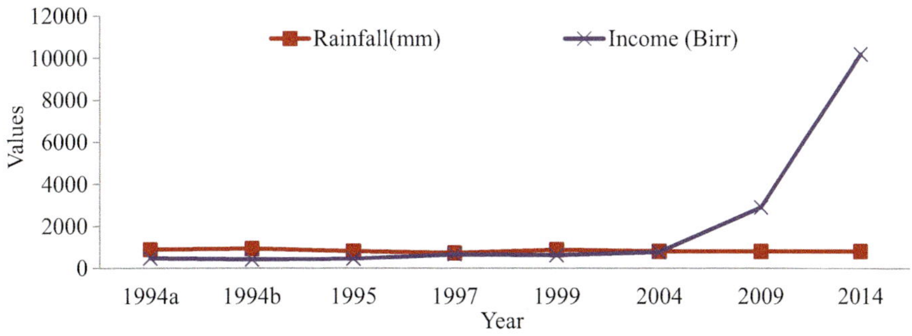

Fig. 1 Rainfall variability and crop income in central highland of Ethiopia. (Source: Ethiopian National Meteorology Agency 2015, and ERHS)

Table 2 Breusch and Pagan LM test

Type of variable	Coefficient	Standard error	P-value
LnCrop income	7.264	2.695	0.000
Overall error component (e)	4.600	2.144	
Fixed or random error component (u)	1.090	1.044	

Note: Ln represents natural logarithm

Table 3 Hausman test result for fixed and random effect model selection

Variable	Coefficient Fixed (FE)	Random (RE)	Difference (FE-RE)	sqrt(diag(V_FE-V_RE)) stan.err.
Ln rain	32.914	−4.873	37.787	3.560
Ln temperature	85.612	−1.787	87.399	8.695
Ln rain X T°	−10.731	1.365	−12.095	1.138
Harvest time rain	−0.518	−0.533	0.016	0.018
Growing time rain	0.748	0.891	−0.143	0.018
Summer rain onset	0.308	0.227	0.081	0.015
(Fe-RE)'[(V_FE-V_RE) ^ (−1)](FE-RE) = $\chi(6)$	195.19	P-value	0.000 accept fixed effect	

Note: Ln represents natural logarithm

factors, climate variability has different effect on smallholder's crop income across time. Once the correlation with individual effects has modeled through an auxiliary regression, random effects specification provides consistent estimators in estimating time-invariant variables (Debarsy 2012). The choice between fixed and random effect model for estimating the panel data dependence on different circumstances like the way of data collection, type of population from which samples drawn, and purpose of analysis (Franco 2001). The author argued that if an individual has an interest in assessing specific difference among small number of regions or countries, it is better to use fixed effect model. Given those ordinary means of identifying method of regressing panel data, this research conducted all the required diagnostic tests of identifying the appropriate model estimating the unbalanced panel data. The Breusch and Pagan test of identifying between random effect and pooled OLS shows that there is significant evidence (P-value = 0.000) to accept the former model of regression (Table 2).

After rejecting pooled OLS, the research just proceed to the next test (Hausman specification test) to identify between random and fixed effect models. If the null hypothesis, *the time-invariant individual factors are uncorrelated with other regressors*, was accepted, then the random effect model would be preferred against its counterpart, the fixed effect. The model selection test in Table 3 considered climate variability elements as an explanatory, and the test result verified to accept fixed effect model. The test result confirmed that time-invariant covariates correlated with the other explanatories, and there was no difference on crop income based on household-related time-invariant dummies.

Table 4 Effect of climate variability on crop income

Description	Coefficients	Robust standard error
Ln rainfall	−11.988***	1.165
Ln temperature	−18.268***	1.444
Ln (rain X temperature)	3.360***	0.342
Shocks	−0.169*	0.095
Idiosyncratic shocks	0.066	0.102
Ln TLU	0.708***	0.083
Pests	−0.367***	0.091
Rain harvest time	−0.203**	0.095
Enough rain at beginning	0.243***	0.092
Off/nonfarm activities	−0.619***	0.084
Credit access	0.388***	0.101
Ln household size	−0.695***	0.151
Ln draft power	0.268***	0.077
Ln cropland	0.178***	0.037
Constant	72.364***	5.173
Sigma_u	1.450	P-value 0.000
Sigma_e	1.985	Corr(u_i, Xb) 0.038
R^2 – overall	0.256	R^2 – within 0.221
Rho(ρ) 0.348 (fraction of variance due to u_i)		R^2 – between 0.333

Note: ***, **, and * refer to 1%, 5%, and 10% level of significance, respectively Ln represents natural logarithm

Regression Result and Discussion

Table 4 shows that more than 34.80% of the variation in the dependent variable was due to panel effect (fraction of variance due to $u_{\_i}$). Interaction between the two dominant climate variability elements (cropping season rainfall and temperature) and crop income was negative and significant as to the model result. This interaction indicated that the partial effect of cropping season rainfall and temperature was negative, which is a finding similar to Félix and Romuald (2014). Those authors argued that rainfall pattern variability is a source of uncertainty and agricultural income fluctuation, which result in significant economic growth reduction for developing countries. Ajay and Pritee (2013) identified positive and statistically significant interaction between rainfall and agricultural produce value, which is a finding contrary to Table 4. Their finding implied that rainfall increment would directly result in drastic quantity supply reduction, which has a positive effect on produce value if smallholders do have marketable quantity. This circumstance could not be acceptable for Ethiopian smallholders who have no marketable amount to collect benefits of crop price increment when there is a supply shortage. A 1% cropping season temperature or rainfall increment, respectively, resulted in 11.99% and 18.27% crop income reduction for a given household as to the model result (Table 4). The covariate generated based on interaction of rainfall and temperature had positive effect on crop income.

Other variables such as flood, frost, and storm affected smallholders' crop income negatively. Those covariates have a devastating effect on crop production that may result in complete loss of produces. Covariates related to rainfall inconsistency like timely onset of the demanded summer rainfall at sowing time significantly affected smallholders' crop income. Availability of sufficient rainfall at the beginning and consistent supply throughout crop growing season may increase the marketable amount of crops through improving the quantity produced. Earlier works conducted in Africa were in support of Table 4, for instance, Ole et al. (2011) found that rain-fed agriculture income reduction largely attributed to rainfall insufficiency in many West African countries. Kaul and Ram (2009) argued that excessive rain and extreme temperature variation adversely affect crop productivity and thereby influence smallholders' income. Some extreme events are associated with very poor harvests of cash crops, which severely affect income of the farming sector (Marc et al. 2013). The other variable sourced from temperature and rainfall inconsistency was pest infestation, which strongly affected crop income (0.15% reduction) (Table 4). The pest commonly known as *faki* highly affects crop production of households in central highlands, and crops planted in areas infested with this pest produce nothing (Dercon and Hoddinott 2011). Drought was the main causal factor triggering poverty through changing the crop production (Yue et al. 2013). Smallholders across the tropics already faced numerous risks including pest and disease outbreaks and extreme weather events that frequently reduced household agricultural income (Celia et al. 2014).

Ajay and Pritee (2013) argued that agricultural labor and mechanization negatively affect the produce value in India because of excess production; the overall crop income effect is positive through increasing the marketable amount. Similarly, draft power allocated for agriculture, which is a critical input for smallholders' crop production, had positive and significant effect on crop income in the study area (Table 4). For a given smallholder, 1% increment in the draft power would increase the crop income by 0.34%. Household member's participation in off-farm income had negative and significant effect on crop income, which was a finding similar to Tesfaye et al. (2017): "each additional earning from off-farm work has negative and statistically significant effect on marketed surplus." The negative effect of off-farm income participation may be due to labor competition between crop production and other activities that would result in crop produce reduction. Having easy access to savings and credit facilities had significant and positive effect on households' crop income in the study area. Smallholders' capacity to purchase agricultural inputs could be easy if they have access to credit, which could enhance production and marketable amount.

Large family size has negative effect on household income in rural Nigeria as of enormous consumption of the produce (Omideyi 1988). Similarly, Tessema (2015) argued that the *number of dependency ratio significantly reduced rural household's income*. Smallholders' family size has two contradictory effects on crop income: the consumption effect that reduced crop income due to large home consumption demand and the production effect that increases crop income through excess supply of marketable amount. Thus, the overall effect of family size increment on crop income always depends on the magnitude of the two effects within the household. Based on Table 4, the variable had negative effect on crop income, which means the consumption effect outweighs the production in the study area. The reduction in household's consumption demand could shift produces to be marketable type that guarantees better crop income. The negative effect of this covariate on crop income implied that large family size had more of consumption effect but not production for smallholders in central highland farming system of Ethiopia.

Conclusions and Recommendations

The study used unbalanced panel data and employed fixed-effect method of data analysis to examine the effect of climate variability on crop income of smallholders in central highland farming system of Ethiopia.

Successive partial increment in either cropping season rainfall or temperature adversely affects smallholders' crop income in the farming system. Similarly, inconsistency in rainfall availability at different stages of the crop production has significant and negative effect on crop income. The ongoing climate variability creates favorable condition for pest and disease outbreak, which have significant and negative effect on crop income as to the model result.

Parallel increment of cropping season rainfall and temperature that has positive effect on smallholders' crop income implies that counterbalance effect of rainfall on temperature increment is strong in the farming system.

Smallholder's family size negatively affects the crop income that may be because of lower marketable amount after the larger demand for home consumption.

Production inputs including access to credit, draft power, and cropland, which enhance smallholder's crop productivity, have positive and significant effect on crop income of the farm households in the study area.

Shocks sourced from climate variability elements had significant effect on crop income of smallholders in the study area.

Based on the above findings, the research forward the following recommendations:

- The concerned party should have coordinated efforts to cope up with the rainfall inconsistency problems that would result in significant crop income variability.
- There should be readiness and awareness of the concerned party to support smallholders when there is unexpected rainfall at harvesting time that would drastically reduce smallholders' crop income in the study area.
- The crop production of smallholders should be protected from pests and disease to reduce negative effect of the problem on the crop income. In doing this, biological methods may be advisable since they are easily affordable for smallholders who have lower purchasing power.

- Rural smallholders should be aware on how to allocate labor to avoid competition between crop production and off-farm participation that significantly reduce crop income.
- Smallholders should have easy access to credit and draft power to improve their crop production practices, which could directly increase smallholders' crop income.

References

Ajay K, Pritee S (2013) Impact of climate variation on agricultural productivity and food security in Rural India discussion paper no. 2013-43. http://www.economicsejournal.org/economics/discussionpapers/2013-43\

Arjan R, Mark de B, Minna K, Vincent L, Nico P (2011) Adaptation to climate variability: the role of past experience and institutions. Conference paper: conference on the economics of adaptation to climate change in low-income countries. Washington DC

Blein R, Longo R (2009) Food price volatility–how to help smallholder farmers manage risk and uncertainty? Discussion paper prepared for the Round Table organized during the Thirty-second session of IFAD's Governing Council

Celia AH, Zo Lalaina R, Nalini RS, Radhika D, Hery R, Rivo HR, Haingo R, James LM (2014) Extreme vulnerability of smallholder farmers to agricultural risks and climate change in Madagascar. Philos Trans R Soc Biol Sci 369:20130089. https://doi.org/10.1098/rstb.2013.0089

Chérif S, Leal Filho W, Azeiteiro UM (2017) The role of farmers' perceptions in coping with climate change in Sub-Saharan Africa. Int J Glob Warm 12(3/4):483

Cline W (2007) Global warming and agriculture: impact estimates by country. Peterson Institute for International Economics, Washington, DC, pp 1–5

Debarsy N (2012) The Mundlak approach in the spatial Durbin panel data model. Spat Econ Anal 7:109–131

Dell M, Jones BF, Olken BA (2008) Climate change and economic growth: evidence from the last half century. Working paper 14132, NBER working paper series. National Bureau of Economic Research, Cambridge, MA

Dercon S, Hoddinott J (2004) The Ethiopian rural household surveys: introduction. International food policy research institute. Addis Ababa. http://www.tammyschirle.org/teaching/Fall2007/IP602/ifpri_ethiopia/OverviewandAcknowledgementsandDataDescription.pdf

Dercon S, Hoddinott J (2011) The Ethiopian rural household surveys 1989–2009: introduction. A report from international food policy research institute. Addis Ababa. https://www.datafirst.uct.ac.za/dataportal/index.php/catalog/244/download/4203

Deressa TT (2007) Measuring the economic impact of climate change on Ethiopian agriculture: Ricardian approach. Policy research working paper, no. 4342. World Bank, Washington, DC. © World Bank. License: CC BY 3.0 IGO https://openknowledge.worldbank.org/handle/10986/7290

Di Falco S, Mahmud Y, Gunnar K, Claudia R (2011) Estimating the impact of climate change on agriculture in low-income countries: household level evidence from the Nile Basin, Ethiopia. Environ Resour Econ 52(4):457–478

ENMA (Ethiopian National Meteorology Agency) (2007) Annual meteorological data. Annual data report. Addis Ababa, Ethiopia, pp 35–45

ENMA (Ethiopian National Meteorology Agency) (2015) Annual meteorological data. Annual data report. Addis Ababa, Ethiopia

FAO (Food and Agriculture Organization) (2008) The state of food insecurity in the world. Food and Agriculture Organization of the United Nations, Rome, pp 20–24

FAO (Food and Agriculture Organization) (2009) Food security and agricultural mitigation in developing countries: options for capturing synergies. FAO, Rome, pp 12–15
Félix B, Romuald KS (2014) Climatic variability and food security in developing countries. Archive ouverte en Sciences de l'Homme et de la Société. Preprints, Working papers, 014.05. 2015
Fischer G, Shah M, Tubiello FN, vanVelthuizen H (2005) Socio-economic and climate change impacts on agriculture: an integrated assessment, 1990–2080. Philos Trans R Soc Biol Sci 360:2067–2083
Franco P (2001) Econometrics. Wiley, Chichester, pp 236–245
IPCC (Intergovernmental Panel on Climate Change) (2001) Climate change: impacts, adaptation and vulnerability. Report of the Working Group II. Cambridge University Press, Cambridge, pp 12–14
IPCC (Intergovernmental Panel on Climate Change) (2007) Agriculture in climate change 2007: mitigation. Working Group III Contribution to the Fourth Assessment Report of the IPCC. Cambridge University Press, Cambridge, pp 13–16
IPCC (Intergovernmental Panel on Climate Change) (2012) Managing the risks of extreme events and disasters to advance climate change adaptation. Cambridge University Press, Cambridge, UK, p 582. Available from June 2012
IPCC (Intergovernmental Panel on Climate Change) (2014) Impacts, adaptation, and vulnerability, vol 8. Cambridge University Press, Cambridge, p 10
Jain S (2007) An empirical economic assessment of impacts of climate change on agriculture in Zambia. Policy research working paper, no. 4291. World Bank, Washington, DC. © World Bank. https://openknowledge.worldbank.org/handle/10986/7478. License: CC BY 3.0 IGO
John MA (2009) Agriculture and the food system: adaptation to climate change. An initiative of the climate policy program at RFF. http://www.rff.org/.../Download/RFF-Rpt-Adaptation-Antle.pdf
Kaul S, Ram G (2009) Impact of global warming on production of sorghum in India. Agric Situat India 66(5):253–256. https://www.cabdirect.org/cabdirect/abstract/20103205344. ref.6
Leal Filho W (2011) The economic, social and political aspects of climate change. Springer, Berlin. ISBN: 978-3-642-14775-3
Marc M, Safietou S, Wolfram L (2013) Impact of changing seasonal rainfall patterns on rainy-season crop production in the Guinea Savannah of West Africa. https://ageconsearch.umn.edu/record/150412/files/AAEAPaper_RainySeasonCrops_MM130603
Mathilde M, Zaneta K (2014) Climate Variability and Migration: evidence from Tanzania. Fondation pour les études et recherchessur le développement international. Development policy working paper, pp 31–33. https://pdfs.semanticscholar.org/5cda/9293d1b14af8ac1dec505206735ba5003a6c.pdf
Mendelsohn R, Nordhaus W, Shaw D (1994) The impact of global warming on agriculture: a Ricardian analysis. Am Econ Rev 84:753–771
Nhemachena C, Hassan R, Kurukulasuriya P (2010) Considered net farm revenues measuring the economic impact of climate change on African agricultural production systems. Clim Change Econ 1(1):33–55
Ole M, Cheikh M, Anette R, Lorenzo G, Eric FL, Sarah D, Malicki Z, Kjeld R, Drissa D, Bruno B, Ibrahim BM, Awa D, Jonas ØN, Inge S (2011) Adaptation strategies and climate vulnerability in the Sudano-Sahelian region of West Africa. Atmos Sci Lett 12:104–108. Published online in Wiley Online Library
Omideyi AK (1988) Family size and productivity of rural households in Nigeria. https://www.ncbi.nlm.nih.gov/pubmed/12315558
Paul AL, Fisher M, Weber B (2011) Do rainfall conditions push or pull rural migrants? Evidence from Malawi. J Agric Econ 43:191–204
Rosenberg NJ, Brown RA, Izaurralde RC, Thomson AM (2003) Integrated assessment of Hadley Centre (HadCM2) climate change projections in agricultural productivity and irrigation water supply in the conterminous United States. Agric For Meteorol 117(1–2):73–96
Sara L (2012) Climate change, weather variability and food consumption: a multidisciplinary study of Rural Uganda. Catholic University of the Sacred Heart, Piacenza

Schlenker W, Lobell D (2010) Robust negative impacts of climate change on African agriculture. Environ Res Lett. http://iopscience.iop.org/article/10.1088/1748-9326/5/1/014010/pdf

Susanne U (2012) The impact of increasing rice and maize prices on household income in north western Vietnam. Scientific world. University of Hohenheim Stuttgart, Germany. Int J Rural Dev 52(2). https://www.rural21.com/uploads/media/rural2012_01-S41-43.pdf

Tesfaye BW, Thomas H, Yves S (2017) Effect of off-farm income on smallholder commercialization: panel evidence from rural households in Ethiopia. The Journal of the International Association of Agricultural Economists 48(2). https://onlinelibrary.wiley.com/doi/abs/10.1111/agec.12327

Tessema U (2015) The determinants of agricultural productivity and rural household income in Ethiopia. Ethiop J Econ 24(2):30–35

UNDP (2007) Climate shocks: risk and vulnerability in an unequal world. In: Human development report 2007/2008. Fighting climate change: human solidarity in a divided world. United Nations Development Program. Palgrave Macmillan, New York, pp 73–107

Yue L, Conway D, Yanjuan W, Qingzhu G, Rothausen S, Wei X, Hui J, Erda L (2013) Rural livelihoods and climate variability in Ningxia, Northwest China. Clim Chang 119:891–904. https://doi.org/10.1007/s10584-013-0765-9. Open access at Springerlink. com

Sustainability of Climate Change Adaptation Measures in Rivers State, South-South, Nigeria

33

A. Henri-Ukoha and O. M. Adesope

Contents

Abstract

Climate change is a multidimensional and sectorial issue all over the world. The incidence of climate change variations has, in recent times, affected global food production, supply, and farm income of the farmers. This study investigated the sustainability of climate change adaptation strategies among Maize-based farmers in Etche Local Government Area of Rivers State with the following objectives: to identify the sustainable climate change adaptation strategies as well as to ascertain the sustainability index of those climate change adaptation strategies practiced in the study area. Data were obtained through the administration of questionnaire to the maize-based farmers. Multistage sampling was used to select 64 maize-based farmers in the study area. Data were analyzed using descriptive statistical tools such as the mean, frequency count, percentages, and sustainability index. Results showed the sustainable and unsustainable climate change adaptation strategies adopted by the farmers. Results further revealed that the climate change adaptation strategies practiced in the area were sustainable. An enlightenment campaign should be organized in the state to educate and expose more

A. Henri-Ukoha (✉) · O. M. Adesope
Department of Agricultural Economics and Extension, University of Port Harcourt, Choba/Port Harcourt, Nigeria
e-mail: adanna.henri-ukoha@uniport.edu.ng

W. Leal Filho (ed.), *Handbook of Climate Change Resilience*,
https://doi.org/10.1007/978-3-319-93336-8_8

farmers to the sustainable adaptation strategies such that they would withstand the devastating effects of the changing climate in the study area.

Keywords
Sustainability index · Adaptation strategies · Greenhouse gases · Maize-based farmers

Introduction

Global warming occurs as a result of accumulation of heat-trapping gases known as greenhouse gases basically caused by natural and anthropogenic activities (United Nations Framework Convention on Climate Change 2007; Crosson 1997). This causes erratic and life-threatening weather events which affect availability of soil water, soil erosion, droughts, floods, sea level rises, and high incidence of diseases and pest infestations (Adejuwon 2004; Zoellick and Robert 2009) resulting in low and unpredictable crop yields which increases the vulnerability of farmers (Ziervogel et al. 2006). Climate change is associated with increasing demand for food, feed, fiber, and fuel and has probable irreversible damage to the natural resource base, with significant consequences for food insecurity (IAASTD 2008). Onyeneke (2010) explained that Sub-Saharan Africa does not make a significant contribution to the changing climate, but the region is mostly affected. Studies have shown that climate change has a negative impact on African agriculture (Pearce et al. 1996; McCarthy et al. 2001; Onyeneke 2010), and Nigerian farmers depend largely on rainfed agriculture (Onyeneke 2010). Adaptation is one of the policy options for reducing the negative influence of the changing climate (Adger et al. 2003; Kurukulasuriya and Mendelsohn 2006).

"Adaptation to climate change refers to adjustment in natural or human systems in response to actual or expected climatic stimuli or their effects, which moderates harm or exploits beneficial opportunities" (Intergovernmental Panel on Climate Change 2001). Common adaptation methods in agriculture include the use of new crop varieties and livestock species that are better suited to drier conditions, irrigation, crop diversification, adoption of mixed crop and livestock farming systems, and changing of planting dates (Bradshaw et al. 2004; Kurukulasuriya and Mendelsohn 2006; Nhemachena and Hassan 2007). Others include portfolio diversification, soil conservation, changing of tillage operations, planting of trees, and irrigation (Onyeneke 2010).

Though farmers reported that their adaptation actions are driven by climatic factor (Maddison 2006; Nhemachena and Hassan 2007), they are often not sure of the sustainability of the adaptation strategies. Sustainable agricultural production systems involve those approaches to food production that ensure continuous increases in productivity without jeopardizing the needs of future generations to provide for themselves (Field and Field 2009). It involves production practices that ensure environmental conservation and no or minimal disturbance to the natural eco support system, hence protects the potentials of the natural regeneration of the flora and fauna (Nwaiwu et al. 2014).

"Sustainable development is humanity's ability to survive by means of the rational use of renewable resources by refraining from disrupting the ecosystem or over-exploiting natural resource and by refraining from activities that destroy cultures or societies and instead allow them to reach their potential" (Ballara 1991). Broadly speaking, sustainable development has to do with environmental protection among other factors (Nwaiwu et al. 2014; Harwood 1987). Agricultural sustainability involves production activities that minimize the use of external inputs and maximize the use of internal inputs, which already exist on the farm (Liebhardt 1987). According to Nwaiwu et al. (2010a, b), "External inputs as those inputs that are artificially manufactured, very capital intensive in procurement, usually purchased, depends on very high skill and technology to produce and use and not readily available to the resource poor farmers; while internal inputs refer to those inputs that are naturally endowed, relatively very cheap to produce, do not require high skill to use, depends on indigenous technology, very readily available and affordable to the farmers." This is vital as some of the adaptation practices may cause a long-term damage to the ecosystem or the environment. Moreover, Agyei (2016) pointed out low-cost strategy, economic equity, flexibility to precipitation, and temperature as ways of determining sustainable climate change adaptation strategies. According to them climate adaptation strategy can be sustainable if it is less costly to establish and flexible to places and seasons. Hopkins (2014) further stressed that for adaptation strategies to be sustainable, they should consider the environment, economy and society.

Several recent studies have been conducted on sustainable agriculture and climate change (Ching 2008; Pimbert 1999; Wall and Smit 2005; Nwaiwu et al. 2014; Henri-Ukoha and Osuji 2017).

Furthermore, Agyei (2016) studied the sustainability of climate change adaptation strategies with experiences from Eastern Ghana; Hopkins (2014) looked at the sustainability of climate change adaptation strategies in New Zealand's ski industry from a range of stakeholder perceptions, but little or none has dwelt on the sustainability of the adaptation strategies to the changing climate in the study area especially as regards maize farming, leaving a void in research which this study intends to fill.

This study is therefore geared toward ascertaining the sustainability of those climate change adaptation strategies used by farmers to adapt to the devastating effects of global warming. This will promote and facilitate the adoption of ecosystem and environmentally friendly sustainable adaptation options which will ultimately favor the farmers and the economy.

Materials and Method

The study was conducted in Etche Local Government Area of Rivers State – Etche, a subsistent agrarian ethnic nation in Rivers State and a typical rainforest with rainfall between 160 and 236 cm. Temperature ranges between 24°C and 38°C, with high humidity of up to 90% during the wet season. The people are mostly farmers and traders; the major crops grown in this local government area are cassava, maize, yam, plantain, etc.

Multistage sampling was used to select maize-based farmers. In the first stage, random sampling was used to select four communities from the local government areas. In the second stage, random sampling was used to select 2 villages from the communities, and in the third stage, 8 maize-based farmers were randomly selected from each village, giving a total of 64 respondents. The selection was done using a list compiled by the extension agent resident in the areas.

Primary data were collected from the respondents using structured questionnaire and scheduled interviews. Data were analyzed using descriptive statistics such as mean, frequency count, percentages cum Likert scales, and sustainability index.

Sustainability index is expressed as:

$$\mathrm{Ss} = \frac{\mathrm{Nsa}}{\mathrm{TNa}} \times \frac{100}{1}$$

(Liebhardt 1987; Nwaiwu et al. 2010a, b, 2013; Henri-Ukoha and Osuji 2017), where:

Ss = sustainability index or sustainable adaptation practice (%)
Nsa = number of sustainable adaptation practices used by a farmer
TNa = total number of adaptation strategies available to all farmers in the study area

Finally, the mean sustainability index or level for the study area was determined to conclude whether the strategy is unsustainable or sustainable as thus:

$$\mathrm{Ss} = \frac{\frac{\mathrm{Nsa}}{\mathrm{TNa}} \times \frac{100}{1}}{\mathrm{N}}$$

(Kadiri et al. 2014).

The criteria of sustainability are.

$\leq$40% unsustainable adaptation practice
>40% $\leq$ 50% sustainable adaptation practice
>50% highly sustainable adaptation practice (Liebhardt 1987; Nwaiwu et al. 2010a, b, 2013; Kadiri et al. 2014)

where:

n = sample size of farmers

Results

Sustainable Adaptation Strategies Practiced by Maize-Based Farmers in the Area

Sustainability Level of the Maize-Based Farmers

Table 1 shows that the maize-based farmers practiced a total of sixteen (16) climate change adaptation practices. The highest sustainable climate change adaptation strategy was relocation to a new farmland (84.38%). This was followed by use of improved tillage practices (81.25%), planting at different locations (67.19%), weed management (62.50%), use of improved crop residue (54.69%), irrigation practices (50.00%), and the least was use of pest- and disease-resistant varieties (18.75%). This is in line with the findings of Agyei (2016) who identified short rotation, mixed species cropping, farming at several locations, and drought-tolerant crop varieties as sustainable adaptation strategies used by farmers in Eastern Ghana.

Relocation of farmland could be a sustainable way to adapt to the recent high incidence of flood and drought arising from the high degree of temperature and rainfall variability as a result of global warming in the study area. They move to a different location entirely that is not prone to flood. They may also move to areas that are close to water bodies where they have access to water for irrigation. Entzinger and Scholten (2015) observed that farmers involved in Mekong River Delta project used relocation of projects as an adaptation strategy to environmental stress in Ho

Table 1 Sustainable adaptation strategies practiced by maize-based farmers in the study area

Sustainable adaptation practices	Frequency[a]	Percentage
Change in planting dates	27	42.19
Use of improved variety	25	39.06
Cultivation of early maturing varieties	18	28.13
Use of crop rotation	35	54.69
Mixed cropping	15	23.44
Irrigation practices	32	50.00
Improved agro-forestry practices	22	34.38
Relocation to new farmland	54	84.38
Diversification	17	26.56
Mulching	20	31.25
Use of pest and disease-resistant varieties	12	18.75
Improved crop residue	35	54.69
Weed management	40	62.50
Improved disease control methods	24	37.50
Use of improved tillage practices	52	81.25
Planting at different locations	43	67.19
Use of chemical fertilizers	24	37.50

Source: Field Survey (2017)
[a]Multiple responses recorded

Table 2 The distribution of maize-based farmers according to the sustainability level in the study area

Sustainability level (%)	Frequency	Percentage (%)
1.00–20.00	5	7.81
21.00–40.00	8	12.50
41.00–60.00	12	18.75
61.00–80.00	22	34.38
81.00–100.00	17	26.56
Total	64	100

Mean = 62.38%
Source: Field Survey 2017

Chi Minh City in Vietnam. In a similar vein, National Climate Change Adaptation Research Plans (2012) advocated the relocation of farmlands as a way to adapt to the changing climate in broadacre farming in South Australia.

From the results also, improved tillage practice is a sustainable climate change adaptation measure. They could adopt practices that would reduce the emission of greenhouse gases. Reduced tillage leads to minimal soil disturbance with little release of carbon into the atmosphere which encourages the less buildup of greenhouse gases.

Conservation tillage methods such as no-till and/or minimum tillage help to protect soil quality and productivity (Govaerts et al. 2009 as cited in Jorge et al. 2013). This ultimately helps the soil in adapting to adverse climatic variables. Minimum tillage used as a climate change adaptation strategy among smallholder farmers in Zambia helped in improving crop productivity levels in smallholder farming systems (Kuntashula and Chabala 2014).

The results in Table 2 showed that the highest sustainability level of maize-based farmers was 61.00–80.00 (34.38%). This was followed by 81.00–100.00 (26.56%), then 41.00–60.00 (18.75%), 21.00–40.00 had (12.50%), and the least was 1.00–20.00 (7.81%). The mean sustainability level of the maize-based farmers was 62.38%. This implied that maize-based farmers in the study area practiced sustainable climate change adaptation strategies. This indicated that farmers use more of climate change adaptation strategies that are not harmful to the environment. This is evident in the adaptation strategies used by the farmers as shown in Table 1. Such adaptation strategies include relocation to a new farmland, use of improved tillage practices, planting at different locations, weed management, use of improved crop residue, and irrigation practices. Sustainable agricultural practices, such as no-tillage, and technologies that mitigate extreme climatic conditions help protect the soil (CropLife International 2015). CropLife International (2015) also stressed that minimal tillage helps to conserve the soil leading to a reduction in greenhouse gas emissions by 90–95% or more compared to conventional tillage practices.

However, there are certain adaptation strategies that damage the soil. These include industrial, chemical-intensive climate change adaptation practices; they degrade soils and destroy resources that are critical to storing carbon, such as forests and other vegetation. For instance, fertilizer application accounts for 0.6–1.2% of the world's total greenhouse gases (Bellarby et al. 2008). Sustain (2013) reiterated that industrialized agriculture leads to such environmental damages as soil erosion and water pollution and damages the wildlife habitats through the use of pesticides and other intensive farming techniques. Similarly, Devarinti (2016) explained that chemical farming practice and indiscriminate use of chemical fertilizers and pesticides are not sustainable adaptation options as they destroy the beneficial soil microflora which changes the soil nature contributing to the high cost of crop production.

Conclusion and Recommendations

The study showed that maize-based farmers in the study area used both sustainable and unsustainable climate change adaptation strategies as evident in the results. However, sustainable adaptation practices were used more than the unsustainable adaptation practices. This is evident in the very high adaptation index and suggests the use of climate change adaptation practices that are not harmful to the environment in the study area. This also implies that those adaptation measures will not lead to the emission of greenhouse gases and other outcomes that will worsen the effect of the changing climate in the study area. Therefore, farmers in the area should be encouraged to upscale these sustainable climate change adaptation strategies so as to encourage others that are yet to adopt these measures since the result is not perfect. Hence, the study recommended that various workshops be facilitated by the government to enhance public awareness and in-depth knowledge on sustainable adaptation strategies to climate change for upscaling of the sustainable climate change adaptation strategies in the study area.

References

Adejuwon SA (2004) Impact of climate variability and climatechange on crop yield in Nigeria. Contributed Paper to Stakeholders Workshop on Assessment of Impact and Adaptation to Climate Change (AIACC): 2–8 held in Columbia

Adger WN, Huq S, Brown K, Conway D, Hulme M (2003) Adaptation to climate change in the developing world. Progress in development studies no 3

Agyei FK (2016) Sustainability of climate change adaptation strategies: experiences from eastern Ghana. Environ Manag Sustain Dev 5(2):84–103

Ballara M (1991) Women and literacy. Women and world development series of the joint. UN/NGO Groups on Women and Development London. Zed Books Ltd. Centre Obafemi Awolowo University, Ile Ife

Bellarby J, Foereid B, Hastings A, Smith P (2008) Cool farming: climate impacts of agriculture and mitigation potential. Greenpeace International, Amsterdam. CBN (Central Bank of Nigeria) (2005). Annual report and statement of accounts

Bradshaw B, Dolan H, Smith B (2004) Farm-level adaptation to climatic variability and change: crop diversification in the Canadian prairies. Clim Chang 67:119–141

Ching LL (2008) National conference on "Sustainable agriculture: moving from grassroots initiatives to mainstream policies". Organised by the consumers' association of penang and held in Petaling Jaya on 24 July 2008. Internet paper accessed on 11/06/2013 @ www.twnside.org.sg; www.biosafety-info.net

CropLife International (2015) Top 5 environmentally friendly farm trends. https://croplife.org/news/top-5-environmentally-friendly-farm-trends/
Crosson P (1997) Impacts of climate change on agriculture. Climate issues brief no. 4. Resources for the Future, Washington, DC
Devarinti SR (2016) Natural farming: eco-friendly and sustainable? Agrotechnology 5(2):147
Entzinger H, Scholten P (2015) Policy brief series volume 1 (6): relocation as an adaptation strategy to environmental stress. Migration, Environment and Climate Change
Field BC, Field MK (2009) Environmental economics: an introduction. McGraw-Hill Publishers, Singapore
Harwood RR (1987) Low input technologies for sustainable agricultural systems. In: Ruttan VW, Pray CE (eds) Policy for agricultural research boulder colorado. Westview Press, Boulder
Henri-Ukoha A, Osuji (2017) Determinants of arable crop farmers' use-levels of sustainable soil management techniques in Imo State, Nigeria. Nigerian J Agric Food Environ 13(2):163–168
Hopkins D (2014) The sustainability of climate change adaptation strategies in New Zealand's ski industry: a range of stakeholder perceptions. J Sustain Tour 22(1):107
IAASTD (2008). International Assessment of Agricultural Knowledge, Science Technology for Development. www.agassessment.org
IPCC (2001) Climate change 2001: impacts, Vulnerability and Adaptation. Contribution of Working Group III to the Third Assessment Report on the Intergovernmental Panel on Climate Change. Cambridge University Press, Cambridge
Jorge AD, Mark AN, Charles WR (2013) Advances in Agronomy in Conservation practices for climate change adaptation. Adv Agron 121:47–115
Kadiri F, Eze C, Orebiyi S, Nwaiwu I (2014) Determinants of sustainability of Paddy Rice production in Niger Delta region of Nigeria. J Econ Sustain Dev 5(14):43–48
Kuntashula E, Chabala LM (2014) Impact of minimum tillage and crop rotation as climate change adaptation strategies on farmer welfare in smallholder farming Systems of Zambia. J Sustain Dev 7:95–110
Kurukulasuriya P, Mendelsohn R (2006) A Ricardian analysis of the impact of climate change on African crop land. CEEPA discussion paper no. 8. Centre for Environmental Economics and Policy in Africa. University of Pretoria, Pretoria
Liebhardt WC (1987) Low input sustainable agricultural production systems. In: Paper prepared for NAS, CSAN joint workshop on agricultural development and environmental research, Ceske Budejovic, 6–16th April
Maddison D (2006) The perception of and adaptation to climate change in Africa. CEEPA. Discussion paper no. 10. Centre for Environmental Economics and Policy in Africa. University of Pretoria, Pretoria
McCarthy J, Canziani OF, Leary NA, Dokken DJ, White C (2001) Climate change 2001: impacts, adaptation, and vulnerability. Contribution of Working Group II to the third assessment report of the intergovernmental panel on climate change. Cambridge University Press, Cambridge
National Climate Change Adaptation Research Plans (2012) Adapting agriculture to climate change. Policy Brief 4. https://www.nccarf.edu.au/sites/default/files/attached_files_publications/AGRICULTURE_A4Printable.pdf
Nhemachena C, Hassan R (2007) Micro-level analysis of farmers' adaptation to climate change in southern Africa, International food policy research institute (IFPRI) discussion paper no. 00714. Environment and Production Technology Division, IFPRI, Washington, DC
Nwaiwu IU, Odii MACA, Ohajianya DO, Nwosu FO, Oguoma NNO, Ben-Chendo NG, Amaechi ECC (2010a) A comparative analysis of returns from cassava farms under externa and interna input use in Imo State, Nigeria. World Rural Obs 2(1):56–60. Marsland Press, Brooklyn, New York
Nwaiwu IU, Ohajianya DO, Ibekwe UC, Amaechi ECC, Emenyonu CA, Onyemauwa CS, Henri-Ukoha A, Kadiri FA (2010b) Comparative analysis of the allocative efficiency of cassava producers that use external and internal inputs in Imo State, Nigeria. Academia Arena 2(11):96–108. Marsland Press, Richmond Hill, New York. http://www.sciencepub.net/academia

Nwaiwu IU, Ohajianya DO, Orebiyi JS, Eze CC (2013) Determinants of agricultural sustainability in South Eastern Nigeria. The climate change debacle. Glob J Agric Res 1(2):1–13

Nwaiwu IUO, Orebiyi JS, Ohajianya DO, Ibekwe UC, Onyeagocha SUO, Henri-Ukoha A, Osuji MN, Tasie CM (2014) The effects of climate change on agricultural sustainability in Southeast Nigeria – implications for food security. Asian J Agric Ext Econ Soc 3(1):23–36

Onyeneke RU (2010) Climate change and crop farmers' adaptation measures in the southeast rainforest zone of Nigeria. Unpublished M.Sc. Thesis submitted to the Department of Agricultural Economics, Imo State University Owerri, 112 pp

Pearce D, Cline W, Achanta A, Fankhauser S, Pachauri R, Tol R, Vellinga P (1996) The social costs of climate change: greenhouse damage and benefits of control. In: Bruce J, Lee H, Haites E (eds) Climate change 1995: economic and social dimensions of climate change. Cambridge University Press, Cambridge

Pimbert M (1999) Sustaining the multiple functions of agricultural biodiversity. In: FAO background paper series of the conference on the multifunctional character of agriculture and land, The Netherlands, September 1999

Sustain (2013) Environmentally-friendly farming, Sustain: the alliance for better food and farming. https://www.sustainweb.org/ethicaleats/friendlyfarming/

United Nations Framework Convention on Climate Change (2007) Climatic change impact, vulnerabilities and adaptation in developing countries UNFCCC Secretariat, Martin-Luther-King-Straat 8 53175, Bonn. www.un fccc.int

Wall E, Smit B (2005) Climate change and adaptation in light of sustainable agriculture. J Sustain Agric 27(1):113

Ziervogel G, Nyong A, Osman B, Conde C, Cortes S, Dowing T (2006) Climate variability and change: implications for household food security. Assessments of impacts and adaptations to climate change (AIACC) Working paper no. 20, January 2006. The AIACC Project Office, International START Secretariat, Washington, DC, pp 678–691

Zoellick S, Robert BA (2009) Climate smart future. The Nation Newspapers. Vintage Press Limited, Lagos, p 18

Climate and Climate Change Scenarios in the Indian Thar Region

34

Surendra Poonia and A. S. Rao

Contents

Abstract

The Thar Desert region of India, which extends in more than 2.0 lakh sq. km area, experiences variable rainfall from 100 mm to 450 mm in a year. Frequent drought, which occurs once in 2 or 3 years in the region, causes extreme stress to fauna due to limited seasonal grazing resources. Besides the xerophytic type of ecosystem, the fauna in the Thar Desert is subjected to extreme diurnal and seasonal variation in temperatures ranging as low as −5 °C in winter to a high of +49 °C in summer,

S. Poonia (✉)
ICAR-Central Arid Zone Research Institute, Jodhpur, Rajasthan, India
e-mail: poonia.surendra@gmail.com

A. S. Rao
ICAR-Central Research Institute for Dryland Agriculture, Hyderabad, India
e-mail: ashivrao@gmail.com

W. Leal Filho (ed.), *Handbook of Climate Change Resilience*,
https://doi.org/10.1007/978-3-319-93336-8_12

causing thermal stress to the fauna. The Intergovernmental Panel on Climate Change (IPCC 2007) projected for hotter days and warm nights and a reduction in rainfall in the Thar region by the twenty-first century. Such projected climate change results in shifting rainfall pattern, higher temperatures, and more demand for water and will be significant driver of biodiversity with changing life cycles, loss, migration, and invasion of new habitat in the Thar region. The present study on annual rainfall and temperature for the Thar region showed, by the end of the twenty-first century, an increase in temperature by +3.3 °C at Bikaner, +3.4 °C at Jaisalmer, +2.9 °C at Jodhpur, and +2.3 °C at Pali, if the present rate of warming continues. Similarly, though there was no significant rise (at 0.56 mm/year) in the annual rainfall of 12 arid districts of western Rajasthan, the annual rainfall is likely to be increased by +100 mm at Bikaner, +124 mm at Jaisalmer, −40 mm at Jodhpur, and +21 mm at Pali. The spatial and temporal variation in potential evapotranspiration requirement of the Thar region ranged from 2.1 mm/day to 12.2 mm/day and on an annual basis between 1500 mm and 2220 mm. Further, due to global warming, if the projected temperatures rise by 4 °C, by the end of the twenty-first century, water requirement in arid Rajasthan increases from the current level, by 12.9% for pearl millet and cluster bean, 12.8% for green gram, 13.2% for moth bean, 17.1% for wheat, and 19.9% for mustard.

Keywords
Climate change · Thar Desert region · Crop water requirement · Drought · Elevated air temperature · Rainfall

Climate Variability

The hot arid zone of India, which is located in northwest India, is spread in 3,17,000 km^2 and is characterized by limited and erratic nature of rainfall, extreme temperatures with large diurnal and seasonal variation, strong solar radiation, and wind regime resulting in the demand for high water requirements (Rao and Roy 2012). The weather conditions, even in normal years, for most part of the year remain too dry and inhospitable for human and livestock.

Rainfall and Its Distribution

The mean annual rainfall in arid Rajasthan varies from 185 mm at Jaisalmer to more than 467 mm at Sikar (Table 1 and Fig. 1). About 80–90% of the annual rainfall is received during southwest monsoon period from west southwesterly moving depressions which are the main source for *kharif* crops. The westerly moist wind from the Arabian Sea during May or early June move in east-northeasterly direction and cause rainfall, but such events occur with a lesser frequency of once in 2 or 3 years. The western disturbances during winter also cause precipitation particularly in the northern districts favoring *rabi* crops. The high interannual variability in rainfall

Table 1 Normal rainfall (mm) and its coefficient of variation (%) in arid Rajasthan

Month	Winter (Dec–Feb)	Summer (Mar–May)	Monsoon (Jun–Sept)	Post-monsoon (Oct–Nov)	Annual mean	Coefficient of variation
Barmer	6.8	12.5	242.6	4.8	266.7	63
Bikaner	13.5	23.5	243.1	8.0	290.6	47
Churu	19.7	21.6	314.4	9.5	365.7	37
Ganganagar	21.0	26.8	204.6	2.8	255.1	53
Hanumangarh	18.9	19.5	207.2	4.9	250.5	56
Jaisalmer	9.1	12.3	161.4	2.9	185.3	65
Jalore	17.9	14.9	349.1	10.1	381.0	52
Jodhpur	7.1	27.9	332.2	12.0	379.2	46
Jhunjhunu	21.5	28.4	340.4	11.7	402.0	36
Nagaur	15.0	20.9	285.1	6.7	327.7	53
Pali	5.8	17.0	384.4	16.9	424.1	50
Sikar	22.0	25.5	410.5	9.4	467.4	42
Wes. Rajasthan	15.3	19.4	290.6	7.7	332.3	50

causes the high degree of coefficient of variation ranging from 36% in eastern parts to 65% in western parts of arid Rajasthan. Thus, the variability in annual rainfall is the most important factor influencing crop and pasture yields in this region.

Solar Radiation and Duration of Sunshine

Solar radiation is generally high in arid Rajasthan throughout the year. During winter, it varies between 15.12 and 17.71 MJ/m^2/day, and in summer months, the values range from 22.79 to 26.50 MJ/m^2/day with a mean of 22 MJ/m^2/day (Table 2). The daily duration of bright sunshine hours in the area remains above 10 h/day in May, reduces to 6.6 h/day in July and August, and is above 8.8 h/day in the winter season.

Air Temperatures

The western Rajasthan experiences high extreme air temperatures considerably influencing the vegetation and crops and demanding higher water requirements. During winter, mean monthly maximum air temperature in the region varies from 22.4 °C to 29.0 °C and minimum temperatures between 4.1 °C and 14.3 °C (Table 3). Air temperatures increase sharply from April onward and stand highest during May till pre-monsoon showers set in the area. Summer air temperatures vary between 31.2 °C and 42.0 °C with peak values as high as 50 °C in summer and – 5.7 °C during winter months. Temperatures fall during the monsoon period (June to September) but however rise after recession of the monsoon by about 3–5 °C and again start falling from December onward due to winter conditions.

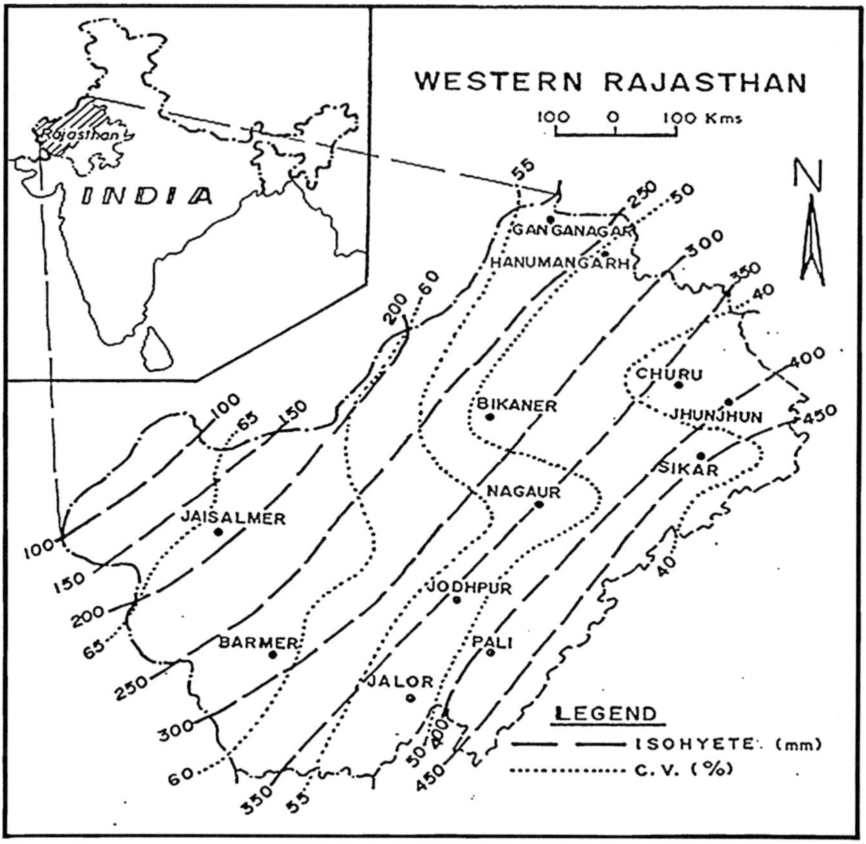

Fig. 1 Spatial distribution of annual rainfall

Relative Humidity

Relative humidity in arid Rajasthan is often less than 30% during the summer months, but gradually increases up to 80% by monsoon season, and then again decreases from October onward following withdrawal of the monsoon. Low humidity combined with strong wind regime leads to advection, a phenomenon which causes evaporation losses more than the energy actually available through solar radiation.

Wind Regime

The winds over arid Rajasthan blow in southwesterly direction during monsoon and in northeasterly direction during winter season. Though winds during winter are low, a strong wind regime builds up along with the temperature regime from April onward. Strong winds of the rate 15–18 kmph are observed during June, but the average winds

Table 2 Solar radiation (MJ/m^2/day) over arid Rajasthan

Station	Winter (Dec–Feb)	Summer (Mar–May)	Monsoon (Jun–Sept)	Post-monsoon (Oct–Nov)	Annual average
Barmer	17.71	26.50	24.55	20.05	22.18
Bikaner	16.81	26.39	26.17	19.37	22.18
Hanumangarh	15.12	22.79	21.96	20.16	19.98
Jaisalmer	17.42	26.49	26.60	19.87	22.57
Jodhpur	17.71	24.94	23.15	19.87	21.31

Table 3 Seasonal variation in air temperatures over arid Rajasthan

Month		Winter (Dec–Feb)	Summer (Mar–May)	Monsoon (Jun–Sept)	Post-monsoon (Oct–Nov)	Annual mean
Barmer	Max	25.4–29.0	34.6–42.0	34.9–40.3	35.9–39.6	34.5
	Min	10.6–13.3	19.0–26.7	24.6–27.3	16.6–22.1	20.7
Bikaner	Max	22.7–26.3	32.3–41.7	36.6–41.7	30.5–35.8	33.8
	Min	05.0–08.6	14.0–26.3	24.9–29.0	11.1–19.0	18.4
Churu	Max	22.4–25.6	31.6–40.9	35.4–41.1	29.4–34.7	32.9
	Min	04.1–07.4	19.4–24.3	23.2–27.9	10.0–17.3	17.1
Ganganagar	Max	21.1–24.2	24.9–40.6	36.6–41.9	29.2–35.1	32.8
	Min	05.3–06.3	13.4–23.9	23.0–28.2	10.8–17.9	17.6
Jaisalmer	Max	23.6–27.3	32.7–41.5	36.2–40.8	30.9–36.0	33.9
	Min	07.1–10.2	16.0–25.2	23.9–26.8	13.0–19.8	18.6
Jalore	Max	25.9–28.1	34.4–41.1	33.4–39.1	32.2–36.2	34.1
	Min	10.1–14.3	20.1–28.5	25.6–28.9	17.8–23.7	21.6
Jodhpur	Max	24.6–27.9	33.3–41.6	33.2–40.1	31.6–35.7	33.6
	Min	09.5–12.0	17.1–27.3	24.1–28.5	13.9–19.6	19.8
Nagaur	Max	23.6–26.7	32.7–41.2	3475–40.6	30.3–35.7	33.4
	Min	06.0–09.6	15.2–25.2	23.6–27.6	13.1–18.7	18.2
Sikar	Max	22.5–25.7	31.2–39.6	33.7–39.2	29.1–33.7	32.0
	Min	05.1–07.9	13.5–23.5	22.4–26.7	10.0–16.6	16.7

during summer vary from 9 to 13 kmph. Peak winds occasionally reach as high as 60–80 kmph during a severe dust storm period. Strong wind regime during May and June cause wind erosion depleting soil fertility particularly during drought years.

Evapotranspiration Rates

The extreme climatic conditions in the region result in high evapotranspiration rates. The potential evapotranspiration rates were 1.3–3.5 mm/day in winter months of December and January; as the summer temperatures increase by May, the values increase to 3.7–9.5 mm/day. During monsoon period, the evapotranspiration rates were 5.2–10.6 mm/day but again decrease from October onward till December

Table 4 Potential evapotranspiration (mm/day) over arid Rajasthan

Station	Winter (Dec–Feb)	Summer (Mar–May)	Monsoon (Jun–Sept)	Post-monsoon (Oct–Nov)	Annual average
Barmer	2.3–3.5	5.2–8.4	5.5–8.0	3.1–4.7	5.1
Bikaner	1.6–2.6	4.3–7.6	5.9–8.6	2.3–4.0	4.9
Ganganagar	1.3–2.3	3.7–7.0	5.4–7.3	2.0–3.7	4.6
Jaisalmer	2.1–3.2	4.9–8.4	6.4–10.6	2.8–4.8	5.7
Jodhpur	2.5–3.5	5.0–8.5	5.2–8.7	2.9–4.3	5.0
Phalodi	2.1–3.0	5.1–9.5	6.2–9.8	3.5–4.3	5.6

Table 5 Frequency of different intensities of meteorological droughts in arid Rajasthan

District	Period	Mild	Moderate	Severe	Total years	% years
Ganganagar	1926–2012	20	13	15	48	57
Churu	1906–2012	34	17	05	58	55
Hanumangarh	1906–2012	17	16	19	52	49
Jalore	1901–2012	31	17	16	64	58
Nagaur	1901–2012	29	23	14	66	60
Jhunjhunu	1901–2012	28	21	09	58	53
Jodhpur	1901–2012	24	23	16	63	57
Sikar	1901–2012	23	21	15	59	54
Barmer	1901–2012	22	18	22	62	56
Bikaner	1901–2012	22	21	13	56	51
Pali	1901–2012	18	30	11	59	54
Jaisalmer	1901–2012	15	23	23	61	55

(Table 4). The annual potential evapotranspiration ranges from 1400 mm in eastern parts to more than 2000 mm/year in the western parts of arid Rajasthan.

Drought Scenarios in the Thar Region

The arid Rajasthan is frequently prone to severe drought due to low and erratic nature of rainfall of the region. Failure of monsoon leaves the arid region totally dependent for food and fodder on buffer stocks to sustain its 19.8 million human and 48 million livestock population.

According to a classification given by Ramana Rao et al. (1981), the frequency of agricultural drought (Table 5) in arid Rajasthan indicated that out of 112 years (1901–2012), the region experienced agricultural drought in one part or the other in 49–60 years, which suggests drought occurs in the region once in 3 years to every alternate year. Jaisalmer district is the most prone region to drought. During 1901–2012, the agricultural drought in the region occurred in 70% of the years, out of which drought in 43% years was of severe nature and in 27% years moderate, thus considerably affecting the crop and fodder production. Next to Jaisalmer district, the Barmer district experienced severe drought in 30% years and moderate drought in 20% years. Bikaner district experienced severe agricultural drought in 23% years and moderate in 25% years, whereas Jodhpur district experienced severe drought in 17% years and moderate drought in 28% years.

Often, drought persists continuously for 2–6 years. Such prolonged droughts were experienced by this region during 1903–1905, 1957–1960, 1966–1970, 1984–1987, and 1997–2002. Century's worst drought conditions were recorded during the years 1918 and 2002. Prolonged drought puts tremendous stress on natural resources and leads to severe scarcity of food, fodder, and water (Rao 1997; Rao et al. 1997; Rao and Boken 2005). However, the western Rajasthan sometimes gets normal to above normal rainfall in subsequent drought years like the years of 1988 to 1996, which followed the 1987 severe drought, leaving ample scope for harvesting and reuse it during scarce periods.

Climate Change Impacts on the Thar Region

The Intergovernmental Panel on Climate Change (IPCC 2007) projected that the likely impact of climate change will be higher in arid regions than in semiarid or subhumid regions of India. Arid Rajasthan, which spreads in 12 western districts of the state covering 19.61 million ha, is very fragile and is subjected to excessive stresses due to frequent drought and low rainfall. The arid phase of northwest India has a history of about 3000 years (Pant and Maliekal 1987). In the northwest India covering Punjab, Haryana, west Rajasthan, and west Madhya Pradesh, there was a marginal increase in the rainfall by 141 mm and fall in air temperature by −0.52 °C in the past 100 years (Pant and Hingane 1988) and more so in irrigated belts of Sri Ganganagar region particularly during the past three decades (Rao 1996). The studies for Jodhpur region showed that the changes in rainfall and air temperatures were not alarming, but the increase in human population (by 400%) and livestock (by 127%) during the twentieth century resulted in a major shift in land use pattern and put tremendous pressure on surface and groundwater resources (Rao 1996; Rao and Miyazaki 1997). The desertification process may continue due to increased biological activity as a result of overgrazing and loss of vegetation cover with consequent more radiant energy loss and reduction in convective activity. Soil degradation and vegetation loss impact the thermodynamic balance in the northwestern India, and expansion of the Thar Desert can lead to a pronounced and large-scale impact on summer monsoon hydroclimate of the northwestern region of India. The Thar Desert is very rich in biodiversity with arid climatic conditions of the region suitable for adaptation of different species in the region. But extreme weather conditions like low and erratic rainfall, high temperatures, strong winds, and low humidity make it inhospitable to different habitats leaving to migration and loss of habitats in the region (Rao 1992, 2005 and 2009).

Climate Change Scenarios for Arid Western Part of India

According to PRECIS model, the simulated future climate of India under A_2 scenario indicates that by the last quarter of the present century, the mean annual temperature in the country will most likely increase by 3–5 °C. The annual spatially average rainfall is increase during the period is expected to be 7–10% (Rupa Kumar et al. 2006). The model also predicted in high variability in the changes in distribution of rainfall and temperature. North India is expected to be warmer than the south, but more importantly, night temperature and winter temperature would register higher of 5 °C increases over the most part. The rainfall trend during the last 100 years revealed that the summer monsoon rainfall, which contributes more than 85% of the total annual rainfall in the region, has increased marginally (< 10%) in the southern and eastern parts of the Thar Desert but has already declined by 10–15% in its northwestern part. Earlier studies on changes in rainfall and air temperatures of northwest India showed that the rainfall increased marginally by 141 mm in the past 100 years (Pant and Hingane 1988), especially in the irrigated belt of Ganganagar region particularly during the past three decades (Rao 1996).

Secular Trends in Annual Rainfall

The long-term rainfall trends in annual rainfall of 12 arid districts in Rajasthan showed that there was no significant change in annual rainfall during 1901 to 2012 (Fig. 2). Severest drought occurred in arid Rajasthan during 1917 and 2002, and the wettest year was in 1918. The rainfall (1960–2008) analysis at tehsil-wise locations of arid Rajasthan shows that the rainfall of June increased at 58 out of 65 locations, whereas the scenario for subsequent monsoon months is different as it decreased during July,

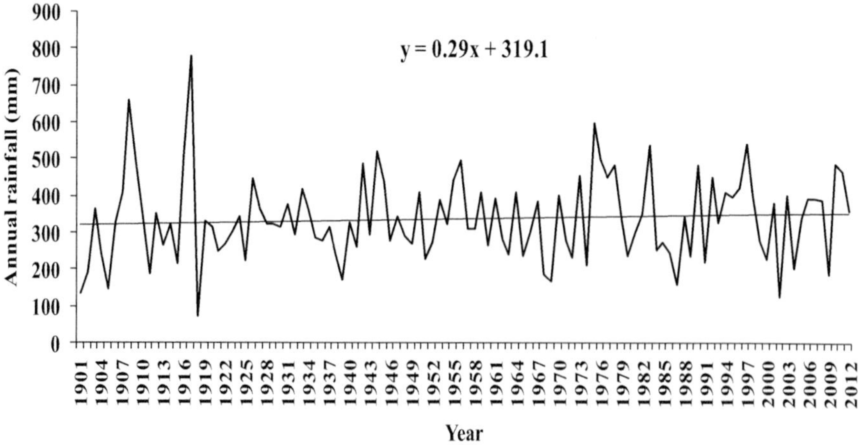

Fig. 2 Long-term annual rainfall trend of arid Rajasthan

August, and September. The rainfall was decreasing at 36 out of 65 locations in July, 49 in August, and 45 in September indicating larger arid area is under reduction in monsoonal rainfall of July, August, and September. The seasonal rainfall was decreasing at 37 locations during monsoon (JJAS), 7 in post-monsoon (ON), 16 in winter (DJF), and 15 during summer (MAM) (Rao and Roy 2012).

Changes and Shift in Rainfall

Twelve arid districts of western Rajasthan constitute 61% area of Indian hot arid zone, where the annual rainfall varies from 100 mm in the extreme west to 400 mm toward eastern part of the region. The coefficient of annual rainfall varies from 40 in the east to 70% in the west of the region, causing larger interannual variability in rainfall influencing crop production. According to a classification given by Ramana Rao et al. (1981), the frequency of agricultural drought in arid Rajasthan indicated that out of 109 years (1901–2011), the region experienced agricultural drought in one part or the other in 54–62 years, which suggest drought occurs in the region once in 3 years to alternate year. Jaisalmer district is most prone to drought. During 1901–2011, the agricultural drought in the region occurred in 70% of the years, out of which drought in 44% years was of severe in nature and in 29% years moderate, thus drought affecting considerably the crop and fodder production. Bikaner district experienced severe agricultural drought in 24% years and moderate in 26% years, whereas Jodhpur district experienced severe drought in 18% years and moderate drought in 29% years.

The overall regional trend in annual rainfall (1960–2011) for Thar showed no significant rise (at 0.56 mm/year) in the rainfall; however, the rainfall trend at different locations showed that the annual rainfall is likely to increase by +100 mm at Bikaner, +124 mm at Jaisalmer, −40 mm at Jodhpur, and +21 mm at Pali (Fig. 3). Thus, the projected rainfall is likely to increase from 252 mm to 308 mm at Bikaner, from 176 mm to 234 mm at Jaisalmer, and from 487 mm to 613 mm at Pali, whereas in Jodhpur the rainfall is likely to be decreased from 325 mm to 275 mm. Thus, in Jodhpur region, long-duration crops like pearl millet and sorghum are likely to be replaced with short-duration and traditional crops like cluster bean, moth bean, and green gram where rainfall is expected to decrease by the twenty-first century (Rao and Purohit 2009). To cope up with the delayed monsoon conditions, crop contingency plans (Joshi and Kar 2009) should be adopted. Adoption of traditional agroforestry systems in arid region buffers against climate variability, improves carbon sequestration, and also provides improved livelihood to the people (Roy et al. 2011 and Poonia and Rao 2013).

Changes in Air Temperatures

Poonia and Rao (2013) studied the long-term trends in mean air temperature (1971–2011) for the Thar region and showed, by the end of the twenty-first

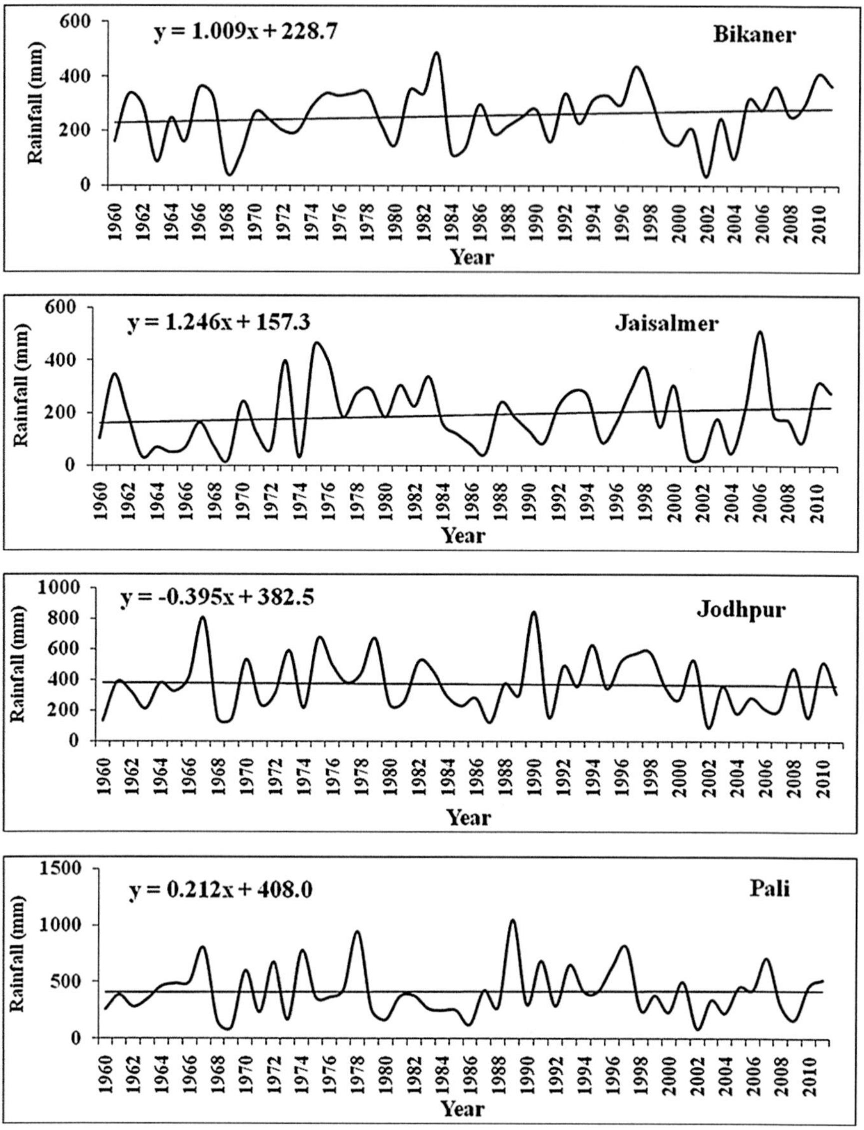

Fig. 3 Long-term trends in annual rainfall in the Thar region

century, an increase in temperature by +3.3 °C at Bikaner, +3.4 °C at Jaisalmer, +2.9 °C at Jodhpur, and +2.5 °C at Pali if the present rate of warming continues (Fig. 4). Jodhpur experienced highest day temperature of 48.3 °C on June 8, 2011, and warmest winter in 2008–2009 surpassing all past 50 years of extreme hot summer and warm winters in the region. Desert fauna suffers causality during severe drought years, reducing population, but multiplies during consecutive good rainfall years when adequate feed is available. The Thar region experienced

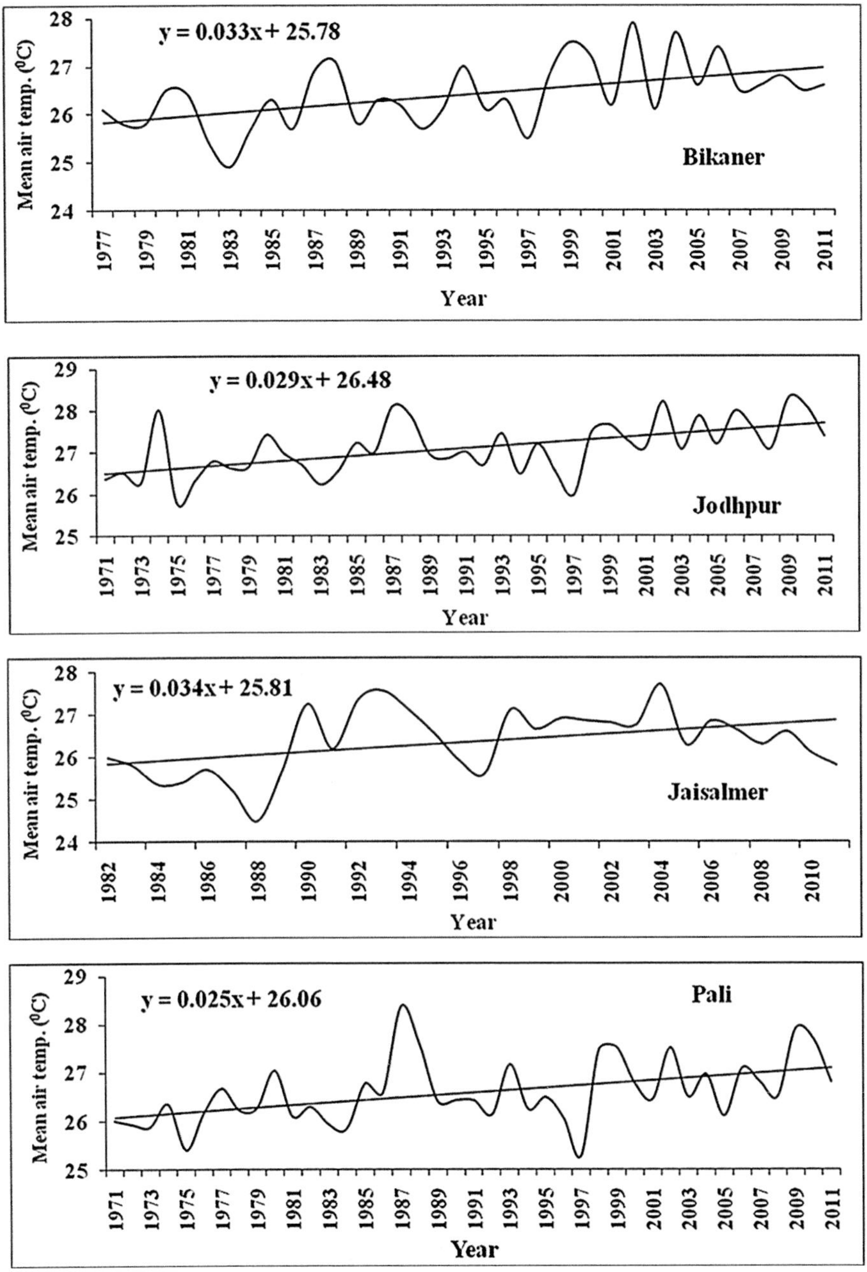

Fig. 4 Long-term trends in annual mean air temperatures in the Thar region

severe drought during 1918, and 1987, 2002, and 2009 were most severe, when rainfall departure from the normal was −81%, −65%, and −70%, respectively. In 2009, a rainfall deficiency of 40% from its normal rainfall caused drought affecting desert fauna due to scarcity for feed and drinking water. Drought followed by high temperatures touching 45–49 °C during late summer period of June 2010 resulted in a casualty of chinkaras and black bucks in Barmer, Churu, and Jodhpur districts of the Thar region. It was reported in local newspapers more than 177 chinkaras and black bucks died in villages of Bhacharna, Guda Vishnoi, Janguwas in Jodhpur district, Chawa in Barmer district, and Tal Chhapar in Churu. Soil fauna is not going to be affected directly by high temperatures due to their habit of living in burrows where the subsurface temperatures are not greatly influenced by high air temperatures (Poonia and Rao 2013).

Impact of Rising Temperatures on Crop Water Requirements

Rao and Poonia (2011) studied the impact of projected climate change by the twenty-first century on water requirements of rain-fed monsoon and irrigated winter crops of arid Rajasthan. Crop water requirements were estimated by them from daily potential evapotranspiration at ambient and projected air temperature by 2020, 2050, 2080, and 2100 using modified Penman-Monteith equation and then by multiplying with crop coefficients. The spatial variability in annual PET over arid Rajasthan shows that high PET prevails in the extreme western districts of Bikaner (2066 mm) and Jaisalmer (2221 mm) and low PET in northern districts of Ganganagar (1712 mm) and Hanumangarh (1736 mm). The northern districts are fed with canal irrigation from Beas and Sutlej rivers and find lower impact due to raising crop water requirements. However, the western districts of Jaisalmer and Bikaner are often prone to drought on an average every alternate year, thus causing frequent crop failures. PET isolines follow reverse trend to the rainfall isohyets in the arid Rajasthan. Crop water requirements in the region varied from 308 to 411 mm for pearl millet, 244 to 332 mm for cluster bean, 217 to 296 mm for green gram, 189 to 260 mm for moth bean, 173 to 288 mm for wheat, and 209 to 343 mm for mustard. Further, due to global warming, if the projected temperatures rises by 4°C, by the end of the twenty-first century, water requirement in arid Rajasthan increases from the current level by 12.9% for pearl millet and cluster bean, 12.8% for green gram, 13.2% for moth bean, 17.1% for wheat, and 19.9% for mustard (Rao and Poonia 2011; Fig. 5). Their study reveals that the impact will be more severe on *rabi* crops than *kharif* crops, the *rabi* crops being dependent on depleting ground water resources in the region.

The western districts of Barmer, Bikaner, and Jaisalmer which are already in low rainfall zone form hot spots due to the impact of projected rise in temperature leading to higher crop water demand, whereas in the districts Sri Ganganagar and Hanumangarh having canal water facility for irrigation, the impact due to climate change in these areas is likely to be less.

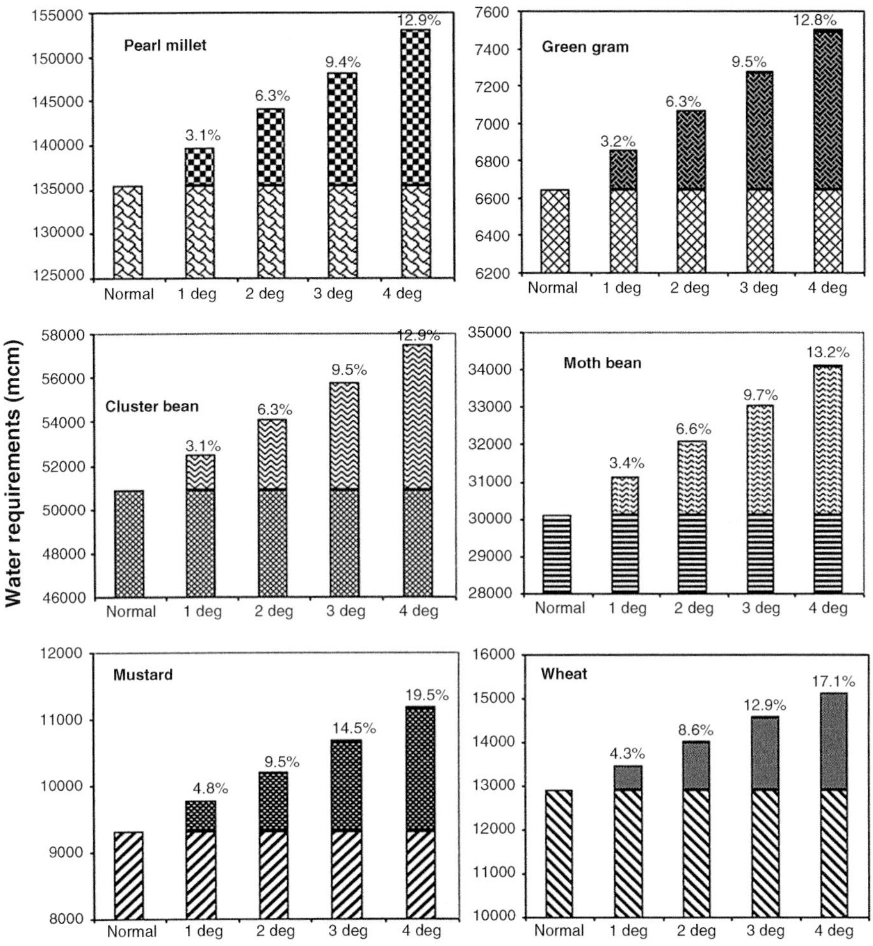

Fig. 5 Seasonal crop water requirements of *kharif* and *rabi* crops at elevated temperatures

References

IPCC, Cambridge (2007) Climate change 2007: the physical science basis. In: Solomon S, Qin D, Manning M, Chen Z, Marquis M, Averyt KB, Tignor M, Miller HL (eds) Contribution of the working group I to the fourth assessment report of the intergovernmental panel on climate change. Cambridge University Press, Cambridge, UK/New York 996 pp

Joshi NL, Kar A (2009) Contingency crop planning for dryland areas in relation to climate change. Indian J Agro 54(2):237–243

Pant GB, Hingane LS (1988) Climatic changes in and around the Rajasthan desert during the 20th century. J Clim 8:391–401

Pant GB, Maliekal JA (1987) Holocene climatic changes over north-west India. An appraisal. Clim Chang 10:183–194

Poonia S, Rao AS (2013) Climate change and its impact on Thar desert ecosystem. J Agric Phys 13(1):71–79

Ramana Rao BV, Sastri ASRAS, Ramakrishna YS (1981) An integrated scheme of drought classification as applicable to Indian arid region. Idojaras 85:317–322

Rao AS (1992) Climate, climatic changes and paleo-climatic aspects of Rajasthan. In: Sharma HS, Sharma ML (eds) Geographical facets of Rajasthan. Kuldeep Publications, Ajmer, pp 38–44

Rao AS (1996) Climatic changes in the irrigated tracts of Indira Gandhi canal region of arid western Rajasthan, India. Ann Arid Zone 38(2):111–116

Rao AS (1997) Impact of droughts on Indian arid ecosystem. In: Singh S, Kar A (eds) Desertification control in the arid ecosystem of India for sustainable development. Agro Botanical Publishers (India), Bikaner, pp 120–130

Rao AS (2005) Impact of introduction of IGNP canal irrigation on micro- and secular changes in climate of Thar desert region. In: Tyagi BK, Baqri QH (eds) Changing faunal ecology in the Thar desert, vol 3. Scientific Publishers, Jodhpur, pp 37–44

Rao AS (2009) Climate and microclimate changes influencing the fauna of the hot indian arid zone. In: Sivaperuman C, Baqri QH, Ramaswamy G, Naseema M (eds) Faunal ecology and conservation of the great Indian desert. Springer-Verlag, Heidelberg, pp 13–24

Rao AS, Boken VK (2005) Agricultural drought monitoring and management in India (Chapter 23). In: Boken VK, Cracknell AP, Heathcote RL (eds) Monitoring and prediction of drought: a global study. Oxford University Press, New York, pp 297–312

Rao AS, Miyazaki T (1997) Climatic changes and other causative factors influencing desertification in Osian (Jodhpur) region of the Indian arid zone. J Arid Land Stud 7(1):1–11

Rao AS, Poonia S (2011) Sensitivity of crop water requirements to elevated temperatures in arid Rajasthan. Ann Arid Zone 50(2):131–138

Rao AS, Purohit RS (2009) Spatial variability and shifts in rainfall patterns of arid Rajasthan, India. In: Proceedings international conference on "Nurturing arid zone for people and the environment: issues and agenda for the 21st century." Central Arid Zone Research Institute, Jodhpur, pp 9

Rao AS, Roy MM (2012) Weather variability and crop production in arid Rajasthan. Central Arid Zone Research Institute, Jodhpur 70p

Rao AS, Gupta JP, Faroda AS (1997) Changing climatic scenarios and strategies for drought management in the Indian arid region. Drought Netw News 9(1):8–11

Roy MM, Tewari JC, Ram M (2011) Agroforestry for climate change adaptations and livelihood improvement in Indian hot arid regions. Int J Agric Crop Sci 3(2):43–54

Rupa Kumar K, Sahai AK, Krishna Kumar K, Patwardhan SK, Mishra PK, Revadekar JV, Kamala K, Pant GP (2006) High-resolution climate change scenarios for India for the 21st century. Curr Sci 90:334–345

Ecosystem-Based Adaptation and Gender Perspectives from a Participatory Vulnerability Assessment in Mountainous Rural Vietnam

35

Tran Thi Kim Lien and Kathryn Brown

Contents

Abstract

Smallholder farmers in rural Vietnam are dependent on ecosystems for their daily subsistence and income. Female and male farmers are differently affected by climate change, and they have different responses to and ability to cope with climate change. Men and women in Son Tho commune, a mountainous rural site in Ha Tinh province, Vietnam, were surveyed to determine what ecosystems and

T. T. K. Lien (✉)
Deutsche Gesellschaft für Internationale Zusammenarbeit (GIZ) GmbH, Ha Noi, Vietnam
e-mail: tranthikimlien@gmail.com

K. Brown
Marine Research Centre, School of Environment, Science and Engineering, Southern Cross University, Lismore, NSW, Australia

W. Leal Filho (ed.), *Handbook of Climate Change Resilience*,
https://doi.org/10.1007/978-3-319-93336-8_14

services they depend on, what losses and damage from climate change they are facing, and identify any appropriate ecosystem-based adaptation (EbA) measures that could be implemented to build resilience. Data were collected at meetings with commune staff and male and female farmers through open-ended and structured questionnaires and focus-group discussions. Both groups acknowledged that livelihoods of people in the study area are ecosystem-based. Perceived changes in the local climate included an increase in heat waves; more intense droughts due to delays in the onset of rain and reduction in total annual rainfall; cold spells; typhoons; and floods. The influences of these phenomena on livelihoods were interpreted differently by female and male farmers. As a result, EbA interventions are defined and selected differently by gender.

Keywords
Smallholder farmers · Climate change · Gender · Ecosystems · Ecosystem-based adaptation · Vulnerability assessment

Introduction

Vietnam is located in Southeast Asia's tropical monsoon belt and commonly affected by floods, droughts, typhoons, and landslides, as well as occasional forest fires. The nation is among the countries most severely affected by climate change. Average temperatures have been rising and the total precipitation has increased, especially during the rainy seasons (Government of Vietnam 2011; The World Bank Group 2011). The Ha Tinh province as well as the Son Tho commune experiences adverse impacts of natural disasters like floods, storms in the rainy seasons, along with hot days and droughts in the dry seasons. Every year, storms and tropical depressions cause floods, tube floods, flash floods, and landslides in the mountainous areas. Hot days and droughts lead to scarcity of water for daily use and irrigation. The changes in temperature, rainfall, and the frequency or intensity of extreme weather events directly affect crop and animal productivity as well as household food security, income, and well-being (Schmidt-Thomé et al. 2015). As a result, great loss of human life and property brings about difficulties for socioeconomic development in the region (Ha Tinh Provincial People Committee 2016).

Climate change in Vietnam affects both men and women in many areas, including livelihoods, health, migration, adaptation capacity, participation, and decision-making. The Intergovernmental Panel on Climate Change (IPCC) Fourth Assessment Report (IPCC 2007) notes that impacts of climate change differ according to gender; however, most studies of climate change impacts tend to group countries together without considering disproportions within groups such as gender.

Vietnam is still an agricultural society, where women are concentrated in agriculture and participate in most production activities as self-employed (United Nations Vietnam n.d.). In addition, Vietnamese culture remains male dominated and women have less access to and control over the resources that they are dependent on for livelihoods. It is suggested that gender equality will likely not be reached if males refuse to share the responsibility of domestic work, and are dominant over decision-making in both private and public domains (Institute for Social Development Studies 2015). Long-term gradual climate change will impact agricultural and ecological systems. Thus, women are more vulnerable to different impacts of climate change (United Nations Vietnam n.d.). Similarly, smallholder farmers are considered to be disproportionately vulnerable to climate change (Parikh et al. 2012).

The National Target Program on Responding to Climate Change to 2020 (2007) reports the need to conduct vulnerability assessments at sectorial, regional, and community levels and recognizes the poor, women, and children as among the vulnerable groups to climate change impacts. However, what this program lacks is an explanation of the different roles of men and women, and concrete guidance on implementation of gender-based strategies (United Nations Vietnam n.d.). The implementation of existing legal frameworks and the development of new legal documents and related action plans tend to have little or no reference to gender equality. In practice, women may perform all climate change adaptation/disaster risk reduction (CCA/DRR) tasks, especially when men migrate for additional work. Women's valuable role in providing support to those affected by disasters is not officially recognized. There is limited research, analysis, and evaluation of gender equality and mainstreaming in CCA/DRR from which appropriate solutions can be formulated (United Nations Development Program Vietnam 2013). These are critical issues as climate change may worsen existing gender inequalities, create additional workloads for women, and lead to higher vulnerability of women in poor households (United Nations Vietnam n.d.).

The national and local governments in Vietnam have introduced and implemented a number of adaptation measures, both hard and soft, to cope with climate change. The measures are in different sectors, including water, agriculture, infrastructure, human health, tourism, transport, energy, and others. Many recent climate change adaptation initiatives have focused on the use of engineering-based construction and the design of climate resilient infrastructure. However, there is growing recognition that maintaining healthy ecosystems can assist people to adapt to climate change.

The advantages of ecosystem-based approaches for climate change adaptation are promoted at the international level, and the necessity to strengthen sustainable approaches has received attention from scientists and governmental bodies alike (Wilkinson et al. 2013; Chong 2014). Among these solutions, ecosystem-based adaptation (EbA) is a relatively new concept, defined as the "use of biodiversity and ecosystem services as part of an overall adaptation strategy to help people adapt to the adverse impacts of climate change" (United Nations Framework Convention on Climate Change 2016). The concept of EbA is to systematically harness the services provided by ecosystems to buffer communities against the adverse impacts of climate change (Foster et al. 2011; Jones et al. 2012). The EbA model relies on ecosystems and services to provide climate change adaptations to enhance sustainable planning in order to tackle the impacts of climatic extremes and variability (Kok and de Coninck 2007, Cowling et al. 2008, Vignola et al. 2009, Preston et al. 2010, Daily et al. 2011 cited in Wamsler 2015).

A participatory study was carried out in the Sơn Thọ commune in rural Vietnam with the aim of assessing vulnerability to climate change, and EbA interventions preferred, according to gender. This commune is an interesting site for this study for at least two reasons. Firstly, it is one of the most vulnerable areas to climate change in the province and is frequently affected by natural hazards (Institute of Strategy and Policy on Natural Resources and Environment – [ISPONRE] 2016). Secondly, a wide range of ecosystem services that benefit people are closely correlated with the communities' resilience to the challenges. This indicates that good management practices may have an important role in changing the landscapes, in addition to generating ecosystem services. The specific objectives of this study are: (i) provide a framework for vulnerability assessment and identify EbA measures through a gender lens; (ii) to explore the advantages and disadvantages of implementing EbA interventions by gender, and; (iii) to show both opportunities and barriers for enhancing the roles of women in reducing vulnerability and increasing adaptive capacity in the socio-ecological system.

Study Area

The study was conducted in Sơn Thọ commune, a mountainous area of Vu Quang district, Ha Tinh province, Vietnam, located about 70 km from the provincial center (Fig. 1). Its natural area is approximately 4591 ha, in which agricultural land takes up about 668 ha

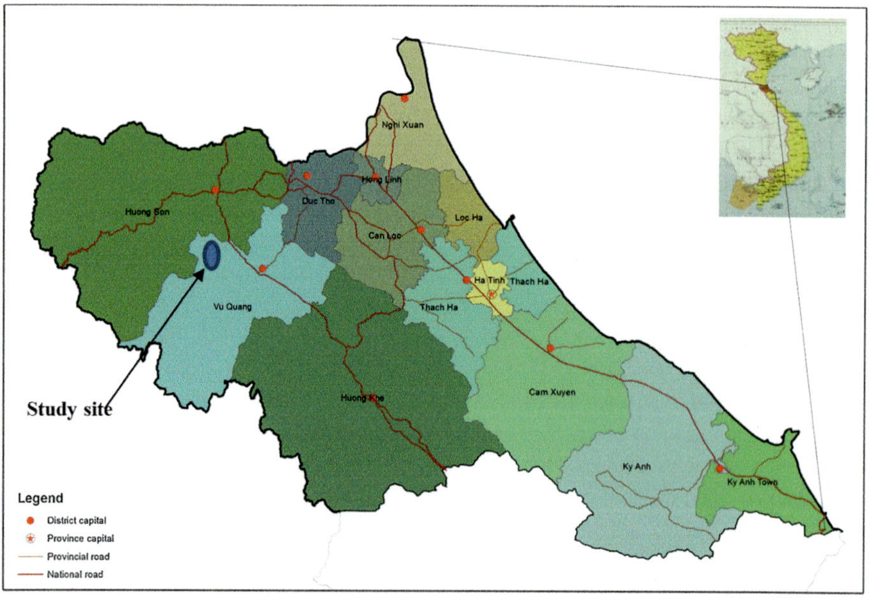

Fig. 1 The study site. (Source: Department of Natural Resources and Environment of Ha Tinh (2015))

(14.5%), forestry land about 3465 ha (75.5%), and nonagricultural land 290 ha (6.3%). Roughly 2715 residents live in 756 households, distributed over 10 villages.

Methods

The participatory vulnerability assessment in Son Tho commune collected data and information via 12 focus group discussions with 240 key informants, divided by gender. A range of methods were applied and combined to obtain a holistic perspective view on vulnerability assessment from smallholder male and female farmers and local authorities. Data was collected in May and June 2017. The research was participatory, requiring the proactive interest and engagement of the smallholder male and female farmers who lived in the area.

In the selection of study respondents, smallholder women and men were sampled at focus groups. Data collection techniques included participatory rural appraisals, focus group discussions, and structured questionnaires with open-ended questions, as well as field observations. Twelve focus groups were comprised of four groups of men, four groups of women, and four groups of mixed gender, with 20 participants in each group. There were two replications to test assumptions for the most frequently offered answers. In all, 287 villagers and 10 commune staff were randomly surveyed from a total of 2715 inhabitants (10.9%). Participants came from different income classes and ages.

The focus group discussions included questions about hazards, exposures, sensitivity, and impacts relating to climate change in the last 10 years (2008–2017). Transect walks across the commune landscape were taken together with five experienced farmers, with stops every 100 m to verify details of points covered in the focus group discussions. Additional details were provided by 42 farmers who were met by chance during these walks. Attention was specifically paid to the description of vegetation, soils, crops, and the signs of impacts from extreme weather conditions, which gave insights into differences between ecosystem services in terms of land use, distribution of vegetation, and soil types.

The selected respondents were personally interviewed using a questionnaire to record individual responses on the importance of ecosystem services in relation to droughts. Interviews were also conducted with 10 key informants, selected from commune cadastral and agricultural staff, village leaders, and elders, to generate broader insights into the constraints faced by farmers and ecosystems. Information from respondents' interviews and key informant interviews were cross-checked for consistencies.

Results

Existing Ecosystems in the Study Area

All 12 groups confirmed that the livelihoods of people in the commune are based on ecological systems. The existing ecosystems in 2017 are shown in Fig. 2. There are five main ecosystems in the area, of which, mono acacia plantation accounted for the

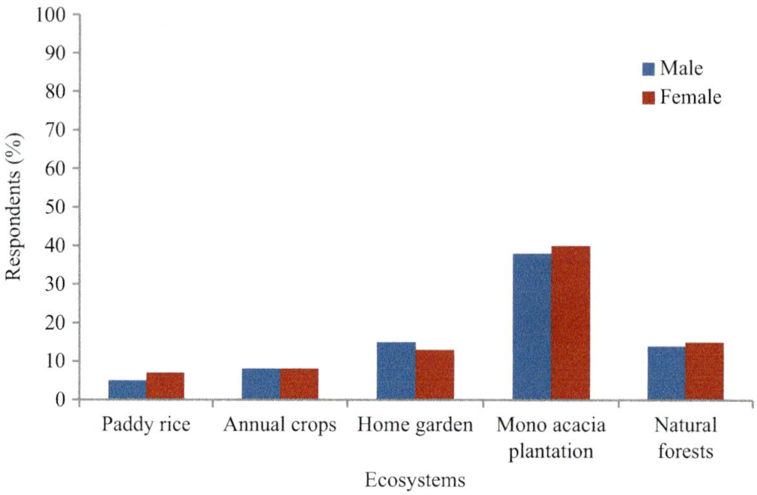

Fig. 2 Proportion of existing ecosystems within the study area in 2017 estimated by men and women in Son Tho commune

largest proportion, followed by relatively similar areas of home gardens and natural forests. Paddy rice and annual crops accounted for the smallest proportion. Men and women provided information with very little difference in estimates of the areas covered by each ecosystem type.

Participatory Vulnerability Assessment by Gender

The results of the focus group discussions and interviews showed that people in the study site have experienced various extreme weather events, including heat waves, droughts, cold waves, typhoons, and floods. According to farmers' observations, changes in the local climate included an increase in the number of hot days, increasing intensity and frequency of droughts (due to delays in the onset of rain and reduction in total annual rainfall), cold spells, typhoons, and floods. The impacts of these phenomena on livelihoods were interpreted differently by men and women. Figure 3 depicts the occurrence and impacts of extreme weather events within the 10 year period 2008–2017 reported by men and women of Son Tho commune.

Droughts were reported as occurring most frequently by both men and women, followed by heat waves, typhoons, and floods (Fig. 3); however, women reported slightly higher frequencies than men for these three extreme events. In contrast, men reported slightly higher frequencies of cold snaps than women (Fig. 3).

Both men and women agreed that droughts caused major loss and damage to ecosystems in Son Tho commune, as did typhoons and floods; however, the greatest impacts on paddy rice and home gardens were from cold snaps. Annual crops and

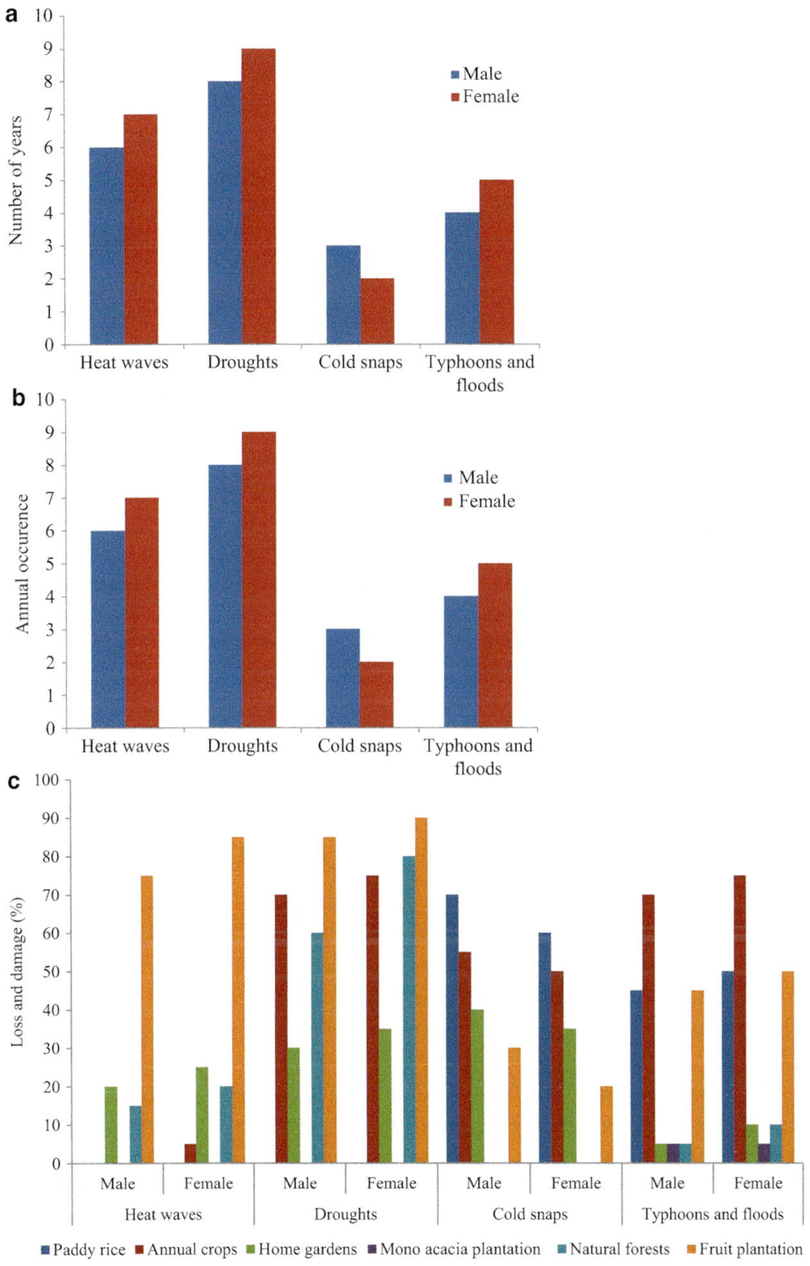

Fig. 3 Vulnerability assessment to extreme weather events in Son Tho commune by gender: (**a**) the number of years extreme events occurred within the 10 year period 2008–2017, (**b**) the annual occurrence of each event, (**c**) the degree of loss and damage suffered by each ecosystem type due to extreme weather events

fruit plantations were affected by all four types of hazards, whereas mono acacia plantations were resilient to extreme events. The damage degree was consistently higher by women's assessment than men's, with the exception of cold snaps. While male groups considered loss and damage by cold snaps to ecosystems as being 5–10% higher than female groups did, negative effects to ecosystems by heat waves, droughts, typhoons, and floods were 5–10% more serious by women's conclusions than men's. The greatest difference in each gender's assessment of loss and damage was in the effects of drought on natural forests, with women's estimates of impacts being 80% compared to men's 60%.

Promising EbA Measures by Gender

Most women (95%) selected farm intercropping planting (grass for animals and vegetable, green tea, fruit trees) plus bee keeping as the most effective EbA measure, whereas men selected forest protection and enrichment plus bee keeping, and water harvesting and drip irrigation facilities for fruit trees as their priority EbA interventions (90% of men interviewed).

The greatest differences between genders in preferred interventions was in soil binding and coverage planting for fruit trees, with 20% higher preference by men.

Discussion

Existing Ecosystems in the Study Area

The ecosystems of Son Tho commune are dominated by agricultural and forestry activities, with monoculture hybrid acacia plantations currently accounting for the largest proportion in the study area. The introduction of hybrid acacia plantations in this region has created many benefits, improving small farmers' livelihoods while providing enhanced environmental values in rural landscapes (CIFOR 2014). This fast-growing species provides marketable commodities, including wood of good quality for paper-making (Sein and Mitlöhner 2011), which helps reduce the pressure on native forests as a source of industrial raw materials (The Australian Centre for International Agricultural Research [ACIAR] 2004). The acacia plantations generate employment, both within the plantations and in associated value-adding activities (Sandewall et al. 2010; Sein and Mitlöhner 2011). Environmental benefits are also gained by the ability of acacia to fix atmospheric nitrogen, facilitated by a symbiosis between the roots and nitrogen-fixing bacteria (Sein and Mitlöhner 2011). Furthermore, acacia offers other ecosystem benefits, such as the conservation or restoration of soils, carbon sequestration, and climate regulation (New Generation Plantations Project 2009; Bauhus et al. 2010). The capacity of acacia hybrids to improve infertile soil could provide an economic benefit to subsequent rotations of acacia or other crops.

In 2017, rice paddy fields in the commune were estimated to occupy an area of 5% and 7% by men and women, respectively. The commune's rice production is consumed by the commune's total population, maintaining a crucial role in the region's food security and providing a foundation for economic and social stability (Yoon 2009; Asian Development Bank 2012). The rice crop also plays a part in the region's ecological and environmental conservation (Chen and Liu 2002) as the lowland paddy fields act as a buffer and subsequently increase the water storage capacity of the basin (Matsuno et al. 2006).

Agricultural lands are under threat to maintain sufficient production for an expanding population in food-deficient mountainous regions of Vietnam. In addition, agricultural activities are important to the restoration and management of natural resources. Maintaining agricultural and forestry activities can reduce the environmental risks and prevent direct economic damage caused by landslides, erosion, forest fires, etc. (Food and Agriculture Organization of the United Nations [FAO] 2006). However, agricultural production must be increased in the future to meet the food and feed demands of rising human populations and increasing livestock production (Maxwell and Fernando 1989; Longhurst 2009). Annual crops, including peanut, green bean, maize, and cassava not only contribute the income source of farmers but also the cultivation of maize and the residuals of rice and peanut crops are important fodder sources to animals. More broadly, crop products play a central role in the livelihood of growers (Poulton et al. 1998; Cuong and Anh 2016). In particular, poorer households depend almost entirely on agricultural products for both food and cash to buy other necessary items (Carr 2008).

Home gardens account for 15% and 13% of the study area according to men and women. The survey informants agreed that their home gardens not only provide income sourced from fruits and livestock but also offer a wide range of rewards for the gardeners, providing a large return in terms of ecosystem services and cultural integrity. Immediate access to fresh food, quality control of the food grown, positive environmental impacts, erosion control, enjoyment, and family satisfaction are recognized as significant benefits by the informants (Agbogidi and Adolor 2013). The importance of ecosystem services from the home garden differs from those provided by other types of agro-ecosystems (Calvet-Mir et al. 2012). Mohri and colleagues argue that home gardens have broader benefits measured across various household scales than abandoning cultivation for alternative incomes (Mohri et al. 2013). Livelihood benefits, beyond providing income, include resilience to food scarcity and security from broad-scale economic and environmental shocks (Landreth and Saito 2014).

In the study site, natural forests produce timber for noncommercial uses at the household level including fencing, animal shelters, equipment, and low-cost building materials (Costanza et al. 1997; Nasi et al. 2002; Turner and Daily 2007; Natural Resources Canada 2016). Non-timber forest products also provide income, fuelwood, and traditional medicines (Lien et al. 2017). Cash income from the sale of non-timber forest products can be highly variable from sources such as honey, bamboo shoots, rattan, and ornamental plants (Nasi et al. 2002; Turner and Daily 2007; Natural Resources Canada 2016). Forests regulate local temperature extremes by providing shade and surface cooling of soils. They can act as a local greenhouse agent to hold heat, block searing winds, and trap warmth (Costanza et al. 1997; Kreg et al. 1997; Nasi et al. 2002; Turner and Daily 2007). The provision of organic contents through leaf and branch fall in forests assists in limiting erosion and preventing soil run off by buffering the direct impacts of heavy rainfall (Costanza et al. 1997; Kreg et al. 1997; Turner and Daily 2007). Another hypothesis is that natural forests improve the quality of water, with purity of drinking water, the efficiency of hydroelectric power plants, and the quality of fish habitats likely to be better in forested catchments than in barren areas, and it can be argued that adverse effects on water quality relate to bad management practices rather than elements within the forests themselves (Nasi et al. 2002; Turner and Daily 2007; Natural Resources Canada 2016). Participants often agree that natural forests will become the key issue for ecosystem biodiversity in the future, in themselves, as well as habitat for other species if they can be well protected (Costanza et al. 1997; Kreg et al. 1997; Turner and Daily 2007).

Participatory Vulnerability Assessment by Gender

Smallholder farmers in Son Tho commune are vulnerable to extreme weather events, with the effects of droughts being the most severe. The most immediate impact of droughts is a reduction and often losses in crop production (paddy rice, peanut, maize, etc.) due to inadequate and poorly distributed rainfall. Informants agreed that small harvests caused by droughts meant there was insufficient food for their family members and to provide for other needs in the home. Another severe impact of drought is water shortages for humans, animals, crops, and trees (Charlotte 1997; Polthanee et al. 2014). Pasture production is also likely to decline, which may lead to a decrease in fodder supplies for animals with consequent weight loss and increased disease among stock. Significant impacts on agricultural crops, livestock, and human beings are also caused by heat waves. Extreme temperatures over several days can lead to reduced pollen fertility and damage to the crop's reproductive parts responsible for producing grain and fruits, reducing reproductive potential, and thus greatly reducing the crop yield (Schlenker and Roberts 2009; van der Velde et al. 2012). High temperatures, in particular when combined with high humidity and low air movement, can exceed the ability of animals to cope, causing a loss of appetite, productivity, reproductive vigor, and sometimes death.

In a policy brief, United Nations Vietnam found that 64% of rural women and 53% of rural men are engaged in crop production and are highly vulnerable to loss from drought and uncertain precipitation. Climate change adds to water insecurity, which particularly increases the work load of rural women involved in small scale farming, as they need to spend more time and effort on land preparation, fetching water, watering, and protecting crops from disease (United Nations Vietnam 2013).

Some of the highest disaster-related agricultural losses occurred in the study area as a consequence of floods, mainly associated with typhoons. Floods destroyed standing crops of paddy rice and fruit trees and also stored food products. Rice and other agriculture products are normally stored in the upper part of the house, so are not damaged by low-level flooding; however, higher floods can destroy the entire house, its appliances, and important food stores (Gupta 1990). Apart from the spoiling of food stores, grains may also have to be discarded if they are too wet to be milled. An additional problem of flooded agricultural land is the proliferation of fungi, which has consequences for the next crop rotation.

Intense floods that persist over a long duration can deliver serious health problems for local people, in particular, disabled and elderly family members or those who live in poor conditions with limited food sources, polluted water sources, and bad sanitation. Epidemic diseases, including dengue fever and diarrhea, appear after floods (Few et al. 2004). The study informants also confirmed livestock were afflicted by health problems as a consequence of floods. The typhoons associated with floods can also uproot crops and trees, break off stems and branches, and defoliate any remaining trees (Few et al. 2004). Typhoons can blow the roofs off houses and animal shelters, and subsequent exposure to the extreme weather can cause death to valuable livestock.

Women and men experienced threats on their ecosystems from floods and typhoons differently. Women, particularly poor women, are actively engaged in agricultural activities, including paddy cultivation, annual crops, home gardens, and fruit plantation, which will be affected by these two hazards. The destruction and damage caused by typhoons and floods will impact livelihoods and increase the vulnerability and marginalization of women. This situation is similar to that observed for women in agricultural regions in Bangladesh and India (Parikh et al. 2012).

Low temperature stress has also been shown to cause considerable agricultural yield loss to sensitive crops such as rice and peanut (Charlotte 1997; Thakur et al. 2010) and to pastures. Unseasonal cold waves kill seedlings of rice and peanut at the germination period, requiring extra resources to regerminate crops, which then reduces productivity or can result in crop failure. Cold also affects livestock digestion and growth, and animals often lose weight and body condition during cold spells. Livestock are often kept confined to avoid cold-related diseases, but with limited fodder their growth is reduced. In extreme cold weather, weaker animals will die and the females will bear underweight offspring (Kerven 2003). Cold extremes also affect human health directly, for example, during annual outbreaks of winter diseases such as influenza. Cold spells can lead to hypothermia and increase morbidity and mortality from cardiovascular disease (McMichael et al. 2003).

Smallholder farmers in agricultural communities are among the most vulnerable people to adverse impacts of climate change (Lambrou and Piana 2006). With warmer temperatures due to climate change, evaporation rates are higher, and water supplies are further strained by greater demand from competing sectors of the economy. Rainfall patterns are also less predictable due to climate change, and drought and flooding are more common. Higher rain intensities can also cause more floods, which in turn tend to cause greater run-off and erosion. This can harm crop yields and human health (Buechler and Mekala 2005).

These climatic events can threaten women's livelihoods more than men's, due both to gendered responsibilities within the household and gendered constraints to labor-market participation. Women are highly dependent on natural resources, such as water and food for household usage and for farming. They undertake farming either together with men or alone as in the case of women's subsistence production, when men migrate or when women are single or widowed (Buechler 2005). Men may increase their migratory activities during adverse periods, and their absence increases women's workloads (Buechler 2009; Djoudi and Brockhaus 2011).

In most cases, climate change impacts are linked to gender issues and women's vulnerability. Women are usually responsible for providing the family with basic nutrition, yet they rarely have access to and control over the resources required to fulfill this task when cultivation conditions deteriorate. Decreasing availability of clean water, decreasing agricultural production, decreasing access to crop residues and biomass for energy, and increasing risk of famine are all areas where women are primarily at risk of experiencing severe negative impacts. Hence climate-induced crop failure also puts the food security of the entire population at risk (Denton 2002). A major factor in women's vulnerability is that they lack important assets such as land rights, ownership rights for the means of production, technology, finances, information and training in climate adaptation, and disaster prevention (Buechler 2009).

Promising EbA Measures by Gender

There is a need to capture the positive effects of choosing ecosystem-based adaptive investments that take into account the different needs of men and women (Buechler 2009). Table 1 shows that women perceived farm intercropping planting (grass for animals and vegetables, green tea, fruit trees) plus bee keeping as the most effective EbA measures. This selection is to reduce the negative impacts of natural hazards on women, in particular in relation to their daily roles in the village of providing food resources for animals. Women understood that their animals are severely affected by

Table 1 Preferred EbA interventions according to gender (n = number of informants)

EbA measures	Men (%) $n = 120$	Women (%) $n = 120$
Forest protection and enrichment + bee keeping	90	85
Farm intercropping planting (grass for animals, vegetables, green tea, fruit trees) + bee keeping	85	95
Water harvesting and drip irrigation facilities for fruit trees	90	80
Soil binding and coverage planting for fruit trees	70	50

cold snaps. This planting is therefore intended to provide more food for animals during cold days, and farmers can take necessary care of plantings as they are nearby. This EbA measure preferred by women might be due to a focus on more immediate and feasible outcomes. Every family member can contribute directly to this intervention, which allows animals to access and receive their fodder without any obstacles or delays.

In contrast, men appear to be considering greater scales and less domestic solutions when they selected forest protection and enrichment plus bee keeping; and water harvesting and drip irrigation facilities for fruit trees as their priority EbA interventions. Men also placed greater emphasis than women on the value of soil coverage to prevent erosion. Again, these trends are potentially connected to different livelihood roles of men and women. Female household members select measures that are highly domestic and immediate in their outputs, and that furthermore do not extensively increase the women's already large workload, since every family member (including children) can directly support this intervention due to its proximity to the house, and fodder delivery remains quick and easy as distances are short. In case cold spells occur, more resistant fodder plants along garden boundaries would constitute a response that would not increase the work load of women who are in charge of supplying fodder to the animals, which in turn would make female household inhabitants less vulnerable to cold spells. Men, whose main tasks are less closely connected to the physical space of the house and direct food provisioning, take on the more long-term and large-scale perspective of creating additional fodder sources on infertile lands. Djoudi and Brockhaus (2011) point out a similar tendency in the preference and perception of forest products, where "[...] women often prefer trees with multiple uses because these trees offer more domestic and supplementary value," while men pay more attention to timber productivity (Djoudi and Brockhaus 2011).

Challenges for Enhancing the Roles of Women in Reducing Vulnerability and Increasing Adaptive Capacity in the Socio-Ecological Systems

Male and female farmers have unlike roles in natural resources management, different access to, and different ways of using resources. As a part of climate change adaptation, traditional gender-based roles are being reduced, which has created opportunities for women to be involved in activities that not only improve livelihoods but allow them to better adapt (Djoudi and Brockhaus 2011).

Women's unpaid jobs, such as creating gardens; cooking and putting food on the table; caring for children, the sick, and elderly; undertaking domestic tasks and collecting fuel and water for the family's maintenance, are some main obstacles to increasing women's participation in minimizing risks and enhancing their adaptive capacity (UNFCCC 2016). However, women play a crucial role in many activities essential for coping with climate change and ability to cope with change (Parikh et al. 2012). Recently, women have reallocated time from domestic chores to productive roles, for example, for creating and selling handicrafts; they become income earners and are perceived in a new light. Women become active economic providers, earning money to improve their livelihoods, to invest in their children's education, and access to health services. This strengthens women's adaptive capacity and encourages gender equality and women's participation in decision-making (UNFCCC 2016).

Women often have less say in decision-making related to natural resources and disaster management, despite their role and experience in these areas. At the household and community levels in Vietnam, while there has been some progress, clear gender-based differences in decision-making processes remain. Though discussion is increasingly common among couples, decision-making in households still reflects a gender stereotype: men make decisions on big purchases and other important issues while women make them for small items and spending. Men have the final say in cases of disagreement (United Nations Vietnam n.d.).

Farmers who did not adapt to climate challenges in the past have given many reasons for their failures to adapt, including lack of information, lack of money, shortage of labor, shortage of land, and poor potential for irrigation (Deressa et al. 2009). Male-headed households are more likely to get information about new technologies and undertake risky businesses than female-headed households (Asfaw and Admassie 2004). Moreover, Tenge and Hella (2004) argue that having a female head of household may have negative effects on the adoption of soil and water conservation measures, because women may have limited access to information, land, and other resources due to traditional social barriers. A study by Nhemachena and Hassan (2007) finds contrary results, arguing that female-headed households are more likely to take up climate change adaptation methods. The authors conclude that women are more likely to adapt because they are responsible for much of the agricultural work in the region and therefore have greater experience and access to information on various management and farming practices (Deressa et al. 2009).

There are several aspects linked to the design and implementation of EbA on-site (e.g., awareness of the problem and potential solutions, local needs, and level of knowledge) as well as within a wider context (e.g., representation of the local population and their needs in political processes), which determine the context and potential for designing and implementing ecosystem-based approaches to adaptation (Naumann et al. 2013).

Conclusion

This study provides an overview of the concept that smallholder female and male farmers in rural Vietnam are differently affected by climate change. Men and women showed different views on the occurrence of extreme weather events and the effect on their livelihoods. In most cases, female farmers considered more events, higher number of occurrence, and more exposure to hazards, while male farmers reported greater frequency and effects from cold snaps than females did. The influences of these phenomena on livelihoods were interpreted differently by female and male farmers, thus EbA interventions are differently defined and selected by men and women. While women focus on more immediate and feasible outcomes, men appear to be considering greater scales. There are numerous challenges that need to be considered for enhancing women's participation in minimizing vulnerability and increasing their capacity for climate change adaptation. We believe that these findings will show some relevance and interest in applying EbA measures as part of the overall adaptation strategy to counteract climate instability in the region, with due consideration for the requirement of the gendered approach.

Cross-References

- ▶ Biodiversity, Ecosystem Degradation, and Climate Change Effects on Livelihoods in the Bitumen Area of Nigeria
- ▶ Early Warning System and Ecosystem-Based Adaptation to Prevent Flooding in Ibadan Metropolis, Nigeria
- ▶ Ecosystem-Based Adaptation in Tigray, Northern Ethiopia: A Systematic Review of Interventions, Impacts, and Challenges
- ▶ Mitigating and Adapting to "Arid" Climate Extremes: Impacts of Locally Prioritized Ecosystem-Based Adaptations at the Water-Energy Nexus

References

Agbogidi OM, Adolor EB (2013) Home gardens in the maintenance of biological diversity. Appl Sci Rep 1:119–125

Asfaw A, Admassie A (2004) The role of education on the adoption of chemical fertilizer under different socioeconomic environments in Ethiopia. Agric Econ 30(3):215–228

Asian Development Bank (2012) The rice situation in Vietnam. Technical Assistance Consultant's report. https://www.adb.org/sites/default/files/project-document/73083/43430-012-reg-tacr-04.pdf

Bauhus J, van der Meer PJ, Kanninen M (2010) Ecosystem goods and services from plantation forests. Earthscan, London/Washington, DC

Buechler S (2005) Women at the helm of irrigate agriculture in Mexico: the other side of male migration. In: Opposing currents: the politics of water and gender in Latin America. University of Pittsburgh, Pittsburgh

Buechler S (2009) Gender, water, and climate change in Sonora, Mexico: implications for policies and programmes on agricultural income-generation. Gend Dev 17(1):51–66

Buechler S, Mekala GD (2005) Local responses to water resource degradation in India: groundwater farmer innovations and reversal of knowledge flows. J Environ Dev 14(4):410–438

Calvet-Mir L, Gómez-Baggethun E, Reyes-García V (2012) Beyond food production: ecosystem services provided by home gardens. A case study in Vall Fosca, Catalan Pyrenees, Northeastern Spain. Ecol Econ 74:153–160

Carr ER (2008) Men's crops and women's crops: the importance of gender to the understanding of agricultural and development outcomes in Ghana's central region. World Dev 36:900–915

Center for International Forestry Research (CIFOR) (2014) Australia acacia tree genus enhances livelihoods in Asia. http://www.cifor.org/forestsasia/role-acacias/

Charlotte B (1997) The economic impact of natural disasters in Vietnam. Working paper 98. Results of ODI research presented in preliminary form for discussion and critical comment, April 1997,

Overseas Development Institute, Portland House, Stag Place, London, pp 1–122. https://www.odi.org/resources/docs/7024.pdf
Chen S-K, Liu CW (2002) Analysis of water movement in paddy rice fields experimental studies. J Hydrol 260:206–214
Chong J (2014) Ecosystem-based approaches to climate change adaptation: progress and challenges. Int Environ Agreements Polit Law Econ 14:391–405
Costanza R, d'Arge R, de Groot R, Farber S, Grasso M, Hannon B, Limburg K, Naeem S, O'Neill RV, Paruelo J, Raskin RG, Sutton P, van den Belt M (1997) The value of the world's ecosystem services and natural capital. Nature 387:253–260
Cuong VN, Anh NT (2016) The role of crop land during economic development: evidence from rural Vietnam. Eur Rev Agric Econ 43:561–582
Denton F (2002) Climate change vulnerability, impacts, and adaptation: why does gender matter. Gend Dev 10(2):10–20
Deressa TT, Hassan RM, Ringler C, Alemu T, Yusuf M (2009) Determinants of farmers' choice of adaptation methods to climate change in the Nile Basin of Ethiopia. Global Environ Change 19(2):248–255
Djoudi H, Brockhaus M (2011) Is adaptation to climate change gender neutral? Lessons from communities depending on livestock and forests in northern Mali. Int For Rev 13(2):123–135
Few R, Ahern M, Matthies F, Kovats RS (2004) Floods, health and climate change: a strategic review. Tyndall Centre working paper 63, November 2014, pp 1–65. https://www.unisdr.org/files/1985_VL206506.pdf
Food and Agriculture Organization (2006) The role of agriculture and rural development in revitalizing abandoned/depopulated areas. Document prepared under the supervision of the Policy Assistance Branch Regional Office for Europe, 2006. Food and Agriculture Organization of the United Nations, pp 1–28. http://www.fao.org/fileadmin/user_upload/Europe/documents/Publications/Abandoned_en.pdf
Foster J, Lowe A, Winkelman L (2011) The value of green infrastructure for urban climate adaptation. Center for Clean Air Policy, Washington, DC. http://www.amwa.net/galleries/climate-change/Green_Infrastructure_FINAL.pdf
Government of Vietnam (2011) National strategy on climate change. http://chinhphu.vn/portal/page/portal/English/strategies/strategiesdetails?categoryId=30&articleId=10051283
Gupta SP (1990) Core shelter assistance project in the Philippines for typhoon victims. Disaster Manag 3:14–20
Ha Tinh Provincial People Committee (2016). Report on the implementation of the CC prevention and response in Ha Tinh province. http://qppl.hatinh.gov.vn/vbpq_hatinh.nsf/4b438b320dbf1cda4725719a0012432c/9D5C25374F934C5147257C84002F3BBE/$file/bao%20cao%2050.pdf
Institute for Social Development Studies (2015) Social determinants of gender inequality in Vietnam. Hong Duc, Ha Noi
Institute of Strategy and Policy on Natural Resources and Environment (ISPONRE) (2016) Vulnerability assessment in Ha Tinh. http://www.climatechange.vn/en/va-ht/
Intergovernmental Panel on Climate Change (2007) Climate change 2007: impacts, adaptation and vulnerability. Contribution of working group II to the fourth assessment report of the intergovernmental panel on climate change. https://www.ipcc.ch/pdf/assessment-report/ar4/wg2/ar4_wg2_full_report.pdf. Accessed 10 Jun 2017
Jones HP, Hole DG, Zavaleta ES (2012) Harnessing nature to help people adapt to climate change. Nat Clim Chang 2:504–509
Kerven C (2003) The influence of cold temperatures the influence of cold temperatures and snowstorms on rangelands and snowstorms on rangelands and livestock in northern Asia. Working paper. School of Government, University of the Western Cape, Bellville, 26–27 March
Kreg L, Brian F, Marily S, Rosemary B (1997) Ecotourism and other services derived from forests in the Asia-Pacific. Asia-Pacific forestry sector outlook study: Asia-Pacific forestry sector outlook study working paper series no. 24, 1997. Food and Agriculture Organization of the United Nations, pp 1–80. http://www.fao.org/3/a-w7714e.pdf

Lambrou Y, Piana G (2006) Gender: the missing component of the response to climate change. Food and Agriculture Organization, Gender and Population Division, Sustainable Development Department, Rome

Landreth N, Saito O (2014) An ecosystem services approach to sustainable livelihoods in the homegardens of Kandy, Sri Lanka. Aust Geogr 45:355–373

Lien TTK, Nichols JD, Brown KE (2017) Firewood extraction and use in rural Vietnam: a household model for three communes in Ha Tinh Province. Agrofor Syst 91(4):649–661

Longhurst R (2009) Cash crops, household food security and nutrition. Inst Dev Stud Bull 19:28–36

Matsuno Y, Nakamura K, Masumoto T, Matsui H, Kato T, Sato Y (2006) Prospects for multifunctionality of paddy rice cultivation in Japan and other countries in monsoon Asia. Paddy Water Environ 4:189–197

Maxwell S, Fernando A (1989) Cash crops in developing countries: the issues, the facts, the policies. World Dev 17:1677–1708

McMichael AJ, Campbell-Lendrum DH, Corvalán CF, Ebi KL, Githeko AK, Scheraga JD, Woodward A (2003) Climate change and human health, risks and responses. World Health Organization, Geneva

Mohri H, Lahoti S, Saito O, Mahalingam A, Gunatilleke N, Irham HVT, Hitinayake G, Takeuchi K, Herath S (2013) Assessment of ecosystem services in homegarden systems in Indonesia, Sri Lanka, and Vietnam. Ecosyst Serv 5:124–136

Nasi R, Wunde R, Campos JJ (2002) Forest ecosystem services: can they pay our way out of deforestation? A discussion paper prepared for the global environment Facility for the Forestry Roundtable to be held in conjunction with the United Nations Forum on Forests II in Costa Rica, 11 March 2002, Costa Rica, pp 1–38. http://www.cifor.org/publications/pdf_files/Books/BNasi0201.pdf

Natural Resources Canada (2016) Forest ecosystem products and services. http://www.nrcan.gc.ca/forests/canada/sustainable-forest-management/13177

Naumann S, Davis M, Munang R, Andrew J, Thiaw I, Alverson K, Mumba M, Kavagi L, Han Z (2013) The social dimension of ecosystem-based adaptation. The United Nations environment programme policy series. Ecosystem Management. Ecologic Institute, United Nations Environment Programme (UNEP), Cornell University

New Generation Plantations Project (2009) Ecosystem integrity and forest plantations. Technical Paper, New Generation Plantations Project, WWF International, June 2009. http://newgenerationplantations.org/multimedia/file/3ab82968-74a2-11e3-92fa-005056986313

Nhemachena C, Hassan R (2007) Micro-level analysis of farmers' adaptation to climate change in Southern Africa. IFPRI discussion paper no. 00714. International Food Policy Research Institute, Washington, DC

Parikh J, Upadhyay DK, Singh T (2012) Gender perspectives on climate change & human security in India: an analysis of national missions on climate change. CADMUS 1(4):180–186

Polthanee A, Promkhumbut A, Bamrungrai J (2014) Drought impact on rice production and farmers' adaptation strategies in Northeast Thailand. Int J Environ Rural Dev 5(1):1–8

Poulton C, Dorward A, Kydd J (1998) The revival of smallholder cash crops in Africa: public and private roles in the provision of finance. J Int Dev 10:85–103

Sandewall M, Ohlsson B, Sandewall RK, Viet LS (2010) The expansion of farm-based plantation forestry in Vietnam. Ambio 39:567–579

Schlenker W, Roberts MJ (2009) Nonlinear temperature effects indicate severe damages to U.S. crop yields under climate change. Proc Natl Acad Sci U S A 106(37):15594–15598

Schmidt-Thomé P, Nguyen TH, Pham TL, Jarva J, Nuottimaki K (2015) Climate change adaptation measures in Vietnam. Springer, Cham

Sein CC, Mitlöhner R (2011) Acacia hybrid. Ecology and silviculture in Vietnam. CIFOR, Bogor. http://www.cifor.org/publications/pdf_files/Books/BCIFOR1104.pdf

Tenge De Graaff J, Hella JP (2004) Social and economic factors affecting the adoption of soil and water conservation in West Usambara highlands. Tanzania Land Degrad Dev 15(2):99–114

Thakur P, Kumar S, Malik JA, Berger JD, Nayyar H (2010) Cold stress effects on reproductive development in grain crops: an overview. Environ Exp Bot 67:429–443

The Australian Centre for International Agricultural Research (ACIAR) (2004) Acacia hybrids in Vietnam. ACIAR Project FST/1986/030. http://ageconsearch.umn.edu/bitstream/113223/2/ias27.pdf

The World Bank Group (2011) Vulnerability, risk reduction, and adaptation to climate change. http://sdwebx.worldbank.org/climateportalb/doc/GFDRRCountryProfiles/wb_gfdrr_climate_change_country_profile_for_VNM.pdf

Turner RK, Daily GC (2007) The ecosystem services framework and natural capital conservation. Environ Resour Econ 39:25–35

United Nations Development Program Vietnam (2013) Gender equality in climate change adaptation and disaster. Policy Brief. http://www.vn.undp.org/content/vietnam/en/home/library/environment_climate/Policy-Brief-Gender-equality-in-climate-change-adaptation-and-disaster.html

United Nations Framework Convention on Climate Change (2016) Opportunities and options for enhancing adaptation actions and supporting their implementation: reducing vulnerability and mainstreaming adaptation. https://unfccc.int/files/adaptation/groups_committees/adaptation_committee/application/pdf/tp_adaptation_2016.pdf

United Nations Vietnam (2013) Gender equality in climate change adaptation and disaster risk reduction in Vietnam. http://www.un.org.vn/vi/publications/cat_view/130-un-viet-nam-joint-publications/209-climate-change-joint-un-publications.html

United Nations Vietnam (n.d.) Gender and climate change impacts in Vietnam. Results of a desk review. http://portal.unesco.org/en/files/44957/12385110459Climate_change_and_Gender.pdf/Climate+change+and+Gender.pdf

van der Velde M, Tubiello FN, Vrieling A, Bouraoui F (2012) Impacts of extreme weather on wheat and maize in France: evaluating regional crop simulations against observed data. Clim Chang 113:751–765

Wamsler C (2015) Mainstreaming ecosystem-based adaptation: transformation toward sustainability in urban governance and planning. Ecol Soc 20:30–47

Wilkinson C, Saarne T, Peterson GD, Colding J (2013) Strategic spatial planning and the ecosystem services concept – an historical exploration. Ecol Soc 18:37–45

Yoon CG (2009) Wise use of paddy rice fields to partially compensate for the loss of natural wetlands. Paddy Water Environ 7:357

Climate Change: Depression in Egg Production in Chickens During the Hot Season with Long-Term Honey Administration

36

M. O. Abioja and M. O. Adekunle

Contents

Abstract

Effect of climate change on poultry production is becoming more evident globally, with resultant elevation on environmental temperature. Honey has been reputed for alleviation of detrimental effects of heat stress in laying pullets. The effect of honey on egg production (EP) and organs in laying pullets during the hot-dry season was determined. Pullets (aged 28 weeks) were randomly assigned to 3 doses of honey: 0 (0H), 10 (10H), and 20 ml honey (20H)/L water for 16 weeks, divided into 4 phases. Honey had a significant ($P < 0.01$) effect on EP in Phase 1. Birds in 20H produced more eggs than birds in CONTROL (0H). EP in 10H chickens was not however ($P > 0.05$) different from those in CONTROL. In Phase 2, the effect of honey on EP was significant ($P < 0.05$). Chickens in 10H had lower EP than in the CONTROL and 20H groups. EP in birds in CONTROL

M. O. Abioja (✉) · M. O. Adekunle
Department of Animal Physiology, College of Animal Science and Livestock Production, Federal University of Agriculture, Abeokuta, Ogun, Nigeria
e-mail: abiojamo@funaab.edu.ng

W. Leal Filho (ed.), *Handbook of Climate Change Resilience*,
https://doi.org/10.1007/978-3-319-93336-8_15

and 20H was not different. Phases 3 and 4 show significantly ($P < 0.05$) lower EP in honey groups than CONTROL. Honey had a significant ($P < 0.05$) effect on the relative weight of the ovary and the whole reproductive tract and the number of yellow and deformed follicles with birds in 20H recording poorer values. Long-term administration of honey during the hot-dry season caused reproductive organ atrophy. Farmers should not extend the use of honey beyond 4 weeks.

Keywords

Egg · Chickens · Follicles · Yolk · Heat stress · Global-warming

Introduction

High environmental temperature above the thermal comfort zone (TCZ) elicits a series of responses in laying chickens generally termed heat stress (Ayo et al. 2011). Heat stress (HS) is a continuous problem faced by poultry production in the tropics, especially in Africa. Most farmers are low- to mid-income people that cannot afford the expensive closed-house, temperature-controlled poultry pens commonly used in developed world. HS results in drastic reduction in egg production, size, and quality (Grieve 2003) and may be lethal if not controlled. The onslaught is more aggravated as the effects of climate change become more evident. Moreover now that it has been reported that the year 2017 was the third warmest on record over the African land masses since 1950 following 2010 (the warmest year) and 2016 (the second warmest year), some steps must be taken. The average temperature over the continent in 2017 was +1.20 °C higher than the 1961–1990 average. Years 2010 and 2016 recorded an average temperature rise of +1.41 and + 1.26 °C, respectively. The warming rate (+3.7 °C per century) over the last 27 years is the most alarming signal, highlighting the need for urgent action on mitigation and climate resilience. Intergovernmental Panel on Climate Change in the Fourth Assessment Report (IPCC 2007) concluded that besides other things, the surface air warming in the twenty-first century by best estimate will range from 1.1 °C to 2.9 °C for a *low scenario* and 2.4–6.4 °C for a *high scenario* (Nardone et al. 2010). Hisas (2011) stated that by 2020, the temperature of the planet would increase by, at least, 2.4 °C above preindustrial times, going by current business-as-usual path. Climate change is advancing much faster than anticipated. However, Gaughan et al. (2009) stated that the negative effects of climate change cum global warming in livestock had not received the same attention given to humans. It should be noted that animal species react to high environmental temperature as do human beings. Any procedure that will arrest further increment in global greenhouse gases (GHGs) must be followed. Agriculture and other sectors are under increasing public pressure to reduce emissions of these gases. Carbon dioxide from livestock respiration accounts for 21% of anthropogenic GHGs worldwide. Industrial production of various chemicals, used in poultry subsector, leads to generation of more GHGs. Layers have higher manure CH_4 emissions than broilers. This is because layers have a greater proportion of their manure managed in anaerobic conditions, which lead to higher CH_4 emissions (MacLeod et al. 2013).

For emissions from egg production, feed production contributes 69% of emissions; direct on-farm energy use, 4%; post-farm processing and transport of meat, 6%; and manure storage and processing, 20%.

Most poultry farmers make use of commercial and synthetic anti-stress and antioxidants to help chickens cope with HS. Alternatives to the use of chemicals such as anti-stress, antioxidants, and antibiotics lie in the discovery and proper utilization of natural plant materials and extracts that have the needed properties (Bedford 2000; Wenk 2003; Vidanarachchi et al. 2005; Ramnath et al. 2008; Zhang et al. 2009; Ali et al. 2010). Various efforts had been geared toward exploration of these materials. One of the promising sources of natural anti-stress/antioxidant is honey (Estevinho et al. 2008; Mohamed et al. 2002; Gheldof et al. 2003; Aljadi and Kamaruddin 2004; Wasagu et al. 2013). In human beings, honey has been used as antibacterial (Adetuyi et al. 2009; Omafuvbe and Akanbi 2009) and antioxidant (Akanmu et al. 2009) and in semen diluents (Oyelowo et al. 2014). It is a thick, viscous, and sweet liquid made by bees from the nectar of flowers, transformed and stored in the honeycombs. Honey is a mixture of many compounds including carboxylic acids, aldehydes, alkynes, nitrites, alkynes, and ethers (Adebiyi et al. 2004). Honey is a good example of natural substance that contains phytochemicals such as vitamin C, thiamine, riboflavin, pyridoxine, pantothenic acid, nicotinic acid, phenolic compounds, and enzymes glucose oxidase, catalase, and peroxidase. It has recently been reported by Osakwe and Igwe (2015) that honey could be included in layers' feed up to the level of 20% without any deleterious effect. Honey was said to elicit positive physiological responses in layer chickens under stress-related conditions and improve egg characteristics. However, it was reported that in order to ensure improvement in egg production and eggshell thickness in laying pullets during the hot season, honey with 20 ml/L of water can be offered for 4 weeks (Abioja et al. 2016). Presently, information on the use of honey in poultry is limited (Hegazi et al. 2013), and its mode of action still needs to be elucidated. Also, some reports calling for caution in the use, misuse, and abuse of honey abound in literature. Chronic consumption of unprocessed Nigerian honey resulted in decreased bile flow, increased bile cholesterol, and decreased plasma cholesterol in albino rats (Alagwu et al. 2009). Therefore, the present study aimed at determining the effect of honey on the improvement of egg production in laying chickens during the hot season.

Materials and Methods

Experimental Location and Meteorological Observations

This experiment was carried out at *Aiyedoto Farm Settlement*, Ojo Lagos, Nigeria (latitude 6° 27′ 25″N, longitude 3°12′ 21″E, and altitude of 36 m above sea level). The experimental site's climate is humid, and it is located in the rain forest vegetation zone of Western Nigeria. Wet- and dry-bulb temperatures and relative humidity at the level of the birds in the pen at 0800 h and 1600 h were monitored throughout the experimental period. The temperature-humidity index was calculated from the relative humidity and wet- and dry-bulb temperature data.

Experimental Animals and Management

ISA Brown layer chickens (n = 120; aged 28 weeks) were used for the experiment during the hot-dry season. The birds were randomly allocated to three (3) treatments consisting 4 replicates and 10 layer birds per replicate. Birds in groups I, II, and III received 0 (CONTROL), 10 (10H), and 20 ml honey (20H) per liter of water, respectively, for a period of 16 weeks, divided into 4 phases. Standard ration was given ad libitum to the birds.

Data Collection

Egg production (EP): Daily records of the number of eggs laid by chickens in each replicate were taken and egg hen-day production calculated as the ratio of the number of eggs laid daily to the number of hens in the pen.

Relative weight of organs: After 16-week honey treatment, 2 laying pullets from each replicate were sacrificed. Fat pad and organs (liver, heart, spleen, gizzard, proventriculus, ovary, and whole reproductive tract) were excised and weighed. Relative weight was determined by calculating as percentage of live weight.

Follicles and eggs: The number of yellow follicles, deformed follicles, and eggs were counted and recorded.

Statistical Analysis

Data collected was subjected to one-way analysis of variance (ANOVA) using SYSTAT (1992) with the model $Y_{ij} = \mu + T_i + \sum_{ij}$, where Y_{ij} is the dependent variable, μ is the population mean, T_i is the ith effect due to the addition of honey to the drinking water ($i = 1,2,3$), and $\sum_{ij}$ is the residual error. Means that are statistically ($P < 0.05$) different were separated with Duncan's multiple range test (DMRT).

Results

Table 1 shows the summary of climatic conditions during the experiment. The average dry-bulb and wet-bulb temperatures were 31.5 and 28.0 °C, respectively. The average relative humidity during the experiment was 79.5%, while the temperature-humidity index of 84.9 was recorded.

The average egg hen-day production in laying chickens in the four phases is presented in Table 2. There was a significant ($P < 0.05$) increase in the average egg production (EP) from Phase 1 (57.0%) to Phase 2 (61.5%). However, EP gradually

Table 1 Summary of climatic conditions during the experiment

	08.00 h	16.00 h	Average
Dry-bulb temperature (°C)	30.4	32.5	31.5
Wet-bulb temperature (°C)	27.8	28.8	28.0
Relative humidity (%)	82.3	76.7	79.5
THI	83.8	86.0	84.9

Table 2 Effect of continuous administration of honey on chicken egg production in four phases of four weeks each

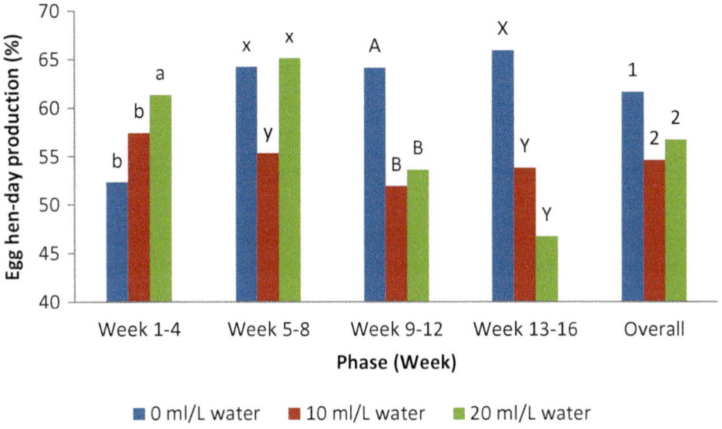

[a,b]Means in weeks 1–4 with different small letters (a and b) differing significantly ($P < 0.05$)
[x,y]Means in weeks 5–8 with different small letters (x and y) differing significantly ($P < 0.05$)
[A,B]Means in weeks 9–12 with different capital letters (A and B) differing significantly ($P < 0.05$)
[X,Y]Means in weeks 13–16 with different capital letters (X and Y) differing significantly ($P < 0.05$)
[1,2]Overall means with different numbers (1 and 2) differing significantly ($P < 0.05$)

decreased in the two consecutive phases (3 and 4). Figure 1 presents the egg hen-day production in laying chickens offered varying dosages of honey in drinking water during the hot season. Addition of honey in drinking water had a significant ($P < 0.01$) effect on EP in Phase 1. Birds that received 20H produced more eggs than birds in the CONTROL group (0H). EP in chickens on 10H treatment was not ($P > 0.05$) different from the 0H. In Phase 2, the effect of honey on EP in laying chickens was significant ($P < 0.05$). Chickens on 10H had lower EP than the CONTROL and 20H groups. The EP in 0H and 20H was not different. Phases 3 and 4 show significantly ($P < 0.05$) lower EP in honey (10H and 20H) groups than the CONTROL group. Combining the data for the 16 weeks experimental period, honey groups had significantly *($P < 0.01$)* lower EP than the CONTROL group.

Administration of honey in drinking water of laying pullets for 16 weeks did not significantly ($P > 0.05$) affect the relative weight of the fat pad, liver, heart, spleen, gizzard, and proventriculus. However, honey had a significant ($P < 0.05$) effect on the relative weight of the ovary and the whole reproductive tract and the number of

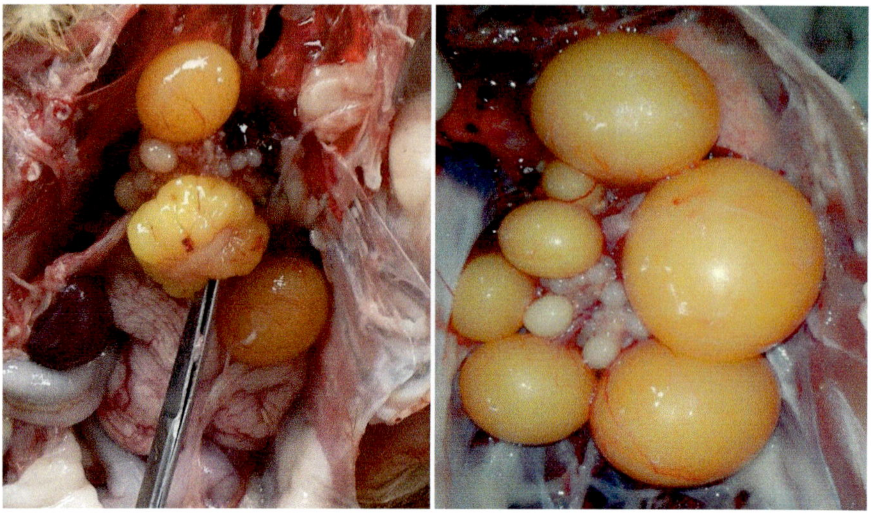

Fig. 1 Egg hen-day production in laying chickens offered varying dosages of honey in drinking water during the hot season

yellow and deformed follicles with birds in 20H recording poorer values. Birds in 20H treated group had relatively lower ovary and whole tract weight compared to birds in the CONTROL group. Those in 10H group recorded similar values to the CONTROL group. Though the number of yellow follicles was higher in 20H group than the CONTROL group, the occurrence of deformed follicles and eggs was observed in birds in 20H group. Deformation of follicles and/or eggs was not present in 0H and 10H groups (Table 3).

Discussion

Climate change, defined as a change in the state of atmospheric conditions that can be identified by using statistical tests, is obvious. It may be observed as deviations in mean and/or the variability of climatic properties that persists for an extended period typically decades or longer (IPCC 2007). Although the length of time it takes the changes to manifest matters, the level of deviation from the normal and its impacts on the ecology are most paramount. Meteorological variables obtained in this study were above the thermal comfort zone for egg-laying chickens. The average temperature recorded during this experiment was 31.5 °C, about 6.5 °C above the upper critical limit. Chickens perform optimally under environmental temperature ranging from 16 to 25 °C (Sahin et al. 2006). Birds tolerate a narrow range in environmental temperature and so are vulnerable to climate change and global warming. Though technology favors the intensification of poultry production in most developing countries, the issues of the environment may pose a hindrance. Besides temperature, various environmental variables play a significant role in the production and

Table 3 Effect of honey on the relative weight of visceral organs of laying chicken during the hot-dry season

Organ	Control	10H	20H	Sem	P-value
Fat pad (%)	0.7	2.1	0.7	0.58	0.204
Liver (%)	1.8	2.1	2.6	0.22	0.080
Heart (%)	0.5	0.5	0.4	0.03	0.660
Spleen (%)	0.09	0.09	0.13	0.023	0.545
Gizzard (%)	2.7	2.6	3.1	0.23	0.362
Proventriculus (%)	0.48	0.51	0.52	0.036	0.640
Ovary (%)	2.0^{a}	2.5^{a}	1.3^{b}	0.28	0.037
Whole reproductive tract (%)	6.6^{a}	6.1^{a}	3.8^{b}	0.53	0.012
Number of yellow follicles	5.0^{b}	8.0^{a}	8.0^{a}	1.55	0.030
Deformed follicles	- - -	- - -	+++		
Deformed eggs	- - -	- - -	+++		

^{a,b}Means with different superscripts in the same row are significantly different ($P < 0.05$)
- - - None
+++ Common

reproduction of chickens. The relative humidity obtained ranged between 76 and 82% during the study. High atmospheric moisture content is characteristic of the experimental location. Combining high temperature with high relative humidity further aggravates the effect of HS in chickens. It has been reported that high temperature in poultry pens, especially during the dry season, does adversely affect egg fertility and hatchability in *avian* broiler-breeder hens under humid tropical conditions (Abioja et al. 2015). The same authors stated that relative humidity further adversely affected the birds.

Attaining the optimal range in the open-sided poultry houses common in the tropics is difficult (Abioja 2010). The average temperature-humidity index during the experiment was such that elicited stress responses from the birds. This accounts for the reduction in egg production over the phases, instead of expected increment. HS, among other adverse effects, causes reduction in egg production (Mashaly et al. 2004; Deng et al. 2012; Lara and Rostagno 2013). Reduction in egg production may result from the behavioral reduction in feed consumption by chickens during heat spell in the bid to reduce metabolic heat production. This causes a low intake of essential metabolites needed for egg formation. Another reason for low egg production may be the diversion of nutrients to thermoregulation instead of production activities. Vasodilatation brings much blood and metabolites into the periphery (areas not covered with feather, skin, comb, and wattle) at the expense of the internal organs at the body core.

In the present study, honey was found to be effective in improving egg production in laying chickens only in the first 4 weeks of administration, while its effect on egg weight was positive only in the first 1 week. Honey contains various vitamins, electrolytes, natural antioxidants, and phenolic compounds (Blasa et al. 2006) that may help chickens overcome the negative demands imposed by HS. Comprehensive report on the effect of honey in drinking water of broiler chickens was given by Abioja et al. (2012). Honey reduced pulse and respiratory rate during heat spell and helped improve calcium metabolism in broiler chickens. By-products of honey such as honey slum gum (Babarinde et al. 2011), propolis (Chen et al. 2009), and bee pollen (Wang et al. 2005; Hašcík et al. 2012) had been used in broiler production and found to be effective.

Long-term administration of honey to laying chickens was causing a decrease in EP. The reason may be adduced to the higher dosage than needed by the birds or longer duration of treatment. Recently, Osakwe and Igwe (2015) reported that dietary inclusion of honey up to 20 ml reduced heart rate, respiratory rate, and rectal temperature but improved egg characteristics. This seems to be in contrast to the findings of the present study. The authors however did not report on the effect on egg production. Analyses of the data were on overall means for 84 days while the present study was phased. As good as honey is, there is a need for caution in its application in poultry production. Avwioro et al. (2012) reported that there were infiltrations of fat cells in the liver tissue of albino rats which was dosage-dependent. This may result in nonalcoholic fatty liver disease or in other unpleasant conditions harmful to health if abused. Alagwu et al. (2009) had earlier stated that chronic consumption of unprocessed Nigerian honey resulted in decreased bile flow, increased bile cholesterol, and decreased plasma cholesterol in albino rats.

In the present study, 16-week administration of honey in drinking water of laying chickens caused hypertrophy of follicle, giving it a deformed shape instead of the usual round shape. The mechanism is yet to be elucidated. In literature, however, deformed follicle in avian is associated with stress. Egg yolk proteins and lipids are synthesized in the liver, and they are transported via the blood to the ovary (Reece 2009). Any disruption in this may lead to deformation of follicles and other components of egg. Wilson et al. (2011) reported in a study involving the effects of honey on liver integrity that histological analysis of the liver showed distortion of the radial arrangement of the sinusoids, hepatic necrosis, and desquamated wall of the central vein in the treated groups, while the control rats appeared normal. The damage noticed was dose-dependent. The authors concluded that chronic consumption of honey may increase the risk of hepatic damage.

Future Directions

Various questions could be raised from the assertion of this study: Would the administration of other known antioxidants such as vitamins A, C, and E elicit similar effect? Which biochemical and metabolic pathways are disrupted by overdose of honey? What components of honey elicited the effects observed? Will administration of lower doses of honey over a long period improve egg production during heat spell? These and many more are object of future research into the use of honey as a local source of antioxidants for ameliorating the detrimental effects of heat stress in chickens in the face of ensuing climate change. Also, the use of other naturally occurring materials containing substances with antioxidant properties such as lime juice and coconut water could be explored.

Conclusion

Any effort leading to increased resilience in the face of the ensuing climate change and variability is worthy. The use of honey, a naturally occurring liquid that contains various antioxidants to combat HS in chickens, is advocated. However, to ensure improvement in egg production in laying chickens during the hot season, duration of honey supplementation up to 20 ml per liter of water should not exceed first 4 weeks. Heat stress has been a menace in chicken egg production. If the predicted and observed rise in atmospheric temperature (IPCC 2007) is not abated, it may spell doom for egg production all over the world, especially in the tropical regions. High ambient temperature causes reduction in egg production, size, and quality. The use of honey by rural farmers, which is relatively cheap, readily available, and easy to use, may be a way of building resilience in the face of climate change.

References

Abioja MO (2010) Temperature-humidity effects on egg fertility and evaluation of vitamin C and cold water on broiler growth in hot season. Ph.D thesis Department of Animal Physiology, University of Agriculture, Abeokuta, Nigeria. 165pp

Abioja MO, Ogundimu KB, Akibo TE, Odukoya KE, Ajiboye OO, Abiona JA, Williams TJ, Oke OE, Osinowo OA (2012) Growth, mineral deposition and physiological responses of Broiler Chickens offered honey in drinking water during hot-dry season. Int J Zool 2012:1–6

Abioja MO, Abiona JA, Williams TJ, Smith OF, James IJ, Oke OE, Daramola JO, Osinowo OA (2015) Egg fertility and hatchability in Avians broiler-breeder hens under humid tropical conditions. Niger J Anim Prod 42:21–27

Abioja MO, Adekunle MO, Abiona JA, Sodipe OG, Jegede AV (2016) Laying performance, survival rate, egg quality and shell characteristics in laying pullets offered honey in drinking water during hot season. Agric Trop Subtrop 49(1–4):12–19

Adebiyi FM, Akpan I, Obiajunwa EI, Olaniyi HB (2004) Chemical/physical characterization of Nigerian honey. Pak J Nutr 3:278–281

Adetuyi FO, Ibrahim TA, Jude-Ojei O, Ogundahunsi GA (2009) Total phenol, tocopherol and antibacterial quality of honey Apis mellifera sold in Owo community, Ondo State, Nigeria. Elec J Env Agricult Food Chem Title 8(8):596–601

Akanmu MA, Echeverry C, Rivera F, Dajas F (2009) Antioxidant and neuroprotective effects of Nigerian honey. In: Proceedings of the Nueroscience meeting planner. Society for Neuroscience, Washington, DC

Alagwu E, Nneli RO, Okwari OO, Osim EE (2009) Bile secretion in albino rat following chronic honey intake. Niger J Physiol Sci 24(2):203–206

Ali MN, Qota EMA, Hassan RA (2010) Recovery from adverse effects of heat stress on slow-growing chicks using natural antioxidants without or with sulphate. Int J Poult Sci 9(2):109–117

Aljadi AM, Kamaruddin MY (2004) Evaluation of the phenolic contents and antioxidant capacities of two Malaysian floral honeys. Food Chem 85:513–518

Avwioro GO, Onyije FM, Atoni AD, Nduku A (2012) Non-alcoholic fatty liver disease following administration of unprocessed Nigerian honey. Adv Biol Res 6(4):141–145

Ayo JO, Obidi JA, Rekwot PI (2011) Effects of heat stress on the well-being, fertility and hatchability of chickens in the northern guinea savannah zone of Nigeria: a review. ISRN Vet Sci 2011:1–10

Babarinde SA, Oladunjoye IO, Ojebiyi OO, Oyedeji SA (2011) Inclusion of honey bee slum gum in broiler chicken feed. Int J Agric Biol 13(5):781–785

Bedford M (2000) Removal of antibiotic growth promoters from poultry diets: implications and strategies to minimise subsequent problems. Worlds Poult Sci J 56:347–365

Blasa M, Candiracci M, Accorsi A, Piacentini MP, Albertini MC, Piatti E (2006) Raw *Millefiori* honey is packed full of antioxidants. Food Chem 97:217–222

Chen YJ, Huang AC, Chang HH, Liao HF, Jiang CM, Lai LY, Chan JT, Chen YY, Chiang J (2009) Caffeic acid phenethyl ester, an antioxidant from propolis, protects peripheral blood mononuclear cells of competitive cyclists against hyperthermal stress. J Food Sci 74(6):H162–H167

Deng W, Dong XF, Tong JM, Zhang Q (2012) The probiotic *Bacillus licheniformis* ameliorates heat stress-induced impairment of egg production, gut morphology, and intestinal mucosal immunity in laying hens. Poult Sci 91:575–582

Estevinho L, Pereira AP, Moreira L, Dias LG, Pereira E (2008) Antioxidant and antimicrobial effects of phenolic compounds extracts of Northeast Portugal honey. Food Chem Toxicol 46:3774–3779

Gaughan J, Lacetera N, Vartorta SE, Khalifa HH, Hahn L, Mader T (2009) Response of domestic animals to climate challenges. In: Ebi KL, McGregor GR, Burton I (eds) Biometeorology for adaptation to climate variability and change. Springer Science + Business Media B.V., Dordrecht, pp 131–170

Gheldof N, Wang XH, Engeseth NJ (2003) Buckwheat honey increases serum antioxidant capacity in humans. J Agric Food Chem 51:1500–1505

Grieve D (2003) Heat stress in commercial layers and breeders. Techn Bull Hy-Line Int Iowa HLST 19(1):1–3

Hašcík P, Elimam I, Garlík J, Kacániová M, Cubon J, Bobko M, Abdulla H (2012) Impact of beer pollen as feed supplements on the body weight of broiler *Ross 308*. Afr J Biotechnol 11(89):15596–15599

Hegazi A, Abdou AM, Abd Allah F (2013) Influence of honey on immune response against Newcastle disease vaccine. Int J Basic and Appl Virol 2(1):1–5

Hisas L (2011) The food gap: impacts of climate change on food production: a 2020 perspective. FEU-US Universal Ecological Fund, USA, p 55

IPCC (2007) Intergovernmental panel on climate change: AR4-The intergovernmental panel on climate change 4th assessment report. Jackson Institute, University College, London

Lara LJ, Rostagno MH (2013) Impact of heat stress on poultry production. Animals 3:356–369

MacLeod M, Gerber P, Mottet A, Tempio G, Falcucci A, Opio C, Vellinga T, Henderson B, Steinfeld H (2013) Greenhouse gas emissions from pig and chicken supply chains – a global life cycle assessment. Food and Agriculture Organization of the United Nations (FAO), Rome

Mashaly MM, Hendricks GL, Kalama MA, Gehad AE, Abbas AO, Patterson PH (2004) Effect of heat stress on production parameters and immune responses of commercial laying hens. Poult Sci 83:889–894

Mohamed A, Ali A, Molham A (2002) Antioxidant activities and total phenolics of different types of honey. Nutr Res 22:1041–1047

Nardone A, Ronchi B, Lacetera N, Ranieri MS, Bernabucci U (2010) Effects of climate changes on animal production and sustainability of livestock systems. Livest Sci 130:57–69

Omafuvbe BO, Akanbi OO (2009) Microbiological and physico-chemical properties of some commercial Nigerian honey. Afr J Microbiol Res 3(12):891–896

Osakwe I, Igwe R (2015) Physiological responses of laying birds fed honey and vitamin C in drinking water. In: Conference on international research on food security, natural resource management and rural development. Humboldt-Universität zu Berlin and the Leibniz Centre for Agricultural Landscape Research (ZALF) Tropentag, Berlin, September 16–18

Oyelowo OT, Adekunbi DA, Dada KA (2014) Protective role of Nigerian honey on sperm indices and testis in sucrose-fed rats. Bangladesh J Med Sci 13(2):180–189

Ramnath V, Rekha PS, Sujatha KS (2008) Amelioration of heat stress induced disturbances of antioxidant defense system in chicken by Brahma Rasayana. Evid Based Complement Alternat Med 5(1):77–84

Reece WO (2009) Functional anatomy and physiology of domestic animals, 3rd edn. Lippincott Williams and Wilkins, USA, pp 403–441

Sahin K, Onderci M, Sahin N, Gursu MF, Khachik F, Kucuk O (2006) Effects of lycopene supplementation on antioxidant status, oxidative stress, performance and carcass characteristics in heat-stressed Japanese quail. J Therm Biol 31(4):307–312

Vidanarachchi JK, Mikkelsen LL, Sims I, Iji PA, Choct M (2005) Probiotics: alternatives to antibiotic growth promoters in monogastric animal feeds. Recent Adv Anim Nutr Australia 15:131–144

Wang J, Jin GM, Zheng YM, Li SH, Wang H (2005) Effect of bee pollen on development of immune organ of animal. Zhongguo Zhong Yao Za Zhi 30:1532–1536

Wasagu RSU, Shehu S, Mode YD (2013) Comparative proximate composition and antioxidant vitamins contents of two honey varieties (light amber and dark amber) from Sokoto State, Nigeria. Bayero J Pure Appl Sci 6(2):118–120

Wenk C (2003) Herbs and botanicals as feed additive in monogastric animals. Asian–Australasian J Anim Sci 16:282–289

Wilson JI, George BO, Umukoro GE (2011) Effects of honey on the histology of liver in adult Wistar rats. Biol Med 3(1):1–5

Zhang GF, Yang ZB, Wang Y, Yang WR, Jiang SZ, Gai GS (2009) Effects of ginger root (*Zingiber officinale*) processed to different particle sizes on growth performance, antioxidant status, and serum metabolites of broiler chickens. Poult Sci 88(10):2159–2166

Climate Variability and Change in Guinea Savannah Ecological Zone, Nigeria: Assessment of Cattle Herders' Responses

37

Ayansina Ayanlade and Stephen M. Ojebisi

Contents

Abstract

Understanding climate variability and change impacts on the livelihood of farming communities, especially the herders, is essential in designing effective technological and policy interventions for increasing the adaptive capacities of the herders. This study aims at assessing the herders' responses to climate variability and changes using the Guinea Savannah region of Nigeria as a case study. The assessment was made through qualitative and qualitative approaches, and the household survey was used to identify the responses of herder to climate variability and change. Palmer drought severity index (PDSI) method was used to assess the drought events in the study area. Results revealed severe drought in recent years. From survey results, 97.5% of the herders identified drought as the major extreme weather event affecting livestock. These scenarios have led to the migration of cattle herders from the northern part of the study area toward the southern part in recent years, in order to adapt to change in climate. During

A. Ayanlade (✉) · S. M. Ojebisi
Department of Geography, Obafemi Awolowo University, Ile-Ife, Nigeria
e-mail: aayanlade@oauife.edu.ng; ojebisimuyiwa@yahoo.com

W. Leal Filho (ed.), *Handbook of Climate Change Resilience*,
https://doi.org/10.1007/978-3-319-93336-8_18

prolonged dry spell and drought, the majority of cattle herders covered several distances. Some migrate between 5 km and 10 km from their sojourn to get water and green pasture. The study recommended that for herders to improve their adaptive capacities during climate change there is an urgent need for adequate support from the government and stakeholders in Nigeria.

Introduction

Climate is not static but dynamic, and its elements have shown significant variability in recent years, which vary with time and space. Based upon the unevenness nature of climate, diverse definitions of climate change exist in the literature. But, the consensus of these definitions is that climate change is the modification in the composition of climate over a long period of time, whether the cause was natural or anthropogenic, resulting in changes of internal processes or external forces. The foremost definition is the one by the Intergovernmental Panel on Climate Change (IPCC) report. In this report, climate change is described as any changes in climate over time due to natural changes or as a result of human activities (McCarthy 2001; Change 2007). The United Nations Framework Convention on Climate Change (UNFCCC) further defined climate change as direct or indirect activities of human, leading to change in global atmosphere components that create changes in natural climate variability observed over a period of time (Protocol 1997). The key point from all these definitions is that climate change manifests in four main ways: slow changes in mean climate conditions, changes in the inter-annual and seasonality/patterns of climatic elements, increased frequency of extreme events, and rapid climate changes causing catastrophic shifts in ecosystems (Tompkins and Adger 2004; O'Brien et al. 2017). Other studies have shown that climate is rapidly changing with disruptive impacts in recent years and that change is progressing faster than any seen in the last 2000 years. The IPCC (2007) ascertains that the global average temperature has increased by at least 0.6 °C in the past few decades. Scenarios from climate models have proven that the global average temperature has increased by 0.3–0.7 °C in the last century with an estimation of an increase of about 1.4 °C, which may likely increase to 5.8 °C by the end of the twenty-first century (Parry et al. 2007). Change in climate patterns affects both natural ecosystems and human activities such as cattle rearing.

Climate change is a threat to cattle sustainability, as climate determines the level of fodder sufficiency and its sustainability in grass-based farming systems (Bellard et al. 2012; Reeves et al. 2017). The effects of climate variability and change on cattle rearing cannot be overstated. The basic effect of the climate variability and change on herders is the loss of livestock, which resulted in a reduction in the food supply of the population. The reductions in number of livestock are due to death caused by disease and starvation and as herders sell livestock to raise money to buy food for the cattle, because there is insufficient green pasture for cattle to graze on. Nonetheless, the response of herders to climate variability and change varies greatly, from almost none to extensive. It has been observed that communities (cattle herders) who are vulnerable to climate variability and change hazards often fail to act, or do very little, to minimize the destructive effects of climate variability and change (Grothmann and Patt 2005; Martin et al. 2008). Other studies have shown that response to climate stress is determined by some factors. There are several factors responsible for herders' failure to carry out coping strategies against potential climate variability and change risk (Leach et al. 1997; Mortimore and Adams 2001; Davies 2016; Sattler et al. 2018). Thus, it is apparent that there is enough background to warrant a study that will assess cattle herders' responses to climate variability and change in Guinea Savannah region of Nigeria. This is because there is a little scientific research information on the subject of this study, that focuses on assessment of herders' responses to climate variability and change in Guinea Savannah ecological zone of Nigeria. Therefore, this study will further contributes to existing knowledge on the subject. The primary effect of climate variability and change in Guinea Savannah region is the incidence of prolonged drought and intervening dry spell which has severely affected water and vegetation composition and distribution, thereby affecting the availability of pasture for feeding cattle. This has caused occasional imbalances between the demand for food and its supply For example, in Nigeria (Lockett et al. 2000; Ayanlade et al. 2018), livestock production, and forestry contributes a substantial share of national income, export earnings, rural employment, and nonfarm enterprises. These sectors are responsible for providing food security for both the rural and urban populations (Abaje et al. 2016; Wossen et al. 2017). The livestock sector has, over time, experienced declining animal productivity and fluctuations in contributions to GDP and export earnings. The study of climate variability and change is indispensable, particularly in the Savannah region, where poor and vulnerable herders have always suffered the extreme climate events. The knowledge of climate-induced extreme events such as droughts and intervening spell is very limited in terms of its characteristics, its impact on farmer livelihood, farmers' coping strategies, and future welfare implications, although there are diverging opinions on how well farmers are coping with climate change and its destructive effects on livestock production (Tambo and Abdoulaye 2012; Olayide et al. 2016; Papaioannou 2016).

Little attention was paid to drought incidence in the Guinea Savannah despite the fact that it is always devastating whenever it occurs because of less anticipation for such event. Climate variability is less noticeable and virtually unpredictable by the local herders. Also, the focus of many researchers was on Sahel and Sudan Savannah because it was seen as the largest cattle-rearing region and it is a zone of recurrent drought. But recently due to the climate perturbation in the Sahel and Sudan Savannah region, there is a serious incursion of cattle herders into the Guinea Savannah ecological zone of Nigeria, thereby making this ecological zone a major cattle-rearing region (Jabbar 1993; Bourn and Wint 1994). However, an understanding of the impacts of climate variability and change on the livelihood of farming communities, especially the herders, is essential in designing effective technological and policy interventions for increasing the adaptive capacities of the herders. Therefore, this study filled the knowledge gap regarding vulnerability toward climate variability at the local level. This study attempts to assess the vulnerability of cattle herders to current climate change and extreme climate events (drought) in major pastoral villages within the Guinea Savannah region. This part of Nigeria is selected for this study because it has become one of the major cattle-producing areas of the country, which has suffered greatly from climate variability and change. Cattle herders have encroached greatly to the Guinea Savannah because of several advantages of this ecological zone to cattle production. Therefore, in this chapter, this study makes an assessment of climate variability and change to determine the spatiotemporal variation and changing trend of the climatic condition in the Guinea Savannah of Nigeria. It will also determine the extreme weather events to know the degree of vulnerability of the herder's community to the climate extreme events. Climate variability has resulted in several extreme climate events such as drought, intervening dry spell, flood, storm, and cyclone among others. But the commonest of them in the Guinea Savannah of Nigeria are drought, intervening dry spell, and global warming. This is because one of the ecological zones in Nigeria with the same climatic characteristics is the Guinea Savannah. Yet, within Guinea Savannah ecological zones, climate also varies with respect to time and space. The herders in this region are going to be most vulnerable to climate variability and climate change due to their heavy dependence on natural pasture which is under the threat of climate variability and change.

Study Area and Data Acquisition

This case study focused on the assessment of extreme weather events through examination of spatiotemporal variations in rainfall and temperature, their impacts on water available for cattle production, and the responses of herders to climate variability/change in the western part of Guinea Savannah ecological zone of Nigeria. The study was conducted among the Fulanis and Bororojoes in the Guinea Savannah ecological zone of Nigeria (Fig. 1). The Fulani and Bororojoes are the largest populations of pastoralist in West Africa (Hukill 2017). Guinea Savannah zone has a unimodal rainfall distribution with the average annual temperature and rainfall of about 27.3 °C and 1051.7 mm, respectively. The climates and soil type of all Guinea Savannah of Nigeria give opportunities for mass agricultural production. The area is endowed with a good climatic condition which favors the mass production of the livestock of which the study will focus on one (cattle). Others include sheep, goat, and poultry. These make the environment unique. Also, the climate and the soil types of the zone give the possibility of production of different types of tuber crop and cereal crop. The Fulanis in the Guinea Savannah have a considerable low population; this may be because they do not usually intermarry with other groups. Unfortunately, the climate variability and change has drastically reduced the suitability of Guinea Savannah for the production of the cattle and affected the livelihood of the herders' communities.

The assessment was made through qualitative and quantitative approaches, and the household survey was used to identify the responses of herder to climate

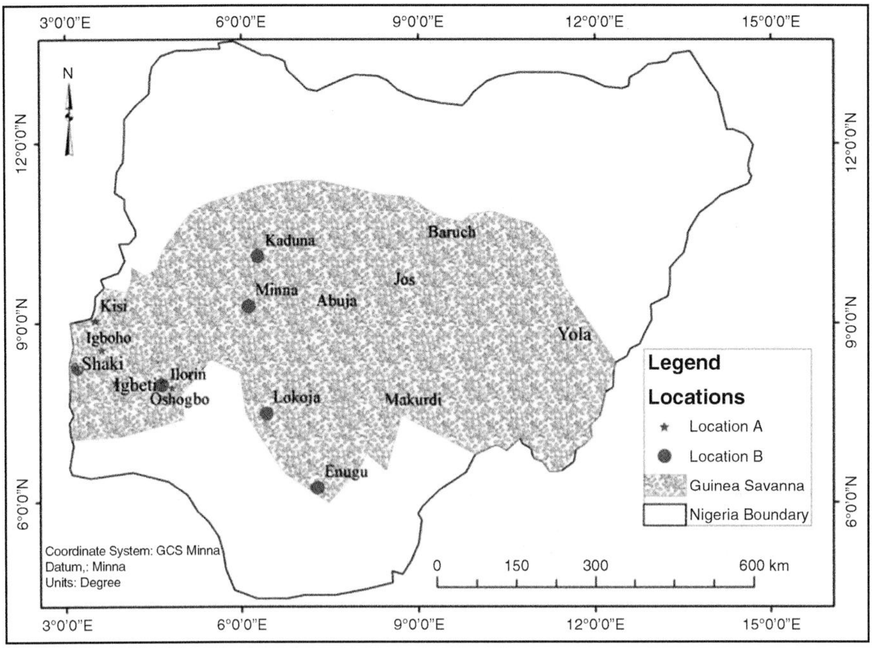

Fig. 1 Map of Nigeria showing the study area and the sites where climate data (circle) and survey data (asterisked) were collected

variability and change. Daily temperature and rainfall data were used. The monthly rainfall and temperature data were used in this study. These data were used to determine the trend of drought in the last three decades. These two climatic data were used in this study because they go a long way in helping us to examine the spatiotemporal variability in rainfall and temperature in the Guinea Savannah of Nigeria and to identify the major drought events in the study area over the past three decades. Studies have shown that the two most important elements of climate that determine the occurrence of drought are temperature and rainfall; the occurrence of drought is evident in high temperature and decline or no rainfall. This gave an overview of the previous pattern of the rainfall and temperature in the past 30 years. It enabled us to know how many occurrences of drought were recorded in the past 31 years. These increased our understanding of the responses of cattle herders to climate variability and change in the Guinea Savannah. According to Camuffo and Jones (2002), historical records have begun to be recognized as valuable sources for investigating climate history and human response to extreme weather events.

To acquire social data, questionnaire and in-depth interviews were used in this study, to understand the herders' response to climate variability in the Guinea Savannah ecological zone of Nigeria. To achieve this, questionnaire surveys were carried out targeting two different groups: the herdsmen and the herder's wife in the field and their hut, respectively. A total number of 100 questionnaires were administered using random sampling method, while the head of herders (Serikis) were interviewed based on purposive sample method. The questionnaires were administered in 6 herders' communities: 12 from Eyenkorin (10 herdsmen and 2 herdsmen's wives), 19 from Igbeti (16 herdsmen and 3 herdsmen's wives), 19 from Igboho (16 herdsmen and 3 herdsmen's wives), 16 from Ilorin (12 herdsmen and 4 herdsmen's wives), 16 from Kisi (12 herdsmen and 4 herdsmen's wives), and 18 from Shaki (12 herdsmen and 6 herdsmen's wives). Information on extreme climate events collected through the survey include water availability and extreme climate event which was the major threat to the herders' cattle, major source of water supply, average amount of water a cattle consume daily, and the distance cover in seeking for water during rainy and dry season and during the prolonged drought. The study also adopted face-to-face interviews with herders' head in each village. The interview is an important data gathering technique – involving verbal communication between the researcher and the subject. Interviews are commonly used in survey designs and in exploratory and descriptive studies. Studies have shown that in-depth interview (unstructured interview), especially face-to-face or personal interview, has the advantage of being one of the best ways of collecting quality data; it is preferable when the subject matter is very sensitive, if the question is very complex, or if the interview is likely to be lengthy. This method was adopted for the study of response to drought and coping strategies of rearing cattle in Guinea Savannah region of Nigeria, to acquire high-quality data, because of the complexity and sensitivity of this study. Face-to-face interviewing is one of the foremost widely used methods in obtaining social survey data. There are many advantages of using face-to-face interviews, such as the use of visual aids and the detection of social cues and body language; through the use of face-to-face interviews, body language and facial expressions are more clearly identified and understood. Face-to-face interviews allow for more in-depth data collection and comprehensive understanding. Another reason for using this qualitative method of data collection lies in its inherent ability to know the mind of the respondent. According to Grover and Singh (2004), the qualitative approach enables a researcher to explore the minds of their informants.

Data Analysis

Palmer drought severity index (PDSI) method was used to assess the drought events in the study area. PDSI is a measurement of dryness based on recent precipitation and temperature. It was developed by meteorologist Wayne Palmer in 1956. In this study the PDSI classifications ranges from 3.0 to 0.50, 0.49 to −0.49, −1.0 to −2.99, and −3.0 to −4 which represent wet, near normal, moderate drought, and severe drought, respectively (Palmer 1965). PDSI was used in this study for the analyses of annual and seasonal rainfall variability, respectively. It was also used to assess the trend of rainfall during dry season early rainfall season and late rainfall seasons, patterns of onset and length of both dry and rainy season, and assessment of the overall intensity and within-season variability of rainfall. Descriptive statistics such as mean, percentage, and tabulation were used to determine herders' adaptation strategies to climate change, especially the distance covered searching for water during drought.

Results and Discussion

Demographic Characteristics of the Respondents

Table 1 shows the demographic characteristics of the respondents which indicated that the sampled population has 80% male and 20% female. This indicated that the herdsmen are directly involved in the rearing of the cattle; they faced the heat of the climate variability and change and always bear the lost due to low productivity of cattle. They also knew more about the reaction of the cattle to the extreme weather events and responded when needed in order to ameliorate the impact of climate variability and change. Notwithstanding, it is important to know the effects of climate variability and change and the herdsmen's wives' responses as well. The age of the respondents was as follows: 18 years of age was 23%, 18–30 years was 42%, 27% were between 31 and 45 years, and 8% were above 45 years of age. This shows that rearing of cattle is not limited to a specific age but cut across different ages. It also ascertained that views of people of different ages were used in the study. The results showed that majority of the respondents were without formal education; 99% were without formal education, with only 8% having primary education and 1% percent with tertiary education. The result indicates a high level of illiteracy among the herders and their wives. It has been reported in the literature that one of the challenges facing smallholder farmers in developing countries is low level of education (Kalirajan and Shand 1985). This has resulted in their inability to use modern methods in their cattle production. This could be as a result of their nomadic life and their choice of cattle rearing and farming as their major occupation. Owing to their high level of illiteracy, the majority of them could not turn to modern adaptation strategies. Also, many of the herders could not benefit from government program on radio and television on climate variability and change.

Figure 2 shows distribution and positions of the respondents to the herders. It was observed that majority of the respondents (30%) were herdsmen, that is, the owner of the cattle. 22% of them were hireling, 28% were herdsmen children; this indicates that they were tending the herd of their father. Majority of the Fulanis engage in cattle rearing from their tender age. This also reflects the choice of cattle rearing as the major occupation of this ethnicity, because they got to know the detail of this occupation from their childhood. Figure 2 also indicates that 20% of the respondents are herders' wives. The population engaged in cattle rearing were adequately represented. This made the sample a good representation of the whole cattle herders in the Guinea Savannah ecological zone of Nigeria, though there was no accurate population of the herders, because of their constant migration and inaccessibility of most of them (Blench 2004).

Table 1 Characteristics of the respondents

Variables	Mean or percentage
Gender of the respondents	
Male	80%
Female	20%
Age of the respondents (in years)	
<18	23%
18–30	42%
31–45	27%
>45	8%
Marital status of the respondents	
Single	22%
Married	78%
Educational status of the respondents	
No formal education	91%
Primary education	8%
Tertiary education	1%
Monthly income of the respondents (in #)	
<5000	44%
5000–15,000	38%
15,001–25,000	5%
25,001–35,000	11%
>35,000	2%
Number of cattle in the herd of the respondent	
<20	28.8%
21–60	42.5%
61–100	10%
>100	18.8%

Source: Field survey

Drought in Guinea Savannah of Nigeria

This study also identified the major extreme climate events through evaluation of the intensity of the dry spells in the study area. Figure 3 shows the annual PDSI over the period of study and the drought severity in the study sites. The results indicated years: 2000, 2007, and 2010, as periods of severe drought (Fig. 3). There are many years with high drought intensity; 17 years out of 31 years are above the normal level. It also can be inferred from the analysis that year 2007 was a drought year; it has the highest dry spell intensity and low amount of rainfall as indicated in. The herders in this region grossly suffer from extreme climate events. Droughts have a lot of effects on cattle production. It determines the quality and quantity of the pasture in the rangeland. Severe drought event reduces the amount of green pasture and its rate of regeneration after being grazed upon by the cattle. During the interview with the herders, the majority of them claim that "prolonged drought reduce the amount of

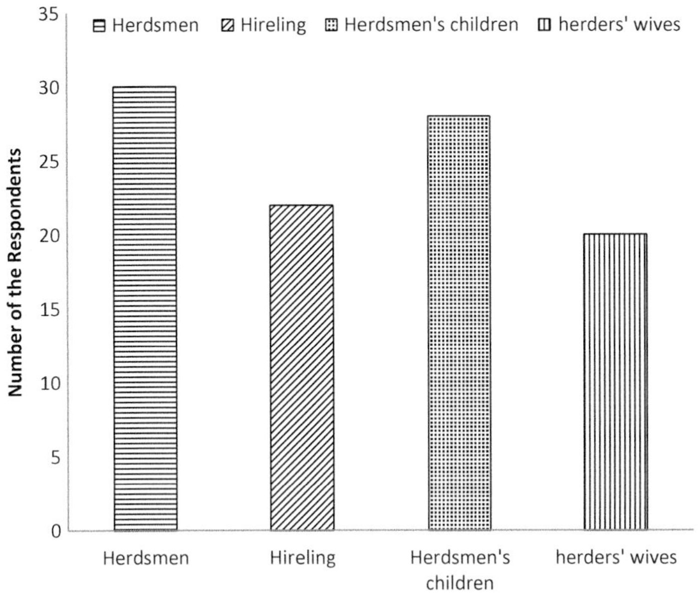

Fig. 2 Distribution and positions of the respondents to the herd

water in the water bodies (rivers) or make some to dry up." This has increased the stress of herders in search for water and food for their cattle. This is because cattle cannot adapt to water restriction. Majority of herders revealed that they "move from one place to another," while some herders "sell their cattle to raise enough money to buy animal feed while other fails to act or do little to ameliorate the effects of increasing dry spell intensity."

Out of 31 years in Ilorin, 23 years are above the normal dry spell intensity with only 8 years which are wet years. This ascertains extreme weather events such as drought. This might be as a result of lesser quantities of water for cattle production during the dry season. It also affects the availability of green pasture for the cattle. Rainfall and soil fertility determine the available green pasture. Lokoja shows a considerable variation in the intensity of dry spell. The analysis indicates that more than 50% of the years under study had high dry spell intensity. The wettest year in the dry season over the period of study was 1986, while the driest years are 1998, 2007, and 2010. This implies drought. The results from the survey show the lesser or no green grasses for the cattle to graze during prolonged drought. And most of the streams and rivers dried up. Out of 31 years, in Minna, there are only 7 years which were below normal of dry spell intensity with about 23 years above the average and 1 year on the normal level. This is a strong indication of the high dry spell in this region. The analysis implied that the climate of this region is fast changing with a negative effect. During the wet season, the grasses grow in abundance, and it regenerates very quickly, but in times of prolonged drought, there are little or no grasses. When the cattle feed on the available pasture, it will not regenerate or

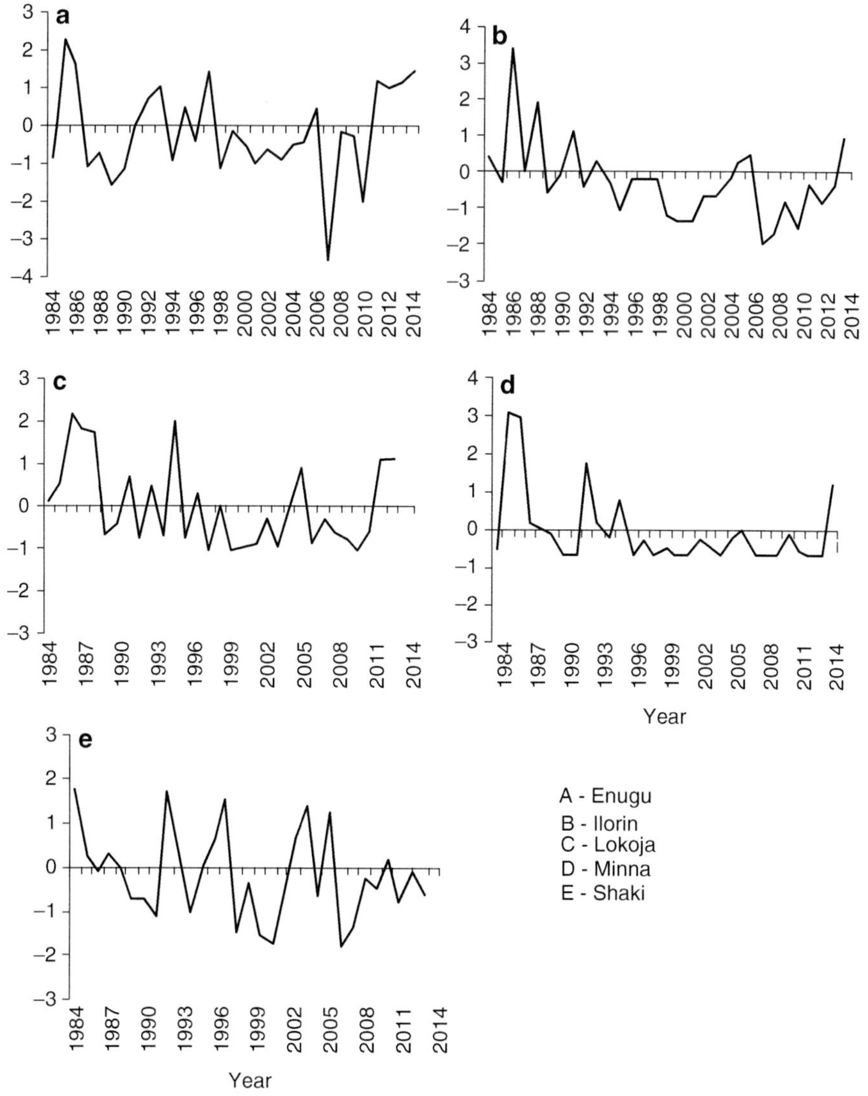

Fig. 3 Severity of drought in the study area between 1984 and 2014 based on recent precipitation

regenerate slowly even in a good soil because of lack of moisture. This has led to a reduction in the efficiency of cattle production in this region, as confirmed by the herders. It has resulted in different responses from the herders, depending on their knowledge and resources at hand. Only the herders with good responses were able to thrive during the time of dryness, while herders with low adaptive capacity suffered greatly. Therefore, it is necessary to help farming communities by promoting adaptation measures and make them more invincible to associated climatic variations (Adger et al. 2003).

From the analysis, it was observed that the region had high dry spell and low amount of rainfall. This also indicated that the region has less amount of surface water which is the major source of water for cattle production. It also indicates that the little available surface water got dried up in time of prolonged dry spell and drought. This result has corroborated previous studies that surface water is on decline in some African countries such as Benin, Burkina Faso, Ghana, Mauritania, Niger, and Nigeria (Dinar et al. 2012). This left the herders with little or no water and green pasture for cattle rearing. These, in turn, have left the cattle with little water and fodder shortage. High temperature also reduces the comfort of the cattle and led to the outbreak of diseases among the cattle, greater incidence of diseases and pest (insects and mites). These were confirmed by the herders and their Seriki. In this context of increasing risks, herders need to identify and implement methods of adaptation. The research will also go a long way in helping the herders in identifying vulnerabilities and ways of reducing them, for example, genetic materials more resistant to disease or water stress, management practices, etc. (Oyhantçabal et al. 2010).

Temperature Trend and Events in the Study Area.

Cattle are sensitive to change in climate. Therefore change in temperature affects the cattle products. Increase in the mortality of young cattle reduces the rate of increase in the number of cattle in a herd. Many herders have suffered from this decrease in temperature to the extent that they cannot regain in time of normal temperature, especially the free range herd. Therefore, for effective production of cattle in this area, semi-free range or padlocking method of production will be suitable. The trends showed an increase in both annual maximum and minimum temperature averages in recent years for all stations (Fig. 4). It also shows that there is a recent increase in both the minimum and maximum temperature of Ilorin (Fig. 4a). This implies that the temperature of the area has increased the problem of water and food availability for cattle production. These have altered the effective production of cattle in this region. Increase in temperature reduces the amount of green pasture available to the cattle to feed on, thereby causing stress. Herders who have little resources will lose a large number of cattle either to death or be sold to raise money to buy animal feed for the remaining cattle. This implies that increased temperature leads to low cattle production. This, in turn, reduces the quantity and quality of cattle products in the country. Though the cattle herders have responded to the problem of increase in temperature in some ways, their responses were unable to adequately ameliorate its effect on cattle production in this region. This is as a result of their low adaptive capacity.

Figure 4 shows the variation in the temperature of Lokoja between 1984 and 2014. The temperature of the area is fast changing with adverse effects on the communities living in the area and their activities. One of such activities that are sensitive to temperature is cattle production. It occupied an integral position in the food security of the country. As stated by the herders and their village head, temperature is one of the major climatic elements affecting the cattle. It determines

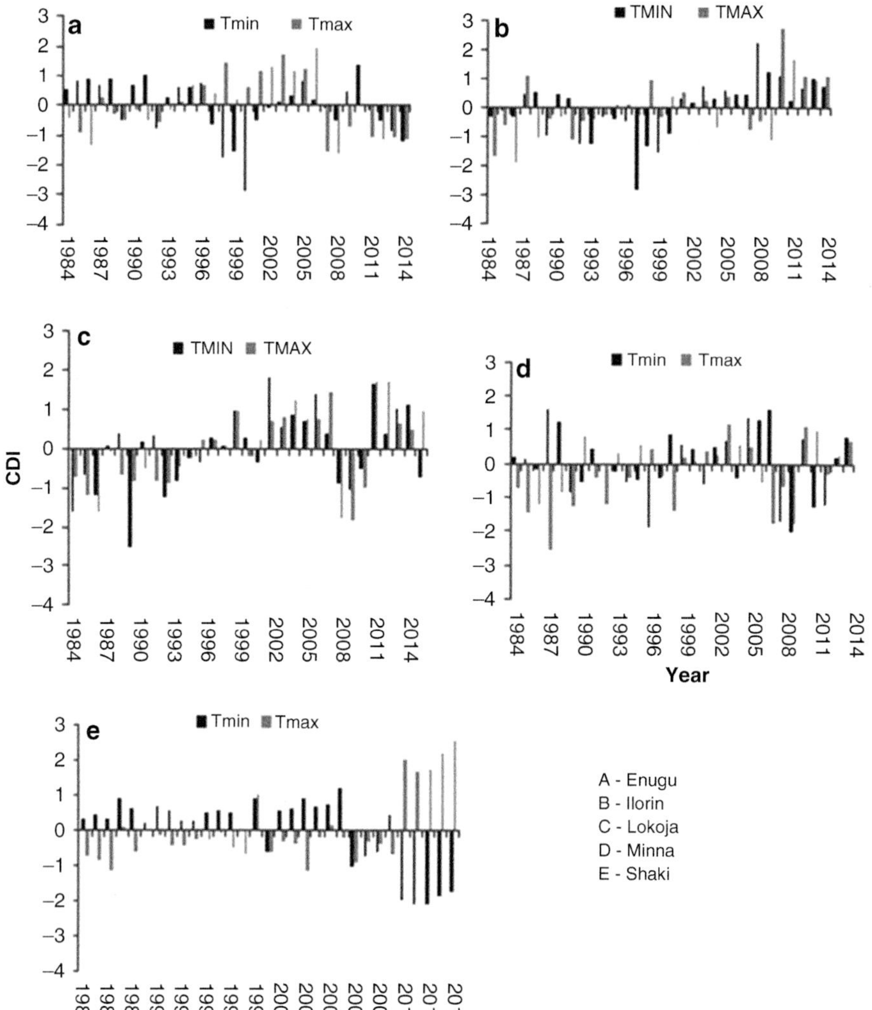

Fig. 4 The average maximum (T_{max}) and minimum (T_{min}) temperature in Enugu, 1984–2014

their comfort, state of health, drinking habit, availability of water, and composition of grasses in grassland (pasture). Increase in temperature consequently increases the rate of evaporation leaving most of the river with little or no water. It also reduces the comfort of cattle making cattle to use more water for evaporative cooling (during hot weather). More so, increase in temperature increases the rate of evapotranspiration which makes most of the green grasses to dry leaving the cattle with no succulent feed but fibrous and less digestible feed needs. They need more water to maintain coarse feed in the cattle gut. Increase in temperature (hot weather) reduces the availability of water, consequently increasing the water needed by effective cattle production. In those years of less fodder combined with the water stress, herders' responses are very sharp to avert or ameliorate this unforeseen doom (Wall and Smit 2006; Ati et al. 2009).

It is evident from Fig. 3 that there is variation in the pattern of both annual maximum and minimum temperature of Shaki. The trends of both maximum and minimum temperature showed that temperature is increasing. This implies that temperature of this area is on the increase. It also implies that there are two extremes in this region. These have altered the effective production of cattle in this region. Therefore herders can only thrive in their production using breeds which are resistant to high temperature. But the majority of the herders have no enough money to buy new breed of cattle. Increase in temperature reduces the amount of green pasture available to the cattle to feed on, thereby causing food stress. This has lots of impacts on the food security in the country (Rigolot et al. 2017; Lopez-Ridaura et al. 2018). Therefore, there is need to help the herders to raise the standard of the cattle production in this region.

Cattle Herders Adaptation Strategies to Climate Change

Table 2 shows how climate variability and change determine the grazing pattern of the cattle in Guinea Savannah. It was observed from the respondents that out of about 87.5% of the respondents' cattle grazing on natural pasture in the rainy season, only 15% graze on natural pasture during the dry season. Majority of the cattle grazing on the natural pasture during the dry season or drought are herders who live close to large water bodies in Guinea Savannah ecological zone of Nigeria. The result from interview further indicates that majority of the cattle that feed on plant residue (fibrous and less digestible feed) need more water. Fibrous and less digestible feed need more water; this is because water is needed to maintain coarse feed in the cattle gut. This widens the mismatch of water supply and water demand of cattle as there is lesser water available with the occurrence of drought to meet increased demand for water caused by drought. This mismatch necessitated the constant migration of Fulani herders from an area with low water availability and dry pasture to where there are perennial water supply and green pasture.

The result in Table 3 shows that majority of the respondents (42.5%) covered less than 1 km in seeking for water when the climate is normal, with only 3.8% covering more than 10 km. But during the drought, majority of the respondents (47.5%) covered between 5.1 km and 10 km, and majority of the respondents (45%) covered more than 10 km during the dry season. This implies that climate variability and change pushed the herders far from their residence in search of water for their cattle. Odjugo (2010) have examined an increase in temperature and a decrease in frequency and intensity of rainfall over the last century. He asserted that there has been an increase of 1.1 degree in the average temperature and a reduction in average rainfall by 81 mm between 1901 and 2005. These are evidence of drought in Nigeria, especially in Savannah region of Nigeria. The causes, duration, and impacts of

Table 2 The grazing pattern of the cattle

Variables	Mean or percentage
During rainy season	
Natural pasture	87.5%
Crop residue	8.8%
Pasture regrowth after bush burning	3.7%
During the dry season	
Natural pasture	15.0%
Crop residue	30.0%
Planted pasture	3.7%
Pasture regrowth after bush burning	51.3%

Source: Field survey

Table 3 Distance covered in seeking water and green pastures

Variables	Mean or percentage
During the rainy season	
<1 km	42.5%
1.1 km to 5km	35.0%
5.1 km to 10 km	18.8%
>10Km	3.8%
During the dry season	
<1 km	2.5%
1.1 km to 5 km	11.3%
5.1 km to 10 km	41.3%
>10 km	45.0%
During the drought	
<1 km	3.8%
1.1 k to 5 km	12.5%
5.1 km to 10 km	47.5%
>10 km	36.3%

Source: Field survey

drought have recently been the focus of a number of publications (Cooke 1985; Glantz and Ausubel 1988). The effects of drought on herders have recently been discussed by Hogg (1980), Oba and Lusigi (1987), and Adams et al. (1998). Tacher (1983), Western and Finch (1986), and Campbell (1999) had earlier observed the effects of drought on livestock populations.

It was obvious from the results of the present study that in time of normal rainfall, water was always available for the cattle within 1 km of the herders' residents. This was as a result of input of water into the rivers from the rainfall. Cattle also need less or no water for evaporative cooling because of cool weather. But in time of drought, intermittent rivers easily dried up through evaporation. Also, increase in temperature reduced the comfort of the cattle, making them need more water for evaporative cooling. This high demand for water with a deficit in water made herders in this region to move more than 5 km away from their residence, in seeking for water. Majority of the herders moved more than 10 km away from their residence in seeking for water during the dry season because only the large bodies of water were able to survive the high rate of evaporation with no input from rainfall. Drought is a threat to livestock farm sustainability, as climate partly determines the level of fodder self-sufficiency and its sustainability in grass-based farming systems.

What is obvious from this study is that pattern of movement, level of education, low resources, and their poor knowledge of climate variability and change influence their adaptive capacity (Brooks et al. 2005; Grothmann and Patt 2005). This is in accordance with the work of Smit and Wandel (2006), Ainsworth et al. (2008), and Cooper et al. (2008) that there is a strong relationship between climate and human migration patterns; it also suggested that migration is the main strategy of people in the rural area in climate change circumstance. The combination of both climate and survey data in the assessment of herders' responses to climate variability and change makes this study a thorough assessment. Climate data ascertained the variability and change. Climate varies greatly across time and space. It also indicated that climate is hard to predict especially by the local farmers because of its inconsistent trend. As also noted by Ayanlade et al. (2017), that farmers in southwestern Nigeria have an adequate perception of climate change and its impacts on livestock agriculture. Also, survey data indicated the degree to which climate perturbation had injurious effects on cattle production in Guinea Savannah ecological zone of Nigeria. The effectiveness of the responses of cattle herders was not adequate enough to raise the standard of cattle production in the Guinea Savannah ecological zone of Nigeria. It has been reported by earlier studies that there are several factors responsible for herders' low adaptation strategies against potential climate variability and change risk. Some of these factors include low level of education, lack of confidence on protective measures or perceived ability to carry out these measures, lack of adequate and timely information, and lack of available resources, while low income for better adaptation innovation or costs of response actions are seen as the main important factors (Leach et al. 1997; Mortimore and Adams 2001; Davies 2016).

Conclusion

The study assessed extreme climate events and the responses of cattle herders to climate variability and change in the Guinea Savannah ecological zone of Nigeria. Analysis of the rainfall data using the Palmer drought severity index (PDSI) within the period of the study showed that there were fluctuations of rainfall both in the dry and rainy season. It varied greatly with the incidence of extreme climate events such as drought, intervening dry spell, and seasonality of rain. Analysis of the intensity of the dry spell and temperature showed that there is an increase in the dry spell in Guinea Savannah of Nigeria. The temperature in the Guinea Savannah also varied greatly with the trend showing an increase in temperature of the area. Increase in the intensity of dry spells and temperature led to high evapotranspiration and high temperature. The PDSI also showed a prolonged change in rainfall pattern, which was also confirmed by the herders in the region. This evidence of climate change is in line with the work of previous scholars (O'Brien and Vogel 2003; Hristov et al. 2017; Rigolot et al. 2017). The analysis further indicated that rainfall declined in its amount, frequency, and intensity.

Herders perceived drought as the major extreme climate event that had critical effects on the production of the cattle in the area. Increased temperature and drought reduced the quantity and quality of water and pasture for cattle production. It also reduced the comfort of both the herders and the cattle (Thornton et al. 2009, 2015; Panthi et al. 2016; Descheemaeker et al. 2017). In a recent paper, Thornton et al. (2009) examined some of the direct and indirect impacts of climate change on livestock and livestock systems. First, one of the most evident and important effects of climate change on animal husbandry is changes in composition of feed resources (Thornton et al. 2006, 2009). Increasing drought leads to change in composition and development capacity of grass and crop feed. According to Thornton et al. (2009), although indirectly, effects on feed resources to a large extent determine cattle productivity, the ability of the ecosystems for the grazing system, prices of stoves and grains, changes in feeding options, and grazing management. Impacts of climate change on the availability of feed resource for livestock are shown in four aspects (Thornton et al. 2009). Herders' principal adaptation strategies include reduction of daily water consumption, migration, and turning to other sources of water. Survey analysis also indicated that the distances covered in search for water and food for the cattle were increased by the occurrences of drought and during the dry season. Majority of the herd which feed on natural pasture during normal rainy season pasture feed on plant regrowth and plant residues during the dry season and in time of drought or related extreme weather events. There is a need for herders to improve their adaptive capacities during climate change, and this can only come from the sufficient support from the government and stakeholders in Nigeria.

References

Abaje I, Sawa B, Iguisi E, Ibrahim A (2016) Impacts of climate change and adaptation strategies in rural communities of Kaduna state, Nigeria. Ethiopian J Environ Stud Manage 9:97–108

dams RM, Hurd BH, Lenhart S, Leary N (1998) Effects of global climate change on agriculture: an interpretative review. Clim Res 11(1):19–30

Adger WN, Huq S, Brown K, Conway D, Hulme M (2003) Adaptation to climate change in the developing world. Prog Dev Stud 3:179–195

Ainsworth EA, Beier C, Calfapietra C, Ceulemans R, Durand-Tardif M, Farquhar GD, Godbold DL, Hendrey GR, Hickler T, Kaduk J (2008) Next generation of elevated [CO_2] experiments with crops: a critical investment for feeding the future world. Plant Cell Environ 31:1317–1324

Ati O, Stigter C, Iguisi E, Afolayan J (2009) Profile of rainfall change and variability in the northern Nigeria, 1953-2002. Res J Environ Earth Sci 1:58–63

Ayanlade A, Radeny M, Morton JF (2017) Comparing smallholder farmers' perception of climate change with meteorological data: a case study from southwestern Nigeria. Weather Clim Extreme 15:24–33

Ayanlade A, Radeny M, Morton JF, Muchaba T (2018) Rainfall variability and drought characteristics in two agro-climatic zones: an assessment of climate change challenges in Africa. Sci Total Environ 630:728–737
Bellard C, Bertelsmeier C, Leadley P, Thuiller W, Courchamp F (2012) Impacts of climate change on the future of biodiversity. Ecol Lett 15:365–377
Blench R (2004) Natural resource conflicts in north-central Nigeria. Mandaras, London
Bourn D, Wint W (1994) Livestock, land use and agricultural intensification in sub-Saharan Africa. In: Pastoral Development Network. Overseas Development Institute, London
Brooks N, Adger WN, Kelly PM (2005) The determinants of vulnerability and adaptive capacity at the national level and the implications for adaptation. Glob Environ Chang 15:151–163
Campbell DJ (1999) Response to drought among farmers and herders in southern Kajiado District, Kenya: a comparison of 1972-1976 and 1994-1995. Hum Ecol 27:377–416
Camuffo D, Jones P (2002) Improved understanding of past climatic variability from early daily European instrumental sources. Clim Chang 53:1–4
Cooke H (1985) The Kalahari today: a case of conflict over resource use. Geogr J 75–85
Change IPCC (2007) Fourth assessment report. IPCC, Geneva
Cooper P, Dimes J, Rao K, Shapiro B, Shiferaw B, Twomlow S (2008) Coping better with current climatic variability in the rain-fed farming systems of sub-Saharan Africa: an essential first step in adapting to future climate change? Agric Ecosyst Environ 126:24–35
Davies S (2016) Adaptable livelihoods: coping with food insecurity in the Malian Sahell. Springer, London
Descheemaeker K, Zijlstra M, Masikati P, Crespo O, Tui SH-K (2017) Effects of climate change and adaptation on the livestock component of mixed farming systems: a modelling study from semi-arid Zimbabwe. Agric Syst 159:282–295
Dinar A, Hassan R, Mendelsohn R, Benhin J (2012) Climate change and agriculture in Africa: impact assessment and adaptation strategies. Routledge, London
Fratkin E, Roth EA (1990) Drought and economic differentiation among Ariaal pastoralists of Kenya. Hum Ecol 18(4):385–402
Glantz MH, Ausubel JH (1988) Impact assessment by analogy: comparing the impacts of the Ogallala aquifer depletion and CO2-induced climate change. Societal responses to regional climate change, 113–142
Grothmann T, Patt A (2005) Adaptive capacity and human cognition: the process of individual adaptation to climate change. Glob Environ Chang 15:199–213
Grover N, Singh KN (2004) Cultural geography, form and process: essays in honour of prof. AB Mukerji. Concept publishing company, New Delhi
Hogg R (1980) Pastoralism and impoverishment: the case of the Isiolo Boran of northern Kenya. Disasters 4(3):299–310
Hristov A, Degaetano A, Rotz C, Hoberg E, Skinner R, Felix T, Li H, Patterson P, Roth G, Hall M (2017) Climate change effects on livestock in the northeast US and strategies for adaptation. Clim Chang:1–13
Hukill JP (2017) International development and West African pastoralism. Analysing conceptions of livestock ownership (Master's thesis, UiT Norges arktiske universitet)
Jabbar MA (1993) Evolving crop-livestock farming systems in the humid zone of West Africa: potential and research needs. Outlook Agric 22:13–21
Kalirajan K, Shand R (1985) Types of education and agricultural productivity: a quantitative analysis of Tamil Nadu rice farming. J Dev Stud 21:232–243
Leach M, Mearns R, Scoones I (1997) Environmental entitlements: a framework for understanding the institutional dynamics of environmental change. Brighton, Institute of Development Studies, University of Sussex
Lockett CT, Calvert CC, Grivetti LE (2000) Energy and micronutrient composition of dietary and medicinal wild plants consumed during drought. Study of rural Fulani, northeastern Nigeria. Int J Food Sci Nutr 51:195–208

Lopez-Ridaura S, Frelat R, Van Wijk MT, Valbuena D, Krupnik TJ, Jat M (2018) Climate smart agriculture, farm household typologies and food security: an ex-ante assessment from eastern India. Agric Syst 159:57–68

Martin WE, Raish C, Kent BM (2008) Wildfire risk: human perceptions and management implications. Earthscan, New York

McCarthy JJ (2001) Climate change 2001: impacts, adaptation, and vulnerability. Contribution of working group II to the third assessment report of the intergovernmental panel on climate changel. Cambridge University Press, Cambridge, UK

Mortimore MJ, Adams WM (2001) Farmer adaptation, change and 'crisis' in the Sahel. Glob Environ Chang 11:49–57

Oba G, Lusigi WJ (1987) An overview of drought strategies and land use in African pastoral systems, Overseas Development Institute Agricultural Administration Unit., pp. 1–33

O'Brien K, Vogel C (2003) Coping with climate variability: the use of seasonal climate forecasts in southern Africa. Ashgate, Burlington

O'Brien MJ, Engelbrecht BM, Joswig J, Pereyra G, Schuldt B, Jansen S, Kattge J, Landhäusser SM, Levick SR, Preisler Y (2017) A synthesis of tree functional traits related to drought-induced mortality in forests across climatic-zones. J Appl Ecol

Odjugo PAO (2010) General overview of climate change impacts in Nigeria. J Hum Ecol 29(1):47–55

Olayide OE, Tetteh IK, Popoola L (2016) Differential impacts of rainfall and irrigation on agricultural production in Nigeria: any lessons for climate-smart agriculture? Agric Water Manag 178:30–36

Oyhantçabal W, Vitale E, Lagarmilla P (2010) Climate change and links to animal diseases and animal production. In: Compendium of technical items presented to the OIE World Assembly of Delegates and to OIE Regional Commissions, pp 169–186

Palmer WC (1965) Meteorological drought, vol 30. US Department of Commerce, Weather Bureau, Washington, DC

Panthi J, Aryal S, Dahal P, Bhandari P, Krakauer NY, Pandey VP (2016) Livelihood vulnerability approach to assessing climate change impacts on mixed agro-livestock smallholders around the Gandaki River basin in Nepal. Reg Environ Chang 16:1121–1132

Papaioannou KJ (2016) Climate shocks and conflict: evidence from colonial Nigeria. Polit Geogr 50:33–47

Parry M, Canziani OF, Palutikof JP, van der Linden PJ, Hanson CE (2007) Climate change 2007: impacts, adaptation and vulnerability, vol 4. Cambridge University Press, Cambridge

Protocol K (1997) United Nations framework convention on climate change. Kyoto protocol, Kyoto 19

Reeves MC, Bagne KE, Tanaka J (2017) Potential climate change impacts on four biophysical indicators of cattle production from western US rangelands. Rangel Ecol Manag 70(5):529–539

Rigolot C, De Voil P, Douxchamps S, Prestwidge D, Van Wijk M, Thornton PK, Rodriguez D, Henderson B, Medina D, Herrero M (2017) Interactions between intervention packages, climatic risk, climate change and food security in mixed crop–livestock systems in Burkina Faso. Agric Syst 151:217–224

Sattler D, Seliger R, Nehren U, de Torres FN, da Silva AS, Raedig C, Hissa HR, Heinrich J (2018) Pasture degradation in South East Brazil: status, drivers and options for sustainable land use under climate change. In: Climate change adaptation in Latin America, Springer

Smit B, Wandel J (2006) Adaptation, adaptive capacity and vulnerability. Glob Environ Chang 16:282–292

Tacher G (1983) Drought and strategy of animal production development in Sahelian countries. Quarterly journal of international agriculture= Zeitschrift fur auslandische Landwirtschaft

Tambo JA, Abdoulaye T (2012) Climate change and agricultural technology adoption: the case of drought tolerant maize in rural Nigeria. Mitig Adapt Strateg Glob Chang 17:277–292

Thornton PK, Jones PG, Owiyo T, Kruska R, Herrero M, Kristjanson P, Notenbaert AMO, Bekele N, Omolo A (2006) Mapping climate vulnerability and poverty in Africa. ILRI, Nairobi (Kenya)

Thornton PK, Van de Steeg J, Notenbaert A, Herrero M (2009) The impacts of climate change on livestock and livestock systems in developing countries: a review of what we know and what we need to know. Agric Syst 101:113–127

Thornton PK, Boone RB, Ramirez-Villegas J (2015) Climate change impacts on livestock. CCAFS Working Paper no. 120. CGIAR Research Program on Climate Change, Agriculture and Food Security (CCAFS), Copenhagen

Tompkins E, Adger WN (2004) Does adaptive management of natural resources enhance resilience to climate change? Ecol Soc 9

Wall E, Smit B (2006) Agricultural adaptation to climate change in the news. Int J Sustain Dev 9:355–369

Western D, Finch V (1986) Cattle and pastoralism: survival and production in arid lands. Hum Ecol 14:77–94

Wossen T, Abdoulaye T, Alene A, Feleke S, Menkir A, Manyong V (2017) Measuring the impacts of adaptation strategies to drought stress: the case of drought tolerant maize varieties. J Environ Manag 203:106–113

Assessment of the Degree of Households' Vulnerability to Climate Variability in Nigeria

38

Abiodun Emmanuel Awoyemi and O. Adeola Olajide

Contents

Abstract

Climate variability is one of the major challenges facing the world today. It is significantly felt on the health, homes, herds, and crops of Africans, including the natural resources upon which many depend to survive, and if adequate measures are not promptly taken, Africa's population might face an urgent crisis. This paper seeks to determine the relationship between crop production and rainfall pattern,

A. E. Awoyemi (✉) · O. A. Olajide
Department of Agricultural Economics, Faculty of Agriculture and Forestry, University of Ibadan, Ibadan, Oyo State, Nigeria
e-mail: enioduntanawoyemi@gmail.com; preciousfunso@yahoo.com

W. Leal Filho (ed.), *Handbook of Climate Change Resilience*,
https://doi.org/10.1007/978-3-319-93336-8_28

vulnerability of households to climate variability, and the effect of climate variability on agricultural productivity across the geopolitical zones of Nigeria. It will increase the level of awareness of respondent's exposure and sensitivity to climate variability in the study area. The findings of this study will help stakeholders to know the relationship between climate variables and their productivity and be able to take appropriate coping strategies. Therefore, it is envisioned that the outcome of this research will contribute in designing program through an elaborate understanding of the degree of household's vulnerability to climate variability which will enable policy makers to formulate policies that will be beneficial to affected persons and areas.

Introduction

Climate variability is the oscillation of the weather condition of a place within a short period of time. Studying of climate variability in agriculture is very crucial because of its adaptability to an average weather condition and vulnerability to unpredictable weather conditions (Risbey et al. 1999 cited in Ajewole 2013). Although the agricultural sector constitutes Nigeria's biggest employer of labor on which over 75% of its citizens depends for their sustenance (Apata 2009; Adebo and Ayelari 2009) and the second largest export earner after petroleum (Ajewole 2013), it stills remain the most vulnerable sector of any economy (Kurukulasuriya and Mendelsohn 2006). The variation in the major climatic parameters (rainfall and temperature) affects the growth and sprouting of major tree crops which often result in yield decline per hectare (Kimengsi 2012).

Climate variability and climate change are significantly felt on the health, homes, herds, and crops of Africans and the natural resources upon which many depend to survive, and if adequate measures are not taken promptly, the African population might face an urgent crisis. According to the slogan of Nature is Speaking Conservation International, "People need nature but nature does not need people." This statement unveils how crucial the effect of environmental degradation could be on human existence. Africa experienced 0.5 °C rise in temperature over the course of the twentieth century with some areas getting warmer than the other (Eriksen 2008).

According to Claudia (2010), Nigeria's experience of climate variability in terms of drier and warmer climates, changes in rainfall pattern, and frequencies of drought in some parts of the country over the years is now a reality. Some areas like the Niger Delta regions receive more than normal rainfall, while other areas, especially the northern region, receive almost no rainfall (Williams et al. 2015). As a result, growing seasons are changing, ecological zones are shifting, and rainfall is becoming more unpredictable and unreliable both in its timing and volume (Brett 2009). Very little has been known on the effect of climate variability across the zones in Nigeria. Thus, this paper will examine among others the relationship between crop production and rainfall pattern, the degree of households' vulnerability to climate variability in Nigeria, and the effect of climate variability on agricultural productivity across the geopolitical zones in Nigeria. It will increase the knowledge of stakeholders on their level of sensitivity and exposure to climate variability and the measure they could take to mitigate its effect.

Vulnerability in this context is the degree to which a system is susceptible to and unable to cope with adverse effects of climate variability (IPCC 2007). According to Notenbaert et al. (2013), there are several factors which limit the capacity of self-protection, diminish access to social protection, and delay recovery or exposure of some households to frequent hazards than others, and these summed up to vulnerability. Thus, vulnerability is a function of adaptive capacity, sensitivity, and exposure. Adaptive capacity is the ability of a system to adjust to the effects of climate variation, while exposure is the extent to which a system is open to climate variation, and sensitivity is the degree to which a system is affected either beneficially or adversely by climate-related stimuli (IPCC 2001).

Agriculture is at the core of environmental vulnerability and major concern of natural resource management, such as land degradation, water scarcity, deforestation, and the threat of biodiversity (Adebo and Ayelari 2009). It is a primary sector that mostly depends on climate variables (Mendelsohn and Dinar 2009). Time of planting, choice of crop to be cultivated, and harvesting time depend directly on the prevailing weather condition in that area (Nastis 2012). According to Easterling (2007) (FAO 2009), there will be an increase in the percentage of the world population that will become more vulnerable to hunger, due to regional agricultural productivity differences caused by climate variability.

In recent time, there was incidence of sudden and unprecedented change in weather condition all over the zones of Nigeria with problems of abnormal rainfall leading to flooding, extreme sunlight during dry season which led to excessive heat, loss of properties, and farms being seriously affected (Adebo and Ayelari 2009).

Methods

Study Area

Nigeria is bordered in the north by Niger Republic, in the west by the Republic of Benin, to the east by the Republic of Cameroon and Chad, and in the south by the Atlantic Ocean. It has a total land area of 923,766 square kilometers. Nigeria's topography ranges from lowland along the coast of the lower Niger to high plateau in the north and high mountains along the eastern border. Nigeria has an ecology which varies from tropical rainforest in the south to dry savannah in the far north, yielding a diverse mix of plants and animal life. It is divided into 36 states and a Federal Capital Territory (FCT). The states are grouped into six (6) geopolitical zones vis-a-vis the North-East, North-West, North-Central, South-East, South-West, and South-South. The southern zones are characterized by high rainfall intensity, long wet season, dense vegetation, rugged topography, and a temperature range of 26–28 °C with small farmholdings. The most common problems of

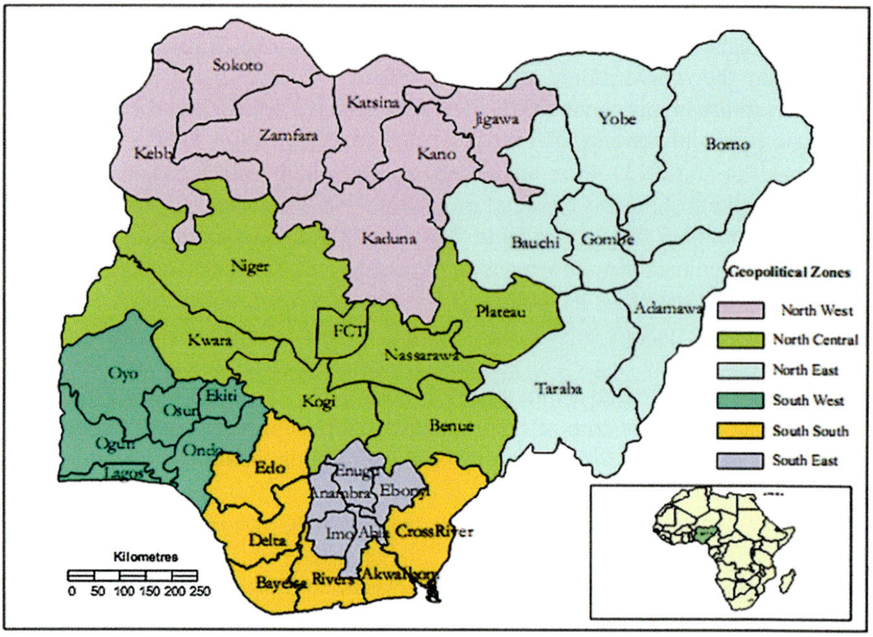

Fig. 1 Nigeria map showing the six geopolitical zones

these zones, i.e., South-West, South-East, and South-South, are flood and water erosion. Maize, cassava, yam, and vegetables are the major crops grown in these zones (Fig. 1).

Type and Sources of Data

The type of data that was used for this study was secondary panel data. The data was sourced from the National Bureau of Statistics (NBS) in conjunction with the World Bank. GHS-Panel data is a nationally representative survey of 5,000 households drawn in a multistage random sampling process across selected enumeration areas in the 36 States. The households were surveyed twice between two waves (2010/2011 and 2012/2013) of an Integrated Surveys on Agriculture (ISA Program).

Analytical Techniques

Data collected was analyzed using STATA 14 software. Method of analysis used in this study includes descriptive analysis, correlation analysis, principal component analysis (PCA), Household Vulnerability Index (HVI), and multiple regression analysis (fixed effect regression analysis).

Table 1 Household Vulnerability Index variables

Adaptive capacity	Sensitivity	Exposure
Household source of income Household size Literacy rate (level of education) (years) Farm size (Ha) Household asset	Temperature variation (°C) Rainfall variability (mm)	Percentage of farmers who used irrigation during rainy season Population density/km^2 Percentage of land managed Consumption of fertilizer/ha (Kg)

Descriptive Analysis

The descriptive analyses such as mean, frequencies, and percentages were used to examine the socioeconomic characteristics of the respondents in the study area.

Correlation Analysis

In other to determine the relationship between crop production and rainfall pattern, correlation analysis was used. This was done against the yield of maize, cassava, sorghum, and rice.

Principal Component Analysis (PCA)

In order to obtain the Household Vulnerability Index to climate change across different zones in Nigeria, the variables listed in Table 1 were first normalized and then principal component analysis (PCA) was performed.

It is essential to normalize the data before carrying out the PCA as this will enable all the variables to have the same standard deviation and the same weight, thereby making it easy for the PCA to calculate the relevant axis so as to get the covariance of all variables so that each of the variables could have the properties of a standard normal distribution with $\mu = 0$ and $\sigma = 1$.

The variables for this study as shown in Table 1 were normalized following the methodology used in Human Development Index (HDI) (UNDP 2006).

Where μ is the mean and σ is the standard deviation

$$z = \frac{x - \mu}{\sigma} \tag{1}$$

PCA is a technique used in extracting some orthogonal linear combinations of variables that most successfully capture common information from a set of variables (Deressa 2008). It is performed in order to assign equal weight to the variables in the data.

The variables were selected based on the signs that come with them. Adaptive variables with positive components were selected. The sensitivity variables and exposure variables with negative signs were also selected. Any variables against the aforementioned rules were dropped. Thus the selected variables enter into Household Vulnerability Index (HVI).

Household Vulnerability Index (HVI)

Household Vulnerability Index (HVI) was used to measure the vulnerability of households in Nigeria to climate variability.

The index was scaled by zones.

The model states that

$$V = (Wa_1X + Wa_2X + Wa_3X + Wa_4X + Wa_5X) \\ - [(Ws_1Y + Ws_2Y) + (We_1Z + We_2Z + We_3Z + We_4Z)] \quad (2)$$

That is,

$$\text{Vulnerability} = (\text{Adaptive Capacity}) - (\text{Sensitivity} + \text{Exposure})$$

where V is vulnerability; X, Y, and Z are adaptive capacity, exposure, and sensitivity, respectively; and W shows that the variables were standardized for normalcy (Madu 2012, 2016) as shown in Table 1.

W_{ai} = Standardized adaptive variables
W_{si} = Standardized sensitivity variables
W_{ei} = Standardized exposure variable

The direction of relationship within the vulnerability indicators always follows that positive is attached to adaptive capacity, while negative value is attached to both exposure and sensitivity. This is justified by Deressa (2008) that areas with households which are highly exposed to issues of climate variability are bound to be more sensitive to damages, assuming constant adaptive capacity. According to Madu (2012), the higher net value indicates lesser vulnerability, and the lower net value signifies greater vulnerability. Thus, it stood as the bases for categorizing the households across the zones into highly vulnerable, moderately vulnerable, and least vulnerable.

The variables used in the HVI analysis were adapted from UNIQUE forestry and land use (2013:12) following Madu (2016).

Hypothesis

H_0: Households in Nigeria are not vulnerable to climate variability.

Multiple Regression Analysis

Fixed effect regression which was selected by Hausman test was used to analyze the relationship between yield and some climatic variables. The Hausman test (1978) was used to decide whether a random effect model or fixed effect model should be used for a panel data model. Hausman test has an asymptotic distribution with a null hypothesis that the fixed effect and random effect do not differ substantially.

From the result of the Hausman test, b which is the fixed effect regression is found to be consistent under the null and alternative hypothesis, while B which is the random effect is found to be inconsistent under the alternative hypothesis and efficient under the null hypothesis. Fixed regression analysis most fits the regression analysis at 1% level of confidence, thus the selection of fixed effect regression.

The model is stated as follows:

$$Y_{it} = \beta_0 + X_{1it}\beta_1 + X_{2it}\beta_2 + U_{it} \tag{3}$$

where

U_{it} is a time-varying residual and i, t denote households and year, respectively. The data is an unbalanced panel of 5,000 households for two waves in Nigeria.

β_1 and β_2 = are the coefficients and slopes of the climatic variables in the model, respectively.

β_0 = is the intercept

X_2 = Climate variables (temperature variation and rainfall variability)

Data Limitation

The data did not include some relevant climate-related variables as such it was impossible to make use of the known variables for measuring exposure such as flooding, erosion, fire outbreak, wind, etc. Thus percentage of farmers who irrigate their land during rainy season, population density/km^2, percentage of land managed, and fertilizer consumption/ha were adopted. Data on climate variables (i.e., rainfall and temperature) and yield of the selected crops for southern zones were also missing.

Results and Discussion

Socioeconomic Characteristics

The result in Table 2 showed that majority (87.37%) of the respondents had one form of education or the other ranging from primary to tertiary institution. The level of education of the respondents might influence the probability of adopting any technology that can improve their level of production and make them resilient to climatic conditions. This is consistent with the result of Oduro-Ofori et al. (2014) that a unit increase in the level of education or years of education of farmers would result in 0.03% increase in production. Also, on the average, it could be seen that the respondents had about 10 years of formal education. This is consistent with Anselm 2011, who said that unemployment might have driven educated young people back to the farm.

Table 2 Socioeconomic characteristics of respondents

Characteristics	Frequency	Percentage
Level of education		
No formal education	631	12.62
Primary education	2,357	47.14
Secondary education	2,383	27.66
Post-secondary education	629	12.58
Sex		
Female	924	18.48
Male	4,076	81.52
Marital status		
Single	184	3.67
Married	4,236	84.72
Divorced	21	0.43
Widowed	559	11.19
Land ownership		
Outright purchase	268	5.35
Rented for cash or in kind	476	9.52
Used free of charge	562	11.25
Distribution by community or family	3,694	73.88
Primary source of income		
Professional	194	3.88
Civil servants	91	1.81
Farming	3,241	64.82
Trading	236	4.73
Artisans	657	13.14
Forestry and mining	581	11.62
Farm size		
0–2.1 ha	4,499	89.98
2.1–4.2 ha	358	7.16
4.2–6.3 ha	69	1.38
Above 6.4 ha	74	1.48

Age

From Fig. 2, most of the households were within the age range of 19–54 years, which suggests they are in their active age. At this age, they are capable of handling tedious farming activities, covering large distances, and will do anything to sustain their existence including adopting adaptation strategies to climate change (Fatuase et al. 2015). This age group is considered by Daniel et al. (2009), as the major stakeholders in agricultural production. They have the strength and energy required to carry out any rigorous activities that are involved in farming. Majority (81.52%) of the respondents were male while the female population was 18.48%. It may be due to the drudgery involved in carrying out farming operations in Nigeria that made the sector to be dominated with male farming households. This, according to Muhammed-Lawal et al. (2012), may be due to the cultural belief of the people which preclude women to go out freely and engage in certain activities such as farming. A good number of the respondents as shown in Fig. 3 have large families,

Fig. 2 Age

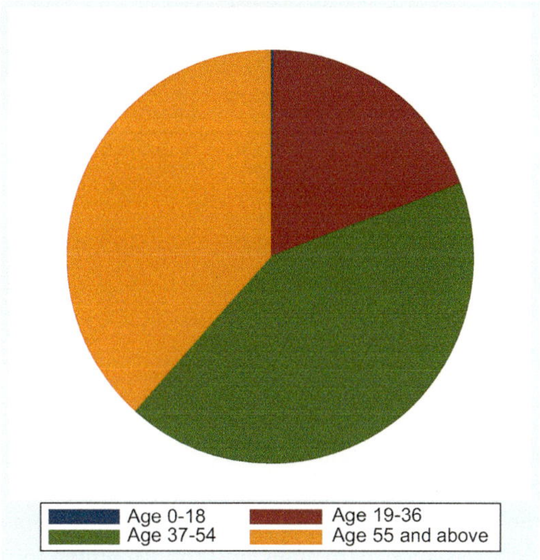

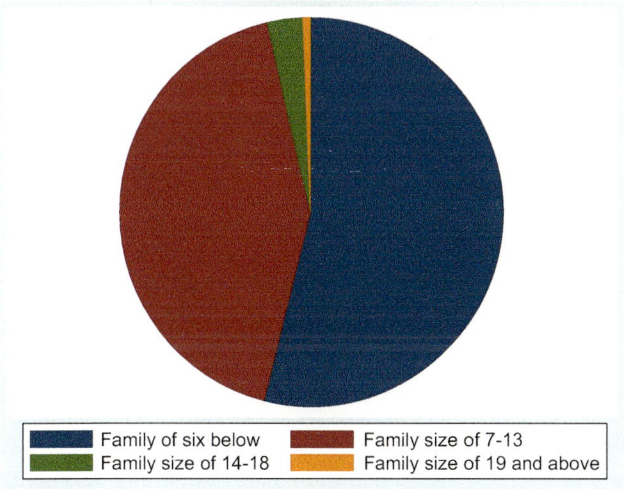

Fig. 3 Household size

i.e., an average of eight members. This according to Abaje et al. (2014) provides cheap labor to the households, increases their farm size, and enables the adoption of some practices which can mitigate the impacts of climate variability. In line with this, Nhemachena and Hassan (2008), Anyoha (2013), and Agbo et al. (2015) said that households with larger sizes have higher likelihood of choosing farming as their means of livelihood compared to households with smaller family size.

Majority (84.72%) of the respondents being married means more support from the wife or wives and more labor to work in the farm which invariably increases

Table 3 Correlation analysis of rainfall variation and crop yield

	Zones					
Variables	North-Central	North-East	North-West	South-East	South-South	South-West
Rice yield (tonnes)	−0.0767	−0.0781	−0.0415	–	–	–
Maize yield (tonnes)	0.1632	−0.0856	−0.0179	–	–	–
Cassava yield (tonnes)	−0.1866	−0.1322	−0.0312	–	–	–
Sorghum yield (tonnes)	0.1137	−0.0170	−0.1667	–	–	–

their levels of production. Anyoha (2013) said that the majority of the respondents being married indicate more responsibility and expectation to meet up with the family demands. It was only a few percentage of the land used in cultivation that was owned by the households under investigation. It was also noted that households made their choice of livelihood depending on the level of technology and access to infrastructures. On the average across the zones, majority of the respondents were into farming compared to other activities. This agrees with Agbo et al. (2015) and Oni and Fashogbon (2013) that farming is the most dominant means of livelihood in Nigeria. The mean farm size was 1.14 ha, with the majority (89.98%) of the respondents planted between 0.1 and 2.1 ha, and agrees with Fatuase et al. (2015) that majority of the farmers in Nigeria had an average farm size of 1.7 ha.

Correlation Analysis of Rainfall Variation and Yield of Selected Crops in Nigeria

Table 3 shows the coefficient of correlation between rainfall variation and yield across the zones in Nigeria. A high correlation exists between rainfall variation and yield of rice in the North-East (−0.0781), rainfall variation and cassava yield in the North-Central (−0.1866), and rainfall variation and sorghum yield in the North-West (0.1667), though they have a negative association. This suggested that the yield of households in this area might have been sensitive to too much rainfall over some times. This might expose them to some hazards by reducing the percentage of land used for farming loss of nutrients through flooding. Fertilizer consumption per hectare (ha) might have increased. These rainfall variation and maize yield in the North-Central (0.1632) also correlate. The relationship between yield and cereal crops and cassava that was observed in the northern part of the country may be as a result of the fact that sorghum is widely consumed in the north and the region has weather and soil characteristics that are suitable for its production (Oni et al. 2009).

Table 4 Principal components (eigenvectors)

Variable	Comp1	Comp2	Unexplained
Household source of income	**0.0715**	0.4372	0.7549
Household size	**0.4738**	0.3044	0.5076
Literacy rate	**0.1875**	0.5138	0.6142
Farm size (Ha)	**0.3222**	0.1127	0.8096
Household asset	**0.1020**	0.4244	0.7596
Mean temperature (°C)	**−0.5657**	0.2770	0.3667
Mean rainfall (mm)	**−0.0706**	−0.1471	0.9648
Population density	**−0.4834**	0.3828	0.3856

Households' Vulnerability to Climate Variability in Nigeria

PCA was run on the variables listed on Table 1 using statistical software (STATA). The PCA of the data set on the vulnerability variables revealed that four components with eigenvalues were greater than one from Table 4 and the table in the appendix below. The tables explain about 48.11% of the total variation in the data set. The first principal components explained 15.95% of the variation, the second components explained 11.72%, the third explained 10.61%, and the fourth explained 9.83%. Based on the definition of PCA, all the indicators that were positively associated with adaptive capacity negatively associated with both sensitive and exposure as adopted by Deressa (2008) and were used for the construction of the vulnerability index. Table 5 shows rural households' vulnerability to climate change across the zones of Nigeria with majority of the households falling in most-highly vulnerable and moderately vulnerable category. It shows that in the North-Central, majority (50.73%) of the households that were interviewed in this area were highly vulnerable to climate variability, while 49.14% were moderately vulnerable and 0.13% were less vulnerable. In the North-East, 66.76% of the households interviewed were highly vulnerable, i.e., about 1,178 households were highly vulnerable and 33.24% were moderately vulnerable. In the North-West, 64.87% households were highly vulnerable, and 35.13% households were moderately vulnerable. From the South-East, 5.49% of the households in this area were highly vulnerable, 94.37 were moderately vulnerable, and 0.14% were less vulnerable. Across the South-South zone, 15.41% of the households were highly vulnerable, and 84.59 were moderately vulnerable. In the South-West zone, 65.81 households were highly vulnerable, and 34.19 were moderately vulnerable.

Thus, 44.81% of the households across the six geopolitical zones were highly vulnerable, 55.14% were moderately vulnerable, while 0.05% were less vulnerable. This shows that the entire houses in this area are vulnerable with the difference depending on the level of vulnerability to climate variability. The level of vulnerability to climate variability showed in Table 5 may be influenced by the choice of variables used as the proxy for the commonly used variables which was a result of the limitation of the data. This might be because of the lower level of literacy in the

Table 5 Vulnerability index of households (HVI) across the zones in Nigeria

Zone	Highly vulnerable	Moderately vulnerable	Less vulnerable	Total
North-Central	496 50.73 22.11	480 49.14 17.40	1 0.13 50.00	977 100.00 19.53
North-East	774 66.76 34.53	385 33.24 13.97	0 0.00 0.00	1,159 100 23.18
North-West	550 64.87 24.57	298 35.13 10.81	0 0.00 0.00	848 100.00 16.97
South-East	49 5.49 2.18	838 94.37 30.41	1 0.14 50.00	888 100.00 17.77
South-South	113 15.41 5.05	621 84.59 22.53	0 0.00 0.00	734 100.00 14.68
South-West	259 65.81 11.56	135 34.19 4.88	0 0.00 0.00	3934 100.00 7.87
Total	2,241 44.81 100.00	2,757 55.14 100.00	2 0.05 100.00	5,000 100.00 100.00

study area, and it aligned with Madu (2016) that the 13 states which experienced high vulnerability were located in the northern geopolitical zones of Nigeria. The table also demonstrates that the households in the northern zones are highly vulnerable compared to the households in the southern zones. This may be due to exposure to drought which has led them to adopt irrigation technique during the raining season, land degradation due to erosion that influences their population per square kilometers, and even percentage of the land managed and depletion of soil fertility which influences the use of fertilizers on their plots.

Effect of Climate Variability on Agricultural Productivity

The result from Table 6 shows that all the variables were significant at 1% confidence interval. Coefficient of variation in rainfall and temperature variability had effect on agricultural yield.

The coefficient of variation in precipitation had a negative significant influence on the yield at 1% confidence interval. This implies that an increase in coefficient of variation (changes) in precipitation will lead to 0.00970 tonnes drop in agricultural yield across the zones of Nigeria. This may be as a result of intense and excess rainfall which might have disrupted the flowering and fruiting of crop across the zones, thus reducing the yield of these crops. This is consistent with Igwe (2013) that very intense rainfall affects the yield of cereal crops especially from the time they

Table 6 Effect of climate variability on agricultural productivity

Variables	Coefficients
Coefficient of variation in rainfall (mm)	−0.00970*** (0.000453)
Temperature variability (°C)	0.01*** (0.00182)
Constant	5.464*** (0.477)
Observations	5,000
R-squared	0.038

start flowering and fruiting because it destroys their inflorescence and fruiting thereby reducing the yield. It is ordinarily expected to have a positive impact on the yield, but where it leads to flooding, erosion, and leaching, its negative influence cannot be outweighed.

Temperature variability increase per degree centigrade will increase the yield of crop by 0.01 tonnes in Nigeria. This may be as result of the fact that some crops are heat-loving crops like millet and sorghum as it helps their fruiting. According to Harold (2015), this is one of the indications of climate variability. Thus, climate variability affects productivity both negatively and positively.

Conclusion and Recommendation

Conclusion

The paper examines the degree of vulnerability of households to climate variability in Nigeria. The study looked at the socioeconomic characteristics of the respondents, the relationship between crop production and rainfall pattern, and the vulnerability of the rural households to climate variability in Nigeria. Though majority of the respondents derive their means of livelihood from agriculture, they could not access enough plot to use for farming as they want because majority of the households were not having plots of their own. It established that there is a relationship between yield of different crops and rainfall in the study area. Considering the result of HVI, 44.81% of the households across the six geopolitical zones were highly vulnerable, 55.14% were moderately vulnerable, while 0.05% were less vulnerable. Thus households in Nigeria are vulnerable to climate variability, and this suggests that all hands need to be on deck as to ensure that the households that are now moderately vulnerable do not become suddenly most vulnerable due to climate vulnerability in the nearest future.

The result also showed that vulnerability decreases down from the north through the southeast. The result of the effect of climate variability on agricultural productivity revealed that rainfall variation and temperature variability had both positive and negative impacts, respectively. This suggests that much effort needs to be concentrated on all the factors that influence agricultural productivity in Nigeria.

Recommendations

Based on the research findings, the following are therefore recommended. There should be more awareness on the influence of climate variability on their productivity. There is need for government to include funding of climate change research and collection of data on climate variables in their budget in order for anyone who wants information on climate variability to be able to gain access to such data in the data bank which collects all the necessary information on climate variability and its related issue. This could also lead to employment generation for young graduates and builds their interest in research and development. Increasing the ability of the people to adapt by strengthening their coping strategy is essential in reducing the vulnerability of households to climate variability. Improved technology on adaptations and mitigations should be encouraged and provided by government and the non-governmental organizations (NGOs) to reduce the households' level of exposure and sensitivity. Policies on social protection scheme which could reduce the vulnerability of households to climate variability are essential at this time. This will reduce their exposure and also increase their adaptive capacity. Incorporating climate variability study into school curriculum in Nigeria at all levels so that there could be more awareness on the exposure to environmental hazards and its implication on agricultural productivity will go a long way in creating awareness on climate change.

Cross-References

▶ Climate Change, Vulnerability, and Adaption Under the Small Farming Households of Konso Community, Southern Ethiopia
▶ Climate Variability and Change in Guinea Savannah Ecological Zone, Nigeria: Assessment of Cattle Herders' Responses
▶ Assessment of Climatic Variability Effect on Millet Production and Yield
▶ Effect of Climate Variability on Crop Income in Central Highlands of Ethiopia

Appendices

Principal component/covariance	**Number of observation = 5000**
	Number of comp = 2
	Trace = 10.55122
Rotation (un-rotated = principal)	**Rho = 0.2767**

Component	Eigenvalue	Difference	Proportion	Cumulative
Comp1	1.68269	0.445701	0.1595	0.1595
Comp2	1.23699	0.117776	0.1172	0.2767
Comp3	1.11921	0.0818093	0.1061	0.3828
Comp4	1.03741	0.0591666	0.0983	0.4811
Comp5	0.978238	0.0951876	0.0927	0.5738
Comp6	0.883051	0.0360326	0.0837	0.6575
Comp7	0.847018	0.053797	0.0803	0.7378
Comp8	0.793221	0.0641303	0.0752	0.8130
Comp9	0.729091	0.0376251	0.0691	0.8821
Comp10	0.691466	0.138633	0.0655	0.9476
Comp11	0.552832	.	0.0524	1.0000

References

Abaje IB et al (2014) Climate Variability and Change, Impacts and Adaptation Strategies in Dutsin-Ma Local Government Area of Katsina State, Nigeria. J Geogr Geol 6(2):103

Adebo GM, Ayelari TA (2009) Climate change and vulnerability of fish farmers in southwestern Nigeria. African J Agric Res 6(18):4230–4238

Agbo FU et al (2015) Effects of climate variability on the choices of livelihood among farm households in Anambra State, Nigeria. African J Agric Res 10(144):413–4141, 29

Ajewole OC (2013) Adoption of crop management options as coping strategies to climate variability among arable crop farmers in Ekiti State, Nigeria. IJAFS 4(1&2):392–400

Anyoha NO (2013) Socio-economic factors influencing climate change adaptation among crop farmers in Umuahia South Area of Abia State, Nigeria. Net J Agric Sci 1(2):42–47

Apata TG (2009) Analysis of climate change perception and adaptation among arable food crop farmers in south western Nigeria. In: Paper presented at the conference of international association of agricultural economics, pp 2–9

Brett H (2009) Food and agriculture, features, climate change threat to food security at: http://www.people and Planet.Net/doc

Chukwuma O (2015) Climate change and conflict in Nigeria: the Boko haram challenge. Am Int J Soc Sci 4(2)181–190

Claudia R (2010) Climate change and hunger: Africa's smallholder farmers struggle to adapt. The Agricultural Economics Society and the European Association of Agricultural Economists 9(3):21

Daniel JD et al (2009) Economics of cotton production in southern parts of Adamawa State, Nigeria. Int J Crop Sci 1(1):73–80

Deressa T (2008) Analysis of the determinants of Farmers' choice of adaptation methods and perceptions of climate change in the Nile Basin of Ethiopia. IFPRI discussion papers no. 798. International Food Policy Research Institute, Washington, DC

Easterling W (2007) Climate change impacts, adaptation and vulnerability. In: Parry M et al (eds) pp 273–313

Eriksen S (2008) Climate change in eastern and southern Africa: impacts, vulnerability and adaptation, global environmental change and human security report. GECHS Project, University of Oslo, Oslo

Fatuase AI et al (2015) Determinants of adaptation measures to climate change by arable crop farmers in Owo local government area of Ondo state, Nigeria. World Rural Obs 7(1). http://www.sciencepub.net/rural

Food & Agricultural Organization of the United Nations (FAO) (2009) The state of food insecurity in the world: economic crises: impacts & lessons learned. Electronic Publishing Policy & Support Branch Communication Division, FAO

Harold M (2015) Cereal crops: rice, maize, millett, sorghum, wheat. Feeding Africa 2015, An action plan for African agricultural transformation. Abdou Diouf international conference centre

Igwe KC (2013) Investigation of the direction of causality and effect of climate change on food grain output in Nigeria. Int J Eng Sci (1970–2010) 2(3):49–54

Intergovernmental Panel on Climate Change (IPCC) (2001) Climate change 2001: impact, adaptation and vulnerability. In: McCarthy JJ et al (eds) Contribution of working Group II to the third assessment report of the intergovernmental panel on climate change. University Press Cambridge, Cambridge

Kimengsi J (2012) Climate variability and cocoa production in meme division of cameroon: Agricultural development policy options. Greener J Agric Sci 3(8):606–617

Kurukulasuriya P, Mendelsohn R (2006) A Ricardian analysis of the impact of climate change on African crop land. CEEPA discussion paper no. 8. Centre for Environmental Economics and Policy in Africa. University of Pretoria

Madu AI (2012) Spatial vulnerability assessments of rural households to climate change in Nigeria: towards evidence-based adaptation policies. In: A paper presented at the 2012 Berlin conference on the human dimensions of global environmental change held at Environmental Policy Research Centre Freie Universitat, Berlin from 5–6th Oct 2012

Madu AI (2016) Rurality and climate change vulnerability in Nigeria: assessment towards evidence-based rural development policy. In: A paper presented at the 2016 Berlin conference on global environmental change held at Freie Universität Berlin from 23–24 May 2016

Mendelsohn R, Dinar A (2009) Climate change and agriculture: an economic analysis of global impacts, adaptation and distributional effects. Edward, Cheltenham, UK

Muhammed-Lawal A et al (2012) Assessment of vulnerability of framing households to climate change in Ekiti State Nigeria. Ethiop J Environ Stud Manag 5(2)

Nastis AS (2012) Climate change and agricultural productivity. African J Agric Res 7(35): 4885–4893

Nhemachena C, Hassan R (2008) Determinants of African farmers' strategies for adapting to climate change: multinomial choice analysis. African J Agric Res Econ 2(1):83–104

Notenbaert A et al (2013) Derivation of a household-level vulnerability index for empirical measures of adaptive capacity and vulnerability. Reg Environ Chang 13:459–470

Oni OA, Fashogbon AE (2013) Food poverty and livelihoods issues in rural Nigeria. Chapter 5: in poverty, price volatility, efficiency and the impacts of population shifts. Accessed on 9/2/2015. Available at: http://www.afjare.org/resources/issues/vol_8_no2/Chapter%205.pdf

Oni O et al (2009) Trends and drivers of agricultural productivity in Nigeria, International Food policy Research Institute. Nigeria Strategy Support Program (NSSP), report 1

Oduro-Ofori E et al (2014). Effects of education on the agricultural productivity of farmers in the offinso municipality. Int J Develop Res 4(9):1951–1960

UNDP (2006) Human Development Report United Nations Development Programme. Available at: http://hdr.undp.org/hdr2006/statistics/

Williams JJ et al (2015) An assessment of impact and adaptation strategies to climate change by local indigenous farmers of Michika local government area of Adamawa State, Nigeria

Impacts of Climate Change on Agriculture, Livelihoods, and Women in Nile Delta, Egypt

39

Marwa R. Hafez

Contents

Abstract
Climate change is a significant and a fundamental threat to global food security in the whole world and has a significant threat to agricultural sector in Egypt.

M. R. Hafez (✉)
Institute of Graduate Studies and Research (IGSR), Alexandria University, Alexandria, Egypt
e-mail: marwa.hafez@yahoo.com

W. Leal Filho (ed.), *Handbook of Climate Change Resilience*,
https://doi.org/10.1007/978-3-319-93336-8_53

Without adaptation to climate change in agriculture sector, it will not be possible to achieve sustainable development especially in food security, and many people will lose their source of living. Putting adaptation and resilience into account before and during planning and national action plans will make great difference.

Egyptian rural communities face other kind of pressures in addition to climate change that reduce their adaptability and affect the agriculture sector and its workers, which must be taken into account when considering the impact of climate change on agriculture, such as the lack of governmental information, government coordination with farmers, and guidance on adaptation to climate change, among others.

Climate change impacts are not experienced in the same way by both genders. Women are suffering in many ways to reach to information and deal with the changing climate either that relate to agriculture or the extreme events. According to literature there is a gender gap in agriculture; addressing this gap before planning and making decisions could lead to better outcomes and butter futures.

Keywords
Climate Change · Agriculture · Women · Nile Delta · Vulnerability · Adaptation

Introduction

The agriculture sector is one of the most critical and influential sectors of climate change because carbon emissions are very much linked to the agricultural system. Agriculture accounts for one-third of global carbon emissions (Kutywayo et al. 2013), making it one of the largest contributors to climate change. High temperatures and humidity, among others, are affecting crops and animals and cause disease transmission, low productivity, and thus food insecurity.

In Egypt, agriculture is the largest sector with about 14% of GDP (gross domestic product) and the largest number of workers (Central Agency for Public Mobilization and Statistics, CAPMAS 2017). However, this vital sector is considered to be vulnerable to the risks of climate change for some reasons. The most important of which is the high reliance on natural resources, which are also under pressure due to climate change, such as in the Nile River, groundwater and rainwater, as well as using traditional agricultural systems and the lack of access to information and technology.

The increase in temperature is expected to reduce the availability of many areas in Egypt (and the world) to produce the crops that are famous in such places, leading to many economic problems for a large number of the population, starting with those working in agriculture (farmers) through the workers in the transportation of goods (drivers), wholesalers, retailers, ending by consumers who may not be able to afford the high costs of this series of economic rises, making the impact of climate change on agriculture and crops very deep especially in developing countries, along with low crop productivity that means that consumers are expected to have to pay more money to buy worse quality than they are used to.

El-Ramady et al. (2013) indicated that countries must increase agricultural production by 70% by 2050 in order to cope with the expected large population growth and change of diets, bearing in mind that agricultural productivity must be increased while conserving natural resources and the environment, such as conserving plants while minimizing the use of toxic pesticides or using plant nutrients without depending on nitrogen or phosphorus, which are harmful to the environment.

As agriculture is considered to be one of the most important sectors of the world, real and rapid solutions must be taken, away from temporary solutions if there is an intention to embark on sustainable agriculture.

Despite the importance of the large agricultural sector in providing food supplies to the population and being the first sector in Egypt, it is considered the most sensitive and affected by the change in temperature and any slight changes in climate. The sensitivity and vulnerability of the agriculture sector depend on the impacts of climate change on its ability to deal with current, future, and projected climatic conditions.

Global climate models suggest that the world will see an increase in average annual temperatures by 2100, which means more challenges in many parts of the world, especially islands, coastal areas, and poor countries.

Climate change has many phenomena, for example:

- Change in rainfall systems, the winter will be more humid and summer will be drier, for example, or the winter will be colder and the summer will be hotter and so on.
- Change in frequency and intensity of storms.
- Change in soil temperature and humidity.

Among these risks, there are some which corrupt the agricultural system:

- Change in productivity of crops, vegetables, and grains
- The negative effects on some crops
- Increased pests, diseases, weeds, and fungi
- Crop damage due to extreme weather events (e.g., heat waves, frosts)
- Increased drought stress and increased demand for water in the future
- Increased evaporation
- Increased pressures on livestock due to extreme weather events (as in hot waves)

The world population is projected to increase from 7 billion in 2011 to nearly 9 billion in 2050, and the growth in developing countries, which already have considerable pressure on their natural resources, will be more concentrated than the developed ones (Abd-El Wahab 2005). In its 2010 report, FAO (Food and Agriculture Organization) indicated that the Green Revolution Technology led to a doubling of food production between 1950 and 2010 (FAO 2010), an increase of only 10% in the area allocated for agriculture and production.

Although climate change may disrupt the ability of green systems to operate and multiply their production as it hits the soil directly, there is a big need to increase production by 2050 by about 70% as response to the increasing in population, rising in living standards, and changes in diet preferences.

Regardless of the fact that climate change became an undeniable reality, there are still a number of skeptics in this matter. When planning, nobody should be left behind, so we will deal with this matter as natural resources (agricultural land, water resources, and soil) depletion, as well as the urban crawling and turning fertile lands to residential cities, so we have to develop plans and strategies to ensure food security in the world.

We must prepare to cope and adapt to climate change from this moment, because if it happens quickly, there will be no enough time to adapt, and the destruction of farmland and famines will enforce people to leave the land and migrate, causing political problems due to overcrowding throughout the world. In case of gradual impacts, the damage will be simple and manageable. In all cases, we must be prepared for any confrontation.

Based on studies of Egypt's situation in the face of climate change, the most affected sectors are water resources and agriculture. The problem with agriculture is very important and more complicated as it needs water resources that are already threatened. As well as the sea level rise due to climate change is also a threat to the densely populated Nile Delta, which includes a large infrastructure and a very large proportion of the country's fertile agricultural land. Coastal agricultural lands in the delta are also low-lying lands, posing an inevitable need to important and rapid steps that should be taken to address the problems of agriculture related to climate change and to ensure Egyptian food security.

Geographical Location of the Study Area

Egypt occupies the northeastern corner of the continent of Africa, bordered on its northern coast by the Mediterranean Sea and on its eastern coast by the Red Sea, Libya from the west, and Sudan in the south. Egypt has an area of about 1 million km^2. However, the Nile Valley and the Delta constitute only about 4% of the country's total area, which is densely populated. Most of the country's economic activity, especially agriculture, is located in the Nile Valley corridor, which is about 1,000 km long from the Aswan High Dam in the south of the country to the Nile Delta in the north (El Sharkawy et al. 2009).

The Delta region (the study area) is particularly important because it accounts for nearly 60% of the country's agricultural land and more than 60% of the population, which is very densely populated and concentrated in the country's largest economic activities. Given that the region has a very large proportion of land below 1 meter, which is at risk of water intrusion and salinity increase, along with the land subsidence of North Africa and the significant rise in sea level, climate change on that region and on agriculture in particular threatens the security of the country as a whole, not only food security but in all respects. All these reasons make the Delta region worthy of study.

Annual temperatures range from 19.2 °C to 21.1 °C in Lower Egypt, from 20.5 °C to 22.5 °C in Central Egypt, and from 22.9 °C to 26.9 °C in southern Egypt. Egypt's natural situation is the low rainfall except for the northern coastal region, where precipitation is moderate, ranging from 100 to 200 mm/year. Within the delta, precipitation ranges from 40 to 60 mm/year and in Central Egypt less than 20 mm/year (Eid et al. 2007).

The agricultural sector in Egypt consists of diverse production systems, as in the "**crop production system,**" which includes field crops, vegetables, and aromatic crops. Egypt's main crops include wheat (basic crop), maize, cotton, rice, sugarcane, sorghum, and soybeans. Crop production contributes about 68% of the total value of agricultural in GDP. The "**livestock production system**" includes raising sheep, goats, cattle, buffalo, and poultry in the agricultural sector. Livestock are raised on farmland for animal fattening on one hand and for soil fertilization by animal manure on the other hand (Fernando et al. 2012).

There are some problems facing Egypt, and among them are the following:

- Egypt suffers from a very big problem as directing the current plans for the reclamation of new land less fertile on the delta margins and in deserts, while the most productive land falls under the urbanization losing fertile land.
- Egypt is one of the most vulnerable countries to sea level rise impacts, which, according to the Medany (2008) report, will affect about 6 million people in Egypt, as well as the loss of up to 12% to 15% of agricultural land in the Nile Delta region. According to the same report, Alexandria, Beheira, Port Said, Damietta, and Suez will be the most heavily affected cities if preventive measures are not taken. The agricultural sector will be the most negatively affected sector. The loss could reach 90% of the total area of the governorates at risk, followed by the industrial sector because the industries in Egypt are concentrated on the coastline with a loss of up to 65% and the tourism sector with 55% loss due to SLR and land subsistence in Egypt The projected impact of sea level rise on agricultural land can be intensified to 80% and 120% of the rise of groundwater (Abd-El Wahab 2005).
- Some studies also estimate the expected population in Egypt to reach 115 to 179 million by 2050 (El-Ramady et al. 2013). High temperatures are expected to reduce agricultural productivity, increase crop demand for water, the loss of some land to its fertility, and the salinity of others and the urbanization, especially in the delta, resulting in a reduction in the production of nutrition and the possibility of exposure of Egypt to the famine in addition to the displacement of many families that depend in one way or another on the agriculture sector.
- There are still some problems with accurately updating the data and making it available and linking different disciplines together. In the case of agriculture, the data that should be available: climatic data (e.g., temperature, humidity, etc.), agricultural data (all relevant to the production, quantity and quality of crops, livestock data, insect studies data, plant diseases, fisheries, etc.), adaptive capacity

for agricultural workers, the capacity of the community as a whole to cope with climate change and its direct and indirect impact on agriculture, the role of women in agriculture, measuring community awareness of climate change issues, etc.), and other economic, historical and geographical data, geomorphology, mathematical and statistical models, and the resulting models to anticipate and manage climate risk.

- In order to get a complete picture to make the right decision; a national database must be completed and ready to use, but this is something that is not currently present in Egypt. In Egypt each discipline works alone, except some of them that started recently in integration. As well as studies that concern with the experiences of other countries suffering from the same problem and presented what has been done and its advantages and disadvantages as well as what can be learned from those lessons and what can be implemented and taken from others and make it available to those who want to learn from our experience later.
- One of the problems Egypt is facing and is expected to worsen in the future the stability of its share of Nile water, which reaches only 55.5 billion m^3 per year (76.7% of the available water resources in the country) with the increase of the population very significantly, especially over the next 50 years. The amount of desalinated seawater is only 0.08%, and the treated groundwater is 20.65 billion m^3 per year (28% of the available water resources), which cannot be added to Egypt's water (El-Ramady et al. 2013).

A Brief Overview of Ancient Egypt

Ancient Egyptian Climate

Ancient Egypt was located in a hot and dry desert climate. This climate affected the lifestyle of Egyptians and influenced the construction of their homes. Some researchers believe that the ancient Egyptians built their homes with flat surfaces for sleeping on them in hot nights. These houses were built with dried mud bricks. Most of the houses were one-floor. Some research also shows that ancient Egyptians wore light clothing because of high temperatures. The men (including the pharaoh himself) were dressed in loose, short-sleeved clothes, while women wore long linen dresses (a natural fabric made of flax that allowed the sweat to evaporate easily). Most of the clothing was white to reflect and reduce heat (Conti et al. 2016).

Egypt was suffering from dust and glare because of the surrounding desert on all sides, and eye infections were common; stone sculptures were found in the tombs showing a group of blind people. A collection of ancient texts on papyrus shows that they used bat blood to treat the eye. Scientists believe that ancient Egyptians, men and women, put on heavy eye makeup (called kohl) to help protect their eyes from dust and glare. The men and women shaved their head hair (Conti et al. 2016) to keep their head moisturized and cooled as they wore wig outdoors and keeping their heads shaved indoors, allowing them to keep their scalp clean.

Agriculture in Ancient Egypt

About 30,000 years ago, the deserts of North Africa were green lands and forests, but they began to dry about 8000 BC, forcing the population to trace the source of water and fertile land and to move away from deserts and dry areas, where the climate emphasizes its ability to move the population and force them to move easily. This has resulted in the entire population of Egypt moving to areas adjacent to Nile and Delta to ensure soil fertility and water suitable for irrigation and other uses.

The Nile begins from Central Africa and flows north to the Mediterranean Sea. The desert environment in Egypt has made the Nile a very important geographical advantage, so the ancient Egyptians worshiped and sanctified it. The annual flood of the Nile was very important for the ancient Egyptians to the extent that they put them in their annual plans for their entire lives, where they leave their land in the period when there is a flood and go to build pyramids or trade or anything away from agriculture, and when the season ends, the Egyptian farmer goes to agriculture in the fertile new land given them by the river.

Egypt was the first country to set dates for agriculture, tillage, irrigation, and harvesting, and they were the first country to set up a number of agricultural sites. They record those dates on the walls as well.

The Main Groups in the Ancient Egyptian Society

Ancient Egyptian society was very well organized, where there was a central government, and they also had a large army at the highest level of training since the fifteenth century BC. People were well aware of their social responsibilities, and the ancient Egyptian society was very interested in education where the learner was placed in a high position of the social pyramid.

Women in Ancient Egypt

The great and important role of women in ancient Egypt was to raise children and care for the family. The ancient Egyptian was concerned about his pregnant wife as this event was very great in the Egyptian family, but the deaths of women during childbirth were very high as well. However, Egyptian women enjoyed more rights and freedoms than other ancient societies such as ancient Greece, for example. The ancient Egyptians believed that the world was dominated by male and female gods with the same number. Women also inherited and owned agricultural lands and started businesses. Children become under guardianship of their mothers if there was separation or divorce, and they had legal rights against men (Conti et al. 2016).

Sustainable Agriculture in Ancient Egypt

Sustainability was a normal part of the stable life of Egyptians; it was presented in the use of techniques that facilitated the construction of pyramids, temples, good medicine systems, irrigation systems, agricultural production techniques, shipbuilding and glass-making, as well as the first peace treaty in the world with Hittites. The sustainability of agriculture was primarily through the annual flooding of the Nile and the resulting sedimentation in fertile clay soil containing phosphorus and other minerals that come with the river on its long journey from the heart of Africa to Egypt. Their environment encouraged them to think about natural processes as they work in predictable cycles.

There were three clear seasons in ancient Egypt, the flood season (Akhet), the planting season (Peret), and the harvest season (Shemu). The flood season lasted from June to September, and the planting season lasted from October to February (El-Ramady et al. 2013).

Food Security in Ancient Egypt

Ancient Egyptians used to store crops for use during the flood, when they stopped farming, so ancient Egypt did not face as much famine as other ancient civilizations, where Egypt was called the bread basket of the Old World where it was the world's leading exporter of wheat in addition to a number of other crops. Because Egypt's soil is fertile and rich in minerals, Egyptians were able to produce an abundance of food, allowing the population to devote more time and resources to cultural, technological, and technical fields. The management of agricultural land was important and critical in ancient Egypt, and there was a system of taxation collected regularly and systematically.

Ancient Egypt suffered from some famine due to climate variability. However, ancient governments managed to overcome this problem by storing food in years when it was surplus in good ways and used in dry periods, which led to the internal balance of the Egyptian market well. However, now that the population is too large to handle, just storing the surplus is not the best solution.

Agriculture in Modern Egypt

Climate is the main determinant of agricultural productivity in Egypt. As food production is necessary to maintain and promote human well-being, agriculture is a source of great interest in Egyptian society at the level of governments and scientific and civil societies.

In 1990 agriculture (including livestock) accounted for nearly 20% of GDP and employed approximately 37% of the labor force. In 2017 agriculture accounted for approximately 14% of GDP, and agricultural workers accounted for 52% of Egypt's labor force (CAPMAS 2017).

Sustainable Agriculture

The importance of sustainable agriculture is that the world population will increase by 2050 with the persistence of the area of cultivated land. Considering the climate changes that are expected to negatively impact agriculture, it could be found out how bad it is and how sustainable agriculture is needed. Sustainable agriculture will not only address the problem of food availability but will also protect the world from wars that may be caused by the scarcity of food and famine, as well as social issues and maintaining balance in societies.

The strategy of sustainable agricultural development aims to develop an intelligent model that coordinates natural, social, and economic aspects with the aim of developing and ensuring agricultural production through long-term and multi-generational programs.

Sustainable Agriculture in Modern Egypt

With regard to sustainable agriculture in modern Egypt, the main challenges to sustainable development facing Egypt include the problem of climate change and its impact on health and food security; sustainable agricultural development issues, which require a national urban development plan; the problem of using non-renewable energy sources; and improved irrigation efficiency.

Agricultural production in Egypt is concentrated mainly in the following areas (El-Ramady et al. 2013):

- The Nile Delta region, where the Nile River water is mainly used for irrigation. The agricultural production in this area is about 6.6 million acre.
- Reclaimed desert areas on the margins of the Nile Valley region, where groundwater is used for irrigation.
- The northern coastal area, which starts from the northwest coast and is heading east to the Sinai Peninsula.
- The area in the Sinai Peninsula and the Eastern Desert, where the main source of irrigation is groundwater.
- Western Sahara, including oases and remote southern regions, where it is entirely dependent on groundwater.

The relationship between sustainable agriculture and rural development and the environment and the promotion of food security and poverty reduction are central issues for achieving economic and social development in Egypt. The contribution of the agriculture sector in Egypt exceeds 14% of GDP and up to 52% of employment. At the same time, about 57% of the population lives in rural areas, which are very poor (CAPMAS 2017). Hence, the promotion of sustainable agriculture and rural development as a means of reducing poverty and food insecurity in light of the expected climate change, which is a prerequisite for sustainable social and economic development, and therefore should be considered a social and political priority for Egypt.

Rural Poverty and Food Security

There is a very strong relationship between economic growth and the reduction of hunger and poverty, together with a strong link between poverty and food insecurity, where a large proportion of the poor people are either undernourished or food insecure due to their vulnerability to food price changes. Poor households spend a large part of their monthly income on food purchases. Globally, nearly 70% of the poor people live in rural areas, where they rely heavily on agriculture for their food supply and income generation.

Pollution of watercourses and groundwater due to domestic and industrial wastewater and solid waste disposal is another problem that reduces the availability of quality water suitable for use. It is known that the number of water irrigation complaints due to lack of water has recently increased, and because of the limited water resources, the lack of irrigation water is exacerbated by a number of causes, including violations of agricultural laws, as well as the cultivation of crops that consume water in huge quantities such as rice cultivation. It is expected that treated wastewater will become the water resources available for future agricultural expansion, especially if Egypt does not increase its share of Nile water (El-Ramady et al. 2013).

Materials and Methods

Field Work and Data Collection

A household survey was conducted among 200 households using a semi-structured questionnaire and face-to-face interviews during February and March 2018 in the areas of Alexandria, El-Beheira, and Kafr El Sheikh, which were selected as study areas to represent the rest of the region. The purpose of this analysis is to examine the degree of awareness of farmers living in different agricultural environments on the issue of climate change and how they intend to deal and cope with them, whether they have actually happened and whether they have been adversely affected by this phenomenon and how farmers have been impacted according to their viewpoints and how livestock have been affected. Farmers were asked to answer some questions, for example, Did the rainfall change in the last 5 years? Have temperatures changed in the last 5 years? And so on.

Most farmers indicated that climate change (e.g., change in temperature and rainfall) has affected farm productivity and profitability. Participants also predict that climate change will affect future generations in a more difficult and complex way. However, many believe they can adapt to climate change through adaptation strategies developed by specialists and governments.

The analysis revealed that many factors such as economic, social, environmental, institutional, and other factors influence farmers' choice of climate change adaptation methods. Household size, income, education, access to climate information through television, radio, etc., the fact that the individual is a member of the local farmer group, the location of the land itself, the type of crops, access to credit, market distance, and access have significantly affected adaptation. For example, farmers who have access to climate information and climate patterns both TV and radio have a higher capacity to adapt to climate change than those who do not have access to information at all.

Results and Discussion

Farmers' Perceptions of Climate Change

Nearly half (49%) of the farmers interviewed for climate change have seen an increase in temperature over the past 20 years, as well as wind and rainfall especially the sudden ones. Extreme weather events such as flash floods in Egypt, especially the North Coast, have been increased in recent years.

There were a number of factors influencing farmers' strategies to deal with climate change, including but not limited to family size, wealth, access to information, size of farm, soil fertility, agricultural experience, and agricultural extension.

Some farmers have also identified a number of problems that could be key constraints to adaptation to climate change, such as inadequate seed availability, poor planning, lack of employment, market problems, water scarcity and irrigation problems, poverty and lack of capital, lack of services, lack of health services, and lack of believing in climate change itself.

The questionnaire included a set of questions intended to measure farmer awareness of the issue of climate change; 60% of respondents admitted they do not know much about it by answering different questions, and most of them said they care about more important problems than atmosphere or air or even the environment.

As well as other questions, for example, a question about the common diseases among family members to try to link them to climate change, but it turned out that most of them were related to the nature of hard work, contaminated water, and other causes. It was a multiple choice question so they choose more than one answer. 60% of respondents confirmed they or any of their family members suffers from allergies, 65% kidney diseases, 54% diabetes, 15% heart diseases, 20% cancer, 40% blood pressure, and 10% other diseases.

In order to try to estimate adaptive capacity among different households, we asked a question, "Do you have or can you practice another profession or job if your current one is affected by climate change?" About 80% of respondents denied having or knowing any other jobs if their jobs get affected by climate change, while about 20% said yes, they could find and work other jobs.

When the question was about the sufficient awareness of these risks and how to deal with them and confront them, nearly 85% denied the presence of any awareness at all. And about their suggestions to confront this danger they were to tell them everything honestly and tell them what to do in different cases and, if it is possible, to try to activate the NGOs' roles.

The last question was about the way they have their information. It was a multiple choices too, so the respondents choose more than one answer. The answers were TV (80%), radio (20%), friends (30%), and mobile services news (10%). And some of them said in the category of others that they have all their information from rumors even if they know it is wrong.

Women in Rural Egypt

The official data of Egypt showed that women's contribution to the labor force is still low at 23.2% compared with men (CAPMAS 2017) due to economic reasons related to the ability of the Egyptian labor market to absorb the supply of women's labor force, including low wages, women's involvement in marginalized employment, and women's entry into the labor market at a later age than men.

However, the year 2017, which was declared the year of women in Egypt, witnessed a high political participation of women in the government. The government currently includes four female ministers and one governor. Article 11 of the Egyptian constitution requires the government to take the necessary measures to empower women politically, in addition to creating equal employment opportunities with men in the Egyptian labor market. The government today sets the conditions for any job without taking into account the gender but only education, age, and other conditions for each job. A plan to empower women was also developed by 2030.

In particular, Egyptian women in rural areas have traditionally been following the male in her family, whether they are fathers or husbands. Their role is often in caring for children, caring for the home, doing housework, raising poultry and animals at home, and other activities that do not show the real role of women.

Since women have a completely different view of men and look at things from a different angle not seen by men, they bring to light solutions and vision from another and new angle so the participation of women in the work can bring an integrated solution because there are two angles of thinking complement each other greater than just looking at a problem or issue from just one angle and just one view the man's view.

When working with women and trying to explore their views on various issues, especially climate change, we can draw up a list of possible solutions that can actually be applied to climate change adaptation.

Climate change can widen the gender gap in rural areas because it is highly related to access to resources such as information, skills, and technology that will determine individuals' ability to cope with new circumstances.

Some women (70%) acknowledged that they did not know much about climate change and could not think of strategies to respond to it, while about 30% said they did not care about climate change at all. Women in Egyptian rural society are receptive to discussing with each other more than men who often do not talk together about problems. Thus, educating women increases their awareness of the issue and improves their ability to find solutions to adapt even if they do not know that they are adapting to a global issue and that they are only adapting to what they can maintain in the small family economy.

Gender relations within society determine the extent and degree of vulnerability of each one (men and women) to climate change on their own. Therefore, it is not possible to generalize even within the Egyptian society as a single unit, because the role and status of women in the countryside are very different from the cities. Egyptian women in the cities have many rights that make them equal to men, so that they can choose to move, migrate, and travel in case of need. Access to information, resources, and health facilities is better, as well as decision-making and the implementation of their views.

This is in contrast to rural women who are still dependent on men in everything, especially in taking and implementing their own fateful decisions, making them more sensitive and vulnerable under the current and future changing circumstances. Climate change is likely to deepen gender inequalities in rural Egypt.

Gender disparity in equality is due to the heaviest burden on women, making them weaker and more vulnerable, especially in disaster and pandemic situations. At a time of disaster, women are not only slower in movement but also responsible for rescuing children, infants, and elderly people at home. After the disaster, it is the responsibility of women to nurse and take all responsibility of others.

The questionnaires also show a problem for women alone if climate change affects agriculture, threatens the local economy, and changes prices. In Egypt men give women a certain amount each day according to their personal and economic ability. Women must adapt to this amount. Therefore, in the case of high prices, women are in the problem of managing family conditions and adapting them to the current situation as much as possible because salaries do not rise but remain as they are to cope with the high costs resulting from problems and damages related to climate and agriculture.

The traditional form of questionnaire was not used with women because of the low level of education among most of them and to ensure that the pencil and paper were not confused. Open discussions were held to obtain the desired information freely and to give women a sense that they were just discussing solutions to climate change that could affect the economy of the state as a whole; there has been a very good response.

Women's Perception of Climate Change

Women have not been particularly interested in climate change in the countryside because of their ability to adapt easily to various changing temperatures, rain, and other conditions and deal with anything very flexibly. The four seasons in Egypt, which are characterized by the transition from cold to hot weather throughout the year, make women's ability to adapt and take care of their family normal throughout the year. There was not enough time to study these traditional adaptation options in more depth but later on.

In general, rural society and women are not particularly concerned about the steps that can be taken to respond to climate change. However, their response to the information and trying to understand what was happening around them was clearly high, and there was some interest in knowing what could happen in the future because of climate change.

Adaptive Capacity Among Women

When asked if there is domestic work for females at home, all responses were yes and varied in poultry and animal husbandry, dairy industry, field assistance to men, and others. All of these businesses fall within the agriculture sector, making adaptive capacity low and vulnerability high.

Conclusion and Recommendations

Agriculture in Egypt is the key to social and economic development and can become a key driver of growth in all sectors if it finds its way into sustainability. The climate of Egypt is subject to three main types of changes, gradual increase in temperature, changes in the pattern of rainfall, and increase in the frequency and intensity of extreme weather events.

The study shows that there are many things that can be linked to agriculture and affect farmers' families as in the case of market fluctuations, changing the supply/demand relationship, low farm income, and other income-related issues affecting agriculture. Such changes are beyond the control of farmers and can pose major challenges and threats to farmers' families.

Farmers in Egypt and even most decision-makers do not consider adaptation to climate change to be a priority in agriculture today compared with other sectors where the impact of climate change is more pronounced. So in order to raise awareness among farmers, there is a need to develop programs about adaption to climate change, to have sufficiently aware of the potential impacts of climate change.

Different scenarios of possible climate changes can be discussed with farmers to measure their adaptability on one hand and to identify traditional ways they are accustomed to on the other hand. Involve these strategies in the decision-making process and ensure farmers' implementation of the laws as they have contributed to develop them with the government and thus to ensure a better future for the agriculture sector.

It is the responsibility of the government, civil society associations, and researchers to inform farmers about the risks and potential opportunities for climate change and related threats to adapt in a measured manner.

The first solutions that can be thought of as the baseline in such studies are the management of the ecosystem, including natural resources in a balanced and good manner through many things as in the decisions, policies, laws, and other elements that must be in harmony without neglecting the human behavior of the population in dealing with the ecosystem.

In the case of agriculture in Egypt as a natural resource in all its components (water and soil), the laws that have been put in place to preserve sustainable agriculture in Egypt should be enforced, as is the case in maintaining the agricultural land, especially that the laws of urban sprawling are among the most famous laws that are breached.

The state has already signed fines to return the agricultural land and already removed the buildings that were built on them, but some of the land has already lost its fertility after being filled with concrete blocks.

Therefore, the second important solution after the government activates the laws, raising the awareness of the population, ensures that agricultural land does not fall in this problem from the beginning, and the principle of reward and punishment can be used not only punishment. For example, the government could buy from farms at better prices than they can get from others or to encourage those who preserve the natural resources.

One of the most important problems facing our world today and agriculture scientists are very interested in is how we will feed future generations and how to ensure sustainable agriculture and food security sufficient for the growing population in the world as a whole, especially in developing countries that already have problems in agriculture and will suffer more as warming continues.

Despite the current number of the world's population and despite the technological and scientific progress currently accepted, there is an anticipated food crisis.

The problem of plants lies in their low adaptability to immobility, migration, and other ways in which animals or humans adapt. The expected problems of rising temperatures, salinity, flood drowning, severe drought, and other climate-related problems greatly reduce plant productivity and thus a food crisis. Putting these things together that climate conditions cannot be accurately predicted is more complicated, especially as predictions take the negative path.

David et al. (1998) estimates Egypt's population in 2060 to be more than double (2.2 times) for the 1990 census. This represents an annual growth rate of 1% over a period of 70 years, based on World Bank estimates as cited in David et al. (1998) and as quoted by Fischer et al. (1994). This increase goes with more efforts in agriculture to increase agricultural production and through land reclamation and technological improvement, which will have a significant impact on the future social and economic conditions in the country.

The above work presented two conflicting views regarding Egypt's long-term economic development. Views of pessimists and optimists were presented. The gloomy group has shown that Egypt is in a very difficult position, as its economic growth is very low and the prospects for uncontrolled population growth will lead to short-term consumption rather than long-term investment. Egypt as a country will remain a major consumer through cheap imports of food commodities, and its external debt will increase as policy will continue to hinder economic growth. Large investments in new technologies will remain very small, and economic reform will be difficult (David et al. 1998).

While the optimists believe that despite Egypt sinking into the current debt and low level of its technological industries, it will invest in new and innovative technologies and will exceed many other countries, and it is waiting for a bright future, where the owners of this view all the privileges enjoyed by Egypt, especially its location on the Mediterranean and Red Sea and the Suez Canal and its workforce, making Egypt a major center for manufactured goods and exports to other countries in the world. However, both types of scenarios do not meet expectations and projections but only mere viewpoints.

In order to adapt to climatic conditions, the agricultural community had to use a wide range of coping mechanisms, such as reducing household consumption, storing grain, renting agricultural tools, working with others, resorting to small trade, immigration to seek employment elsewhere in our out the state, work in the field of handicrafts, obtaining loans from banks, and others.

The farmers in the study area made some adjustments in the agricultural calendar (planting and harvesting dates) compared to what was followed 15 or 20 years ago. A relatively small number of farmers have found that the current agricultural assessment of most crops is shorter than before, and the result implicitly showed a steady trend toward shortening the growth season. Some farmers (very small percentage did not exceed 10%) changed the crop type in some cases, reduced the number of animals due to the death of some of them, reduced the use of water to maintain it, reduced the cultivated area so that they could control it (not exceeding 15%), as well as engaged in nonfarm activities (the large percentage was for young farmers).

In addition, the Egyptian rural communities are facing other pressures other than climate change that reduce their adaptability and affect the agriculture sector and its workers. Therefore, consideration should be given to studying the impact of climate change on agriculture. As is the lack of government information, government coordination with farmers, and guidance on adaptation to climate change and lack of adequate financial support from the state, infrastructure degradation, and poor technology and supplies, sometimes unevenly distributed to farmers.

Recommendations

The story of Joseph, mentioned in different books (Bible and Quran) and religions, is one of the most telling stories of the ability to avoid disasters through perfect and correct predictions (mentioned here because they occurred in Egypt in an agricultural crisis). By anticipating the next 7 years of famine (as his Lord told him), he skillfully collected surplus food in the seven fruitful years and was able, through an exemplary agricultural management plan, to overcome the famine that struck Egypt for 7 years as documented in history books. This is another proof that Egypt has been facing climate change for a long time and that the impact of these changes is on agricultural production and the agriculture sector in a large and direct manner.

Although Egyptian planners today do not have the luxury of ideal predictions, climate change must be treated as a reality that requires special strategies to deal with them and with the potential problems and impacts of climate change on the agricultural system in a large and specialized way. So here we will list a set of recommendations that can be worked on in order to reach the closest form of adaptation and avoid the damage caused by climate change to agriculture.

The real solution lies in changing our agricultural practices to help ourselves and the environment. Man-made practices of natural resource depletion, fuel combustion for agricultural machinery operation, crop transportation, tillage, mono-crop production, use of chemical pesticides, herbicides, fungicides, fertilizers, etc. have caused soil erosion, deforestation, and ultimately climate change and food insecurity at risk. That is, our actions have a devastating impact on the health of our environment and our bodies.

However, the roles will be divided into the three main parties of any society:
First: The role of science

- The importance of improving scientific capacity in various areas related to climate change should be considered as a top priority among specialists of all disciplines. Current and future research should be directed to make plants more capable of adapting and enduring by developing mechanisms that allow them to transfer qualities or characteristics that adapt from the strong plants to those that are sensitive and have a role both in food and economics through the separation and transfer of genes and so on.
- Integration of research on food security to complete the picture for researchers because it is of great importance in clarifying the problem and developing real and effective solutions, especially since not only agriculture is concerned with food security or agricultural production but agriculture, plant biology, soil science, chemistry, and many other field of biological science.
- The interest in studying biodiversity and linking it to agriculture because of its many benefits is not sufficiently clear in Egypt because of the unifying of each discipline by itself.
- Diversity cultivated crops and livestock, as well as change in planting and harvest times, to commensurate with current and future weather conditions and temperatures.
- There should be an ongoing dialogue between the scientific community and the stakeholders in agriculture to provide sound input to decision-making while ensuring that climate change is included in the list of management practices at the local level.

Second: The role of Government

- The government can contribute to addressing the impacts of climate change on agriculture by increasing investment in agricultural research, drought-resistant varieties and diseases, and other varieties that grow early or faster and shorter than their current counterparts.
- Enhancement of technologies such as modern agricultural machinery to cope with floods and promote water filtration in the soil during and after floods, as well as the development of irrigation technologies, market expansion, and education (establishing and increasing the number of training centers for farmers, in addition to their formal education).
- The establishment of local meteorological stations and the dissemination of climatic data obtained and training farmers to read, understand, and deal with them.
- Develop climate forecasts and make recommendations with each forecast for farmers and stakeholders and formulate well-planned adaptation strategies.
- Create jobs by expanding nonagricultural sectors so that the dependence on one sector is not already threatened in the near future.
- Develop an early warning system and train farmers to use, understand, and deal with it.
- Direct further economic growth toward rural development to eradicate poverty and malnutrition and job creation, to overcome the double trap of poverty for small farms and poor soils by unifying farms to larger sizes and commercial entities that can be more productive and sustainable.
- Governments, banks, and companies must support smallholders to expand agricultural production projects, particularly in developing countries. Such support must include legal land tenure, equal access to employment, access to capital, markets, organized training, and investment in technology and infrastructure.
- A policy should be developed to deal with the negative impacts of climate change on agriculture, with emphasis on the three most important areas: crop management, water management, and land management.
- Policies that ensure sustainability in agriculture must be concerned with economic diversification of the farming family, promotion of agricultural and non-agricultural development, and assistance to poor people suffering from food insecurity through access to land and water that will enable them to move forward on sustainable development and ensure a decent life especially marginalized groups of the poor, such as women.
- Information and education campaigns should be carried out on an ongoing basis to encourage farmers to develop positive attitudes toward climate change. A positive attitude may ultimately lead to constructive behavior in the selection and implementation of adaptation options.
- The potential increase in extreme weather events should also be important and relevant to farmers, and they are of concern at all levels as well as at the level of governments.

Third: The role of the farmers

- It is possible to cultivate leafy vegetables that do not require much space or effort on the roofs of houses, especially that the countryside is characterized by the fact that each house is independent and private for one family and not shared so this may be considered a small-scale economic project or insurance for home supply of small things useful to reduce pressure on markets and land.
- Switching crops is the future for other crops that do the same but are less affected by climate change – temperatures, humidity, and so on – as in Mexico, which is looking to replace coffee crops that may not be suitable for growth by in 2025 for some types of cocoa, for example, due to high temperatures resulting from climate change. It is also necessary to work on the principle of modifying the dates of agriculture and crops and their places based on the recommendations of scientific studies.
- Finally, the gap between scientists and specialists, on the one hand, and decision-makers, on the other hand, and public people in the community must be worked on in a way so that the picture is clear and everyone is a partner in decision-making, which ensures its implementation.

References

Abd-El Wahab HM (2005) The impact of geographical information system on environmental development. MSc thesis, Faculty of Agriculture, Al Azhar University, Cairo

Central Agency for Public Mobilization and Statistics, CAPMAS (2017) Egypt in figures. Central Agency for Public Mobilization and Statistics (CAPMAS), Cairo

Conti L, Easton M, Saldais M et al (2016) Humanities and social sciences 7: Western Australian curriculum. Oxford University Press, South Melbourne, pp 216–226

David N, Yates K, Strzepek M (1998) An assessment of integrated climate change impacts on the agricultural economy of Egypt. Clim Change 38:261–287

Helmy M. Eid, Samia M. El-Marsafawy, Samiha A. Ouda (2007) Assessing the economic impacts of climate change on agriculture in Egypt: a Ricardian approach. In: The World Bank, Development Research Group, policy research working paper 4293, Sustainable rural and urban development team

El Sharkawy H, Rashed H, Rached I (2009) Climate change: the impacts of sea level rise on Egypt. In: 45th ISOCARP congress, Porto

El-Ramady HR, El-Marsafawy SM, Lewis LN (2013) Sustainable agriculture and climate changes in Egypt, Sustainable agriculture reviews, vol 12. Springer, Dordrecht. https://doi.org/10.1007/978-94-007-5961-9_2

FAO (2010) Challenges and opportunities for carbon sequestration in grassland systems, Integrated Crop Management, vol 9. FAO, Rome

Fernando HJS, Klaić Z et al (2012) National security and human health implications of climate change, NATO science for peace and security series C: environmental security. Springer, Dordrecht

Fischer G, Frohberg K et al (1994) Climate change and world food supply, demand, and trade: who benefits, who loses? Glob Environ Chang 4(1):7–23. https://doi.org/10.1016/0959-3780(94)90018-3

Kutywayo D, Chemura A, Kusena W et al (2013) The impact of climate change on the potential distribution of agricultural pests: the case of the coffee white stem borer in Zimbabwe. PLoS One 8(8):e73432

Medany M (2008) Impact of climate change on Arab countries: chapter nine report of The Arab forum for environment and development. Technical Publications and Environment & Development magazine, Beirut. ISBN: 9953-437-24-6

Impacts of Climate Change on Food Security: An Appraisal

40

Damilola Grace Ogunrotimi and Joshua Kayode

Contents

Abstract

Climate change is a serious threat to agriculture, one of the major backbones of Nigeria's economy. There has been a drastic shift from the actual climate patterns over a long period with attending impacts on human, animal, and plants. This is attributed to the increase in the emission of greenhouse gases which raise the Earth's temperature, a phenomenon referred to as global warming. In some regions of the world, the melting of polar ice caps and glaciers has led to a rise in sea level, increased precipitation, and flooding. Nigeria is no doubt undergoing a shift in its dry and wet seasons with agricultural lands becoming arid as a result of incessant drought. This alternation in temperature and rainfall pattern due to global warming has its attendant consequence on plant yield. The rise in global temperature has led to the increased proliferation of pests and diseases which

D. G. Ogunrotimi (✉)
Department of Plant Science and Biotechnology, Faculty of Science, Ekiti State University, Ado Ekiti, Nigeria
e-mail: damilolarotimi91@gmail.com

J. Kayode
Department of Plant Science and Biotechnology, Ekiti State University, Ado Ekiti, Ekiti State, Nigeria

W. Leal Filho (ed.), *Handbook of Climate Change Resilience*,
https://doi.org/10.1007/978-3-319-93336-8_70

thrive successively in hot climate. This chapter is a review of the impacts of climate change and the threat it poses to food security and sustainable development. To this end, the chapter suggests adequate public sensitization on the dangers inherent in climate change both on the regional and global scale. Biotechnological advancement that will promote crops that are well able to adapt to adverse weather conditions and produce high yields is advised.

Keywords
Agriculture · Climate change · Crop · Food security · Hunger · Weather

Introduction

Agriculture feeds and clothes the world; nevertheless, several millions of people suffer from poverty and hunger globally. Rapid improvements in agricultural productivity and economic growth over the second half of the twentieth century brought food security to broad swaths of the developing world. Some other regions in Africa did not share in that success and remain no better off today and in some cases are worse off than they were decades ago (Lobell and Burke 2010); these however can be attributed to poor economy and low coping strategies. Countries in Africa are likely to suffer the most because of their geographical location, low incomes, low institutional capacity, as well as their greater reliance on climate-sensitive renewable natural resources sectors like agriculture (Eboh 2009). The climate change is a phenomenon that is a threat to human and natural resources. Several disasters have led to the destruction of biodiversity especially agricultural lands that produce major foods for consumption. Human influence on the climate system is clear, and recent anthropogenic emissions of greenhouse gases are the highest in history with widespread impacts on human and natural systems. About 76% of all disaster events are hydrological, meteorological, or climatological in nature, and these accounts for 45% of the death and 79% of the economic losses (ISDR 2008). In the last three decades, the Earth's surface has been successively warmer than in any preceding decade since the year 1850 (IPCC 2014). Several flooding cases where houses and farmlands have been submerged by water have occurred. Anthropogenic greenhouse gas (GHG) emissions since the pre-industrial era have driven large increases in the atmospheric concentrations of carbon dioxide (CO_2), methane (CH_4), and nitrous oxide (N_2O). Of the total GHG emissions, CO_2 from fossil fuel combustion and industrial processes have contributed about 78% (Adejuwon 2006). Economic and population growths remain the most important drivers of CO_2 emissions, for example, the increased use of coal has reversed the long-standing trend of gradual decarbonization of the world's energy supply (IPCC 2014).

The release into the atmosphere of high concentrations of these greenhouse gases (GHGs) has led to the increase in the Earth's temperature, a phenomenon referred to as global warming. About 40% of these emissions have remained in the atmosphere, and some are removed from the atmosphere and stored on land (in plants and soils), while the ocean has absorbed about 30% of the emitted anthropogenic carbon

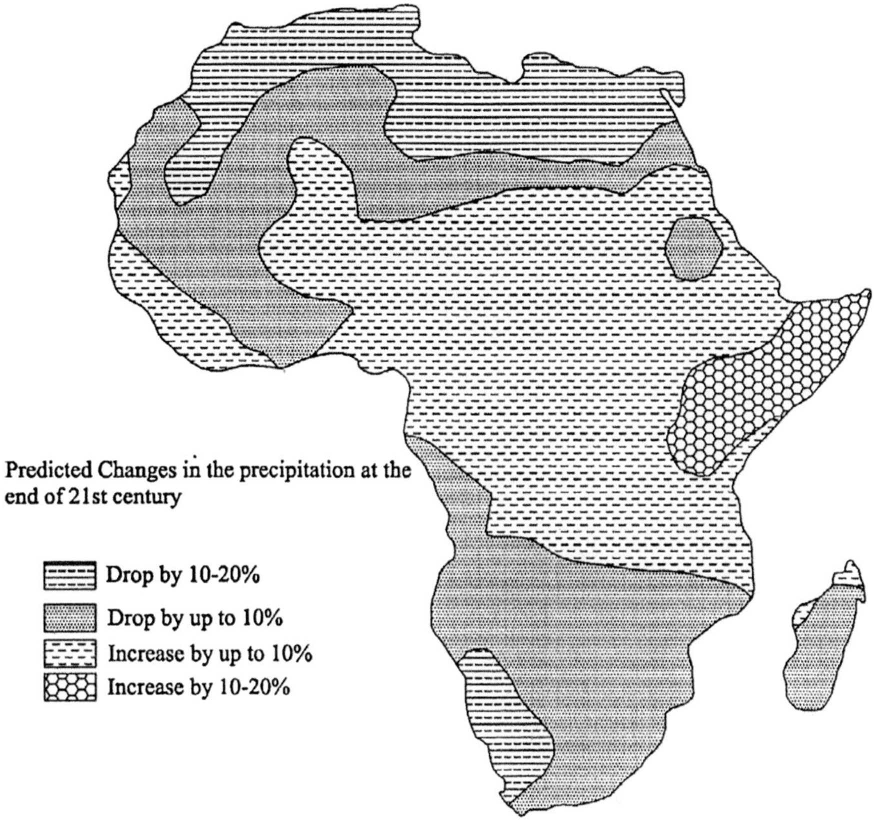

Fig. 1 Predicted changes in the precipitation in the African continent due to climate change at the end of the twenty-first century. (Source: de Wit and Stankiewicz 2006)

(IV) oxide, causing ocean acidification (IPCC 2007b). The Intergovernmental Panel on Climate Change also reported that climate change may affect natural systems so much that dry areas may become drier, leading to drought and wet areas wetter leading to floods. This will greatly impact agricultural lands posing a threat to food security and combating hunger (IPCC 2007b) (Figs. 1 and 2).

Globally, climate change has been observed in rising sea levels and inundating coastal lands. In Nigeria, floods, strong winds, fierce storms, and high temperature are currently being experienced with high annual rainfall variability (Enete 2008; Adefolalu 2000). Climate change affects yearly harvest as precipitation and temperature are major determinants of plant growth and development. Carbon (IV) oxide in the atmosphere drives the process of photosynthesis in plants; however, fluctuations in its atmospheric value may be detrimental to optimum plant growth and development. Extremely high temperature affects the soil moisture content, increases evapotranspiration rate, affects photosynthetic activity, and consequently reduces crop yield. As a result, developing countries of the world may experience severe hunger and malnutrition if food demands are not met.

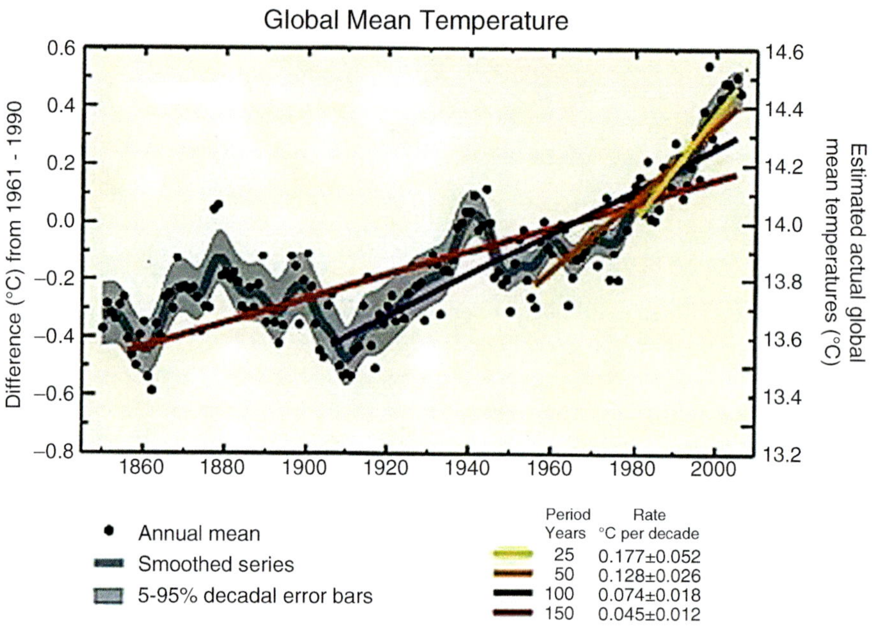

Fig. 2 Global mean temperature during the last 100 years. (Source: IPCC 2007b)

The focus of this chapter therefore is to understand much better the dynamics of climate change as it affects food security and to propose appropriate adaptation and mitigation measures.

The Concept of Food Security

Several scholarly definitions have been assigned to food security; however, the most appropriate is that which was given by the Food and Agricultural Organization:

> Food security is a situation that exists when all people at all times have physical, social, and economic access to sufficient, safe and nutritious food that meets their dietary needs and food preferences for an active and healthy life. (FAO 2009)

This definition underscores the basic parts of food security – food availability (physical presence of food), food accessibility (ability to produce or monetary power to purchase), and food utilization (processing food into edible forms and effective use by the body) – and in addition to these is food stability (availability over a long range of time as it relates to sustainability). Measuring food security may be complex as food availability does not equal accessibility, as accessibility is determined by income in addition to market value.

It is important to therefore ask the following questions:

Table 1 Crop production in Nigeria from 2003 to 2005 (million tons)

Crop	2003	2004	2005	% Growth
Maize	8,685.1	9,503.4	10,369.6	19.4
Millet	6,561.1	6,963.3	7,394.7	12.7
Sorghum	9,460.8	9,994.4	10,593.6	12.0
Rice (paddy)	3,520.3	3,713.9	3,929.4	11.6
Plantain	1,096.0	1,161.5	1,246.7	13.8
Potato	1,442.1	1,528.3	1,640.4	13.7
Yam	25,073.3	26,700.2	28,521.8	13.8
Cassava	31,698.1	33,393	36,057.8	13.8
Cocoyam	2,350.5	2,407.0	2,479.2	0.5
Cowpea	4,210.7	4,328.3	4,462.2	0.6

Source: National Bureau of Statistics (in Bello et al. 2012)

1. Do we have the capacity to produce enough food to feed the teeming Nigerian population?
2. Does the economy support access to food by the poor?
3. Is the agricultural technology currently employed in food production sustainable enough to ensure food stability?
4. Are there adequate storage and management techniques to ensure proper food utilization?

Providing appropriate answers to these questions may be the way forward in tackling the challenge of food insecurity.

The table below shows crop production in Nigeria from 2003 to 2005 by Bello et al. (2012). A careful analysis of the table shows that plantain was the lowest crop produced in 2003, though with higher growth rate (13.8%). Cassava production on the other hand was the highest with about 31, 698.1 million tons and highest growth rate of 13.8% to year 2005, followed by yams, millet, maize, cowpea, and rice. Ibrahim (2004) opined that changes in temperature and rainfall are likely to result in spatial shifts in patterns of agriculture and in crop yields, which may have considerable socioeconomic impacts. This phenomenon has been observed in the shift in crops cultivated by farmers in Northern Nigeria with the most preferred crops grown to be cereals and legumes such as guinea corn, groundnut, and maize (Odjugo 2009) when compared to vegetables and fruits (Table 1).

Climate Change and Its Impacts

Climate change is a drastic turn from actual weather patterns over a long period of time, and this can be natural or anthropogenic. Several human activities are responsible for this change; however, it is usually triggered by the release into the environment of harmful gases known as greenhouse gases (GHGs) which raise the Earth's temperature. Total anthropogenic GHG emissions have continued to increase over 1970–2010 with larger absolute increases between 2000 and 2010, despite a growing number of global climate change mitigation policies (IPCC 2014). Weather shifts have resulted in longer growing seasons as a result of reduced rainfall and high temperature which reduce soil moisture content and lower the amount of water available for plant growth. In many regions, changing precipitation or melting snow and ice are altering hydrological systems, affecting water resources in terms of quantity and quality. An assessment of studies covering a wide range of regions and crops shows that negative impacts of climate change on crop yields have been more common than positive impacts. The Intergovernmental Panel on Climate Change report of the year 2014 states that impacts from recent climate-related extremes, such as heat waves, droughts, floods, cyclones, and wildfires, reveal significant vulnerability and exposure of some ecosystems and many human systems to current climate variability.

The World Meteorological Organization also opines that flooding remains the most prevalent and extreme weather event especially in the last decade. The world's mean temperature has increased by 0.74 °C in the last century, and this has resulted in the melting of polar ice caps and glaciers and a concurrent sea level rise (IPCC 2007a). The massive flood of water erodes agricultural lands washing away soil nutrients which otherwise should be available for plant growth. In addition, drought has led to a reduction in quantity and quality of essential food commodities such as fruits, vegetables, and cereals as warmer temperatures have been reported in various parts of the world (Aggarwal and Singh 2010; Hocking and Meyer 1991; Ziska et al. 1997). Studies have shown that drought, desert encroachment, and coastal inundation are currently affecting the country's ecosystem leading to ecological destabilization due to climate change impact in the semiarid region of Northern Nigeria (Odjugo and Ikhuoria 2003; Ayuba et al. 2007). Therefore, limiting climate change would as a matter of fact require substantial and sustained reductions in greenhouse gas emissions which, together with adaptation, can limit climate change risks (IPCC 2014).

Extreme temperatures result in heat stresses on crops. Although warmer temperatures may favor the growth of some crops such as melon, tomatoes, and grapes, other species like corn, wheat, potatoes, and oats may experience lesser yields (Aggarwal and Singh 2010). Hot temperatures affect important crop development stages, thereby reducing quality and yield (Edame et al. 2011). Such staple crops as grains and cabbage may become difficult to grow, thereby reducing availability. For example, apple is known to thrive successfully in cool weather, and as such a hot weather may prevent its fruiting stage.

Pest and weed invasion is triggered in hot climatic conditions. Warm weather condition and increased carbon (IV) oxide (CO_2) spread in the atmosphere promote the growth and rapid development of invasive weed species over crops. Seeds of weeds exhibit resilience in all environments and so compete with crops of economic value. High temperatures also favor breeding of insect populations. Severe windstorms have destroyed large expanse of agricultural lands leading to crop losses. As climate change continues, agricultural yields are expected to reduce over the next hundred years (Thornton et al. 2011).

Adaptation and Mitigation Measures in Climate Change

Adaptation and mitigation are two complementary strategies for reducing and managing the risks associated with climate change. Adaptation is the process of adjustment to actual or expected climate and its effects in order to either lessen or avoid harm or exploit beneficial opportunities. Mitigation on its own is the process of reducing emissions or enhancing sinks of greenhouse gases (GHGs), so as to limit future climate change. Both adaptation and mitigation are capable of reducing and managing the risks of climate change impacts. Substantial emission reductions over the next few decades can reduce climate risks in the twenty-first century and beyond, increase prospects for effective adaptation, reduce the costs and challenges of mitigation in the longer term, and contribute to climate-resilient pathways for sustainable development (IPCC 2014). As stated by the IPCC reports, without additional efforts to reduce GHG emissions beyond those in place today, global emission growth is expected to persist, and it will continually be driven by growth in global population and economic activities. Climate change adaptation is also a process of adjusting to a changing environment either by man or natural systems that exploits beneficial opportunities and moderates negative effects. Developing and low-income countries of the world have been shown to have poor adaptive capacities to climate change (Apata et al. 2009), and this may be due to their economic level. The continued emission of greenhouse gases will cause further warming and long-lasting changes in all components of the climate system, increasing the likelihood of severe, pervasive, and irreversible impacts for people and ecosystems.

Limiting climate change would require substantial and sustained reductions in greenhouse gas emissions which, together with adaptation, can limit climate change risks (IPCC 2014). Climate change has the characteristics of a collective action problem at the global scale; therefore, effective mitigation will not be achieved if individual agents advance their own interests independently. Cooperative responses, including international cooperation, are therefore required to effectively mitigate GHG emissions and address other climate change issues as the effectiveness of adaptation can be enhanced through complementary actions across levels. A first step toward adaptation to future climate change is reducing vulnerability and exposure to present climate variability. Transformations in economic, social, technological, and political decisions and actions can enhance adaptation and promote sustainable development (IPCC 2014). Agricultural productivity in Nigeria is weather sensitive and highly vulnerable to climate change, and the southern ecological zone of Nigeria largely known for high rainfall is currently confronted by irregularity in the rainfall pattern, while Guinea Savannah is gradually experiencing increasing temperature (Bello et al. 2012). The northern zone faces the threat of desert encroachment at a very fast rate per year occasioned by fast reduction in the amount of surface water and flora and fauna resources on land (Obioha 2008). Some authors have opined that Nigeria is already being plagued with diverse ecological problems which have been closely linked to the ongoing climate change (Adefolalu 2007; Ikhile 2007) and some farmers have even abandoned farming for non-farming activities (Apata et al. 2010). Destruction of mangrove ecosystems, contamination of water, and transmission of waterborne diseases, leading to displacement and communal crisis, have been observed in flood-prone and coastal regions (Odjugo 2010). All of these have led to tragic crop failures, reduced agricultural productivity, increased hunger, poverty, malnutrition, and diseases (Zoellick and Robert 2009; Obioha 2009).

Over 90% of agricultural production in Nigeria is rain-fed with about 79 million hectares of arable land, of which 32 million hectares are cultivated (Nwajiuba 2012, cited in FAO 2012). Recent drought experiences in northern states and flooding in southern parts are evidences of disasters that have occurred due to climate change. Adaptation and mitigation strategies are not mutually exclusive. There is a need to work toward mitigating and adapting to best meet the challenges of climate change. Green infrastructures, such as urban wetlands and urban forests, reduce storm water runoff and could help cities adapt to increased flood risks due to climate change. Eme et al. (2014) stated that the Nigerian government is apparently inconsistent in its policy intervention and implementation strategies which further compound the problem of food security. It is therefore imperative to correctly apply adaptation and mitigation strategies that will moderate the negative effects of climate change. Rapid changes in temperature and precipitation patterns may have further disastrous consequences if not properly planned for.

Current Food Security Challenges in Nigeria

Severe hunger currently affects over a billion people against the 800 million people recorded in 1996 (FAO 2009). A large percentage of the population of sub-Saharan Africa (SSA) is being affected by chronic hunger; SSA has the highest prevalence of undernourishment with some 32% of the total population deprived of access to food. Furthermore, the number of undernourished individuals in the developing world is estimated at 820 million, equivalent to 17% of the total population. Agriculture though a climate-sensitive sector still remains necessary for survival. Changes in the climate pattern and variability in weather are major hindrances to achieving food security and eradicating poverty (Nyong 2005). A study by International Food Policy Research Institute (IFPRI) concluded that 25 million more children will be malnourished in 2050 due to the effects of climate change such as decreased crop yields, crop failures, and higher food prices (Stephen 2009). By the year 2080, agricultural output in developing countries may decline by 20% due to climate change, while output in industrial countries is expected to decrease by 6%. Also due to climate change, yields in developing countries could further decrease by 15% on average by 2080 (FAO 2008).

Global temperature increases of about 4 °C or more above late twentieth-century levels, combined with increasing food demand, would pose large risks to food security all over the world. Rural areas are expected to experience major impacts on water availability and supply, food security, infrastructure, and agricultural incomes, including shifts in the production areas of food and non-food crops around the world (Edame et al. 20111 Ibrahim 2004). Extreme temperature impacts plant growth and productivity through heat stress, high evapotranspiration, and increased irrigation requirements. An assessment by the Nigeria Meteorological Agency (NIMET) states that from 1941 to 2000, climate in Nigeria has demonstrated dramatic changes in weather patterns such as irregular rainfall pattern leading to fewer rainy days (Adelekan 2009). The assessment report also stated that late onset and early cessation of rains led to a shorter rainy season in most parts of the country from 1971 to 2000, annual rainfall dropped by 81 mm, and sharp temperature increases in most parts of the countries except for coastal areas such as Delta, Bayelsa, and Akwa-Ibom states.

In Nigeria, several government administrations have made interventions toward increasing food production and reducing food insecurity. This includes:

- Farm Settlement Scheme
- National Accelerated Food Production Programme (NAFPP) by Gen. Yakubu Gowon
- Operation Feed the Nation (OFN) by Murtala/Obasanjo administration
- River Basin and Rural Development Authority (RBRDA)
- Green Revolution (GR)
- World Bank-funded Agricultural Development Project (ADP) by Shehu Shagari
- Directorate for Food, Road and Rural Infrastructure (DFRRI) by Ibrahim Babangida
- Fadama Development Project in flooded regions which is still ongoing

However, to date, agriculture has been constrained by numerous challenges like rural-urban migration, wavering policy formulations, implementation, insufficient infrastructural support, poor input distribution system, focus on oil economy, pricing system, overdependence on rain-fed farming, poor capacity utilization, low investor's confidence, environmental degradation, poor access to funds, poor socioeconomic status of farmers, insufficient technological transfer system, corruption, and poor commitment to implementation of agricultural policies (Eme et al. 2014). The authors further stated that the Yar'Adua/Goodluck administration took a proactive measure by outlining agriculture as part of his agenda and vision 2015 of curbing hunger and poverty to improve the lives of 140 million Nigerians. Over N134 billion approved for agriculture in the 2008 budget is a clear testimony. The federal government constituted a practical and positive intervention toward eliminating hunger; this intervention resulted in the release of N80 billion from the Natural Resource Development Fund for importation of 500,000 mt of rice from abroad and 11,000 mt of grains to complement the local output. The government even proceeded to approve the release and distribution of 65,000 mt of assorted food from the Strategic Food Reserve to lessen the effect of low yield during 2007 season while stocking food items in the reserves to guarantee the required level of food security. Odjugo (2009) stated that climate change led to a shift in crops cultivated by farmers in Northern Nigeria. The preferred crops grown include guinea corn followed by groundnut and maize.

The National Adaptation Strategy and Plan of Action on Climate Change for Nigeria (NASPA-CCN) was also constituted to manage climate change-related issues, and it does this by incorporating stakeholders at the national and subnational levels.

Higher temperatures result in decreased agricultural productivity and production, high evaporation rates and reduced soil moisture, lowering of the groundwater table, and shrinking of surface water. Heat stress reduces human labor use on farms, lowers labor productivity, and leads to rapid deterioration and wastage of farm produce.

Achieving Resilience in the Nigeria Food System

To combat climate change and ensure food security in Nigeria, increased awareness of the causes and effects of climate change must be promoted through incessant public enlightenments and seminars on climate change. Farmer's education through agricultural extension services to rural farmers is of utmost importance as there is a dearth of information on weather forecast to ensure that farmers are adequately prepared for impending climate-induced disasters. Providing this information in farmer's native language will facilitate their understanding. This effort will immensely improve their knowledge of climate change and its management. This chapter also suggests increased endowments to farmers to facilitate coping strategies and provision of alternative means of sustenance especially in poor regions during wind and flood disasters.

There is also a need for agricultural organizations to provide drought-tolerant varieties and improved crop varieties to farmers at little or no cost. A wake-up call to the government is of utmost necessity so as to engender commitment to land tenure security and land access in the face of changing population dynamics, migration, and potential competition between agribusiness and smallholder and subsistence producers.

Furthermore, the following mitigation measures can be employed during incidences of climate change: reduction in greenhouse gas emission through enforcement by environmental agencies, mixed cropping for effective management of available land space, carbon reduction through soil carbon, and storage facilities to prevent waste of perishable agricultural produce, in addition to establishment of local panel on food, agriculture, and climate change in every state and local councils.

Conclusions and Recommendations

Global climate is projected to continue to change over this century and beyond, but there is still time to act to limit the amount of change and the extent of damaging impacts. Drought-prone areas should adopt irrigation capacities to ensure that crops are well fed with water to promote adequate growth. The constructions of small dugout tanks and ponds may be adopted in arid and semiarid regions of sub-Saharan Africa. These structures will be able to meet the demands for water for agricultural and domestic purposes during droughts. "New crops for a new climate" is a phenomenon that may be applied, for example, drought-tolerant, disease-resistant, and high-yield crop varieties and shift in planting dates are advised to mitigate crop losses. Pest and weed should be controlled once detected by farmers before infestation occurs. In rural poor communities, integrated pest management is advised as it is more economical and health and environmentally friendly. In flooded farmlands, reduced tillage and improved soil organic matter through composting and manure are advised. For large-scale agriculture in flood-prone areas, farmlands should be designed with drainage patterns to allow flow of water through rows and ridges. Agricultural productivity is affected by unpredictable climatic conditions as most crop failures are due to either lack or excess rainfall and extreme temperature. Climate forecasting may provide adequate knowledge to farmers to make proper and informed decisions that will support profitable agricultural yield. The IPCC reports have shown that GHGs remain the major cause of climate change; efforts must therefore be put in place to ensure the reduction of these gases in the environment through burning of fossil fuels. Switching to renewable energy sources like solar or wind and electricity will help reduce carbon dioxide emissions.

References

Adefolalu DO (2000) An introduction to the challenges of climate change and its impact in Nigeria. A workshop on climate change and natural disaster in Nigeria held at FUT, Minna, 3 Apr 2000

Adefolalu DO (2007) Climate change and economic sustainability in Nigeria. Paper presented at the international conference on climate change, Nnamdi Azikiwe University, Awka 12–14 June 2007

Adejuwon J (2006) Food security, climate variability and climate change in Sub Sahara Africa. In: A final report submitted to Assessments of Impacts and Adaptations to Climate Change (AIACC), Project No. AF23, USA: The International START Secretariat Washington, DC 20009. http:www.aiaccproject.org/finalreport

Adelekan IO (2009) Vulnerability of poor urban coastal communities to climate change in Lagos, Nigeria. Fifth Urban Research Symposium 2009

Aggarwal PK, Singh AK (2010) Implications of global climatic change on water and food security. In: Ringler C et al (eds) Global change: impacts on water and food security. Springer, Berlin, pp 49–63. https://doi.org/10.1007/978-3-642-04615-5_3

Apata TG, Samuel KD, Adeola AO (2009) Analysis of climate change perception and adaptation among Arable Food Crop Farmers in South Western Nigeria. Beijing, 16–22 Aug 2009

Apata TG, Ogunyinka A, Sanusi RA Ogunwande S (2010) Effects of global climate change on Nigerian Agriculture: an empirical analysis. Paper presented at the 84th. Annual conference of Agricultural Economics Society held in Edinburgh, pp 345–351

Ayuba HK, Maryah UM, Gwary DM (2007) Climate change impact on plant species composition in six semi-arid rangelands of Northern Nigeria. Niger Geogr J 5(1):35–42

Bello OB, Ganiyu OT, Wahab MKA, Afolabi MS, Oluleye F, Ig SA, Mahmud J, Azeez MA, Abdulmaliq SY (2012) Evidence of climate change impacts on agriculture and food security in Nigeria. Int J Agric For 2(2):49–55

de Wit M, Stankiewicz J (2006) Changes in surface water supply across Africa with predicted climate change. Science 311:1917–1921

Eboh E (2009) Implications of climate change for economic growth and sustainable development in Nigeria. Enugu forum policy paper 10. African Institute for Applied Economics, Nigeria

Edame GE, Ekpenyong AB, Fonta WM, Duru EJC (2011) Climate change, food security and agricultural productivity in Africa: issues and policy direction. J Humanit Soc Sci 1 (21):205–223. http://www.ijhssnet.com/journals

Eme OI, Onyishi T, Uche OA, Uche IE (2014) Challenges of food security in Nigeria: options before government. Arab J Bus Manag Rev (OMAN Chapter) 4(1):15

Enete IC (2008) Potential impacts of climate change on coastal settlements of Nigeria. A. Paper presented at the 15th Annual Conference of Environment and Behavior of Nigeria (EBAN) held at Babcock University, Ogun State from 6–9 May 2008, Nigeria
Food and Agricultural Organization (FAO) (2012) Food security statistics- Nigeria. FAOSTAT. FAO, Rome
Food and Agricultural Organization, FAO (2008) Climate change, bioenergy and food security. Options for decision makers identified by expert meeting. Prepared for the high-level conference World Food Security: The Challenges of Climate Change and Bioenergy, Rome, 3–5 June
Food and Agricultural Organization, FAO (2009) The state of food security in the world. Food and Agriculture Organization of the United Nations, Rome
Hocking PJ, Meyer CP (1991) Carbon dioxide enrichment decreases critical nitrate and nitrogen concentrations in wheat. J Plant Nutr 14:571–584
Ibrahim EY (2004) Climate change and food security in Nigeria. Paper presented at the scientific session of the international conference on possible impacts of climate change on Africa, organized by the institute of African research and studies, Cairo University, Cairo, 18–20 May
Ikhile CI (2007) Impacts of climate variability and change on the hydrology and water resources of the Benin-Owena River Basin. PhD thesis submitted to the Department of Geography and Regional Planning, University of Benin, Benin City, Nigeria. pp 234–236
Integrated Strategy for Disaster Reduction (2008) Integrating disaster risk reduction into the fight against poverty: global facility for disaster reduction and recovery. World Bank, Washington, DC
IPCC (2007a) Fresh water resources and their management climate change Impact, adaptation and vulnerability. Contribution of working group-II to the fourth assessment report of the intergovernmental panel on climate change. Cambridge University Press, Cambridge, UK
IPCC (2007b) Climate change, the physical science basis; working group contribution to the fourth assessment report of the intergovernmental panel on climate change; summary for policymakers, technical summary and frequently asked questions
IPCC (2014) Climate change 2014: synthesis report. Contribution of working groups I, II and III to the fifth assessment report of the intergovernmental panel on climate change [Core Writing Team, R.K. Pachauri and L.A. Meyer (eds.)]. IPCC, Geneva, Switzerland, pp 151
Lobell D, Burke M (2010) Climate change and food security. Advances in global change research 37. © Springer Science Business Media, B.V. Springer Netherlands. https://doi.org/10.1007/978-90-481-2953-9_2
Nwajiuba C (2012) Nigeria's agricultural and food security challenges. http://ng.boell.org/downloads/Agriculture_Green_Deal_Nigeria_study.pdf
Nyong A (2005) Abstract in impacts of climate change in the tropics: the African experience. Avoiding Dangerous Climate Change, Meteorological Office, Exeter. www.stabilisation2005.com
Obioha E (2008) Climate change, population drift and violent conflict over land resources in North Eastern Nigeria. J Hum Ecol 23(4):311–324
Obioha E (2009) Climate variability, environmental. J Hum Ecol 26(2):107–121
Odjugo PAO (2009) Quantifying the cost of climate change impact in Nigeria: emphasis on wind and rainstorm. J Hum Ecol 28(2):93–101
Odjugo PAO (2010) General overview of climate change impacts in Nigeria. J Hum Ecol 29(1):47–55
Odjugo PAO, Ikhuoria AI (2003) The impact of climate change and anthropogenic factors on desertification in the semi-arid region of Nigeria. Glob J Environ Sci 2(2):18–126
Stephen L (2009) Climate change, food supply hangs in the balance, IPS NEWS, 2 Oct 2009. http://ipsnews.net/news.asp?idnews=48689
Thornton PK, Jones PG, Ericksen PJ, Challinor AJ (2011) Agriculture and food systems in sub-Saharan Africa in a 40C+ world. Philos Trans R Soc A Math Phys Eng Sci 369(1934):117–136
Ziska LH, Namuco O, Moya T, Quiland J (1997) Growth and yield response of field grown tropical rice to increasing carbon dioxide and air temperature. Agron J 89:45–53
Zoellick S, Robert BA (2009) Climate smart future. The nation newspapers. Vintage Press Limited, Lagos, p 18

Climate Change, Vulnerability, and Adaption Under the Small Farming Households of Konso Community, Southern Ethiopia

41

Yohannes GebreMichael

Contents

Abstract

The overall aim of this case study is to understand climate change, vulnerability, and adaptation at small farming household scale. A total of 244 household heads were randomly selected from the clusters of wealth rank groups in the two farming practices of ox-plough and hoe cultivation. The findings indicate that both the community perceptions and meteorological evidences imply an increasing trend of drought occurrence and rainfall intensity accounting for the continuous vulnerability of the agrarian community with climate change. Moreover, the root causes of the vulnerability of the community to the impacts of climate change are attributed to the combined effects of environmental and socioeconomic and

Y. GebreMichael (✉)
Department of Geography and Environmental Studies, Addis Ababa University, Addis Ababa, Ethiopia
e-mail: yohannesgmichael@gmail.com

W. Leal Filho (ed.), *Handbook of Climate Change Resilience*,
https://doi.org/10.1007/978-3-319-93336-8_71

political factors, where the poor and the landless youth are affected the hardest. The study also found that the communities are practicing a wide range of ingenious land husbandry practices, innovations, and early warning systems as strategy of adaption to climate change. Finally, it concludes with the need of supporting local experimentation and innovation in developing resilience to the impact of climate change and food insecurity.

Keywords
Climate change · Vulnerability · Perceptions · Local innovation · Adaptation · Resilience

Introduction

Scientific evidence has indicated that climate change is occurring and will be the fundamental challenge of achieving millennium development goals in developing countries (IPCC 2014; Adger et al. 2007; Haile 2005). Moreover, the rural communities with an agrarian livelihood base will be the victims of the impacts of climate change, including drought, famine, flood, land degradation, and decline of land productivity (Okonya et al. 2013; Fosu-Mensah et al. 2010; Mertz et al. 2009).

Today, more than 70% of the population in Ethiopia depends on agriculture, which is influenced by the impacts of climate change and numerous other pressures, including population pressure and policy changes. The Ethiopian government has developed a Climate Resilient Green Economy (CRGE) strategy with a vision of becoming a middle-income state by 2025. Also, achieving the Millennium Development Goals is a challenge in Ethiopia, as annually up to eight million people are chronically food insecure (DFID 2010). The problem has been compounded by the anticipated climate change, with a projected increase in rainfall variability, frequent drought, and more extreme flooding (World Bank 2008; Green Forum 2008; EEA 2008; Aragie 2013). It is also predicted that an increase in temperature and in frequency of drought will lead to increased famine, spread of human and livestock disease, rural-urban migration, and dependency on external food aid (Ringer 2008).

The history of famine and drought in Ethiopia is recorded back to at least the sixteenth century. From 1540 to 1800 AD, 26 major droughts and famines occurred, and in 1889–1892 the great Ethiopian famine led to the spread of diseases and deaths of humans and livestock (McKee 2008; Pankhurst 1985). The famines were due not only to shortage of rain but also to excessive rain. Thus, climate variability is not a new phenomenon to Ethiopian farmers, and, because local people have a long experience with climatic risks, they have developed different adaptation strategies and innovations that could also be relevant for adaptation to current climate change. Factors compounding the impact of climate change in Ethiopia are rapid population growth, land degradation, widespread poverty, dependency on rained agriculture, and lack of appropriate policies and legislation (NMA 2007).

Currently, the global and national development strategies and plans are preoccupied with addressing poverty; fostering economic growth; developing appropriate technologies in insuring food security, equity, and sustainability and good governance; and mainstreaming sustainability. Accordingly, indigenous knowledge (IK) can contribute to achieving the strategy. The recognition of IK is not only based on the deep-rooted experience of rural communities but also support and stimulates marginalized and poor people to solve their own problems (Agrawal 1995; Senanayake 2006; Briggs 2005; Finucane 2009). Accordingly, there is a growing awareness that scientific knowledge alone is not solving global problems and that IK has a fundamental role to play in filling the gap (Briggs 2005; IPCC 2007; Finucane 2009; Yohannes and Waters-Bayer 2007). IK comprises the knowledge and experience accumulated by local people over generations. It is dynamic, flexible, and adaptable to environmental and policy changes and is embedded in the social culture of the community (Berkes 2012; Nakashima et al. 2012).

However, in the course of modernization, IK has been marginalized in Africa. This has led to a decline in IK and a weakening of the values and identities of local communities (Briggs 2005). The "Green Revolution" with its top-down approach to agricultural development has replaced local knowledge and practices and made smallholder farmers more vulnerable to famine and drought (Chossudovsky 2000).

Therefore, the overall aim of this case study is to understand the spatial and temporal climatic changes, its inherent causes, vulnerability, and adaption under the differentiated community. It is assumed that a better understanding of the vulnerability and practices at grassroots level is rational entry point to develop appropriate climate resilience strategy and extension services in the livelihoods of the rural community (Elizabeth et al. 2009).

Materials and Methods

Sample Design

The approach used for data collection for this study was based on interviews and focus group discussions (FGDs) with different stakeholders and complemented by field observations.

Currently, the Konso Special *Woreda* has 50 *kebele* (lowest administrative unit of government in rural areas); see also Fig. 1. With the help of resource persons in the district, two *kebeles* were identified as study sites on the basis of the following combination of factors:

- Chronic food insecurity and dependency on the Productive Safety Net Program (PSNP)
- Practicing different farming practices (ox-plough and hoe cultivation)
- Prevalence of different indigenous soil and water conservation (SWC) and local innovations

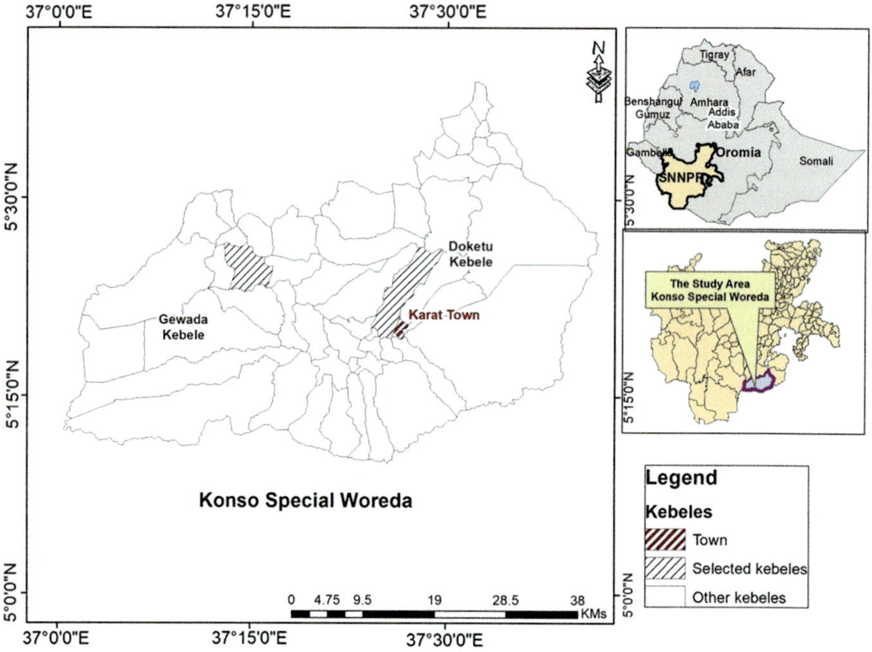

Fig. 1 Study area

The two *kebeles* selected were Gewada and Doketu. Gewada represents the *kolla* agroecological zone (500–1500 masl) and ox-ploughing system. In Doketu both *kolla* and *weyna dega* agroecological zones (1500–2300 masl) are dominated by hoe cultivation and wide range of indigenous SWC practices. Both *kebeles* are under the PSNP.

As a framework for data collection and analysis, wealth ranking of the community was applied. The hypothesis is that vulnerability and resilience to the risk of climate change is highly determined by access to resources. Similarly, the agroecology and the farming practices also have an influence on perceptions and adaptation to climate change.

Sample Size

As shown in Table 1, a total of 244 household heads were randomly selected from the list of wealth categories in the respective *kebeles* for an in-depth interview survey. The main issues covered in the survey were demography, agricultural production, and sources of livelihood and perceptions of climate change, vulnerability, impact, and adaptation. In addition, a series of focal group discussions (FGDs) were held with elders, local leaders, women and youth, and individual stakeholders

Table 1 Sample population with different wealth ranks

	Sample *kebeles*			
Category of household	Gewada	Doketu	Total	%
Better-off	10	15	25	11
Medium	23	30	53	23
Poor	60	50	110	46
Ultra-poor				
Woman-headed	25	20	45	19
Landless	6	5	11	1
Total no.	124	120	244	100

such as local healers, artisans, development agents (DAs) and staff of schools, local administration, and government line offices.

Moreover, the metrological data from the weather station of the Konso district from 1987–2011 has been accessed from the National Metrological Agency (NMA).

Study Area

Konso Special *Woreda* is located in the Rift Valley of southwestern Ethiopia in the Southern Nations, Nationalities, and Peoples' Region (SNNPR). It covers a total area of 200,000 ha and has 50 *kebeles*. It lies about 600 km south of Addis Ababa.

By and large, Konso is a semiarid area, and in terms of the Ethiopian agroecological zones, Konso has a dry *kolla* agroecology in about 70% of the total land area and dry *weyna dega* in the remaining 30%. The landscape in some parts of the area is very steep and highly dissected, creating microclimatic niches. Konso has bimodal rainfall: the main rains are locally known as *Hagaya* (March–May) and the small rains as *Katana* (September–November). The average annual rainfall is about 800 mm, ranging from 519 to 1094 mm.

Based on the individual and group discussions, the fundamental sources of livelihood in the rural areas of Konso are crops and livestock (mixed farming). In the ox-ploughing areas, different types of cereals, pulses, and oilseeds are produced, and cattle and small ruminants are reared. In the hoe-farming areas, cereals, pulses, oilseeds, and tubers are produced as components of agroforestry systems with multipurpose trees for food, fodder, medicine, and timber. Fattening of livestock is a complementary activity in all wealth-rank groups; some are also engaged on traditional beekeeping. Many of the poor farmers, particularly women household heads, are engaged on petty trade: selling local drinks and attending periodical market days to sell different crops and fruits. Some youth groups (female and male) also generate income by working as a group in different farming activities, while other youth groups migrate seasonally to neighboring areas to work as farm laborers. Many of the poor farmers are highly dependent on the food aid under the productive safety net. There are also community members who complement their livelihoods with some craftwork (see Table 2). Firewood selling from private and

Table 2 Sources of livelihood

	Doketu (%)				Gewada (%)			
Category of household	MF	T	LC	PSNP	MF	T	LC	PSNP
Better-off	100	20	10	4	100	47	7	5
Medium	62	9	11	18	91	100	12	20
Poor	47	22	14	37	64	72	18	70
Woman-head	60	45	10	72	44	60	5	63

Key: *MF* mixed farming, *T* trade, *LC* labor and craft, *PSNP* Productive Safety Net Program

Table 3 Firewood selling as means livelihoods

	Doketu (%)			Gewada (%)		
Category of household	Frequently	Sometimes	No	Frequently	Sometimes	No
Better-off	7	25	69	10	40	51
Medium	20	10	72	10	15	75
Poor	20	5	76	60	20	20
Woman-head	5	0	95	40	7	53

communal lands is also a source of income among some of the different wealth-rank groups, mainly during drought periods. Wood is the only source of energy for cooking and heating in the rural areas, and cow dung is used for soil fertility improvement (Table 3).

Results and Discussions

Resource Potential at Local Level

Soils

Based on the discussions with the farmers in the study areas, different soil categories were identified. In Gewada, under the ox-plough farming system, the soil types are classified into four major categories based on their fertility, soil color, and slope (Table 4A). Usually, fertile soils (*borober*) have a potential for a wide and diverse range of crops and are less vulnerable to drought and erosion. Unfortunately, such soils are not widespread (15%) and are usually localized in flat areas and exposed to erosion by flooding and riverbank erosion. The poor and very poor soils (*tahak* and *mekorko)* account for up to 35% of the farmland and have serious problems of crop failure and soil erosion. The farmland with poor soil is usually on plots far from homestead and used by the poor, who have no livestock to provide manure for improving the soil fertility and usually assigned for sharecropping with the medium or better-off farmers. The medium-fertile soils account for up to 50% of the farmland and are favorable for different species of crops. However, they need frequent ploughing to control weeds. The poor with shortage of labor and oxen lease the land or made sharecropping agreements.

Table 4A Major local soil types of farmlands in Gewada

Soil types	Area (%)	Color	Slope	Fertility	Major crops
Borober	15	Red	Flat/sloping	Fertile	All
Tahak	23	White	Sloping	Poor	Sorghum + maize
Mekorko	12	Yellow	Sloping	V. poor	Sorghum + finger millet
Shelhakoh	50	Yellow	Flat/sloping	Medium	All

Table 4B Major local soil types of farmlands in Doketu

Soil type	Area (%)	Color	Slope	Fertility	Major crops
Borober	20	Black	Flat	Fertile	All
Klkayta	50	Red	Sloping	Medium	All
Dogolata	10	Red	Flat	Medium	All
Achaita	20	White	Sloping	Poor	All

In the hoe-farming system in Doketu, crop diversity is not strongly correlated with soil fertility because the physical conservation, mixed cropping, and agroforestry practices create a synergy for soil and water improvement. Compared to Gewada, Doketu has a lower proportion of poor soil coverage (Table 4B). However, the diversity of plants per plot under hoe farming varies with distance from homestead and farm size. Usually the youth who inherited land from family have plots far from their homestead and of small size.

Soil fertility varies greatly from place to place on account of microclimate and cropping patterns. Overall, the fertile soils account for up to 20% of total soils, while the medium-fertile and poor soils account for up to 70% and 30%, respectively. In other words, the soil fertility tends toward medium. This suggests that its better status is due to the indigenous physical, biological, and agronomic SWC measures (see also section "Indigenous Soil and Water Conservation Practices"). It also implies local existence of wide range of options for fertility improvement.

Land Use and Cover

Based on the DAs' estimates in the study *kebeles*, the dominant form of land use is crop farming, which covers more than 60% of the total area (see Table 5). The badlands are usually scattered rocky outcrops. The other land uses are grazing land (8%) and forests (16%); forest cover is smaller in the ox-ploughing than in the hoe-farming system. This has at least two fundamental implications: (i) low potential for further land distribution with population increase; and (ii) the poor who complement their livelihood from the potential communal land for firewood, pasture, and wild foods will be diminishing and may be more vulnerable to any changes including the impact of climate change and food insecurity.

Landholdings and Fragmentation

In 1994, the total population of Konso Special *Woreda* was about 152,106; this grew to 235,087 in 2007, which means that the population is growing by about 5% per

Table 5 Land use and cover in the study areas (%)

Land use/cover	Gewada	Doketu	Mean
Cultivated land	65	60	63
Grazing land	10	5	8
Forest area	12	20	16
Badlands (rook outcropped)	5	5	5
Settlements	5	5	5
Others	3	5	4

Table 6 Household landholdings in ha (%)

Category of household	Gewada				Doketu			
	0.25–0.75	1.00–1.50	1.75–2.25	2.50+	0.25–0.75	1.00–1.50	1.75–2.25	2.50 +
Better-off	40	60	–	–	38	38	19	6
Medium	100	–	–	–	76	24	–	–
Poor	100	–	–	–	9	8	–	–
Woman-headed	100	–	–	–	75	5	–	–

Table 7 Household ownership of the number of fragmented plots (%)

	Gewada (no. of plots)				Doketu (no. of plots)			
Category of household	1–2	3–4	5–6	7+	1–2	3–4	5–6	7+
Better-off	40	53	7	–	25	63	6	6
Medium	90	10	–	–	48	38	14	–
Poor	100	–	–	–	84	12	2	2
Woman-headed	80	20	–	–	75	5	–	–

year. The average landholding per household in the study areas is about 0.5 ha, but this varies with wealth rank and between *kebeles*.

As shown in Table 6, the medium-wealth and poor households have less than 0.5 ha, while some of the better-off households use more than 3 ha, using sharecropping with their oxen, while the better-off under the hoe with more than 2 ha are usually the community leader (*poqalla*). Hence more land implies more diversity of crops and production.

Fragmentation of plots is a rational adaptation strategy to access different farm features (soil fertility, microclimate), allowing crop diversity and increasing resilience to risks and uncertainties (drought and crop diseases). Moreover, remote plots can be assigned for inheritance by children or for sharecropping or for sale. However, the number of fragmented plots owned by a single household has been diminishing mainly because of inheritance by children and expansion of settlements. As shown in Table 7, the average number of plots per household in the study areas is three, but the better-off farmers have up to 11 fragmented plots. Moreover, the number of fragmented plots per household is higher in the hoe-cultivation system than in the ox-plough system. More land fragmentation implies access to different microclimate resource potential, spreading of risks of climate change, and diversity of production.

Community Perceptions of Changes in Their Locality

Local Perceptions on Changes in Their Locality

Speaking in general terms, the interviewees in the study areas have observed in the span of their life change in the rainfall and temperature distribution and amount across the seasons and months. For example, when the two wet seasons (*Hagaya* and *Katana*) have some rain, it is locally considered a good year. Other times, one of the rainy seasons fails, and it is considered a normal year, while when both rainy seasons fail, it is considered as bad year. However, the trend is skewed to bad years due to the frequent and prolonged droughts. The communities have traditional early warning systems to determine the good and bad years, such as the position of stars, the direction of the winds, natural vegetation growth (flowering and bearing fruits), and wild and domestic animal behavior on specific seasons. Usually, the occurrence of some rainfall in the dry seasons leads to pests and diseases in crops, livestock, and humans and coincides with shortage of food and fodder and low resistance of livestock and human beings to the different risks. Most farmers are experiencing more frequent shortage of rain (*enodae*), while some having excessive rain (*ferayn*).

Regarding the monthly temperatures, the local people perceived that most of the wet months are hotter than before, and in recent years, there has been no frost in the months of March and April, as used to occur previously. As shown in Table 8, the different wealth groups, including women household heads, have underlined that, over time, the amount of rainfall is becoming less and hot months are predominating.

"The current drought in the locality might not be [as/more] serious than in the past. In the past, we used a wide range of strategies to adapt, such as soil and water conservation, diversified crops and agro-forestry, livestock and wild plants. However, today, with the loss of livestock, diminishing of farming lands and communal areas, feeding the increasing population become difficult. Hence, vulnerability to any risks and uncertainties becomes evident." Elderly farmer in Doketu

Table 8 Perceptions on indicators of climate change and variability

	Gewada (%)				Doketu (%)			
Classification	RDU	TmI	SoD	DCY	RDU	TmI	SoD	DCY
Better-off	48	30	11	11	32	37	5	26
Medium	69	8	8	15	26	40	9	25
Poor	58	17	8	17	28	35	17	20
Woman-headed	62	22	4	12	25	43	3	30

Key: *RDU* rainfall decline and unpredictability, *TmI* temperature increase, *SoD* spread of disease, *DCY* decline in crop yield

"Malaria was with us since our ancestors but less people were affected, as we had good resistance because we had access to sufficient milk, meat, butter and different kinds of local food types. Today, the only food is cheka drinks (Local beer made from sorghum or maize) *and quarkufa food (*Food made from flour of sorghum and *Moringa* leaves*). Hence, many people are easily affected by malaria because of low resistance."* Elderly women in Gewada

"Historically our parents cultivated the remote plots (fora) in the lowlands with sorghum using the river floods. They visit such plots only during seeding and harvesting due to high prevalence of malaria and very hot temperature. However, today, due to population pressure and shortage of land many of the youth are inheriting such plots which are permanently cultivated and settled. Hence the youth and the fora plots are becoming highly vulnerable to the impact of climate change." Youth farmers from Doketu

"In the past, the traditional huts had a narrow door to make the house warm, and the cloths we used were relatively thick, but now all months are becoming hot, the houses are with big doors and the thick cloths are no longer demanded and we prepare the thin cloths for our locality and outsiders." A weaver in from Doketu

During the FGDs, the elders, youth, and women reflected on the overall changes they observed in their locality. The challenges or threats farmers mentioned included the increasing trends in human and animal disease, resource-based conflicts and dependency on food aid, and the decreasing trend in landholdings, crop, and tree diversity (see Table 9). The opportunities or potentials they identified were the better possibilities for livelihood diversification, including tree plantations, and the increasing attention given to SWC. To adapt to the challenges, some local people are practicing intensification of agriculture, sharecropping arrangements beyond the local boundary, and revitalization of customary institutions for conflict management and social safety nets. Furthermore, many are practicing agronomic and conservation measures in land husbandry (see section "Indigenous Soil and Water Conservation Practices").

Events of Climate Changes

The outcomes of the FGD with a group of elderly people about past events of severe drought and famine are shown in Table 10 and can be summarized as follows:

- Historically, several severe drought events led to the loss of human life and livestock. No events were mentioned of serious floods although some elders said that heavy rainfall was causing some damage on the floodplain around riverbanks and physical SWC structures.
- The intervals between severe droughts and famines gradually declined from 50 to 30 years and, more recently, to 10-year and 1-year intervals, while the duration of severe drought ranged from 8 years to 1 year – still a serious drought from which a poor household cannot easily recover.
- One of the fundamental causes for the community's vulnerability was the cumulative effect of the injustice and damage of their resource base by the past regimes as in the case of the territorial expansion of emperor Menelik II, in the eighteenth

Table 9 Local people's perceptions of different changes in the study area

No	Indicators	Trends	Remarks
1	Family size	Increasing	Earlier marriage Limited use of contraceptives
2	Landholding	Declining	Inheritance by children Expansion of settlements Enclosure of communal land Riverbank erosion
3	Malaria infestation	Increasing	Permanent settlement in the lowlands Diminishing access to balanced diet
4	Diversity of crops and landraces	Declining	Decline in farm size and fragmentation of farming plots Introduction of "improved" seed by extension Prolonged droughts
5	Livelihood diversity	Increasing	Involved in cropping, livestock fattening, petty trade, crafts, and hiring out labor
6	Planting trees	Increasing	Multipurpose uses Drought resistance and synergetic effects Opportunity to engage in off-farm activities
7	Dependency on food aid	Increasing	Prolonged droughts Dependency syndrome Incentive to introduce new technologies Marginalization in decision-making
8	Out migration	Declining	In normal season, working on farm In bad season, seeking seasonal labor Voluntary resettlement in other areas
9	Power of customary institution	Declining	Introduction of government structures External development interventions by NGOs Expansion of new religion (Christianity)
10	Use of physical SWC measures	Increasing	Inherited by children Maintained to avoid loss of seed and soil and to harvest water (immediate benefit) Incentives by the PSNP to build SWC structures
11	Conflict over resource use	Increasing	Population pressure Prolonged droughts Defective land tenure regulations *Kebele* leaders less accountable to community

century (Table 10). This historical event in the Konso area suggests the following: (i) climate change and irregular drought events are not new phenomena for the Konso people. This has triggered local experimentation and innovation in land husbandry. (ii) The cumulative effects of the frequency and prolonged drought compounded with and external injustice have contributed to the slow recovery and increase vulnerability. To cope with the different risks and uncertainties, the Konso elders have mentioned they used to collect wild food sources and migrate to big riverbanks to exercise traditional irrigation. Moreover, the external emergency support from missionaries, government, and NGOs was present since the 1950s, but it was on and off and small scale.

Table 10 Events of severe drought and famine in the Konso area

Period	Drought event (local name)	Major features of the event
1891–1899	*Lota Koksh* (drying up of the evergreen trees)	For 9 years, there was no rain, and all trees including the drought-resistant evergreen tree locally known as *koksh* (hop bush, *Dodonaea angustifolia*) dried up. The event, given the name *Lota Koksh,* was the worst event in the history of Konso, when people were forced to eat the dry skin of livestock and when dogs attacked weak people. A significant number of people died, almost all livestock were lost, and wild animals migrated to other areas. At that time, the Konso area was inaccessible to external support
1896–1899	*Kormo Kaleto* (slaughter of young cattle)	Because the Konso people resisted Emperor Menelik II, in 1896 his army set fire to the houses and crops in the field and stored in the house. Nothing remained for sowing. This attack was followed by a severe drought and death of humans and livestock. People were forced to slaughter young cattle (*kaleto*) against the cultural value of the community, as young cattle are normally not slaughtered for any purpose
1950–1951	*Lota Kore* (grain market center)	There was a total failure of rains over 2 consecutive years. The Konso people travelled to the neighboring Gamo District of Kore to buy maize. The event is named after the market center (Kore) that contributed to their survival
1984–1986	*Lota Kaws* (long beared)	A 3-year drought led to serious loss of lives of people and livestock. For the first time, the Konso received external food aid (through the Norwegian Lutheran Church). The drought was given the name *Kaws* (long beard) after the driver of the truck that transported the food. This also marked the introduction of an external religion into the Konso area
1992/3–1994	*Lota Farm* (NGO support)	A 2-year drought led to little human death but high livestock losses because of shortage of pasture and spread of livestock diseases. The NGO FARM-Africa intervened with humanitarian relief and some development activities
2004/5–2008		There was no rain for about 4 years. Despite the humanitarian relief, there was loss of livestock and huge migration of people to the neighboring areas including riverbanks of Segen and Woyito. No specific name was given to this event
2009		The first wet season *Katana* (March–May) was good, but the second and main wet season (*Hagaya)* totally failed. Days of light rain at wide intervals led to the loss of sown seed and dependency on food aid
2010	Good year	Both *Katana* and *Hagaya* were very good
2011	Bad year	Both *Katana* and *Hagaya* were failing
2012	Normal year	Katana was very good but *Hagaya* was totally failed

(*continued*)

Table 10 (continued)

Period	Drought event (local name)	Major features of the event
2013	Normal year	Katana totally failed but *Hagaya* was very good
2014	Good year	Both *Katana* and *Hagaya* were not bad
2015	Normal year	The first season *Katana* was not good, but the second season (*Hagaya*) was good

Metrology Evidence

Based on the secondary data of the weather station of Konso (1987 to 2011) from the National Metrological Agency, the following analyses were made.

Distribution and Trends

The Konso traditionally divide the year into four seasons: two wet seasons, the big rains *Hagaya* (March–May) and the small rains *Katana* (September–November), and two dry seasons, *Masana* (June–August) and *Tona* (December–February). On average, 359 mm or 45% of the total precipitation falls in the *Hagaya* season, 229 mm or 29% of the total precipitation in the *Katana* season, and 26% of the total precipitation in the two dry seasons: 96 mm or 12% in *Masana* and 113 mm or 14% in *Tona* (Fig. 2). Thus, rainfall is spreading across the year, which is favorable for perennial crops and pastures but also for the infestation of malaria.

The variability within each season is high (Fig. 3). For example, in *Hagaya*, the percentage of the total annual rainfall ranges from 29 to 63%, for *Katana* it ranges from 14 to 49%, for the *Masana* it ranges from 2 to 27%, and in *Tona* from 1 to 37%. Thus, the coefficient of variation is relatively high in all seasons and increases as one moves from the main wet season, *Hagaya* (30%), to the short wet season *Katana* (37%) and the dry seasons *Masana* (54%) and *Tona* 73%. This also fits to the community perception of irregularity of rainfall and frequency of drought in their locality.

The two rainy seasons are characterized by high fluctuations in rainfall distribution, which is typical for arid and semiarid areas. Moreover, as indicated in Fig. 4, the rainfall in the *Hagaya* rainy season has a pattern of fluctuation every two season with some degree of predictability. However, in the *Katana* rainy season, the pattern of fluctuation is between an interval of 2 and 5 years. There is also a trend of increasing of rainfall in *Katana* and a slight decreasing trend in *Hagaya*. When rainfall is higher than average in both wet seasons, farmers regard it as a good year; the probability of this is about 26% considering the 1987 to 2011 rainfall data. When one wet season has higher than average rainfall and the other has lower than average rainfall, it is overall an average year in Konso in terms of annual rainfall; such years account for 37% of the total years measured. A lower than average rainfall in both wet seasons is regarded as a bad year and accounts for about 37% of all years (Fig. 4). Generally, the rainfall in the past decades has tended to be "normal," as 63% of the seasons are "good" or "average." However, in reality, the good and bad seasons are not only

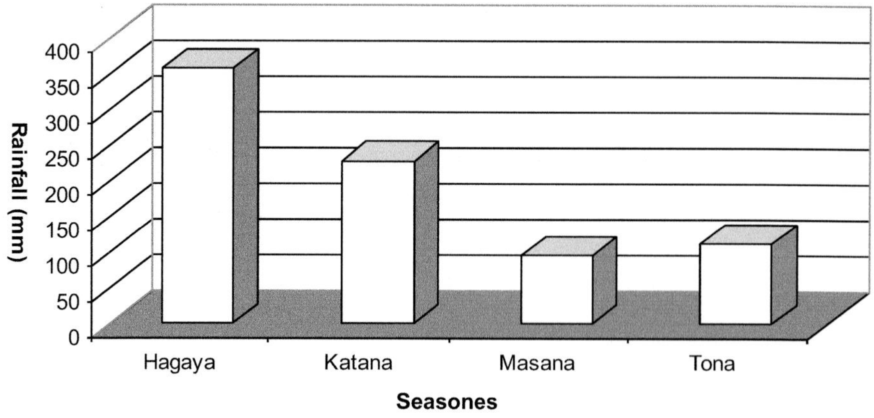

Fig. 2 Annual average rainfall distribution across the seasons (1987–2011)

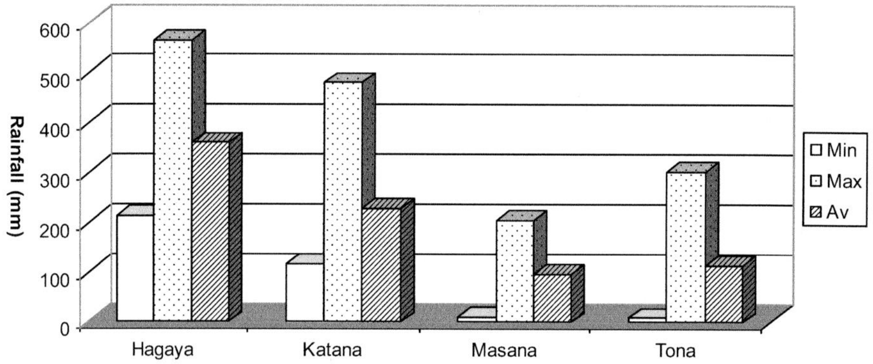

Fig. 3 Range of rainfall variability across seasons (1987–2011)

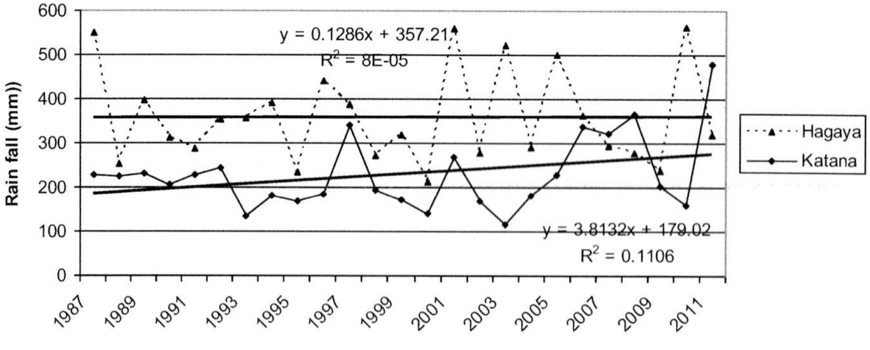

Fig. 4 Trends of rainfall distribution during Hagaya and *Katana* seasons

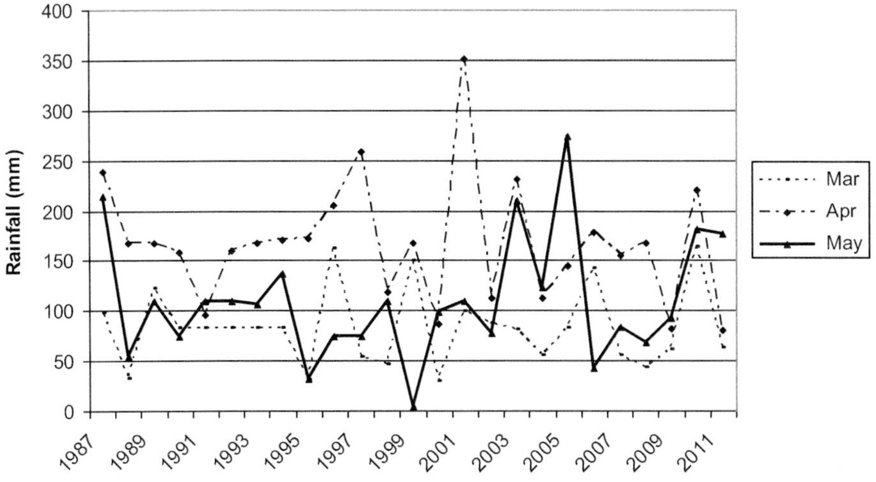

Fig. 5 Monthly rainfall distribution in the *Hagaya* season (1987–2011)

influenced by environmental factors; the socioeconomic condition of the community is also equally important in the determination of the vulnerability and resilience to the impact of climate change.

Also, the monthly rainfall within each rainy season is highly variable. For example, in the *Hagaya* season, the rainfall in the month of March varies from 28 to 183 mm, in April from 80 to 352 mm and in May from 5 to 273 mm (Fig. 5). Similarly, in *Katana*, the rain in the month of September varies from 15 to 182 mm, in October from 41 to 193 mm, and in November from 14 to 230 mm (Fig. 6). The high variability between and within the rainy seasons has a strong impact on the farming practice and household economy based on agriculture.

The metrological data from the weather station may indicate some tendencies in the variability of rainfall and climate change, but they do not seem to reflect the whole truth for at least two fundamental reasons: (i) the rainfall data are highly aggregated on a monthly and annual basis and do not reflect the distribution and intensity of rainfall, which are fundamental in plant growth, splash effect, and erosion; and (ii) a single metrological station in Konso cannot represent the micro-climatic variations in the study area.

Intensity of Rainfall

Regarding the intensity of rainfall during the *Hagaya* season, the average maximum rainfall per day is 32% of the total rainfall. This also varies within the *Hagaya* months. For example, in March, the average maximum rainfall per day was 35% of the total, while for April and May, it was 27% and 34%, respectively. The maximum rainfall per day with 40% and above from the total rainfall on the specific month was 35% for March, 5% for April, and 20% for May. This indicates that the rainfall in March is more intensive than in the other *Hagaya* months (Fig. 7). Similarly, in the

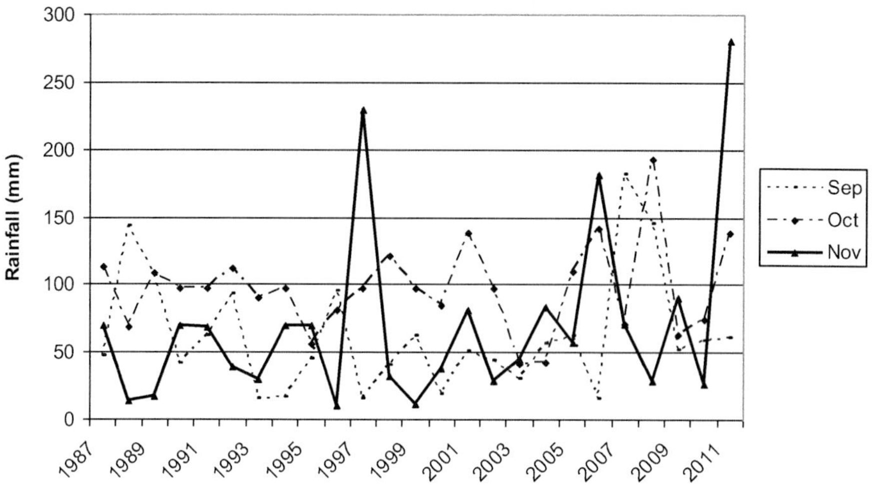

Fig. 6 Monthly rainfall distribution in the *Katana* season (1987–2011)

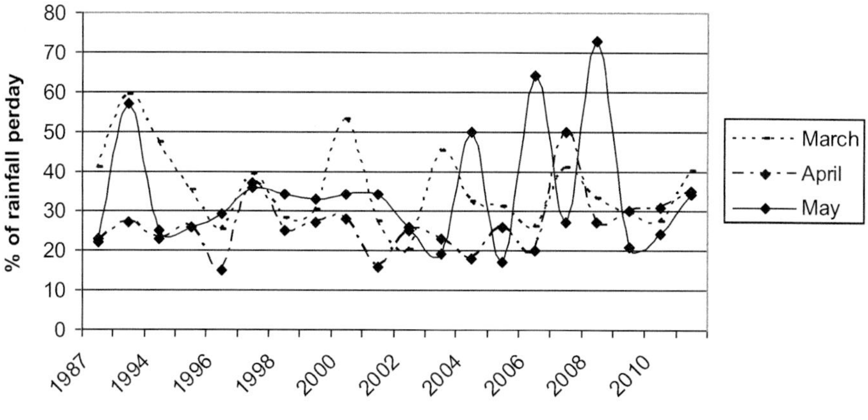

Fig. 7 Percentage of maximum rainfall per day per month in the *Hagaya* season

Katana season, the average percentage of maximum rainfall per day was 37%, and this varied within the months. For example, the average maximum rainfall per day for September was 41%, while for October and November, it was 25% and 46%, respectively (Fig. 8). This indicates that November is the month of most intensive rainfall in the *Katana* season and that rainfall in *Hagaya* is less intensive than in *Katana*. With such intensive rainfall patterns, splash effects and runoff will be inevitable. Hence the farming practices need to have good land husbandry including agroforestry and some soil and water conservation measures.

The distribution of rain days also resembles that of rainfall intensity: in March, rainfall is concentrated in only 5–10 days, while in April, it is spread over up to 20 days (Fig. 9). Similarly, in the short wet season (*Katana*), the rainfall in the month

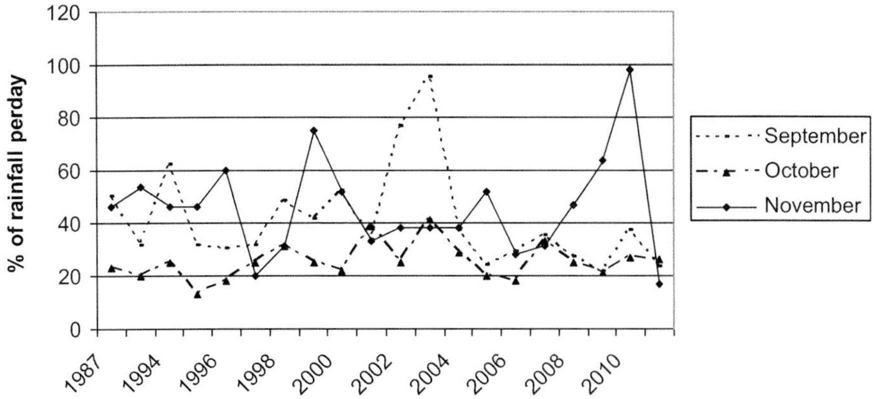

Fig. 8 Percentage of maximum rainfall per day per month in the *Katana* season

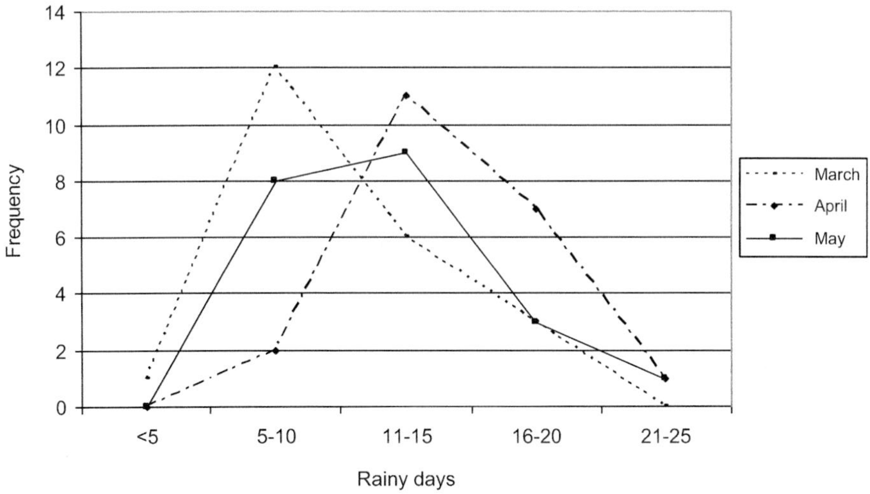

Fig. 9 Distribution of rain days in the *Hagaya* rainy season (1987–2011)

of November is concentrated in only 5–10 days (Fig. 10). It is also important to examine the onset and end of the rains, which often start early but then stop at the flowering or maturing stage of the crops.

Causes of Vulnerability to Climate Change

Various documents have highlighted the causes for chronic food insecurity in Konso, including drought, population pressure, soil erosion, deforestation, traditional farming practices, and marginalization of the community in decision-making (Foerch 2003; Beshah 2003; Tadesse et al. 2008; Tadesse 2010; Meron 2012). Based on the

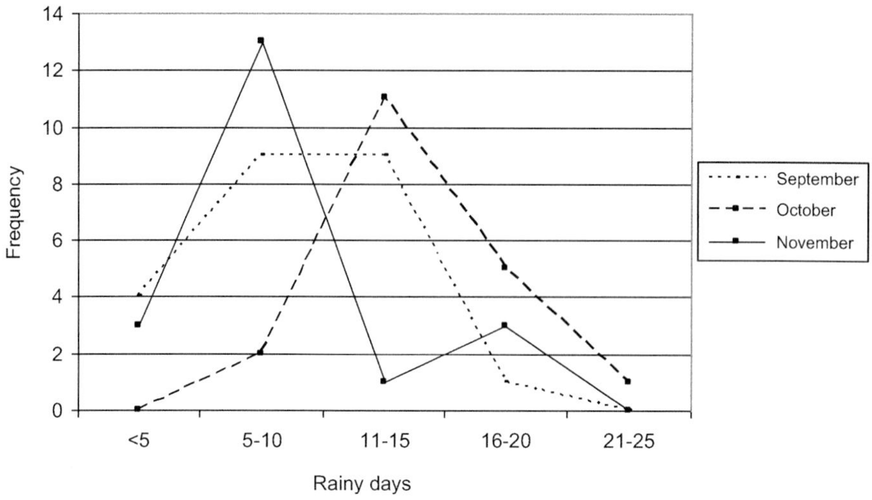

Fig. 10 Distribution of rain days in the *Katana rainy* season (1987–2011)

household survey and group discussions, the root causes of vulnerability to climate changes appear to be very diverse and complex, as indicated below.

Environmental factors: Over the past decades, rainfall distribution in the big and small rainy seasons is erratic. Sometimes, the rain comes too early or too late, and it often stops in the middle of the growing season. This situation provides favorable conditions for infestation by pests and diseases in both crops and livestock. Consequently, crop failure, livestock losses, seasonal outmigration of people, and dependency on food aid have become more common.

Reduction of access to resources: Population pressure has generally led to a decrease in farm size and in a number of fragmented plots per household. Many farmers lost their livestock during the prolonged drought 2004–2008 and found it difficult to recover. With the expansion of cultivation and land enclosure by the government, farmers have less access to communal land with its multiple functions of providing pasture, fuelwood, beekeeping areas, and medicinal plants as complementary sources of livelihood. Because they have fewer and narrower options, the farmers – and especially poor women and the youth – have become more vulnerable to risks of any kind.

Introduction of inappropriate technologies and polices: With the Ethiopian government policy of the Growth and Transformation Plan (GTP), food security through the green revolution approach has been given a top priority. This included the introduction of "improved seed" and chemical fertilizer and the modernization of small-scale irrigation schemes. However, the introduced crop varieties are less resistant to drought and disease, demand more labor, and encourage sole cropping, thus replacing the indigenous farming practices. Similarly, the indigenous terraces are being replaced by the standard modern terraces with the support of incentives from the PSNP. Generally, the extension services have many undesired outcomes attributing to the vulnerability of the poor farmers. On the other hand, for the Konso people, the big rivers are emergency-reserve areas for traditional irrigation during prolonged drought and for collecting wild fruits from the riverbanks. However, many of these areas have been allocated to private investors, where the community resilience to different risks has been diminishing.

Erosion of cultural practices: The Konso people are known for their deep-rooted ritual practices, indigenous land husbandry skills and grassroots customary institutions that tried to ensure the livelihood of the community during bad and good years. These cultural practices and associate landscape have been recognized by UNESCO as a global heritage. The forest-based rituals and annual festivals had reinforced the values and multifunctionality of the forest and strengthened social safety nets and community solidarity (Otto 2004). However, during the Derg regime (1974–1990), the ritual houses were set on fire, the ritual drums were destroyed, and the Konso people were told not to practice such "backward culture." Many of the elderly and middle age farmers, irrespective of sex, have believed that this destruction of their cultural practices and weakening of their values were root causes of their vulnerability to climate change until today.

Adaptation to Climate Change

Indigenous Soil and Water Conservation Practices

Based on the group and individual discussions with the farmers and field observation of the different land use and cover, the community is using a wider range of SWC practices. The indigenous practices are by and large designed as adaption and resilience strategies to the impacts of climate change. Some of the indigenous practices are widespread, while others are declining or localized. Generally, the indigenous SWC measures can be categorized as agronomic, biological, or physical as indicated below.

As shown in Table 11, the agronomic measures include mixed cropping, timing of sowing, seed rates, manuring, mulching, and crop rotation. Thinning of finger millet and sorghum land races are also widely practiced as a way to deal with poor germination or to prevent disease in the crops but also as a source of livestock feed and to adapt to climate variability. After harvesting crops such as finger millet and sorghum, the roots are left on the ground and resprout up to three times as a component of ratooning. However, under the ox-ploughing, the range of agronomic measures is very limited to manuring, crop rotation, and thinning. Many of the agronomic measures are very similar across the wealth rank except the manure use of wider areas by the better-off.

The biological measures applied in Konso include living fences, agroforestry, enclosures, afforestation, cut-and-carry feeding, grass strips, stall-feeding, and controlled grazing mainly under the hoe practice. However, under the ox-plough practice, the grass strips are widely occupied by free livestock grazing.

The physical measures applied in Konso include stone terraces, micro-basins, trashlines, and check dams. Some of the physical structures, including terraces and

Table 11 Statues of land husbandry at different wealth ranks

Indicators	Trends	Remarks
Agronomic measures		
Mixed cropping	Wide spreading	Different cereal, pulse, oilseed, root, and vegetable crops are mixed
Manuring	Widespread	The better-off farmers use manure over larger areas; the middle-ranked and poor farmers use it only around the homesteads
Crop rotation	Wide spreading	Complementary species of cereals and legumes are sown together
Thinning	Widespread	Thinning of maize and sorghum stands is widespread to obtain feed for livestock and for fodder sale and to reduce competition between plants under moisture stress
Double cropping	Declining	In normal years, Konso has bimodal rainfall, which permits double cropping, but greater frequency of drought has led to a decline in this practice
Trashlines	Declining	Trashlines are common practice in hoe farming, but this practice is declining because the sorghum and maize straw is used for fodder and fuelwood
Fallow	Declining	Fallowing in Konso is targeted within a plot; a specific spot with low soil fertility is left under grass cover, which also serves as a conservation measure and source of cash, when the grass is sold as fodder and cottage construction
Ratooning	Medium	Ratooning is still widespread in hoe farming, mainly in crops like sorghum and millet, although slightly declining with greater frequency of drought
Biological measures		
Agroforestry	Wide spreading	Agroforestry is widespread in hoe farming and is expanding quickly from homestead to field stead. As diversification, intensification, and conservation strategy
Grass strips	Localized	Irrespective of wealth rank, making grass strips is common in the ox-ploughing system on gentle slopes. The grass strip serves as conservation measure and source of cash when the grass is sold as fodder
Stall-feeding and tethering	Wide spreading	Zero grazing is essential in the hoe-farming system; free grazing might damage the stone terraces and trees. Many farmers fatten animals to generate some income
Physical measures		
Stone terraces	Wide spreading	In the hoe-farming system, terraces are integral components on both steep and gentle slopes; in the ox-ploughing system, terraces are found only on the field boundaries on steeper slopes
Diversion ditches	Localized	On land near grazing areas, woodland, roads, and settlements, drainage ditches are often dug to harvest water and lead it to the plots. This is more common under the hoe than the ox-plough
Micro-basins	Wide spreading	Micro-basins are integral components of the hoe-farming system, mainly to harvest water and conserve soil. Hoe farming is not possible without making micro-basins, and the ridges provide a larger soil surface for planting crops

Note: Widespread (>50%), medium (40–50%), localized (<10%)

check dams, are permanently established, while micro-basins and trashlines are semipermanent (Kruger et al. 1997). The hoe practice has more complex physical conservation practices than the ox-plough, and some of the physical practices are widespread, while others are localized and diminishing.

Generally, the indigenous farming practices in Konso are characterized by multipurpose, complementarily with synergetic effect, flexibility, and site specific. Moreover, the agronomic, biological, and physical measures are very limited under the ox-ploughing system than hoe practice.

Indigenous Innovation in Production

Indigenous innovation is a broad term with a wide range of definitions. It refers not only to technologies but also to husbandry and management practices and new ways of social organization. It includes discovery of completely new ideas in a given area but also to modification of existing technologies and practices. It is a dynamic social process involving different actors and institutions (Waters-Bayer et al. 2006). Identifying innovation within small-scale farming households in a heterogeneous landscape (slope, soil, relative location) and socioeconomic setting (wealth rank, family size, age and sex composition of households) is a big challenge. Each farmer is an innovator to some extent, as s/he attempts to solve specific problems in specific areas. Two fragmented plots worked by the same farmer cannot be treated equally in terms of designing SWC structures or combining crops and trees, let alone plots worked by different farmers. A technology cannot be applied in exactly the same way in different plots; some site-specific modifications will be necessary, and only the basic principles or functions of the technology will remain the same. In any case, local innovation is not what is inherited from parents or grandparents nor is it the practice of a model farmer who does what is told by the extension worker. Rather, a farmer innovator is someone who explores and experiments to discover something new or to modify an existing technology (Yohannes 2001). Local innovation is dynamic and flexible according to season. For example, during the small rains, new SWC structures may serve to harvest water, while during the main rains or short periods of intensive rainfall, other measures such as drainage ditches may be added. If the livelihoods of small-scale farmers in Africa are to be improved, their farming and land use need to be diversified and intensified with the integration of local innovation.

In the study areas, the following major innovations by groups and individuals in Konso that are relevant for improving land productivity and food security were identified:

- ***Conversion of cropland to tree plots:*** Many farmers in Doketu and a few in Gewada have converted their farm plots from cropping to planting multipurpose trees because of a combination of factors: (i) adaptation to recurrent drought and poor soil fertility that makes the plots easily vulnerable to shortage of rain; (ii) trees demand less labor, thus releasing household labor to be engaged on complementary activities such as crafts, petty trade, and daily wage labor in towns; and (iii) timber is in high demand at a good price in both urban and rural areas. For example, from a small plot (0.2 ha), up to 30,000 Ethiopian Birr, which is about 1100 US dollar (July 2018 exchange rate was 1 US dollar = 27.58 Ethiopian Birr), can be earned by selling trees that can grow in 3–4 years, while family members are doing other off-farm activities to generate income. This practice is slowly spreading in the study areas.

- ***Modification of farming tools:*** Almost all tools used in ploughing, hoeing, and cutting have been gradually improved by the artisans. The iron traditionally used has been modified in size and sharpness. This has helped the farming community to dig deeper into the soil for better access to soil nutrients and for better moisture-holding capacity. The modified tools thus help reduce vulnerability to drought.
- ***New design of water-harvesting technologies:*** Some farmers harvest water from roadways and footpaths and make water diversion ditches to private plots and communal water ponds in a zigzag form to minimize erosion on their plots and to reduce siltation of the ponds.
- ***Use of micro-basins to gain more land:*** Within their stone terraces, many Konso farmers use a series of micro-basins to harvest more water. They gradually raise the height of the micro-basin ridges to harvest more water and plant the sides of the ridges with shallow-growing root, legume, oilseed, and vegetable crops. Depending on the size of the plot and the number of micro-basins within the plot, this can increase the size of the plot up to 30%. Thus, the micro-basin has multiple functions in preventing soil erosion, maximizing water harvesting, and providing space on the heaps of soil for more crop production. This innovation is spreading mainly in the hoe-farming areas.
- ***Reclaiming rocky outcrops:*** Some farmers have covered the outcropped rocks with some soil and made it productive by planting fruit trees.

Generally, the different local innovations work toward harmonizing land uses, water and soil conservation, improving soil fertility and agricultural tools, reclamation of land, and efficient use of energy as a component of the integrated farming system.

Conclusions

The trend of increasing frequency of drought and high intensity of rainfall implies the vulnerability of the rural community depending on farming in general and the poor in particular. Hence, the development of a strategy of intensification and diversification of production with the integration of soil and water conservation remains fundamental.

Moreover, in order to develop proactive early warning adaption strategies, the intermarriage of the traditional early warning systems with the modern metrological evidence is more effective and reliable.

Both the community perception and metrological evidence have some similarities, indicating a trend of increasing drought. However, to the community, the traditional early warning systems (complex and diverse) are more reliable than the modern metrological measurements. Hence, the intermarriage of both approaches is fundamental in developing proactive early warning adaption strategies. Similarly, the root causes of vulnerability to climate change are by and large the result of the cumulative effect of the environmental and socioeconomic and political factors. Hence, in order to develop resilience to climate change and assure food security, the strategic integration of the environmental and human parameters remains valid.

Lastly, the rural community applies a wide range of indigenous practices and innovations in adaption to climate change. Hence, supporting the local initiatives and creating policy environment for the community to experiment and innovate is an integral part of developing resilience to the impact of climate change and assuring food security.

Cross-References

- ▶ Assessment of the Degree of Households' Vulnerability to Climate Variability in Nigeria
- ▶ Biodiversity, Ecosystem Degradation, and Climate Change Effects on Livelihoods in the Bitumen Area of Nigeria
- ▶ Building Resilience to Climate Change: Water Stewardship in Rainfed Agrarian Villages in Maharashtra, India
- ▶ Climate Change and Variability in the Mixed Crop/Livestock Production Systems of Central Ethiopian Highland
- ▶ Climate Change Vulnerability Among Pastoralists and Farmers in Ethiopia
- ▶ Ecosystem-Based Adaptation in Tigray, Northern Ethiopia: A Systematic Review of Interventions, Impacts, and Challenges
- ▶ Effect of Climate Variability on Crop Income in Central Highlands of Ethiopia
- ▶ Tradition Versus Climate Change: Cultural Importance of Indigenous Fruit Tree and Adaptation in Benin, West Africa
- ▶ Understanding Kenyan Farmers' Perceptions of and Responses to Climatic Variability to Build Their Resilience
- ▶ Working Together to Build Climate Resilience in Hudson Riverfront Communities

References

Adger N, Agrawala S, Mirza MMQ, Conde C, O'Brien K, Pulhin J, Pulwarty R, Smit B, Takahashi T (2007) Assessment of adaptation practices, options, constraints and capacity. In: Parry ML, Canziani OF, Palutikof JP, van der Linden PJ, Hanson CE (eds) Climate change 2007: impacts, adaptation and vulnerability. Contribution of working group II to the fourth assessment report of the intergovernmental panel on climate change. Cambridge University Press, Cambridge, pp 717–743

Agrawal A (1995) Dismantling the divide between indigenous and scientific knowledge. Dev Chang 26:413–439

Aragie EA (2013) Climate change, growth and poverty in Ethiopia. Working paper no.3, June 2013, The Robert S. Strauss Center for International Security and Law at The University of Texas
Berkes F (2012) Sacred ecology, 3rd edn. Routledge, New York
Beshah T (2003) Understanding farmers: explaining soil and water conservation in Konso, Wolaita and Wello. Ethiopia tropical resource management paper no. 41, Wageningen University and Research Centre, Wageningen
Briggs J (2005) The use of indigenous knowledge in development: problems and challenges. Prog Dev Stud 5: 99, SAGE Publisher, http://pdj.sagepub.com/content/5/2/99
Chossudovsky M (2000) Genetically modified seeds imposed on farmers in developing countries triggered famine and social devastation, sowing the seed of famine in Ethiopia, *The Ecologist*, September 2000, Centre for Research on Globalization, Montreal
Department for International Development (DFID) (2010) Concept note: reducing hunger in Ethiopia: support to the government's food security programme – DFID's approach to environment. Department for International Development, London
Elizabeth B, Temesgen TD, Glwadys AG, Claudia R (2009) Adaptation to climate change in Ethiopia and South Africa: options and constraints. Environ Sci Pol 12(4):413–426
Ethiopian Economic Association (EEA) (2008) Climate change and development adaptation measures. Econ Focus 11(1). Addis Ababa
Finucane ML (2009) Why science alone won't solve the climate crisis: managing climate risks in the Pacific. Asia Pacific issues no. 89. East-West Centre, Honolulu
Foerch W (2003) Case study: the agricultural system of the Konso in South-Western Ethiopia. Published 2003 online: FWU Water Resources Publications 2003:1, University of Siegen, http://fwu.fb10.uni-siegen.de/bkd/FWU_WRP.htm
Fosu-Mensah BY, Vlek PLG, Manschadi AM (2010) Farmers' perception and adaptation to climate change; a case study of Sekyedumase district in Ghana. Paper presented at Tropentag 2010 "world food system-a contribution from Europe", 14–16 September, 2010, Zürich
Green Forum (2008) Climate change – a burning issue for Ethiopia: proceedings of the 2nd green forum conference held in Addis Ababa, 31 October–2 November 2007, Green Forum, Addis Ababa, 145pp
Haile M (2005) Weather patterns, food security and humanitarian response in sub-Saharan Africa. Philos Trans R Soc B Biol Sci 360:2169–2182
Intergovernmental Panel on Climate Change (IPCC) (2007) Impacts, adaptation and vulnerability working group II contribution to the intergovernmental panel on climate change fourth assessment report. Summary for policymakers, IPCC, Brussels
Intergovernmental Panel on Climate Change (IPCC) (2014) Summary for policymakers. In: Climate change 2014: impacts, adaptation, and vulnerability
Kruger HJ, Berhanu F, Yohannes GM, Kefeni K (1997) Inventory of indigenous soil and water conservation measures on selected sites in the Ethiopian highlands. SCRP Research Report 34. University of Berne, Berne
McKee J (2008) Deconstructing some myths about climate change adaptation and mitigation. In: Green Forum (ed) Climate change – a burning issue for Ethiopia: proceedings of the 2nd Green Forum conference held in Addis Ababa, 31 October–2 November 2007. Green Forum, Addis Ababa, pp 111–135
Meron A (2012) The impact of climate change and adaptation through agro ecological farming practices: a case study of the Konso area in Ethiopia. Master's thesis, Swedish University of Agricultural Sciences (SLU), Uppsala
Mertz O, Cheikh M, Reenberg A, Diouf A (2009) Farmers perceptions of climate change and agricultural adaptation strategies in rural Sahel. Environ Manag 43(5):804–816
Nakashima DJ, Galloway McLean K, Thulstrup HD, Ramos Castillo A, Rubis JT (2012) Weathering uncertainty: traditional knowledge for climate change assessment and adaptation. UNESCO/United Nations University, Paris/Darwin
National Meteorological Agency (NMA) (2007) Climate change national adaptation program of action (NAPA) of Ethiopia. NMA, Addis Ababa

Okonya JS, Syndikus K, Jürgen Kroschel J (2013) Farmers perception of and coping strategies to climate change: evidence from six agro-ecological zones of Uganda. J Agric Sci 5(8). https://doi.org/10.5539/jas.v5n8p252

Otto S (2004) Traditional Konso culture and the missionary impact. Annales d'Éthiopie 20:149–180

Pankhurst R (1985) The history of famine and epidemics in Ethiopia prior to the twentieth century. Relief and Rehabilitation Commission, Addis Ababa

Ringer C (2008) Food and water under global change: developing adaptive capacity with a focus on rural Africa. Paper presented in the workshop "How can African agriculture adapt to climate change? 11–13 December, Nazareth

Senanayake SGJN (2006) Indigenous knowledge as a key to sustainable development. J Agric Sci 2 (1):87–94

Tadesse M (2010) Living with adversity and vulnerability: adaptive strategies and the role of trees in Konso, Southern Ethiopia. PhD thesis, SLU, Uppsala

Tadesse M, Kassa H, Sriskandarajah N, Powell N (2008) Do conservation and intensification have their limits? The challenges of food security and vulnerability amongst the Konso people of southern Ethiopia. Proceedings, meeting global challenges in research cooperation, 27–29 May, SLU, Uppsala, pp 229–239

Waters-Bayer A, van Veldhuizen L, Wongtschowski M, Wettasinha C (2006) Recognising and enhancing local innovation processes. PROLINNOVA working paper 13, PROLINNOVA International Secretariat, ETC Foundation, Leusden

World Bank (2008) Ethiopia: a country study on the economic impacts of climate change. World Bank, Addis Ababa

Yohannes GM (2001) Community assessment of local innovators in northern Ethiopia. In: Reij C, Waters-Bayer A (eds) Farmer innovation in Africa: source of inspiration for agricultural development. Earthscan, London, pp 171–177

Yohannes GM, Waters-Bayer A (2007) Trees are our backbone: integrating environment and local development in Tigray region of Ethiopia. IIED issue paper 145, IIED, London

Role of Climate Justice in Strengthening Adaptive Capacities in Developing Countries

42

Ruchi Sachan

Contents

Abstract

Climate change intensifies the pre-existing vulnerabilities for already-struggling communities. It disproportionately affects the developing countries. Further, adaptation to climate change is becoming a significant issue within the climate change agenda, yet little attention is paid to the need to build on poor people's own coping strategies. Local practices and indigenous knowledge to adapt and build resilience needs to be incorporated so that those most affected do not just become "passive" victims of state/global forces. When natural resource-dependent societies respond to climatic or non-climatic disturbances, the concerns exist regarding equity and social justice issues in adaptation to these changes. Thus, it becomes

R. Sachan (✉)
School of International Studies, Centre for African Studies, Jawaharlal Nehru University, New Delhi, India
e-mail: iruchisachan@gmail.com

W. Leal Filho (ed.), *Handbook of Climate Change Resilience*,
https://doi.org/10.1007/978-3-319-93336-8_74

important to see how addressing these issues of climate justice can strengthen the local level adaptive capacities in the developing countries.

This paper emphasizes on a need to tap into the local knowledge systems of communities for a more comprehensive understanding to make effective adaptation policies and strategies, as "one-fits-all" approach will not work and yield the desired results in adapting to climate change. It aims to highlight climate change equity and justice issues relating to adaptation strategies in the developing world, specifically at the subnational levels, with a focus on Africa. The paper will also look into how climate justice and sustainable development both complement each other, as both the concepts emphasize the need to develop local citizen participation, the processes of community capacity building, more responsive governance, and greater empowerment of citizens.

Keywords
Climate justice · Adaptation · Local level strategies · Sustainability

Introduction

Climate change imposes fundamental challenges to the ecosystem in terms of food insecurity, environmental degradation, depletion of necessary resources, and conflict over natural resources. Human actions and decisions have always been made in response to environmental pressures and changes. Adaptation responds to the risks while building future resilience. Nevertheless, these challenges imposed by climate change cannot be resolved without an acknowledgment of ecological realities and commitment to environmental justice. It encourages inclusivity and wider social responsibility in the face of ecological realities, along with value-based perspectives to utilize and transform recognized legal frameworks toward addressing sustainability (Popke et al. 2016).

In Africa, 75% of the population lives in the rural areas, with most of it involved in agriculture, which makes more than 30% of the continent's economy (IPCC 2007). It is predicted that the extent of arid and semi-arid areas will expand by almost 5–8% in the future under the current climatic scenario (Stringer et al. 2009), with noticeable changes in the timing of rainfall and frequency of droughts. Similar conditions are noticeable in the other developing regions like South Asia, Southeast Asia, and Caribbean regions. It has to be understood that the challenges experienced are not only natural but also accentuated due to the dry policies of the governments. However, few attempts have been made to look into this issue in terms of equity and social justice at the subnational level, and there remains a relative lack of attention to climate change equity and social justice issues relating to natural resource users in developing areas. This makes it essential to look into the implications of adaptation policies among vulnerable groups at subregional levels and develop appropriate policy instruments to address the challenges; therein lies the importance of the paper. For the paper, the debate of climate justice is restricted to national and subnational levels and does not include the issues at the global level (Popke et al. 2016).

Differential Impact of Climate Change

The climate system changes with increased warming, decreasing precipitation, seasonal changes, and increase in extreme climate events are likely to have a more devastating impact on tropical and sub-tropical agricultural systems in the developing world (Thomas and Twyman 2005). "The causes and consequences of climate change are deeply intertwined with global patterns of inequality" (Popke et al. 2016: 75). Differential vulnerability is observed in the impacts of climate change with developing countries being disproportionately affected by its impact than the developed countries. This inequitable distribution of negative impacts is attributed to the low adaptive capacity in the developing world. Adaptive potential is linked to the levels of vulnerability, which itself is dependent on a variety of factors of exposure, sensitivity, adaptive capacity, and resilience, which is further dependent on levels of economic development and advanced technological infrastructure (Stringer et al. 2009; Paavola and Adger 2006; White 2011). A considerable amount of research has been conducted to suggest that the negative effects of climate change would disproportionately affect the poorest and the most vulnerable groups, with a high natural resource dependency, limited financial and institutional mechanisms, low levels of economic development, and high levels of poverty, along with a lack of safety nets (Thomas and Twyman 2005; Muigua and Kariuki 2015).

The impacts are projected to rise both in frequency and asperity, with severe and prolonged droughts, flooding, loss of arable land due to desertification and soil erosion, and loss of livestock creating critical socioeconomic issues and posing a challenge to stability of a region in terms of food and water security and resource availability (Tadesse 2010). For this reason, it is appropriate to characterize climate change as both an urgent policy concern and a matter of social and environmental justice.

In Africa, particularly, millions will suffer from climate change impacts on agriculture, water availability, ecosystem services, and biodiversity, in varied ways, depending on the spatiality of impacts and features of the local system (Paavola and Adger 2006). According to leading experts, around 250 million people will be exposed to increased water stress due to climate change by the year 2020. In some countries, yields from rain-fed agriculture could drop by at least 50% (IPCC 2007). As observed by farmers in East Africa, the rainfall has declined in its quantity and quality, thereby increasing the uncertainties at the local level (Eriksen and Lind 2009). Since the year 2000, there has been a continuous decline in the *maguna ng'ombe* (short rains), which ensures that the grass would grow to sustain the livestock until the start of rainy season in March every year (Ogalleh et al. 2012). The impact on local and national economies of such harvest failures would be devastating, especially to those who have the least women and children.

Thus, climate change intensifies the pre-existing vulnerabilities. The first response to climate change is to adapt; however, it impacts the ability of a household, a community, or a region to withstand shocks and stresses. The burden of disruptions caused by climate change is expected to fall disproportionately on those who are already socially and geographically disadvantaged, particularly poor communities with low adaptive capacity. Limited resources and lack of alternatives further increase environmental degradation that leads to household and community conflict. Alongside undermining the coping strategies, especially of the natural resource-dependent population, it may also result in internally displaced population or environmental migrants. Despite this, there are no clear provisions in the domestic laws for addressing the problem of internal displacement or human mobility in response to climatic and non-climatic drivers. Here it is also important to highlight the lack of an accepted definition of people displaced as a result of climatic changes. As defined by the International Organization for Migration (2014: 13), "Environmental migrants are persons or groups of persons who, predominantly for reasons of sudden or progressive changes in the environment that adversely affects their lives or living conditions, are obliged to leave their habitual homes, or choose to do so, either temporarily or permanently, and who move either within their country or abroad." Internally displaced persons, on the other hand, are "persons or group of persons, who have been forced or obliged to flee or to leave their homes or places of habitual residence, in particular as a result of or to avoid the effects of armed conflict, situations of generalized violence, violations of human rights or natural or human-made disasters, and who have not crossed an internationally recognized state border" (IOM 2014: 12). However, environmental change and climate variability do not act in isolation in creating such fragile conditions; other factors of conflict, political instability, demographic changes, prevalence of diseases, and colonialism play an important role in shaping the situation (Bob and Bronkhorst 2010; Barrett 2013). The exposure and sensitivity of marginal environments with climate variability to extreme weather events such as floods, droughts, storms, etc. are also driven by manifestations of poverty and underdevelopment.

Adaptation to Climate Change

Adaptation measures are required to counter the consequences of perceived and predicted climate change. Adaptation is an important aspect of the global response to climate change and is an effective tool in reducing climate risks. Changing weather conditions have prompted communities to adapt, cope, and survive throughout history. The adaptation mechanisms vary according to the ability, resources, and situations. Adapting to changing weather patterns is crucial if people have to secure livelihoods and increase resilience to environmental hazards. However, many have few resources and low adaptive capacities for risk reduction or response.

For the natural resource-dependent societies in the developing countries, it is not new to diversify their livelihoods in adaptation to climatic or non-climatic factors (Yamin et al. 2005; Thomas and Twyman 2005). Some practices are applied even before the climatic event and others after the event has struck. The response also varies depending on whether the impact is slow (drought, desertification) or sudden (floods, hurricanes). Adaptation mechanisms are both reactive (triggered by past or current events) and also anticipatory (assessment of conditions in the future) in nature. The strategy implemented depends on the need of the hour. Stringer et al. (2009) call agricultural adjustments like selling off livestock as tactical while changes in land use, land management, crop types, and livelihood alteration as strategic. Diversification, mixed farming, adjusting the farm sizes, and migration have been identified as some of the potential adaptation strategies (Fankhauser 2016; Ogalleh et al. 2012). Farmers also mix long-cycle crops with short-cycle crops on their farms to cope with decreasing rainfalls. Diversification is noticed also in the activities not dependent on the environment. The fact that diversification occurs indicates a level of responsiveness toward the factors of change, which may or may not be driven by climate and may be significant in terms of their capability to adapt.

However, it is feared that the pace of climate change could exceed the limits of adaptations and coping strategies, in many parts of the world, especially among vulnerable population in the developing countries (Stringer et al. 2009). Today climate change is causing changes faster than the vulnerable populations and ecosystem can cope with and adapt to.

There are also signs that people's responses to climate risks are not always effective; sometimes they are even insufficient or inefficient (Fankhauser 2016). It depends on the availability of extension services and access to credit and household size, at the individual level, along with institutional, financial, or technological capacity of a region or nation. A commonly expressed concern is that adaptation to climate change is hampered by dependence on natural resources, which is further complicated by a lack of equity in terms of access to the natural resource base (IPCC 2001). The kind of policies governing resource access and utilization plays a huge role in determining the coping capacity of the communities affected. For example, "in developing countries, partly maintaining subsistence production in rural communities provides greater flexibility and insurance than do more rigid and risky monetary strategies that are influenced by the global markets" (Fazey et al. 2010: 416). All communities generally have their own, formal or informal, ways to adapt. The local level adjustments to climate change do not act in isolation from the other seasonal changes and pressures. In the dry-land eastern Africa, the interconnected regional resource system has allowed the different identity groups to develop pluralistic production techniques (Eriksen and Lind 2009). The diversity of response options must therefore be generated by encouraging innovations and creating spaces from which new responses can emerge.

Adaptation Policies as a Challenge to Notion of Climate Justice

Developing effective mitigation and adaptation strategies presents some of the most significant public policy challenges of our time (Kaswan 2008). To date, the debate on climate change policies has given insufficient attention to their climate and environmental justice implications. To understand the justice issues presented by climate change policies, it is critical to understand the environmental and climate justice movement and the nature of its claims.

Climate Justice

Climate justice highlights the ways in which climate change is an ethical issue and how the causes and effects of climate change relate both to environmental and social justice (Huntjens and Zhang 2016). It relates to the questions of equity and fairness and simply an ethical act or response. One of the crucial components of climate and environmental justice is that it seeks to tackle social injustices and environmental problems through an integrated framework of policies, along with recognition of cultural identity, especially of the poorest and least powerful. "Environmental and climate justice movements share many core principles, including, but not limited to, a respect for indigenous conceptions of the nature, the importance of self-determination of all peoples, and opposition to the role trans/multi-national corporations play in the environmental decimation of communities" (Schlosberg 2014: 02). Equity in the context of climate change also relates to relationships between climate change impacts and other factors that affect and disturb livelihoods.

Addressing the question of climate justice, at the local and global level, requires taking into account both the parties, one who is immediately affected by climate-related problems and one who is responsible primarily when designing climate-related policies (Running 2015; Popke et al. 2016), both relevant within a generation and between generations. Policies addressing climate change need to have the potential to address long-standing societal problems like distributional inequities.

Climate justice ensures that there is fair and equal distribution of environmental burdens and benefits, and all have an opportunity to participate democratically in the decision-making process – who decides and who responds (Paavola and Adger 2006). The former refers to the distributive justice while the latter to the procedural justice. These two tenets of climate justice are co-dependent on each other. Distributive justice refers to the distributional consequences of environmental decisions and actions. Nevertheless, equity in the context of climate change relates to a wide range of issues, including decision-making processes and frameworks for taking and facilitating actions, and also to relationships between climate change impacts and other factors that affect and disturb livelihoods. It therefore has a strong procedural dimension too, which can affect the opportunities that are available for adaptation. In terms of climate change and its effect on natural resources, there are clear questions of procedural justice at stake in the frameworks and institutions shaping climate change policy and action at both local and global scales (Popke et al. 2016). There is an administrative reluctance to recognize any obligations toward affected communities.

The debate of climate justice has also been majorly on fair sharing of the burden of mitigation by reduction of greenhouse gases. The decision-making processes, which would make burden sharing fair, have been ignored. "Techno-fixes, which focus on the symptoms rather than the causes, are important for short-term mitigation and coping, but for effective, long-term adaptation, greater emphasis should be placed on strategies that reduce the behavioral causes of undesired environmental change" (Fazey et al. 2010: 416). However, distributive justice is likely to be insufficient for climate justice, and procedural justice is needed to maintain the legitimacy of climate change regime (Paavola and Adger 2006). "Distributive justice is fulfilled when there is procedural justice and when the distribution of conservations costs and benefits among actors are perceived as just" (Gustavsson et al. 2014). Hence, when the adaptation plans do not consider both of them, they are more likely to induce inequality, rather than reducing it. It ensures that the parties affected have a chance to voice their concerns and be an active participant in the decision-making process, which affects the opportunities available for adaptation. Passive participation where people are told what has already been decided and decisions are made and announced without listening to people's responses would be ineffective.

Even at the international policy level, it is seen that historically, the major emphasis is on the mitigation measures. It was only after the 2001 Marrakesh Accords that mitigation and adaptation were both considered as important in tackling future climate change. Stringer et al. (2009: 03) reason that the unwillingness to focus on adaptation has been "linked to discussions on liability, compensation, equity and fairness which developed countries were keen to avoid." More significantly, the discussions on climate justice though focused on the causes ignore the climate change consequences and its adaptation (Huntjens and Zhang 2016). This unbalanced approach is unjust and unfair to those who have no other choice but to adapt to climate change (keeping in mind that the most vulnerable have contributed the least to building up of the situation). It is to be understood that alongside mitigation, adaptation is urgent. Paavola and Adger (2006: 02) highlight "four important principles that would help resolve the dilemmas of justice in fair adaptation – avoiding dangerous climate change, forward-looking responsibility, putting the most vulnerable first and equal participation for all."

Gaps in Adaptation Policies

Natural resource policies offer significant means of addressing the needs and concerns of climate change adaptation. Nevertheless, introduction of new policies and interventions with limited/insufficient institutional capacity, without the active and sustained participation of communities and local resource users, may end up supplanting, undermining, or eroding community institutions governing resource use or adapting to climate variables.

Climate change regime can be understood as "the collection of principles, norms, rules, and decision-making procedures around which actor expectations converge in a given issue-area, such as climate change" (Paavola and Adger 2006: 05). The focus of climate change regimes at national and global scales leaves formidable dilemmas of justice at subnational and local levels, in regard to the equitable nature of adaptive strategies to climate change (Thomas and Twyman 2005). Little attention is paid to the issues of equity and justice at smaller (local or subnational) scales than at larger (national and global) scales. This is prevalent more acutely in the natural resource-dependent communities in the developing countries. No explicit stance for responsibility toward climate change impacts is taken by the climate change regime, in terms of dedicated responsibility of developed countries to assist developing countries in adapting to climate change (Paavola and Adger 2006).

Nevertheless, the UN convention on climate change does not ignore the responsibility completely. The parties to the convention, under Article 2, are responsible for ensuring that the adaptive capacities of the ecosystem, as a whole, are not surpassed by climate change and its impacts. The preamble of the convention adds that the most vulnerable, i.e., small island nations; low-lying areas; arid or semi-arid areas; areas prone to floods, droughts, and desertification; and areas with fragile mountain ecosystem, must be given priority. There are other responsibilities with regard to adaptation such as UN convention Article 4. Article 4 commits the parties to cooperate in adaptation planning and to incorporate climate change considerations into their economic, social, and environmental policies so as to minimize adverse effects on public health, environmental quality, and mitigation and adaptation measures (IPCC 2001). It asks of developed countries to particularly assist vulnerable developing countries in adaptation and participation, by financing, technology transfer, and insurance. It provides a reason for putting the most vulnerable first when it comes to distributive justice, which would eventually help them move up the vulnerability ladder. However, gap persists in the amount of assistance and no binding agreement. The contributions are fairly voluntary, which brings in negligible funds for assistance. Further, the UN convention does not pay enough attention to the concept of vulnerability and sensitivity of vulnerable groups to climate change impacts and their adaptive capacity, which is central to the notion of climate justice as "it helps to tie the primary concerns of adaptation scholarship to those of moral philosophy" (Paavola and Adger 2006: 11).

One of the tenets of the United Nations Convention to Combat Desertification (UNCCD) is creation of National Adaptation Plans, which requires engaging with local stakeholders in strengthening the existing coping mechanisms by the means of community participation. Impacts of climate change are felt everywhere; however, the visibility of impacts is more in the developing world, with low adaptive capacity and high vulnerability. National Adaptation Plans are at infant stage for many developing countries and even in the scenarios where climate change adaptation policies have been developed; the major challenge remains of "real" local participation and recognition of household-level strategies (Eriksen and Lind 2009). In terms of participation, at the national and subnational levels, politico-economic factors along with institutional rules affect the participation and representation of non-state actors and other ethnic groupings. Also, there is a lack of will to formulate any solution to facilitate participation across levels. When this is absent, planning for tackling climate change is no more than mere rhetoric, leading to unsustainable development. Participation and representation thus remain contested issues in the climate change debate (Eriksen and Lind 2009). Even at international conventions, developing countries will not have the same levels of effective participation as developed countries (Paavola and Adger 2006), with less capacity to back up their negotiation teams. The use of English as the primary language for negotiations is a further impediment, and the assistance to developing countries does not fully address inequalities in participation.

There are two important issues that arise here. One, several vulnerable countries are yet to formally define their adaptation priorities (Stringer et al. 2009). Second, for those who have formulated it, "one-fits-all" approach cannot work and will not yield the desired results. Local-level climate impacts are likely to affect different sections of society differently (Thomas and Twyman 2005; Yamin et al. 2005). Appropriate adaptation strategies need to be assessed on both temporal and spatial scales. What may be effective for a region might have a negative effect on the other, which makes it imperative to understand the specificity of factors affecting local practices. Policies need to promote divergent adaptation interests for different groups to minimize inequality and promote justice in local adaptation. The socio-economic vulnerability of the people of the developing world requires that equity and justice be included in all climate change debates if it has to remain relevant to the affected regions. This does not mean that equity and justice should be ignored in other livelihood-disturbing factors (Thomas and Twyman 2005).

For instance, Kenya's policy for arid and semi-arid lands tends to be a "one-fits-all" solution across geographical regions (Roden et al. 2016). The policies/solutions at the macro-level are not favorable at the micro-levels. The integration of climate change and variability into strategies is also limited. The government of Kenya itself recognized in a document on arid and semi-arid lands (GoK 2007) that its policy to promote sedentary agriculture without providing sufficient resources for many decades, despite the climatic conditions of the region, has contributed to the food insecurity and vulnerability of the region (Schilling et al. 2015). The reasons given for the failure were "mainly top-down, discriminative, and not consultative" nature of the policies. Instead giving access to credits and investments in infrastructure would have given better security to the pastoralists, along with increase in income security with integration into the market structure (ibid., 2015). This shows that without inclusivity, institutional linkages are less effective and susceptible to manipulation (Robinson and Berkes 2011).

With the climate change projections, the agricultural policies in Kenya, which aim to improve farmer's livelihood, fall short of mechanisms that promote farmer's adaptation (Ogalleh et al. 2012). Moreover, the smallholder's adaptive role in agriculture is overlooked, as seen in the national climate change response strategy. This policy recognizes the smallholder's as the most vulnerable however suggests nothing in regard to integration of smallholders' knowledge into national climate policies. At the policy level, desertification and climate change adaptations overlap within broad development approaches, but they are insufficient. Similarly, mutual supportive links are missing between policy-driven and autonomous local adaptations, though common ground exists between them (Stringer et al. 2009). "Access to justice in Kenya for the poor and the marginalized groups of persons is still a mirage," which results into "feelings of bitterness, marginalization, resentment and other negative feelings that also affect the stability and peace of the country" (Muigua 2015: 8), which makes effective management of natural resource conflicts a necessary ingredient in the quest of climate justice.

In Swaziland, though there are efforts being made to develop and enact policies with participatory approaches, there still remains a gap between the policy proposals and their local acceptability (Stringer et al. 2009). Even in Botswana, in contrast to the local level, which tried to address the interlinked challenges, simultaneously, the National Adaptation Plan had some priority areas for impeding land degradation. In Malawi, though the common ground is apparent in the national plans and local level adaptation, and in finding a degree of acceptability amongst them, but the plan encourages the cultivation of groundnuts and sweet potato, which rarely covers 10% of country's cultivated area. Similar to Swaziland, smallholder agriculture relies heavily on maize, which loses its significance as a subsistence crop. In addition, the cultural context of such changes is not recognized within policy proposals. For example, under the national plan, in Swaziland, microcredit was to be made available for a small collateral such as cattle or other assets. But in much of southern Africa, cattle ownership bestows a social status upon the people; hence they were reluctant to do so, even when they could not afford the collateral by any other means (Stringer et al. 2009).

Another example can be cited from Zanzibar, Tanzania, where top-down interventions have not increased participation or justice, nor sustainable resource use and conflict resolution (Gustavsson et al. 2014). The government's assumption of distributing equipment and alternative income-generating projects instead created conflict and injustice within and between villages owing to insufficient resources which did not target those in need. This is suggestive of the fact that interactive and effective participation by all local actors is a must to create just trade-offs.

Taking a Cue from Climate Justice

Given the concerns that exist regarding equity and justice issues in adaptation to climate change, it is necessary to explore the issues of equity and justice that arise at the subnational or local level when natural resource-dependent societies respond to climatic or non-climatic disturbances with adaptation strategies and coping mechanisms. Moreover, it becomes important to see how addressing these issues of climate justice strengthens the local-level adaptive capacities in the developing countries. The incorporation of environmental procedural rights in emerging international and regional instruments can secure access to justice in environmental decision-making. It should be understood that most of the development interventions do not yield the desired result due to incomplete understanding of how people live and build their lives (Eriksen et al. 2005).

Stallworthy (2009: 55) has clearly stated that the "urgent process of addressing climate related threats, through a meaningful degree of mitigation and adaptation, and multiple levels of decision-making and response, must be informed by the precepts of environmental justice." Any comprehensive assessment must consider local perceptions of change and adaptation, for the social well-being and equity. It is necessary so that the vulnerable people do not just remain at the receiving end of the first climate change and then to its adaptation mechanisms (Foster 1998). The groups disadvantaged in terms of income should not be disadvantaged in other spheres of justice (Paavola and Adger 2006). There is a need to encourage community empowerment and wider recognition for equitable considerations in decision-making, seeking fair means of reconciling apparent conflicts (Foster 1998). Necessary attention needs to be given to justice-related arguments in relation to climate mitigation and adaptation.

Empowerment is regarded as a key element of creating equity in decision-making; however, it needs to be coupled with notions of procedural and distributive justice, addressing distributional inequities, within all levels of decision-making in emerging international and regional instruments (Thomas and Twyman 2005). It is thus critical in promoting greater access to justice and public participation in environmental matters and other livelihood disturbing factors, as the effect of even best legal instruments can have varied impacts on diverse groups in a given community and context (Kameri-Mbote 2013). Thus, there is a need to recognize the interrelationship between climate change policies and broader economic and environmental issues, which is indeed one of the hallmarks of the climate justice movements (integration of environmental and economic justice).

There are institutions that one finds in most parts of and among most communities in Africa, whose mandate increasingly includes environmental matters, as most of the conflicts in pastoral communities revolve around livestock and pasturelands. For example, each clan of the Akamba ethnic group, living in the vicinity of Endau hilltop forest in Kenya, has its own mechanism for land management and regulation. The traditional resource governance structure governs natural resources use, management, and conservation. The council of elders from the different groups resolves serious conflicts using traditional negotiating mechanisms during the times of drought (Owuor et al. 2005). It is also seen that the community-based natural resource management finds more supporters among the local population than that of centralized state-centered or private-managed provision like in rural dry-land Africa (Thomas and Twyman 2005). In a situation where people need help, they find effective solutions from justice systems that are accessible, affordable, and comprehensible to ordinary people (Stringer et al. 2009). While climate change is global, adaptation has to be local. An example of this can be seen in the Ethiopian constitution which gives much power to the ethnic groups in governing their own affairs, although all the land belongs to the state. The constitution also gives the regions power to recognize customary dispute resolution mechanisms (Unruh 2005). Although in parts of a region there is significant interaction between local customary dispute resolution institutions and regional and state authorities in regard to access to commons, the legitimacy, spatial applicability, and utility of traditional resource management and dispute resolution institutions are yet to be ascertained. One of the problems faced is that the local government personnels can be much younger than the elders, creating a traditional conflict. Thus, there is a need to link these informal institutions to the formal institutions; otherwise the policies developed might promote measures far removed from the local livelihood security needs and in turn aggravate vulnerability, rather than diminishing it (Owuor et al. 2005; Yamin et al. 2005; Muigua and Kariuki 2015). Moreover, the local people's perception of what entails environmental justice should be effectively incorporated in any government measure, aimed at achieving the same.

While inclusion, formal or informal, with procedural rights can allow some to reclaim rights, this is not universal as seen in the Directorate of Rural Water Supply Program in Namibia, which aimed to put all rural-water points under community management. The scheme meant to decentralize resource management by empowering local communities that witnessed cases of exploitation of marginalized and vulnerable people by the rural elite (Thomas and Twyman 2005). Another example can be cited from the research in Machakos district in Kenya where the reinstatement of traditional practices, leading to more intensive market-oriented production, aimed at local empowerment (Tiffen et al. 1994). It, however, led to higher polarization with some households emerging as winners and marginalizing the others (Murton 1999). This polarization has its seeds in the diminishing resource base as well, which do not necessarily lead to conflict, but do create a space for "winners" and "losers." Thus, understanding the issues of fairness and equity would require supporting strategies, which enhance secure livelihoods and reduce poverty (Thomas and Twyman 2005). The presence of procedural justice attributes in the decision-making process does not ascertain that distributive justice would prevail too, in terms of access to resources and actual local-level decision-making.

Further, for contemporary understanding of adaptation, it is necessary to look into its historical context; otherwise, the impact of the policy might turn regressive. For instance, as seen in the Financial Assistance Programme (FAP) in Botswana, where a policy shift from food sufficiency to food security in post-independence times, had a major impact on farming and non-farming households in rural areas. This same policy which was successful in Southeast Botswana opened up spaces for adaptation by livelihood diversification, reduced sensitivity and increased resilience to drought and climate change, and restricted adaptation options in the Southwest. This not only affected the national and regional marketing systems but also had an adverse impact on local credit systems, ploughing practices, and food security (Thomas and Twyman 2005). The less well-off could not make the investment in terms of both money and time to access the Financial Assistance Programme benefits. It also reflects that such interventions, which though well intended, are not being evenly applied in the regions due to lack of financial infrastructure. Such a situation increases the vulnerability of the poor, instead of reducing it, and by at large limits local adaptation and leads to unsustainable resource use (ibid., 2005).

All adaptive responses have justice implications, both distributive and procedural. It is important to identify how decisions on adaptive responses are made and what is the incidence of consequences of adaptation decisions (Paavola and Adger 2006) to understand the justice implications of adaptation. The justice concerns to adaptive responses require a change in multi-level responses, both proactive and reactive, as climate change certainly affects the feasible adaptive responses (ibid., 2006). Successful adaptation strategies are those which decrease the vulnerability from multiple threats and increase overall resilience (Stringer et al. 2009), especially for the population whose livelihoods are directly tied to natural resource base and subsistence like in semi-arid, dry land, or flood-prone areas. Adaptation exacerbates injustice when policy actions render disadvantaged groups even more vulnerable (Huntjens and Zhang 2016). The most commonly discussed adaptive responses include policy changes like grants or subsidies for new lines of agricultural production so that food production could adapt well to changing climate. What can also be discussed and redistributed are the burdens of climate change impact and adaptation (Paavola and Adger 2006). This involves looking into the collective actions and cross-scale interactions at the local, national, and international levels, as adaptation does not take exclusively at one level (Huntjens and Zhang 2016).

A policy space is required which allows local and regional level innovations and adaptation options to evolve, alongside retaining the principles of equity and justice, as the ability is clearly present in the developing world, both at the community and the household levels. There is a greater role for multi-level adaptation approaches incorporating local innovation. Moreover, there can never be one single right level of decision-making. This could yield multiple benefits and can help address land degradation, drought, and climate change, simultaneously. Thus, policies need to promote divergent adaptation interests for different groups to minimize inequality and promote justice in local adaptation.

Importance of Local Level Participation in Climate Adaptation Policies and Its Challenges

Climate change impact, though discussed and projected at the global or national scales, is ultimately felt at the local levels. Policy proposals for climate change and conflict management without peoples' participation is futile.

Local practices and indigenous knowledge need to be incorporated in the adaptation policies so that those most affected do not just become "passive" victims of state/global forces (Eriksen and Lind 2009). It is also required for a more comprehensive understanding for effective adaptation policies and strategies. The firsthand knowledge of the communities makes sure that the adaptive mechanisms are effective and do increase resilience. Their participation in the decision-making process and actions that affects their own livelihoods and well-being needs to be significant and ingrained in their capacity to adapt and build elements of resilience. The local picture raises issues of within and between community equity and justice that require consideration. Even the 1992 Rio Declaration provided that environmental issues are best handled with participation of all concerned citizens, at the relevant level, and an informed public. However, public participation should be meaningful and not just "cosmetic" (Huntjens and Zhang 2016; Muigua and Kariuki 2015), for the public to feel that their concerns are addressed and consequently for them to have trust and support the decisions of the government relating to the particular natural resources and environmental concerns. Sharing lessons and identifying best practices are essential for adaptation and building a long-term vision (Yamin et al. 2005).

It needs to be highlighted that the marginalized groups remain vulnerable to the ill effects of climate change because they cannot participate in or influence the structures that determine the range of alternative options made available to them (Adano et al. 2012). Moreover, the local adaptive capacity, in regard to access to required capital and adjustment options under different contexts, depends on the ability to promote interests in the decision-making process (Eriksen and Lind 2009). Thus, power and access to social capital becomes an important determinant of adaptation outcomes. Promoting community-led adaptation would require new ways to incorporate vulnerabilities and the perceptions of communities into the relevant policy process (Yamin et al. 2005).

An interdisciplinary, multi-stakeholder approach incorporating the techniques of community-based vulnerability and capacity assessment and participatory rural appraisal is needed for a holistic policy (Stringer et al. 2009), as the local adaptations undertaken are not just a response to climate change, instead relate to broader issues of desertification, drought, and climate. For climate change adaptation to work, we need to believe that environmentally vulnerable communities have the agency to act. Moreover, there can never be one single right level of decision-making. The interplay of local, national, and international drivers is to be considered in building a complete picture for measuring the adaptation outcomes. The results are to be learned from in exposing any policy gap at each level.

Addressing the questions of equity is important so that the most vulnerable do not get trapped in the cycle of impoverishment and destitution at the hands of climate events. While addressing the equity issues, the distributional dimension is considered, more commonly, under the pretext that people in the developing world are the "passive victims," with little role to play in the decisions affecting their own well-being and livelihood (Thomas and Twyman 2005). This kind of an approach of vulnerability ignores the people's own capacity to adapt and build elements resilience. It undermines and undervalues the actions of people most significantly affected by the impacts of climate change. Thus, equity in context of climate change impacts would not be achieved by just providing a buffer from the unduly burdens of the impacts. It would rather need to address the procedural dimension of the decision-making process by recognizing the importance of participation of the affected communities at the local levels, which can affect the opportunities available for adaptation. It must be emphasized that while managing resources sustainably, states must have an environmental policy that takes account of those who depend on the resources for their livelihoods (Muigua and Kariuki 2015). This highlights the importance of recognition and participation of affected communities in decision-making and planning for fairness in regard to community adaptation strategies.

The perception needs to change that local level adaptation is traditionally reactive, while the policy-driven adaptation mechanisms are planned. The challenges of agricultural production can be tackled with inclusion of local knowledge with the policies. The farmers with their experiences and long-term observations have developed an in-depth understanding of climate variability, which helps them formulate their coping and adaptation strategies. Local knowledge, mostly location-specific, also gets passed on from one generation to another and helps them adapt to environmental changes or extreme events.

The integration of local knowledge also tends to increase the legitimacy of the decision-making process for the most vulnerable (Thomas and Twyman 2005; Paavola and Adger 2006). Solutions are looked for with the farmers, rather than just prescribing what should be done. It is also seen that the community-based natural resource management finds more supporters among the local population than that of centralized state-centered or private-managed provision like in rural dry-land Africa (Thomas and Twyman 2005).

An example can be seen in India (Banda District, Uttar Pradesh), where agriculture is being revolutionized with an alternative way of farming, called as *Avartanasheel Kheti*. This is a dry, mostly semi-arid area which is severely affected by drought-like conditions and the farmer's woes are never ending. This approach makes use of mixed farming for sustainable production, dividing the land into three parts and using them for three different purposes. One part of the land is dedicated to fruit trees and plants and wood cultivation, the other for cattle farming and livestock breeding, and the third for agricultural farming and the different types of crops. The produce left after consumption is processed and value-added and then forwarded to the markets through farmer's cooperative. This alternative farming is climate change resistant and weatherproof, as the choice of crops and trees depends on the demand of agro-climatic zone. Majority of the crops grown require less water and are drought resistant. The policy-makers promoted sesame cultivation, which is not natural to the region, and hence the crop was affected by drought conditions. This initiative started by one farmer is now being brought to farmers through farmer's learning center and is an example of how farmers can adapt to climate change and prosper at the same time. It also helps in maintaining the environmental and ecological balance of the area, along with providing healthy and nutritious food for the farmers. There are also other effective watershed management examples for drought adaptation in various other parts of the country, through the usage of common pools for water storage.

When policies affect the social institutions that govern key natural resources, especially at the local level, some fundamental questions relating to equity and justice need to be raised, while incorporating the fundamentals of procedural "legitimate" justice (Thomas and Twyman 2005; Paavola and Adger 2006). There are several procedural justice dynamics that impact communities negatively due to the power relations at the community level (Barrett 2013). Moreover, formal attempts at distributive justice can be well placed but may not always have the intended outcomes. The political dynamics of the local level adaptation, having a fundamental impact on livelihood adjustments, need to be considered for climate change policies to be successful or minimize inequities, social and economic (Eriksen and Lind 2009; Barrett 2013). The political agenda diverts away from the issues of landlessness, resource inequality, and development of basic infrastructure (Owuor et al. 2005). "Powerful legal, institutional and political obstacles stand in the way of making disaster relief more development orientated, in particular of shifting disaster management to disaster risk reduction frameworks – as seen in Kenya and Bangladesh case studies" (Yamin et al. 2005: 09). This may require correcting imbalances in the distribution of powers and resources in a system, which leads to development inequalities. Thus, policies addressing climate change, mitigation and adaptation, need to be enabling local or regional options, rather than restraining them.

While resource shortages may not necessarily create conflict, they may well create different spaces in which it is more likely for winners and losers to polarize. Thus, the room for equity is likely to be reduced through the diminishment of resource base availability. The issues of scale raised by adaptation to climate change itself generate a range of complexities for the processes necessary to engender equity and justice. This may require correcting imbalances in the distribution of powers and resources in a system, which leads to development inequalities, in addition to putting communities at the center stage in determining how the vulnerabilities are addressed.

However, burdening the local governments with the responsibility of just and sustainable adaptation without strengthening their financial and technical capacity only furthers the differences in their ability to adapt (Eriksen and Lind 2009). Therefore, ensuring climate justice in climate change adaptation requires fair distribution and access to resources, alongside meaningful participation of marginalized and vulnerable groups in decisions directly affecting their livelihoods and well-being. This is to ensure that inequality and injustice are not exacerbated with absence of disadvantaged groups in adaptation planning outcomes (Shi et al. 2016). Thus, a need for institutionalization of most vulnerable groups in development processes arises.

Acknowledging and incorporating the local stakeholders in the process does not mean that little attention should be paid to the ill effects of certain adaptation practices. There is a need for vulnerability and risk assessments to avoid maladaptation practices, so that adaptation actions reduce vulnerability and not enhance it. “In the northern Kenyan drylands, where livestock raiding is a widespread practice among pastoralists, the use of violence is common and part of the group-specific accepted codes of conduct and cultural institutions” (Adano et al. 2012: 68). Though during the times of drought the parties patch up and reconcile to use limited scarce resources and settle differences, the use of accepted violence during the other times cannot be justified. Moreover, the lack of access to resources is compounded by social relations and paucity of representation leading to exclusion of certain groups, especially women, as seen within smallholders in Kenya and Tanzania (Eriksen et al. 2005). Further, it is important to recognize how the adaptation responses taken by one group affect the vulnerability context of the other groups, reducing vulnerability and resource sustainability, in regard to different interests and potential value-based conflicts. As seen in the case of Akamba ethnic group, only in areas with fair representation for all groups were the clan committees were found to be effective. Further, with increased government control of the hilltop forests, the power of Mwamba, group of elderly women regulating the use of an access to the hill, has dwindled (Owuor et al. 2005). In Northwest Namibia, there have been cases of exploitation where individuals have to work for a select individuals controlling water points to gain access to water resources, if unable to fulfill monthly payments to water point committees (Thomas and Twyman 2005). Climate action policies could only be effective when implemented in convergence with vulnerability assessment, mainstreaming local-level techniques with climate change responses (Eriksen et al. 2011; Huntjens and Zhang 2016), which was one of the main drawbacks in the case of Durban. This integration enhances the understanding of the implications of climate change in the local context, along with building local interest in policy actions. Having an umbrella approach integrating the tenets of social inclusion, justice, and sustainable development yields better results.

Sustainable Development and Climate Justice

Climate change adaptation has been understood as the adjustments which people make to mitigate the negative effects of climate change on their livelihoods, well-being, and security (Eriksen et al. 2011). However, little attention is paid to the synergy between adaptation and sustainable development, and they have been differentiated by differences in discourse and the effect of these development and adaptation mechanisms on the poor is seldom assessed.

In terms of participation, at the national and subnational levels, politico-economic factors, along with institutional rules, affect the participation and representation of non-state actors and other ethnic groupings. Under such conditions, planning for tackling climate change is no more than mere rhetoric, leading to unsustainable development.

It is evident that climate justice has an important role to play in sustainable development and social inclusion. Moreover, there is a conceptual and practical nexus between them. Both the debates of sustainable development and climate change question the efficacy of the underlying development pathways in addressing the problems of poverty and environment. Climate justice is not just about equally distributing risks but also about preventing them in the first place (Agyeman and Evans 2004). It would be justified to say that environmental problems are vested disproportionately on the poor, though are caused by actions of the affluent. It would be insufficient to simply reject environmental "bads" and not do anything about it. One of the most significant barriers to sustainable development is inequality – economic, social, or environmental – and climate justice addresses the similar aspects of equity and fairness, and hence, both complement each other. Moreover, Local Agenda 21, which is a nonbinding, voluntarily implemented United Nations action plan for sustainable development, talks about local-level empowerment of people through greater citizen partnership and more responsive governance with community capacity building.

Policy practices need to be mutually supportive with broader development goals and adaptation mechanisms in order to be successful. Responses to policy development will be limited if the local-level strategies are not absorbed. This could moreover develop into negative externalities, both spatially and temporally. Integrating local knowledge-based community adaptation initiatives becomes an essential for sustainable adaptation to climate change. Local-level communities have time and again used their understanding of their environment and vulnerability to survive a cataclysmic event, rather than depending solely on technical approaches prescribed by the scientific discourse. Isolating the local responses from the global changes would only further worsen the problem.

The pillars of sustainable adaptation should be based on recognition of the context for vulnerability from multiple stressors, acknowledging the differences in the adaptation outcomes due to the values or interests involved and integrating local indigenous knowledge into the adaptation mechanisms, along with a continuous feedback loop of response between the local and the global. In fact, sustainable adaptation practices and policies are those that are qualified in terms of their effects on social justice and environmental integrity (Eriksen et al. 2011; Huntjens and Zhang 2016). Sustainable adaptation is not just about readjustment of the current practices and development paths, instead focuses on novel and dynamic development ways. The practices adopted are more structural and contextual, for reducing vulnerability and greenhouse gases emissions, and less concerned about spatial or temporal uniformity. There are multiple stressors in these areas, and unless the socioeconomic dimension is not integrated with other extreme climate-related events, the adaptation measures will not be successful in terms of livelihood viability and long-term vulnerability reduction. Thus, sustainable adaptation also works toward building social capital and community empowerment through "strong citizen participation, local identity and local organization" (Eriksen et al. 2011). It should be noted that just economic development will not make the economies climate-proof (Fankhauser 2016). Longer-term adaptation strategies and development plans need to be harmonized with other socioeconomic development goals, such as girl's education and women and youth empowerment (Hardee and Mutunga 2009). Effective adaptation would require knowledge, planning, coordination, and foresight. Adopting the principles of sustainable adaptation is a step forward in guiding approaches toward social justice and environmental integrity.

Conclusion

As climate change worsens the stresses faced in the present times, it is essential to strengthen the links between policy and practice. The political, social, and ethical dimensions need to be integrated, both at the analysis and policy front. The policy framework on adaptation needs to focus on the local realities and the practical implications of adaptation, understanding the underlying causes of vulnerability, social differentiation, and conflicting interests between different groups to efficiently target the most vulnerable. Additionally, participatory assessment frameworks would help in the improvement of environmental decision-making, making it more suitable to the local sociocultural and biophysical contexts.

Reducing vulnerability is an effective precautionary step toward adaptation as it is easier to create generic resources which would have wide potential for application across a variety of areas. However, these will not be able to target services, resources, or regions in a focused manner. The approach, of collaborative resource management and decision-making, is extremely pertinent to developing climate change adaptation strategies that reduce vulnerability, increase resilience, and improve well-being among natural resource-dependent groups. Adaptation techniques should incorporate peace and conflict resolution along with environmental and socioeconomic developmental measures and access to justice for the marginalized. In addition, adaptation finance needs to be decentralized with an understanding of what climate-vulnerable communities perceive as their climate risk, to enable communities sustainably address climate risks and not just through informal measures and community-based conservancies alone.

Addressing the issues of fairness, equity, and justice is significant to the climate change mitigation and adaptation strategies, as climate change is likely to impact the development processes and the national and local responses could potentially exacerbate inequalities by creating winners and losers. The adaptation mechanisms vary according to the ability, resources, and situations, and resilience is eroded by political decisions. A justice perspective questions whether adaptation is backed by a collective commitment and some notion of fairness or rather an ad hoc process initiated by those with the ability and resources to do so. The first step to increase community's resilience is lifting people out of poverty through local knowledge matched with innovation. It is important to link the individual vulnerabilities to the ability to access the environmental services in a way which is socially more inclusive. This would help in realizing environmental justice by integrating the participation of the society and providing fair intra- and intergenerational access to environmental services. A climate-just future would be achieved when individuals have equal access to safe and sustainable livelihoods in their communities without having to compromise access to clean food and water or energy security. If the processes are well-designed, it would lead to capacity building of the communities and societies in terms of effective and proactive response to the changes in the climate.

Although the research was carefully prepared, it is likely that there are still a lot of gaps in the knowledge base that need to be filled. The lack of quantitative information constitutes a shortcoming of the study. The scope of discussions might also be compromised due to lack of experience in conducting researches, compared to the works of experienced scholars. Local level participation and community resilience can be best measured with an exposure to the field itself; thus, the next step could be to build a stronger overall empirical evidence base. The research can also be enriched with a regional focus on the affected areas. Further, exploring the role of international climate change regime for climate justice would make for an interesting work.

References

Adano WR, Dietz T, Witsenburg K, Zaal F (2012) Climate change, violent conflict and local institutions in Kenya's drylands. J Peace Res, Special Issue: Climate Change and Conflict 49(1):65–80

Agyeman J, Evans B (2004) 'Just sustainability': the emerging discourse of environmental justice in Britain? Geogr J 170(2):155–164

Barrett S (2013) Local level climate justice? Adaptation finance and vulnerability reduction. Glob Environ Chang 23:1819–1829

Bob U, Bronkhorst S (2010) Environmental conflicts: key issues and management implications. Afr J Con Res 10(2):9–30

Eriksen S, Lind J (2009) Adaptation as a political process: adjusting to drought and conflict in Kenya's drylands. Environ Chang 43:817–835

Eriksen S, Aldunce P, Bahinipati CS, Martins RD, Molefe JI, Nhemachena C, O'Brien K, Olorunfemi F, Park J, Sygna L, Ulsrud K (2011) When not every response to climate change is a good one: identifying principles for sustainable adaptation. Clim Dev 3(1):7–20

Eriksen SH, Brown K, Kelly PM (2005) The Dynamics of vulnerability: locating coping strategies in Kenya and Tanzania. Geogr J 171(4): 287–305

Fankhauser S (2016) Adaptation to climate change. [Online: web]. Accessed 16 July 2017. http://www.lse.ac.uk/GranthamInstitute/wp-content/uploads/2016/11/Working-Paper-255-Fankhauser.pdf

Fazey I, Gamarra JGP, Fischer J, Reed MS, Stringer LC, Christie M (2010) Adaptation strategies for reducing vulnerability to future environmental change. Front Ecol Environ 8(8):414–422

Foster S (1998) Justice from the ground up: distributive inequities, grassroots resistance, and the transformative politics of the environmental. Calif Law Rev 86(4):775–841

GoK (2007) National policy for the sustainable development of arid and semi arid lands. Government of Kenya, Nairobi

Gustavsson M, Lindstrom L, Jiddawi NS, de la Torre-Castro M (2014) Procedural and distributive justice in a community-based managed Marine Protected Area in Zanzibar, Tanzania. Mar Policy 46:91–100

Hardee K, Mutunga C (2009) Strengthening the link between climate change adaptation and national development plans: lessons from the case of population in national adaptation programmes of action (NAPAs). Mitigation and Adaptation Strategies for Global Change 15(2):1–14

Huntjens P, Zhang T (2016) Climate justice: equitable and inclusive governance of climate action – Policy and governance recommendations for advancing climate justice. Working paper 15. Institute for Global Justice, The Hague

IOM (2014) Migration, environment and climate change: assessing the evidence. International Organization of Migration, Geneva

IPCC (2001) Climate change 2001: impacts, adaptation, and vulnerability. Contribution of Working Group II to the Third Assessment Report of the Intergovernmental Panel on Climate Change. Cambridge University Press, Cambridge

IPCC (2007) Climate change 2007: impacts, adaptation, and vulnerability. Contribution of Working Group II to the Fourth Assessment Report of the Intergovernmental Panel on Climate Change. Cambridge University Press, Cambridge

Kameri-Mbote P (2013) Climate change and gender justice: international policy and legal responses. In: Ruppel OC, Roschmann C, Ruppel-Schlichting K (eds) International Climate Law and Global Governance: Legal Responses to a. Changing Environment. Nomos Publishers, Baden-Baden, pp 323–348

Kaswan A (2008) Environmental justice and domestic climate change policy. Environmental Law Reporter, News and Analysis, pp 1–10

Muigua K (2015) Empowering the Kenyan people through alternative dispute resolution mechanisms. [Online web] Accessed 30 Jan 2018. https://www.ciarb.org/docs/default-source/centenarydocs/speaker-assets/kariuki-muigua.pdf?sfvrsn=0

Muigua K, Kariuki F (2015) Towards environmental justice in Kenya. [Online: web]. Accessed 20 July 2017. http://www.kmco.co.ke/attachments/article/140/Towards%20Environmental%20Justice%20in%20Kenya-January%202015.pdf

Murton J (1999) Population growth and poverty in Machakos District, Kenya. Geogr J 165:37–46

Ogalleh SA, Vogl CR, Eitzinger J, Hauser M (2012) Local perceptions and responses to climate change and variability: the case of Laikipia district Kenya. Sustainability 4:3302–3325

Owuor B, Eriksen S, Mauta W (2005) Adapting to climate change in a dryland mountain environment in Kenya. Mt Res Dev 25(4):310–315

Paavola J, Adger WN (2006) Fair adaptation to climate change. Ecol Econ 56:594–609

Popke J, Curtis S, Gamble DW (2016) A social justice framing of climate change discourse and policy: adaptation, resilience and vulnerability in a Jamaican agricultural landscape. Geoforum 73:70–80

Robinson LW, Berkes F (2011) Multi-level participation for building adaptive capacity: formal agency-community interactions in northern Kenya. Glob Environ Chang 21:1185–1194

Roden P, Bergmann C, Ulrich A, Nüsser M (2016) Tracing divergent livelihood pathways in the drylands: a perspective on two spatially proximate locations in Laikipia County, Kenya. J Arid Environ 124:239–248

Running K (2015) Towards climate justice: how do the most vulnerable weigh environment–economy trade-offs? Soc Sci Res 50:217–228

Schilling J, Akuno M, Scheffran J, Weinzierl T (2015) On arms and adaptation: climate change and pastoral conflict in Northern Kenya. Working paper, CLISEC-15. [Online web] Accessed 1 Feb 2018. https://www.researchgate.net/publication/263242776

Schlosberg D (2014) From environmental to climate justice. [Online web]. Accessed 25 Jan 2018. http://sydney.edu.au/environment-institute/blog/from-environmental-to-climate-justice/

Shi L, Chu E, Anguelovski I, Aylett A, Debats J, Goh K, Schenk T, Seto KC, Dodman D, Roberts D, Roberts JT, VanDeveer SD (2016) Roadmap towards justice in urban climate adaptation research. Nat Clim Chang 6:131–137

Stallworthy M (2009) Environmental Justice Imperatives for an Era of Climate Change. J Law Soc, Economic Globalization and Ecological Localization: Socio-legal Perspectives 36(1):55–74

Stringer LC, Dyer JC, Reed MS, Dougill AJ, Twyman C, Mkwambisi D (2009) Adaptations to climate change, drought and desertification: local insights to enhance policy in southern Africa. Environ Sci Policy 12:748–765

Tadesse D (2010) The impact of climate change in Africa, vol 220. Institute for Security Studies, ISS Paper 220, [Online web] Accessed 1 Feb 2018. https://www.files.ethz.ch/isn/136704/PAPER220.pdf

Thomas DSG, Twyman C (2005) Equity and justice in climate change adaptation amongst natural-resource-dependent societies. Glob Environ Chang 15:115–124

Tiffen M, Mortimore M, Gichuki F (1994) More people, less erosion: environmental recovery in Kenya. ACTS Press, Nairobi

Unruh JD (2005) Changing conflict resolution institutions in the Ethiopian pastoral commons: the role of armed confrontation in rule-making. GeoJournal 64:225–237

White G (2011) Climate change and migration- security and borders in a warming world. Oxford University Press, New York

Yamin F, Rahman A, Huq S (2005) Vulnerability, adaptation and climate disasters: a conceptual overview. IDS Bull 36(4):1–14

Toward Climate-Resilient African Indigenous Vegetable Production in Kenya

43

Winifred Chepkoech, Nancy W. Mungai, Hillary K. Bett, Silke Stöber, and Hermann Lotze-Campen

Contents

W. Chepkoech (✉) · S. Stöber
Center for Rural Development (SLE), Humboldt Universität zu Berlin, Berlin, Germany
e-mail: sangwinfred@gmail.com; silke.stoeber@agrar.hu-berlin.de

N. W. Mungai
Department of Crops, Horticulture and Soils, Njoro, Kenya
e-mail: nmungai@egerton.ac.ke

H. K. Bett
Department of Agricultural Economics and Agribusiness Management, Egerton University, Njoro, Kenya
e-mail: hkbett@gmail.com

H. Lotze-Campen
Potsdam Institute for Climate Impact Research (PIK), Potsdam, Germany

Department of Sustainable Land Use and Climate Change, Humboldt-Universität zu Berlin, Berlin, Germany
e-mail: lotze-campen@pik-potsdam.de

W. Leal Filho (ed.), *Handbook of Climate Change Resilience*,
https://doi.org/10.1007/978-3-319-93336-8_76

Abstract

Climate change presents a global environmental threat to all economic sectors and particularly to the agricultural sector. Kenya is one of the countries negatively affected by climate change due to its high exposure to extreme events and the low adaptive capacity of smallholder farmers. Farmers are facing water scarcity, unpredictable weather patterns, dry spells, droughts, and rising temperatures. The effects of high temperatures, drought, and dry spells lead to serious losses in vegetable yields. Smallholder farmers involved in vegetable production are most at risk due to the sensitivities of vegetable production and their high vulnerability. Drought and water stress have been identified as important limiting factors in vegetable production. This paper examines the climate change adaptation strategies of farmers of African indigenous vegetable (AIV) in three agro-climatic zones (ACZs) in Kenya. Data from 18 focus group discussions (FGDs) and 269 interviews with farmers were analyzed. This study showed that AIV farmers have responded to climate change with a wide range of farm-level adaptation measures and all of the respondents use a combination of these strategies. Farm production practices, such as the application of manure, frequent weeding, and watering of vegetables, were most widespread, while migration to urban areas and buying insurance were the strategies adopted least across all zones. The results revealed a significant association between particular adaptation strategies and ACZs, particularly in soil and water management practices and land-use adjustments. This study offers policy recommendations for accelerating AIV farmers' resilience by supporting opportunities for livelihood diversification.

Keywords

Climate change · Adaptation strategies · African indigenous vegetables · Agro-climatic zones · Kenya

Introduction

Climate variability and change are expected to adversely affect crop production, with the negative effects of this being more common than positive ones (IPCC 2014). The frequency, intensity, predictability, and spatial distribution of extreme climatic conditions, such as temperature extremes, floods, droughts, heat waves, and storms, are likely to be altered by climate change (FAO 2013). These risks to agricultural production directly translate into additional risks for the food security and nutrition of the people who directly depend on agriculture for their food and livelihood (FAO 2016a). Furthermore, the world's poorest and most vulnerable countries, including developing states, arid and semi-arid areas, and countries in which people depend on rain-fed agriculture and natural resources, are disproportionately affected by climate change (FAO 2016b). Furthermore, scientific evidence suggests that climate variability and change are expected to further exacerbate the variability in rainfall and temperatures, particularly in arid and semi-arid areas (Bobadoye et al. 2016a). Africa's agricultural production is also expected to be greatly affected by climate change (Bryan et al. 2013; Elum et al. 2017). Low adaptive capacity, greater reliance on climate-sensitive sectors such as agriculture, and the geographic location of most African countries on the lower latitudes make climate change one of the most important challenges these countries face (Belloumi 2014).

Smallholder farmers make up about 75% of the agricultural sector in Kenya and are heavily dependent on weather predictability and reliability for their rain-fed farming systems (Kuyiah et al. 2013). Agriculture is not only the major sector driving Kenya's economy but also the means of livelihood for the majority of the population (GoK 2009). Extreme weather variability and climate change are already affecting agricultural production in Kenya. The National Climate Change Action Plan 2013–2017 (NCCAP) reveals high vulnerability to climate variability and change. Increases in the frequency and intensity of extreme events, such as drought, heavy rainfall, flooding, and high maximum temperatures, have already occurred and are expected to accelerate (GoK 2013). According to the IPCC (2014), these effects will not be evenly distributed. Water scarcity, dry spells, and drought are likely to increase by the end of the century, particularly in arid regions (GoK 2016). The problems in Kenya are exacerbated by a high level of poverty combined with a rapidly increasing population, with about half of the country living below the national poverty line, often resulting in malnutrition and poor health (KNBS and SID 2013). Furthermore, it is estimated that over 10 million people suffer from chronic food insecurity and poor nutrition. Notably, 30% of children aged 5 and under are undernourished (Mohajan 2014; UNICEF 2013; Famine Early Warning Systems Network (FEWSNET) 2013).

Environmental extremes such as high temperatures and soil moisture stress make vegetables sensitive to climate change (Ayyogari et al. 2014). Vegetable crop growth and development as well as the incidence of insect pests and diseases are influenced by the concentration of the greenhouse gas carbon dioxide (CO_2) in the atmosphere (Naik et al. 2017). African indigenous vegetables (AIVs) for example spider plant (Cleome gynandra), cowpea (Vigna unguiculata) and nightshade (Solanum villosum) are significant contributors to food and nutritional security for smallholder farmers in sub-Saharan Africa (Abukutsa-Onyango 2010). They are rich in vitamins (A and C), dietary fiber, minerals such as iron and calcium, and protein compared to other widely consumed vegetables such as tomatoes, cabbages, and onions (Abukutsa-Onyango 2007; Luoh et al. 2014). According to Luoh et al. (2014), some indigenous vegetables are considered to have stronger adaptation mechanisms to water stress than exotic vegetables. In contrast, some research has revealed that even though they are known for their relative resistance to climate change, some of the species are also partly constrained by extreme climatic conditions (Ngugi et al. 2006; Stöber et al. 2017).

Although local farmers are adjusting to harsh weather conditions and developing strategies to lessen the adverse impacts of climate change, more rigorous adaptation measures are necessary (Ngugi et al. 2006). Given the heterogeneity of the agroecological and socioeconomic environments, it is not possible to provide one solution that fits all (Giller et al. 2011). According to Bryan et al. (2013), adaptation is vital to enhance the resilience of the agriculture sector, protect the livelihoods of the poor, and ensure food security. Adaptation is defined as "the process of adjustment to actual or expected climate and its effects. In human systems, adaptation seeks to moderate or avoid harm or exploit beneficial opportunities. In some natural systems, human intervention may facilitate adjustment to expected climate and its effects" (IPCC 2014 p. 5). Climate change adaptation issues have become the subject of intense global discussions in the past few decades (Elum et al. 2017). While some of the problems associated with climate change are emerging gradually, action is urgently needed in order to allow enough time to build resilience into agricultural production systems (FAO 2016a) because adaptations are manifestations of adaptive capacity and offer ways of reducing vulnerability (Smit and Wandel 2006). According to the IPCC (2014), the decisions regarding adaptations can be undertaken by private individuals, local communities or institutions, national governments, or international organizations. The main categories of adaptation strategies discussed in this paper are autonomous and planned adaptation. Planned adaptation follows a specific program, ranging from awareness-based, evidence-based, and conscious policy decisions (IPCC 2007). In contrast, autonomous adaptation strategies are developed independently from climate change policies and programs. Instead autonomous adaptation is initiated by changes in natural or human systems, such as new products due to market demand and on-farm investments made out of welfare or ecological necessity. According to Forsyth and Evans (2013), autonomous adaptation might include practices such as altering agricultural inputs, introducing on-farm water management technologies, changing cropping cycles, and diversifying economic activities. The examples of planned adaptation measures include programs on crop breeding and promotion in different agro-climatic zones, crop substitution, fertilizer subsidies, insurance promotion, and infrastructure development, such as the expansion of rainwater harvesting, water storage, irrigation, and conservation techniques, in order to increase water-use efficiency (IPCC 2007).

Little is known about the range of both categories of adaptation strategies employed by small-scale AIV farmers. With the intention of filling this gap, the present study aimed to provide farm-level evidence of existing adaptation measures adopted by AIV farmers in three ACZs in Kenya. This study answered the following research questions: (1) Which on-farm adaptation measures are practiced by AIV farmers in the three different ACZs? (2) How context-specific are the adaptation strategies and do adoption rates differ by ACZ? (3) How can this information feed into local agricultural extension service systems with the aim of making location-specific adaptation strategies more mainstream? Acquiring a better understanding of farmers' adaptation practices is relevant in order to support the government's plan to incorporate location-specific climate-resilient strategies into agricultural extension services delivered by the government and its partners across Kenya (GoK 2010). While a comprehensive adaptation program covers questions of governance and livelihood management, the present study concentrated on farm-level adaptation measures that are highly complementary to the sustainable intensification of vegetable production (Campbell et al. 2014).

Description of the Study Areas and Methodology

Study Areas

Kenya has a wide range of climate and ecological conditions (Bryan et al. 2013). The study area was made up of three purposively selected agro-climatic zones in Kenya: Kakamega, Nakuru, and Kajiado counties.

Kakamega County (humid zone) is located in the former Western Province (Mulinya et al. 2016). It comprises 12 sub-counties and lies between longitudes of 34° and 35° east and latitudes of 0° and 10° north of the equator at an altitude of 1250–2000 m. It covers an area of about 3,017.66 km^2 (GoK 2009). Kakamega County mainly experiences hot and wet conditions most of the year, with mean annual rainfall between 1800 and 2000 mm and an average temperature of 22.5 °C (Barasa et al. 2015).

Nakuru County (semi-humid zone) is in the Great Rift Valley and comprises 11 sub-counties. The county covers an area of 7235.3 km^2 and is located between longitudes of 35° and 36° east and latitudes of 0° and 1° south (GoK 2014b). The county is between 1520 and 2400 m above sea level (Ogeto et al. 2013). The temperature varies between 12 °C and 29.6 °C, and annual rainfall ranges between 500 and 1800 mm (GoK 2014b).

Kajiado County (semi-arid zone) is located in the southern tip of the former Rift Valley Province. It lies between longitudes of 36°5 and 37°5 and latitudes of 10° and 30° south. The altitude ranges between 500 and 2500 m above sea level (GoK 2014a; Bobadoye et al. 2016b). Annual rainfall is between 450 and 900 mm, with an average temperature of 22 °C. It comprises five sub-counties and covers an area of 19,600 km^2 and has a population of 136,482 (GoK 2009).

Research Methods

Study Site Selection and Sampling Methods

This study employed a multistage purposive sampling procedure. In the first stage, the three counties of Kakamega, Nakuru, and Kajiado were selected to represent humid, semi-humid, and semi-arid zones. In the second stage, two sub-counties were purposefully selected from each of the selected counties. In the third stage, a sample of farmers in each sub-county was identified. The sample size was determined proportionately. For pretesting, 30 farmers of African indigenous vegetable from Kiambu County were purposefully identified and interviewed prior to the actual interview of the target sample. This allowed questions to be restructured prior to comprehensive data collection. Based on the limitations identified in the pretest, the questionnaire was amended and enriched for the actual interview.

Data Sources and Collection Methods

Primary data presented in this paper were sourced through individual farmer interviews and focus group discussions. Structured questionnaires were administered to 269 farmers of African indigenous vegetables to gather information on adaptation strategies. One hundred interviews were held in the humid zone and 107 and 62 in the subhumid and semi-arid zones, respectively. For the adoption of adaptation practices, farmers were requested to select from a list of 26 practices that had been identified following the pretest.

To triangulate the data obtained from the questionnaire, a total of 18 focus group discussions (FGDs) were conducted to collect in-depth information on the scope and quality of adaptation strategies and the adaptation gaps as perceived by vegetable growers. Each focus group discussion comprised both men and women. The focus group discussions were organized by agricultural extension officers who had been trained in advance on the research purpose and methodology. Each focus group discussion had between 7 and 16 participants and lasted approximately 90 min.

Data Analysis

The study followed a mixed method approach combining both qualitative and quantitative approaches in a parallel design in which both sets of primary data are analyzed independently and used to triangulate the results (Fetters et al. 2013). Data obtained from the questionnaire were analyzed using the Statistical Package for Social Sciences (SPSS) by descriptive statistics. Ranking based on percentages was performed in order to select the strategies that were adopted most. The strategies that were adopted most by farmers in the three zones were considered for further analysis. A chi-square test was then used to determine whether there was an association between the selected adaptation strategies and the ACZs. This hypothesis was tested at the $p<0.05$ level of significance. The tested hypothesis stated that:

(H_0) there is no statistically significant relationship between adaptation strategies and ACZs.
(H_1) there is a significant relationship between adaptation strategies and ACZs.

Qualitative data collected during the FGDs were analyzed using qualitative content analysis. The field notes and videos were studied repeatedly and transcribed in order to develop an analysis structure. This allowed the content of text data to be classified systematically. This classification process included a series of coding cycles to identify themes or patterns and followed the process described by Hsieh and Shannon (2005), Morgan (1993), and Altheide (1987).

Results and Discussion

African indigenous vegetable production in Kenya is facing a number of challenges associated with climate change and therefore requires the implementation of appropriate adaptation measures (Stöber et al. 2017). The categories of adaptation strategies presented in this paper were adapted and modified from various sources (Smit and Skinner 2002; Below et al. 2010; Ngo 2016). The categories discussed in this paper focus on short-term incremental farm-level adaptation practices and include farm financial management, water-use management, farm production practices, planting calendar, and crop portfolio strategies.

Adjustment of Farm Financial Management Strategies

The adaptation strategies categorized as farm financial management included migrating to an urban area, purchasing insurance, leasing extra land, borrowing money, and obtaining off-farm employment. From the results shown in Fig. 1, it can be observed that in all three zones fewer than 33% of farmers took out loans. According to Mukasa et al. (2017) stringent loan requirements marginalize farmers from credit services. The results also revealed that more than half (65%) of the farmers in the humid zone leased extra land. This could be attributed to the rural nature of the humid zone as opposed to the semi-arid and semi-humid zones, which are peri-urban in nature. Furthermore farmers may not have enough land to be able to lease some of it to others. A lower proportion, less than 8%, of farmers from the semi-humid and semi-arid zones leased extra land.

The purchase of insurance has been found to build financial resilience as it helps farmers access credit assistance more easily, allowing them to innovate and invest in technologies that boost productivity (Zwane and Montmasson-Clair 2016). However

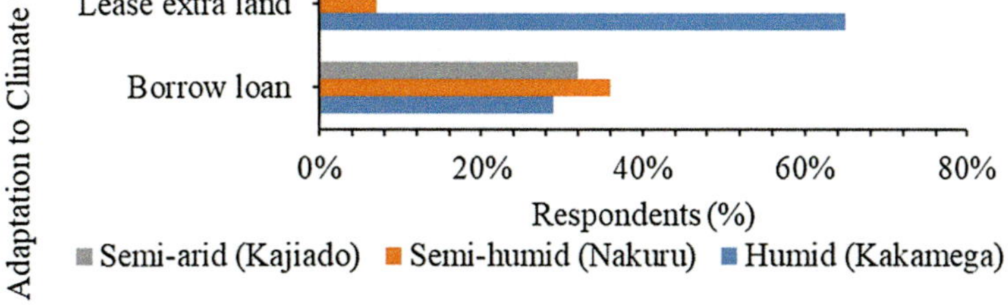

Fig. 1 Distribution of farm financial management strategies by ACZ

fewer than 5% of the AIV farmers in the three zones considered crop insurance as an adaptation measure that they would adopt. In Kenya, a lack of knowledge about the mechanism of agriculture insurance has been identified as the single greatest hindrance to farmers considering agriculture insurance as a risk mitigation measure. Furthermore, existing informal systems of managing risks, such as income diversification, multi-cropping, and scaling down on production in seasons perceived to be unfavorable for farming, seem to be working (Association of Kenyan Insurers 2016). Similarly in a farm-level study in Bangladesh, Uddin et al. (2014) found that crop insurance was ranked as the least important adaptation strategy by farmers.

Only 1% of farmers in the semi-humid zone considered migration to an urban area as a measure for adaptation to climate change and variability. None of the farmers from the humid and semi-arid zones migrated to urban areas. According to Warner (2010), people move to urban areas when they are driven by sudden or progressive changes in climate. Migration can be temporary, permanent, or seasonal. In Nepal, declining agricultural production due to climate change has been identified as one of the major drivers of migration (Dhakal et al. 2016). Migration is an important strategy that is discussed when families are struggling with rainfall variability and food insecurity. This is often motivated by attempts made by households dependent on rain-fed agriculture to stabilize household income and food consumption. Similarly, Warner and Afifi (2013) noted that migration is often the last option considered when farmers are unable to access stable agriculture-based incomes and food consumption in situ.

Finding an off-farm job was adopted by 31% of farmers in both the humid and semi-humid zones and by 24% of farmers in the semi-arid zone. Bezabih et al. (2010) note that even though off-farm employment can be seen as an opportunity to cope with anticipated risks, poorer households will not necessarily opt for it because the investment needed may be beyond their reach. Overall, it is important to note that adaptation strategies in the category of farm financial management were the measures least adopted by farmers. A higher number of farmers in the three zones borrowed money, while leasing extra land was practiced more in the humid zone only. It appears evident that the limited use of these adaptations in the category of farm financial management could be attributed to the need for sufficient financial capital to implement them. A cross-tabulation of farm financial management strategies and ACZs was performed and is shown in Table 1, which also gives the

Table 1 Contingency table of farm financial management strategies and ACZs

			Agro-climatic zones					
Adaptation strategies	Kakamega n(%)	R	Nakuru n (%)	R	Kajiado n (%)	R	Total n (%)	R
Borrow money	29(29)	16	38(36)	12	20(32)	11	87(24)	17
Find off-farm job	20(20)	18	25(23)	19	19(31)	12	64(24)	17
Lease extra land	6(6)	22	7(7)	22	2(3)	19	15(2)	23
Buy insurance	1(1)	23	1(0.9)	23	3(5)	18	5(6)	22
Migrate to urban areas	0(0)	24	1(9)	21	0(0)	20	87(32)	14

Values in parentheses are percentages. R = rank

ranking of adaptation strategies based on percentage distribution per zone. The adaptation strategies in this category were ranked low in all three ACZs. They were not considered for chi-square test analysis since they were not adopted by many farmers in the three ACZs.

Adjustment of Water-Use Management Strategies

The adaptation strategies categorized as water-use management included irrigation and water-harvesting schemes. The results showed that only 15% of farmers in the humid zone had a water-harvesting scheme, while they had been installed by 40% and 60% of farmers, respectively, in the semi-humid and semi-arid zones. In response to long-term climate change, over 60% of AIV farmers chose to irrigate their vegetables in the humid, semi-humid, and semi-arid zones. Notably, most of the farmers relied on hand watering with buckets and watering cans to irrigate their vegetables, as shown in Table 2. The majority of farmers cited that they had insufficient water to irrigate their vegetables and added that they often completely abandoned vegetable farming during long dry seasons. Drip irrigation was adopted by fewer than 7% of farmers in the humid, semi-humid, and semi-arid zones. Watering by bucket/watering can was adopted by 29.8%, 26.6%, and 7.8% in the humid, semi-humid, and semi-arid zones, respectively. This could be attributed to the minimum costs associated with hand watering as compared to other irrigation methods. Overhead/sprinkler irrigation was adopted by 4.6% and 6.4% of farmers in the humid (Kakamega) and semi-humid (Nakuru) zones, while a slightly higher percentage (13.3%) in the semi-arid zone of Kajiado had adopted this practice. Some of the most commonly used smallholder irrigation innovations include drip irrigation, gravity irrigation, pipe irrigation, furrow irrigation, and sprinkler irrigation (AGRA 2014).

Farmers have also been reported to harvest water from rooftops and divert water from natural springs into tanks to ensure they have a substantial amount of water stored up for the dry season. Interestingly this water has been used for the supplementary irrigation of vegetables (Altieri and Koohafkan 2008; Nakuja 2012).

Table 2 Irrigation methods practiced in the ACZs (in %)

Type of irrigation	Kakamega (humid zone) n = 100	Nakuru (semi-humid zone) n = 107	Kajiado (semi-arid zone) n = 62	Total
Drip	1.8	0	6.9	8.7
Bucket/ watering can	29.8	26.6	7.8	64.2
Overhead/ sprinkler	4.6	6.4	13.3	24.3
Furrow	2.3	0.5	0	2.8
Total	–	–	–	100

Source: Survey data

The results from the FGDs showed that majority of the farmers in the semi-arid zone harvested rainwater, while very few of them did in the humid (7%) and semi-humid (1%) zones. Farmers cited a lack of funds for purchasing or building water storage tanks as the main constraining factor. The irrigation methods adopted least were sprinkler and drip irrigation with pipes. The limited use of these irrigation methods could be attributed to their high initial costs and the technical skills required to maintain and operate these systems. The use of buckets or watering cans was a reasonably well-adopted measure, used by 44%, 15%, and 21% of farmers in the semi-arid, semi-humid, and humid zones, respectively (Stöber et al. 2017). Farmers in the FGDs noted that it is easy to water vegetables with a bucket using wastewater from the kitchen. Research has shown that manual irrigation with watering cans is time-consuming. According to Danso et al. (2002), manual irrigation takes 13% of the total cost (excluding family labor) or 38% of a farmer's time and high water application rates (640–1600 mm $year^{-1}$) in year-round irrigation of peri-urban vegetable production in Ghana. Even though it is the least adopted method, research shows that enhanced irrigation has the potential to help communities access, manage, and convey scarce water resources and therefore conserve water in times of drought (Kuang-Idba et al. 2016). According to Elliott et al. (2014), crop water demands are likely to increase with warming, necessitating the use of irrigation practices. Irrigation could also improve farm productivity by allowing diversification in production whenever climate-related changes occur (Barasa et al. 2015). Table 3 shows that irrigation (hand watering) was ranked highly and is widespread, while the construction of a water-harvesting scheme was not common in these three zones. A statistically highly significant association between water-use management strategies was reported at $p<0.001$ (Fig. 2).

Adjustment of Farm Production Practices

Figure 3 shows the adaptation strategies that are classified as the adjustment of farm production practices. The common strategies in this category include soil management practices such as the use of manure, terracing, fallowing, crop rotation, agroforestry, shifting cultivation, intercropping, the use of cover crops, increased weeding, and increased use of chemical pesticides.

Results showed that more than half of the farmers in the three agro-climatic zones adopted mulching as an adaptation to climate change. These results could be attributed to the fact that the application of mulching may not require specific skills and often the materials used are readily available from plant residues that cannot be used further for livestock feeding. As learned in the focus group discussions, extension agents from the government and community-based organizations often train farmers about the importance of land management practices such as mulching. Research shows that plant residues left on the soil surface can have several effects on weed germination and growth. They help suppress weeds by serving as a physical barrier to normal plant growth and development (Altieri et al. 2011). Farmers use

Table 3 Contingency table of the ranking and distribution of water-use management strategies by ACZ

	Agro-climatic zones								
Adaptation strategies	Kakamega n(%)	R	Nakurun (%)	R	Kajiado n(%)	R	Total n(%)	*R*	χ^2
Irrigation	81(81)	3	69(65)	5	60(97)	2	210 (78)	5	24.70****
Invest in water-harvesting schemes	15(15)	20	43(40)	11	40(65)	6	98 (36)	13	41.60****

****indicates significance at a level of $p<0.001$. Values in parenthesis are percentages. R = rank

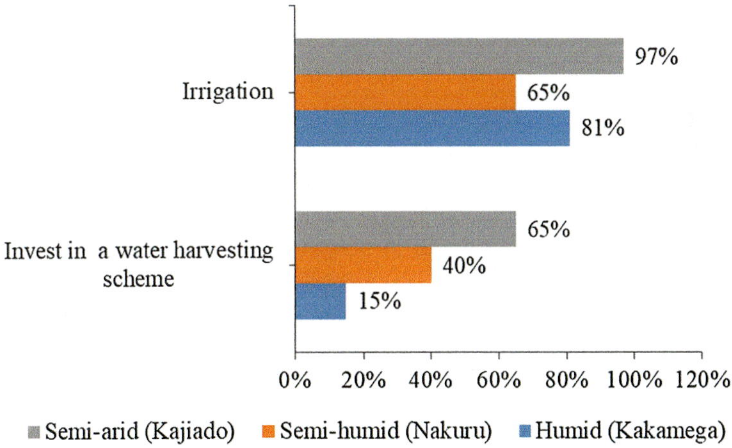

Fig. 2 Distribution of water-use management strategies

composting and mulching to conserve soil moisture and improve soil fertility to increase their crop production (Ndamani and Watanabe 2016; Asayehegn et al. 2017). Additionally, conventional as well as organic vegetable production often employs some type of mulch system to help suppress in-row weed emergence (Price et al. 2011). An overwhelming percentage of farmers (over 90%) in all three zones applied manure on their vegetables farms. These results may be attributed to the contribution made by manure to increasing yields in AIV production. Farmers mentioned that they apply manure because it is readily available, is retained longer in the soil compared with inorganic fertilizers, and therefore increases soil fertility for a longer period. Farmers also mentioned that it improves the soil's water-retention capacities, therefore supplying moisture to the vegetables during unexpected dry spells. Furthermore the results revealed that over 80% of the farmers in the humid and semi-arid zones practiced terracing, while 48% in the semi-humid zone did so. The extensive adoption of this strategy can be explained by soil erosion and waterlogging issues that farmers in the FGDs confirmed they had experienced.

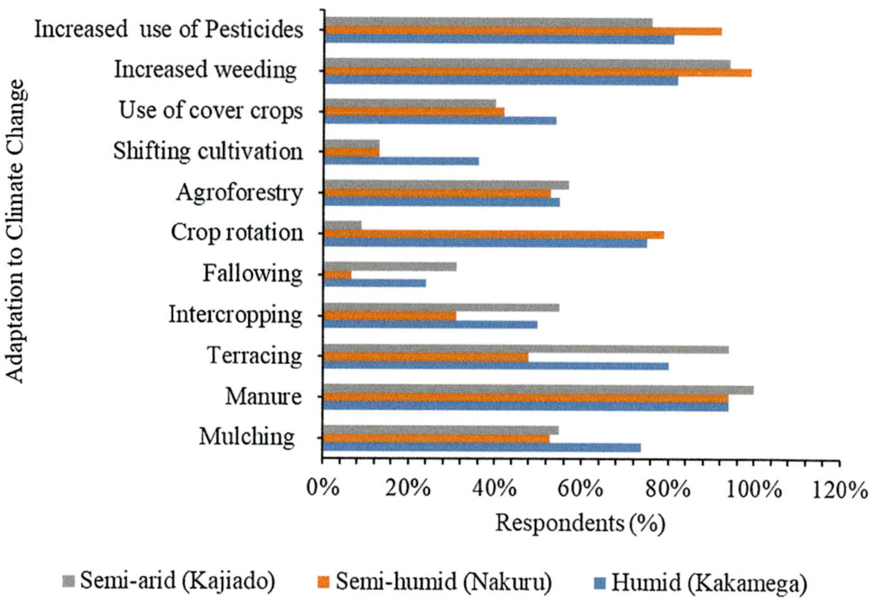

Fig. 3 Adjustment of farm production practices

Fallowing was not only carried out by at most 34% of farmers in the three zones. This could be explained by land fragmentation, which has left farmers with very small areas of land and therefore little or no space for fallowing. The results further revealed that the majority (75%) of farmers in both the humid and semi-humid zones practiced crop rotation, while only 9% did so in the semi-arid zone. Typical vegetable crop rotations in the humid zone included:

1. Slenderleaf (*Crotalaria Brevidens)* – cowpea (*Vigna unguiculata*) or jute mallow (*Corchorus olitorius*) or maize (*Zea mays*)-cowpea *or* jute mallow – nightshade (*Solanum villosum*).
2. Slender leaf – nightshade/cowpea – jute mallow – amaranth (*Amaranthus* sp.).
3. Cowpea – pumpkin (*Cucurbita moschata*) – nightshade.
4. Cowpea or jute mallow – slenderleaf – spider plant (*Cleome gynandra*) – maize – kale (*Brassica oleracea var. sabellica*) – nightshade – maize – nightshade.

Intercropping was practiced by 50% of farmers in the semi-arid zone and by 31% and 55% of farmers in the semi-humid and humid zones, respectively. In the humid zone, farmers practiced intercropping mostly with cowpea and maize (FGD 9). These cropping systems (crop rotation and intercropping) provide many ecosystem services, including nitrogen fixation and carbon sequestration, breaking the life cycle of pests, improving weed suppression, and smoothing out the impacts of price fluctuations. This can further save farmers the cost of fertilizers and pesticides by minimizing the use of inputs (Teklewold et al. 2013).

The adoption of agroforestry was popular with more than half of the farmers in all the zones. Agroforestry uses both traditional and modern land-use systems in which trees are managed together with crop production systems in agricultural settings (FAO 2016b). Many farmers grow crops in agroforestry designs and shade tree cover to protect crop plants from extremes in the microclimate (Sujakhu et al. 2016) and soil moisture fluctuation (Altieri and Koohafkan 2008). Besides cushioning farmers from the effects of climate change, agroforestry systems simultaneously contribute to reducing greenhouse gases because of their high potential for carbon sequestration. Other benefits include soil fertility improvement and soil moisture retention through increasing soil organic matter and nitrogen fixing, stabilizing soils, increasing infiltration rates, arresting land degradation and soil porosity, and reducing runoff (Luedeling et al. 2014; Mbow et al. 2014).

Shifting cultivation was employed by 36% of the farmers in the humid zone. Some mentioned that they moved their production plots next to rivers or wells in order to minimize the distance water needed to be carried or pumped to irrigate vegetables. The low adoption of shifting cultivation in the semi-humid and semi-arid zones could be attributed to their peri-urban location, where rapid urbanization causes stiff competition for private ownership of land that in turn results in the subdivision of land into very small plots/portions. Evidence from the Eastern Cape Province in South Africa shows that shifting cultivation is commonly used by farmers in response to climate change. Farmers report shifting their cultivation from one field to another, with a general trend of moving away from uplands toward wetlands in search of a high moisture content and fertile soils (Taruvinga et al. 2016). The results also showed that more than half of the farmers in the humid zone planted cover crops, while 42% and 40% did so in the semi-humid and semi-arid zones. According to Price and Norsworthy (2013), weed suppression can be achieved in vegetable production by a variety of cover crops grown prior to the vegetable crop. Cereal grains, legumes, and mustards are common cover crops shown to be beneficial in vegetable systems. In the FGDs, farmers mentioned that they planted sweet potatoes, lucerne, and Napier grass as cover crops (FGD 10). Reed-Jones et al. (2016) argue that cover cropping also has multiple purposes such as suppressing nematodes, serving as a nitrogen source, improving soil quality, controlling weeds, and providing seasonal protection against soil erosion. They have also been known to increase crop productivity, possibly by the suppression of growth of a nearby plant by releasing toxins and by having a number of effects on soil quality, fertility, and moisture (Altieri et al. 2011).

The majority (more than 80%) of the farmers in the three zones weeded their vegetable fields more frequently as an adaptation strategy. Although labor-intensive, it is a common practice in AIV production, as learned in the FGDs. Farmers noted that a good yield can only be achieved on a weed-free AIV plot. More than 75% of the farmers were starting to apply more chemical pesticides/herbicides/fungicides than before in the study area. According to the farmers, this was because the vegetables were more greatly affected and on occasion new pests and diseases were arriving as a result of climate change. According to Abang et al. (2014), pests and diseases have been identified as important constraints to vegetable production. As noted by Gatto et al. (2016), climate change may increase pest populations, including weeds, invasive species, insects, and insect-borne diseases, which is likely to lead to large increases in the use of pesticides. In the prevalence of pest diseases and weeds, however, the frequent application of chemicals will have negative implications for the environment in the long run (Juroszek and Tiedemann 2011). Furthermore, Sujakhu et al. (2016) argue that excessive use of chemical pesticides produces a good yield for the first year but reduces the natural resistance of plants to harmful microorganisms. It also increases the resistance of the target pests to pesticides in subsequent years. Therefore, the ability to respond to the impacts of climate change calls for a major shift from input-intensive approaches to more sustainable and resilient food systems (FAO 2016b). As identified in the FGDs, farmers sometimes also use organic remedies to fight pests and diseases in AIVs. For example, a mixture of onions, garlic, ash, and Mexican marigold (*Tithonia diversifolia*) is quite commonly used in the semi-arid zone (see Stöber et al. 2017). Due to less exposure to organic farming practices, only 12% and 14% used organic remedies in the humid and semi-humid zones. The use of ash was reasonably well adopted by 50% of the farmers in both the humid and semi-humid zones as compared to those from the semi-arid zone (ibid). Furthermore, the results also showed that very few (less than 14%) of the farmers in the study areas changed from crop production to livestock as a strategy to adapt to climate change. This may be because, like vegetable production, livestock keeping is equally vulnerable to the effects of climate change and farmers may lack the necessary capital for restocking and purchasing animal feed.

Overall, the adoption of strategies categorized as adjustments in farm production practices were reasonably well adopted in the three agro-climatic zones, except for crop rotation which was the least adopted strategy in the semi-arid zone and shifting cultivation which was the least practiced method in the semi-humid and semi-arid zones. Notably the majority of activities in this category were commonly practiced by more than 40% of the farmers, with the exception of fallowing, shifting cultivation and crop rotation (in the semi-arid zone), and intercropping (in the semi-humid zone), which were the least practiced methods (Table 4). The uptake of these strategies by the majority of AIV farmers could be associated with the lower expense and ease of access by farmers than that of adjustment of water-use management, adjustment of farm financial management, and adjustment of planting calendar and crop portfolio.

The findings from the focus group discussions also showed that the most common strategies included crop rotation, intercropping, use of organic pesticides (especially in the semi-arid zone), use of manure, composting, and crop residue mulching. These strategies were ranked highly as the most practiced in the three ACZs. The results in Table 4 show that frequent weeding, terracing, shifting cultivation, and fallowing are highly significantly associated with the ACZs at $p<0.001$.

Table 4 Ranking and distribution of farm production practices and difference per ACZ

			Agro-climatic zones						
Adaptation strategies	Kakamega n(%)	R	Nakuru n(%)	R	Kajiado n(%)	R	Total n(%)	R	χ^2
Manure application	94(94)	1	101(94)	2	62(100)	1	257 (96)	1	3.78^{*}
Frequent weeding	82(82)	2	106(99)	1	58(94)	3	246 (91)	2	19.71^{****}
Increased use of pesticides, herbicides, fungicides	81(81)	3	98(91)	3	47(76)	5	226 (84)	3	8.36^{***}
Terracing	80(80)	4	51(48)	9	58(94)	3	189 (70)	6	46.78^{****}
Crop rotation	75(75)	6	85(79)	4	55(89)	4	215 (80)	4	4.51^{*}
Mulching	74(74)	7	57(53)	7	34(55)	8	165 (61)	7	10.80^{*}
Agroforestry	55(55)	9	57(53)	7	35(57)	7	147 (55)	9	0.17
Cover cropping	54(54)	10	45(42)	10	25(40)	9	124 (46)	11	4.05^{*}
Intercropping	50(50)	11	33(30)	15	34(55)	8	117 (44)	12	11.94^{**}
Shifting cultivation	36(36)	13	14(13)	20	8(13)	16	58 (22)	18	19.62^{****}
Fallowing	24(24)	17	7(7)	22	19(31)	12	50 (19)	20	18.15^{****}
Change from crop to livestock	7(7)	21	14(13)	20	6(10)	17	27 (10)	21	2.13

****, ***, ** and * indicate significance at $p<0.001$, $p<0.01$, $p<0.05$, and $p<0.1$, respectively. Values in parentheses are percentages. R = rank

Adjustment of Planting Calendar and Crop Portfolio

The adaptation strategies in this study classified as planting calendar and crop portfolio included changing the AIV variety, changing the amount of land under AIVs, the cessation of planting some varieties, the use of improved seeds, crop diversification, and changed planting dates. The results in Fig. 4 show that more than 36% of the farmers in the humid and semi-humid zones changed their AIV variety in order to adapt to the changing climate. This could be because a diverse range of vegetables is produced in the two zones as compared to the semi-arid zone, where only 10% of the farmers adopted this strategy. The moderate adoption of this strategy in the humid and semi-humid zones could also be attributed to farmers clearly understanding the sensitivity of the vegetables to different conditions. As observed in the FGDs, farmers were already aware of the kind of AIVs that do well in different

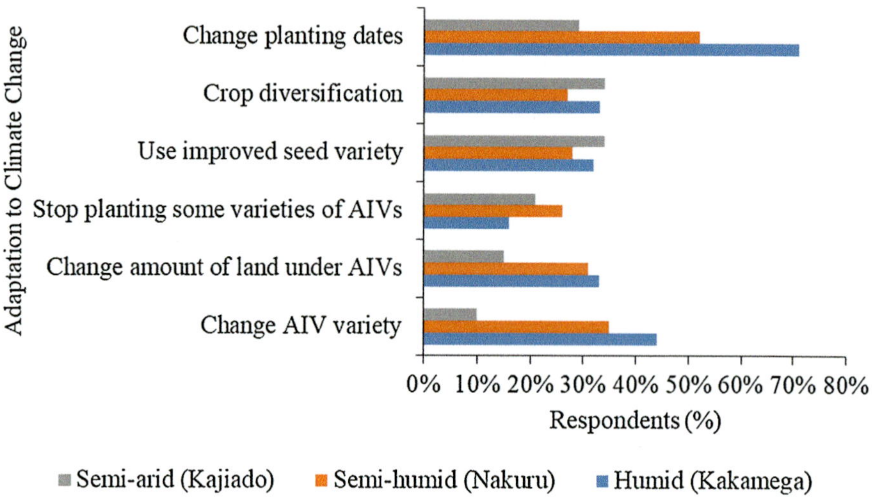

Fig. 4 Distribution of adjustment of planting calendar and crop portfolio

seasons and did not have to switch from one variety to another when seasons change. Fisher et al. (2015) noted that switching to crop varieties that are less sensitive to climatic stress is one of the preferred strategies of farmers in SSA.

Slightly more than 30% of the farmers in the humid and semi-humid zones changed the amount of land under AIVs, while only 15% did so in the semi-arid zone. This meant that very few farmers were either expanding or reducing the plot area on which the vegetables were grown. Interestingly, less than 22% of the farmers in the study areas stopped planting some varieties of AIVs in order to adapt to the changing climate. In the FGDs, many farmers mentioned that AIVs formed part of their food systems, and it may not be easy to abandon production even in the face climate change. Certified seed varieties were used by a similar number of farmers in the humid, semi-humid, and semi-arid zones (32%, 28%, and 34%, respectively). The majority of farmers noted that in most cases, they prepared their own seeds saved from the previous harvest because they could not afford the certified seed varieties. According to Teklewold et al. (2013), the adoption of improved seeds is likely to be an important strategy in adapting to future climate change. Reportedly, the majority of farmers in developing countries have been found to access their uncertified seeds, which have often been saved from the previous harvest, from informal seed markets, and there is widespread exchange within social networks. Research has found that these local crop varieties are often very well adapted to prevailing climate conditions and pests (GIZ 2013). Other results show that, in order to lessen the impacts of climate change, 27–34% of farmers diversified their crop production in the three zones. Multi-cropping has been found to have advantages over more labor-intensive single cropping because it is more resistant to pest attacks, diversifies the diet, and reduces crop failure risk (Dhakal et al. 2016). Furthermore, many adaptations to climate variability made by farmers concentrate on crop diversification to take advantage of the differential effects that a given climate event or condition might have on different crops and activities during the year (Bryan et al. 2013).

The majority of the farmers in the humid and semi-humid areas (52% and 72%, respectively) adapted to climate change by changing their planting dates. However in the semi-arid zone, this was not a common practice, as reported by only 29% of farmers. The uptake of this adaptation by more than 50% of the farmers overall could be attributed to the fact that most of the farmers in the study areas reported that they were experiencing changes in rain onset and cessation. This was perhaps done to shift planting dates to match the seasons in order to minimize losses and take advantage of the opportunities. Rurinda et al. (2015) demonstrate that a moderate delay in planting to avoid early season dry spells can be beneficial to plant growth. Changing planting dates is relatively inexpensive and easy to implement and is particularly important for farmers in rain-fed systems. The onset of the rainy season is crucial to the timing of rain-fed crops: if a farmer plants too early, soil moisture will be insufficient for seed germination, but if left too late, intense rain might wash seeds away (Reason et al. 2005; Bryan et al. 2013). Furthermore, climate information has been identified as an important determinant for changing planting dates in response to changing weather patterns (GIZ 2013). However, according to Fisher et al. (2015), the main limitation is the availability of accessible and accurate weather forecasts, which would complement indigenous knowledge about the weather. In all three agro-climatic zones, the strategies categorized as crop portfolio and planting calendar were the strategies with the second lowest adoption rate after farm financial management (Table 5), except for changing planting date, which was adopted by more than half of the farmers in the humid and semi-humid zones.

The chi-square tests in Table 5 show that adaptation strategies, planting dates, changing AIV varieties, shifting cultivation, and changing the amount of land under AIVs were statistically significantly associated with the ACZs at the $p<0.05$ level of significance. The results revealed that the computed p-value was lower than the significance level alpha, so the null hypothesis H0 was rejected and the alternative hypothesis H1 was accepted [$x^2(50, N = 269) = 185.675 p = 0.0001$]. The risk of rejecting the null hypothesis while it is true is less than 0.01%. The results therefore showed that there is an association between adaptation strategies and ACZs.

Adaptation Strategies per Farm by Agro-Climatic Zone

The mean distribution of the number of adaptation strategies per farm by ACZ was computed, as shown in Table 6. Humid Kakamega has the highest mean number (10.03) of adaptation strategies while also having the greatest variation, with some farmers applying only 2 and others up to 19 strategies. Nakuru (semi-humid zone) had an average of 9 adaptation strategies, with a minimum of 3 and a maximum of 18 adaptation strategies per farm. The semi-arid Kajiado also had an average of 9 adaptation strategies, with a minimum of 4 and a maximum of 19 strategies per farm.

Table 5 Ranking and distribution of adjustments in the planting calendar and crop portfolio by ACZ

			Agro-climatic zones						
Adaptation strategies	Kakamega n(%)	R	Nakuru n (%)	R	Kajiado n (%)	R	Total n(%)	R	χ^2
Changing planting dates	77(77)	5	56(52)	8	18(29)	13	151 (56)	8	36.80****
Change AIV variety	44(44)	12	37(35)	13	6(10)	17	87 (32)	14	21.01****
Shifting cultivation	36(36)	13	14(13)	20	8(13)	16	58 (22)	18	19.62****
Crop diversification	33(33)	14	29(27)	17	21(34)	10	83 (31)	15	1.19
Change amount of land under AIVs	33(33)	14	33(31)	14	9(15)	15	75 (28)	16	1.28**
Use improved AIV seed varieties	32(32)	15	30(28)	16	21(34)	10	83 (31)	15	0.72
Stopped planting some AIVs	16(16)	19	28(26)	18	13(21)	14	57 (21)	19	3.20

****, ** indicate significance at $p<0.001$ and $p<0.05$, respectively. Values in parentheses are percentages. R = rank

Table 6 Mean distribution of adaptation strategies per farm by agro-climatic zone

County/ACZ	Mean number of adaptation strategies	N	Std. deviation	Min	Max
Kakamega (humid zone)	10	100	3.89	2	19
Nakuru (semi-humid zone)	9	107	3.45	3	18
Kajiado (semi-arid zone)	9	62	3.47	4	19
Total	9	269	3.63	2	19

Conclusions and Recommendations

African indigenous vegetable farmers in Kenya are not ignoring the consequences of climate variability and change. They are aware of the need to adapt in order to reduce the risks associated with climate change. This paper provides evidence of the wide range of adaptation strategies practiced by African indigenous vegetable farmers in three ACZs. The uptake of certain adaptation strategies significantly differed between the ACZs, indicating the location-specific nature of farm-level adaptation.

Water management practices were most widely disseminated in semi-arid areas. Humid and subhumid ACZs lagged behind in improved water management. Climate change adaptation programs need to consider the limited experience farmers have with irrigation and water-harvesting practices as these grow in importance due to the increased variability in rainfall in the subhumid and humid ACZs. Crop calendar and portfolio strategies, such as shifting planting dates and changing AIV varieties, were more common in the humid ACZ compared to the subhumid and semi-arid ACZs, which implies a relative advantage for farmers in humid ACZs.

The results also showed that even though the majority of farmers irrigated their vegetables, hand watering by buckets or watering cans was most commonly used compared to advanced irrigation methods such as drip and sprinkler irrigation. Rainwater-smart, low-tech, or low-input water management technologies are therefore a relevant field for climate change adaptation programs by vegetable farmers in Kenya.

The five adaptation strategies that were most widespread across all three zones included the application of manure, frequent weeding, the increased use of pesticides, crop rotation, and irrigation by hand watering. Typical for these widely adopted strategies is that they can be adopted with minimal risk and little financial investment but are associated with higher labor input.

In contrast, the prominent features of the five least adopted strategies were that they are not easily accessible and require financial inputs or knowledge and networks. In detail these were migrating to urban areas, purchasing insurance, and leasing extra land. A very small percentage of AIV farmers were looking for off-farm jobs. Two other strategies least adopted were stopping planting AIVs and fallowing land. This means that farmers operate within the limited context of making maximum use of their farm land.

There were also similarities across the ACZs, for example, the mean number of adaptation strategies adopted per farm, which overall ranged between 2 and 19 strategies. This indicates that the number of adaptation strategies was influenced more by other factors, such as the socioeconomic state of the farm household, than by the ACZ itself.

Adaptation strategies categorized as farm production practices were the most widespread across the ACZs. The lowest adoption rates were in the categories of adjusting the planting calendar and crop portfolio and farm financial management strategies.

Although AIV farmers have responded to climate change by adopting various strategies, the study showed that most of the adaptation strategies that had been adopted were largely autonomous adaptations rather than part of deliberate climate change policies. As climate change adaptation will probably require a broader rural transformation in Kenya, alternative livelihood programs will be needed to overcome the present narrow adaptation strategy portfolio of AIV farmers, which mainly rely on low-barrier and easy-to-implement farm production activities. Therefore, programs need to be implemented that help farmers adopt transformative and livelihood diversification adaptation practices, which require technical and financial support. A more resilient AIV production system could also be part of the program by ensuring accessibility to and the availability of improved AIV seed varieties, money lending institutions, climate change extension services, and reliable sources of weather information but not just limited to these.

Furthermore, farmers should be made aware of the need to implement sustainable adaptation strategies. Given the high percentage of AIV farmers implementing harmful practices through the increased use of chemical pesticides, comprehensive training programs on the sustainable use of pesticides, cultural practices to prevent pests and diseases, and natural pest control should be carried out to ensure environmental protection and safer production of healthy vegetables.

References

Abang AF, Kouamé CM, Abang M, Hanna R, Fotso AK (2014) Assessing vegetable farmer knowledge of diseases and insect pests of vegetable and management practices under tropical conditions. Int J Veg Sci 20(3):240–253

Abukutsa-Onyango MO (2007) The diversity of cultivated African leafy vegetables in three communities in western Kenya. Afr J Food Agr Nutr Develop 7(3). www.ajfand.net

Abukutsa-Onyango MO, Kavagi P, Amoke P, Habwe FO (2010) Iron and Protein Content of Priority African Indigenous Vegetables in the Lake Victoria Basin. J Agr Sci Technol 4(4)

AGRA (2014) Africa agriculture status report: climate change and smallholder agriculture in sub-Saharan Africa. CCAFS, Nairobi, Kenya

Altheide DL (1987) Reflections: ethnographic content analysis. Qual Sociol 10:65–77

Altieri MA, Koohafkan P (2008) Enduring farms: climate change, smallholders and traditional farming communities third world network 131 Jalan Macalister 10400 Penang, Malaysia. ISBN: 978–983–2729-55-6. www.twnside.org. Accessed 24 Sept 2018

Altieri MA, Lana MA, Bittencourt HV, Kieling AS, Comin JJ, Lovato PE (2011) Enhancing crop productivity via weed suppression in organic no-till cropping systems in Santa Catarina, Brazil. J Sustain Agric 35(8):855–869

Asayehegn K, Temple L, Sanchez B, Iglesias A (2017) Perception of climate change and farm level adaptation choices in Central Kenya. Cahiers Agric 26(2):25003

Association of Kenya Insurers (2016) Situational analysis of agriculture insurance landscape in Kenya: consumer options, research and beyond August, 2016. http://www.akinsure.com/images/pdf/AgriculturalInsuranceSituationalAnalysisPresentation.pdf. Accessed 24 Sept 2018

Ayyogari K, Sidhya P, Pandit MK (2014) Impact of climate change on vegetable cultivation – a review. Int J Agric Environ Biotechnol 7(1):145. https://doi.org/10.5958/j.2230-732X.7.1.020

Barasa MOB, O'tengi SB, Wakhungu JW (2015) Farmer adaptation strategies to the impacts of climate variability in Kakamega County, Kenya. Researchjournali's J Agric 2(5):1–18

Belloumi M (2014) Investigating the linkage between climate variables and food security in ESA countries AGRODEP working paper 0004

Below T, Artner A, Siebert R, Sieber S (2010) Micro-level practices to adapt to climate change for African small-scale farmers a review of selected literature. In: IFPRI discussion paper 00953 February 2010. https://www.preventionweb.net/files/12863_IFPRIfeb2010.pdf. Accessed 24 Sept 2018

Bezabih M, Gebreegziabher Z, GebreMedhin L, Köhlin G (2010) Participation in off-farm employment, rainfall patterns, and rate of time preferences: the case of Ethiopia environment for development. Discussion paper series August 2010 EfD DP 10–21. http://www.rff.org/files/sharepoint/WorkImages/Download/EfD-DP-10-21.pdf. Accessed 24 Sept 2018

Bobadoye AO, Ogara WO, Ouma GO, Onono JO (2016a) Pastoralist perceptions on climate change and variability in Kajiado in relation to meteorology evidence. AJIS 5(1): E-ISSN 2281-4612). https://doi.org/10.5901/ajis.2016.v5n1p37

Bobadoye AO, Ogara WO, Ouma GO, Onono JO (2016b) Assessing climate change adaptation strategies among rural Maasai pastoralist in Kenya. Am J Rural Dev 4(6):120–128. https://doi.org/10.12691/ajrd-4-6-1

Bryan E, Ringler C, Okoba B, Roncoli C, Silvestri S, Herrero M (2013) Adapting agriculture to climate change in Kenya: household strategies and determinants. J Environ Manag 114:26–35

Campbell BM, Thornton P, Zougmoré R, van Asten P, Lipper L (2014) Sustainable intensification: what is its role in climate smart agriculture? Curr Opin Environ Sustain 8:39–43

Danso G, Drechsel P, Wiafe-Antwi T, Gyiele L (2002) Income of farming systems around Kumasi, Ghana. Urban Agric Mag 7:5–6

Dhakal S, Sedhain G, Dhakal S (2016) Climate change impact and adaptation practices in agriculture: a case study of Rautahat District, Nepal. Climate 4(4):63

Elliott J, Deryng D, Müller M, Frieler K, Konzmann M, Gertlotter M, Flörkeg M, Wada Y, Best N, Eisner S, Fekete BM, Folberth D, Foster I, Gosling NS, Haddeland I, Khabarov K, Ludwig F, Masaki Y, Olin S, Rosenzweig C, Ruane CA, Satoh Y, Schmid E, Stacke T, Tang Q, Wisser D (2014) Constraints and potentials of future irrigation water availability on agricultural production under climate change. PNAS 111(9):3239–3244

Elum ZA, Modise DM, Marr A (2017) Farmer's perception of climate change and responsive strategies in three selected provinces of South Africa. Climate Risk Manage 16:246–257. https://doi.org/10.1016/j.crm.2016.11.001

FAO (2013) Resilient livelihoods: disaster and risk reduction. FAO, Rome

FAO (2016a) Climate change food security: risks and responses. FAO, Rome

FAO (2016b) FAO'S work on climate change. In: Paper presented at the United Nations climate change conference 2016. FAO, Rome

Fetters MD, Curry LA, Creswell JW (2013) Achieving integration in mixed methods designs – principles and practices. Health Serv Res 48(6 Pt II):2134–2156

FEWS NET (2013) Kenya food security brief. United States Agency for International Development (UASID) Famine Early Warning Systems Network (FEWS NET). December 2013. http://www.fews.net/sites/default/files/documents/reports/Kenya_FoodSecurity_In_Brief_2013_final_0.pdf. Accessed 24 Sept 2018

Fisher M, Abate T, Lunduka RW, Asnake W, Alemayehu Y, Madulu RB (2015) Drought tolerant maize for farmer adaptation to drought in sub-Saharan Africa: determinants of adoption in eastern and southern Africa. Clim Chang 133:283–299

Forsyth T, Evans N (2013) What is autonomous adaption? Resource scarcity and smallholder agency in Thailand. World Dev 43:56–66. https://doi.org/10.1016/j.worlddev.2012.11.010

Gatto MP, Cabella R, Gherardi M (2016) Climate change: the potential impact on occupational exposure to pesticides. Ann Ist Super Sanità 52(3):374–385

Giller KE, Tittonell P, Rufino MC, van Wijk M T, Zingore S, Mapfumo P, Adjei-Nsiah S, Herrero M, Chikowo R, Corbeels M, Rowe EC, Baijukya F, Mwijage A, Smith J, Yeboah E, van der Burg WJ, Sanogo OM, Misiko M, de Ridder N, Karanja S, Kaizzi C, K'Ungu J, Mwale M, Nwaga D, Pacini C, Vanlauwe B (2011) Communicating complexity: integrated assessment of trade-offs concerning soil fertility management within African farming systems to support innovation and development. Agric Syst 104(2):191–203

GIZ (2013) Report on rural development and adaptation to climate change: what do we know? Freiburg, Germany, March 2013. https://www.giz.de/fachexpertise/downloads/giz2013-en-adaptation-on-climate-change.pdf. Accessed 24 Sept 2018

GoK (2009) The national population census report. National Bureau of Statistics, Kenya. https://www.knbs.or.ke/publications/. Accessed 24 Sept 2018

GoK (2010) Agricultural sector development strategy, 2010–2020. Ministry of agriculture livestock development quarterly bulletin. Government Printers, Nairobi http://www.gafspfund.org/sites/gafspfund.org/files/Documents/Kenya_strategy.pdf. Accessed 24 Sept 2018

GoK (2013) National climate change action plan 2013–2017. Executive summary, Nairobi. https://cdkn.org/wp-content/uploads/2013/03/Kenya-National-Climate-Change-Action-Plan.pdf. Accessed 24 Sept 2018

GoK (2014a) Kajiado county integrated development plan 2013–2017. https://cog.go.ke/images/stories/CIDPs/Kajiado.pdf. Accessed 10 Mar 2017

GoK (2014b) Nakuru County integrated development plan 2013–2017. http://www.nakuru.go.ke/wp-content/uploads/2014/03/Nakuru-county-intergrated-dev-plan-2013-2017.pdf. Accessed 24 Sept 2018

GoK (2016) Kenya national adaptation plan 2015–2030. Enhanced climate resilience towards the attainment of Vision 2030 and beyond. http://www4.unfccc.int/nap/DocumentsNAP/Kenya_NAP_Final.pdf

Hsieh HF, Shannon SE (2005) Three approaches to qualitative content analysis. Qual Health Res 15(9):1277–1288

IPCC (2007) In: McCarthy JJ, Canziani OF, Leary NA, Dokken DJ, White KS (Eds.), Climate Change 2007: Impacts, Adaptation, and Vulnerability, Contribution of Working Group II to the Fourth Assessment Report of the Intergovernmental Panel on Climate Change. Cambridge University Press, Cambridge

IPCC (2014) Impacts, adaptation, and vulnerability. Part B: regional aspects. Contribution of working group II to the fifth assessment report of the intergovernmental panel on climate change. In: Barros VR, Field CB, Dokken DJ, Mastrandrea MD, Mach KJ, Bilir TE, Chatterjee M, Ebi KL, Estrada YO, Genova RC, Girma B, Kissel ES, Levy AN, MacCracken S, Mastrandrea PR, White LL (eds) Climate Change 2007. Cambridge University Press, Cambridge, UK/New York, 688 pp

Juroszek P, Tiedemann AV (2011) Potential strategies and future requirements for plant disease management under a changing climate. Plant Pathol 60:100–112

KNBS and SID (2013) Exploring Kenya's inequality: pooling apart or pooling together? (978–9966–029-19-5). Kenya National Bureau of Statistics (KNBS) and Society for International Development (SID), Nairobi

Kuang-Idba K, Kisyombe V, Dobardzic S (2016) Climate adaptation for rural livelihoods and agriculture in Malawi "Time to adapt: insights from the GEF's experience in adaptation to climate change". https://medium.com/@theGEF/climate-adaptation-for-rural-livelihoods-and-agriculture-in-malawi-393646d9fe06. Accessed 24 Sept 2018

Kuyiah JW, Kwena K, Esilaba AO, Nyamwaro, SO, Matere SJ, Ndegwa W, Wamae DK, Ruttoh R and Kibue AM (2013) Climate change adaptation and agricultural development. In: Paper presented at the transforming rural livelihoods in Africa: how can land and water management contribute to enhanced food security and address climate change adaptation and mitigation? 20–25 October 2013, Nakuru

Luedeling E, Kindt R, Huth NI, Koenig K (2014) Agroforestry systems in a changing climate – challenges in projecting future performance. Curr Opin Environ Sustain 6:1–7

Luoh JW, Begg CB, Symonds RC, Ledesma D, Yang R-Y (2014) Nutritional yield of African indigenous vegetables in water-deficient and water-sufficient conditions. Food Nutr Sci 05(09):812–822

Mbow C, Smith P, Skole D, Duguma L, Bustamante M (2014) Achieving mitigation and adaptation to climate change through sustainable agroforestry practices in Africa. Curr Opin Environ Sustain 6:8–14

Mohajan HK (2014) Food and nutrition scenario of Kenya. Am J Food Nutr 2(2):28–38

Morgan DL (1993) Qualitative content analysis: a guide to paths not taken. Qual Health Res 3:112

Mulinya C, Ang'awa F, Tonui K (2016) Constraints faced by small scale farmers in adapting to climate change in Kakamega County. IOSR J Human Soc Sci (IOSR-JHSS) 21(10):08–18

Mukasa AN, Simpasa AM, Salami AO (2017) Credit constraints and farm productivity: Micro-level evidence from smallholder farmers in Ethiopia. ADB Working Paper Series No. 247. Abidjan, Côte d'Ivoire: African Development Bank

Naik PS, Singh M, Ranjan JK (2017) Impact of climate change on vegetable production and adaptation measures. In: Minhas P, Rane J, Pasala R (eds) Abiotic stress management for resilient agriculture. Springer, Singapore

Nakuja T (2012) Water storage for dry season vegetable farming as an adaptation to climate change in the upper east region of Ghana. Afr J Agric Res 7(2):298–306. https://doi.org/10.5897/AJAR11.1601

Ndamani F, Watanabe T (2016) Determinants of farmers' adaptation to climate change: a micro level analysis in Ghana. Sci Agric 73(3):201–208

Ngo TQ (2016) Farmers adaptive measures to climate change induced natural shocks through past climate experiences in the Mekong River Delta, Vietnam. Afr J Agric Res 11(15):1361–1372

Ngugi IK, Gitau R, Nyoro JK (2006) In: IIED (ed) Access to high value markets by smallholder farmers of African indigenous vegetables in Kenya, Regoverning markets innovative practice series. IIED, London

Ogeto RM, Cheruiyot E, Mshenga P, Onyari CN (2013) Sorghum production for food security: a socio-economic analysis of sorghum production in Nakuru county. Kenya Afr J Agric Res 8(47):6055–6067

Price AJ, Norsworthy JK (2013) Cover crops for weed management in southern reduced-tillage vegetable cropping systems. Weed Technol 27(1):212–217

Price AJ, Balkcom KS, Culpepper SA, Kelton JA, Nichols RL, Schomberg H (2011) Glyphosate-resistant palmer amaranth: a threat to conservation tillage. J Soil Water Conserv 66(4):265–275

Reason CJC, Hachigonta S, Phaladi RF (2005) Interannual variability in rainy season characteristics over the Limpopo region of southern Africa. Int J Climatol 25(14):1835–1853

Reed-Jones NL, Marine SC, Everts KL, Micallef SA (2016) Effects of cover crop species and season on population dynamics of Escherichia coli and listeria innocua in soil. Appl Environ Microbiol 82(6):1767–1777

Rurinda J, van Wijk MT, Mapfumo P, Descheemaeker K, Supit I, Giller KE (2015) Climate change and maize yield in southern Africa: what can farm management do? Glob Chang Biol 21(12):4588–4601

Smit B, Skinner MW (2002) Adaptation options in agriculture to climate change: a typology. Mitig Adapt Strateg Glob Chang 7:85–114

Smit B, Wandel J (2006) Adaptation, adaptive capacity and vulnerability. Glob Environ Change 16 (3):282–292

Stöber S, Chepkoech W, Neubert S, Kurgat B, Bett H, Lotze-Campen H (2017) Adaptation pathways for African indigenous vegetables' value chains. In: Filho L et al (eds) Climate change adaptation in Africa, fostering resilience and capacity to adapt, Climate Change Management. Springer International, Cham

Sujakhu NM, Ranjitkar S, Niraula RR, Pokharel BK, Schmidt-Vogt D, Xu J (2016) Farmers' perceptions of and adaptations to changing climate in the Melamchi Valley of Nepal. Mt Res Dev 36(1):15–30

Taruvinga A, Visser M, Zhou L (2016) Barriers and opportunities to climate change adaptation in rural Africa: evidence from the eastern Cape Province of South Africa. Int J Develop Sustain 5(11):518–535

Teklewold H, Kassie M, Shiferaw B, Köhlin G (2013) Cropping system diversification, conservation tillage and modern seed adoption in Ethiopia: impacts on household income, agrochemical use and demand for labor. Ecol Econ 93:85–93

Uddin M, Bokelmann W, Entsminger J (2014) Factors affecting farmers' adaptation strategies to environmental degradation and climate change effects: a farm level study in Bangladesh. Climate 2(4):223–241. https://doi.org/10.3390/cli2040223

United Nations Children's Fund (UNICEF) (2013) Improving child nutrition. The achievable imperative for global progress UNICEF, New York, USA. https://www.unicef.org/publications/files/Nutrition_Report_final_lo_res_8_April.pdf. ISBN: 978–92–806-4686-3. Accessed 24 Sept 2018

Warner K (2010) Global environmental change and migration: governance challenges. Global environmental change, special issue focusing on resilience and governance. Glob Environ Chang 20:402–413

Warner K, Afifi T (2013) Where the rain falls: evidence from 8 countries on how vulnerable households use migration to manage the risk of rainfall variability and food insecurity. Climate Develop 6(1):1–17

Zwane M, Montmasson-Clair G (2016) Climate change adaptation and agriculture in South Africa: a policy assessment. In: Report compiled for WWF-SA. South Africa

Agricultural Impacts of the 2015/2016 Drought in Ethiopia Using High-Resolution Data Fusion Methodologies

44

James M. Warner and Michael L. Mann

Contents

Abstract

Integrating satellite data into GIS and traditional agricultural survey data remains relatively understudied, particularly in developing countries. This paper seeks to partially fill this gap by creating a 6 year, panel data that incorporates the relatively severe 2015/2016 Ethiopian drought. A variety of interesting results emerge, including the ability to better understand impacts on productivity from

J. M. Warner (✉)
International Food Policy Research Institute, Addis Ababa, Ethiopia
e-mail: j.warner@cgiar.org

M. L. Mann
Department of Geography, George Washington University, Washington, DC, USA
e-mail: mmann1123@gwu.edu

W. Leal Filho (ed.), *Handbook of Climate Change Resilience*,
https://doi.org/10.1007/978-3-319-93336-8_79

remote-sensed, GIS and farm management perspectives on maize productivity. In addition, the study explores potential impacts at initial stages in the crop cycle which better understand productivity and production changes earlier in the production season. The goal is to facilitate providing timelier and more accurate, information for potential policy action in the face of extreme weather events such as the recent Ethiopian drought.

Introduction

The ability to monitor and predict crop yields in developing countries is critical to the successful adaptation to changes in our climate. Increased temperatures and variability have already been linked to losses in major crop yields, such as maize and wheat, and subsequent impacts on crop prices. Although much effort has been placed on modeling the spatial distribution of these shifts, less effort has been placed on how yields vary across space and time. Advances in remote sensing provide avenues to monitor agricultural crop health at high spatial and temporal resolution, especially in large-scale more developed economies. However, our ability to monitor changes in plant productivity is still limited in the more complex environments common in many developing countries. In other words, the small fragmented parcel agricultural systems, typical in most parts of Africa, have remained a challenge for coarser remote sensing products.

Here the authors develop a nonspatial panel regression model to estimate crop output per hectare using both Stata and R quantitative software. This model integrates historical data on plant phenology statistics obtained from the Moderate Resolution Imaging Spectroradiometer (MODIS) satellite as well as the annual Ethiopian agricultural household questionnaire generally referred to as the Agricultural Sample Survey (AgSS). The objective is to develop a handful of metrics that can be used to better predict interannual variability in yields at a somewhat finer resolution (median area approximately 24 km^2).

Considering the relatively small scale of agriculture in this region, the authors utilized 250 m vegetation products from the MODIS satellites. Vegetation indices are obtained from two 16-day MODIS products (MOD13Q1, MYD13Q1) from the Aqua and Terra satellites. Due to the staggered nature of acquisition, these products are treated as partially overlapping windows representing 8-day periods. In particular, the authors examine the predictive power of the Normalized Difference Vegetation Index (NDVI) using panel econometric techniques. NDVI is sensitive to the amount of chlorophyll in any given pixel and is commonly used to estimate plant productivity and health in agricultural applications. After eliminating cloud cover and other flagged low-quality cells, the authors removed all nonagricultural cells through the use of a custom-made land use map generated for this project.

The authors combined the satellite data with both GIS and the AgSS data to provide a more comprehensive assessment of the impacts of a variety of different weather, geographic, and farm management strategies to better understand the effects of a major weather event, the 2015/2016 drought, on maize productivity in Ethiopia.

Research Question

By using data fusion techniques, the models developed here seek to explain extreme weather events and their effects on crop productivity in a more accurate and timelier fashion. (For our purposes, the authors define data fusion as combining quantitative data from a wide variety of relevant sources to augment traditional survey-based economic analysis.) The ultimate goal is to develop new methodologies that will help both researchers and policymakers understand the variety of variables that impact production, where in the crop production cycle these variables become predictive and be able to take appropriate action based on the development of these models. Specifically, for our purposes here, the goal is to develop a system that can reduce uncertainty of prediction concerning Ethiopian staple crop yields and do so earlier than other current techniques being employed in Ethiopia. More specifically, predicting major crop production and yields in a timelier manner will help reduce the gap between knowledge of a potential shock and actionable policy. (Actionable policy can include such things as price floors, in the event of bumper crop yields, or distribution of food aid in the case of production shortfalls.) In addition, by creating a more localized model, better estimates of the extremely localized effects of weather events can be performed.

Recent climatic variations across the Sahel provide an opportunity to estimate the impacts of variability on rain-fed smallholder farmers in Ethiopia. Estimates of the national impact of the 2015–2016 drought in Ethiopia ranged from −18.6% according to a USAID rapid assessment performed during the crop season (AKLDP (Agriculture Knowledge, Learning, Documentation and Policy Project) 2016) to −1.3% according to government-collected statistics also collected during and after harvest (The Federal Democratic Republic of Ethiopia – Central Statistical Agency 2015). Of importance here is the trade-off between time and relative accuracy. Of course, as it should be noted, both the rapid assessment and AgSS were taken before crop harvesting, and sales were finished and did not have the benefit of hindsight if they were performed later in the crop cycle. The question the authors pose here is whether improved accuracy and, equally importantly, timeliness can be improved using data fusion methodologies.

As the authors have stressed elsewhere (Mann and Warner 2017), we do not believe that any one data methodology (e.g., survey, remote sensing, etc.) is superior to another, and many kinds of relevant data are important for understanding crop production and productivity. Relying exclusively on satellite data has many weaknesses including the exclusion of crop management techniques (farmer demographics as well as application of improved seed, fertilizer, etc.) and weaknesses with the data itself (cloud cover, identifying crop fields, lack of granular study, etc.). However, survey data has shortfalls as well and includes cost considerations, spatial coverage, the need to deploy enumerators, data processing, and the dissemination of findings. The authors argue for a synergy of remote-sensed variables (e.g., NDVI), along with other near real-time drivers of productivity such as water availability and evapotranspiration rates, which can augment current yield assessments in a more accurate and timely fashion. Overall, the authors argue for integration of data from a variety of sources to better model the effects of weather, climate, policy, and farm management on agricultural productivity.

The General Depiction of Ethiopia's 2015–2016 Drought

The 2015–2016 drought, typically ascribed to the effects of El Nino, has been depicted by several international and Ethiopian agencies, and a typical representation is provided in Fig. 1. Broadly speaking, the hardest hit areas are located in the northeast and southeast part of the country. Or, as depicted in an OCHA situational report issued in October of 2016, *Southern and southeastern parts of the country are being highly affected by poor performance of the 2016 seasonal rain which normally starts in late September* (Disaster Risk Management Technical Working Group (DRMTWG) 2016). Visually, the areas in orange were found to have a failed Belg

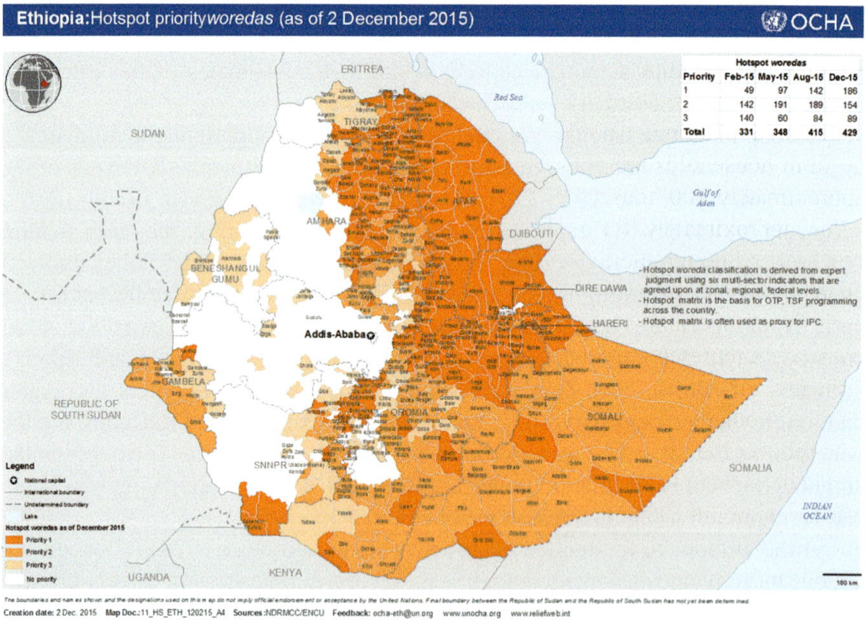

Fig. 1 OCHA's drought affected areas of the 2015/2016 drought in Ethiopia. 1. Map provided courtesy of the UN Office for the Coordination of Humanitarian Affairs; 2. The boundaries and names shown and the designations used on this map do not imply official endorsement or acceptance by the United Nations (OCHA 2015, accessed August 16, 2017 (https://reliefweb.int/map/ethiopia/ethiopia-hotspot-priority-woredas-2-december-2015))

(spring) and Kiremt (summer) rains in 2015. For our purposes here, the authors focus on Kiremt, also referred to as the Meher cropping season, as it provides the vast majority of production, and spatial planting coverage, for the country. The next section provides a detailed review of the variety of data sources used for our data fusion model.

Data and Methods

Survey Data: Agricultural Sample Survey Data (2010–2016)

Survey data was obtained from Ethiopia's Central Statistical Agency's (CSA) Agricultural Sample Survey (AgSS) and was chosen for its annual collection, spatial coverage, and unique sampling structure through these six crop years. The AgSS is an annually collected government-administered large-scale survey tasked with measuring agricultural production in Ethiopia at the zonal level. (Beyond the nation, Ethiopia has four official levels of administrative areas. These include, in order of geographic size, regions, zones, woredas, and kebeles. There are approximately 12 regions, 88 zones, 690 woredas, and 15,000 kebeles. For our purposes, the authors use the sub-kebele, not recognized as a geographic region for administrative purposes, but commonly used for statistical sampling and are roughly based on population. There are over 75,000 sub-kebeles in the country.) The survey interviews approximately 45,000 farmers per year on a range of farm management questions covering some basic demographics of the household as well as a range of questions concerning planting, harvesting, and selling at the plot level. Typically, about 20 farm households are randomly sampled from small local village-level areas of approximately 200 households (the sub-kebele level). On average, based on population, approximately five of these sample areas exist at the kebele-level administration area. Although the figures vary widely, sub-kebeles are meant to approximate about 200 households per unit, and the average kebele has around 1,000 households each. From this sampling frame, a random selection of about 2,200 sub-kebeles are chosen as a representative subsample for zonal-level production projections of major agricultural production areas. Population weights are then applied to project agricultural production at the zonal level. For this study, the authors construct a panel data set over the six crop seasons identified because approximately 75% of the same sub-kebeles were sampled by CSA over the six Meher crop seasons of interest. This effectively builds a base of five relatively favorable crop seasons, from 2010 to 2014, and allows for the drought effects of 2015–2016 as a study in contrast in terms of this significant weather event. The sample is with replacement and therefore representative of the sub-kebele but not a true panel because the same farmers were not chosen. In addition, no sampling weights are applied because the authors have omitted approximately 25% of the sample and are generally most interested in the localized effects. However, when the authors do aggregate for zonal analysis, the authors take simple farmer averages which are an acknowledged shortcoming and not a true prediction of actual zonal figures typically projected by CSA.

While this study collected data on the five principal Ethiopian field crops (teff, wheat, maize, barley, sorghum), because of their predominant importance in value terms for Ethiopia as well as their geographically widespread adoption, the analysis should therefore be placed within this context. This study covers the drought effects of only these five crops and not all agricultural production. In other words, when the authors outline percentages of crop losses, the authors are referring to losses of only these five crops and not total output in the country. The authors have no reason to believe that these are not generally representative of all crop production, in terms of the drought, but a more detailed analysis would be needed to make this assertion. Finally, our models, presented in section "Preliminary Model," reflect maize as it serves as a principal food crop for much of Ethiopia.

The principal unit of analysis is the sub-kebele and all relevant CSA data is aggregated to this level. The survey data represents an amalgamation of all 20 households as a single representative farmer, the authors refer to as a "super-farmer." This was done for a variety of reasons including the fragmented plot farming system common in Ethiopia as well as CSA data collection methodology. More specifically, CSA data collection methodology relies on crop cuts to estimate productivity at the local level. Depending on the actual number of farmers growing the specific crop, CSA collects up to five different individual farmer crop cuts, averages the yields, and projects this figure onto all plot areas for that crop in the sub-kebele. In this way, individual plot-level yields are not able to be determined, and, consequently, household level estimates are not possible. Additionally, while the level of the pixel used for the remote sensing data is relatively granular (6,250 m^2), cost and logistic considerations make plot-level estimations prohibitive in this research. Therefore, the data from 20 farmers is aggregated to a single super-farmer resulting in 1,780 observations sampled from the sub-kebele area (approximately 23,900,000 m^2) over a 6-year period. The total number of observations collected is about 10,680 but will vary in sample size depending on the crop choice (i.e., some areas do not grow all five designated crops). A breakdown of each super-farmer is done with aggregated non-crop-related information combined with the specificity of the individual crop chosen. More precisely, several variables are consistent across all five crop choices (elevation, proportional numbers of male and female household heads, total land area, etc.), and others are crop specific (use of fertilizer, seed, etc.). Each crop year was checked for consistency, cleaned, and aggregated to the sub-kebele level. The 6 years were then aggregated and merged with other data to develop the panel data set. Importantly, for our analysis, the authors chose specific variables to reflect farmer management decisions and other actions affecting yields.

Remotely Sensed Data

For remotely sensed data, the Normalized Difference Vegetation Index (NDVI) data from the MODIS satellite is used. *NDVI* is sensitive to the amount of chlorophyll in any given pixel and is commonly used to estimate plant productivity and health in agricultural applications (Mann and Warner 2017). NDVI measurements are collected every 16 days from the MYD13Q1 MOD13Q1 data products. Collectively, these data cover each cell every 8 days. After removal of snow, cloud, and other flagged low-quality cells, the authors remove all nonagricultural cells using a custom classification of land use for Ethiopia at a 250 m resolution (see Annex A for specific classification methodology). After cleaning, the authors are left with measurements of NDVI for each 8-day period for all agricultural pixels in Ethiopia over the 2010–2016 study period.

In order to estimate the effects of precipitation (PPT) on our crop production models, the Climate Hazards Group InfraRed Precipitation Station (CHIRPS) data was included (Funk et al. 2015). Data is collected as total precipitation by dekad. A dekad represents three periods each month, the first two periods being 10 days each and the third being any remaining days of that month.

Hydrological availability of water and available energy for plant growth have been shown in previous research to be important factors in crop models and are therefore included here. For this reason, the authors include monthly estimates of potential evaporation (PET) and actual evapotranspiration anomaly (*ETA*). Actual evapotranspiration (AET) is the sum of transpiration of water through plants plus evaporation from soils and water surfaces. AET is a correlated with vascular plant productivity and correlates with the biomass accumulation and regrowth (Major 1967; Mann et al. 2016). The ETA variable used in this study is the current AET compared to the 2003–2013 mean. ETA therefore is a proxy for drought as low current values of AET would correspond to reductions either in available energy to move water or water itself. Potential evapotranspiration (PET) is an estimate of the total amount of water that could be moved through the system if water was not a limiting factor. As such it is an ideal indicator of available (or excessive) energy in the form of light and heat, but also includes the impact of wind speed, pressure, and humidity among others. The standard deviation of PET reflects variability at the local level of these weather conditions. Both ETA and PET are available for download through FEWSNET (Monthly Evapotranspiration (ETA) Anomaly, Daily Potential Evapotranspiration (PET)).

Summarizing Spatial Data over Time

Properties of the time signature of NDVI, PPT, PET, and ETA must be obtained to characterize and identify important components correlated with crop yields. These summary statistics take two forms: first, growing season statistics, spanning the estimated planting date of designated crops until harvest date and second annual season statistics, spanning the Meher growing season. (The Meher crop season is the dominant crop season and the focus of this study. A second, relatively minor crop season (Belg) is not included because of its relatively limited spatial adoption as well as low production compared to national output of cereal crops during Meher.) For both annual and growing season statistics, three classes of statistics are estimated. First, basic statistics (e.g., mean, median), second integrated statistics (e.g., total area under the curve and divided areas under both the growing (up to maximum NDVI point) and grain-filling (from maximum point to harvesting dates)), and third comparisons to the norm (e.g., difference of the current year from the 90th percentile of the historical record). For our purposes here, the authors chose to use the first two types of variables in the model presented in this paper.

The mean of all pixels within the administrative boundary is used to represent that area. In this study, sub-kebele mean values of NDVI, PPT, PET, and ETA are used to represent estimated effects of agricultural productivity for each period.

Annual and Meher Growing Season Metrics

In order to determine the length of the crop season at the local level, estimated dates of planting and harvest for the Meher growing season are estimated using an iterative search algorithm finding the date of the global minimum NDVI value nearest to the a priori estimated date for both planting and harvest. These dates are then adjusted for each administrative area to match the estimated planting and harvest patterns for that particular area.

Summary statics including minimum, maximum, mean, and standard deviation were estimated for each sub-kebele area. The authors also calculate a series of integrated metrics to evaluate the compounding influence of good or poor growing conditions. These include the total area under the NDVI curve (total AUC), the increasing portion of the NDVI curve (from estimated planting date to growing season maximum – referred to as leading), and the AUC of the declining portion of the NDVI curve (from growing season maximum to estimated harvest date – referred to as trailing). The timing of events (i.e., planting, maximum NDVI, and harvest) is also determined for estimation use. For instance, precipitation PPT can be summed across the entire growing season or for the first or second half of the growing season (after the maximum value). The same can be done with other variables like PET and ETA, to look at the impact of available water and energy for important parts of the phenological cycle.

The current NDVI conditions are also compared to those of "ideal" years by comparing them to the estimated 90th percentile of historic observations. As crop production was generally considered good over the 2010–2015 period, this can form a base to analyze the exceptional 2015–2016 weather shock in particular parts of the country. Most functions described in this section can also be used to calculate annual metrics. Annual metrics, which include seasons preceding the main growing season, may be of importance because they are indicative of plant health during early phases of plant growth (when the NDVI signal is minimal) and may also be an indication of water availability for plants leading up to planting. As such, annual metrics can be an indicator of the effects of longer-term phenomenon such as drought.

Table 1 Model variable names and descriptions

Name	Description
Survey questionnaire data	
Maize_OPH	Maize crop productivity (quintal/hectare)
Improved seed	Crop area using improved seed (%)
Fertilizer	Amount of chemical fertilizer applied (quintal/hectare)
Extension area	Maize area receiving extension services (%)
Irrigation area	Maize area irrigated (%)
Erosion prevention	Land area where erosion prevention implemented (%)
Female household heads	Number of female heads of household (absolute number)
Area of reported damage	Farmer reported damage to maize area (%)
GIS data	
Area of sub-kebele	Land area of the sub-kebele (sq. kms.)
Elevation	Average elevation of sub-kebele (meters)
Distance to pop. 50 k	Euclidean distance from the centroid of the sub-kebele to the nearest population center of at least 50,000 people (kms.)
Remote sensed data	
NDVI max	Difference between NDVI maximum and NDVI minimum (%)
NDVI_leading	Area under the curve of NDVI for the ascending part of the curve during the Meher season
NDVI_trailing	Area under the curve of NDVI for the decreasing part of the curve during the Meher season
Potential evapotranspiration	Estimates the total amount of water that could be moved through the system if water was not a limiting factor
Precipitation_leading	Amount of precipitation under the curve of NDVI for the ascending part of the curve during the Meher season
Precipitation_trailing	Amount of precipitation under the curve of NDVI for the descending part of the curve during the Meher season

Variable Definitions

Table 1 provides the name and basic definitions of the remote sensed data specifically determined for the models presented below, and Table 2 provides a statistical summary of their values.

Results

Statistical Review of the Data

This section provides a brief overview of AgSS farmer-reported damage before the drought and then contrasts this baseline data to the drought experience in 2015/2016. The data reveal the relative spatial concentration of the drought with traditionally lower productivity areas particularly adversely affected. Overall, this means that

Table 2 Variable summary statistics

Variable	Mean	Std. dev.	Min	Max
Maize_OPH (Qtl./H)	25.8	12.6	0.3	64.3
Improved seed (%)	20.0	31.0	0	100
Extension area (%)	28.9	35.2	0	100
Irrigation area (%)	2.3	11.6	0	100
Erosion prevention (%)	20.5	33.1	0	100
Female heads (number)	3.9	2.3	0	16
Reported damage – maize (%)	21.0	22.8	0	100
Area of sub-kebele (sq. km.)	23.9	15.8	0.7	287.4
Elevation	1,889.0	441.9	341.3	3545.2
Distance to pop. 50 k (km.)	68.0	46.8	2.4	306.7
Net NDVI (max.-min.) (%)	1.7	0.7	0.3	24.7
Area under NDVI – leading	102,863	40,455	4,700	306,656
Area under NDVI – trailing	109,819	45,174	11,634	319,014
Potential evapotranspiration (std. dev.)	54.4	15.8	26.8	118.3
Precipitation – leading	1,033.2	425.5	1.7	2659.6
Precipitation – trailing	400.0	351.9	9.6	2316.8

These statistics represent only maize growing areas ($n = 8977$)

while the effects where significant in some areas, national production figures were not as affected because higher productive field crop areas experienced normal or above normal rain. In addition, the authors found large intrazonal effects of the drought which suggests the need for higher-resolution studies.

Figure 2 depicts the AgSS annual average baseline values (2010–2014 Meher planting) of farmer self-reported amount of cropland lost, relative to the five represented crops planted, due to a variety of natural and human causes aggregated to the zonal level. (The AgSS first asked a filtering question if the farmer experienced loss of cropland on the plot during the season. A follow-up question asks to identify the source of the damage. While the choices were slightly expanded over the 6 years, the 2015 survey listed a total of 16 categories. Principal categories included crop disease, frost, too little rain, too much rain, hail, weeds, depletion of soil, and others. For the purposes of the pie charts in Figs. 2 and 3, responses were aggregated into five categories that include pests, management, weather related (drought), weather related (not drought), and an "other" category.) Potential values range from 0%, where no land was damaged, to 100% or total cereal crop loss. A simple, non-weighted, zonal average of 12.4% was reported over the five baseline years which indicates substantial losses due to weather, pests, and management techniques. During the base years, the minimum zonal average was 4.6%, and a maximum of 26% crop loss was reported. Interestingly, values tended to "cluster" in certain area and suggest areas where higher, and relatively lower, losses can be found. The representative pie charts indicate the source of zonal damage with yellow being weather-related damage, not associated with too little rain (e.g., hail, flooding, etc.), identified as the principal determinant of crop damage.

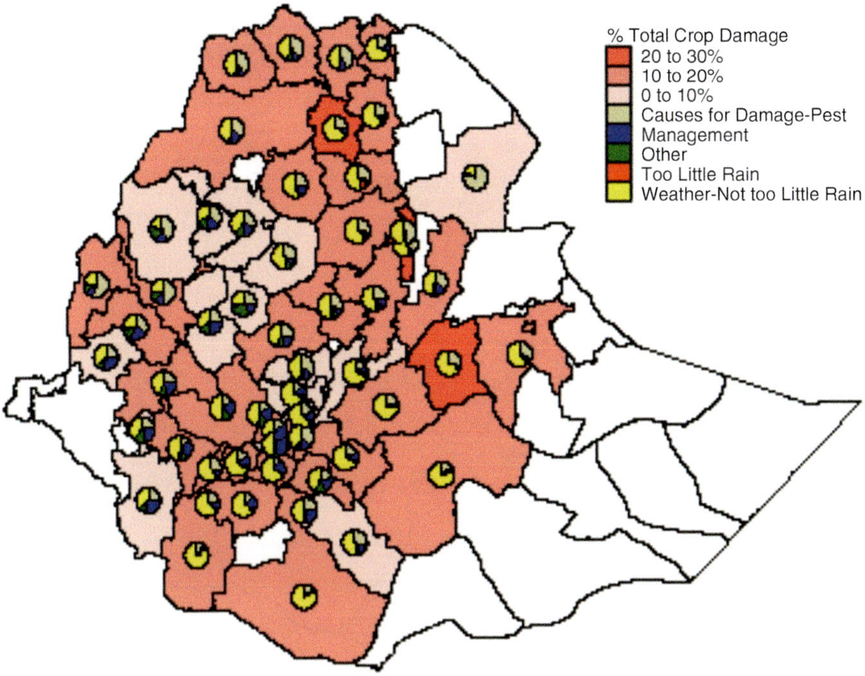

Fig. 2 Total crop damage: average from 2010 to 2014

Figure 3 graphically depicts the unique characteristics of the 2015–2016 drought using the same methodology. Crop loss is significantly higher and principally attributed to too little rain and is depicted by red in the pie graphs. The simple zonal average was 24.8% with a high of 76.4% and a low of 2.5%. The band of higher crop losses is generally in the north and southeastern parts of the country and is consistent with OCHA's map presented earlier. The southwestern parts of the country had relatively little crop loss and, causes of crop loss were generally not weather related.

In Fig. 4 the authors contrast the relative change of 2015 with the baseline figures. The concentrated effects, at the zonal level, demonstrate that northeastern and eastern parts of the country disproportionately saw drought conditions, while other parts of the country had lower than average crop losses. Contrasting 2015, to the previous 5 years' results, demonstrates a significant shock in certain areas but relatively good weather in many areas of the country and confirms a very uneven effect of the 2015 drought in Ethiopia. While the net increase in zonal damage was about 12.4%, individual zonal averages ranged from a high of an additional 57.9% damage to an actual reduction of damage of −2.8% as contrasted to the 2010/2014 baseline average. It is important to emphasize that simple zonal averages do not reflect national production losses as zones do not have equal levels of production of the five designated crops. As this research is not weighted for administrative area

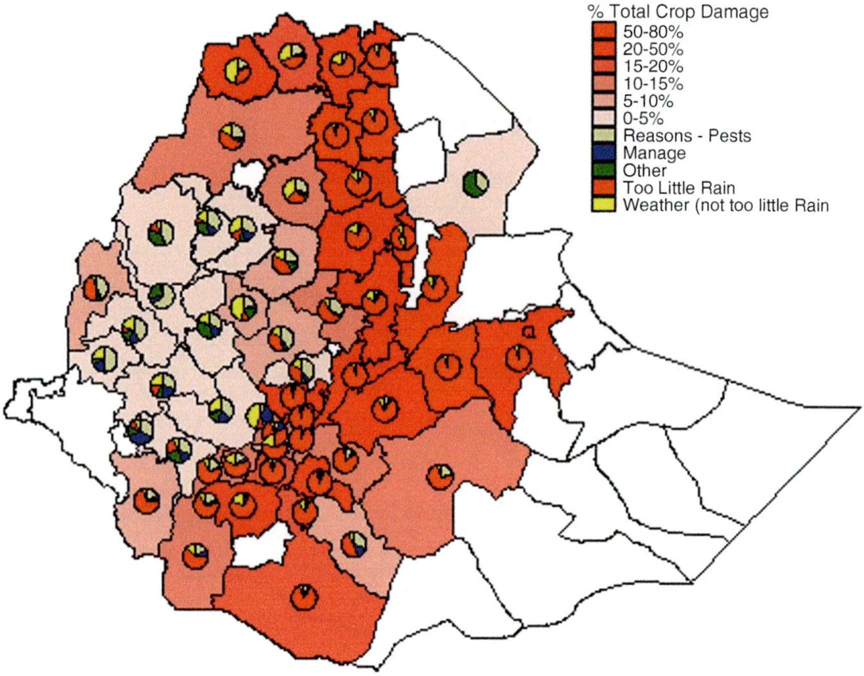

Fig. 3 Total crop damage: 2015

production figures (i.e., zonal, regional, or national figures), and only simple damage averages are taken, the relative production crop loss figures are not explicitly determined by this analysis. However, based on expert opinion, and some basic analysis performed next, the drought disproportionately affected lower production areas for our five cereal crops.

In order to understand the potential drought impacts on the national economy, estimated effects on production were undertaken. As would be expected, production of cereal crops is not uniformly distributed throughout the country. The following analysis attempts to provide a basic review of the drought from a production perspective and finds that the greatest impacts of the drought were in the relatively less productive areas. Figure 5 depicts the drought effects relative to overall cereal production in Ethiopia. Using AgSS data, cereal production is projected at the zonal level, and the figures are indexed to represent the percent of total production in the 2014/2015 Meher crop year. (Aggregation of all cereal crops was performed by weight, and this information was obtained from AgSS annual bulletins.) For example, 1.4 indicates that this particular zone contributed 1.4% of total cereal crop production in the 2014/2015 crop year. The choropleth mapping represents the farmer-reported relative percentage crop loss or crop gain in 2015/2016 relative to their contribution to total production in 2014/2015. Consistent with other works presented here, areas in the southwest did not suffer in terms of weather-related

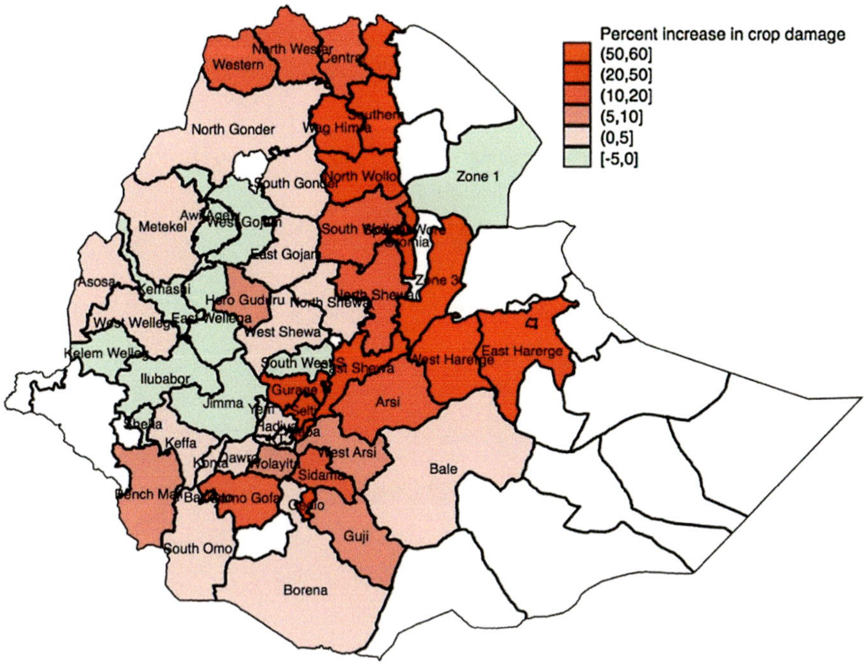

Fig. 4 Net difference in total crop damage (2015– (2010–2014 Avg.))

impacts and are represented by green to represent crop production gains relative to the 2014/2015 base year. This graph represents, along with the following table, that the relative production contributions are higher in areas that were not impacted by the extreme effects of the drought. More specifically, Table 3 creates five categories of zones based on relative crop gain or loss. Of the 54 zones represented, almost half (25) experienced relative increases in their national production contributions. Some locations experienced significant gains of 10% or more. (It should be noted that these increases could have come from crop switching, increased input use, or other non-weather-related events. The important thing to remember is that there were gains in production of these five principal crops and not significant losses as would be expected during negative weather shocks.) On the other hand, about a third of the zones (20/54) had losses of up to 10%. Of critical importance, and significantly affecting the overall averages, were the top 9 worst affected zones which contributed an average of less than 0.7% each to national cereal production. This indicates that lower cereal production zones suffered larger crop losses than more productive areas and national total production effects were not as severe. However, while zonal disaggregation indicates spatial variation, moving to finer resolution reveals even greater intra-area variation. The next section reviews the drought effects at, what the authors believe to be, the more accurate sub-kebele unit of analysis. In other words, using the data at the smaller sub-kebele level (on average about 24 km^2 area) reveals

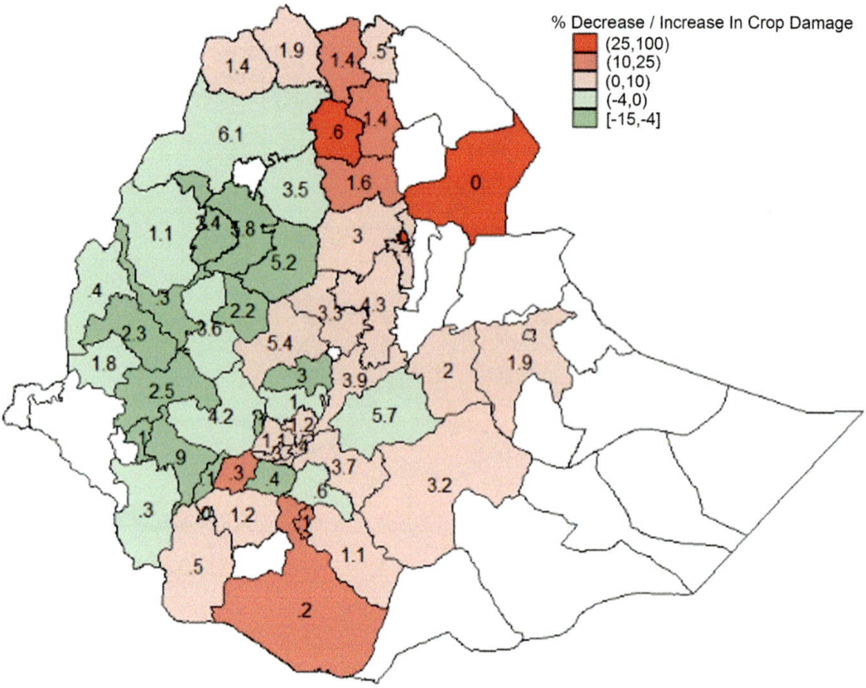

Fig. 5 Relative production effects of the drought: contrasted to 2014

Table 3 Zonal percentage contribution of cereal crops

Range of crop gain/ loss	Sample size (n)	Mean (% of 2015 prod.)	Std. dev.	Min.	Max.
Gain: 4–14%	13	1.95	1.9	0.1	5.8
Gain: 0–4%	12	2.36	2.2	0	6.1
Loss: 0–10%	20	2.04	1.5	0.3	5.4
Loss: 10–25%	4	0.80	0.7	0.1	1.4
Loss: 25–100%	5	0.48	0.7	0	1.6
Total	54	1.85	1.6	0	6.1

significant variation of impacts even in areas most affected by the drought and suggests the need for higher-resolution analysis for potential policy interventions.

This section reviewed drought impacts at the zonal level, but, for our purposes, the authors believe that the strength of our analysis rests at the local level where geographically specific data can better capture the effects of weather. For example, anecdotes concerning the effects of the 2015/2016 drought reveal that crop failure can occur in one village with relative abundant harvests in neighboring villages. The next section performs analysis at a more localized level and reveals significant intrazonal variation.

Statistical Analysis at the Sub-Kebele Level

Figure 6 A demonstrates the distribution of crop area damage for both the baseline (average 2010–2014) and drought period (2015) at the sub-kebele level. At the sub-kebele level, the overall average reported for 2010–2014 is 12.8% with an almost 10% increase in the 2015 drought year (22.9%). Clearly the baseline, represented by the blue distribution, had significantly lower overall damage with less than 5% of sub-kebeles reporting an average of 20% crop loss or higher. In 2015, about 25% of all sub-kebeles indicated significant crop loss (i.e., greater than 20%). Interestingly, about 40% of 2015's sub-kebeles reported relatively less damage than the distribution in the baseline. However, it should be remembered that these figures are not direct comparisons at the sub-kebele level but ranked overall distributions of damages. The two distributions do reflect, however, that many of the sampled sub-kebeles did relatively better in 2015 as compared to the 2010–2014 averaged baseline. Figure 6b depicts that the major source of the crop area loss was too little rain and is consistent with the zonal depiction in Fig. 3. Almost half of all zones indicated at least some crop loss due to a shortage of rain in 2015. This stands in sharp contrast to the base year average where almost no sub-kebeles reported losses from too little rain. Put another way, the average impact of too little rain, at the sub-kebele level, was about 0.26% for the baseline and about 16% for 2015. These figures represent an approximate 6,200% increase in drought-reported impact. However, this should be tempered with the actual overall increase in crop damage which is a much more modest increase of 79%. While significant, the impacts also have to be contextualized into where the effects where most severe. Additional research indicates that the level of variation is relatively significant even at the zonal level. The authors explore intrazonal effects of the drought in the following section.

Intrazonal Effects of the Drought

Most African agriculture is critically dependent on rainfall for production, and rainfall differentials can be highly localized with no respect for administrative boundaries. This suggests varied productivity, even at relatively smaller administrative areas (for Ethiopia, this can mean at the zonal, woreda, or even kebele level). Traditional methods of measuring rainfall were designated at meteorological stations. However, typical meteorological information suffers from sparse geographic coverage which makes local rainfall projections relatively imprecise. By using remote sensed data, coupled with sub-kebele survey data, the authors believe that the authors have much more accurately captured localized weather and farm management effects.

Exploring the data, from this perspective, reveals that many sub-kebeles themselves, within the zone, were not affected by crop loss with only a select few suffering more catastrophic results. This is important for several reasons and, most notably, complicates potential policy response to local areas. While the OCHA map and our subsequent zonal level maps indicate general areas of drought, it is important to realize that there is significant reported variation in cereal crop loss. Maps

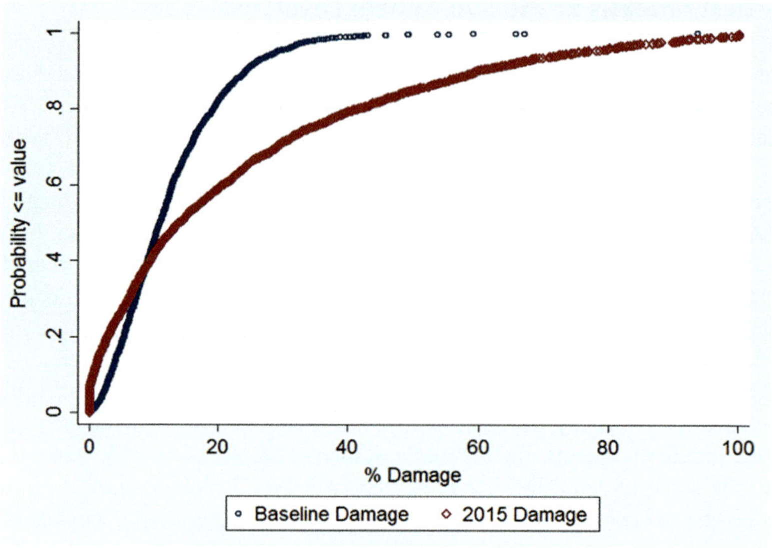

Graph A - Cumulative Total Damage

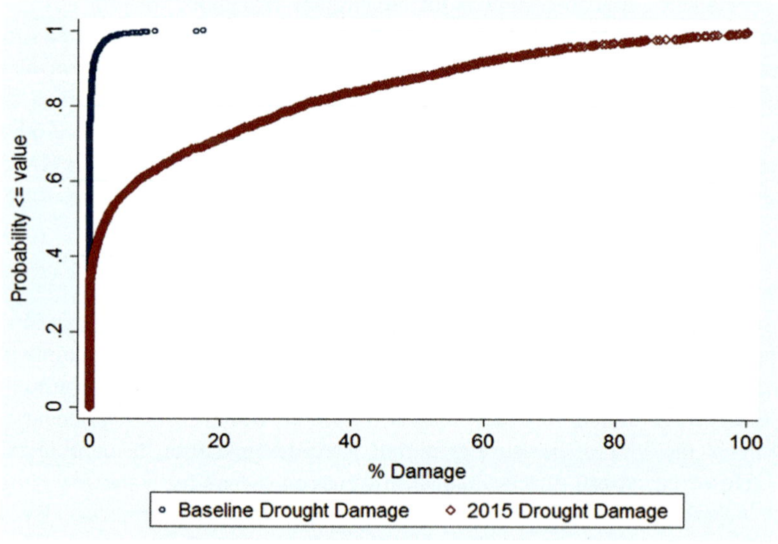

Graph B - Cumulative Drought Damage

Fig. 6 Cumulative total damage curves

depicted by emergency response institutions (like OCHA) may paint a broad brush when it comes to the distribution of food aid, but this research demonstrates that effects are not experienced equally across local areas. In addition, early rapid assessment of production, such as the USAID project (AKLDP (Agriculture

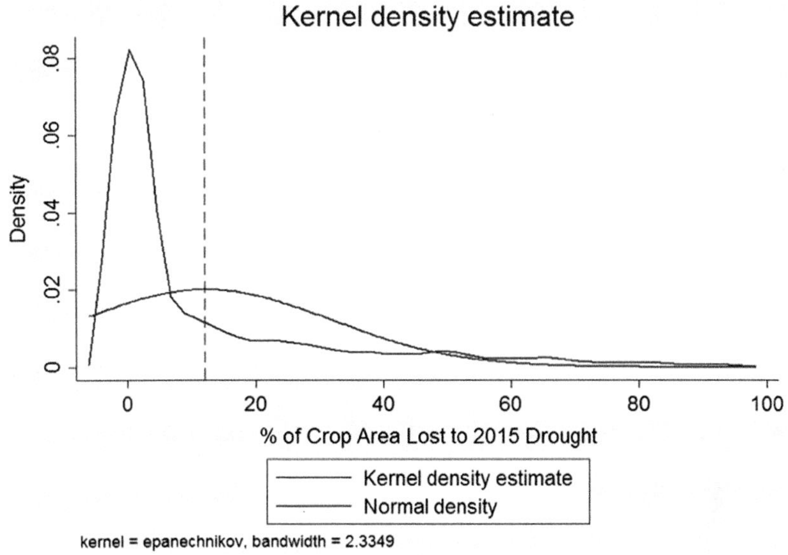

Fig. 7 Total sub-kebele percentage area crop loss

Knowledge, Learning, Documentation and Policy Project) 2016), which projected broad impacts based on local interviews, could be assisted by remote sensing data that can better capture local weather events and therefore production and productivity estimates. In this way, care can be given to avoid supplying food aid in local areas with relative robust harvests. The goal would be to target those that actually need aid and avoid those who have had relatively normal production. Obviously, providing aid inappropriately could have deleterious effects on both crop planting and market activity.

The net relative impact of reported crop loss due to drought in 2015, compared to the baseline, was about 12% over the entire sample. However, the distribution of sub-kebele AgSS-reported cropland loss reveals relatively non-normal distribution that skews far left. In other words, while the mean loss was 12%, the median reported loss was only 1.5%. Figure 7 visually depicts this distribution and indicates that a large majority of sub-kebeles suffered little effect, and only a select few had large impacts. Importantly, while there was some obvious concentration of affected sub-kebeles in particular zones, the distribution at the intrazonal level also bifurcates among those affected and those that were not. For example, one of the more drought affected zones in Tigray is the South Tigray zone which exhibits significant intrazonal variation.

In the South Tigray zone, the average effect of the drought, taken across the sample of 31 sub-kebeles, is almost 41% and indicates an area where considerable crop loss was reported. However, the data bifurcates around low values, with little crop loss, and higher values where significant crop loss was experienced. Figure 8 reveals a relative "missing middle" where few values approximate the simple

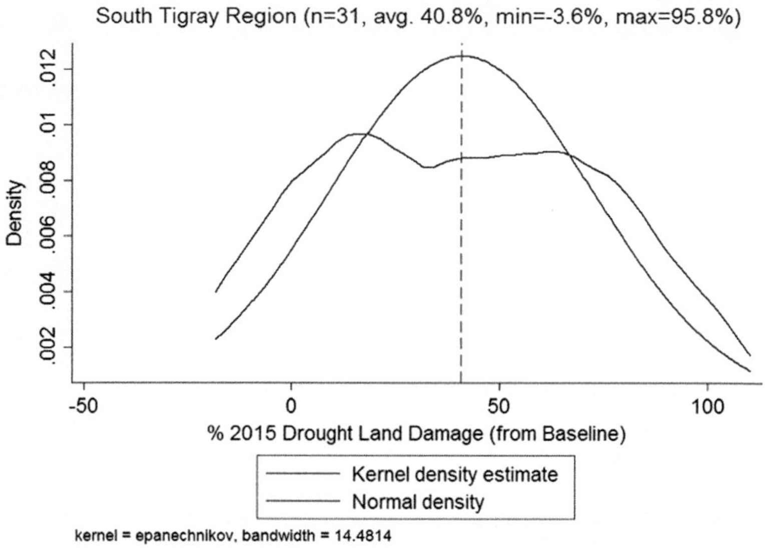

Fig. 8 Crop loss distribution in South Tigray region

average of 40% and indicates a non-normal distribution within the zonal administrative area. Individually reported crop losses in South Tigray zone range from a reported reduction of crop loss due to too little rain (−3.6%) to nearly complete crop loss in another sampled sub-kebele (96%). This analysis reveals the extremely localized effects of drought and suggests that augmenting survey data with remote sensed data may help in providing a more complete picture of drought effects. Relying on coarser levels of aggregation may obfuscate local effects and create problems with policy response. Overall, this suggests that the sub-kebele may be a better unit of analysis and augmenting survey data with remote sensed data could improve both in and out of sample projections. Of course, there have been many expressed doubts concerning the effectiveness of remote sensed data, and the authors perform a brief data verification exercise below to build confidence in both sets of data before outlining the econometric model.

Data Verification: Does Remote Sensing Data Confirm Farmer-Reported Crop Loss?

As with any data, increased confidence is obtained by triangulating the data both within a data set, using several variables that measure similar attributes, and between data sets to measure consistency and reliability. In this case, by contrasting variables that measure somewhat similar effects of weather in both the survey and the remote sensed data, confidence would be increased in the confidence of both data sets. For purposes here, the authors estimated the effects of the drought using both farmer-reported crop loss in the AgSS with the relative area under the NDVI and performed simple correlations to see if they were similar. The results were highly significant and indicate that these vastly different data collection methods roughly approximate similar predictive measures. This is critical for in-survey sampling as well as out-of-sample predictions and approximates ground-truthing for augmenting drought effect predictions. In other words, crop loss could potentially be estimated at the sub-kebele level for the entire country using remote sensing and accepting a degree of error.

Both the survey variable and the remote sensed NDVI were constructed as differences, contrasting the 2015 drought effects to the aggregated baseline values (2010–2014). In the case of the AgSS data, the relative change of reported effects of too little rain in 2015 was subtracted from the averaged baseline responses from the 2010–2014 period at the sub-kebele level. This created a percentage variable of reported cropland loss due to lack of rain relative to the baseline. In the case of the remote sensing variable, the authors hypothesized that a reduction in rain would reduce the area under the NDVI curve. This percentage measures, in the case of drought affected areas, the relative shortfall of chlorophyll detected by satellites at the sub-kebele level. The smaller the area, the relative less chlorophyll, and the greater the effects of the drought. A strong negative correlation was hypothesized because as the percentage of NDVI increases, indicating greater plant growth, a reduced amount of farmer cropland loss should be reported. This hypothesis is visually depicted in Fig. 9. In fact, the results strongly collaborated this proposition and are discussed below.

A simple correlation value of −0.53 was obtained between the two variables (NDVI and farmer-reported crop loss) and is depicted in Fig. 10. (Given their relative extreme values, two observations were dropped in this analysis. While they did

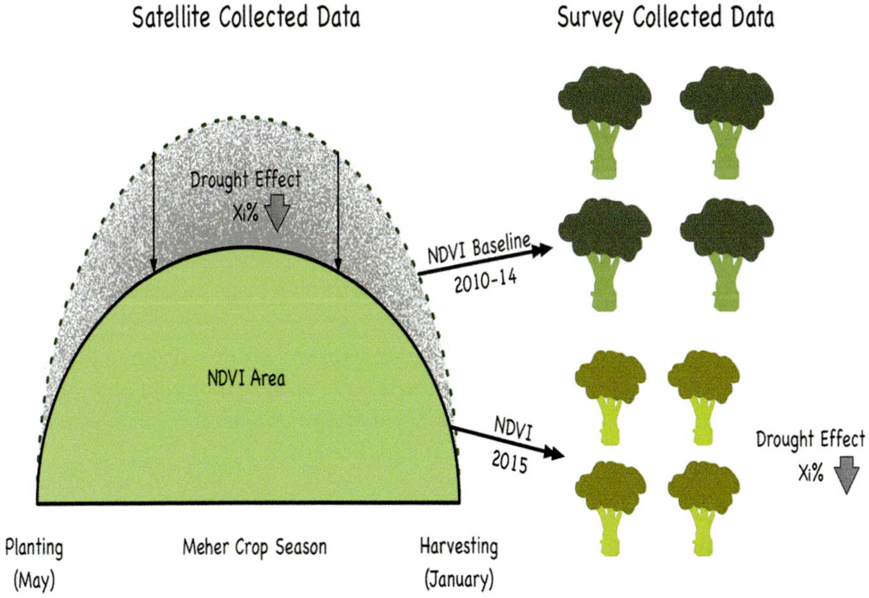

Fig. 9 Data verification – contrasting satellite and survey collected data

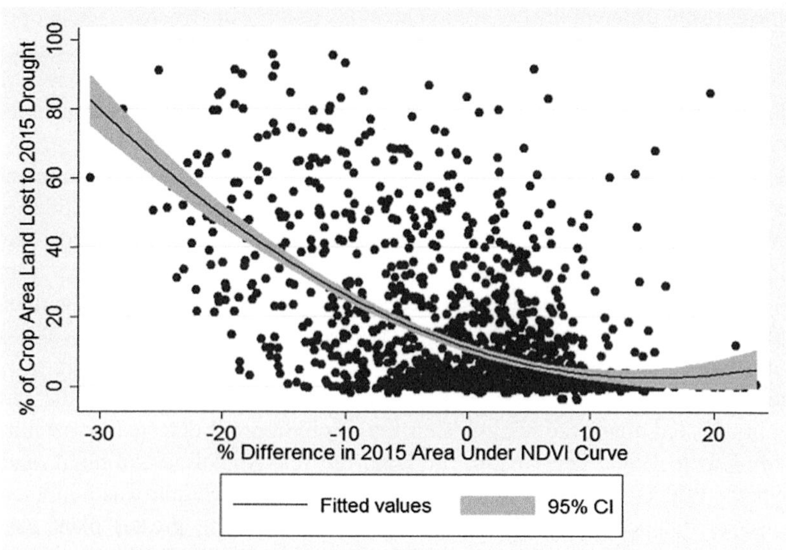

Fig. 10 Plot of % crop area land lost to 2015 drought versus % difference in NDVI area

affect the graphical component of the results, these observations did not significantly affect the statistical analysis.) In addition, the logs of both variables were taken, and a simple regression was performed to measure relative elasticity (area under the NDVI curve was assumed to be the independent variable). The elasticity of −1.03 suggests that for every 1% relative increase in NDVI, a reduction of 1.03% loss in cropland was reported. The overall explanatory power of the model is a little over 30% (R^2) and indicates a high degree of explanatory power based on this single variable. Of course, these variables are not identical and should not be considered as such. A perfect correlation would not be feasible for both measurement errors but also because these variables do not measure the same effects on crop productivity.

The result is important for a variety of reasons including the verification of data question, but also, given the detailed level of remote sensing data, future research could use out-of-sample predictions using the NDVI data based on the training data of the AgSS questionnaire data. In this way, detailed granular maps could be generated for a variety of research or policy purposes and create a better understanding of the localized effects of extreme weather events on agricultural productivity. In the next section, the authors provide two econometric models that seek to explain maize productivity using our data fusion methodologies.

Regression Methods and Diagnostic Results

Traditional ordinary least squares (OLS) approaches look at cross-sectional or time-series data, exploiting variance in one dimension. However most social and physical processes occur over both space and time. For this reason, the authors focus on the use of panel data sets which increases the amount of observed heterogeneity by including information about individuals i over time t. Critically, despite its widespread use in the geography field, cross-sectional analysis should generally *not* be used for forecasting or for modeling of phenomenon with strong temporal components.

Compared to cross-sectional approaches, panel analysis also substantially increases the degree of observed variance over both space and time. When pooled together, the integration of multiple data sets provides significantly more observations ($n = 1{,}780$) over the 2010–2016 sample period ($t = 6$). In this case, our panel provides approximately ($N = 10{,}680$) total observations.

The authors use panel data to model crop yields over time at the sub-kebele level for the planting seasons starting in 2010 and ending in the drought year of 2015/2016. The use of panel data in this study helps to alleviate two key problems, unobserved spatial and temporal dynamics and homogeneity (lack of variance). Now i is used to designate the sub-kebele sample area, rather than individual observations, and individual observations are the unique combination of i and t. From both theoretical and quantitative diagnostic perspectives, the random effects model was deemed appropriate. In addition, heteroskedasticity was present and therefore corrected for using White-Huber standard errors.

Preliminary Model

The authors believe that this data set provides a comprehensive balanced panel data set, collected from a variety of relevant sources that indicate the effects of drought at both local and regional impact levels in Ethiopia. The importance of understanding climatic variability and farmer response to these shocks is critical for policy formulation. In addition, many of the contemporaneous variables collected provide insights into more timely and accurate predictions of production earlier in the crop cycle. Logs of all variables were taken, with the exception of the *female head* variable which is a limited discrete variable. Table 4 presents two basic maize models.

Both models preformed relatively well with most variables exhibiting correct signs as well as statistical significance. Overall the first model used about 83% of all observations and reflects the relatively widespread use of maize throughout the country. Model significance was good with R^2 of 0.27 and 0.32, respectively. The variables are divided into four color-separated sections to reflect various aspects of our data fusion methodology. General control variables, too numerous to be listed here, are presented in white. Survey data from AgSS was included in orange, GIS information in blue, and variables derived from remote sensed data in green. A lagged value of the dependent variable was included, in the second model, to capture local dynamic effects on productivity. While some research has suggested potential bias coefficients downward when including the lagged dependent variable, it should be noted that the coefficients appear to be relatively stable when contrasted in both models. Most of the included variables are of the hypothesized sign and statistically significant. The following briefly discusses each of the variables in turn.

Table 4 Maize regression models

Dependent variable:	***Maize productivity***	
Control variables: years, agroecological zones, zones, regions		
Lag Maize OPH(t-1)		0.1961** (13.01)
Constant	-1.5618* (-2.14)	-2.7922** (-3.65)
Improved seed	0.0111** (8.10)	0.0121** (8.23)
Fertilizer	0.0018 (1.56)	0.0024 (1.82)
Extension area %	0.0058** (3.69)	0.0060** (3.64)
Irrigation area %	0.0025 (1.04)	0.0041 (1.50)
Erosion prevention—(terracing %)	0.0033* (2.48)	0.0052** (3.64)
Maize planted area	-0.0058** (-2.68)	-0.0066** (-2.95)
% of female heads	-0.1124* (-2.22)	-0.1266* (-2.26)
Area reported damage	-0.0170** (-10.36)	-0.0201** (-10.18)
Area of sub-kebele	0.0383** (2.96)	-0.0376** (2.63)
Elevation	-0.1045* (-2.33)	-0.0518** (-1.17)
Dist. To 50k city	-0.0247 (-1.77)	-0.0228 (-1.73)
NDVI max	0.1887** (7.30)	0.2183** (7.42)
NDVI area leading	0.2953** (6.56)	0.2946** (5.95)
NDVI area trailing	0.2336** (4.80)	0.2360** (4.54)
Pot. evapotranspiration (st. dev.)	-0.1307** (-4.24)	-0.0794* (-2.03)
Precipitation-leading	-0.0230 (-0.48)	-0.0127 (-0.22)
Precipitation-trailing	0.0416* (2.22)	0.0465* (2.21)
Precipitation (st. dev.)	0.0572 (1.24)	0.0287 (0.53)
N	8818	7007
Overall R^2	0.27	0.32

t statistics in parentheses
$^{*}p < 0.05$ $^{**}p < 0.01$

The zonal coefficients are omitted for brevity but are included to capture time-invariant spatial differences. The year coefficients are meant to estimate general time-sensitive productivity changes. All, but 2011, are statistically significant at the 99% level relative to the omitted 2010 value. It is interesting to note that the 2015 coefficient has a relatively large drop in value when compared to the previous year and may reflect some of the effect of the drought.

The orange-colored variables, or the AgSS questions, were all of the correct sign, and most were statistically significant. Maize is one of the few crops in Ethiopia that has widespread adoption of improved seed, and it is therefore not surprising that the coefficient is positive and highly significant. The government has placed a lot of emphasis on extension services, and it is encouraging that the coefficient is positive and has 95% statistical significant. The *terracing* variable is positive and significant and provides some insight into preventative farm management techniques. While the *irrigation* variable is positive, it is not statistically significant and may reflect the limited amount of land currently irrigated (about 2.3% of the total sampled area). As expected, farmer-*reported damage* was negative and highly significant. *Percent of female heads* measures the number of sampled female head of households in the sub-kebele. While the literature is mixed in Ethiopia, this analysis shows a reduction in productivity with increased number of women heads and is significant at the 95% level.

The GIS variables, highlighted in blue, exhibited the correct signs, but *distance to 50 k city* was not statistically significant. The variable is meant to proxy relative remoteness and access to markets. Therefore, the expected sign on this coefficient should be negative as the farther the distance the more remote the sample area and relative less access. *Elevation* is negative, and statistically significant, and reflects that maize tends to grow better in lower elevation areas while controlling for all other factors. The following discusses elements that were derived from remote sensing or other crop models.

The green-highlighted variables depict remote sensed or remote sensed derived variables. All of the NDVI variables performed as expected and were statistically significant. The authors took the percentage difference between the minimum and maximum point of the NDVI in the Meher season to measure the relative change of greenness (i.e., chlorophyll). As expected, the larger this difference, the greater positive impact on maize productivity. Concomitant to this measurement, the area under the curve, both in the run up to the maximum and from the maximum to harvest, also contributed positively to productivity. Further research is warranted, but the leading NDVI variable, coupled with survey data, may help to provide earlier crop productivity estimates for possible policy action. The standard deviation of *potential evapotranspiration* is meant to estimate variation in weather-related stress on the sub-kebele. As expected, this variation has a negative impact on productivity. The PPT measurements reflect relative rainfall, and an attempt was made to divide them into early (leading) rainfall before the maximum NDVI signature and post-maximum (lagging) or the grain-filling period. The later rains variable was significant at the 95% level of significance and suggests that late rains may be critical to prevent crop failure and enhance productivity. The early rains exhibited the incorrect sign and were statistically insignificant. This is disappointing, and more research is warranted here. The variation of rainfall also had the incorrect sign and was statistically insignificant. Overall, the NDVI variables performed extremely well, but the rainfall variables did not. As there may be some degree of correlation between the two, additional research of all of these variables is needed.

These findings provide a detailed analysis of the determinants of crop yields at the sub-kebele level. The AgSS data, coupled with GIS and remote sensing information, provides a unique panel data that includes a major drought. When AgSS is combined with other data, the synergies are apparent both within-sample and the potential for out-of-sample estimation is important. Beyond an improved understanding of various effects on cereal crop productivity, the leading variables are meant to provide initial estimates of productivity earlier in the crop cycle. However, they exhibit mixed results, and additional research is needed.

Overall, this data explores the influence of water availability, farm management practices, GIS variables, as well as other variables and its relative impact on maize productivity. The report reviewed each determinant's observed impact on yields with implications for policy choice and improved targeting of interventions to be discussed going forward.

Conclusion

Many important variables, derived from GIS or remote sensing, are usually omitted from traditional household survey analysis. This research demonstrates the importance of incorporating data fusion techniques for enhanced explanatory analysis. In addition, the potential for earlier, time-sensitive, derived data which can be incorporated into models for understanding current production impacts for appropriate policy action has been prototyped here. The leading variables are meant to be predictive variables that are available earlier than other survey-based questions impacting crop yields. Here the authors see that simple metrics, such as change in area under the curve of NDVI, can describe meaningful variation in crop yields for maize. This research is in its early stages and certainly needs refinement and further study.

The authors also find that spatial heterogeneity is a critical component of crop modeling in an environment as complex as Ethiopia's. In particular, the authors find that the effects of the 2015 drought were highly geographically specific, where traditional sampling methods might tend to over(under)state crop losses. The authors find significant variability below the regional and even zonal level. With this in mind, the authors urge the application of data integration techniques that can better match the temporal and spatial variability of drought.

Moving forward, there are several elements of this research that need to be developed. First and foremost, a more comprehensive model for maize, as well as other cereal crops, should be developed for improving predictions of productivity earlier in the crop cycle. This represents a possible impact for improving government action. In addition, the comprehensive coverage of remote sensed data means universal coverage of the country. Using survey data, such as AgSS, could provide a base for out-of-sample predictions for the entire country. Given our preliminary results of significant intrazonal variations in weather impacts, this could mean a far better understanding of the impacts of extreme weather events on local areas. Of course, additional collection of other variables, from other sources, would further enhance these models.

In conclusion, the authors believe that these types of data fusion methodologies are important for better understanding of crop productivity and production at a variety of scalable levels. They also have the potential for more accurate predictions earlier in the crop cycle. If this can be shown to be the case, in addition to the overall enhancements to crop productivity predictions, the impacts for both more accurate and preventative policy action in the context of extreme weather events should be significant.

Cross-References

- ▶ Assessment of the Degree of Households' Vulnerability to Climate Variability in Nigeria
- ▶ Climate Finance as a Catalyst for Initiating an African Green Revolution to Enhance Drought Resilience: Policy Prospects and Challenges
- ▶ Determinants of Household Food Security Status and Challenges of Building Resilience to Climate Variability and Change Posed by Drought in Tharaka Nithi, Kenya
- ▶ Effect of Climate Variability on Crop Income in Central Highlands of Ethiopia
- ▶ Impacts of Climate Change on Food Security: An Appraisal
- ▶ Managing Vulnerability to Drought and Enhancing Smallholder Farmers Resilience to Climate Change Risks in Zimbabwe

References

AKLDP (Agriculture Knowledge, Learning, Documentation and Policy Project) (2016) Food security in Ethiopia in 2016: analyzing crop production and market function after the main Meher agricultural season. AKLDP, USAID, Addis Ababa

Central Statistical Agency (2010). Agricultural Sample Survey 2009/2010, Vol. 1: Area and Production of Major Crops, Central Statistical Agency, Federal Democratic Republic of Ethiopia, Addis Ababa

Central Statistical Agency (2011). Agricultural Sample Survey 2010/2011, Vol. 1: Area and Production of Major Crops, Central Statistical Agency, Federal Democratic Republic of Ethiopia, Addis Ababa

Central Statistical Agency (2012). Agricultural Sample Survey 2011/2012, Vol. 1: Area and Production of Major Crops, Central Statistical Agency, Federal Democratic Republic of Ethiopia, Addis Ababa

Central Statistical Agency (2013). Agricultural Sample Survey 2012/2013, Vol. 1: Area and Production of Major Crops, Central Statistical Agency, Federal Democratic Republic of Ethiopia, Addis Ababa

Central Statistical Agency (2014). Agricultural Sample Survey 2013/2014, Vol. 1: Area and Production of Major Crops, Central Statistical Agency, Federal Democratic Republic of Ethiopia, Addis Ababa

Disaster Risk Management Technical Working Group (DRMTWG) (2016) Ethiopia: humanitarian response situation report no. 06 (as at 31 October 2016), 1–7

FEWSNET Monthly Evapotranspiration (ETA) Anomaly https://earlywarning.usgs.gov/fews/downloads/index.php

FEWSNET Daily Potential Evapotranspiration (PET) https://earlywarning.usgs.gov/fews/downloads/index.php

Funk C, Peterson P, Landsfeld M, Pedreros D, Verdin J, Shukla S, Husak G, Rowland J, Harrison L, Hoell A (2015) The climate hazards infrared precipitation with stations – a new environmental record for monitoring extremes. Sci data 2:150066

Major J (1967) Potential evapotranspiration and plant distribution in western states with emphasis on California. In: R. H. Shaw (ed.), Ground Level Climatology, pp. 93–126. American Association for the Advancement of Science, Washington, D.C.

Mann ML, Warner JM (2017) Ethiopian wheat yield and yield gap estimation: a spatially explicit small area integrated data approach. Field Crop Res 201:60–74

Mann ML, Moritz MA, Berck P, Batllori E, Waller E, Krawchuk M (2016) California wildfire past, present, and future: ecological and anthropogenic determinants in California's diverse ecosystems. PLoS One 11:e0153589

Central Statistical Agency (2015) Agricultural Sample Survey 2014/2015, Vol. 1: Area and Production of Major Crops, Central Statistical Agency, Federal Democratic Republic of Ethiopia, Addis Ababa

Intensifying Maize Production Under Climate Change Scenarios in Central West Burkina Faso

45

Omonlola Nadine Worou, Jérôme Ebagnerin Tondoh, Josias Sanou, Thomas Gaiser, Pinghouinde Michel Nikiema, Jules Bayala, Paulin Bazié, Catherine Ky-Dembele, and Antoine Kalinganiré

Contents

O. N. Worou (✉)
The International Crops Research Institute for the Semi-Arid Tropics (ICRISAT), Bamako, Mali
e-mail: n.worou@cgiar.org; omonlola@yahoo.fr

J. E. Tondoh
WASCAL and UFR des Sciences de la Nature, Nandjui Abrogoua University, Abidjan, Côte d'Ivoire

J. Sanou · P. Bazié
Institut de l'Environnement et de Recherche Agricole (INERA), Ouagadougou, Burkina Faso

T. Gaiser
Institute of Crop Science and Resource Conservation, University of Bonn, Bonn, Germany

P. M. Nikiema
Agence Nationale de la Météorologie du Burkina Faso, Ouagadougou, Burkina Faso

J. Bayala · C. Ky-Dembele · A. Kalinganiré
World Agroforestry Centre, West and Central Africa Regional Office – Sahel Node, Bamako, Mali

W. Leal Filho (ed.), *Handbook of Climate Change Resilience*,
https://doi.org/10.1007/978-3-319-93336-8_81

Abstract

Combination of poor soil fertility and climate change and variability is the biggest obstacle to agricultural productivity in Sub-Saharan Africa. While each of these factors requires different promising adaptive and climate-resilient options, it is important to be able to disaggregate their effects. This can be accomplished with ordinary agronomic trials for soil fertility and climate year-to-year variability, but not for long-term climate change effects. In turn, by using climate historical records and scenario outputs from climate models to run dynamic models for crop growth and yield, it is possible to test the performance of crop management options in the past but also anticipate their performance under future climate change or variability. Nowadays, the overwhelming importance given to the use of crop models is motivated by the need of predicting crop production under future climate change, and outputs from running crop models may serve for devising climate risk adaptation strategies. In this study we predicted yield of one maize variety named Massongo for the time periods 1980–2010 (historical) and 2021–2050 (2030s, near future) across agronomic practices including the fertilizer input rates recommended by the national extension services (28 kg N, 20 kg P, and 13 kg K ha^{-1}). The performance of the crop model DSSAT 4.6 for maize was first evaluated using on-farm experimental data that encompassed two seasons in the Sudano-Sahelian zone in six contrasting sites of Central West Burkina Faso. The efficiency of the crop model was evidenced by reliable simulations of total aboveground biomass and yields after calibration and validation. The root-mean-square error (RMSE) of the entire dataset for grain yield was 643 kg ha^{-1} and 2010 kg ha^{-1} for total aboveground biomass. Three regional climate change projections for Central West Burkina Faso indicate a decrease in rainfall during the growing period of maize. All the three scenarios project that the decrease in rainfall is to the tune of 3–9% in the 2030s under RCP4.5 in contrast to climate scenarios produced by the regional climate model GCM ICHEC-EC-Earth which predicted an increase of rainfall of 25% under RCP8.5. Simulations using the CERES-DSSAT model reveal that maize yields without fertilizer show the same trend as with fertilizer in response to climate change projections across RCPs. Under RCP4.5 with output from the climate model ICHEC-EC-Earth, yield can slightly increase compared to the historical baseline on average by less than 5%. In contrast, under RCP8.5, yield is increased by 13–22% with the two other climate models in fertilized and non-fertilized plots, respectively. Nevertheless, the average maize yield will stay below 2000 kg ha^{-1} under non-fertilized plots in RCP4.5 and with recommended mineral fertilizer rates regardless of the RCP scenarios produced by ICHEC-EC-Earth. Giving the fact that soil fertility improvement alone cannot compensate for the adverse impact of future climate on agricultural production particularly in case of high rainfall predicted by ICHEC-EC-Earth, it is recommended to combine various agricultural techniques and practices to improve uptake of nitrogen and to reduce nitrogen leaching such as the splitting of fertilizer applications, low-release nitrogen fertilizers, agroforestry, and any other soil and water conservation practices.

Keywords
DSSAT 4.6 · Climate change projection · Site-specific fertilization · Validation

Introduction

As the planet warms, climate change predictions increasing mean surface air temperatures from 2 °C to 5 °C by the end of the century are considered robust for West Africa, whereas no consensus has emerged on how precipitation is likely to evolve (Christensen et al. 2007; Cooper et al. 2008). Meanwhile, there seems to be an agreed perception about the exacerbation of intra-seasonal variability and rainfall patterns shift that future climate is likely to be characterized by Sylla et al. (2016) and Roudier et al. (2011). This may lead to lowering or unpredictable crop yield resulting in food insecurity especially in Sudanian zone of West Africa where farmers are the most vulnerable (Zoellick 2009). Although great efforts have been deployed in the research on the nexus climate change-agriculture, these were focused on evaluating the sensitivity of various cropping systems to pest, diseases, weeds, etc. (Enete and Amusa 2010). Less has been done on anticipating the effects of rainfall variability in order to cope with the negative effects of climate change and variability.

Globally, the national development programs tend to focus on more "high-tech" solutions to increasing food production such as increasing subsidies of mineral fertilizers and promoting improved seeds and equipment. However, the potential benefits of these approaches will be undermined unless the depletion of soil fertility is reversed and climate change impacts are overcome (Winterbottom and Reij 2013). Therefore, it is important to evaluate cropping systems under various environmental and climatic conditions including drought, nitrogen, and phosphorus-depleted soils as our study site located in Central West Burkina Faso. Soils in this area are usually relatively well weathered with low fertility which combined with high rainfall variability induce poor crop yields. Like in other Sub-Saharan countries, the low use of fertilizers especially the most important nutrients (nitrogen, phosphorous, and potassium) is considered as a major constraint to increasing productivity (Bationo et al. 1998; Samaké et al. 2005; Tadele 2017). As a consequence the extension staff tend to only promote the use of fertilizers, ignoring adaptive measures to climate change or variability for high yield crop production. Indeed, there is a need for information about the potential impact of climate change on agricultural systems in the future, and this is currently tackled by applying physiology-based crop simulation models (Asseng et al. 2013). As those crop models are used to predict variability of crop yield and related variables as well as natural resource use from driving factors such as short- and long-term variation of weather and climate conditions, linking climate-induced risk with yield gap analysis might seem relevant to predict crop yields in the future. To do so, these models need to be thoroughly tested and validated for given site/region to establish their credibility (Boote et al. 1996). Gaiser et al. (2011) recommended detailed site-specific soil input parameters (e.g., organic matter pools, water retention) for reliable prediction of scenario effects for individual sites. Bassu et al. (2014) found that increasing the level of input information (i.e., high information in calibration) significantly reduced the uncertainties in the models' outputs. In the present study, we utilized DSSAT model which has been successfully tested under water- or nutrient-limited production conditions in West Africa over a wide range of crops and management options (Fatondji et al. 2010). Most of the past trials of this model have used on-station data, thus unable to capture the real conditions of the farmers, except few cases in agroforestry with other models (Bayala et al. 2008; Coulibaly et al. 2014; Luedeling et al. 2016). Our study is a contribution to fill in that gap by using on-farm data as they better represent farmers' real conditions.

The objective of this study was to assess the impacts of climate change on agronomic productivity of maize (*Zea mays*) by using crop growth model simulation based on historical climate records (1981–2010) and bias-corrected near-future (2021–2050) climate projections for the Central West Burkina Faso. In this study, the key research questions were to find out (1) whether DSSAT crop model can reliably predict maize production under smallholder farming conditions in the Sudan Savanna of West Africa with and without fertilizer application and (2) what are the impacts of climate change on the production risk of maize for these cropping systems when fertilizer is applied or omitted. The hypothesis tested is that conventional intensification (high mineral N supply), reported to sustain yield under various climate scenarios (Folberth et al. 2014) at the continental scale, is negatively affected by the projected future climate change in Burkina Faso.

Materials and Methods

Study Sites, Experiment Description, and Data Collection

The agronomic trials were carried out in farmers' fields in three villages, Cassou, Dao, and Kou, from the departments of Cassou, Gao, and Bakata, respectively, located in Ziro Province of the Central West Burkina Faso (11°42′N, 2°03′W) (Fig. 1). The average altitude of Ziro Province is about 300 m a.s.l., and this province lies within the South-Sudanian ecological zone (Fontes and Guinko 1995) and receives 900–1200 mm rainfalls annually. Its unimodal rainy season lasts for about 6 months, from May to October. According to the FAO's soil classification system (Driessen et al. 2001), the most frequently encountered soil type in the area is *Lixisol* (tropical ferruginous soil), which is poorly to fully leached, overlying on sandy, clayey-sandy, and sandy-clayey materials.

The trials were set up to identify the nutrients limiting crop productivity with the ultimate ambition to be able to recommend site-specific sustainable land management options. In this chapter we focused on the response of maize crop to nutrients as NPK at 200 kg ha^{-1} which is the recommended fertilizer dose by the extension services containing 28 kg N ha^{-1}, 20 kg ha^{-1} P_2O_5 ha^{-1}, and 13 kg K ha^{-1}. The fertilizer was applied in broadcast during sowing. Two seeds per hole were sown to

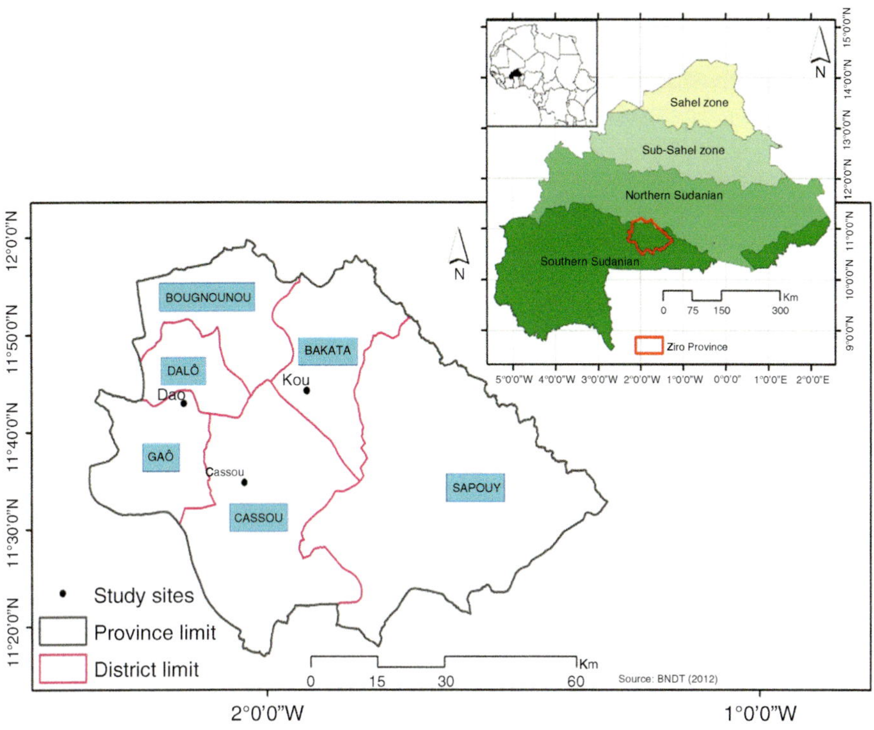

Fig. 1 Location of Ziro Province and study sites in Burkina Faso, West Africa

ensure germination and good stands of the crop and then thinned to one plant per hole, 5 days after emergence. The trial was carried out with an improved variety of medium maturing cycle maize named *Massongo*. Twelve experimental plots were established in Dao, Kou, and Cassou (Cassou 1, Cassou 2, Cassou 3, and Cassou 4), while data of Cassou 1 was used for calibration.

Each experimental plot covers an area of 100 m^{-2} (10 m × 10 m). Prior to trial establishment, initial soil status was characterized from composite sample (0–20 cm and 0–50 cm depths) analysis. Concentrations of total organic carbon and total nitrogen were determined using the Walkley-Black (1934) method and Kjeldahl method, respectively. The pH-H_2O measurements were made on a soil/water solution at a ratio of 1: 2.5. Available P was extracted according to the Bray I method. The cation exchange capacity (CEC) was obtained from Kjeldahl distillation. Information on soil textural properties were obtained from literature (BUNASOLS 2001).

In order to estimate maize crop yield and total aboveground biomass, hills were harvested within a sampling plot size of 50 m^2. On the harvested maize plants, grains and shoots were separated, oven-dried at 60 °C for 72 h, and weighted. The dry matter of stover weight (kg ha^{-1}) and the grain weight (kg ha^{-1}) was determined by scaling-up from the harvested plot area to 1 ha.

DSSAT Crop Growth Model

Crop Growth Model

CERES (Crop Environment Resource Synthesis)-Maize module within the DSSAT version 4.6 (Jones et al. 2003) was used. This model can be classified as a dynamic cropping system model with mechanistic and functional components. In the model, crop development is controlled by temperature, in terms of thermal degree days, while crop biomass formation depends on the simulation of LAI (leaf area index) and its direct relation to light interception. The model owns specificities when fine-tuning to simulate the phenological and physiological traits of specific crops including the processes of assimilation, respiration, crop development, and growth with regard to crop type. A set of reduction factors such as water, nitrogen, and phosphorus stresses influence the plant growth, senescence, and crop yield under unfavorable conditions.

Model Calibration and Validation Procedures

The DSSAT model was evaluated using data from 2014 to 2015 growing seasons at the site of Cassou 1. The model was calibrated for simulation of days after sowing to reach maturity for 100 days with respect to available information in the documentation for *Massongo* maize variety (Nitiema 2009). Then the adjustment of the other parameters consisted of making the best fit of the simulated outputs to the observed total aboveground biomass and grain yield within both growing seasons. The model was validated using site inputs and observation data from Dao, Cassou 2, Cassou 3, Cassou 4, and Kou sites. Farm operation dates and types that were implemented in all the sites and used as inputs in crop management file are described in Table 1.

Model Inputs

Soil Data

An accurate description of the soil profile is important for water-limiting simulations such as rainfed crops. The characteristics of the layers within the soil profile are integrated into the framework available within the crop models. They included limits of water content for each layer (lower limit, field capacity, etc.), pH-H_2O, organic matter, and nitrogen content (Table 2). From a depth of 50 cm to 120 cm, all soil parameters were set to the same value.

Climate File Inputs for Model Calibration and Validation

Weather data for running crop simulation models were obtained from the nearest weather station of Po. This includes daily rainfall and maximum and minimum air temperatures. Solar radiation was computed from Angstrom-type equation. The rainy period started in June and ended in October. The total precipitation for the year of experimentation was typical for the Sudanian regions of Central West Burkina Faso. The rainfall during the growing season was 638 mm and 670 mm in 2014 and 2015, respectively (Fig. 2). In DSSAT, FAO 56 method was chosen to estimate potential evapotranspiration.

Table 1 List of field operations used in the DSSAT model and derived for maize cropping for the two growing seasons in 2014 and 2015 in Ziro Province in Burkina Faso, West Africa. TBD is the total aboveground biomass, GY is the grain yield. Jul, July; Oct, October

Year	Treatment	Sowing date	Fertilizer application		Amount of N, P (kg ha^{-1})			Harvest date	Model outputs
			Type	Date 1	N	P	K		
2014	Control	10-Jul	–	–				31-Oct	TBD and GY at harvest
	+Fertilizer	10-Jul	NPK	10-Jul	28	20	13	31-Oct	TBD and GY at harvest
2015	Control	15-Jul	–	–				31-Oct	TBD and GY at harvest
	+Fertilizer	15-Jul	NPK	10-Jul	28	20	13	31-Oct	TBD and GY at harvest

Crop/Cultivar Input Files

Crop file was calibrated by varying the phenology parameters of the crop cultivars until the simulated phenology dates matched the observed dates. The focused dates were those associated with physiological maturity. After the phenology parameters were set, biomass assimilation and yield component coefficients were adjusted to represent as accurately as possible the measured biomass and yield. This includes the parameters in the cultivar (CUL) file such as P1 affecting anthesis day, PHINT value with leaf appearance rate and phyllochron interval, P2/P2R affecting photoperiod sensitivity, P5 affecting grain filling duration and yield components (G2: maximum possible number of kernels per plant; G3: kernel filling rate during the linear grain filling stage and under optimum conditions (mg day^{-1})) (Table 3).

Climate Projection Information

Post-processing of the raw regional climate model (RCM) (CLMcom and RCA4) data derived from 3 global climate models (GCM), (CNRM-CERFACS's CNRM-CM5 labeled CNRM, ICHEC-EC-Earth labeled ICHEC, MPI-M-MPI-ESM-LR labeled MPI) from the Coordinated Regional Downscaling Experiment (CORDEX) (Giorgi et al. 2009) was applied to the precipitation and temperature data. The precipitation data were bias-corrected at the Rossby Centre, SMHI, in Sweden using the distribution-based scaling method (DBS, Yang et al. 2010) and WATCH-Forcing-Data-ERA-Interim (WFDEI, Weedon et al. 2014) as reference dataset. WFDEI is a meteorological forcing dataset covering the period 1979–2014. Minimum and maximum temperatures were statically bias-corrected using baseline data from 2006 to 2015. For the provided climate data, two scenarios were used: RCP (Representative Concentration Pathway) 4.5, the mitigation scenario where the

Table 2 Soil profile horizon for DSSAT model calibration and validation of data from Cassou, Dao, and Kou villages in Burkina Faso, West Africa

Sites	Soil depth (cm)	Sand content %	Silt content %	Field capacity	Wilting point	Organic carbon %	Nitrogen %	Available phosphorus	pH-H_2O
Calibration									
Cassou 1	0–20	68	7	7	2	1.11	0.08	21.30	6.5
	0–50	54	7	7	2	1.04	0.07	10.64	6.5
Validation									
Dao	0–20	68	7	7	2	0.68	0.05	7.10	5.8
	0–50	54	23	19	13	0.66	0.05	6.87	5.8
Cassou 2	0–20	68	7	7	2	1.42	0.10	7.87	6.3
	0–50	54	23	19	13	1.22	0.09	7.21	6.3
Cassou 3	0–20	68	7	7	2	1.11	0.08	7.43	5.8
	0–50	54	23	19	13	1.07	0.08	7.76	5.4
Cassou 4	0–20	68	7	7	2	1.31	0.08	8.10	5.7
	0–50	54	23	19	13	1.31	0.08	7.30	5.9
Kou	0–20	68	7	7	2	0.67	0.05	6.32	6.3
	0–50	54	23	19	13	0.67	0.04	5.99	6.2
All sites	50–120	60	10	19	13	0.04	0.04	0.30	3.6

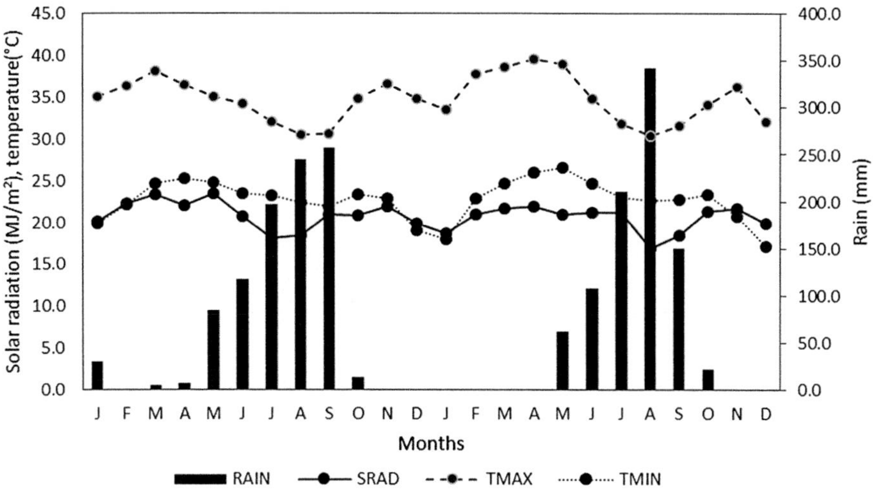

Fig. 2 Monthly rainfall, minimum and maximum temperatures, solar radiation (SRAD) distribution in 2014 and 2015 in the study sites (Cassou, Dao, and Kou) in Burkina Faso, West Africa

Table 3 Values of *Massongo* maize cultivar parameters as calibrated in the CERES-Maize crop model of DSSAT for the maize experiment in Ziro Province, Burkina Faso

P1 (°C day)	P2 (°C days)	P5 (°C days)	G2 (number)	G3 (mg day^{-1})	PHINT (°C day)
260	0.750	800	750	5.30	49

P1: Thermal time from seedling emergence to the end of the juvenile phase (expressed in °C day, above a base temperature of 8 °C) during which the plant is not responsive to changes in photoperiod. P2: Extent to which development (expressed as days) is delayed for each hour increase in photoperiod above the longest photoperiod at which development proceeds at a maximum rate (which is considered to be 12.5 h). P5: Thermal time from silking to physiological maturity (expressed in °C day above a base temperature of 8 °C). G2: Maximum possible number of kernels per plant. G3: Kernel filling rate during the linear grain filling stage and under optimum conditions (mg day^{-1}). PHINT: Phyllochron interval, i.e., the interval in thermal time (°C day) between successive leaf tip appearances

global greenhouse gas emissions culminate in year 2040 and then decrease, and the RCP8.5, the "business-as-usual" scenario where the emissions continue to accelerate. Historical scenario concerns the observed 30-year period data from 1981 to 2010 at Po station.

Model Evaluation and Statistical Analysis

In the model calibration, the observed and simulated means were compared using two-sample t-test for paired data in R software across the treatments and season during the 2 years (2014–2015) of the experiments. Statistics used for the performance evaluation of DSSAT model were the root-mean-square error (RMSE) (Eq. 1) and the coefficient of determination (R^2) which were calculated from observed and simulated variables. The value of RMSE equal to zero indicated the goodness of fit between predicted and observed data and evaluated model performance and accuracy in prediction. The coefficient of determination's value close to 1 indicates better prediction, while R^2 value of zero indicates no predictability.

$$\mathrm{RMSE} = \sqrt{\frac{1}{n}\sum_{i=1}^{n}(S_i - O_i)^2} \tag{1}$$

The root-mean-square error (RMSE), observation (O), simulations (S)

Scenario Analysis

With the seasonal analysis tool of DSSAT, predictions of grain yield over the climate scenario periods were done. Yield frequency plots with the cumulative probability distribution (CPD) for the base period (1981–2010) as historical and near future period (2021–2050) were developed. This allowed the visualization of trends and effects of climate change scenarios and resilience attributes of recommended fertilizer inputs (yield stability and ranges). In CPD, the basic idea is to look at mean yield and variance at 0.5 cumulative probabilities. For a given minimum yield level, the risk of each treatment was defined by looking at the cumulative probability of obtaining such a yield level (Ngwira et al. 2014).

Results and Discussion

Model Calibration and Evaluation for Maize Cultivar *Massongo*

The crop production variables used for the study were maturity date, total aboveground biomass, and maize grain yield based on six genetic coefficients of cultivar *Massongo* of maize, which were estimated in the present study (Table 2). The range of those coefficients lies within the values reported by other researchers for medium maturing maize varieties in Africa (Ngwira et al. 2014).

The calibration process revealed that simulated total aboveground biomass gave 1335 kg ha^{-1} greater yield than observed although the averaged biomass showed no significant ($p > 0.05$) differences between observation and simulation (Table 4). Observed grain yields displayed higher standard deviation compared to simulated ones. Also, the simulation overpredicted the mean yields as a result of general overprediction in both rainfall seasons and soil fertility options (Table 4). Overall, comparison between observed and simulated yields using two-sample t-test for unpaired data showed that the mean values were not statistically different ($p > 0.05$) as indicated in Table 3. This shows that the model was successfully calibrated for the cultivar *Massango* as well as for the two treatments and two

Table 4 Observed and simulated maize *Massongo* total aboveground biomass (TAB) and grain yields (GY) in 2014 and 2015

Crop parameters	Observed yields (kg ha^{-1})		Simulated yields (kg ha^{-1})		Two-sample t-test for paired data	R^2	RMSE (kg ha^{-1})
n = 4	Mean	std	Mean	std			
TAB	6738	3498	7706	559	NS	0.87	2186
GY	750	575	990	206	NS	0.74	617

n, number of plots; std, standard deviation; NS, not significant at the 0.05 level

seasons of study. Moreover, in the years of calibration, DSSAT was able to capture the relative decrease in grain yield between the recommended and non-fertilizer treatments, thus generating the best level of accuracy with its high R^2 (0.74) which is specifically used to assess the extent to which magnitudes of observed means are related to the simulated ones, and allows for sensitivity toward differences between them as well as the proportionality changes.

The CERES-Maize model was evaluated by comparing simulated and observed yield data for sites of Cassou 2, Cassou 3, Cassou 4, Dao, and Kou for 2014 and 2015 growing seasons that are considered to be an independent dataset to Cassou 1. Although, the evaluation experiments were conducted at the different sites, the management operations and in particular the fertilizer inputs were the same as on the calibration site (Table 1). However, soil and climate conditions were different (Table 2, Fig. 2). Validation of DSSAT model for yield simulation was challenging because the stable percentage of organic matter factor, which is one of the soil parameters that had significant effect on yield, should be adjusted. Unique value of stable organic matter at 0–20 cm that provided simulated yields comparable to observed yields was 20%.

Graphically and in overall, regression plot shows that simulated aboveground biomass was well represented in comparison with the observations using the modified parameters for all the experiments (Fig. 3). The model simulated maize grain over all treatments with difference of 9% for total aboveground biomass and grain yield. Ngwira et al. (2014) and Bakhsh et al. (2013) estimated that the error in predicting yield for all treatments was below 12% which was considered to be "good." Overall, the RMSE was found to be 2010 kg ha^{-1} and 643 kg ha^{-1} for total aboveground biomass and grain yield, respectively. In the present study, the DSSAT simulations were of good quality for the mean grain yield and mean aboveground biomass across the two seasons and four sites; such conclusion is based on moderate parameterization efforts and statistics for on-farm growing conditions. It can be outlined that the model has been shown to simulate maize growth under strongly contrasting environmental conditions in the tropics, while it has functions that describe the changes in system states in response to external drivers (e.g., weather and management practices).

However, the variation between the individual plots was quite high, resulting in an R^2-value of 0.74 showing that at field level and for total aboveground biomass, the mean overall treatment value was a good predictor, whereas the individual

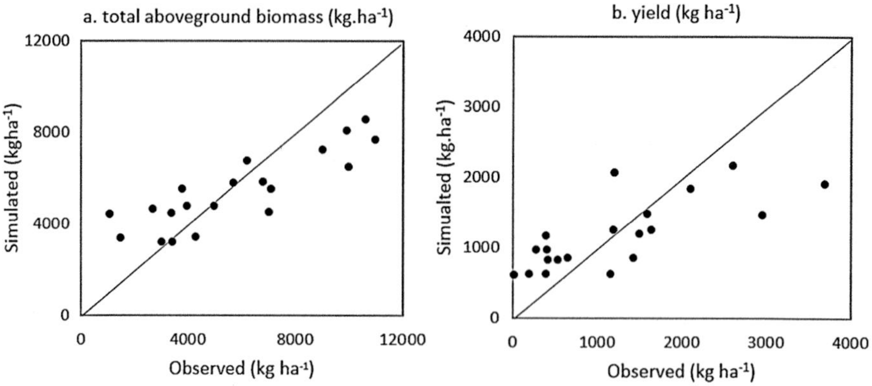

Fig. 3 Regression of simulation for measured maize *Massongo* total aboveground biomass (**a**) and grain yields (**b**) (kg ha^{-1}) after model validation

Table 5 Model validation, observed and simulated maize *Massongo* total aboveground biomass (TAB), and grain yields (GY) between 2014 and 2015 in five study sites (Cassou 2, Cassou 3, Cassou 4, Dao, Kou) in Burkina Faso, West Africa

$n = 10$	Observed yields (kg ha^{-1})	Simulated yields (kg ha^{-1})	Two-sample t-test for paired data	R^2	RMSE (kg ha^{-1})
TAB-Control	4879.3	4836.4	NS	0.66	3019
TAB-NPK	6682.4	5959.2	NS	0.75	3154
GY-Control	1104.5	1076.5	S	0.34	1054
GY-NPK	1324.0	1297.9	NS	0.76	1057

NS, no significant difference at the 0.05 level; S, significant difference at the 0.05 level

plot-wise yield predictions may be more uncertain. Indeed, in Table 5 it is shown that the model simulates grain yield and total aboveground biomass with higher reliability for fertilized plots than non-fertilized plots where R^2 was below 0.5. Some errors related to the gap between the simulated and observed value are mainly due to the level of variability between plots. An experiment under on-farm conditions with various treatments over several cropping seasons may be enough complex for crop production estimation because of the involvement of several factors like pests; heterogeneity within the crop management intensity, e.g., plant density; various interacting nutrient stresses such as micronutrients (Folberth et al. 2014; Voortman et al. 2003); and the soil physical discontinuities with the consequence of a huge yield gap. For the sites in Africa, recorded yield reached only 20% of the attainable yield confirming the large yield gap which weakens the use of favorable seasonal weather conditions (Hoffmann et al. 2017). Nevertheless, the evaluation of the simulation model is on a good level being used as a decision support tool to evaluate the impact of usual fertilizer input rates on farmers' fields.

Interannual and Seasonal Variability in Rainfall for Projected Climate Scenarios

With reference to the analysis of future climatic change as projected by regional climate models around the 2030s (2021–2050), mean rainfall in the maize growing season is predicted to decrease slightly in the Sudanian Savanna environments of Central West Burkina Faso for all the climate scenarios as compared to the simulated value for the historical scenario (1980–2010) except for ICHEC under RCP8.5 (Fig. 4a). Specifically, three regional climate change projections for Central West Burkina Faso indicate a decrease in rainfall during the growing period of maize to the tune of 3–9% in 2030s under RCP4.5. ICHEC predicted an increase of rainfall of 25% under RCP8.5. Similar small magnitude of variation in RCP4.5 and higher magnitude of variation in RCP8.5 were also obtained for mean annual rainfall which is predicted to increase in the Sudanian Savanna by about 1.7% and 1.9% in 2030s and by 4.4% and 16.1% in 2030s under the respective scenarios RCP4.5 and RCP8.5 (Mohammed et al. 2016). With reference to CMIP5 (Coupled Model Intercomparison Project Phase 5), the majority of the climate model outputs we used seem to be contrary to the positive precipitation trend which is predicted by 50% of the models in CMIP5 model ensemble. The results of our climate models are probably closer to the 25% of the models in the CMIP5 archive showing robust decreasing precipitation trend although only one of our scenarios was in agreement with these previous studies (Sultan and Gaetani 2016). In Fig. 4a, b, higher warming is expected for all climate scenarios with RCP8.5 and CNRM 8.5 at the highest.

For analysis of intra-seasonal rainfall distribution, the monthly rainfall variability across the years within the growing season is shown in Fig. 5. Across the months, more consistent trends of rainfall in the growing period among the climate scenarios were obtained under RCP4.5 than under RCP8.5. There will be significant reduction in the distribution of rainfall at the end of the growing season (October) especially in the output of the MPI climate model with RCP4.5 and RCP8.5 (Fig. 5d). Although our study is in disagreement with Ngwira et al. (2014) who used RegCM4 in Malawi projecting more rainfall at planting than in the historical weather data, our conclusions confirm the findings of these authors that there will be a reduction in the length of the growing season in the future. In RCP4.5, lower variance of rainfall across July adding to the lower water volume in this month is in contrast with Tachie-Obeng et al. (2013) who reported from projections an increase in rainfall at the onset of the wet season in Northern Ghana, with a decline in mid-season rainfall in June, followed by a significant shift in the distribution of rainfall toward the tail end of the season, especially in November. In our case and for RCP8.5, the increase in rainfall in the future in the middle season seems to be in favor and in benefit of the normal crop growth and development as this period corresponds to critical stages of maize growth, i.e., flowering and grain filling.

In summary, the use of multiple RCM models that was expected to reduce uncertainty in the model projections seems to show disparities in the climate warming particularly for maximum temperature, while the precipitation projections for only one model (ICHEC) out of the three show an increasing trend. Future rainfall distribution seems to be favorable over the growing season of maize.

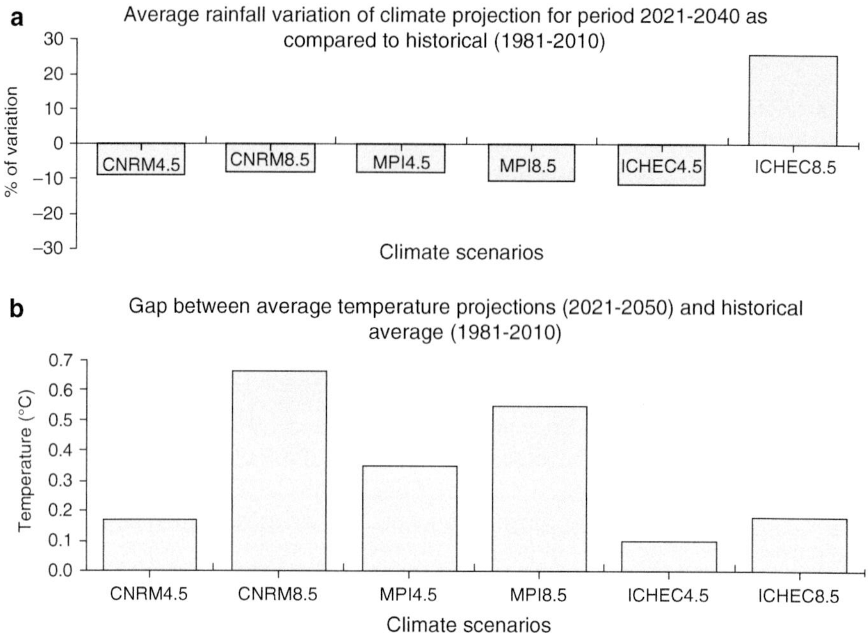

Fig. 4 Comparison of rainfall during the growing period of maize in the baseline (1981–2010) and future climate periods (2021–2050) under RCP4.5 and RCP8.5 climate scenarios for the study sites (Cassou, Dao, and Kou) in Burkina Faso, West Africa. (**a**) Percentage change in rainfall during the growing period in climate projections as compared with historical data average; (**b**) gap between average temperature projections and historical average temperature during the growing season

Risk Analysis

The seasonal analysis program of DSSAT 4.6 was used to compare two management options with recommended fertilizer inputs (Fig. 7) and no fertilizer (Fig. 6) in a soil with medium soil fertility status represented by the soil at Cassou 3 (Table 2). The simulations were carried out for a 30-year period with daily climate data consisting of rainfall derived from three climate models under RCP4.5 and RCP8.5 scenarios and from the historical time series. About 960 runs were derived from the set of treatments of fertilizer inputs tested and encompassed variable rainfall profiles (dry, normal, and wet seasons) while allowing to explore and to generate potential impact of climate seasonality on the risk in the cropping systems.

Figure 6 shows yield changes over 30 years and 3GCMs for RCP4.5 and RCP8.5 when no fertilizer was applied on a soil with moderate fertility status. Under RCP4.5, the plots of cumulative probability distribution (CPD) at the level of 0.5 showed that the lowest mean yield was predicted under CNRM and MPI climate scenarios, while the highest yields were both with historical and ICHEC climate scenarios (Fig. 6a). This pattern was strongly contrasting for the predicted yield with RCP8.5 (Fig. 6b). In RCP8.5, yield will be negatively affected by higher amounts of rainfall in the ICHEC

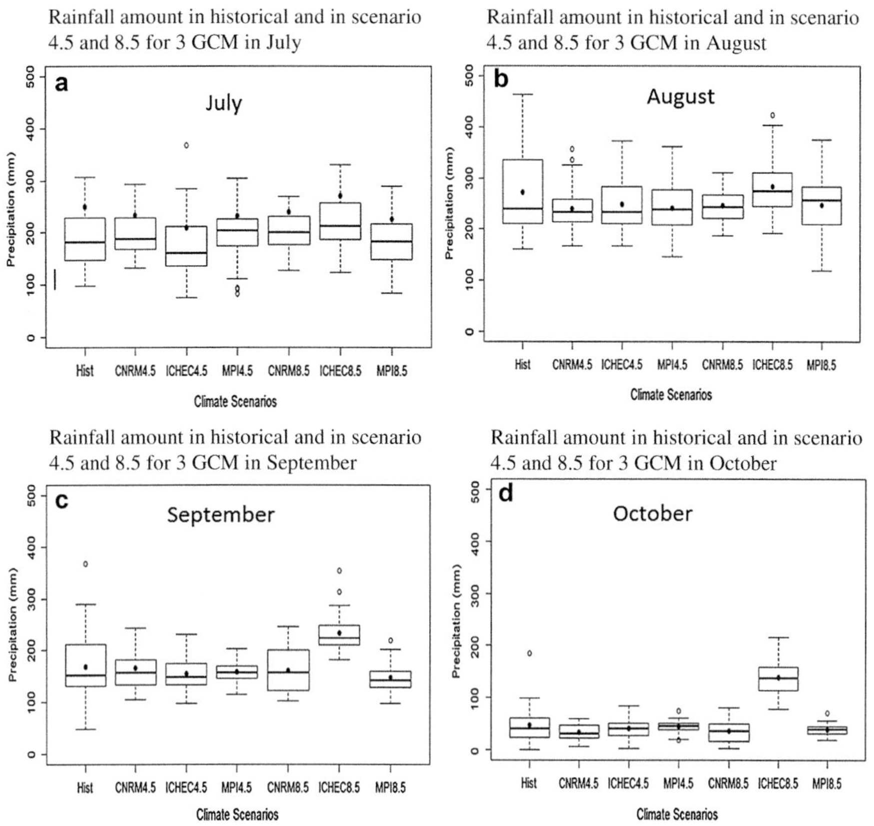

Fig. 5 Seasonal variability of growing period rainfall under different scenarios over the simulation period (30 years) in both RCP4.5 and RCP8.5. Each box in the graph shows the distribution of rainfall over the simulation period. The boundary of the box closest to zero indicates the 25th percentile, the broken line within the box marks the mean, the solid one marks the median, and the upper boundary of the box indicates the 75th percentile. Whiskers above and below the box indicate the 95th and 5th percentiles. Black spot represents the mean. CNRM refers to rainfall amount under climate scenarios using CNRM-CERFACS-CNRM-CM5 global climate model; ICHEC refers to rainfall amount under climate scenarios using ICHEC-EC-Earth global climate model; and MPI refers to MPI-M-MPI-ESM-LR global climate outputs. (**a**) Rainfall amount in historical and in scenario 4.5 and 8.5 for 3 GCM in July. (**b**) Rainfall amount in historical and in scenario 4.5 and 8.5 for 3 GCM in August. (**c**) Rainfall amount in historical and in scenario 4.5 and 8.5 for 3 GCM in September. (**d**) Rainfall amount in historical and in scenario 4.5 and 8.5 for 3 GCM in October

climate scenario as compared with historical yield. Similarly responses of maize yields to the outputs of nine GCMs without adaptation options showed general tendency toward diminishing future maize yields in the single cropping season of the savanna zone, ranging from −6.3% to −42.6% in the near future (Tachie-Obeng et al. 2013). Meanwhile, in our results for RCP8.5, mean-variance of yield and distribution of yield along with maximum and minimum values were higher for the two other climate scenarios (CNRM and MPI) than for historical climate scenario.

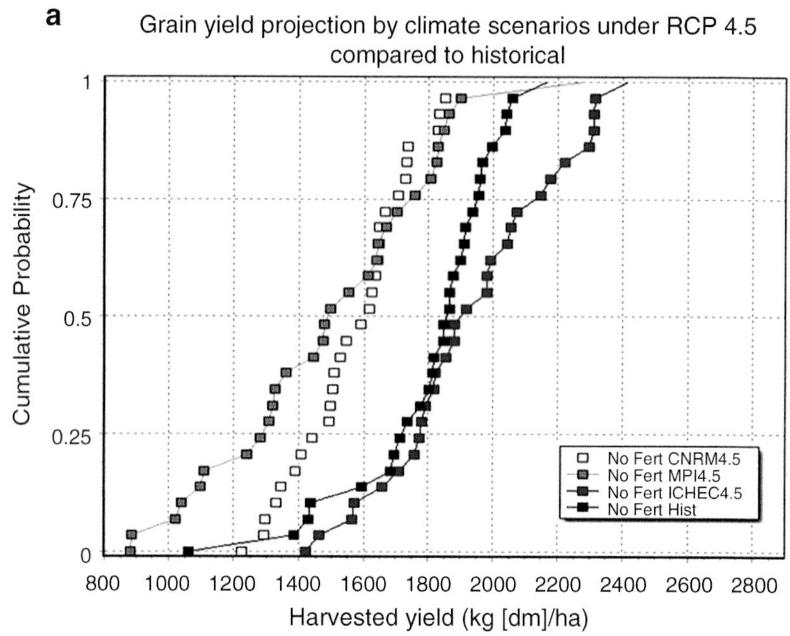

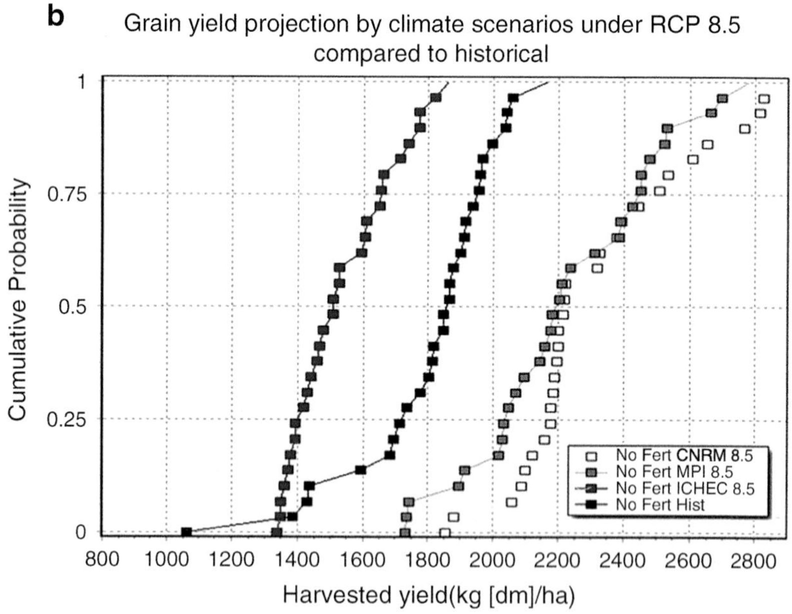

Fig. 6 Cumulative probability distribution plots of maize yields for no fertilizer treatment under medium soil fertility (Cassou3) with the historical baseline (1981–2010), and 2030s time periods under RCP4.5 and RCP8.5 climate scenario with 3 GCM. No fert Hist is the treatment without fertilizer under historical climate. No fert CNRM 4.5 and CNRM 8.5 are treatment without fertilizer

Figure 7 shows yield changes over 30 years for historical and 3GCMs in RCP4.5 and RCP8.5 when applying the recommended fertilizer rates. Under RCP8.5, future average yields (at CDP = 0.5) were projected to increase by 13% (for both CNRM and MPI) during the near future compared with historical climate, whereas there was an 18% decline in maize yield in the ICHIEC scenario when compared with the historical scenario. Under RCP4.5, the decline in simulated yield was also observed for the other climate model scenarios but less pronounced.

Comparing the two RCP scenarios, the results in Figs. 6 and 7 revealed that globally maize grown under nutrient stress conditions (moderate soil fertility and without fertilizer) has favorable conditions under RCP4.5 for the ICHEC scenario, but not for the MPI and CNRM scenarios, whereas under RCP8.5, the conditions are more favorable for the MPI and CNRM scenarios. The inconsistency in the outcomes is similar to the findings by Araya et al. (2015) who found out that there will be an increase and decrease in the yield of chickpea varieties in central highlands of Ethiopia in the upcoming periods depending on the projected climate change under both RCPs by 2050s. Nevertheless, the projections of maize yields confer the same trend for both treatments (without fertilizer and with fertilizer) in response to climate change projections in each of the RCPs. The results contrast with findings from Waongo et al. (2015) who reported that with 8 RCMs and a regional crop model, irrespective of the RCP and period, a higher positive change (>40%) of maize mean yield is expected in the Central West Burkina Faso. In our study, the main contrast exists between RCP4.5 and RCP8.5, where only the ICHEC scenario does not benefit the maize crop under RCP8.5 by resulting in reduced yield within the range of the simulations (yield between 0% and 100% of CDP lower than historic) although its projection of rainfall amount was the highest with this scenario (Figs. 4 and 5). Maize grown under nutrient stress condition (moderate soil fertility without fertilizer) may suffer from the potential nutrient depletion of arable land including micronutrients due to nutrient leaching. Water erosion is also higher as a result of intense rainfall events (Folberth et al. 2012) which have been shown to lead to significant mining of soil nitrogen reserves. Effective conclusions can be drawn up after making more complex analysis from climate change scenarios while taking into account important climatic parameters such as elevated temperatures as they might shorten the growth duration and be a stress factor. Under low nutrient stress (with fertilizer application), maize yields are slightly higher (approx. 200 kg ha^{-1}) irrespective of the climate model and RCP scenarios.

Fig. 6 (continued) under RCP4.5 and RCP8.5, respectively, for regional climate CNRM-CERFACS-CNRM-CM5. No fert MPI 4.5 and MPI 8.5 are treatment without fertilizer under RCP4.5 and RCP8.5, respectively, for regional climate MPI-M-MPI-ESM-LR. No Fert ICHEC4.5 and ICHEC8.5 are treatment without fertilizer under RCP4.5 and RCP8.5, respectively, for regional climate ICHEC-EC-Earth

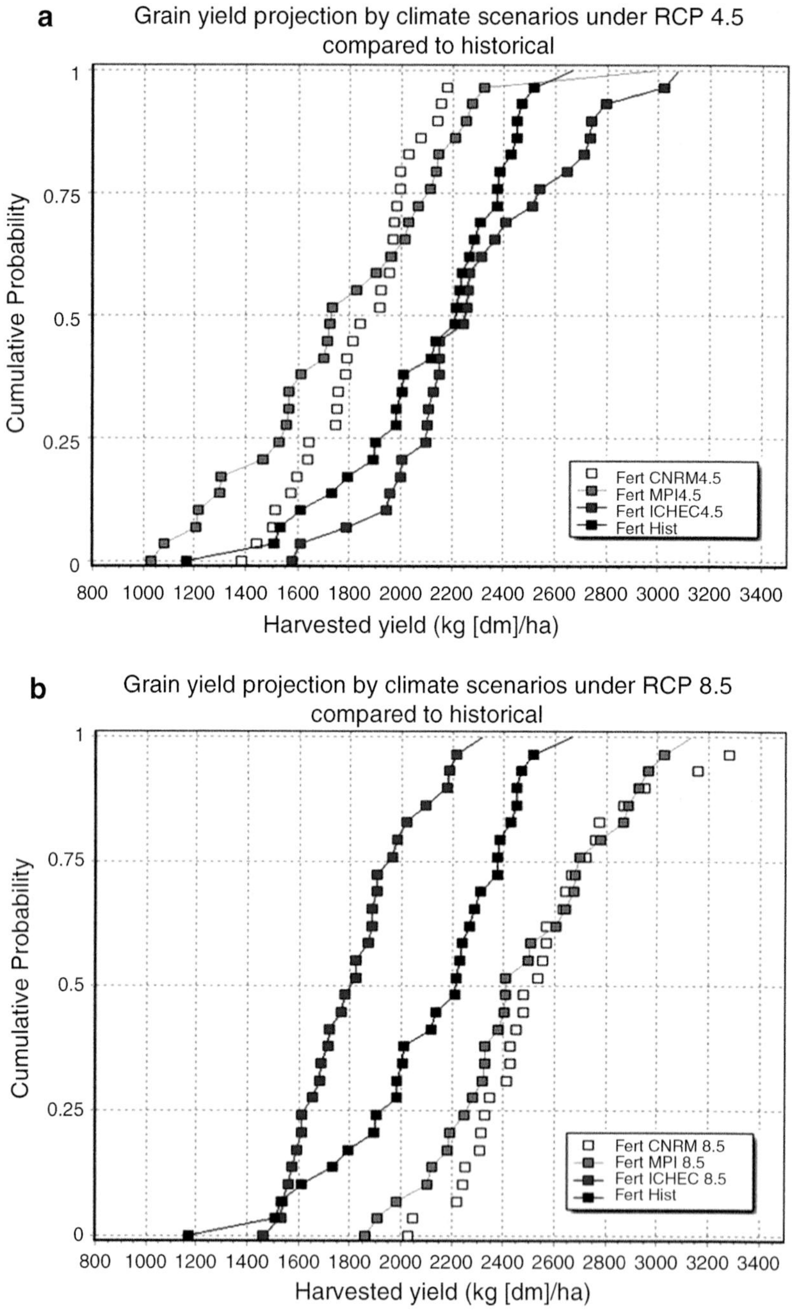

Fig. 7 Cumulative probability distribution plots of maize yields for fertilized treatment under medium soil fertility (Cassou3) with the historical baseline (1981–2010) and 2030s time periods under RCP4.5 and RCP8.5 climate scenarios with 3 GCM. Fert Hist is the treatment with fertilizer

Our study confirms that modifying the fertilizer rate according to seasonal weather forecasting could help better realize the potential for intensification. However the validity of our findings may suffer from the lack of cumulative benefits of crop rotation systems, thereby underestimating maize grain yield as the DSSAT model uses the same initial soil conditions for the entire simulation period. Hence refinements of the algorithms to simulate changes in soil properties in DSSAT are recommended in order to sufficiently and accurately predict yield of crop rotation systems in the seasonal analysis module (Ngwira et al. 2014). As rainfall appeared to consistently meet crop water demand, our study is likely to support the conclusion by Araya et al. (2015) that there was no substantial difference in yield across the future scenarios. The study may be extended to assess the compensation by effects of increased CO_2. Nevertheless, at the current status of knowledge, evidence was made that the positive CO_2 effect is less significant on C_4 crop (e.g., maize, millet, sorghum) for the region (Roudier et al. 2011).

Policy-Induced Recommendations

Given a population growth rate of 2.9% per annum in 2015 and taking into account a maize yield target of 4000 kg ha^{-1} calculated to satisfy a healthy diet of an average smallholder family of 5.4 members (Ngwira et al. 2014), none of the scenarios even with recommended fertilizer rate has reached these yield levels within the limits of the CPD (Figs. 6 and 7). Indeed our climate change impact assessment shows that if current and historical levels of grain yield do not meet the demand, any further reduction in yield, as indicated by the predicted yield losses even with recommended fertilizer rates, will entail some risks of leaching for farmers. The model also highlighted the influence of inorganic fertilizer on increasing the average maize yields on a moderately fertile soil irrespective of the climate model and RCP. Based on these findings, policy direction and support for potential measures to increase maize yield can be reinforced with integrated soil fertility management (combination of mineral fertilizer and crop rotations with legumes, splitting of N fertilizer applications to avoid excessive leaching, low-release N fertilizers) and soil conservation with inputs from agroforestry and water and soil conservation practices that would preserve natural resources while increasing yields.

Fig. 7 (continued) under historical climate. Fert CNRM 4.5 and CNRM 8.5 are treatment with fertilizer under RCP4.5 and RCP8.5, respectively, for regional climate CNRM-CERFACS-CNRM-CM5. Fert MPI 4.5 and MPI 8.5 are treatment with fertilizer under RCP4.5 and RCP8.5, respectively, for regional climate MPI-M-MPI-ESM-LR. Fert ICHEC4.5 and ICHEC8.5 are treatment with fertilizer under RCP4.5 and RCP8.5, respectively, for regional climate ICHEC-EC-Earth

Conclusions

DSSAT model parameterized with site-specific inputs allows evaluating the impacts of projected rainfall variability and temperature on maize production using information from regional downscaling and bias correction of the output of 3 GCMs and the interaction with recommended fertilizer application rates. Total aboveground biomass and grain yield were calibrated and validated with a dataset derived from contrasting sites in Central West Burkina Faso with two soil fertility management options: recommended fertilizer dose and no fertilizer. Overall acceptable $R^2 > 0.5$ and RMSE values were obtained from those exercises. With the validated DSSAT model, the risks associated with future climate change scenarios from 3 GCMs were assessed as well as the effectiveness of fertilizing options to mitigate the effects of rainfall variability on maize yield in the near future under 2 RCPs (4.5 and 8.5). Both non-fertilized and recommended fertilizer treatments responded similarly to the impacts of future climate change, but projections under RCP4.5 contrast to the ones under RCP8.5, and there are also inconsistencies depending on the GCMs. Under RCP8.5, two out of three GCMs showed that maize yield under future climate may increase slightly compared to historical conditions by an average of 17%. However, in RCP8.5, when DSSAT is run with the output of the ICHEC-EC-Earth model, grain yield is projected to decrease by a maximum of 16%, whereas under RCP4.5, maize yield slightly increased by less than 5%. As there is no overall trend of gain or loss in maize yield in the different scenarios (currently taking into account only future changes in rainfall and temperature), there is a need to add the CO_2 effect in future climate impact studies in Burkina Faso.

Acknowledgments This study relies partly on the outputs of the Work Package 1.2 (agroforestry and farm interventions) of the BIODEV Project (2012–2016) funded by the Finnish Government whose significant contribution to combat climate change in West Africa is greatly appreciated.

The assistance provided by the Swedish Meteorological and Hydrological Institute (SMHI) in providing and processing downscaled, bias-corrected regional climate information for climate impact studies is gratefully acknowledged.

References

Araya A, Hoogenboom G, Luedeling E, Hadgu KM, Kisekka I, Martorano LG (2015) Assessment of maize growth and yield using crop models under present and future climate in southwestern Ethiopia. Agric For Meteorol 214–215:252–265

Asseng S, Ewert F, Rosenzweig C et al (2013) Quantifying uncertainties in simulating wheat yields under climate change. Nat Clim Chang 3:827–832

Bakhsh A, Bashir I, Farid HU, Wajid SA (2013) Using CERES-wheat model to simulate rain yield production function for Faisalabad Pakistan, conditions. Exp Agric 9(03):461–475

Bassu S, Brisson N, Durand JL, Boote K, Lizaso J et al (2014) Do various maize crop models vary in their responses to climate change factors? Glob Chang Biol 20:2301–2232

Bationo A, Sivakuman MVK, Acheampong K, Harmsen K (1998) Technologies de lutte contre la dégradation des terres dans les zones soudano-sahéliennes de l'Afrique de l'Ouest. In: Breman H, Sissoko K (eds) L'intensification agricole au Sahel. Karthala, Paris, pp 709–725

Bayala J, van Noordwijk M, Lusiana B, Kasanah N, Teklehaimanot Z, Ouedraogo SJ (2008) Separating the tree-soil-crop interactions in agroforestry parkland systems in Saponé (Burkina Faso) using WaNuLCAS. Adv Agrofor 4:296–308

Boote KJ, Jones JW, Pickering NB (1996) Potential uses and limitations of crop models. Agron J 88:704–716

BUNASOLS (Bureau national des sols) (2001) Etudes morphopédologiques des provinces de Sissili/Ziro. Echelle 1/100000e. Rapport technique n°20. Bunasols, Ouagadougou. 83 pp

Christensen JH, Hewitson B, Busuioc A, Chen A, Gao X, Held R, Jones R, Kolli RK, Kwon W, Laprise R (2007) Chapter 11, Regional climate projections. In: Climate change, the physical science basis. Contribution of Working group I to the Fourth Assessment Report of the Intergovernmental Panel on Climate Change. Cambridge University Press, Cambridge, pp 847–940

Cooper P, Dimes J, Rao K, Shapiro B, Shiferaw B, Twomlow S (2008) Coping better with current climatic variability in the rain-fed farming systems of sub-Saharan Africa: an essential first step in adapting to future climate change? Agric Ecosyst Environ 126:24–35

Coulibaly YN, Mulia R, Sanou J, Zombré G, Bayala J, Kalinganire A, van Noordwijk M (2014) Crop production under different rainfall and management conditions in agroforestry parkland systems in Burkina Faso: observations and simulation with WaNuLCAS model. Agrofor Syst 88:13–28

Driessen P, Deckers J, Spaargaren O (2001) Lectures notes on the major soils of the world. FAO World Soil Resources. Report-94. Food and Agriculture Organization of the United Nations, Rome

Enete AA, Amusa TA (2010) Challenges of agricultural adaptation to climate change in Nigeria: a synthesis from the literature. "Challenges of agricultural adaptation to climate change in Nigeria: a synthesis from the literature". Field Actions Sci Rep 4. http://factsreports.revues.org/678

Fatondji D, Tabo R, Jones JW, Adamou A, Hassane O (2010) Water use and yield of millet under the Zai system: understanding the processes using simulation. In: Humphreys D, Bayot RS (eds) Increasing the productivity & sustainability of rainfed cropping systems of poor smallholder farmers. Proceedings of the CGIAR Challenge Program on Water and Food International Workshop on Rainfed Cropping Systems, Colombo, p 125

Folberth C, Gaiser T, Abbaspour KC, Schulin R, Yang H (2012) Regionalization of a large-scale crop growth model for sub-Saharan Africa: model setup, evaluation, and estimation of maize yields. Agric Ecosyst Environ 151:21–33

Folberth C, Yang H, Gaiser T, Liu J, Wang X, Williams J, Schulin R (2014) Effects of ecological and conventional agricultural intensification practices on maize yields in sub-Saharan Africa under potential climate change. Environ Res Lett 9:044004. https://doi.org/10.1088/1748-9326/9/4/044004. 12 pp

Fontes J, Guinko S (1995) Carte de la végétation et de l'occupation du sol du Burkina Faso. Note explicative. Ministère de la cooperation française, projet Campus, Toulouse. 68 p

Gaiser T, de Barros I, Sereke F, Lange F-M (2011) Validation and reliability of the EPIC model to simulate maize production in small-holder farming systems in tropical sub-humid West Africa and semi-arid Brazil. Agric Ecosyst Environ 135:318–327

Giorgi F, Jones C, Asrar GR (2009) Addressing climate information needs at the regional level: the CORDEX framework. WMO Bull 58:175–183, 2

Hoffmann MP, Haakana M, Asseng S, Höhn JG, Palosuo T, Ruiz-Ramos M, Fronzek S, Ewert F, Gaiser T, Kassie BT, Paff K, Rezaei EE, Rodríguez A, Semenov M, Srivastava AK, Stratonovitch P, Tao F, Chen Y, Rötter RP (2017) How does inter-annual variability of attainable yield affect the magnitude of yield gaps for wheat and maize? An analysis at ten sites. Agric Syst. https://doi.org/10.1016/j.agsy.2017.03.012

Jones JW, Hoogenboom G, Porter CH, Boote KJ, Batchelor WD, Hunt LA, Wilkens PW, Singh U, Gijsman AJ, Ritchie JT (2003) The DSSAT cropping model. Eur J Agron 18:235–265

Luedeling E, Smethurst PJ, Baudron F, Bayala J, Huth NI, van Noordwijk M, Ong CK, Mulia R, Lusiana B, Muthuri C, Sinclair FL (2016) Field-scale modeling of tree-crop interactions: challenges and development needs. Agric Syst 142:51–69

Mohammed A, Tana T, Singh P, Korecha D, Molla A (2016) Management options for rainfed chickpea (*Cicer arietinum* L.) in Northeast Ethiopia under climate change condition. Clim Risk Manag. https://doi.org/10.1016/j.crm.2016.12.003

Ngwira AR, Aune JB, Thierfelder C (2014) DSSAT modelling of conservation agriculture maize response to climate change in Malawi. Soil Tillage Res 143:85–94

Nitiema WJ de D (2009) Contribution des opérations d'urgence de facilitation de l'accès des producteurs a des semences améliorées à l'accroissement du rendement du maïs dans la commune Rurale de Tiefora (province de la Comoe). Graduate Diploma in Agricultural Extension, Institute of Rural Development, Polytechnic University of Bobo-Dioulasso, 70 p

Roudier P, Sultan B, Quirion P, Berg A (2011) The impact of future climate change on West African crop yields: what does the recent literature say? Glob Environ Chang 21:1073–1083

Samaké O, Smaling EMA, Kropff MJ, Stomph TJ, Kodjio A (2005) Effects of cultivation on spatial variation of soil fertility and millet yields in the Sahel of Mali. Agric Ecosyst Environ 109:335–345

Sultan B, Gaetani M (2016) Agriculture in West Africa in the twenty-first century: climate change and impacts scenarios, and potential for adaptation. Front Plant Sci 7:1262. https://doi.org/10.3389/fpls.2016.01262

Sylla MB, Nikiema PM, Gibba P, Kebe I, Kluts NAB (2016) Climate change over West Africa: recent trends and future projections. In: Yaro JA, Hesselberg J (eds) Adaptation to climate change and variability in rural West Africa. Springer International Publishing, pp 25–40. https://doi.org/10.1007/978-3-319-31499-0

Tachie-Obeng E, Akponikpe PBI, Adiku S (2013) Considering effective adaptation option to impacts of climate change for maize production in Ghana. Environ Dev 5:131–145. https://doi.org/10.1016/j.envdev.2012.11.008

Tadele Z (2017) Raising crop productivity in Africa through intensification. Agronomy 7:22. https://doi.org/10.3390/agronomy7010022

Voortman RL, Sonneveld BGJS, Keyzer MA (2003) African land ecology: opportunities and constraints for agricultural development. Ambio 32:367–373

Walkley A, Black IA (1934) An examination of the Degtjareff method for determining organic carbon in soils: effect of variations in digestion conditions and of inorganic soil constituents. Soil Sci 63:251–263

Waongo M, Laux P, Kunstmann H (2015) Adaptation to climate change: the impacts of optimized planting dates on attainable maize yields under rainfed conditions in Burkina Faso. Agric For Meteorol 205:23–39

Weedon GP, Balsamo G, Bellouin N, Gomes S, Best MJ, Viterbo P (2014) The WFDEI meteorological forcing data set: WATCH Forcing Data methodology applied to ERA-Interim reanalysis data. Water Resour Res 50. https://doi.org/10.1002/2014WR015638

Winterbottom R, Reij C (2013) Farmer innovation: improving Africa's food security through land and water management. World Resources Institute WRI. Retrieved from https://www.environmental-expert.com/articles/farmer-innovation-improving-africa-s-food-security-through-land-and-water-management-397859

Yang W, Andréasson J, Graham LP, Olsson J, Rosberg J, Wetterhall F (2010) Distribution-based scaling to improve usability of regional climate model projections for hydrological climate change impact studies. Hydrol Res 41:211–229

Zoellick RBA (2009) Climate smart future. The Nation Newspapers. Vintage Press Limited, Lagos, p 18

Fertilization Strategies Based on Climate Information to Enhance Food Security Through Improved Dryland Cereals Production

46

Komla Kyky Ganyo, Bertrand Muller, Aliou Guissé, and Myriam Adam

Contents

K. K. Ganyo (✉)
Centre d'Etude Régional pour l'Amélioration de l'Adaptation à la Sécheresse (CERAAS), BP, Thiès, Sénégal

Institut Togolais de Recherche Agronomique (ITRA), Lomé, Togo
e-mail: desireganyo@gmail.com

B. Muller
Centre d'Etude Régional pour l'Amélioration de l'Adaptation à la Sécheresse (CERAAS), BP, Thiès, Sénégal

Centre de Coopération Internationale en Recherche Agronomique pour le Développement (CIRAD), UMR AGAP/PAM, Montpellier, France

A. Guissé
Département de Biologie Végétale, Université Cheikh Anta Diop de Dakar (UCAD), Dakar, Sénégal

M. Adam
Centre de Coopération Internationale en Recherche Agronomique pour le Développement (CIRAD), UMR AGAP/PAM, Montpellier, France

Institut de l'Environnement et de la Recherche Agricole (INERA), Bobo Dioulasso, Burkina Faso

International Crops Research Institute for Semi-Arid Tropics (ICRISAT), Bamako, Mali

W. Leal Filho (ed.), *Handbook of Climate Change Resilience*,
https://doi.org/10.1007/978-3-319-93336-8_90

Abstract

Rainfall uncertainty and nutrient deficiency affect sorghum production in Sahel. This study aimed at (i) determining the responses (varieties*water*nitrogen) of various West-African sorghum (Sorghum bicolor L. Moench) varieties to the application of fertilizer (NPK and urea) at selected growing stages according to water regime (irrigated or not, different rainfall patterns) and (ii) simulating them to define alternative fertilization strategies. This chapter proposes alternative fertilization strategies in line with rainfall patterns. Split plot experiments with four replications were carried out in two locations (Senegal), with four improved sorghum varieties (Fadda, IS15401, Soumba and 621B). Treatments were T1, no fertilizer; T2 = 150 kg/ha of NPK (15-15-15) at emergence +50 kg/ha of urea (46%) at tillering +50 Kg/ha of urea at stem extension; T3 = half rate of T2 applied at the same stages; T4 = 150 kg/ha of NPK + 50 kg/ha of urea at stem extension +50 kg/ha of urea at heading, and T5 = half rate of T4 applied at the same stages. Plant height, leaf number, grain yield, and biomass were significantly affected by the timing and rate of fertilizers. Grain yield were affected by water*nitrogen and nitrogen*variety interactions. It varied from 2111 to 261 kg/ha at "Nioro du Rip" and from 1670 to 267 kg/ha at "Sinthiou Malème." CERES-Sorghum model overestimated late fertilizer grain yields. To achieve acceptable grain yield, fertilizers application should be managed regarding weather.

Keywords

Fertilization strategies · Climate information · Food security · Sorghum · Sahel · Modeling

Introduction

Climate is the most important factor that governs food production and is considered as the most weather dependent of all human activities (Hansen 2002) that impacts food security (Tubiello et al. 2007). The severity of climate uncertainties is particularly strong in Sahel where rainfed agriculture is the main source of food and income (Defrance et al. 2017; Ingram et al. 2002). Cereals such as sorghum (*Sorghum bicolor* L. Moench) and millet (*Pennisetum glaucum* (L.) R. Br.) are the main staple foods for millions of people in tropical arid and semi-arid regions of Africa, Asia, and South America (Gueye et al. 2016). They are multipurpose cereals, with grain, forage, and sweet types (Almodares et al. 2007; Gnansounoua et al. 2005).

Although, sorghum and millet are described as rustic and well-adapted crops to tropical and subtropical zones (Legwaila et al. 2003; Bezançon et al. 1997; Kulkarni et al. 1995), cultivating these crops nowadays in Sahel comes with problems. The main limiting factors are nutrient deficiency, short rainy season, and interannual and intraseasonal rainfall variability. A consequence of this rainfall variability is the appearance of drought periods during the growing season (Casenave and Valentin 1989). According to the Intergovernmental Panel on Climate Change (IPCC) fourth assessment report (AR4), climate change resulting in an increase of greenhouse gas emissions will accentuate rainfall uncertainty and extremes events such as extreme droughts, flooding, temperature increase in some parts of world (IPCC 2007). Both variability and change in climate affect food production availability, stability of food supplies, food utilization, access to food, and food prices everywhere in world (Schmidhuber and Tubiello 2007). Dealing with the uncertainty of climate is a challenge for agriculture in general and farmers in particular. Farmers in the Sahel perceiving risks in inputs investments due to unpredictable weather are still using little or no fertilizer, and local varieties, hence maintaining low production level. However, with population growth and increasing demand for food, maximizing crops productivity while minimizing losses and maintaining resilience is a complicated obligation.

To mitigate the effect of rainfall uncertainty and to adapt to droughts during the growing season, it is proposed to manage fertilizer application according to climate information and forecasts. Recent studies (Sultan and Gaetani 2016; Roudier et al. 2012; Sultan et al. 2010) indicated that seasonal forecasts could help to provide advises about varieties choice, sowing date, and inputs use. Roudier et al. (2012) showed that (i) predictions of long rainy season duration (onset and offset) and humid years exclude the use of millet short cycle cultivars because of damages occurring on early maturation and (ii) impact of fertilizer in humid years is positive on yields in South-West Niger. Therefore, forecasts about future climate might help African farmers to take crucial strategic decisions to reduce their vulnerability and increase farm profitability (Sultan et al. 2010).

Field studies on late fertilization showed promising results. Bodson et al. (2003) studied the split of N fertilizer on wheat and found that applying N fertilizer twice at stem elongation and at flag leaf stages gave similar yields with higher grain quality compared to traditional practice (three applications at tillering, stem elongation, and leaf flag stages). Perez et al. (1996) reported an increase rough rice yield by 6% and head rice yield by 17% with an additional N application of 30 kg of N/ha at flowering. Wuest and Cassman (1992) compared the uptake efficiency of preplant versus late-season application of irrigated wheat and observed a significant increase in yield with late application. Brassard (2007) reported that grain maize above-ground biomass yields in Quebec region varies according to year (rainfall), area (soil) and nitrogen (N) rate applied. In general, aboveground biomass increased with an increasing rate of N fertilizer and was more important under low rainfall amount on loam or loam-sandy soil than higher rainfall amount on loam or sandy soil. Rötter and Van Keulen (1997) reported that the response of different maize genotypes (short, medium, long, and short to medium cycle durations) to fertilization in their study varied according to different rainfall regime in the eight study zones. Finally, Keating et al. (1993) and McCown et al. (1991) suggested that farmers would benefit by adjusting cropping practices to rainfed conditions early in the season, i.e., applying fertilizers only in "good" (favorable) seasons.

The objectives of this research were (i) to determine the responses of various West-African sorghum varieties to the application of fertilizer (NPK and urea) at selected stages of development according to water regime (irrigated or not, different rainfall patterns), i.e., varieties*water*fertilizer interaction and (ii) to simulate them to define alternative fertilization strategies in line with climate information. It is hypothesized that (1) late fertilization would be beneficial under rainfed conditions with a well rainfall distribution after a slow beginning of the rainy season; that (2) Fadda (hybrid) is going to reply to fertilization differently from other sorghum varieties, and (3) that long-cycle varieties will benefit to late fertilization compared to short-cycle ones because a catastrophic beginning of rainy season might affect more short-cycle varieties due to their early maturity. This chapter is important because it offers the possibility to diversify fertilization strategies in areas with erratic rainfall and to minimize losses in crop production that could be related to droughts or dry spells in the season.

Description of the Study Areas and Methodology

Study Area

Three experiments were carried out during 2015–2016 growing season at "Nioro du Rip" and "Sinthiou Malème," two experimental stations of Senegalese Institute of Agricultural Research (ISRA) representing the Peanut Bassin and the Eastern Senegal agroecology. Nioro du Rip (13° 44' N; 15°46'W) and Sinthiou Malème (13°46' N; 13°40'W) are located about 260 km and 440 km south-east of Dakar (14°46' N; 17° 21' W), respectively.

Climate in both locations are Soudano-Sahelian with a mono modal rainfall from June to October and with maximum precipitation in August. The two locations differed mainly in total rainfall that occurred during the sorghum growing season. During this period, the total rainfall at Nioro was 1045 mm while it was about 525 mm at Sinthiou Malème. In Nioro du Rip, the rainfall has been complemented by irrigation when needed (early and late season). In Sinthiou Malème, there was no access to irrigation facilities. The mean annual maximum and minimum temperatures for Nioro du Rip and Sinthiou Malème were 35.7 °C and 20.3 °C; 37 °C and 22.2 °C, respectively. Figure 1 shows temperatures, rainfall, irrigation, and timing of fertilization for the 2015–2016 growing season at the two locations.

The soils were similarly sandy at Nioro du Rip and Sinthiou Malème. For the determination of initial soil status parameters, samples were collected from 0–10 cm, 10–20 cm, and 20–30 cm. The initial soil conditions are presented in Table 1, with Sinthiou Malème having a relatively richer soil.

Research Methods

Experimental Design

The experiments were conducted in a split plot design with completely randomized blocks and four replications. Treatments consisted of four different levels and timing of fertilizers application in addition to the control (T1, no fertilizer applied). These

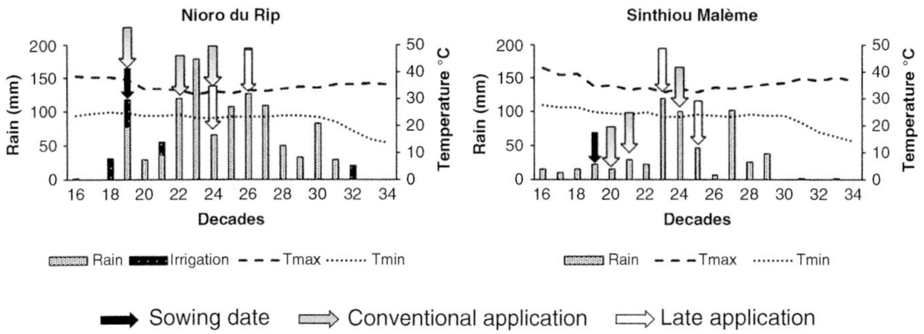

Fig. 1 Weather data, irrigation water, and fertilization dates at Nioro du Rip and Sinthiou Malème, for 621B sorghum variety (applications vary according to variety, due to the relation to growth stages)

Table 1 Initial soil conditions in Nioro du Rip and Sinthiou Malème at three different depths before sowing

	Nioro du Rip			Sinthiou Malème		
	0–10 cm	10–20 cm	20–30 cm	0–10 cm	10–20 cm	20–30 cm
pH	6.7	7.7	7.3	5.7	5.6	5.7
Soil nitrate (N ‰)[a]	3.01	3.51	2.73	4.15	2.49	2.21
Total carbon(C ‰)[b]	0.12	0.14	0.11	0.41	0.37	0.29
C/N	25.85	24.28	23.83	10.03	6.69	7.64

[a]Kjeldahl method
[b]Walkley & Black method

were 150 kg/ha of NPK (15-15-15) at emergence +50 kg/ha of urea (46%) at tillering +50 Kg/ha of urea at stem extension (T2, practice recommended by ISRA); T3 half rate of T2 applied at the same stages (i.e., at emergence, tillering, and stem extension, Fig. 1); T4 = 150 kg/ha of NPK + 50 kg/ha of urea at stem extension +50 kg/ha of urea at heading; and T5 half rate of T4 applied at the same stages (i.e., stem extension and heading, Fig. 1). These were assigned to the main plots. Four improved sorghum varieties (Fadda, IS15401, Soumba and 621B) chosen for their contrasting characteristics (phenology, architecture, response to inputs) were assigned to the subplots. Each subplot consisted of five rows of 6.8 m long with and an interrow spacing of 0.8 m. The inter-hole spacing was 0.4 m. Six seeds per hole were sown and the final plant populations were 62,500 plants/ha.

Experiments were conducted under rainfed and irrigated conditions at Nioro du Rip. Irrigated crops were irrigated four times according to dry spells (5–10 days without rainfall in addition to the visual appearance of plants) with a total of about 110 mm of water (Fig. 1). It was under rainfed conditions at Sinthiou Malème (because of lack of irrigation facilities). After emergence, the cultural operations like weeding, control of pest, and abutment were done as and when necessary for better growth and development of sorghum plants.

Data Collection

Development and growth parameters: Four plants from two center rows were randomly selected and development parameters including leaf number (Nb. Leaf), plant height (PH) were monitored every 15 days from tillering to flowering. Leaf area index (LAI) and aboveground biomass were measured from four hills of three center rows for each subplot. Data presented in this chapter were recorded at flowering for Nb. Leaf, PH, and LAI and at stem extension for biomass before late application (BBLA) and at maturity for biomass after late fertilizer application (BALA).

Yield and biomass: At maturity, 12 hills from the 3 center rows of each subplot were harvested for measuring grain yield and aboveground biomass. Plant material (grains and biomass) was dried first under greenhouse during 2 weeks and in an oven at 65 °C for 24 h. Oven dry weight of the samples was converted to dry matter per ha. To evaluate fertilizers treatment especially plants response to late fertilization, a response index (RI) and a harvest index were calculated as follow:

$$\mathrm{RI} = (\mathrm{Yij}/\mathrm{Y2j})$$

where Yij grain yield with ith fertilization treatment and j the variety and Y2j being the yield for T2 (considered as reference fertilization treatment).

$$\mathrm{HI} = \mathrm{Yij}/\mathrm{YBij}$$

where Yij = grain yield and YB = aboveground biomass yield.

Only harvest data were presented for Sinthiou Malème because growth parameters were not recorded. For the remainder of the chapter, the terms experiment, fertilizer, and variety will be designated by W (meaning water), N (meaning nitrogen), and G (meaning genotype), respectively.

Modeling: Calibration was carried out in the CSM-CERES-Sorghum model of Decision Support System for Agrotechnology Transfer (DSSAT) (Jones et al. 2003) aiming at correctly simulating crop cycles duration and grain yields for Fadda and IS15401 under all treatments. This is a first step to assess model performance in order to define fertilization strategies according to climate information.

Data Analysis

Statistical analyses were performed using R software (R Core Team 2017). First, the statistical analysis was focused on data from each experiment to evaluate the effect of fertilization strategies per site (G*W and G*N). Second, all the data were pooled to assess the effect of total water received on fertilization strategies according to the varieties over experiments (G*W*N). Analysis of variance (ANOVA) of the measured parameters was performed and the treatment means were compared using least significant difference (LSD) at the 5% level of probability.

Results and Discussion

Development and Growth Parameters

Parameters by Environment

Table 2 shows results of the analysis of variance (ANOVA) for development (leaf number, plant height, leaf area index) and growth (biomass before and after late fertilizer application) parameters. The ANOVA showed that, no matter the environment, all parameters are significantly different due to the variety. Guinea varieties (Fadda and IS15401) gave the greatest growth parameters. 621B (caudatum) produced the lowest.

Regarding fertilization, Nb. Leaf and PH were affected by fertilization irrespective of the environment, while LAI and BBLA were affected significantly by fertilization only under rainfed conditions. Conversely, BALA was only affected by fertilization treatments under irrigation. G*N interactions affected significantly LAI under rainfed conditions and Nb. leaf and PH under irrigated ones.

Considering each environment, ANOVA indicated significant effects of N on BBLA under rainfed conditions (Table 2). Plants were stressed experiencing lower biomass under T4, T5, and T1 (882, 685, and 645 kg/ha, respectively). However, BALA was affected only under irrigated conditions with lower dry matter under T1 and T4 (5941 and 5031 kg/ha, respectively). These results indicate that applying reasonable amount of fertilizers later in the season (T5) gives better response, with the plant recovering from water (and nitrogen) stress.

Table 2 Effect of fertilization and varieties on number of leaf, height, leaf area index, biomass before and after late application according to experimental conditions

		Development parameters		Growth parameters		
Experiments (W)[a]	Source	Nb. Leaf (P > F)	PH (P > F)	LAI (P > F)	BBLA (P > F)	BALA (P > F)
Nioro Du Rip Rainfed	**Varieties (G)**	< 0.0001	< 0.0001	< 0.0001	0.027	< 0.0001
	Fertilization (N)	0.013	< 0.0001	0.04	0.011	0.067
	N*G	0.534	0.051	< 0.0001	0.587	0.007
Nioro Du Rip Irrigated	**G**	< 0.0001	< 0.0001	0.0001	0.041	< 0.0001
	N	0.047	<0.0001	0.524	0.13	0.038
	N*G	< 0.008	0.048	0.832	0.897	0.479

G varieties, *N* fertilization, *N*G* interaction between fertilization and varieties, *Nb. Leaf* number of leaf, *PH* plant height, *LAI* leaf area index, *BBLA* biomass before late application, *BALA* biomass after late application

[a]Data at Sinthiou Malème are not presented because not measured

All Environments Included

Irrespective of the environment, the application of 150 kg/ha of NPK at emergence, 50 kg/ha of urea at tillering, and 50 kg/ha of urea at stem extension (T2) produced more leaves (27.78) than the control with no fertilizer (26.47). This highest leaf number is similar to the one produced under T3, T4, and T5. Furthermore, leaf number varied significantly across experiments. Nb. leaf and LAI were neither affected by experiments × fertilization (W*N), nor by experiments × variety (W*G), nor by variety × fertilization (G*N), nor by experiments × fertilization × varieties (G*W*N) interactions.

There was a significant interaction between experiments and fertilization treatments (W*N) and between variety and fertilization (G*N) for plant height. G*W*N interactions for this parameter were not significant. Plant height affected by W*N interaction is presented in Fig. 2. Plant height varied from 2.5 to 1.5 m with an average mean of 2.0 m. High plant size were obtained under T2 in both rainfed and irrigated conditions. Plant height under T4 and T5 was statistically similar to plant height with application of half amount of T2 (T3) in both conditions except T5 under rainfed experiment. Moreover, application of 150 kg/ha of NPK + 50 kg/ha of urea at stem extension +50 kg/ha of urea at heading (T4) and application of its half amount at same stages (T5) in rainfed condition gave similar plant height compared to T4 and T5 under irrigated conditions.

BBLA was not affected by W*N interactions while BALA was (Table 3). Dry matter under T4 in rainfed conditions was higher than T4 under well-watered condition. However, dry matter was lower with T5 under rainfed conditions compared to T5 in irrigated conditions (Fig. 3). As most of the irrigations were applied before late application to avoid dry spells recorded in the beginning of growing season, varieties better response with T4 under rainfed conditions suggested that fertilizer application late-in-season is suitable when rainfall seems to be good after a catastrophic beginning, while in favorable conditions (i.e., irrigated), T5 is recommended rather than T4.

Yield, Plant Biomass, Harvest Index, and Response Index

Yield, Biomass, Harvest Index, and Response Index by Environment

Table 4 shows the summary of ANOVA for grain yield, plant biomass at harvest, harvest index (HI), and response to fertilization index (RI) for the different fertilization and variety treatments. Considering each experiment, variety affected yield, biomass, HI and RI except for yield, and RI at Sinthiou. The fertilization affected significantly yield except at Sinthiou. It affected also biomass and response index in Nioro irrigated and Sinthiou rainfed experiments, respectively.

The splitting pattern of fertilization × variety interactions varied according to environments. Sinthiou Malème and Nioro du Rip (irrigated) experiments which received the lowest and greatest water, respectively were affected significantly by G × N interactions (Table 4) for yield, while the rainfed Nioro du Rip experiment was significantly affected by G*N for biomass.

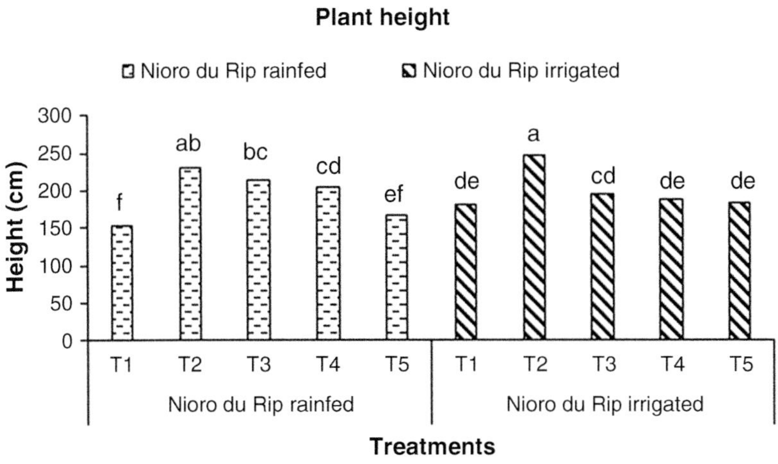

Fig. 2 Interaction between experiment (total water received) and fertilization on plant height. Values are means of four replications. Column followed by same letters are not significantly different at $P < 0.05$. T1 = no fertilizer; T2 = 150 kg/ha of NPK (15-15-15) at emergence +50 kg/ha of urea (46%) at tillering +50 Kg/ha of urea at stem extension; T3 = 75 kg/ha of NPK (15-15-15) at emergence +25 kg/ha of urea (46%) at tillering +25 Kg/ha of urea at stem extension; T4 = 150 kg/ha of NPK + 50 kg/ha of urea at stem extension +50 kg/ha of urea at heading; T5 = 75 kg/ha of NPK (15-15-15) + 25 kg/ha of urea (46%) at stem extension +25 Kg/ha of urea at heading

Table 3 Effect of fertilization, varieties and environments on number of leaf, height, leaf area index, biomass before and after late application

		Development parameters			Growth parameters	
		Nb. Leaf	PH	LAI	BBLA	BALA
Experiments (W)	Source	(P > F)	(P > F)	(P > F)	(P > F)	(P > F)
ANOVA over experiments						
	G	< 0.0001	< 0.0001	< 0.0001	0.004	< 0.0001
	W	< 0.0001	0.527	0.799	0.002	0.709
	N	0.009	<0.0001	0.146	< 0.0001	< 0.0001
	W*N	0.995	0.001	0.232	0.249	0.009
	G*W	0.384	0.737	0.636	0.587	0.221
	G*N	0.091	0.041	0.053	0.964	0.466
	G*W*N	0.237	0.08	0.068	0.79	0.372

G varieties, *W* environment, *N* fertilization, *W*N, G*W, G*N* and *G*W*N* interactions between environment and fertilization; varieties and environment; varieties and fertilization; varieties, environment and fertilization, respectively, *Nb. Leaf* number of leaf, *PH* plant height, *LAI* leaf area index, *BBLA* biomass before late application, *BALA* biomass after late application

Figure 4 displays grain yield for the fertilization × variety (N*G) interactions at Nioro du Rip in irrigated conditions. It varied from 2111 to 261 kg per hectare with a mean value of 1195 kg/ha. Significantly higher grain yield was produced under T2 × Fadda, T5 × Fadda, T2 × Soumba, and T2 × 621B treatments as compared to

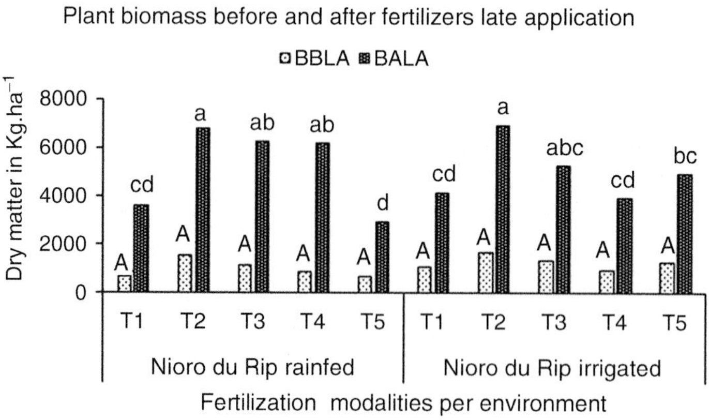

Fig. 3 Interaction between total water received and fertilization on plant biomass before and after late application. Values are means of four replications. Upper case compares BBLA columns and low case comparison between BALA columns. Column followed by same letters are not significantly different at $P < 0.05$. T1 = no fertilizer; T2 = 150 kg/ha of NPK (15-15-15) at emergence +50 kg/ha of urea (46%) at tillering +50 Kg/ha of urea at stem extension; T3 = 75 kg/ha of NPK (15-15-15) at emergence +25 kg/ha of urea (46%) at tillering +25 Kg/ha of urea at stem extension; T4 = 150 kg/ha of NPK + 50 kg/ha of urea at stem extension +50 kg/ha of urea at heading; T5 = 75 kg/ha of NPK (15-15-15) + 25 kg/ha of urea (46%) at stem extension +25 Kg/ha of urea at heading

Table 4 Effect of fertilization and varieties on yield, biomass, harvest index, and response index according to the environments

Source	Yield (P > F)	Biomass (P > F)	HI (P > F)	RI (P > F)
Nioro du rip Rainfed experiment				
G	< 0.0001	< 0.0001	<0.0001	< 0.0001
N	0.04	0.067	0.699	0.129
G*N	0.257	0.007	0.863	0.231
Nioro du rip irrigated experiment				
G	< 0.0001	< 0.0001	< 0.0001	0.026
N	0.033	0.038	0.603	0.196
G*N	0.004	0.479	0.245	0.7
Sinthiou Malème rainfed experiment				
G	0.099	< 0.0001	0.003	0.206
N	0.58	0.79	0.817	0.017
G*N	0.026	0.998	0.501	0.86

G varieties, *N* fertilization, *G*N* interactions between varieties and fertilization, *HI* harvest index, *RI* response index

T4 × Soumba and T5 × Soumba treatments, giving the lowest yield. The grain yield reduction under the T4 and T5 × Soumba treatments was 78% and 88%, respectively, compared to T2 × Fadda treatment which produced the highest grain yield. Intermediate grain yield was obtained with T3 and T4 for Fadda, T5 for IS15401, and with T3 for 621B. Late application gave better grain yield with Fadda under T4

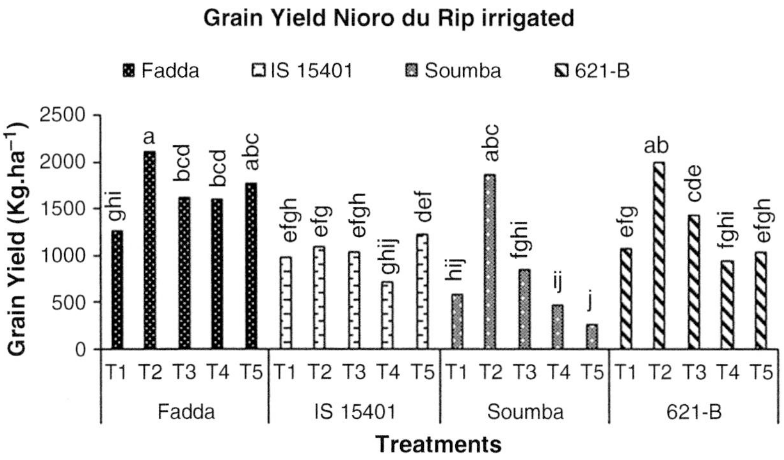

Fig. 4 Grain yield as affected by fertilizers treatments on four sorghum varieties at Nioro du Rip under rainfed with irrigation. Values are means of four replications. T1 = no fertilizer; T2 = 150 kg/ha of NPK (15-15-15) at emergence +50 kg/ha of urea (46%) at tillering +50 Kg/ha of urea at stem extension; T3 = 75 kg/ha of NPK (15-15-15) at emergence +25 kg/ha of urea (46%) at tillering +25 Kg/ha of urea at stem extension; T4 = 150 kg/ha of NPK + 50 kg/ha of urea at stem extension +50 kg/ha of urea at heading; T5 = 75 kg/ha of NPK (15-15-15) + 25 kg/ha of urea (46%) at stem extension +25 Kg/ha of urea at heading

and T5 treatments and with IS15401 under T5. The late application under irrigated conditions seems to give better results with only long cycle varieties. Moreover, this result indicates that late application with low fertilizer rate is more suitable than the total amount applied when we have optimal water conditions.

Grain yield at Sinthiou Malème is represented in Fig. 5. Yield varied from 1670 to 267 kg per hectare with an average value of 888 kg/ha. The grain produced was higher under T2 × Fadda, T3 × Fadda, T4 × IS15401, T5 × IS15401, T4 × 621B, while the lowest yield was obtained with T1 × Fadda. The grain yield reduction under the T1 × Fadda treatment was 84% compared to the T2 treatment for Fadda (highest grain producing treatment). In contrast, T1 × IS15401 was also high (996 kg/ha). The application of fertilizer late-in-season (T4 and T5) increased grain yield for IS 15401 and 621B compared to the actual recommendation in cereal fertilization in Senegal (T2). This result suggests that late application is beneficial for both varieties cycles, under low rainfed conditions.

G*N interactions were significant for two experiments over three. Probably, rainfall distribution in rainfed conditions at Nioro du Rip induced water stress. Sometimes, there was dry spells of 8–9 days up to 10 days before recording rainfall again. Water deficit might reduce nitrogen effect and jeopardize grain filling. The rate and distribution of fertilizer (NPK and urea) considerably affected the grain yield with high production under T2, T4, and T5. The better grain yield recorded with the late fertilizer application treatments (T4 and T5 for some varieties in both conditions) might be due to high nitrogen uptake and water availability (sufficient water uptake for grain filling). The results are

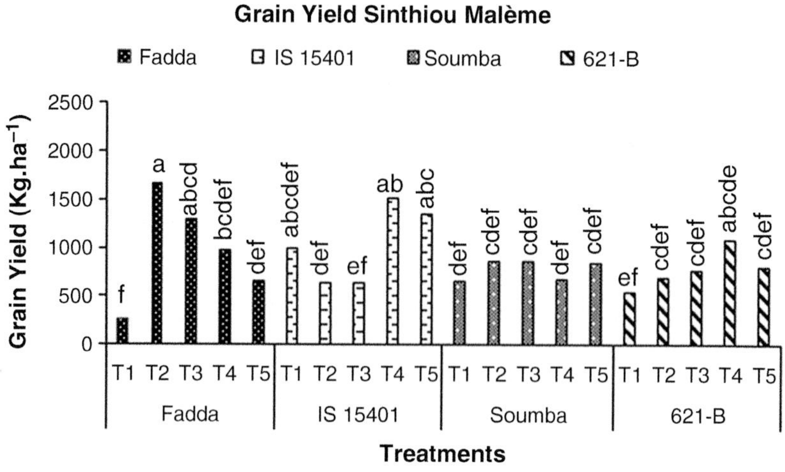

Fig. 5 Grain yield as affected by fertilizers treatments on four sorghum varieties at Sinthiou Malème. Values are means of four replications. T1 = no fertilizer; T2 = 150 kg/ha of NPK (15-15-15) at emergence +50 kg/ha of urea (46%) at tillering +50 Kg/ha of urea at stem extension; T3 = 75 kg/ha of NPK (15-15-15) at emergence +25 kg/ha of urea (46%) at tillering +25 Kg/ha of urea at stem extension; T4 = 150 kg/ha of NPK + 50 kg/ha of urea at stem extension +50 kg/ha of urea at heading; T5 = 75 kg/ha of NPK (15-15-15) + 25 kg/ha of urea (46%) at stem extension +25 Kg/ha of urea at heading

consistent with the findings of Amiri et al. (2014). These authors studying different levels and time of nitrogen application on grain sorghum in Iran reported that the highest sorghum grain yield was obtained with 25% of N applied at planting stage +50% N at stem elongation +25% N at reproductive stage. On the other hand, Blandino et al. (2015) reported granular N soil fertilization at the beginning of heading and foliar liquid N fertilization at anthesis did not influence yield of common wheat but these late-season N fertilizations increased grain yield of durum wheat. However, Fageria and Baligar (1999) observed low grain yield on lowland rice with two late-season N fertilization defined as (i) one-third N applied at planting + one-third at panicle initiation + one-third at flowering and (ii) the application of zero N at sowing + one-third N at the initiation of tillering + one-third N at booting + one-third N at flowering.

All Environments Included

Irrespective of the environment, the fertilization treatment affected significantly grain yield and plant biomass whereas variety affected significantly grain yield, biomass, and HI. Experimental conditions affected grain yield, HI, and RI significantly over experiments. However, neither harvest index nor response to fertilization index varied according to fertilization. W*G interactions affected only yield. Experiment × fertilization × variety (G*W*N) interactions neither affected grain yield nor plant biomass, harvest index, and response to fertilization index (Table 5). Nonsignificant G*W*N interactions indicates that varieties behave similarly in the three experiments.

Table 5 Effect of fertilization, varieties and environments on yield, biomass, harvest index, and response index

Source	Yield (P > F)	Biomass (P > F)	HI (P > F)	RI (P > F)
ANOVA over experiments				
G	< 0.0001	<0.0001	<0.0001	0.403
W	0.0007	0.281	0.001	0.002
N	< 0.0001	< 0.0001	0.817	0.749
W*N	0.045	0.002	0.929	0.476
G*W	0.001	0.084	0.079	0.056
G*N	0.011	0.681	0.622	0.922
G*W*N	0.784	0.828	0.592	0.884

G varieties, *W* environment, *N* fertilization, *W*N, G*W, G*N,* and *G*W*N* interactions between environment and fertilization; varieties and environment; varieties and fertilization; varieties, environment, and fertilization, respectively, *HI* harvest index, *RI* response index

The interactive effect of fertilizers rate and distribution and total water received (experiments) on HI and RI was not significant confirming the potential of late application. The experiment × fertilization (W*N) interactions affected yield and biomass (Fig. 6). T2 irrigated gave the highest grain yield. Better yield with late application were obtained with T5 irrigated and T4 rainfed at Sinthiou. The greatest dry matter was obtained with T2 irrigated, T2, T3, and T4 Nioro rainfed.

Plant biomass at harvest varied according to fertilizers rate and distribution and environment by fertilization interactions. Comparison of means via least significant difference (LSD) showed that the highest dry matter yield (6031Kg/ha) belonged to T2 and the lowest (3739 kg/ha) belonged to the treatment with no fertilizer (T1). T4 and T5 gave identical dry matter but less than T2 one. As nutrients were early available to plant under T2, the N uptake increased and more assimilates were produced.

Grain yield over experiments is significantly affected by G × N interactions (Table 5, Fig. 7). The highest grain yield was produced under application on Fadda with 150 kg of NPK 15-15-15 per hectare at emergence and 100 kg of urea in two equal split at tillering and stem elongation (T2) and the lowest under application of 75 kg/ha of NPK 15-15-15 + 25 kg of urea at stem extension and 25 kg/ha of urea at heading (T5) for Soumba. The grain yield reduction under the T5 × Soumba treatments was 75% as compared to the T2 × Fadda treatment which produced the highest grain yield. Similar and intermediate grain yields were obtained with T2 for Soumba and 621B, T3 for Fadda, and 621B and T4 for Fadda. Late fertilizer applications (T4 and T5) gave better grain yields with IS15401 and Fadda. Variety replied differently to fertilization. Fadda showed a higher grain yield than IS15401, Soumba and 621B with 1403, 842, 770, and 1027 $kg.ha^{-1}$, respectively. These results confirmed the second and third assumptions, that Fadda is going to reply to fertilization differently from other varieties and long-cycle varieties will benefit to late fertilization compared to short-cycle ones.

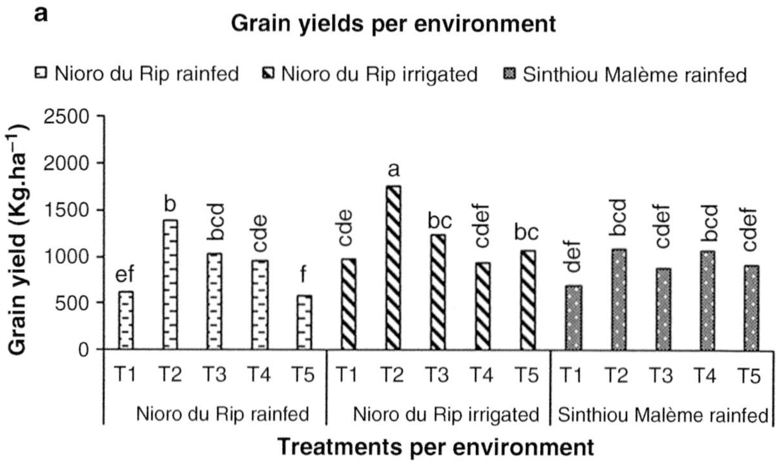

Fig. 6 Grain yield and biomass as affected by fertilization × experiment interaction experiments. Values are means of four replications. T1 = no fertilizer; T2 = 150 kg/ha of NPK (15-15-15) at emergence +50 kg/ha of urea (46%) at tillering +50 Kg/ha of urea at stem extension; T3 = 75 kg/ha of NPK (15-15-15) at emergence +25 kg/ha of urea (46%) at tillering +25 Kg/ha of urea at stem extension; T4 = 150 kg/ha of NPK + 50 kg/ha of urea at stem extension +50 kg/ha of urea at heading; T5 = 75 kg/ha of NPK (15-15-15) + 25 kg/ha of urea (46%) at stem extension +25 Kg/ha of urea at heading

Grain Yield Simulations

Results presented here (Fig. 8) are the first simulations on Fadda and IS15401 at the two sites. Simulations gave good results for T1, T2, and T3 but overestimated yields for T4 and T5. DSSAT gives the same results for T2 and T4 and for T3 and T5. Hence, the delay in fertilizer application is not yet simulated and the model needs to be improved.

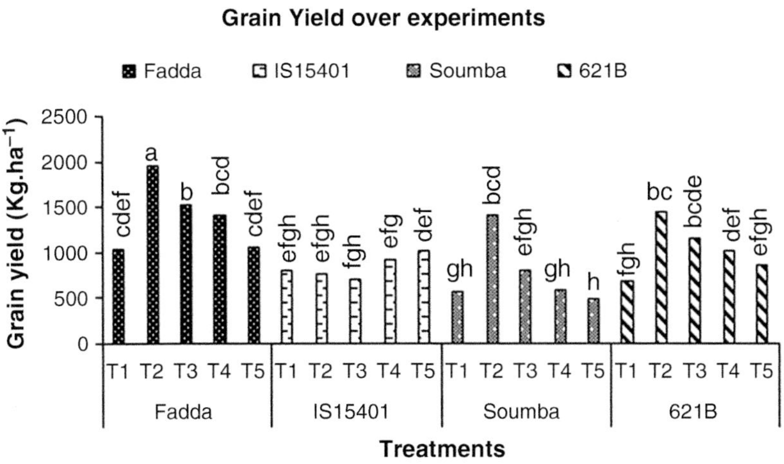

Fig. 7 Grain yield as affected by fertilization x variety interaction over experiments. Values are means of four replications. T1 = no fertilizer; T2 = 150 kg/ha of NPK (15-15-15) at emergence +50 kg/ha of urea (46%) at tillering +50 Kg/ha of urea at stem extension; T3 = 75 kg/ha of NPK (15-15-15) at emergence +25 kg/ha of urea (46%) at tillering +25 Kg/ha of urea at stem extension; T4 = 150 kg/ha of NPK + 50 kg/ha of urea at stem extension +50 kg/ha of urea at heading; T5 = 75 kg/ha of NPK (15-15-15) + 25 kg/ha of urea (46%) at stem extension +25 Kg/ha of urea at heading

Results suggest the need for crop model improvements to better simulate the effect of late-season fertilization on grain yield and thus help to define fertilization strategies based on climate information.

Conclusion

In this research, the effect of different levels of fertilizers and its split application (late-season application) based on climate information (rainfall distribution) was investigated. The results showed that rate and distribution of fertilizers affected significantly plant height, leaf number, biomass, and grain yield. However, the response index was not statistically different when comparing T2 (recommended practice) to other fertilization strategies. Variety*nitrogen (G*N) interactions effects on crop growth and development depend on experimental conditions. Results revealed that under rainfed conditions a late application of 100% of the fertilizers enables the plant biomass, initially stressed, to get back to the same level as if the application was done earlier. However, under irrigated conditions we concluded that T5 was better than T4, for plant biomass as well as for grain yield. Fertilizers late application in-season was beneficial in rainfed condition and Fadda, a hybrid, replied to fertilization differently from other varieties. Both short- and long-cycle varieties replied to the late application. Varieties behaved similarly no matter water and fertilizer received. The findings about a single year study and three water regime

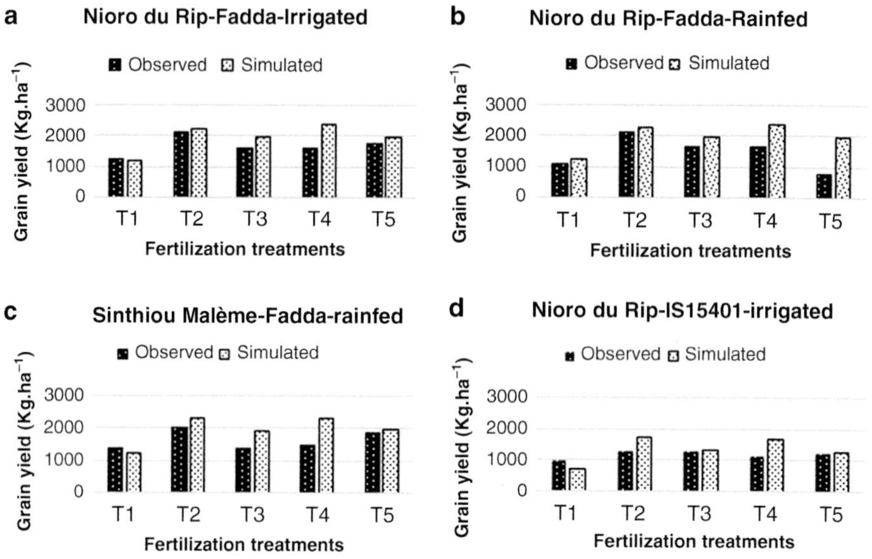

Fig. 8 Comparison of observed grain yield and yield simulated with CERES sorghum for Fadda and IS15401. Values of observed data are means of four replications. T1 = no fertilizer; T2 = 150 kg/ha of NPK (15-15-15) at emergence +50 kg/ha of urea (46%) at tillering +50 Kg/ha of urea at stem extension; T3 = 75 kg/ha of NPK (15-15-15) at emergence +25 kg/ha of urea (46%) at tillering +25 Kg/ha of urea at stem extension; T4 = 150 kg/ha of NPK + 50 kg/ha of urea at stem extension +50 kg/ha of urea at heading; T5 = 75 kg/ha of NPK (15-15-15) + 25 kg/ha of urea (46%) at stem extension +25 Kg/ha of urea at heading

models would be more relevant with more water regime patterns at different soil texture site in long-term study and this will be presented at the conference. Furthermore, the crop model CERES-sorghum seems not to take into account the effects of late fertilizers application. Model improvements will be investigated to broaden decision-making tools that can help to face climate change effects by developing fertilizers recommendation based on rainfall patterns. Once the model is improved, it will define fertilization strategies for sorghum based on rainfall forecasts in Sahel conditions, essential to improve crops production and to maintain resilience in Africa.

References

Almodares A, Hadi MR, Ranjbar M, Taheri R (2007) The effect of nitrogen treatments, cultivars and harvest stages on stalk yield and sugar content in sweet sorghum. Asian J Plant Sci 6:423–426. https://doi.org/10.3923/ajps.2007.423.426

Amiri M, Mojaddam M, Shokouhfar A, Bakhtiarinejad N (2014) The effect of different levels and time of nitrogen application on grain yield, some physiological traits and nitrogen use efficacy in grain sorghum. Indian J Fundam Appl Life Sci 4:223–227

Bezançon G, Renno J-F, Kumar KA (1997) Le mil. In: Hamon S (ed) L'amélioration des plantes tropicales. CIRAD/ORSTOM, Paris, pp 457–482

Blandino M, Vaccino P, Reyneri A (2015) Late-season nitrogen increases improver common and durum wheat quality. Agron J 107:680–690. https://doi.org/10.2134/agronj14.0405

Bodson B, Vancutsem F, Destain J et al. (2003) Evolution du fractionnement de la fumure azotée. In: CRA (ed) Livre Blanc "Céréales." FUSAGx, Gembloux (Belgique), pp 1–8

Brassard M (2007) Développement d'outils diagnostiques de la nutrition azotée du maïs-grain pour une gestion optimale de l'engrais azoté. Université de Laval, Québec

Casenave A, Valentin C (1989) Les états de surface de la zone sahélienne : influence sur l'infiltration, ORSTOM. Institut français de recherche scientifique pour le développement en Coopération, Paris

Defrance D, Ramstein G, Charbit S et al (2017) Consequences of rapid ice-sheet melting on the Sahelian population vulnerability. Proc Natl Acad Sci U S A 114:6533–6538. https://doi.org/10.1073/pnas/.1619358114

Fageria NK, Baligar VC (1999) Yield and yield components of lowland rice as influenced by timing of nitrogen fertilization. J Plant Nutr 22:23–32. https://doi.org/10.1080/01904169909365603

Gnansounoua E, Dauriat A, Wyman C (2005) Refining sweet sorghum to ethanol and sugar: economic trade-offs in the context of North China. Bioresour Technol 96:985–1002. https://doi.org/10.1016/j.biortech.2004.09.015

Gueye T, Sine B, Cisse N et al (2016) Characterization of phenotypic diversity of Sorghum collection for developing breeding material. Int J Sci 5:38–48. https://doi.org/10.18483/ijSci.931

Hansen JW (2002) Realizing the potential benefits of climate prediction to agriculture: issues, approaches, challenges. Agric Syst 74:309–330. https://doi.org/10.1016/S0308-521X(02)00043-4

Ingram KT, Roncoli MC, Kirshen PH (2002) Opportunities and constraints for farmers of west Africa to use seasonal precipitation forecasts with Burkina Faso as a case study. Agric Syst 74:331–349. https://doi.org/10.1016/S0308-521X(02)00044-6

IPCC (2007) Climate change 2007: impacts, adaptations and vulnerability. Contribution of working group II to the fourth assessment of the Intergovernmental Panel on Climate Change. Cambridge

Jones JW, Hoogenboom G, Porter CH et al (2003) The DSSAT cropping system model. Eur J Agron 18:235–265

Keating BA, Wafula BM, Watiki JM, Karanja DR (1993) Dealing with climatic risk in agricultural research – a case study modelling maize in semi-arid Kenya. In: Craswell ET, Simpson J (eds) Soil fertility and climatic constraints in dryland agriculture. ACIAR, Harare, pp 105–114

Kulkarni DP, Almodares A, Somani RB (1995) Sweet sorghum – supplementary sugar crop in Iran. Ann Plant Physiol 9:90–94

Legwaila GM, Balole TV, Karikari SK (2003) Review of sweet sorghum: a potential cash and forage crop in Botswana. J Agric For 12:5–14. https://doi.org/10.4314/uniswa.v12i1.4631

McCown RL, Wafula BM, Mohammed L et al (1991) Assessing the value of a seasonal rainfall predictor to agronomic decisions: the case of response farming in Kenya. In: Muchow RC, Bellamy JA (eds) Climatic risk in crop production. Models and management for the semi-arids tropics and subtropics. CAB Internatioanl, Wallingford, pp 383–409

Perez CM, Juliano B, Liboon SP et al (1996) Effects of late nitrogen fertilizer application on head rice yield, protein content, and grain quality of rice. Cereal Chem 73:556–560

R Core Team (2017) R: a language and environment for statistical computing. R Foundation for Statistical Computing, Vienna

Rötter R, Van Keulen H (1997) Variations in yield response to fertilizer application in the tropics: II. Risks and opportunities for smallholders cultivating maize on Kenya's arable land. Agric Syst 53:69–95. https://doi.org/10.1016/S0308-521X(96)00037-6

Roudier P, Sultan B, Quirion P et al (2012) An ex-ante evaluation of the use of seasonal climate forecasts for millet growers in SW Niger. Int J Climatol 32:759–771. https://doi.org/10.1002/joc.2308

Schmidhuber J, Tubiello FN (2007) Global food security under climate change. Proc Natl Acad Sci U S A 104:19703–19708. https://doi.org/10.1073/pnas.0701976104

Sultan B, Gaetani M (2016) Agriculture in West Africa in the twenty-first century: climate change and impacts scenarios, and potential for adaptation. Front Plant Sci 7:1–20. https://doi.org/10.3389/fpls.2016.01262

Sultan B, Barbier B, Fortilus J et al (2010) Estimating the potential economic value of seasonal forecasts in West Africa: a long-term ex-ante assessment in Senegal. Am Meteorol Soc 2:69–87. https://doi.org/10.1175/2009WCAS1022.1

Tubiello FN, Soussana J-F, Howden SM (2007) Crop and pasture response to climate change. Proc Natl Acad Sci U S A 104:19686–19690. https://doi.org/10.1073/pnas.0701728104

Wuest SB, Cassman KG (1992) Fertilizer-nitrogen use efficiency of irrigated wheat: I. Uptake efficiency of preplant versus late-season application. Agron J 84:682–688

Drip Watering Regimes on Growth Performance, Yield, and Water Use Efficiency of Sorghum in Semi-arid Environment of Tanzania: Effects

47

Athuman Juma Mahinda, Method Kilasara, and Charles K. K. Gachene

Contents

A. J. Mahinda (✉)
Department of Agricultural Engineering, University of Dar es Salaam, Dar es Salaam, Tanzania

Department of Land Resource Management and Agricultural Technology (LARMAT), University of Nairobi, Nairobi, Kenya

Environmental Science and Technology, Kyoto University, Kyoto, Japan

Tanzania Agricultural Research Institute (TARI)-Makutupora Centre, Dodoma, Tanzania
e-mail: athumanmahinda@gmail.com

M. Kilasara
Department of Soil and Geological Sciences, Sokoine University of Agriculture, Morogoro, Tanzania
e-mail: methodkilasa@gmail.com

C. K. K. Gachene
Department of Land Resource Management and Agricultural Technology (LARMAT), University of Nairobi, Nairobi, Kenya

Kenya Agriculture and Livestock Research Organisation, Nairobi, Kenya
e-mail: gachenecharles@gmail.com

W. Leal Filho (ed.), *Handbook of Climate Change Resilience*,
https://doi.org/10.1007/978-3-319-93336-8_92

Abstract

The inadequacy of water for crop production due to low and unreliable rainfall is the characteristic of dryland areas, including Tanzania, whose more than 60% of its arable land is prone to drought and has high frequencies of crop failure. In enhancing climate change resilience for crop production in these areas, field trials were conducted in the semi-arid environment of central Tanzania to assess the effectiveness of three drip watering regimes on the growth performance, grain yield, and water use efficiency (WUE) of sorghum. The irrigation treatments, irrigating early in the morning (EM), late in the evening (LE), and both early in the morning and late in the evening (ELE), were replicated thrice in a randomized complete block design for the two seasons. The results showed that sorghum grain yield and growth parameters were significantly higher ($p < 0.05$) by almost two-fold when the crop was irrigated twice a day compared to the single drip watering regimes. The maximum grain yield (13.12 t/ha) and biomass (24.91 t/ha) were recorded in sorghum irrigated twice a day in the dry and dry-wet season, respectively. While the highest WUE (4.53) was registered from the sorghum irrigated twice a day in the dry-wet season, the lowest WUE (1.97) was registered in sorghum irrigated late in the evening in the dry season. The findings suggest that, although irrigating early in the morning or late in the evening in dry-wet season results in nine times more grain yield than under rain-fed condition (0.9 t/ha), it is recommended to irrigate sorghum twice a day, not only in the dry season but also during the rainfall season by supplementing hydric deficit for a higher growth performance, grain yield, and efficient water productivity.

Keywords

Climate change · Water scarcity · Crop failure · Drought resilience · Dryland areas · Drip irrigation · Irrigation in the morning · Irrigation in the evening · Hydric deficit · Sorghum performance

Introduction

Climate change has had altered hydrological cycles and weather patterns, resulting in the increased intensity and frequencies of extreme weather conditions (Gregory et al. 2005; Jarvis et al. 2010; Thorton et al. 2011; IPCC 2014). Although its impacts cut across different sectors in all regions around the world, it has been observed to have more adverse effects in developing countries particularly in the arid and semi-arid areas (ASALs) of Sub-Saharan Africa (Shemdoe 2011; Jerry et al. 2012; Mahinda et al. 2016). This is mainly because inhabitants of these regions rely on natural resources which are more susceptible to the impacts of the climate change.

The problem of water scarcity, which is heavily influenced by the effect of climate change and its variability, is expected to worsen more on crop production in the ASALs if farmers continue to rely on rain-fed agriculture. According to IPCC (2014), there will be increasingly failure of crops in the ASALs, which constitute most countries of Sub-Saharan Africa. Since agricultural activities in this region rely on rain-fed condition, any adverse effect on climate change could have a devastating effect on food security and livelihood of the majority (Challinor et al. 2005, 2009, 2010; Rodima-Taylor 2012; Mahinda et al. 2016). This serves as a caution for farmers in the ASALs of Sub-Saharan Africa not to solely depend on rain-fed agriculture in this era of higher frequencies of weather and climate variability.

Like the rest of Sub-Saharan Africa countries, ASALs of Tanzania are characterized by low rainfall which is unreliable for food and commercial crop production. The areas are prone to harsh drought condition almost throughout a year. The insufficient amount of rainfall received is poorly distributed to meet crop water requirement of the major crops grown. Farmers in the ASALs of Tanzania have adopted different soil and water conservation measures to improve crop production. For instance, pits have been used as the traditional sowing method to collect water in-situ during the early growing stage of the crop (Hatibu and Mahoo 1999; Shemdoe 2011). Traditional terraces (Matuta), hillside sheet, and runoff utilization, which exploits valley bottoms and plains where runoff collects, have been widely used by farmers for growing high-water demand crops such as maize and grapes. However, due to the high rate of evapotranspiration, the systems have been unsuccessful (MAFC 2013; Mahinda et al. 2016).

Dodoma region is typical of ASALs of Tanzania. The area receives very low (300–600 mm), erratic, and unpredictable rainfall to meet crop water requirement. For several years, there has been persistent crop failure even for drought-tolerant crops such as sorghum, which is the main staple food and one of the cash crops in ASALs of Tanzania and the world in general. Currently, sorghum is demanded by brewing industry for making clear-based alcoholic and nonalcoholic drinks. It also needed in the baking industry for making bread, biscuit, cakes, and pops. Although highly demanded, its supply to meet daily calorie requirement and raw materials for brewing and baking purpose is limited. The farming communities have been maintaining their traditional subsistence farming for sorghum production under flat cultivated land. This farming practice has resulted in low crop yield production per unit area, food insecurity, low-income, and low purchasing power of agricultural inputs (Mbwaga et al. 2007; Pages et al. 2010; Ahmed et al. 2011; Kangalawe and Lymo 2013; IPCC 2014; Kilasara et al. 2015; Mahinda et al. 2016).

In the context of enhancing drought resilience that would improve crop production and contribute to the stabilization of food security and livelihood in the dryland areas, while sustaining scarce water resource and the environment, drip irrigation could be one of the best viable options in exploiting underground water resources, intermittent rivers, and harvested rainwater for sorghum production. However, there is no scientific information on how frequently the crop should be irrigated, neither the amount nor the appropriate time for irrigating sorghum. This paper, therefore, is important in bridging some gaps in knowledge that guide practicability application of the drip irrigation for enhancing sorghum growth performance, grain yield, and water use efficiency in the dryland areas of Tanzania and many other in Sub-Saharan Africa.

Description of the Study Area and Methods

Study Area

The study was conducted in the peri-urban area of Dodoma district at Tanzania Agricultural Research Institute (TARI)-Makutupora center, located between 05°58′S and 35°57′E. The area is characterized by a monomodal rainfall pattern. The rainfall lasts in April, followed by a long dry season that extends from May to the end of November/December. The annual rainfall in the area ranges from 300 to 600 mm with a mean of 450 mm, marked with high variation in amount and distribution. The soil temperature regime is hyperthermic, and the soil moisture regime oscillates between ustic and aridic depending on the rainfall amount, distribution, and pattern in a year under consideration. The area has the soil originating from an alluvium derived from a granitic rock that well developed into a soil profile. The physiographic position of the area is foot slope (piedmont) with flat surrounding landform with a slope of about 1–2%.

Soil and Water Sampling

A systematic soil sampling was done at a 20 cm depth, followed by analytical soil testing to determine the physical and chemical properties of the soil. The soil physical property, soil texture, and chemical properties mainly nitrogen (N), phosphorus (P), potassium (K), potential hydrogen (pH), and percentage organic carbon (%OC) were determined following standard laboratory procedures compiled by Okalebo et al. (2002).

The water for drip irrigation was obtained from the house water piping system from Mzakwe ground water reservoir in Dodoma. The suitability of water quality for irrigation was tested using appropriate laboratory procedures as described by Katerji et al. (2003). The irrigation water was analyzed for its electoral conductivity (EC), potential hydrogen (pH), sodium adsorption ratio (SAR), chloride ion (Cl^-), sodium ion (Na^+), potassium (K^+), calcium (Ca^{2+}), and magnesium (Mg^{2+}). The chemical properties of the tested water for drip irrigation are shown in Table 1. The water quality assessment and interpretation was based on the rating suggested by Park et al. (2014).

Experimental Layout

A randomized complete block design (RCBD) was used to assign treatments in a random manner. Treatments assigned were three drip irrigation regimes, namely, irrigation early in the morning (EM), late in the evening (LE), and both early in the morning

Table 1 Water quality analyzed for irrigating sorghum in the study area at TARI-Makutupora center

Water quality parameter	Value measured	Differential criteria	Severity
K^+ (ppm)	14.9	<20	Not a problem
Ca^{2+}(ppm)	50	20–250	Increased problem
Mg^{2+}(ppm)	30	20–40	Increased problem
Na^+(ppm)	60	<70	Not a problem
Cl^- (ppm)	40	<70	Not a problem
EC (Siemens/cm)	0.75	0.5–0.75	Not a problem
pH	7.9	6.5–8.0	Not a problem
SAR meq/L	9.5	<10	Not a problem

Water quality assessment was based on the rating suggested by Park et al. (2014)

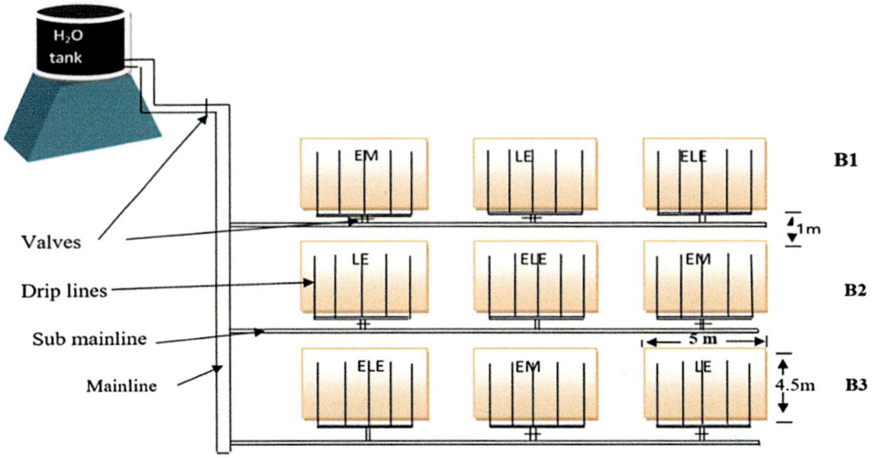

Fig. 1 Experimental field layout and drip irrigation system set up

and late in the evening (ELE). The three drip irrigation regimes were replicated three times (Fig. 1) for the two consecutive seasons (dry season and dry-wet season).

There was no experimental control plot in this study. This is because sorghum was grown during the dry and dry-wet season. The sorghum in the experimental control plots could not have survived at that time. In the dry season, sorghum was sown on 8 August 2013 and harvested on 9 December 2013, while in the dry-wet season, sowing was done on 11 October 2013, 2 months before the onset of the rainfall, and harvested on 7 February 2014.

The size of each experimental plot was 22.5 m^2 spaced by a 1 m path, thus giving the experimental area a total of 214.5 m^2. Seeds of a high-yielding and drought-tolerant sorghum variety, Macia (SDS 3220), were sown at a spacing of 25 × 90 cm. Each experimental plot was having a total of 100 plants, giving 900 plants population in an entire experimental area. The amount of water supplied to every plant was 7 mm on a daily basis (McWilliams 2002; Stichler and Fipp 2003; Assefa et al. 2010;

Table 2 Amount of water used to irrigate sorghum during dry and dry-wet season

				Dry-wet season			Dry season	
Crop stage	Crop Kc	Growing days	Total rainfall (mm)	Suppl. irrig. (mm/crop)	Gross irrig. mm/crop/day	Gross irrig. (L/ha/cropping season)	Gross irrig. mm/crop/day	Gross irrig. (L/ha/cropping season)
Initial growth	0.35	20	0	18.0	18.0	16000.0	18.0	16000.0
Develp.	0.75	30	0	93.9	93.9	125200.0	93.9	125200.0
Mid season	1.1	40	128.3	101.8	230.1	180977.8	234.8	417422.2
Late stages	0.65	30	58.7	100.3	159.0	133733.3	164.0	218666.6
Total		120	187.0	314.0	501.0	455911.1	510.7	777288.8

Develp. development, *Suppl.* supplemental, *Kc* crop coefficient, *Irrig.* irrigation

Seleshi et al. 2009; FAO 2013). Some adjustments were made depending on the sorghum growth stage, weather, and soil moisture condition as indicated in Table 2. The daily crop coefficient adjustment was numerically determined following an equation suggested by Allen et al. (1998)

$$\mathrm{Kc\ i} = \mathrm{Kc_{prev}} + \left[\mathrm{i} - \sum \left(\mathrm{L_{prev}}/\mathrm{L_{stage}}\right]\left(\mathrm{Kc_{next}} - \mathrm{Kc_{prev}}\right)\right. \tag{1}$$

whereby

i: day number within the growing season [1... length of the growing season]
Kc: crop coefficient on day i
L_{stage}: length of the stage under consideration [days]
$\sum (L_{prev}$: sum of the lengths of all previous stages [days]

The pan-evaporation method was used to compute the daily crop water requirement (ETcr, mm). Field installed tensiometers and gravimetric water content measurement methods were used to monitor the amount of water in the sorghum root zone. Three tensiometers were inserted at the three depth zones: above, at the active root zone, and below the root zone at depths of 10, 25, and 40 cm, respectively. Augering was done to ascertain the tensiometers installed at different depths. The augered soil samples were immediately packed in the separate plastic bags and transported to the laboratory. The soil samples were oven-dried at 105 °C to a constant weight for the determination of gravimetric water content and hence volumetric water content.

Determination of Water Use Efficiency (WUE)

In determining sorghum water use efficiency, two parameters were measured, i.e., sorghum biomass in kilogram (kg) and the amount of water used by sorghum (ET_{cr})

in mm. The biomass was obtained by harvesting the whole mature sorghum crop at the soil surface level. The harvested sorghum plant was oven-dried at 75 °C to a constant weight. The dry matter weight in kg was weighed using digital electronic weighing balance. The water used by sorghum was obtained by summing up the recorded daily consumptive water use (ET_{cr}) of the sorghum at every growth stage, from the initial growth stage to physiological maturity. Water use efficiency was determined by dividing the total amount of dry matter (kg) by the total amount of water utilized by the crop in a season (mm).

$$\mathrm{WUE} = \mathrm{DM}/\mathrm{ET}_{cr} \tag{2}$$

whereby, WUE, sorghum water use efficiency; DM, biomass in kg, ET_{cr}, amount of water transpired (mm).

Growth Parameters Measurement

Plant height, stem thickness, leaf width, leaf length, and leaf numbers were measured at every phenological stage to determine the sorghum growth performance under different irrigation regimes. All the crop growth parameters were recorded when a sorghum phenological stage in question was at $\geq$50%. Leaf width and length were lately used to compute leaf area index (LAI).

$$\mathrm{LAI} = \mathrm{LA}/\mathrm{GA} \tag{3}$$

whereby, LAI, leaf area index; LA, leaf area; GA, ground area.

Plant height, leaf width, and leaf length were measured using a measuring tape. Stem thickness was measured using a Vernier caliper.

Grain Yield Measurement

The harvested sorghum grains were sun-dried, and its moisture content was analyzed using a digital grain moisture meter (SINAR grain moisture analyzer). The grains were then weighed with an electronic weighing balance, and its yield response to different watering irrigation regimes was correlated to the safe grain moisture storage level of 12% as recommended by FAO-Food Security Department (1999).

Statistical Data Analysis

GenStat Discovery 14.2 version was used as an aiding software tool to analyze the data at 95% level of significance. The means were separated by employing Duncan's new multiple range test (Wim et al. 2007).

Results and Discussions

Physicochemical Properties of the Soil and Climatic Condition of the Study Area

Laboratory analysis results for the physical and chemical properties of the soil are shown in Table 3. The pH was 7.9, and total nitrogen was 0.13, while the extractable phosphorus was 39.17 mg/kg soil. The soil had a narrow C/N ratio of 6:1. The texture of the soil was sandy clay loam with 23% clay, 7% silt, and 70% sand. Taxonomically, the soil is classified as Arenic Haplustalf, and in the World Reference Base (WRB), the soil is classified as Haplic Luvisol (Aridic, Profondic). The soil temperature regime of the study site is hyperthermic, and during the experimentation period, the soil moisture regime was aridic.

The pattern and amount of the rainfall on monthly basis, and in 10 days interval during the experimentation period, are indicated in Fig. 2a and b, respectively. There was no rainfall when the experiment was started in August and October. The rainfall started on 11 December 2013 and ended on 30 April 2015 with a total rainfall amount of 369.8 mm a year for the 2013/2014 cropping season. The highest amount of rainfall received in a month and a day were 125.6 and 30.5 mm, respectively.

The average temperature throughout the research period was 24.04 °C. While the highest daily temperature recorded was 29.4 °C in 13 November 2013, the lowest daily temperature registered was 17.2 °C on 3 January 2014. The daily variation in air temperature and rainfall in the study area during the research period is shown in Fig. 3.

Effect of Drip Irrigation Regimes on the Growth and Development of Sorghum

The parameters whose data are reported below are for the evaluation of sorghum growth performance, biomass, water use efficiency, and grain yield. The parameters

Table 3 Some chemical and physical properties of the soil in the study area at TARI-Makutupora center

Parameter	Value
Soil pH (H_20)	7.9
% total nitrogen	0.13
Olsen-extractable P (mm/kg)	39.17
Exchangeable K (Cmol/kg)	0.71
% organic carbon	0.74
% organic matter	1.3
C/N	6:1
% clay	23
% silt	7
% sand	70
Textural class	Sandy clay loam

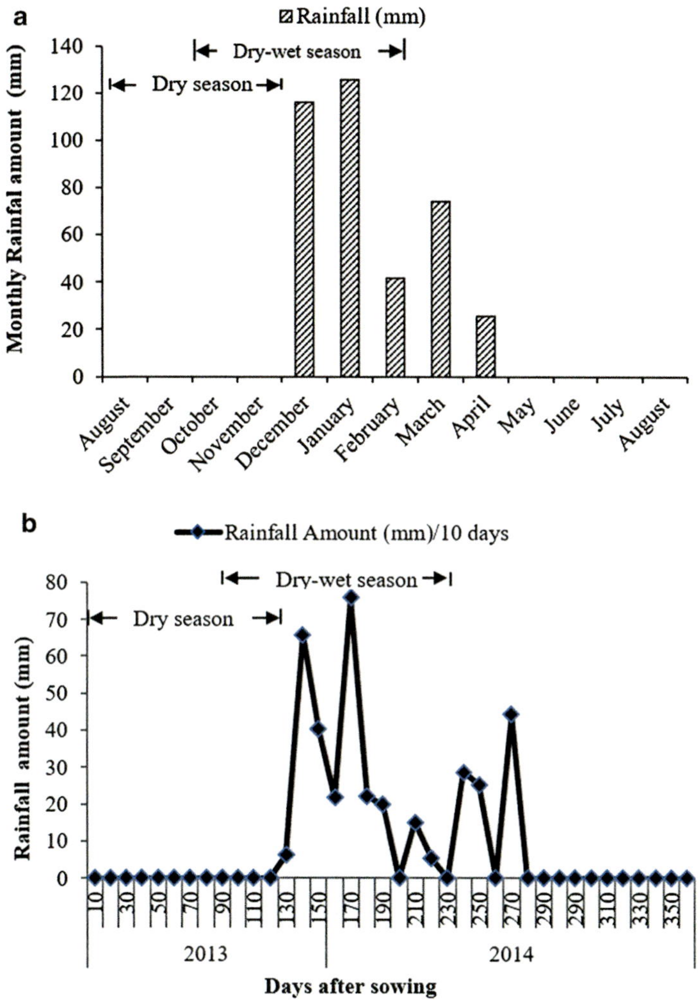

Fig. 2 (**a**) Monthly rainfall amount (mm). (**b**) Line graph of rainfall pattern and amount in 10-days interval (mm)

for sorghum growth and development are seedling emergence, plant height, number of leaves, stem thickness, and leaf area index.

Effect of Different Irrigation Regimes on Sorghum Seedlings Emergence

The results for the seedling emergence are shown in Fig. 4. There were no significant differences (P >5%) in sorghum seedlings emergence between the treatments in both seasons. This is because the amount of water that was irrigated in all the plots

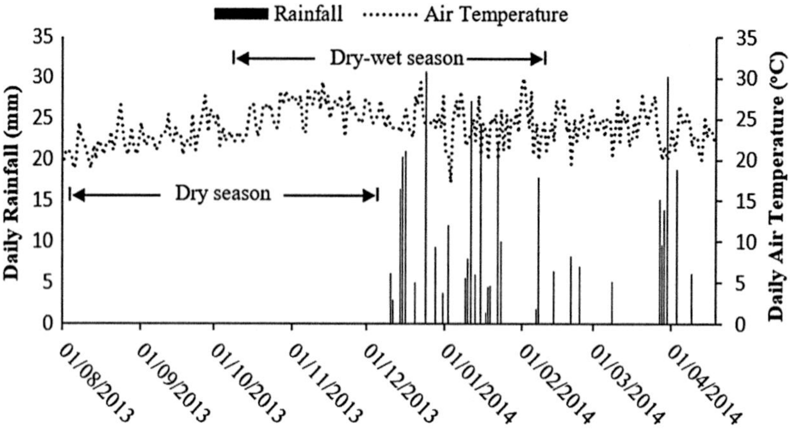

Fig. 3 Recorded daily rainfall and air temperature during the experimentation period of 2013/2014

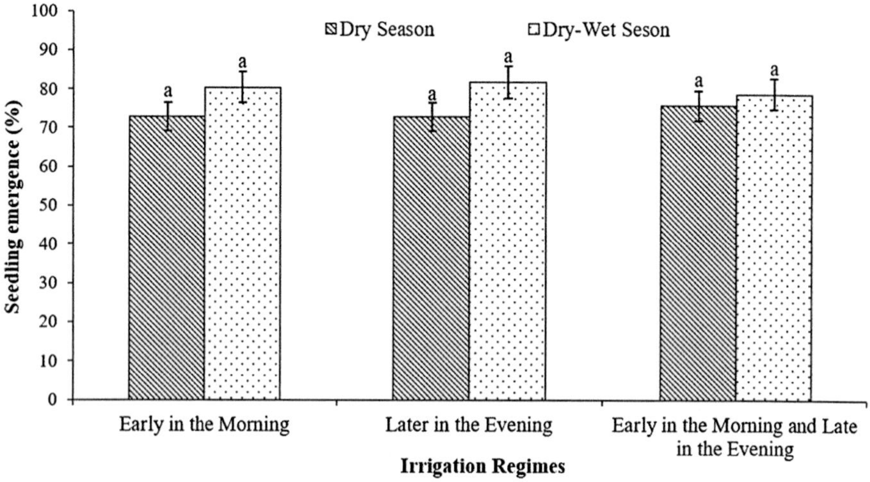

Fig. 4 Response of sorghum seedlings emergence to different irrigation regimes

after sowing was optimal for seed germination and hence seedlings emergence. The result agrees with those obtained by Hossein et al. (2009), who reported uniform seedlings emergence in seeds sown under optimal soil moisture conditions. The lack of the differences between the seasons can be explained by the similarity in favorable germination condition influenced by the optimal soil temperature and the presence of adequate soil moisture.

Although statistical analysis showed no significant differences in seedlings emergence between the two seasons, there was a higher numerical performance of seedling emergence in the dry-wet season by 7.66%, 9%, and 3.02% for EM, LE, and ELE, respectively. The numerical percentage increment in seedling emergence in the dry-wet season was due to more conducive weather condition and soil temperature that accelerated seedling emergence.

Effect of Different Irrigation Regimes on Sorghum Height

The response of sorghum height to different irrigation regimes and seasons was significantly different ($P < 0.05$) in some phenological stages as indicated in Table 4. The performance of sorghum plant height in the dry-wet season was significantly higher ($P < 0.001$) compared to the dry season. This was due to more favorable weather conditions for crop growth and low incidences of pests and diseases in the nearby and during the rainfall period of December and January. It might also be influenced by negligible un-effective rainfall that was falling in the last 2 months of the crop growth.

The highest plant height (138 cm) was measured in dry season at soft dough stage in the plots irrigated twice a day, while the shortest plant height (36.6 cm) was recorded in sorghum irrigated late in the evening in the dry season. The highest sorghum plants heights in the two seasons were 138 cm and 118 cm, for the dry-wet and dry season, respectively. In the dry-wet season, the shorter plant height recorded in LE compared to EM plot was attributed to fact that during the evening and night hours; the rate of transpiration drops down and plant water uptake decreases to

Table 4 Means separation table for the percentage of sorghum seedling emergence and plant height grown in two seasons under different irrigation regimes, at different phenological stages

			Plant height (cm)				
Season	Irrigation regime	Germination percentage	Vegetative	Flag leaf	Flowering	Soft dough	Hard dough
1	EM	72.67a	41.3a	54.21a	55.75a	74.6a	75.9a
1	LE	72.67a	36.6a	61.26b	69.0ab	78.2a	79.6a
1	ELE	75.65a	45.85a	82.78d	98.37c	115.9c	118c
2	EM	80.33a	66.15b	78.5 cd	88.05c	97.6b	98.2b
2	LE	81.67a	64.69b	75.54c	83.69c	91.9b	92.2b
2	ELE	78.67a	83.56c	104.9e	121.3d	138d	136d
LSD, $P < 0.05$	Season	11.33	8.97	5.996	10.42	6.17	6.25
	Irrigation	13.88	10.99	3.462	12.76	7.56	7.65
	Season*irrigation	19.63	15.54	4.240	18.04	10.69	10.82
CV(%)		14	15.2	4.3	11.5	5.9	5.9
F-STAT	Season	ns	***	***	***	***	***
	Irrigation	ns	*	***	***	***	***
	Season*irrigation	ns	ns	ns	ns	ns	ns

EM early in the morning, *LE* late in the evening, *ELE* early in the morning and late in the evening $P > 5\%$, nonsignificant (ns); $P = 5\text{–}1\%$, significant (*); $P = 1\text{–}0.1\%$, very significant (**); $P < 0.01\%$, very highly significant (***)
[1]Dry season
[2]Dry-wet season

negligible level, as relative humidity is higher and atmospheric demand for water is very low. In night-time, there was readily available moisture in the soil, but the plant could not absorb it, as soil-water-atmospheric continuum process ceased. As time passed, the lateral and vertical movement of the available soil moisture at the effective root zone of the crop decreased. This finding agrees with Guang-Cheng et al. (2008), who reported that when the irrigated water is not taken up by the plant in the root zone due to low atmospheric demand, water starts to move down the gradient until it is equally distributed. However, in the dry season, this supporting evidence was only true at the vegetative growth stage; suggesting that, apart from irrigation regimes, plant height is also influenced ($P < 0.001$) by other compounded season's climatic weather variations. When a single drip irrigation was done once a day in the dry season, sorghum plant height in the LE plot increased in height than those in EM at flag leaf ($P < 0.001$), flowering ($P > 0.05$), soft dough ($P > 0.05$), and hard dough stage ($P > 0.05$). During the dry-wet season, EM irrigation regime consistently resulted to taller ($P > 0.05$) plant height than LE across all the sorghum phenological stages.

During the daytime water was highly utilized for photosynthesis, irrigated water was taken up by the plant for transpiration purpose. Therefore, when the sorghum was irrigated both early in the morning and late in the evening, it took advantage of the available soil moisture to equilibrate with atmospheric moisture demand. On the process, the manufactured assimilates were translocated for the normal growth and other physiological activities of the crop, which resulted into a significant ($P < 0.001$) increment in plant height as captured in Fig. 5.

Fig. 5 A side view look of the sorghum plant height performance in the twice double irrigation regime (ELE) versus the single drip irrigation (EM) regime in the study area at Tanzania Agricultural Research Institute (TARI)-Makutupora centre

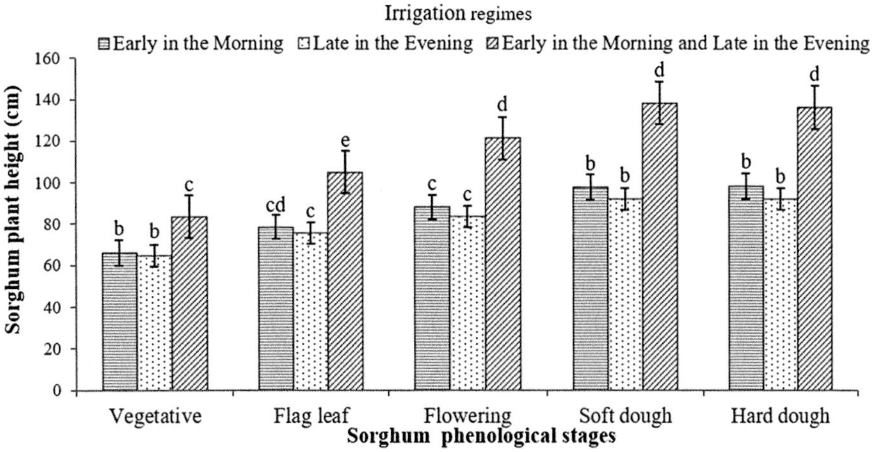

Fig. 6 Average response of sorghum plant height to different irrigation regimes in dry and dry-wet season, at different phenological stages

The results showed that ELE significantly (P <0.001%) increased sorghum plant height compared to EM and EL. This indicates that irrigating twice a day maintains an appreciable amount of readily available soil moisture in the root zone that is easily absorbed by the crop for its optimal primary growth rate (Fig. 6).

Effect of Different Irrigation Regimes and Season on Leaves Number

The response of leaves number per plant to different irrigation regimes, at different phenological stages in the dry season was not significantly different (P >0.05). However, in the dry-wet season, a significant difference (P <0.05) between LE and ELE and EM and ELE was noted at soft dough and hard dough stage, respectively. At this season, EM had fewer number of leaves (P <0.05) than the rest of the treatments. The maximum average number of leaves in dry season was at the flag leaf stage as depicted in Table 5. The leaves number was observed to increase from vegetative to flag leaf stage, thereafter started to decrease due to withering. This implies that flag leaf is the last leaf the sorghum plant produces apart from those arising from tillers and abnormal sorghum growth. In contrast to the dry season, after the maximum number of leaves was attained at flag leaf stage, in the dry-wet season, LE and ELE consistently maintained the same number of leaves until the harvest time. Exception was noted in the EM, whereby the leaves number kept increasing up to the soft dough stage. This increase was due to the formation of tillers in the former treatment.

The results showed a significant difference (P <0.001) in leaves number between the two seasons at vegetative, soft dough, and hard dough stages. In both seasons, sorghum which was irrigated twice a day registered numerically few number of leaves (P >0.05) at vegetative and soft dough stage. The well-irrigated sorghum had long internodes and took a relatively long time to grow, resulting in a few leaves.

Table 5 Means separation table for sorghum plant leaves number grown in two seasons under different irrigation regimes, at different phenological stages

Season	Irrigation regime	Number of leaves				
		Vegetative	Flag leaf	Flowering	Soft dough	Hard dough
1	EM	8.67a	10.4ab	10.4a	10.2abc	10.2abc
1	LE	8.593a	10.6ab	10.5a	9.9ab	9.93ab
1	ELE	8.44a	10.07a	10.3a	9.7a	9.7a
2	EM	9.29ab	10.5ab	10.5a	11.0cd	10.3bc
2	LE	10.15b	10.9ab	10.9a	10.9d	10.9d
2	ELE	10.1b	10.7ab	10.7a	10.7c	10.7d
LSD5 <0.05	Season	0.560	0.3990	0.468	0.3229	0.3085
	Irrigation	0.662	0.4887	0.573	0.3955	0.3779
	Season*irrigation	0.963	0.6911	0.810	0.5592	0.5344
%CV		5.6	3.6	4.2	3.0	2.9
F-STAT	Season	***	ns	ns	***	***
	Irrigation	ns	ns	ns	ns	ns
	Season*irrigation	ns	ns	ns	ns	*

EM early in the morning, *LE* late in the evening, *ELE* early in the morning and late in the evening
P >5%, nonsignificant (ns); P = 5–1%, significant (*); P = 1–0.1%, very significant (**); P <0.01%, very highly significant (***)
[1]Dry season
[2]Dry-wet season

Physiologically, when the plant is grown under constantly replenished appreciable soil moisture condition, it grows to its genetic potential while extending out greenish, broad, and long leaves for active photosynthesis (Salih et al. 1999). At vegetative and soft dough stages, the counted leaves suggested a different finding from those of Monsour (2013), who reported a decrease in some sorghum leaves as a result of decreased soil moisture. However, at flag leaf stage in the dry season, the sorghum double irrigated twice a day led to similar findings to his work. At this growth stage, the double irrigated sorghum had relatively many leaves number per plant compared to single irrigated sorghum. When sorghum was irrigated in the dry-wet season, the leaves number neither complemented the work of Salih et al. (1999) nor the findings of Monsour (2013). This shows that the ability of the sorghum plant to produce leaves number is so versatile to be guaranteed by either irrigation regimes or the growing season.

Most leaves were maintained up to the flowering stage, thereafter started to decrease at the soft dough and hard dough stage. The decrease of leaves number after flag leaf stage was mainly due to senescence. Similar results were also reported by Unlu and Steduto (2000), Moseki and Dintwe (2011), and Monsour (2013).

Effect of Different Irrigation Regimes on Stem Thickness

The effect of different irrigation regimes to plant thickness varied significantly ($P = 0.01$) at vegetative and soft dough stage and highly significant ($P < 0.001$) at

Table 6 Means separation table for sorghum plant thickness grown in two seasons under different irrigation regimes, at different phenological stages

Season	Irrigation regime	Plant stem thickness (cm)				
		Vegetative	Flag leaf	Flowering	Soft dough	Hard dough
1	EM	2.02ab	2.74ab	2.9a	2.4a	2.55a
1	LE	1.989a	2.65ab	3.0a	2.5a	2.6a
1	ELE	2.161b	3.280c	3.1a	3.0b	3.13c
2	EM	2.374c	2.66ab	2.7a	2.8b	3.18c
2	LE	2.589d	2.576a	2.8a	3.1b	3.14c
2	ELE	2.726d	2.813b	2.9a	2.9b	3.0bc
LSD, $P<0.05$	Season	0.0874	0.1120	0.30	0.19	0.214
	Irrigation	0.1071	0.137	0.36	0.24	0.262
	Season*irrigation	0.1514	0.194	0.51	0.32	0.37
%CV		3.6	3.8	9.7	6.4	6.9
F-STAT	Season	***	**	ns	**	**
	Irrigation	**	***	ns	*	ns
	Season*irrigation	ns	*	ns	*	*

EM early in the morning, *LE* late in the evening, *ELE* early in the morning and late in the evening
$P > 5\%$, nonsignificant (ns); $P = 5\text{–}1\%$, significant (*); $P = 1\text{–}0.1\%$, very significant (**); $P < 0.01\%$, very highly significant (***)
[1]Dry season
[2]Dry-wet season

flag leaf stage (Table 6). The highest plant thickness (3.28 cm) was measured in flag leaf stage of the sorghum irrigated twice a day in the dry season, while the lowest plant thickness (1.99 cm) was registered from the sorghum, which was irrigated late in the evening (LE) within the same season. The thickest sorghum was recorded in ELE because application of water both early in the morning and late evening ensures a constant supply of easily available soil moisture, which lowers coincidence of soil moisture deficit that might lead to plant moisture stress. Plant moisture stress affects the bio-physiochemical process for optimal growth of the plant (Zhang et al. 2004; Morison et al. 2008; Mahinda 2014). Irrigating the sorghum only late in the evening provided little room for the light reaction of the photosynthesis process as most of the stomata pores were closed during the night, transpiration ceased; and the favored respiration resulted in much utilization of assimilates synthesized during the daytime.

The sorghum thickness in the plot which was irrigated early in the morning appeared to be thicker at vegetative (2.02 cm) and flag leaf stage (2.74 cm) compared to the sorghum which was irrigated late in the evening ($P > 0.05$). In the daytime, the rate of evapotranspiration was very high. At this time, stomata pores were opened, and more water was taken up by the plant for its bio-physiochemical activities including photosynthesis and transpiration.

The performance of sorghum stem thickness between the seasons appeared to be significantly different ($P < 0.05$) at vegetative, soft dough, and hard dough stage for the plots that were only irrigated once a day. The difference might have been brought by the more favorable climatic condition and less incidence of pest in the dry-wet seasons. For the ELE treatments, the significant differences across the seasons were registered only at vegetative and flag leaf stage, respectively.

In the dry season, the stem thickness of sorghums in plots irrigated once a day was observed to increase up to flowering stage and then decreased at soft dough stage. This decrease could be explained by the fact that accumulated assimilates were being translocated from the stem to fill up grains; this resulted in the decrease in stem thickness at soft dough stage. These results resemble those of Kiniry et al. (1992), Younis et al. (2000), and Bahashti and Behbood (2010) who attributed the phenomenon to the translocation of assimilates from the sink to fill the grains soon after the flowering stage to milking stage. However, the same trend was not clearly observed in the sorghum double irrigated within the same season and those grown in the dry-wet season, whereby the flag leaf stage could not mark the thickest plant girth. In the dry-wet season, the plant thickness steadily increased from vegetative to hard dough stage. This could probably be explained by the occurrences of the lower rate of translocation of photosynthetic materials from the stem for grain filling. Interestingly, the stem thickness of the sorghum that was double irrigated in the dry season sharply decreased in its thickness at the flowering stage, while the sorghum plants in other irrigation regimes were still expanding in stem size. Regardless of the seasons and irrigation regimes, all the sorghum plants increased in thickness at hard dough stage. This un-expected phenomenon of the complex elasticity behavior of sorghum under different irrigation regimes and seasons warrants further studies.

The result suggested that different irrigation regimes and seasons have different influence in plant thickness. As the soil moisture was increased and maintained at the appreciable amount by irrigating twice a day, the plant thickness also increased. This is an indication that an additional increment in water to a certain point did influence the expansion of the plant cells, which increased plant thickness as was also reported by Stone and Schlengel (2006). In both seasons, the stem thickness of the sorghum at flag leaf stage was in the following order: ELE>EM>EL. This trend was influenced by the different watering regimes that were supplied to the plant to meet crop water requirement.

Effect of Different Irrigation Regimes on Leaf Area Index

The influence of the three drip irrigation regimes on the leaf area index (LAI) in the two seasons appeared to be significantly different ($P < 0.001$) at vegetative and soft dough stages (Table 7). There was no significant difference ($P > 0.05$) in LAI between the treatment at flag leaf, flowering, and hard dough stage. The LAI of the sorghum irrigated twice a day outperformed other irrigation regimes in almost all the phenological stages, with significant differences during vegetative and soft dough stage in the dry-wet and dry season, respectively.

The results showed that different drip irrigation regimes and seasons have different influence on the LAI at some phenological stages of the sorghum plant.

Table 7 Means separation table for sorghum leaf area index grown under different irrigation regimes, at different phenological stages

Season	Irrigation regime	Leaf area index				
		Vegetative	Flag leaf	Flowering	Soft dough	Hard dough
1	EM	0.08ab	0.13a	0.133a	0.119a	0.122a
1	LE	0.075a	0.13a	0.148a	0.12a	0.121a
1	ELE	0.101b	0.17ab	0.17ab	0.16b	0.167a
2	EM	0.132c	0.145a	0.148a	0.15b	0.148a
2	LE	0.144c	0.15ab	0.16ab	0.17bc	0.158a
2	ELE	0.168d	0.295b	0.238b	0.188c	0.129a
LSD, $P < 0.05$	Season	0.0128	0.0806	0.0442	0.0129	0.037
	Irrigation	0.0157	0.987	0.0541	0.0158	0.046
	Season*irrigation	0.0222	0.1395	0.077	0.0224	0.0654
% CV		10.4	45	25.4	8.2	25.6
F-STAT	Season	***	ns	ns	***	ns
	Irrigation	**	ns	ns	***	ns
	Season*irrigation	ns	ns	ns	ns	ns

EM early in the morning, *LE* late in the evening, *ELE* early in the morning and late in the evening
$P > 5\%$, nonsignificant (ns); $P = 5–1\%$, significant (*); $P = 1–0.1\%$, very significant (**); $P < 0.01\%$, very highly significant (***)
[1]Dry season
[2]Dry-wet season

In the dry season, the highest LAI (0.17) was measured from the sorghum irrigated twice a day, while the lowest LAI (0.075) was registered from the plots with sorghum irrigated late in the evening. The overall highest LAI (0.295) was recorded at the flag leaf stage from the sorghum irrigated twice a day in the dry-wet seasons, without significant differences from the similar irrigation regime, during the dry season. This indicates that irrigating twice a day maintained appreciable soil moisture condition and ameliorate the rhizosphere, which results in an increment of the sorghum leaf area.

Effect of Different Drip Irrigation Regimes on Biomass and Water Use Efficiency

The response of sorghum biomass and water use efficiency under the three drip irrigation regimes and seasons appeared to be significantly different ($P < 0.001$) across the seasons and irrigation regimes. However, there was no any statistical difference revealed when the former and the latter were interacted. The results showed that, when sorghum is grown in the dry-wet season, it generates more biomass than when is grown in the dry season (Fig. 7a). The highest biomass (24.91 t/ha) was recorded from sorghum irrigated twice a day (ELE) in the dry-wet season, while the lowest biomass (10.85 t/ha) was obtained in the dry season

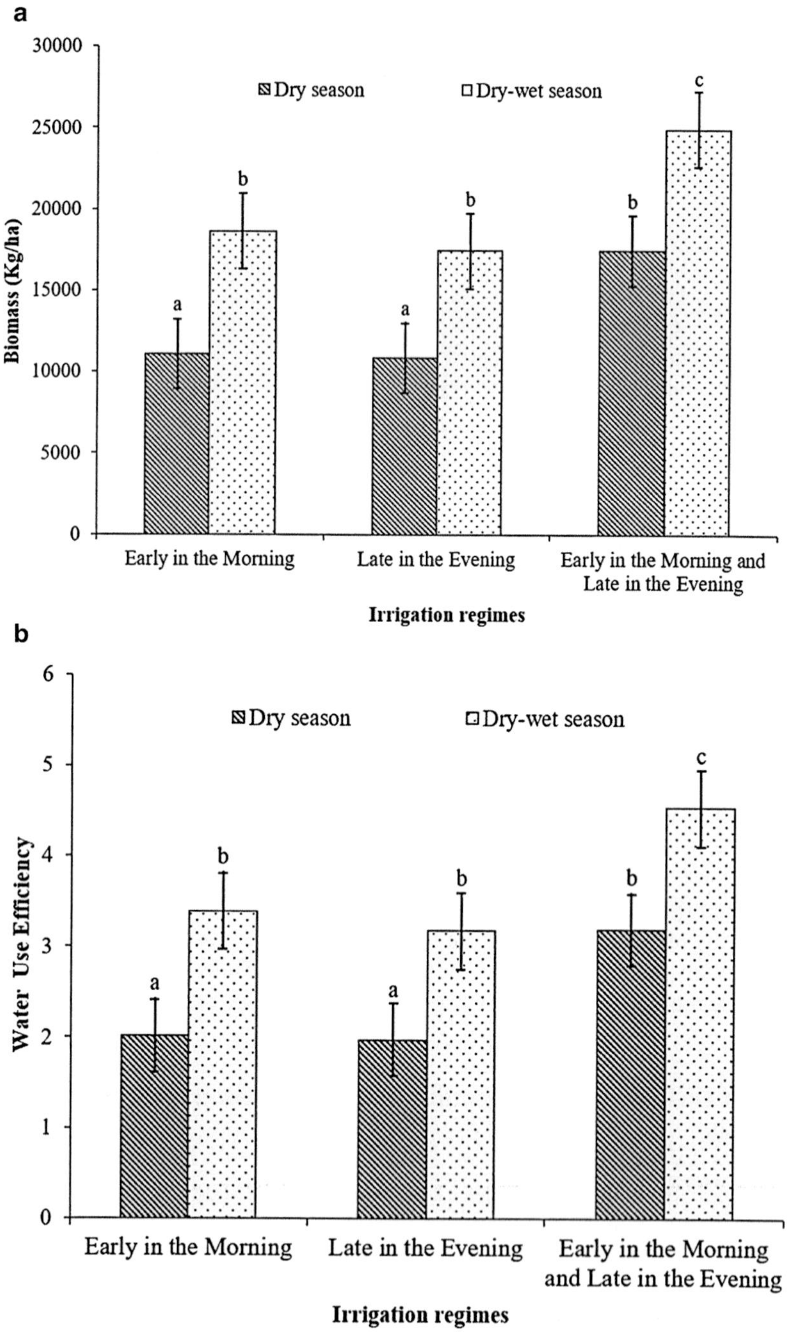

Fig. 7 (**a**) Effect of drip irrigation regimes on sorghum biomass. (**b**) Effect of drip irrigation regimes on sorghum water use efficiency

when sorghum was irrigated only late in the evening (LE). Similarly, the highest water use efficiency (4.53) was recorded from the sorghum irrigated twice a day in the dry-wet season, and the lowest water use efficiency (1.97) was measured in the sorghum irrigated late in the evening in the dry season (Fig. 7b). Both biomass and water use efficiency trends in the dry and dry-wet seasons were in the order of ELE>EM≥LE. This showed that biomass accumulation is not only a function of water but also the time of irrigation.

Although there was no statistical difference ($P > 0.05$) between EM and LE within the season, watering late in the evening produced the lowest numerical biomass indicating that the contribution of irrigation in night hours was comparatively low to accumulate the same biomass quantity, for the same amount of water supplied. A little amount of water used by the sorghum plant for its physiological activities during night hours resulted in the lowest biomass accumulation. Since less biomass (1.97 kg) was produced for every 1 mm of water used by the crop, LE appeared to use water less efficiently. The highest WUE (4.53) was registered in sorghum irrigated twice a day in the dry-wet season. This is because a significant amount of biomass was produced in the dry-wet season, therefore increasing the power of the numerator over the denominator water used by the crop. The result agrees with Zhang et al. (2004); Abdel-Motagally (2010) and Morison et al. (2008) who reported that the less water is used by a plant to produce a gram of biomass, the more efficient it is. It also conforms to the work of Adzemi and Ibrahim (2014), who reported that crops transpire little amount of water during the night hours compared to daytime.

Sorghum Grain Yield Response in Different Drip Irrigation Regimes

The performance of the sorghum grain yield at 12% equilibrium grain moisture content under different drip irrigation treatments appeared to be significantly different ($P < 0.001$) for the plot which was irrigated twice a day compared to those irrigated once a day, independently of the season tested (Table 8). Although the grain yield for the sorghum irrigated only in the morning or evening within and across the seasons did not differ significantly ($P > 5\%$), the overall lowest grain yield (6.8 t/ha) was harvested from the sorghum irrigated late in the evening during the dry season. The highest grain yield (13.2 t/ha) was weighed in the sorghum irrigated twice a day during the dry season. However, there was no significant difference ($P > 5\%$) in grain yield for the sorghum drip irrigated twice a day between the two seasons.

The response of the grain yield in the dry-wet season appeared to be higher than that of the dry season by 16.4% and 18.6% for EM and LE irrigation regimes, respectively. This increment might be due to the influence of precipitation which rained after irrigation regimes were assigned. Contrary to the single drip irrigation regimes, the performance of grain yield for the sorghum irrigated twice a day in the dry season was higher than the dry-wet season by 0.509 t/ha ($P > 0.05$). It is interesting to note that this additional yield was probably due to a higher rate of

Table 8 Means separation table for sorghum grain yield grown under different drip irrigation regimes

Season	Irrigation regime	Yield (t/ha) at 12% EGMC
1	EM	6.833a
1	LE	6.822a
1	ELE	13.212b
2	EM	8.383a
2	LE	8.170a
2	ELE	12.703b
LSD, P <0.05	Season	1.090
	Irrigation regime	1.335
	Season*irrigation regime	1.887
%CV		11.1
F-STAT	Season	ns
	Irrigation regime	***
	Season*irrigation regime	ns

EM early in the morning, *LE* late in the evening, *ELE* early in the morning and late in the evening, *EGMC* equilibrium grain moisture content

P >5%, nonsignificant (ns); P = 5–1%, significant (*); P = 1–0.1%, very significant (**); P <0.01%, very highly significant (***)

[1]Dry season

[2]Dry-wet season

photosynthesis triggered by long time exposure of the sorghum plant to the solar radiation in the dry season under the appreciable soil moisture condition. It is also because of the efficient ability of the sorghum plant to translocate photosynthetic assimilates from the shoot to the grains during the dry season under double irrigation regime as observed in this research study.

The results showed that within the dry season, irrigating twice a day (ELE) increased grain yield significantly (P <0.001) by 1.9-fold compared to the sorghum irrigated once a day. During the dry-wet period, the increase in grain yield in ELE over EM and LE was 1.5 and 1.6-fold, respectively. However, similarly to LE and EM irrigation regime treatments, the yield performance in ELE plot did not differ significantly (P >0.05) between the dry and dry-wet seasons. This proved that sorghum grain yield is more dependent on irrigation regime, which influences the constant distribution of the soil moisture based on the crop water requirement, rather than the total amount of water applied in a single point of time. Application of water twice a day maintained additional easily available soil moisture to sorghum water requirement that created a favorable microclimatic condition for the sorghum growth (Mahinda et al. 2017). Irrigating once a day might have caused a partial root drying and hidden water stress that reduced leaf area and therefore affected the overall photosynthetic activities of the plant which led to the reduction of the final grain yield production. The hidden water stress affects enzymatic reaction and regeneration that are involved in the function of the photosystem II (Shangguan et al. 1999) and, therefore, reduced the rate of photosynthesis and the overall grain yield production as it was captured in Fig. 8.

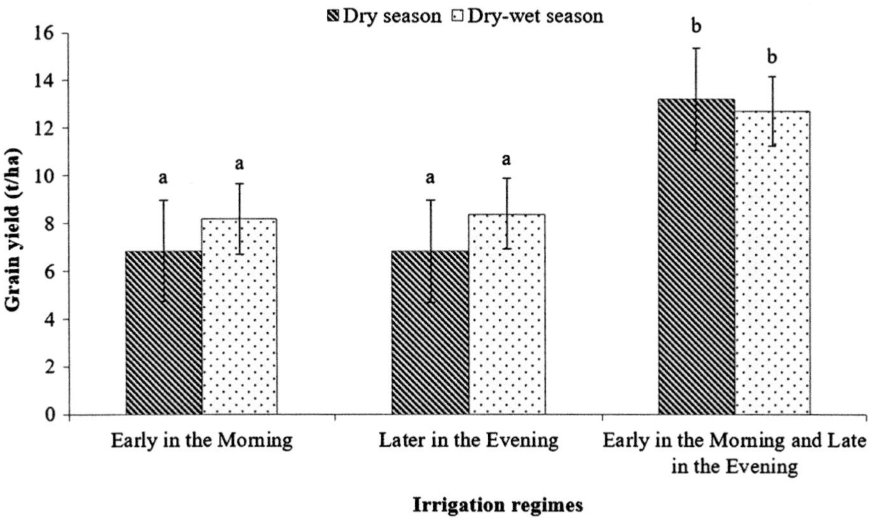

Fig. 8 Effect of different drip irrigation regimes on sorghum grain yield production

Comparing to the average global sorghum production of 1.37 t/ha under rain-fed condition (FAO 2013), the total yield obtained from this study appeared to be ≈ 1000% and 900% times more for the sorghum irrigated twice a day in the dry and dry-wet season, respectively. The result showed that the lowest grain yield (8.38 t/ha) in the dry-wet season was more by 3.83 bags of 100 kg compared to the maximum grain yield which was harvested under a full irrigated condition in the USA as reported by Blum (1996). It is also nine times more of the average yield of 0.9 t/ha which is harvested in Tanzania under the rain-fed condition as was reported by Mbwaga et al. (2007) and FAO (2014). Holding constant the grain yield produced in Tanzania, the highest grain yield (13.21 t/ha) attained from this study is 1400% times more. This implies that by ensuring appreciable constant soil moisture supply throughout a day, in the dry season, under fully solar radiation exposure, results in a higher assimilates accumulation and enhances translocation and subsequently storage of assimilates to the sink, resulting in a higher grain yield. In the dry-wet season, drip supplementing twice a day plays a similar role in ameliorating rhizosphere for a higher yield production but with a less grain yield ($P > 0.05$) than in the dry season, due to a reduction of sorghum plant exposure to the sun radiation. However, irrigating in the dry-wet season is likely to be economically viable, as it saves money as less water is needed to supplement hydric deficit of the unreliable and insufficient rainfalls. Nevertheless, there might be some offset in profitability driven by demand and supply in the season of interest.

Conclusion

In mitigating climate change and stabilizing adaptive capacity of drought resilience for crop production in the dryland areas, effective drip irrigation regime is the viable option to exploit underground water resources, intermittent rivers, and harvested rainwater. Although drip irrigation is highly recommended in ASALs, the study revealed that it is not just the drip watering irrigation that optimizes crop performance and water productivity, but the effective drip watering regime that constantly supplies appreciable soil moisture at the active feeding root zone of the crop. This means pinpointing an effective drip irrigation regime for a specific crop variety is of crucial importance in sustaining scarce water resources and increase drought resilience for crop production against inter and intra-annual rainfall variability, shifting rainfall patterns, and low rainfall amount in ASALs.

From the findings of this study, when sorghum is double irrigated, its plant height, LAI, biomass, grain yield, and water use efficiency increased significantly by almost two-fold compared to the single drip watering regimes. Sorghum that is drip irrigated late in the evening results in the lowest grain yield of 6.82 t/ha and 8.17 t/ha for the dry and dry-wet season, respectively. The double irrigation regimes (early morning and late evening) led to the highest grain yield of 13.21 t/ha in the dry season and 12.7 t/ha in dry-wet season. These yields are astonishingly higher by more than 1400% compared to the average sorghum grain yield of 0.9 t/ha under the rain-fed agriculture in Tanzania. The water use efficiency and biomass significantly varied following the order of ELE>EM≥LE for both dry and dry-wet seasons.

Regardless the significant variations arose in plant height and biomass between the two seasons, the study showed no remarkable differences in the grain yield when sorghum crop is either drip supplemented in the dry-wet season or purely irrigated twice a day in the dry season. However, this does not set a benchmark conclusion on economic implications between the two seasons and irrigation regimes. Therefore, it is imperative to consider economic viability when a drip irrigation regime in question is to be practiced in either of the seasons. To maximize grain yield production, the study suggests drip watering twice a day by practicing pure irrigation in the dry season and supplementing the sorghum crop for hydric deficit during the rainfall season. This will ensure sorghum grains availability throughout a year, thereby, significantly, contributing to the stabilization of food security and livelihood of the majority in the region. To achieve this, the study recommends more work to be conducted to evaluate the rationale of putting these results into practice by smallholder or progressive farmers by hinging on the socioeconomic dimension. Finally, the study proposes a hydrological mapping to identify suitable areas in the ASALs of Tanzania, where drip irrigation will be further practiced.

References

Abdel-Motagally F (2010) Evaluation of water use efficiency under different water regimes grain sorghum (*Sorghum bicolor*, L.). World J Agric Sci 6(5):499–505. IDOSI Publication. Faisalabad, Pakistan

Adzemi M, Ibrahim W (2014) Effect of regulated deficit irrigation on photosynthesis, photos synthetic active radiation on the yield of Sorghum cultivar. J Biol Agric Healthc 4(2): 2224–3208. International Institute for Science, Technology and Education, USA

Ahmed S, Diffenbaugh N, Hertel T et al (2011) Climate volatility and poverty vulnerability in Tanzania. Glob Environ Chang 21(1):46–55. Pergamon Press, United Kingdom

Allen R, Pereira L, Raes, D et al (1998) Crop evapotranspiration-Guideiline for cimputing crop water requirement, FAO irrigation and drainage paper no. 56, vol 56, no 97. Food and Agriculture Organization of the United Nations, Rome, pp 1–300

Assefa Y, Staggenborg S, Prasad V (2010) Grain sorghum water requirement and responses to drought stress: crop management. A review. Crop Manage Plant Management Network, United Kingdom. https://doi.org/10.1094/CM-2010-1109-01-RV

Bahashti A, Behbood F (2010) Dry matter accumulation and remobilization in grain sorghum genotypes (*Sorghum bicolor*, L.) under drought stress. Aust J Crop Sci 4(3):185–189. Southern Cross Journal, Australia

Blum A (1996) Crop responses to drought and the interpretation of adaptation. Plant Growth Regul 20(2):135–148. Springer, Netherlands

Challinor A, Wheeler T, Craufurd P et al (2005) Simulation of the impact of high-temperature stress on annual crop yields. Agric For Meteorol 135(1):180–189. Elsevier BV, Netherlands

Challinor A, Ewert F, Arnold S et al (2009) Crops and climate. Environ Sci Policy 11(7):642–654. Elsevier Inc, United States

Challinor A, Simelton E, Fraser E et al (2010) Increased crop failure due to climate change: assessing adaptation options using models and socio-economic data for wheat in China. Environ Res Lett 5(3):034012. Institute of Physics Publishing Ltd, United Kingdom

FAO (2013) Yield response to water: the original FAO water production function. FAO, Rome. Visited at http://www.fao.org/docrep on 15 May 2017

FAO (2014) FAOSTAT, DATA: production/yield quantities of Sorghum in Word+(Total), item sorghum, 1994–2014. FAO, Rome. Visited at http://www.fao.org/faostat/en/#data/QC/visualize on 2 Aug 2017

FAO- Food Security Department (1999) Sorghum post-harvest operations. In: Danilo M, Beverly L (eds) INPho: post-harvest compendium. Natural Resource Institute, FAO, Ghana, pp 1–33. Visited at www.fao.org/3/a-ax443e.pdf on 16 June 2017

Gregory P, Ingram J, Brklacich M (2005) Climate change and food security. Philos Trans R Soc B Biol Sci 360(1463):2139–2148. The Royal Society Publishing, United Kingdom

Guang-Cheng S, Zhan-Yua Z, Nac L et al (2008) Comparative effects of deficit irrigation (DI) and partial rootzone drying (PRD) on soil water distribution, water use, growth and yield in greenhouse grown hot pepper. Sci Hortic 19(1):11–16. Elsevier BV, Netherlands

Hatibu N, Mahoo H (1999) Rainwater harvesting technologies for agricultural production. A case for Dodoma, Tanzania. In: Kaumbutho P, Simalenga T (eds) Conservation tillage with animal traction. A resource book of the Animal Traction Network for Eastern and Southern Africa (ATNESA). Animal Traction Network for Eastern and Southern Africa (ATNESA), Harare, p 173. Institute for Strategic Studies, South Africa

Hossein N, Siddique K, Palta J (2009) Effect of soil moisture content on seedling emergence and early growth of some chickpea (*Cicer arietinum* L.) genotypes. J Agric Sci Technol 11(4): 401–441. University of Tarbiat Moderres-Faculty of Agriculture, Iran

IPCC (2014) Climate change 2014: impacts, adaptation, and vulnerability. Working group II of IPCC report on 31th, March 2014. Imperial College Press, London. Visited on 5 Apr 2017 at http://www.ipcc.ch/pdf/ar5/pr_wg2/140330_pr_wgII_spm_en.pdf

Jarvis A, Ramirez J, Anderson B et al (2010) Scenarios of climate change within the context of agriculture. In: Reynolds M (ed) Climate change and crop production. CAB International Publishers, Wallingford, pp 9–17

Jerry K, Tim H, Andre D et al (2012) Climate change impacts on crop productivity in Africa and South Asia. Environ Res Lett 7(3):034032. Reading and Cranfield Universities, United Kingdom

Kangalawe R, Lymo J (2013) Climate change, adaptive strategies and rural livelihood in semi-arid, Tanzania. Nat Resour 4:266–278. Scientific Research Publishing Inc. United Kingdom

Katerji N, Van-Hoorn J, Hamdy A (2003) Salinity effect on crop development and analysis of salt tolerance according to several classification methods. Agric Water Manag 62(1):37–66. Elsevier BV, Netherlands

Kilasara M, Boa M, Sway E et al (2015) Effect of in situ soil water harvesting techniques and local plant nutrient sources on grain yield of drought-resistant Sorghum varieties in semi-arid zone,

Tanzania. In: Lal R, Singh B, Mwaseba D et al (eds) Sustainable intensification to advance food security and enhance climate resilience in Africa. Springer, Cham/Heidelberg, pp 255–271

Kiniry J, Tischler C, Rosenthal W et al (1992) Nonstructural carbohydrate utilization by Sorghum and maize shaded during grain growth. Crop Sci 32:131–137. Crop Science Society of America, United States of America

MAFC (2013) Irrigation water management. Technical handbook no. 6. Ministry of Agriculture and Food Security, MAFC, Dar-es-salaam, pp 12–64

Mahinda A (2014) Effect of drip irrigation on the production and economic returns of sorghum (*Sorghum bicolor*) in semi-arid areas of Tanzania. Master dissertation, University of Nairobi, Department of Land Resource Management and Agricultural Technology (LARMAT), University of Nairobi Research Archive, Nairobi

Mahinda A, Gachene C, Kilasara M (2016) Economic impact of drip irrigation regimes on Sorghum production in semi-arid areas of Tanzania. In: Lal R, Kraybill D, Hansen D et al (eds) Climate change and multi-dimensional sustainability in African agriculture. Springer International Publishing, Cham, pp 227–240

Mahinda A, Kilasara M, Gachene C (2017) Effect of drip irrigation regimes on the growth performance, yield and water use efficiency of *Sorghum bicolor* in the semi-arid areas of Tanzania. In: Simó I, Poch R, Pla I (eds) Extended abstracts of the 1st world conference on soil and water conservation under global change – CONSOWA. 12–16 June 2017, Lleida, p 263. ISBN: 978-84-697-2909-0

Mbwaga A, Riches C, Ejeta G (2007) Integrated Striga management to meet Sorghum market demands in Tanzania. In: Ejete G, Gressel J (eds) Integrated new technologies for striga control towards ending the witch- hunt. World Scientific, Singapore, pp 253–264

McWilliams D (2002) Drought strategies for corn and grain Sorghum. Cooperative extension services, 580. New Mexico State University, Las Cruces, p 77

Monsour F (2013) Effect of irrigation regimes and polymer on dry matter, yield and several physiological traits of forage sorghum variety. 'Speedfeed'. Afr J Biotechnol 12(51): 7074–7080. Academic Journal, Nigeria

Morison J, Baker N, Mullineaux P et al (2008) Improving water use in crop production. Philos Trans R Soc Lond B 363:639–658. The Royal Society Publishing, United Kingdom

Moseki B, Dintwe K (2011) Effect of water stress on photosynthetic characteristics of two sorghum cultivars. Afr J Biotechnol 5(1):89–91. Academic Journal, Nigeria

Okalebo J, Gathua K, Woomer P (2002) Laboratory methods of soil and plant analysis; a working manual, 2nd edn. TSBF-CIAT and SACRED Africa, Nairobi

Pages S, Karanja D, Mbwaga AM et al (2010) The underlying cause of the 2009 sorghum failure in Kongwa district and its implication for Tanzania's vulnerability to climate change. Food Secur 2(2):157–167. Springer Publishing, Netherlands

Park D, White S, McCarty L et al (2014) Interpretation irrigation water quality report. Water Resour CU-14-700. Clemson University Cooperative Extension Services: South Carolina Counties, United States

Rodima-Taylor D (2012) Social innovation and climate adaptation: local collective action in diversifying Tanzania. Appl Geogr 33(1):128–134. Pergamon Press, United Kingdom

Salih A, Ali I, Lux A et al (1999) Rooting, water uptake, and xylem structure adaptation to the drought of two sorghum cultivars. Crop Sci 39(1):168–173. Crop Science Society of America, United States

Seleshi BA, Philoppe L, Tafa T (2009) Soil- Plant-Water relationship: Module 3: Improving Productivity and Market Success (IPMS) of Ethiopian farmers project, and Livestock Research Institute (ILRI), Addis Ababa, Ethiopia

Shangguan Z, Shao M, Dyckmans J (1999) Interaction of osmotic adjustment and photosynthesis in winter wheat under soil drought. J Plant Physiol 154(5–6):753–758. Elsevier GmbH-Urnan un Fischer, Germany

Shemdoe R (2011) Tracking effective indigenous adaptation strategies on impacts of climate variability on food security and health of subsistence farmers in Tanzania. Working research paper. African Technology Policy Studies Network, Nairobi. Visited on 14 June 2017 at http://www.atpsnet.org/Files/wps51.pdf

Stichler C, Fipp G (2003) Irrigation in south and south-central Texas. AgricLife communication and marketing, Texas coop Ext. Pub. L-54342-03. Texas A and M University, College Station, pp 1–6

Stone L, Schlegel A (2006) Yield-water supply relationship of grain sorghum and winter wheat. Agron J 98(5):1359–1366. American Society of Agronomy, Inc. United States

Thorton P, Jones P, Ericksen P et al (2011) Agriculture and food systems in Sub-Saharan Africa in a 4 °C+ world. Philos Trans R Soc A Math Phys Eng Sci 369(1934):117–136. The Royal Society Publishing, United Kingdom

Unlu M, Steduto P (2000) Comparison of photosynthetic water use efficiency of sweet sorghum at canopy and leaf scales. J Agric 24(4):519–525. Scientific and Technical Research Council of Turkey, Turkey

Wim B, Roger S, Rice C et al (2007) GenStat discovery edition for everyday use. ICRAF newsletter: 117. ICRAF, Nairobi

Younis M, El-Shahaby O, Abo-Hamed S et al (2000) Effect of water stress on growth, pigments and CO_2 assimilation in three Sorghum cultivars. J Agron Crop Sci 185(2):73–82. Wiley-Blackwell Velerg GmbH, Germany

Zhang Y, Eloise Q, Liu Y, Shen Y et al (2004) Effect of soil water deficit on evapotranspiration, crop yield and water use efficiency in the North China plain. Agric Water Manag 64(2):107–122. Elsevier BV, Netherlands

Managing Vulnerability to Drought and Enhancing Smallholder Farmers Resilience to Climate Change Risks in Zimbabwe

48

Obert Jiri and Paramu L. Mafongoya

Contents

Abstract

Smallholder farmers experience a myriad of climate change risks and weather variability challenges. The effects of climate change threats are expected to worsen, and smallholder farmers' vulnerability to these threats is projected to increase. While the research community has been highly engaged in detecting global climate change and attributing this change to anthropogenic causes, data-driven analysis of adaptation options has been weak. In Southern Africa, smallholder farmers have been coping and adapting to climate threats, using their own

O. Jiri (✉)
Agricultural Practice, Faculty of Agriculture, University of Zimbabwe, Harare, Zimbabwe

School of Agricultural, Earth and Environmental Sciences, University of KwaZulu-Natal, Pietermaritzburg, South Africa
e-mail: obertjiri@yahoo.co.uk

P. L. Mafongoya
School of Agricultural, Earth and Environmental Sciences, University of KwaZulu-Natal, Pietermaritzburg, South Africa
e-mail: mafongoya@gmail.com

W. Leal Filho (ed.), *Handbook of Climate Change Resilience*,
https://doi.org/10.1007/978-3-319-93336-8_94

indigenous knowledge systems. As climate change effects increase, smallholder farmers are failing to cope effectively.

This paper focuses on the role of indigenous knowledge systems and their use in agricultural adaptation by smallholder farmers. It also reviews the role of IKS-based adaptation strategies in the development of locally relevant adaptation policies. The review concludes that policy development should not be distant from community concerns and through integration of IKS, policies become relevant to community needs and concerns; hence they become part of the community.

Keywords
Vulnerability · Climate change adaptation · Smallholder farmers

Introduction

Climate change is adversely impacting rural households that are mostly dependent on rain-fed agriculture, and this has led to calls for the adoption of strategies that can assist farmers to adapt to the phenomenon. Climate change is defined by the Intergovernmental Panel on Climate Change (IPCC), as "a change of climate which is attributed directly or indirectly to human activity that alters the composition of the global atmosphere and which is in addition to natural climate variability observed over comparable time periods" (IPCC 2013). Climate change refers, therefore, to the effect of human-induced increase in the concentration of greenhouse gases in the atmosphere, enhancing the natural greenhouse effect. The impacts of climate change are uncertain. The effects that a global mean change in temperature may have on local and regional scales are very variable and uncertain (Selvaraj et al. 2013). In addition to changes in the mean climatic conditions (such as temperature and rainfall), the frequencies of irregular seasons and extreme events, including fires, floods, and droughts, are likely to change and in some places increase (Mills 2009). Climate patterns may change in terms of timing of rainy seasons and the distribution in time and space of rainfall. Given the potentially dramatic effects on local climate, natural resources, infrastructure, and economic activities, Africa may be particularly physically vulnerable and at risk from climate change. The impact of climate change is an issue that needs to be addressed urgently in Africa. Despite the possible global warming being a problem largely caused by anthropogenic emissions in the North, Africa may be on the receiving end of many of its effects. Africa contributes only a third of global greenhouse gas emissions, yet it has the highest vulnerability in the world. This is accompanied by low adaptive capacity due to the following:

- It has variable climates most of them which are semiarid and arid.
- Low technology development capacity.
- High levels of poverty.
- Poor infrastructure.
- Majority of the population, about 70%, rely on rain-fed agriculture which is climate sensitive (Charles et al. 2014).

Agriculture and Climate Change in Zimbabwe

Reports produced by the Intergovernmental Panel on Climate Change (IPCC) (2001, 2007, 2012) conclude not only that greenhouse gas emissions are already beginning to change the global climate but also that Africa will experience increased water stress, decreased yields from rain-fed agriculture, increased food insecurity and malnutrition, sea-level rise, and an increase in arid and semiarid land as a result of this process (Seipt et al. 2013). Extreme weather events, notably flood, drought, and tropical storms, are also expected to increase in frequency and intensity across the continent (IPCC 2007). These projections are consistent with recent climatic trends in Southern Africa, including Zimbabwe. The effects of this exposure to changes in climate are exacerbated by the high levels of sensitivity of the social and ecological systems in the region and the limited capacity of civil society, private sector, and government actors to respond appropriately to these emerging threats.

It is widely recognized that Africa is one of the most vulnerable regions in the world due to widespread poverty, limited coping capacity, and its highly variable climate (Medany et al. 2006). Zimbabwe is particularly vulnerable due to its heavy dependence on rain-fed agriculture and climate-sensitive resources (Shumba et al. 2012). Agriculture's sensitivity to climate-induced water stress is likely to intensify the existing problems of declining agricultural outputs, declining economic productivity, poverty, and food insecurity, with smallholder farmers particularly affected. Extreme weather events, notably drought, food, and tropical storms, are also likely to threaten development gains across a variety of sectors and intensify existing natural hazard burdens for at-risk populations in both rural and urban areas. Consequently, climate change presents risks to lives and livelihoods at the individual level and to the economy and infrastructure at the regional and national levels (Cairns et al. 2013). Climate change adaptation is therefore a principal development challenge in Zimbabwe.

Zimbabwe lies in a semiarid region with limited and unreliable rainfall patterns and temperature variations. Rainfall exhibits considerable spatial and temporal variability characterized by shifts in the onset of rains, increases in the frequency and intensity of heavy rainfall events, increases in the proportion of low rainfall years, decreases in low intensity rainfall events, and increases in the frequency and intensity of mid-season dry spells (Unganai and Murwira 2010). Extreme weather events, namely, tropical cyclones and droughts, have also increased in frequency and intensity (Unganai and Murwira 2010). Moreover, according to the Zimbabwe Meteorological Service, daily minimum temperatures have risen by approximately 2.6 °C over the last century, while daily maximum temperatures have risen by 2 °C during the same period.

Climate records demonstrate that Zimbabwe is already beginning to experience the effects of climate change, notably rainfall variability and extreme events. These conditions, combined with warming trends, are expected to render land increasingly marginal for agriculture, which poses a major threat to the economy and the livelihoods of the poor due to Zimbabwe's heavy dependence on rain-fed agriculture and climate-sensitive resources. It is expected that farmers, who represent approximately 62% of the total population, will bear disproportionate impacts due to their limited adaptive capacity. Consequently, climate change poses a major threat to sustainable development at the micro- and macro-levels.

Vulnerability of Smallholder Farmers

While poor people generally possess the lowest levels of adaptive capacity, the IPCC (2012) emphasizes the importance of recognizing the differences in vulnerability within groups and between communities (IPCC 2013). The failure of development practice to consider diversity and difference by defining populations in categorical terms (e.g., the poor), which has led to the exclusion of vulnerable groups, needs to be questioned. In reality, vulnerability is a function of a variety of social, economic, and political factors that sequence the distribution of risk across society (Smit and Wandel 2006).

Rising temperatures and increasing rainfall variability, notably drought, are also expected to exacerbate declining agricultural outputs, further compromising economic growth and stability, employment levels, food insecurity, demand for other goods, and poverty reduction. In particular, climate change is expected to lead to the expansion of marginal lands (IPCC 2007), which is already beginning to occur in Zimbabwe. If changing climatic conditions continue to expand these regions, traditional agricultural systems will become increasingly unsustainable.

Even diversified livelihood systems with a livestock component are expected to become more vulnerable. Mixed crop-livestock systems are a traditional livelihood strategy of smallholder farmers in semiarid rural areas (Thornton et al. 2007a). These systems tend to be well adapted to climatic conditions characterized by erratic rainfall patterns. However, climatic variability in semiarid areas poses major threats to natural processes that sustain fodder production for livestock and moisture for rain-fed crop production (Ericksen et al. 2011). Pasture and crop production in the absence of appropriate management practices are at risk of frequent failure with predicted future rainfall expected to be reduced or punctuated by concentrated heavy events separated by prolonged dry spells.

Studying vulnerability is a means of operationalizing locality by explaining the welfare and specific social conditions of people at a local or village level at a point in space and time, as a product both of household ability to secure a livelihood, as well as of environmental processes that interact with social processes on higher levels. Precisely combining the two components – exposure and capacity – allows the concept of vulnerability to link across problem areas and geographical scales. Exposure and the nature of particular events are the focus of much literature concerning vulnerability to environmental change (Luers 2005). Within the concept of exposure, processes at several geographic levels are linked to create local risk. Locally derived changes in an environmental system (which may be global in incidence) are a source of exposure (Luers 2005). Exposure to external environmental events and sensitivity to these are the basis for arguing that certain places may be more or less vulnerable than other areas due to their natural geography. Africa's environment has been described as fragile, and Africa as a region has been portrayed as sensitive in its water resources, wetland habitats, biodiversity, fisheries, and agriculture (Ringler et al. 2011). The concept of exposure is, nevertheless, intrinsically linked to social aspects.

Exposure to global environmental change, as filtered through the physical and social circumstances, leads to exposure at a local level. An additional aspect of the link of exposure to humans relates to human reliance on resources that are affected by environmental events. A broader understanding of exposure employed by Luers (2005) incorporates an internal structure side to exposure. High dependence on agriculture may make a society more exposed to drought. Due to the dynamism of human-environment relationships, vulnerability is not easily quantified. Vulnerability is characterized by many different relationships, some of which are difficult to measure. Exposure cannot be accurately determined unless the exact probability and risk of an event taking place can be established; however, a characteristic of climatic events in dryland Africa is a high degree of uncertainty. In addition, famines often have little to do with drought and more to do with underlying societal factors (Shiferaw et al. 2014). Quantitative analyses of social vulnerability have been performed in the past by measuring, for a country or a particular geographic area, indicators of processes that define vulnerability. Shiferaw et al. (2014) argue that human ecology, expanded entitlements, and political economy and the place and time-specific configuration of these three key analytic variables define the dimensions of vulnerability. Downing et al. determine the level of vulnerability or ability to secure sufficient food (food security) by assessing food availability kcal/day per capita (a measure of human ecology), GNP per capita (a measure of expanded entitlements), and childhood mortality per 1000 (a measure of political economy).

Coping and Adaptation by Smallholder Farmers

Coping strategies can be defined as "the bundle of poor people's responses to declining food availability and entitlements in abnormal seasons or years" (Thornton et al. 2007b). They purposively evaluate the costs and consequences associated with different coping strategies (Thornton et al. 2007b). Valdivia et al. (2010) argue that coping strategies reflect careful forward planning. A response refers to each individual action undertaken within a strategy.

Adaptation concerns the set of livelihood strategies that is available to households when faced with a natural event, such as drought (Table 1). While coping is a short-term response to an inhabitual decline in food entitlements, adapting is a permanent change in the mix of ways in which food is acquired (Table 1). The former entails actions to survive within the prevailing system, whereas adaptation involves changes

Table 1 Coping and adaptation strategies used by farmers

Scale	Timing		
	Before the season	During the season	After the season
Crop	Variety selection for stress tolerance	Replanting with early maturing varieties	
Field	Staggered planting dates	Changing crops when replanting	Grazing of failed field for animal maintenance
	Low density planting	Increasing plant density when replanting	
	Intercropping		
	Soil and water management strategies		
	Delayed fertilizer use	Split application of topdressing fertilizer	Spreading of anthill soil to fields
Farm	Diversified cropping	Shifting crops between land types	
	Plot fragmentation		
Community level	Social networks	Matching weeding labor inputs to expectations of the season	Asset sales for cereal purchases
	Off-farm employment networks		Food transfers
	Increase livestock assets		Migration employment
	Assess cereal stocks		

in the rule system to meet livelihoods. Hungry season strategies used for part of most years are "normal" activities, while strategies to survive particularly bad years represent "abnormal" activities and coping. Hungry season strategies are sometimes included as the first stage coping mechanisms (Valdivia et al. 2010). Coping mechanisms becoming a more permanent phenomenon, however, imply a change in livelihood systems. Coping strategies hence become part of a process of adaptation of the food production system.

Adaptation and coping represent a link between household vulnerability and government policies because social and economic conditions affect household need and ability to engage in coping strategies in given situations. Examining coping in adaptation is particularly important as much adaptation may be implicit rather than planned (Table 1). According to Smit and Wandel (2006), "adaptation can be spontaneous or planned, and can be carried out in response to or in anticipation of changes in conditions." Adger and Barnett (2009) point out that relating to social responses, adaptation can occur in two forms. The first is spontaneous impacts or individual or household behavior when faced with climate variability. The second is the planned adaptation policy that can occur on a longer time scale and as collective action. The first perception of adaptation implies that strengthening environmentally and economically sustainable coping strategies is a form of adaptation. The second perception concerns macroeconomic and political adjustments that affect coping. There are multiple catalysts of the need to cope.

A principal challenge facing adaptation projects in Zimbabwe is the uncertainty associated with climatic variability. Most farmers in drylands, such as Chiredzi District, have experienced multiple climate risks and have confronted them with a diverse range of coping mechanisms (Jiri et al. 2015). Rural communities in particular already possess good understandings of local climate patterns, including inter-/intra-annual rainfall variability, and are accustomed to coping with them. However, an important consequence of climate change is that future climate will be less familiar, more uncertain, and, possibly, more extreme (Vincent and Cull 2013). The implication is that current local practices, processes, systems, and infrastructure, which have been more or less adapted to current climate conditions, could become increasingly unsuitable as the climate changes. This situation is aggravated by varying local perceptions and interpretations of climate variability, which can be broad and diverse within different social groups and between communities. The Zimbabwean experience represents a growing recognition in many parts of the developing world that autonomous adaptation strategies that were effective in the past are increasingly ineffective for coping with emergent climate change. Climatic variability and the increasing frequency and magnitude of extreme events means that past climate conditions will become decreasingly useful for predicting future climate (Tubiello et al. 2000). In addition, despite Africa's vulnerability, few regional and subregional climate change scenarios based on regional climate models or empirical downscaling exist to inform local action (Ericksen et al. 2011). The lack of downscaled climate data is particularly problematic in countries like Zimbabwe where climate impacts and vulnerability are highly regionally differentiated. As a result, strengthening national and local capacities to manage climate risks, as they are currently understood, is imperative for dealing with climate risk in the future (Adger and Barnett 2009). Dryland communities tend to rely on sectors highly sensitive to climate change and variability, e.g., rain-fed agriculture and pastoralism (Adger and Barnett 2009). Climate change is already affecting drylands in Sub-Saharan Africa (SSA), where increased droughts due to climate change are intensifying the conditions of poverty (Vincent and Cull 2013). Current knowledge about climate change and variability impacts in drylands does not allow accurate predictions; but as the consequences of climate change – increased climate variability, environmental degradation, and cultural change – are predicted to intensify, dryland communities are expected to be least resilient to climate change (IPCC 2013). Vulnerability to climate change is important to address in drylands dependent on scarce resources, as poor people have limited capacity to cope with everyday stresses and climate risk adds to existing vulnerabilities. Africa is one of the most vulnerable regions to climate change due to multiple stresses coupled with high population growth, increasing pressure on resources, and low adaptive capacity (IPCC 2007). Consequently, SSA dryland populations may be severely affected by food insecurity in the future and should play a central role in global poverty reduction strategies.

Vulnerability involves both physical and social aspects of human-environment interaction, conceptualized as the exposure to an event and the capacity to cope with that event. In addition to the natural event itself, individual or household ability to secure a livelihood is central to descriptions of vulnerability. Entitlement theory explains the concept capacity in terms of an individual's ability to command commodities or food, through ownership of endowments and exchange of these. Extended interpretations of entitlement emphasize the wide variation in time and space in sources of food and income, both formal and informal. Social vulnerability approaches focus on the choices and constraints to livelihood strategies posed by social and political structures. Coping strategies are related to specific ways of securing entitlement or livelihoods during natural events (Table 1). Identifying characteristics of coping strategies and processes that affect these is a means of examining determinants of vulnerability. Coping strategies seek to balance present consumption with future livelihoods and are generally sequenced reflecting increasing present adversity and compromising of future livelihoods as a drought progresses.

Responses to climate change risk have to consider existing narratives. Perceptions, and misperceptions, of drylands are built largely on degradation narratives (Luers 2005). The coexistence of high levels of poverty and food insecurity with high rates of land degradation has led to the understanding that these factors are linked (Kandji et al. 2006). Since the 1930s there have been claims that small-scale farmers are degrading their land, with predictions well into the 1990s that Africa is heading toward environmental disaster (Roberts 2010).

Use of IKS

Subsistence rain-fed agriculture underpins rural livelihoods in the Sub-Saharan Africa. The overdependence on rainfall suggests the need for more reliable climate and weather forecasts to guide farm-level decision-making. Traditionally, African farmers have used indigenous knowledge (IK) to understand weather and climate patterns and make decisions about crops and farming practices (Table 2). Local farmers are concerned with decisions as to when the rains would come, about the start of the rainy season, and about how long the season would be. Indigenous knowledge or local knowledge is generally defined as the understanding of the local environment by local communities and the practices, techniques, and technologies they use to cope and adapt to climate change and variability (Briggs and Service 2005). Most climatic models lack localized climate data and scenarios. Climate scientists can benefit from local observations of weather (Kalanda-Joshua et al. 2011). Understanding the basis of indigenous peoples' coping and adaptation strategies is critical if climate change research and development efforts aimed at these communities are to be successful. Consideration of local IKS will enhance decision-making at local levels as well as influence policy processes and policy choices at the

Table 2 IKS indicators used by smallholder farmers for agricultural season quality predictions

Name	Characteristics
Indicators for whole season quality (rainy or drought)	
Colophospermum mopane (mopani) sprouting	Level of sprouting and color of leaves during spring months, initial heavy flowering indicates good rains, subsequent heavy flowering indicates poor rains
Sclerocarya birrea (mupfura) fruiting	Heavy fruiting
Chochomela birds	When these are seen in summer, it is a sign of a good rainy season
Kwarakwara birds	If this bird is seen in summer, it is a sign of a good season
Madzetse (big frogs)	If these are heard in dry streams in summer, it is a sign of a good rainy season
Ciconia ciconia (stork)	Appearance of these birds indicate a good rainy season
Time of winds	If at the end of October there is too much wind, it indicates a poor rainy season to come
Moon	Profuse halo around the moon
Grasshoppers	Abundances and hatching of grasshoppers in mid-season indicates a good rainy season
Indicators of the start of the rainy season	
Lone baboon	If a lone baboon crosses the area in early summer, it indicates onset of the rains
Millipedes	Big millipedes producing sounds in the soil indicate onset of the rains
Indicators of very short predictions (within season)	
Cuckoo bird (kohwera bird)	Whenever this bird produces its crying sound, rains will fall within 2 days
Bucorvus leadbeateri (southern ground hornbill) – Dendera bird	When a sound of this bird is heard, it rains the following day
White frog	This small frog stays on trees. If it croaks in summer, rains will fall within a day
Wind direction	Winds that come from the south eastern indicated it would rain that week
Cloud types	Dark clouds preceding strong winds means rainfall in a few hours

national level (Adger et al. 2009). Indigenous people have ways of predicting weather within a season and from season to season. These predictions help them to decide what and when to plant for a particular season as well as when to perform certain operations within a season. However, increased rainfall variability in recent years associated with climate change has reduced their confidence in indigenous knowledge, hence reducing their adaptive capacity and increasing their vulnerability to climate change. To address this problem, researchers are advocating the integration of indigenous knowledge into scientific climate forecasts at the local level, where it can be used to enhance the resilience of communities vulnerable to climate change.

Resilience and Agricultural Development

The concepts of resilience, vulnerability, and adaptive capacity become relevant in addressing the social dimensions of climate change and variability. Yet, they emerged from different intellectual traditions, and the challenge is to integrate these concepts to enhance our knowledge of the consequences and response options of communities to climate change and variability (Brenkert and Malone 2005). The concept of resilience is a central element of how a society adapts to change. The use of the concept of resilience has proliferated in the academic literature and in various development contexts, though different definitions and perceptions of the concept are in use (Borron 2006). An improved understanding of resilience to climate change and variability may contribute to forging a better linkage of understanding among different disciplines. The ability of society to sustain desirable features of our environment is increasingly compromised by unprecedented climate change and variability. This requires an enhanced understanding of how people respond to risk and change (Borron 2006). In this context, resilience, vulnerability, and adaptive capacity become crucial in explaining human consequences of climate change and variability (Adger et al. 2011). However, definitions have emerged from different intellectual traditions and thus have diverse meaning (Berkes 2007). A bibliometric analysis of the knowledge domains of resilience, vulnerability, and adaptive capacity illustrates an increase in publications over the last decades, while integration has been slow (Marshall et al. 2010). Multidimensional concepts offer opportunities to integrate knowledge to better comprehend the complex interactions of natural and human systems and to respond to climate change and variability more effectively (Marshall et al. 2010). Resilience is central to understanding the dynamics of social-ecological systems and their vulnerability to shocks and is a crucial element of how societies cope with uncertainty and adapt to change (Adger et al. 2011). The concepts of resilience and vulnerability have been used in diverse ways by various disciplines, e.g., ecological theory, systems analysis, and disaster studies (Luers 2005). While resilience has its roots in the study of ecosystems, vulnerability is used by social scientists in hazards and poverty research (Luers 2005). The study of vulnerability tends to focus on predicting problems while ignoring that people are not passive victims of disasters but active agents of their own lives and resourceful in their strategies to reduce vulnerability (Kelly and Adger 2000). Incorporating the concept of resilience recognizes people as active agents in decision-making rather than passive victims of change. It recognizes the ability to cope with and adapt to change and the capacity to self-organize and learn, i.e., the capacity to seize opportunity for innovation and development.

Will Smallholder Farmers Survive in Light of Climate Change?

For dryland communities that are chronically food insecure, vulnerability to complex disasters is part of the daily struggle to survive (Solh and Van Ginkel 2014). The persistence of structural vulnerability to food insecurity is largely a result of the combination of livelihood insecurity, poverty, and vulnerability to hazards (Annan 2013). Differential vulnerability to climate change and variability means that people are affected differently, making it important to address their resilience and adaptive capacity. The concept of resilience can contribute to both, more sustainable development pathways and climate change and variability responses. In drylands, increasing resilience remains a challenge, despite growing recognition of existing adaptive capacities of communities and inherent resilience of social-ecological systems in the light of adversity (Jat et al. 2012). Stringer et al. (2009) analyze the linkages between climate change, drought, and desertification in SSA and conclude that populations dependent on natural resources for their livelihoods most acutely experience these. They call for an enabling policy environment that helps to build the resilience of social-ecological systems and strengthens local adaptations (Stringer et al. 2009). This situation merits a broader perspective that considers resilience to climate change and variability as integral component of long-term livelihood security.

The emphasis on building resilience to unprecedented climate change and variability in rural livelihoods illustrates the interdependence between environmental risk, social-ecological systems dynamics, social resilience, institutional context, and the broader political economy of development (Adger et al. 2013). It can be applied to understand the challenges and opportunities associated with climate change and variability and help build capacity to support overall development objectives. Important attributes of resilient systems are also critical for development: community capacity to withstand shocks, adapt to change, and capitalize on opportunities for development (Vetter 2009). Social, political, and economic trends reinforce, transform, or weaken existing patterns of resilience.

Local drought coping strategies are key for responding to the patterns of spatial and temporal variability that exist at different scales in the small-scale farming sectors. The ability to respond to this variability is dependent on a range of socioeconomic factors. These include various modes of responses to immediate and future risk such as communal pooling of resources and labor contribution (*nhimbe*), herding labor availability, kin support networks, cattle loaning and sharing relationships, cash income for labor hire or fodder purchase, land tenure restrictions on herd movement, migrations, and diversification of incomes and livelihoods.

Modes of Smallholder Farmers' Adaptation to Climate Change

Impacts of climate change and variability can be classified in a variety of ways: short-term or long-term, those resulting from sudden disasters or those resulting from slow changes in trends, and predictable or unpredictable (Biagini et al. 2014). However, at household level, it is important to examine how climate-related risks affect livelihood capabilities over time, across space, across asset classes, and across households. These four types of risks to livelihood define the major conceptual categories of the ways variability threatens the ability of households to adapt and secure livelihoods.

Households and communities have developed strategies to adapt to climate variability. Smallholder farming communities in Southern Africa have already experienced many forms of extreme climate events. Over time, they have developed a range of adaptive responses to cope with climatic threats to their livelihoods (Biagini et al. 2014). When climate hazards are repeated, the distinction between short-term and long-term adaptation (that is, coping vs. adaptation) breaks down, as does that between proactive and reactive adaptation. Thus, there is need to address adaptation strategies as they assist coping. Local coping and adaptation responses to climate change and variability can be classified into five categories:

1. *Mobility* addresses movements of various types in response to climatic risks and climate-induced scarcities. It is a common adaptation strategy used by smallholder households and communities, particularly in dry areas, such as in Southern Africa (Mertz et al. 2009). It pools risks across space and is successful in combination with clear information about the spatial and temporal distribution of precipitation. It is especially important as an adaptation practice for agropastoralists in Sub-Saharan Africa and most dry regions of the world. In the context of climate change, mobility sometimes has been viewed as a maladaptation, in which climatic stresses lead to involuntary migrations. However, mobility also is a way of life for large groups of people in semiarid regions and a long-standing mechanism to deal with spatiotemporal variations in rainfall and range productivity (Marin 2010). Therefore, whether mobility is considered a desirable adaptation often depends on the status of the social groups in question. For agricultural populations, mobility often can be the last resort in the face of climate risks and disruption of livelihoods. To address the needs of mobile populations, the role of information in tracking human and livestock movements and the mobile provision of basic services such as health, education, credit, and marketing of animal products are especially important to reinforce adaptive capacity.
   Temporary, circular, and internal migration is an important aspect of spatial mobility in drought-affected areas and is often a component of household survival strategies for coping with drought and high levels of production uncertainty (Black et al. 2013). Much of the migration that occurs in areas of drought or increasing degradation is believed to be economically motivated. The destinations of these migrants are often proximate urban areas or, increasingly, other rural areas within a state. However, sudden-onset disasters and prolonged chronic hazards lead to "distress migration." These migrations are composed largely of impoverished people seeking aid until they may be able to return to their homes or communities (if that is possible). Two characteristics of distress-induced migration deserve emphasis: the first characteristic is forced migration as a result of ecological disaster, which mainly results in internal rather than international displacement (Chappell 2011). The second characteristic is the temporary rather than permanent nature of the displacement caused by such stressors. Living mainly in poor countries, victims have little mobility and few options, and the

majority of displaced people return as soon as possible to reconstruct their homes in the disaster zone (Chappell 2011).

The movement of cattle to alternative grazing areas dominates the management strategies used by farmers to ensure cattle survival. Movement is an important component of herd management in "normal" years; this involves localized movements to "key resource" grazing patches within the landscape (Thompson and Scoones 1994). Movement in response to drought-induced shortages of fodder occurs at different scales and in relation to the availability of different pasture and feed sources. Under most drought conditions, flexible movement appears an effective strategy in offsetting mortality.

2. *Storage* of past surpluses is an effective measure against future livelihood failures. Agricultural households, especially in dry areas, have created indigenous storage infrastructure for seeds and harvested crops and have developed time-tested procedures for drying fruits for storage. When combined with well-constructed infrastructure, low levels of perishability, and a high level of coordination across households and social groups, it is an effective measure against drought (Meinke et al. 2006). As an adaptation practice to address risks, storage is relevant to individual farmers and communities to address food and water scarcities. It is critical, in rural communities that, for this strategy to work, improvements in storage technologies and institutions are looked at to improve rural livelihoods. In Zimbabwe, for instance, the *Zunde raMambo* concept envisaged the storage of grain by the paramount chief for distribution among his subjects in the event of a drought.
3. *Diversification* can occur in relation to on- and off-farm employment opportunities, productive and nonproductive assets, and consumption strategies. Diversification into different farm management practices and crop varieties and using a combination crop and livestock farming are common diversification responses in risky environments (Paavola 2008). This would also include mixing herds by keeping cattle, goats, poultry, sheep, and donkeys to hedge against climate risk. Highly varied in form, it can occur in relation to productive and nonproductive assets, consumption strategies, and employment opportunities, e.g., engaging in carpentry (using locally obtained trees), gold panning (risking environmental degradation), or use of river reeds to making sleeping mats. It is reliable to the extent that benefit flows from assets are subject to uncorrelated risks. Diversifying households typically give up some returns in exchange for the greater security provided by diversification. Individuals using this strategy would be willing to live with some level of poverty in exchange for reduction in vulnerability.
4. *Communal pooling* refers to adaptation responses involving joint ownership of assets and resources; sharing of wealth, labor, or incomes from particular activities across households; and/or mobilization and use of resources that are held collectively during times of scarcity. It pools risks across households (Berkes 2007). This practice is most effective when the benefits from assets owned by different households and the livelihood benefit streams are uncorrelated. When a group is affected in a similar manner by adverse climate hazards, for instance, communal pooling is less likely to be an effective response. Although communal

pooling can occur in combination with mobility, storage, and diversification, its hallmark is joint action by members of a group with the objective of pooling their risks and resources. Joint action increases the range of impacts with which collected households might have to deal, compared with the range of impacts individual households could have encountered. It also requires functioning and viable institutions to coordinate activities across households. It is one way for social groups, particularly those natural resources dependent, to enhance their capacity to adapt to the impacts of future climate change. Communities in dryland areas, for example, increase water rationing and/or often prohibit the consumption of certain foods and forest products except during times of famine or long-term rainfall failure.

5. *Market exchange* is perhaps the most common and versatile mechanism for adaptation. To be fair and effective, it requires well-developed markets, exchange instruments, and widespread access; weather-related insurance schemes for agricultural populations (although very scarce) are an example of market-based adaptation to climate change. Market exchange-based adaptation practices can substitute for the other four practices when rural poor people have access to markets (Paavola 2008). However, they are likely to do so mainly when there are well-developed institutions to facilitate market access. Furthermore, equity in adaptation practices based on market exchanges typically requires significant attention to the institutional means through which access to markets and market products becomes available to households. In the absence of institutional mechanisms that can ensure equity, the smallholder farmers are less likely to benefit from purely market exchange-based adaptation.

The cattle loaning system (*kuronzera*) enables grazing pressure within the communal areas to be redistributed. Cattle may be loaned on a temporary basis for the duration of a local crisis or on a longer-term arrangement. Cattle are loaned to both relatives and friends and often at some distance from the owner's home (e.g., 10 km), in order to relieve grazing pressure. Large herd owners rarely keep more than 10–15 animals in their own kraal and prefer to loan out to a number of sites (*miraga*). Loaning lowers the risk of local overgrazing, reduces herding and management requirements, and assists stockless relatives and friends.

Conclusion

Resilience is important as communities are allowed to build capacity to govern livelihood adaptations or transformations in an environment characterized by change and uncertainty. By refocusing development policy and practice toward maximizing adaptive and transformative capacities, resilience can contribute to creating new development pathways that are sensitive to climate change and variability. An improved use of existing adaptive knowledge and skills of dryland communities to respond to uncertainty is important. Framing resilience as the general capacity of a community to change may be an effective strategy to mainstream GEC responses into development. Thus, resilience in drylands should be pursued to decrease vulnerability to multiple threats and increase general resilience in response to immediate needs and future risks and aspirations within a development framework. Weather and climate information should be readily available to the intended recipients through reliable dissemination channels and in a user-friendly language. Platforms for meteorological information dissemination should be increased. The meteorological service departments should collaborate with developmental organizations and national extension services in climate change adaptation programs and research on indigenous knowledge systems and how it can be linked with climate science. There is also a need for weather insurance schemes at the local community levels.

References

Adger WN, Barnett J (2009) Four reasons for concern about adaptation to climate change. Environ Plan A 41(12):2800–2805. https://doi.org/10.1068/a42244

Adger WN, Dessai S, Goulden M, Hulme M, Lorenzoni I, Nelson DR, . . . Wreford A (2009) Are there social limits to adaptation to climate change? https://doi.org/10.1007/s10584-008-9520-z

Adger WN, Brown K, Nelson DR, Berkes F, Eakin H, Folke C, . . . Tompkins EL (2011) Resilience implications of policy responses to climate change. Wiley Interdiscip Rev Clim Chang 2(5):757–766. https://doi.org/10.1002/wcc.133

Adger WN, Quinn T, Lorenzoni I, Murphy C, Sweeney J (2013) Changing social contracts in climate-change adaptation. https://doi.org/10.1038/nclimate1751

Annan K (2013) Growing disruption: climate change, food, and the fight against hunger. Retrieved from www.oxfam.org

Berkes F (2007) Understanding uncertainty and reducing vulnerability: lessons from resilience thinking. Nat Hazards 41(2):283–295. https://doi.org/10.1007/s11069-006-9036-7

Biagini B, Bierbaum R, Stults M, Dobardzic S, Mcneeley SM (2014) A typology of adaptation actions: a global look at climate adaptation actions financed through the Global Environment Facility. Glob Environ Chang 25:97–108. https://doi.org/10.1016/j.gloenvcha.2014.01.003

Black R, Adger WN, Arnell NW (2013) Migration and extreme environmental events: new agendas for global change research. Environ Sci Pol 27(January 2016):S1–S3. https://doi.org/10.1016/j.envsci.2012.09.010

Borron S (2006) Building resilience for an unpredictable future: how organic agriculture can help farmers adapt to climate change. Food and Agriculture Organization of the United . . ., (August), pp 1–25. Retrieved from http://www.mapaq.gouv.qc.ca/SiteCollectionDocuments/Regions/CentreduQuebec/INPACQ2009/ah617e.pdf

Brenkert AL, Malone EL (2005) Modeling vulnerability and resilience to climate change: a case study of India and Indian states. Clim Chang 72(1–2):57–102. https://doi.org/10.1007/s10584-005-5930-3

Briggs J, Service, P (2005) The use of indigenous knowledge in development: problems and challenges. Prog Dev Stud 5:99–114. https://doi.org/10.1191/1464993405ps105oa

Cairns JE, Hellin J, Sonder K, Araus JL, MacRobert JF, Thierfelder C, Prasanna BM (2013) Adapting maize production to climate change in sub-Saharan Africa. Food Security 5:345. https://doi.org/10.1007/s12571-013-0256-x

Chappell L (2011) Drivers of migration in household surveys. The Government Office for Science, UK, pp 1–11

Charles N, Rashid H, James C (2014) Analysis of determinants of farm-level adaptation measures to climate change in Southern Africa. J Dev Agric Econ 6(5):232–241. https://doi.org/10.5897/JDAE12.0441

Ericksen P, Thornton P, Notenbaert A (2011) Mapping hotspots of climate change and food insecurity in the global tropics: Appendix 1 SOFI Country Group Composition, (June). Retrieved from www.ilri.org

IPCC (2001) Climate change 2001: The scientific basis. Contribution of working group I to the third assessment report of the intergovernmental panel on climate change. Houghton JT, Ding Y, Griggs DJ, Noguer M, van der Linden PJ, Dai X, Maskell K, Johnson CA (eds.). Cambridge University Press, Cambridge, UK, and New York, NY, USA, 881pp

IPCC (2007) Summary for policymakers. Climate change 2007: impacts, adaptation and vulnerability. Contribution of Working Group II to the Fourth Assessment Report of the Intergovernmental Panel on Climate Change, (April 2007), pp 7–22. https://doi.org/10.2134/jeq2008.0015br

IPCC (2012) Summary for policymakers. In: Managing the risks of extreme events and disasters to advance climate change adaptation. Field CB, Barros V, Stocker TF, Qin D, Dokken DJ, Ebi KL, Mastrandrea MD, Mach KJ, Plattner GK, Allen SK, Tignor M, Midgley PM (eds.). A Special Report of Working Groups I and II of the Intergovernmental Panel on Climate Change. Cambridge University Press, Cambridge, UK, and New York, NY, USA, pp 1–19

IPCC (2013) Climate change 2013: the physical science basis. Summary for policymakers. IPCC, pp 1–29. https://doi.org/10.1017/CBO9781107415324

Jat RA, Craufurd P, Sahrawat KL, Wani SP (2012) Climate change and resilient dryland systems: experiences of ICRISAT in Asia and Africa. Curr Sci 102(12):1650–1659

Jiri O, Mafongoya PL, Chivenge P (2015) Indigenous knowledge systems, seasonal "quality" and climate change adaptation in Zimbabwe. Clim Res 66(2):103–111. https://doi.org/10.3354/cr01334

Kalanda-Joshua M, Ngongondo C, Chipeta L, Mpembeka F (2011) Integrating indigenous knowledge with conventional science: enhancing localised climate and weather forecasts in Nessa, Mulanje, Malawi. Phys Chem Earth 36(14–15):996–1003. https://doi.org/10.1016/j.pce.2011.08.001

Kandji ST, Verchot L, Mackensen J (2006) Climate change and variability in the Sahel region: impacts and adaptation strategies in the agricultural sector. World Agroforestry Centre (ICRAF) and United Nations Environment Programme (UNEP). UNEP, pp 1–48

Kelly PM, Adger WN (2000) Theory and practice in assessing vulnerability to climate change and facilitating adaptation. Clim Chang 47(4):325–352. https://doi.org/10.1023/A:1005627828199

Luers AL (2005) The surface of vulnerability: an analytical framework for examining environmental change. Glob Environ Chang 15(3):214–223. https://doi.org/10.1016/j.gloenvcha.2005.04.003

Marin A (2010) Riders under storms: contributions of nomadic herders' observations to analysing climate change in Mongolia. Glob Environ Chang 20:162–176. https://doi.org/10.1016/j.gloenvcha.2009.10.004

Marshall N, Marshall P, Tamelander J, Obura D, Malleret-King D, Cinner JE (2010) A framework for social adaptation to climate change sustaining tropical coastal communities and industries. Communities. IUCN, Gland

Medany M, Niang-Diop I, Nyong T, Tabo R, Vogel C (2006) Background paper on impacts, vulnerability and adaptation to climate change in Africa for the African workshop on adaptation implementation of decision 1/CP.10 of the UNFCCC convention, pp 21–23

Meinke H, Nelson R, Kokic P, Stone R, Selvaraju R, Baethgen W (2006) Actionable climate knowledge: from analysis to synthesis. Clim Res 33(1):101–110. https://doi.org/10.3354/cr033101

Mertz O, Mbow C, Reenberg A, Diouf A (2009) Farmers' perceptions of climate change and agricultural adaptation strategies in rural Sahel. Environ Manag 43(5):804–816. https://doi.org/10.1007/s00267-008-9197-0

Mills DM (2009) Climate change, extreme weather events, and US health impacts: what can we say? J Occup Environ Med 51:26–32. https://doi.org/10.1097/JOM.0b013e31817d32da

Paavola J (2008) Livelihoods, vulnerability and adaptation to climate change in Morogoro, Tanzania. Environ Sci Policy 11(7):642–654. https://doi.org/10.1016/j.envsci.2008.06.002

Ringler C, Bryan E, Hassan RM, Alemu T, Hillesland M (2011) How can African agriculture adapt to climate change? Insights from Ethiopia and South Africa. IFPRI – Research Brief, 45 pp. Retrieved from http://www.ifpri.org

Roberts D (2010) Prioritizing climate change adaptation and local level resilience in Durban, South Africa. Environ Urban 22:397. https://doi.org/10.1177/0956247810379948

Seipt C, Padgham J, Kulkarni J, Awiti A (2013) Capacity building for climate change risk management in Africa: encouraging and enabling research for informed decision-making. Environ Dev 5:1. https://doi.org/10.1016/j.envdev.2012.11.006

Selvaraj S, Ganeshamoorthi P, Pandiaraj T (2013) Potential impacts of recent climate change on biological control agents in agro-ecosystem: a review. Int J Biodivers Conserv 5(x):845–852. https://doi.org/10.5897/IJBC2013.0551

Shiferaw B, Tesfaye K, Kassie M, Abate T, Prasanna BM, Menkir A (2014) Managing vulnerability to drought and enhancing livelihood resilience in sub-Saharan Africa: technological, institutional and policy options. Weather Clim Extremes 3:67–79. https://doi.org/10.1016/j.wace.2014.04.004

Shumba EM, Wallgren VL, Carlson A, Kuona M, Moyo N (2012) Community climate change vulnerability assessment in miombo woodlands. WWF, Harare

Smit B, Wandel J (2006) Adaptation, adaptive capacity and vulnerability. Glob Environ Chang 16:282–292. https://doi.org/10.1016/j.gloenvcha.2006.03.008

Solh M, Van Ginkel M (2014) Drought preparedness and drought mitigation in the developing world's drylands. Weather Clim Extremes 3:62–66. https://doi.org/10.1016/j.wace.2014.03.003

Stringer LC, Dyer JC, Reed MS, Dougill AJ, Twyman C, Mkwambisi D (2009) Adaptations to climate change, drought and desertification: local insights to enhance policy in southern Africa. Environ Sci Policy 12(7):748–765. https://doi.org/10.1016/j.envsci.2009.04.002

Thompson J, Scoones I (1994) Challenging the populist perspective: rural people's knowledge, agricultural research, and extension practice. Agric Hum Values 11(2–3):58–76. https://doi.org/10.1007/BF01530446

Thornton PK, Boone RB, Galvin KA, BurnSilver SB, Waithaka MW, Kuyiah J, ... Herrero M (2007a) Coping strategies in livestock-dependent households in East and Southern Africa: a synthesis of four case studies. Hum Ecol 35:461–476. https://doi.org/10.1007/s10745-007-9118-5

Thornton P, Herrero M, Freeman A, Mwai O, Rege E, Jones P, Mcdermott J (2007b) Vulnerability, climate change and livestock – research opportunities and challenges for poverty alleviation. Open Access Journal Published by ICRISAT 4(1):1–23. http://www.dfid.gov.uk/research/climate-change.asp

Tubiello FN, Donatelli M, Rosenzweig C, Stockle CO (2000) Effects of climate change and elevated CO_2 on cropping systems: model predictions at two Italian locations. Eur J Agron 13:179–189 Retrieved from www.elsevier.com/locate/eja

Unganai LS, Murwira A (2010) Challenges and opportunities for climate change adaptation among small-holder farmers in Southeast Zimbabwe. In: Second international conference: climate, sustainability and development in semi-arid regions

Valdivia C, Seth A, Gilles JL, García M, Jiménez E, Cusicanqui J, ... Yucra E (2010) Adapting to climate change in Andean ecosystems: landscapes, capitals, and perceptions shaping rural livelihood strategies and linking knowledge systems. Ann Assoc Am Geogr 100(4):818–834. https://doi.org/10.1080/00045608.2010.500198

Vetter S (2009) Drought, change and resilience in South Africa's arid and semi-arid rangelands. S Afr J Sci 105(1–2):29–33. https://doi.org/10.1590/S0038-23532009000100017

Vincent K, Cull T (2013) Climate change and food security in Southern Africa: implications of theoretical development for the promotion of sustainable equitable development. Kulima Integrated Development Solutions, Johannesburg, p 1

Building Pastoral and Agro-pastoral Community Resilience Against Drought in the Context of the Paris Agreement: The Case of Isiolo County, Kenya

49

David Walunya Ong'are and Anne Nyatichi Omambia

Contents

Abstract

Kenya is a climate-vulnerable country already experiencing adverse sector-wide climate change impacts with significant implications on socioeconomic development. In recent years, the frequency and severity of climate change-induced droughts in the country have had significant adverse effects on the natural resources in the fragile arid and semi-arid lands (ASALs). This has inhibited natural regeneration and recovery of the ecosystems and human communities. Recurrent droughts have contributed to crop and livestock losses, water scarcity, resource-use conflicts, and the resultant threats to livelihood support systems. In view of this, the country has established climate change legislative and policy measures, made institutional strides to access climate finance, and developed and submitted its first Nationally Determined Contribution (NDC) under the Paris Agreement. Further, there are ongoing initiatives in disaster preparedness at both national and county levels in spite of the myriad challenges.

D. Walunya Ong'are (✉) · A. N. Omambia
National Environment Management Authority, Nairobi, Kenya
e-mail: ongare@gmail.com; tichiomambia@gmail.com

W. Leal Filho (ed.), *Handbook of Climate Change Resilience*,
https://doi.org/10.1007/978-3-319-93336-8_95

This chapter expostulates that enhancing the resilience of pastoral communities in Kenya's rangelands to droughts can be achieved through identification and implementation of relevant actions in Kenya's NDC and the National Adaption Plan in tandem with county and community actions, measures, and strategies. Experiences in some counties show that resilience building is a multi-stakeholder collaborative process and up-scaling is possible. This chapter is important because it reinforces the huge potential for ASALs in meeting Kenya's Nationally Determined Contribution with a particular focus on linking science, policy, and practice with a special emphasis on gender and conflict management.

Keywords
Climate change · Adaptation plan · Nationally determined contribution · Medium term plan · Mitigation · Paris Agreement · County Integrated Development Plan · ASAL · Resilience

Background

Kenya, a country located in the Great Horn of Africa, is a climate-vulnerable nation and thus not exempt from the vagaries and adverse effects of climate change (IPCC 2007). Projected observable changes in climate in Kenya include increase in temperature, changes in rainfall distribution and intensity, sea level rise, increase in frequency and severity of extreme events of severe droughts and floods, spread of vector-borne diseases such as malaria in humans and Rift Valley Fever in livestock, sea level rise, coastal erosion, and resultant coral bleaching along the Kenyan coast, among others (Republic of Kenya 2010; Omambia et al. 2015). The main climate observation for Kenya is that the country might get wetter like most of East Africa especially in the highlands West and East of the Rift Valley but its seasonality and intensity will vary across the country. Further, 80% of Kenya is classified as arid and semiarid lands (ASALs), and these areas are predicted to have periodic and severe droughts that will not only threaten the community livelihood support systems but also hinder socioeconomic development, jeopardize the coping and recovery rate, thus affecting their resilience (Republic of Kenya 2015a; Filho et al. 2017). Kenya's ASALs are often climate vulnerable because they are already climatically stressed with high temperatures, low rainfall, and long dry seasons, and the areas also exhibit ecological constraints (Filho et al. 2017).

Kenya has fully perceived the economy-wide implications of climate change and has made tremendous progress in policy, legislation, and activities to ensure that the country's development proceeds along a low-carbon, climate-resilient pathway (Republic of Kenya 2013, 2015a). However, due to the frequency and severity of drought in the recent past and the related impacts to the GDP, the country does not have sufficient resources to face the combined reinforcing impacts of drought and climate change. According to the Kenya Post Disaster Needs Assessment report for 2008–2011 (Republic of Kenya 2012), the Great Horn of Africa drought cost the country in damages, losses, and needs an estimated 12.1 billion US Dollars which included 805.8 million US Dollars for destruction of physical and durable assets and 11.3 billion US Dollars for losses in flows of the economy across all sectors (Republic of Kenya 2012). The said drought had wide-scale impacts across all sectors of the economy with the degree of impact varying across sectors with the livestock sector experiencing 72% of the total sectoral damages and losses (Republic of Kenya 2012). This drought increased the vulnerability of the country to drought especially in the arid and semiarid regions and jeopardized recovery efforts and resilience to future occurrences (Republic of Kenya 2012). The 2008–2012 drought is regarded as one of Kenya's worst droughts in recent times which caused severe prolonged impacts as follows:

- Rainfall deficit persisted leading to lower than normal precipitation duration and intensities at various times scales which rendered it a meteorological drought
- Agricultural drought due to inadequate soil moisture to sustain even drought tolerant crops over time
- Hydrological drought with deficiencies in availability of groundwater
- Socioeconomic drought with physical water shortage affecting health, well-being, and quality of life of communities across the country especially in the ASAL areas

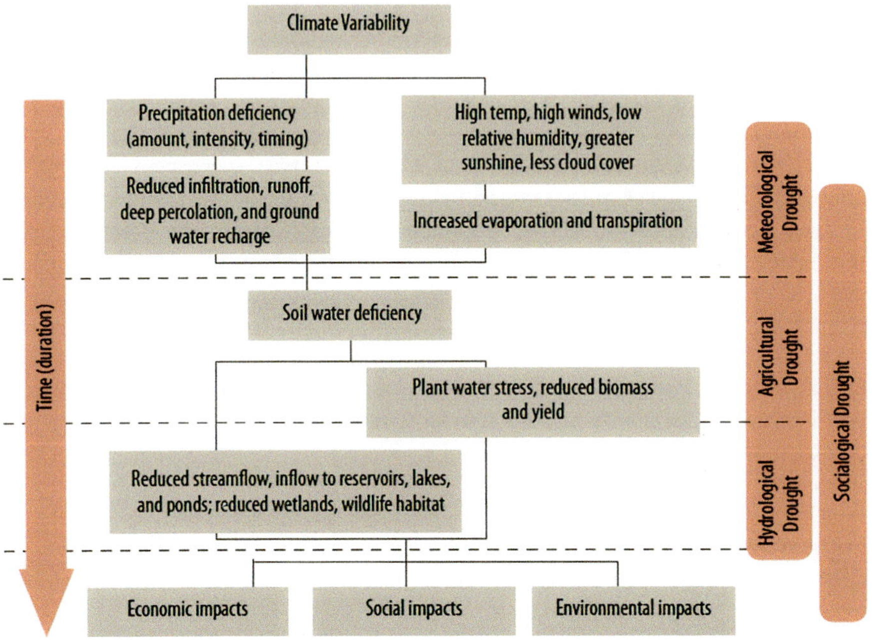

Stages of Drought. (Source: Republic of Kenya 2012; Kenya Post Disaster Needs Assessment 2008–2011 Drought)

It is against this background, among other climate change impacts, that Kenya as a member state of the United Nations Framework Convention on Climate Change

(UNFCCC) has actively participated in international climate change negotiations, domesticated measures and actions to address climate change, and also ratified the Paris Agreement (PA) on enhanced action to climate change. The Paris Agreement was designed and adopted as a Conference of Parties (CoP) decision at the CoP 21 in Paris, France, in 2015 with the overall objective to enhance the global response to climate change by strengthening implementation of the UNFCCC. The Paris Agreement aims at holding the increase in global average temperatures at well below 2° C above the preindustrial levels and pursuing efforts to limit temperature increase even further to 1.5° C in order to significantly reduce the risks and impacts of climate change. Its other objectives were to increase adaptive ability to climate change and to enhance financing flows towards climate resilient, low Green House Gas (GHG) development (UNFCCC webpage). Under article 3 of the Paris Agreement, all parties are to undertake and communicate ambitious efforts with a view to achieving the purposes of the agreement. Further, according to Article 4 paragraph 2 of the Paris Agreement, each party shall prepare, communicate, and maintain successive Nationally Determined Contributions (NDCs) that it intends to achieve (UNFCCC, webpage). Parties shall pursue domestic mitigation measures with the aim of achieving the objectives of such contributions.

Kenya prepared and submitted her first Intended Nationally Determined Contributions (INDC) to the UNFCCC in 2015. Kenya's NDC which consists of both adaptation and mitigation actions is based on her national circumstances and commits to implement it through both domestic and international support (Republic of Kenya 2015a). In its NDC, Kenya seeks to undertake an ambitious mitigation contribution towards the 2015 agreement with a goal to abate its GHG emissions by 30% by 2030 relative to the business as usual (BAU) scenario of 143 $MtCO_2$ equivalent, and it covers all sectors including agriculture and other land use sectors. Further, Kenya's NDC prioritizes adaptation and proposes various priority adaptation actions aimed at mainstreaming climate change adaptation across all sectors such as energy, research and technology development, education, provision of climate information services, agriculture and livestock resilience, and water and irrigation (Republic of Kenya 2015a). Implementation of these priority actions is at various formative stages and this chapter expostulates that if fully implemented, it will increase the country's resilience to slow-onset climate events such as drought that has now become more recurrent and prolonged.

Legislative and Institutional Arrangements for Addressing Climate Change and Drought

Kenya is the first African country and among the first nations globally to enact a Climate Change Act. The Climate Change Act 2016 is a framework law which provides for institutional and climate change measures to ensure that the Kenya develops along a low-carbon climate resilient pathway (Republic of Kenya 2016a). Other policy documents for addressing climate change and drought include the National Climate Change Policy Framework; the National Climate Change Action

Plan 2013–2017; the Kenya National Adaptation Plan 2015–2030; the National Climate Change Strategy 2010; the Environmental Management and Coordination Act Cap 387; the Sessional Paper on Drought Management; and other sectoral laws and policies.

The institutional arrangements for addressing climate change as outlined in the Kenya Climate Change Act 2016 has established the Climate Change Directorate (CCD) in the Ministry of Environment and Natural Resources to perform various duties including coordinating adaptation actions across the country and also doubling up as the national focal point of the UNFCCC. It also works with the climate change coordinating units in different ministries, counties, state departments, and agencies. Further, the CCD works with the Ministry of Devolution and Planning to ensure integration of climate change into the Medium Term Plans (MTPs). The National Environment Management Authority (NEMA) is the Designated National Authority (DNA) for the Clean Development Mechanism, the National Implementing Entity for the Adaptation Fund, and the Accredited Entity for the Green Climate Fund. The National Drought Management Authority oversees adaptation and resilience building in drought prone areas which are mainly the arid and semi-arid lands (ASALs). It is also the secretariat of the Common Programme Framework in ending drought emergencies in Kenya (Republic of Kenya 2016a). The National Climate Change Council is chaired by the President of Kenya and has a mandate to ensure mainstreaming of climate change functions by both the national and county governments. Furthermore, County governments are expected to integrate climate change actions into their County Integrated Development Plans (CIDPs) and County Environmental Action Plans and designate a County Executive Council (CEC) member to coordinate climate change affairs. They are also expected to submit a report on the implementation progress of climate-change actions to the county assembly for review and debate with a copy to the climate change directorate for information (Fig. 1).

MDAs: Ministries, Departments and Agencies; this is an umbrella acronym for all government entities in Kenya. From the illustration, the multiple roles played by a number of MDAs are outlined below.

NCCC: National Climate Change Council; chaired by the president of Kenya

Ministry of Environment (an MDA); the national focal point for the UNFCCC and also the climate change secretariat

The National Treasury (part of the MDAs); the National Designated Authority for the green climate fund

Ministry of Planning (part of the MDAs); this has a role of mainstreaming climate change into the country's medium term plans of vision 2030

National Environment Management Authority (part of the MDAs); the National Implementing Entity for the adaptation fund and the green climate fund; DNA for CDM projects under the Kyoto protocol

The National Drought Management Authority (An MDA); overall coordination over all matters relating to drought management

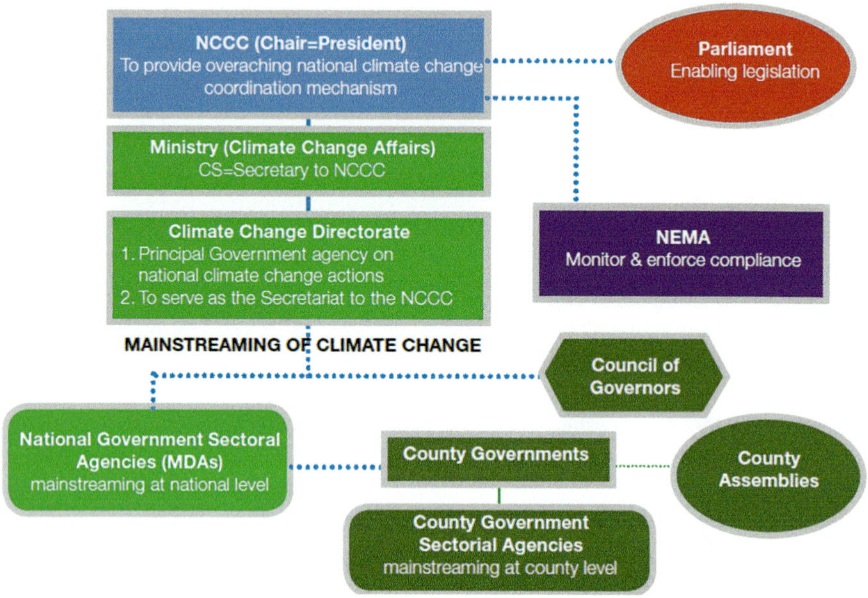

Fig. 1 Diagrammatic representation of the institutional arrangements for climate change management in Kenya. (Source of illustration: The Kenya National Adaptation Plan 2015–2030)

A number of barriers have been identified to the use of climate services including absence of climate data because of limited observational networks, low investments in climate services, low consideration of user needs, low levels of the socioeconomic benefits of a decision based on climate services and good quality information (Smith 2016)

Objectives of the Study

The objectives of this chapter are twofold, namely:

- To review the alignment of national and county government policies and strategies for climate change resilience using Isiolo as a case study.
- To assess opportunities and risks for climate change interventions in Isiolo county

Methodology

The methodology adopted was a desk study approach involving a review and cross-reference of climate change policies, strategies, and plans starting at the international commitments that Kenya has made to national and county levels for coherence and/or gaps. The document then moves on to look at efforts, synergies, and challenges at implementation between the various tiers.

The key documents analyzed included;

Global Level

Paris agreement; this contains the globally agreed intervention areas of focus to help countries fashion, their commitments, and reporting frameworks in a more consistent manner

Country Level

At the Kenyan country level, we analyzed policies and strategies at the national level and therefore cut across all counties and the sectors of the economy (upstream policies/strategies) and those that were more specific to the Isiolo County context (downstream policies/strategies)

(a) *Upstream policies/strategies*
   (i) Kenya's intended Nationally determined contribution 2015; this is the key country link to the Paris agreement that then guides the local level country strategies and actions at the national and county levels.
   (ii) Kenya National Climate Change Response Strategy 2010; the first of a series of documents directly addressing climate change in Kenya. It set the stage for subsequent actions within the country.
   (iii) The Kenya vision 2030 development blueprint. The key document guiding Kenya's development until 2030. Considering its importance, all other sector policies need to be aligned to it.
   (iv) Kenya climate change act 2016 (Republic of Kenya 2016a); the latest document to come into force with regard to climate change. It sets out the institutional arrangements, duties, and obligations of the various entities with respect to climate change interventions in Kenya. It also establishes the climate change fund and prescribes its funding mechanisms and management.
   (v) Kenya National Adaptation Plan of 2016 (Republic of Kenya 2016c); this forms the key platform for delivering on Kenya's intended nationally determined contribution to the Paris agreement. Its robustness or lack thereof is therefore key in the realization of Kenya's commitments under the said agreement.
   (vi) The Kenya post disaster needs assessment report of 2012; this report looked at the impacts of the 2008–2011 drought acknowledging that droughts are recurrent in Kenya (more so in ASAL regions like Isiolo). It gave recommendations for recovery and long-term resilience building.
   (vii) The environmental management and coordination Act CAP 387/(Republic of Kenya 1999); a cross-cutting and overarching piece of legislation covering all aspects of environmental management and natural resource use within the country.

(b) *Downstream policies/strategies – County-level*
   (i) County fiscal strategy paper for Isiolo: This sets out the county's priority programs for implementation in the medium term expenditure framework
   (ii) Isiolo County integrated development plan 2013–2017: This is the key planning document for a county. It outlines the county vision and policy direction for the next 5 years. All critical issues for a county are captured here.
   (iii) Isiolo County climate change fund act, 2016 (Republic of Kenya 2016a).

With ASALs comprising 80% of Kenya's land mass, the focus was to see how climate resilience can be enhanced in these regions which are also prone to disaster risk. Isiolo County, one of Kenya's 33 counties that are ASALs has been used as a case study. There was also analysis of key elements that should be targeted if resilience interventions are to be successful in such a setting. These were compared to current independent studies being carried out in this region by other independent researchers.

Study Area

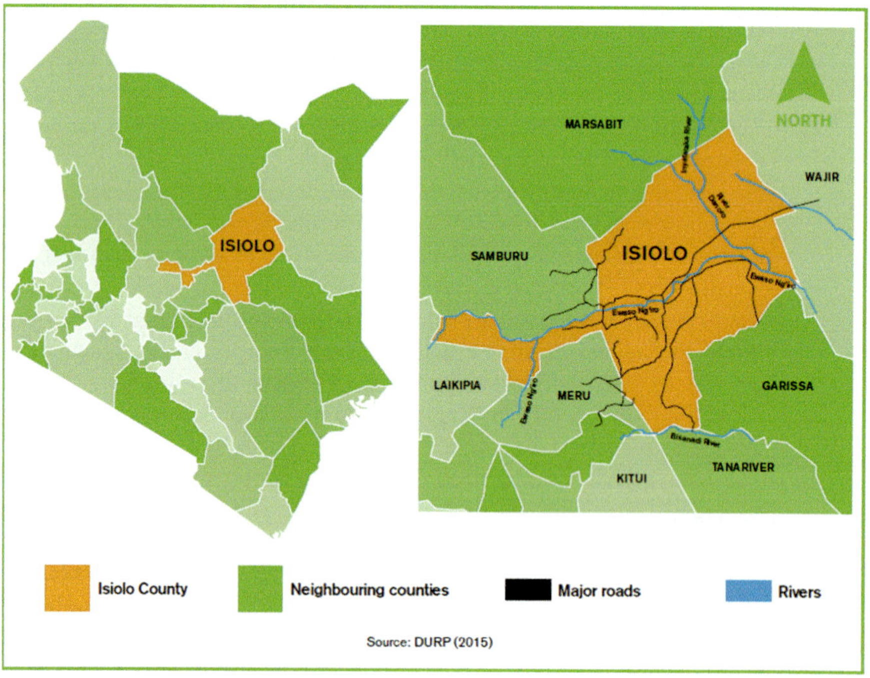

Source: Caroline King-Okumu et al. 2016

Isiolo County, one of the counties in the lower Eastern region of Kenya has a land area of 25, 382 km^2. It borders Marsabit to the North, Samburu, and Laikipia to the West, Garissa to the South East, Wajir to the North East, Tana River and Kitui to the South, and Meru and Tharaka-Nithi to the South West (Republic of Kenya 2013). Isiolo is classified into three ecological zones as follows;

- Semi-arid: Covers 5% of county area. Rainfall is between 400–650 mm annually.
- Arid: Covers 30% of total area of county; rainfall is between 300–350 mm annually; mostly grassland and some sparse shrubs.
- Very arid: 65% of total land area; rainfall of 150–250 mm annually; mostly barren.

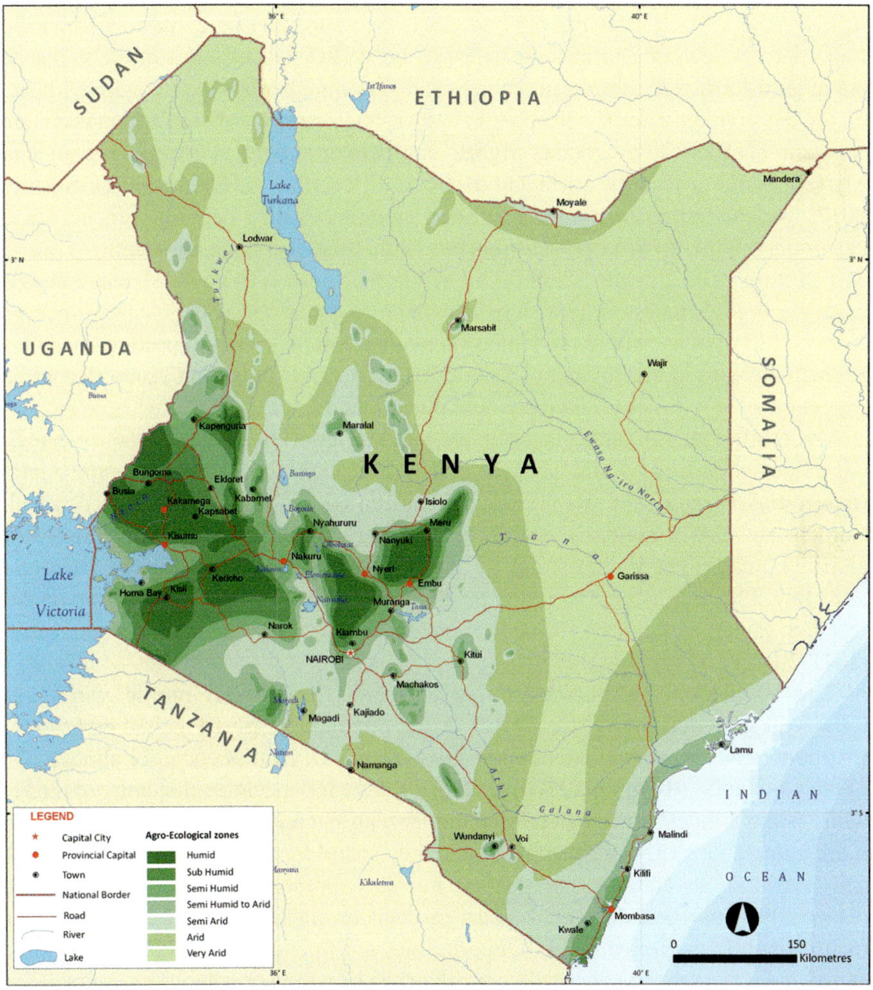

Ecological zones of Kenya. (Source: National Environment Management Authority; SoE Report 2010)

According to the 2009 population census, the county has a population of 143,294 almost evenly distributed between men and women. The population is projected to rise steeply mainly fueled by the construction of an international airport and the on-going Lamu Port Southern Sudan and Ethiopia Transport (LAPSSET) corridor.

Pastoralism is an important activity in Isiolo County with 80% of inhabitants relying on livestock for livelihoods (Republic of Kenya 2013). Livestock hold a special place in the hearts of pastoralist communities; they are a form of social identity beyond their market value (NRT 2012). Drought is the major and frequent disaster in this area that increases the vulnerability of the populace by increasing livestock mortality while generally reducing livestock productivity (Republic of Kenya 2013). As a perhaps contradictory insurance measure, pastoralists keep large herds knowing that at least half will be wiped out during a drought season but at least some will survive and regenerate. The large herds have a negative impact on the environment thereby creating an endless vicious cycle (NRT 2012). There are frequent conflicts over grazing rights; conflicts in Isiolo, Turkana, Pokot, and Marsabit Counties can be attributed to drought (Republic of Kenya 2012). In early 2017, violent skirmishes were witnessed in this locality between ranchers and pastoralists who invaded private ranches with their livestock in search of water and pasture. The likelihood of future recurrence and escalation of resource-use conflicts is high unless urgent steps are taken to arrest the situation (Republic of Kenya 2013). As a result of the migration by men folk in search of pasture, there has been an increase in the number of female-led households in Isiolo County (Republic of Kenya 2012); this is a relatively new phenomenon.

Over 80% of the land is used as grazing land by pastoralists; livestock comprises mostly of Zebu and Borana cattle. Goat breeds include the Galla (main) and Saanen. The black head Persian breed is the dominant sheep breed. Camels are also a major investment area. The county also hosts a number of private ranches. The main market for the county's livestock is Nairobi. There are very few livestock cooperatives to cushion pastoralism from the vagaries of the weather. Sheep and goats have been found to be more drought resilient than cattle (NRT 2012).

The population distribution of Isiolo depicts a youthful profile with ages 0–14 year olds accounting for 44.4% of the population; 14–64 year olds account for about 52% of the population; while those above 65 years account for about 3.6% of the population (Republic of Kenya 2013). Kenya has a gender inequality of 0.651, while Isiolo's index is estimated to be higher than the national average. Women are more vulnerable to poverty than men, and they spend more time searching for water and firewood. In addition, women do not own or control assets like livestock (especially cattle) and rarely take part in decision-making as it is a preserve for men (Republic of Kenya 2013). Opportunities are opening up in the ownership of sheep and goats (shoats) generally considered to be of lower value. The boy child is generally more valued amongst pastoralist communities (Republic of Kenya 2013). Work by the Northern Range Trust has identified several elements used to subjugate women despite the critical role they play in household management and climate

Table 1 Summary of Isiolo county Profile; extract from the Isiolo County Integrated Development Plan

Isiolo county profile
Located in Eastern Kenya
Area: 25,382 km^2
Population: 143,294 of which 60% is youthful
65% of total area classified as very arid with annual rainfall of 150–200 mm
Nomadic pastoralism is the main economic activity
Resource conflict over grazing areas and watering points common during dry spells
First county to establish a climate change fund

resilience (NRT 2016a). The issues of gender and youth have been identified as critical to climate resilience in the ASAL areas (Republic of Kenya 2015a). The county suffers poor communication network, with a road network of 975.5 km of which only 3% are bituminized. Around 77% of the roads have earth surfaces and impassable during the rainy season. Only 7% of the county is under mobile network coverage; there are only 164 landline telephone connections in the entire county (Republic of Kenya 2013). The county has a spatial planning framework in place (Table 1).

The Kenya Constitution creates a devolved system of government comprising of the national government and 47 counties; each county government consists of a county assembly (a parliament) and county executive with the governor serving as the head of the executive. It is on the institutional strength of this devolved system of governance that in 2016 Isiolo County established the Isiolo County Climate Change Fund Act 2016. The Isiolo County Climate Change Fund Act of 2016 was established to facilitate and coordinate financing of climate change adaptation and mitigation activities (Republic of Kenya 2016b). Its specific purposes are the following:

- Financing of climate change programs
- Initiating and coordinating climate change adaptation and mitigation activities at the community level
- Facilitating incorporation of climate finance in the county planning and budgetary frameworks
- Enhancing climate finance mechanisms
- Coordinating support from national governments climate change policy and legislative frameworks
- Coordinating collection and dissemination of climate change information to the public to create awareness and preparedness

The fund has fairly elaborate institutional structures including a board, an administrator, and linkages to the county wards. The county planning committee acts as the technical vehicle for county climate change planning. It identifies its funding sources as:

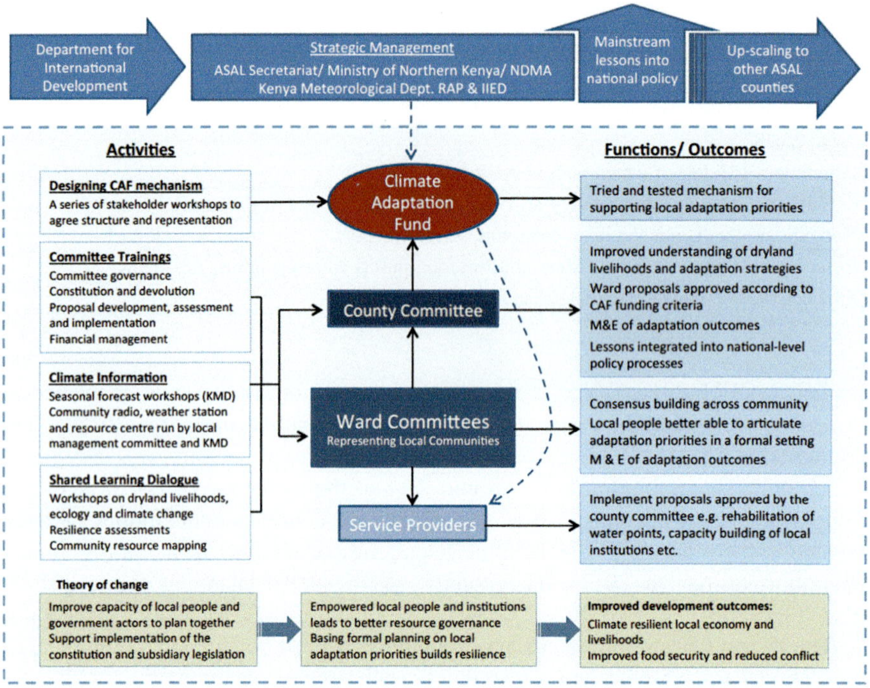

Fig. 2 Summary of the Isiolo Climate Adaptation Fund Process; Kenya National Drought Management Authority (KNDMA) 2014 Report

- Appropriation by county assembly not less than 2% of the development expenditure in a financial year
- Grants and loans from the national climate fund
- Climate finances from national and international entities
- Monies from public benefit organizations
- Fees and charges from climate finance activities
- Grants and donations

In addition, county wards have been identified as the bases for both planning and disbursements based on eligibility of proposals submitted. Amongst key criteria for proposal success are elements such as sustainability, vulnerable groups and gender, and public participation (Republic of Kenya 2016b). Further, Isiolo's CIDP has amongst others prioritized improved weather forecasting, better adapted livestock, and better food processing and preservation techniques as part of climate change adaptation and mitigation areas (Fig. 2).

Since 2010 with primary funding from the UK Department for International Development (DFID; £ 900,000) and additional funding from the Catholic Organization for Relief and Development Aid (Cordaid; Eur 152,327), the county went about setting up the Isiolo County Adaptation Fund (ICAF) with the various officials undergoing public vetting. The ICAF was to assist the county access global climate finance. The initial success led DFID to award an additional £ 6.5 million (2013–2016) to strengthen the Isiolo process and to supported an additional four other ASAL counties – Kitui, Makueni, Wajir, and Garissa (KNDMA 2014).

Findings and Discussion

The Paris Agreement has three main areas, namely, (a) holding the global average temperatures well below 2° C; (b) increasing ability to adapt to adverse impacts; and (c) financing mechanisms for climate change. The national government policies and strategies have captured and to some extent domesticated the three focus areas. The weak area is mostly in the financing area which is still unclear and mostly outward looking without adequate elaboration of internal financial flows to support adaptation and mitigation efforts. This risk is also manifested in the Isiolo county policies and strategies. When we move to the county level policies and strategies, the issue of keeping the temperature rise to below 2° C does not come out, and by extension the ability of counties to contribute to the national and global aspirations contribution gets lost (Table 2).

The Vision 2030 is the key political document guiding Kenya's development pathway until the year 2030; it has three pillars, i.e., the economic, social, and political. It is implemented through 5-yearly Medium Term Plans (MTPs). The environment, and by extension climate change, are a subset of the social pillar of this document and therefore somewhat less-visible; It has only two flagship programs with a direct bearing on climate change, i.e., the solid waste management initiative and the plastics bags initiative. This is somehow odd considering that climate change cuts across virtually all sectors of the country's enterprises, including but not limited to agriculture, construction, energy, transport, etc. One of the challenges arose because the Vision 2030 blueprint preceded the country's

Table 2 Alignment of national and (Isiolo) county policies and strategies with the three thrusts of the Paris Agreement. (Source: Ong'are and Omambia 2018)

Paris Agreement	National government policies	County government policies
a) Holding the increase in global average temperatures to below 2° C	Issue of global and country commitments to lower temperatures well elaborated except in the vision 2030	Issue of country and global commitments to lower temperatures very weakly elaborated; possible county contributions not spelt out
b) Increasing ability to adapt to adverse impacts on climate change	Priority sectors for interventions identified	Priority sectors for interventions identified
c) Making finance flows towards low GHG emissions and climate resilient development – General statement	Statements on financing mechanisms general and outward looking	Statements on financing mechanisms general and outward looking

commitment to the Paris Agreement; in some way this has been cured by some more deliberate write ups in the 5-yearly MTPs through which the vision 2030 aspirations are delivered. How much opportunity the MTPs provide for these provisions without deviation from the original focus areas in the mother vision 2030 document would then be the real question.

Kenya National Climate Change Response Strategy 2010 (Republic of Kenya 2010) was one of the first in a series of policy documents in Kenya to directly address the subject of climate change. It identified key areas of action as being agriculture and food security, livestock and pastoralism, water, health, forestry, energy, Rangelands, social infrastructure, physical infrastructure, fisheries, and coastal and marine resources. A fairly forward-looking document laid the foundation for subsequent policy-legal climate change products in Kenya. It can be credited with proposing the setting up of the climate change secretariat and a National Adaptation Facility (the current climate change fund).

All the policies (national and county based) have resulted in production of relevant implementation tools including Acts and regulations. The main Acts for delivery of the climate change agenda in Kenya include the Climate Change Act of 2016 (Republic of Kenya 2016a) and the Environmental Management and Coordination Act CAP 387 (Republic of Kenya 1999).

The Environmental Management and Coordination Act (Republic of Kenya 1999): This Act has made provisions for management of a wide array of environmental and natural resource assets. Most significantly it, among others, establishes the National Environmental Trust Fund to facilitate environmental research; The National Environment Restoration Fund to serve as supplementary insurance for mitigation against environmental degradation and deposit bonds as security against good environmental practice. With specific regard to climate change, it establishes the County Environment Committee (CEC) to not only be in charge of all environmental matters in a county but to also serve as a link between the national and county government. The CEC comprises of officials from both the county and national government. This committee is appointed by a county governor and serves for a period of 3 years. The EMCA also states that the cabinet secretary in charge of environmental matters shall prescribe measures on climate change. To date, for reasons ranging from financing to reporting lines, less than 10 out of the envisaged 47 county committees have been established thereby putting the whole aspect of environmental management within a county in jeopardy. Isiolo County is among those yet to put their county committees in place.

The Climate Act of 2016 (Republic of Kenya 2016a) is the latest and main legal document that seeks to anchor climate change action in all sectors of Kenya's economy, including national and county governments. It also extends to spheres such as disaster risk reduction. It has spelt out the duties of different players in climate interventions.

The institutional arrangements for climate management (as set out in the act) are a bit complex; there are several ministries playing different and sometimes multiple functions as the different policy documents came into force at different times. Apart from the climate council at the apex, the overall communication and reporting lines amongst the other institutions are not that clear. Moreover, counties being independent governments would be looking for a different engagement process rather than receiving instructions from national government institutions, and reporting back. This is because the constitution recognizes them as being equal to the national government.

The Climate Act also establishes a climate fund with a description of how it should be administered. The fund targets three main sources of financing, i.e.,

(i) Monies appropriated from the consolidated fund by an act of parliament
(ii) Monies received by the fund in the form of donations, endowments, grants, and gifts
(iii) Monies under an act payable to the fund

The climate fund is yet to be activated to date. The Government of Kenya, through the National Treasury, has launched the Green Climate Fund Readiness Project. Acceleration of this process is key as almost all sectors of the economy will be seeking to draw from this fund.

Within the Climate Act, NEMA is supposed to regulate, enforce, and monitor compliance on levels of GHGs as set out by the council. The council has yet to set out the levels; of interest though is the fact that NEMA is already implementing its air quality regulations through which it has been able to bring a number of enterprises in the manufacturing sector into compliance.

The Kenya National Adaptation Plan of 2016 (Republic of Kenya 2016c) is the key delivery mechanism for the country's commitments under the Paris Agreement. It has identified key sectors and earmarked a total budget of USD 38, 255,496, 051 (agriculture- USD 375, 116, 329; livestock- 299, 759, 329) to run till the year 2030. The main risk is a focus on external financing considering the magnitude of the budget in relation to Kenya's GDP. This therefore poses a risk to the actual capacities to deliver on the interventions earmarked across various sectors.

The private sector's role in the adaptation plan is quite marginal at national and county level documents with government entities given mandate to *"impose climate change obligations"* on the private sector (Republic of Kenya 2016c). Bearing in mind the prominent role, this sector plays in national development its perception as a passive player might rob the country of the opportunity to tap into the knowledge and innovation base resident in this sector. More opportunities should be explored for various contributions from this sector including financing mechanism, public private arrangements, research and eco-innovation. Ways to better engage and partner with this sector should be worked out. An important starting point could be the private-sector geared sustainable development goals (SDGs).

The Kenya Post Disaster Needs Report was an important product as gave opportunity to scientifically analyze and project the impacts of climate change related droughts and to facilitate planning and undertake various scenario analyses. It is instructive that it deliberately calls for the institutionalization of data and information to facilitate future adaptation and mitigation efforts in Kenya.

For Isiolo County, there are constants that the county had to deal with in climate planning and interventions which were the demographics, the arid conditions, and the (current) prevalent agro-pastoral practices. Local context is therefore important in county planning for adaptation and mitigation. Any interventions that don't take into account the arid conditions of the study site, together with the youth/gender angle, and the dominant livelihood activities, would probably not find easy take-up within the community and might be costlier to implement.

There is very good effort at linking the county and national government policy instruments such as the county climate change fund act 2016 (Republic of Kenya 2016b), the County Integrated Development Plan, the Climate Act of 2016 (Republic of Kenya 2016a), and the County Environment Action Plan (at final draft stage) amongst others. The County Integrated Development Plan (CIDP) is the chief planning document within a county. However, the Isiolo CIDP, while acknowledging the high percentage of the population made up of youth (60% between 0–24 years), fails to highlight this, together with gender, as an important intervention area within its climate change adaptation and mitigation strategies. The same document goes on to identify conflict over grazing rights as a major challenge in the county but the funding stage only singles out floods, drought, and diseases as priority areas. It is important to do a proper linkage between the huge proportion of the population comprised of unemployed youth and the frequent grazing conflicts in order to design appropriate strategies.

The dilapidated or in many areas absence of appropriate infrastructure is a matter of concern especially from the context of disaster management and livelihood enhancement. The completion of the Nairobi–Isiolo–Moyale road is a major milestone in the opening up of this region. However, the low ICT penetration in the county still poses a major challenge.

The Isiolo CIDP acknowledges that 80% of the land in Isiolo County is used as grazing land and that climate impacts will be worsened by intensified grazing. This is a matter that must be central in the county planning process going forward

ASAL counties have benefited significantly by the presence national government institutions such as the National Drought Management Authority and development partners in the management of drought and its resultant impacts. Further, there is enhanced national support for research, awareness, and generation of policy briefs with respect to climate change and drought. However, there is a need to operationalize the National Climate Change Fund established under the Climate Change Act 2016 in order to fund a county such as Isiolo whose climate change and drought readiness process is fairly well developed. Isiolo like other counties was constrained financially to respond to the recent 2008–2011 and 2016–2017 droughts. Further, like many counties in Kenya, the environment docket is underfunded thereby jeopardizing preventive natural resource conservation actions that are vital to mitigate drought and the resultant impacts. In the 2015 county fiscal strategy paper for Isiolo, the vast sector of livestock, veterinary, agriculture, fisheries, and physical planning is only apportioned 7.3% of the budget (Republic of Kenya 2015b) with no direct mention of specific allocations towards climate resilience. An interesting development is the fact that climate change is starting to find a political voice as evidenced in that fact it has found debate time in the Isiolo county assembly with local leaders looking for external financing to support adaptation and mitigation actions.

Even though pastoralism is a major socioeconomic activity in the county, there is limited mention of strategies for pastoralism aimed at enhancing long-term resilience in the County Climate Change Fund Act of 2016 (Republic of Kenya 2016b). The latter Act also advocates for proposal development as an avenue for accessing climate financing; part of this money should be ring- fenced for activities to be carried out by the county government itself such as the disaster early warning system and or conflict management. However, the establishment of a gazetted county environment committee as envisaged in the Environmental Management and Coordination Act (EMCA CAP 387) has been complementary in facilitating climate change resilience in the county. There is further need to clarify and enhance the role of women and youth; the census results point to the significant proportion of youth to the total population which can therefore not be ignored. In the absence of productive engagement, they could be negatively deployed in the frequent community conflicts. Four critical areas for conflict management in ASALs have been identified as the Kenyan state, the political leadership, local communities, neighboring states, and communities (Republic of Kenya 2015a). More specific interventions are needed to address the increasing number of female-led households.

The process of setting out the Isiolo County Adaptation Fund (ICAF) demonstrated possibility of an ecosystems approach to climate change adaptation since the Isiolo process was also cascaded to other ASAL counties of Kitui, Makueni, Wajir, and Garissa; this opens up further possibilities for joint interventions and other synergies where possible.

As pointed out, Isiolo was hitherto a relatively rural county that relied heavily on pastoralism as a way of life. The massive and rapid development of huge infrastructural projects such as the Lamu Port Southern Sudan and Ethiopia Transport (LAPSSET) corridor have increased pressure on natural resources like water and biomass fuel that are already scarce in this region fueled directly by infrastructural projects and an influx of people raising concerns by locals and creating a sudden inter-section between agriculture-based climate change-related challenges and those that accompany urbanization. This therefore calls for new management approaches. This is a new reality for the country as many counties are rapidly urbanizing due to commissioning of various types of mega-projects spread in many different counties (Table 3).

Conclusion

This chapter set out to look at linkages between the policy/legislative issues at national and county levels in Kenya using Isiolo County as a case study and how these have translated into climate actions at the local level. There is a major challenge under the current circumstances where environment and climate change matters are categorized under the social pillar of the country's key development plan

Table 3 showing summary of climate change policy documents at national and county levels. (Source: Ong'are and Omambia 2018)

Item/policy	Remarks
Overall institutional arrangements	Several institutions with multiple functions; at some levels, one institution could be performing a coordinating role at another level the same institution is reporting to another. Reporting and collaboration lines between MDAs a bit complex
Kenya National Climate Change Response Strategy 2010	First policy document directly addressing climate change in Kenya. Identified country's vulnerable sectors. Proposed the setting up of a climate change secretariat and a National Adaptation Facility
Kenya's intended nationally determined contributions (2015)	Quite ambitious targets. Focuses on both adaptation and mitigation; livestock identified as an important sector
The Kenya post disaster needs assessment report of 2012	Examined impacts of a real drought and how it plays out in terms of impacts. Recommended integration of new programs with the vision 2030 and the need to institutionalize assessments
Vision 2030	Key planning document for Kenya till 2030. Organized into social, economic, and political pillars. Developed before Kenya's commitments to the Paris agreement. Environment and climate change a subset of the social pillar of this document. Efforts at mainstreaming in the 5-yearly plans of vision 2030.
The Environmental Management and Coordination Act (EMCA CAP 387)	The overarching legislation on environmental management in Kenya Less than 10 out of 47 county environmental committees established to date. Puts in jeopardy efforts of environmental management at county level (including climate change)
Kenya Climate Change Act 2016	Has set out the institutional arrangements for climate change governance in Kenya
Kenya National Adaptation Plan 2016	Huge budget estimated at USD 38, 255, 496, 051 for various sectors to run till 2030. It has earmarked priority sectors for intervention. No clear mechanisms for financing-focus is mostly on external financing which poses a risk
Isiolo County Integrated Development Plan (CIDP) 2013–2017	Rapidly growing population 80% of inhabitants reliant on pastoralism Inter-ethnic conflict an ever present threat Role of women in climate adaptation needs higher profiling
County fiscal strategy for Isiolo	Sets out the county's priority programs for implementation within the Government's medium term expenditure framework
Isiolo County Climate Change Fund Act, 2012	Has outlined elaborate management structures Identified ways to incorporate climate finance with county planning and budgeting Most internal sources non-financed and therefore unlikely to be realized

(the vision 2030) in terms of profiling the whole debate around environment and climate change. ASALs are important ecosystems in Kenya and therefore have an important role to play in climate mitigation and adaptation in Kenya. As pointed out in the Kenya post disaster needs assessment report of 2012, drought will continue to be recurrent problem in Kenya (especially ASALs) for the foreseeable future. Coordination at various levels (across institutions and between national and county governments) seems to be one major challenge with regard to climate change action in Kenya. The second key factor is international financing at both national and county levels (currently estimated at 40 billion USD across all the policy documents examined in forms finance, investment, technology, and capacity-building till 2030); with the government facing pressure from other sectors for limited resources, an inordinate amount of attention has been directed at external actors to fund climate change activities in Kenya.

Enhanced and harmonized efforts by relevant national, county, and NGO stakeholders are required to bring to the fore the importance of the environment as the foundation for sustainable development and not a hindrance to the same. It is unequivocal that climate change is real, and its reinforcing impact on Kenya's drought prone arid and semi-arid areas is not only increasing community and ecosystem vulnerability but also threatening both national and county socioeconomic development. Commendable efforts in addressing both climate change and drought are ongoing in Kenya, and the Isiolo case is not only emulative but can be replicated and up-scaled. There is also need for innovative resource mobilization approaches from internal and external sources, technical skills enhancement, and adoption of environmentally sustainable technologies enhancement in order to effectively implement county and national plans aimed at increasing resilience of communities in drought-prone areas. It is important that continuous lesson learning around institutional and governance arrangements around climate change are adopted. It is recommended that efforts to address both drought and climate change should be inclusive and participatory with all stakeholders in order to yield multiple benefits from all players. There is also need to translate resilience plans into priority action especially in the gender and conflict-management arenas.

It is noted that ASALs constitute over 80% of Kenya's land mass and therefore their potential cannot be ignored with regard to the country's development plans. Unfortunately, this study was only limited to one out of several ASAL counties. It might be prudent to collate information from a number of ASAL counties for purposes of informing policy at the national level.

Recommendations

These should be implemented as complementary and not stand-alone actions;

- There is a need to expedite the activation of the climate fund as spelt out in the climate act of 2016; within the said fund explore opportunities to upscale internal financing to complement external sources.

- Urgent discussions should be initiated between the national and county governments to speed up the establishment of county environment committees to be in charge of coordination of all environment and climate functions within the counties.
- Institutionalize assessments as recommended in the Kenya post disaster needs assessment report of 2012.
- The National Climate Change Response Strategy Needs to be reviewed as the aspirations of the old one (2010) have largely been met with regard to the setting up of the legal and institutional frameworks for climate change management in Kenya.
- There is a need to simplify the institutional arrangements around climate change management in Kenya; the roles and inter-plays between ministries, departments, and agencies need to be clearer and complementary.
- More avenues needed to collaborate in multiple ways with county governments which are the actual implementation grounds for climate change interventions in the county. Debate on this issue should be initiated with the secretariat to the council of governors.
- Upscale the role of private sector to that of a critical partner and create the space for eco-innovations within the climate change arena. Potentials exist for these in the agro-pastoral, conflict- prone settings such as Isiolo. These could for instance be in early-warning systems, marketing, pasture, and feed management, etc.
- There is a need to further strengthen the science-policy-action nexus. This should be done through the existing institutional arrangements between the Kenya meteorological department (data collection and analysis that incorporates indigenous knowledge and practice), the National Drought Management Authority, and the county government in order to achieve science led decision-making.
- Avenues (including public-private partnerships) should be explored to urgently improve the cellular telephony penetration in the county as this is a cross cutting enabled for various other sectors of the county's economy from markets access, weather forecasting, and disaster management.
- Financing mechanisms for the environment and natural resources sector, though provided for in the county finance act, the actual realization of the same has been a tall order. New strategies are needed as the environment and natural resource sectors don't seem to be able to attract the required resources when compared to other sectors like infrastructure.
- Support is needed to diversify livelihood options as well as shift pastoralist investments to the more drought tolerant livestock breeds especially the shoats (sheep and goats), while at the same time establishing and sustaining food and fodder banks. The youth should be supported to invest in these areas to avoid them being used to fan conflicts.
- There is a need to strengthen the place of women and girls in society through the development and implementation of a county gender policy. Other actors such as religious institutions, conservancies, etc. could also assist. Their capacities should be built in strategic areas and their education (formal and informal) enhanced.

- There is need to designate a docket in the county specifically for climate change to spearhead development of an elaborate county adaptation action plan that is in line with the county spatial plan; a key element of this plan could entail a grassland and grazing strategy based on an approach developed for Laikipia county by the northern rangelands trust; the latter holistically addresses pasture and livestock management practices as a drought-coping strategy (NRT 2016b). Adoption of an ecosystem based adaptation (EBA) is a good method of avoiding maladaptation since it is it is localized to the ecosystem within which it is practiced and also provides other co-benefits such as job creation (EC 2013).
- There is a need to enhance dialogue at the national and county government levels to synchronize planning processes through the medium term planning process and the county integrated development planning process.
- For both national and county governments and all non-state actors in drought management, disaster risk reduction actors and climate change in the ASALs, there is a need for periodic joint measurement, reporting, and verification of actions to track progress and reinforce benefits.

References

Caroline King-Okumu et al. (2016) Direct use values of climate-dependent ecosystem services in Isiolo County, Kenya; http://pubs.iied.org/1014 IIED; ISBN 9781784312480

EC; (European Commission) (2013) Science for environment policy; Ecosystem-based adaptation provides promising approach. Thematic Issue 37. Ecosystem-based adaptation

Filho LW, Nzengya GM, Chemuliti J, Kalungu WJ (2017) Climate change responses among the Maasai Community in Kenya. Climate Change 145:71. https://doi.org/10.1007/s10584-017-2087-9

IPCC AR4 WGII (2007) Climate change 2007 – impacts, adaptation and vulnerability. Contribution of working group II to the fourth assessment report of the intergovernmental panel on climate change, 2007. http://www.ipcc.ch/pdf/glossary/ar4-wg2.pdf. 3

KNDMA (Kenya National Drought Management Authority) (2014) Isiolo County adaptation fund; activities, costs, and impacts after the 1st investment round

Northern Rangelands Trust (2016a) Gender analysis report, Nairobi

Northern Rangelands Trust (2016b) Grassland and grazing strategy 2016–2026. Said Business School, University of Oxford

NRT (Northern Rangelands Trust) and Kenya Markets Trust (2012) Rethinking the shoat market. SpardAfrica, Nairobi

Omambia AN, Shemsanga C, Sanchez Hernandez Y (2015) Climate change impacts, vulnerability and adaptation in East Africa (EA) and South America (SA). In: Chen W-Y, Seiner J et al (eds) Handbook of climate change mitigation. Springer Science + Business Media, LLC, New York

Republic of Kenya (1999) Environmental Management and Coordination Act (EMCA CAP 387)

Republic of Kenya (2010) Kenya national climate change response strategy. Ministry of Environment and Natural Resources, Nairobi

Republic of Kenya (2012) Kenya Post-Disaster Needs Assessment (PDNA) 2008–2011: Government of Kenya

Republic of Kenya (2013) Isiolo County Integrated Development Plan 2013–2017; Government of Kenya

Republic of Kenya (2015a) Kenya's Intended Nationally Determined Contribution. Ministry of Environment and Natural Resources. 23 July 2015. http://www.environment.go.ke/wp-content/uploads/2015/07/Kenya_INDC_20150723.pdf

Republic of Kenya (2015b) County fiscal strategy paper, Isiolo. County Government of Isiolo. http://kebudgetdocs.ipfkenya.or.ke/docs/IsioloCounty/2015-2016/IsioloCountyFiscalStrategypaper2015.pdf

Republic of Kenya, Kenya Climate Change Act (2016a) Government Printers. Nairobi

Republic of Kenya (2016b) Isiolo County Climate Change Fund Act, 2016; Government of Kenya

Republic of Kenya (2016c) Kenya National Adaptation Plan 2015–2030; Enhanced climate resilience towards the attainment of vision 2030 and beyond

Smith G (2016) Technical paper: opportunities and options for enhancing adaptation actions and supporting their implementation: reducing vulnerability and mainstreaming adaptation. https://unfccc.int/files/adaptation/groups_committees/adaptation_committee/application/pdf/tp_adaptation_2016.pdf

UNFCCC (2015) The Paris Agreement (FCCC/CP/2015/10/Add.1). http://unfccc.int/files/essential_background/convention/application/pdf/english_paris_agreement.pdf

Climate Change Adaptation Among Female-Led Micro, Small, and Medium Enterprises in Semiarid Areas: A Case Study from Kenya

50

Joanes Atela, Kate Elizabeth Gannon, and Florence Crick

Contents

Abstract

This chapter contributes to the literature on private sector adaptation by empirically exploring how female-led micro, small, and medium enterprise (MSMEs) in Kenya's semiarid lands (SALs) experience and respond to climate risk. The chapter argues that strong sociocultural orientations around gender roles and resource use and access not only confine female-led MSMEs to sectors that experience higher exposure to climate risk – most notably agriculture – but also trigger more pronounced barriers to building resilience within their businesses, including reduced access to land, capital, markets, new technology, and educational opportunities. Faced by these barriers, female entrepreneurs may pursue unsustainable forms of coping, as part of which business activity is scaled back through reduced profits, loss of business, and the sale of valuable business assets.

J. Atela
African Centre for Technology Studies, Nairobi, Kenya
e-mail: joanes.atela@gmail.com

K. E. Gannon (✉) · F. Crick
Grantham Research Institute on Climate Change and the Environment, London School of Economics and Political Science, London, UK
e-mail: k.e.gannon@lse.ac.uk; kateelizabethgannon@gmail.com; F.D.Crick@lse.ac.uk

W. Leal Filho (ed.), *Handbook of Climate Change Resilience*,
https://doi.org/10.1007/978-3-319-93336-8_97

Such strategies may help enterprises to cope in the short term but may undermine longer-term MSME adaptive capacity. Social networks, such as women's groups and table banking initiatives, appear to be crucial adaptation tools. Additionally, a strong dependency exists between household resilience and business resilience, implying that building resilience at the household level could support adaptive capacity among female-led MSMEs. Supporting the adaptive capacity of women in business should be a policy priority.

Keywords
Adaptation · Small and medium enterprises (SMEs) · Women · Semiarid lands · Private sector · Narok

Introduction

The private sector is increasingly gaining attention as a means of furthering global climate change mitigation and adaptation responses. "The private sector *for* adaptation" (Pauw and Pegels 2013:258) is often afforded greatest attention, as national and international governance fora look to the private sector as a means of harnessing resources, especially in pursuit of low-carbon development. In this chapter, however, we focus on what Pauw and Pegels label "domestic private sector adaptation": The processes through which firms institute strategies to manage climate risk within their own operations. Within domestic private sector adaptation, we focus specifically on private sector adaptation among micro, small, and medium enterprise (MSME) in SALs.

There are several reasons for adopting this focus. Firstly, "the private sector" constitutes a wide-ranging term that covers all types of formal and informal businesses, ranging from microenterprises, such as local entrepreneurs and smallholder farmers, through to multinational companies operating at global scales. As well as focusing primarily on companies based in developed countries, as noted by Crick et al. (2018a), the private sector adaptation literature has, to date, tended to focus on larger-sized companies (Averchenkova et al. 2016; PWC 2010; Tompkins and Eakin 2012). Yet adaptive capacity varies widely, being intimately connected with wider social vulnerabilities and heavily constrained by local institutional, regulatory, infrastructural, and technological conditions (Ackerman et al. 2012). As a result, it is important to disaggregate the private sector when considering adaptation (Crick et al. 2018a).

We focus on adaptation among MSMEs, since MSMEs are considered highly vulnerable to climate change, to be among the most affected by extreme weather events and to typically have low adaptive capacity (Crick et al. 2018a; Wedawatta et al. 2010). Yet MSMEs form a critical part of developing country economies. Accounting for about 80% of total employment (Dougherty-Choux et al. 2015), they provide most livelihood opportunities in sub-Saharan Africa (SSA). Further, MSMEs are considered to be fundamental to more inclusive and equitable development, since they are strongly integrated into communities and hold the potential to make an important contribution to female employment and the social integration of marginalized groups (Crick et al. 2018a). MSMEs are also seen as important drivers for societal adaptation and for realizing the opportunities of climate change.

Our research focus on MSMEs in Kenya's semiarid lands (SALs) stems from the recognition that SALs in developing countries are climate change "hot spots," areas where climate change is already having significant and documented impacts, such as longer and more frequent droughts. The IPCC's Fifth Assessment report (IPCC 2014) has recognized SALs as among the areas that are especially vulnerable to the impacts of climate change. Indeed, in what has been described as the "triple whammy" of semiarid regions (Mountfort 2015), in SALs this high exposure to climate risk is often coupled with lack of development, including poor infrastructure, poor access to markets, and high levels of poverty (De Souza et al. 2015; Gannon et al. 2018; Jobbins et al. 2016; Tucker et al. 2015). An estimated two billion people live in the world's drylands; 90% of which live in developing countries (FAO 2016). The livelihood challenges faced in SALs will be further exacerbated by anthropogenic climate change, so, if implementation of the 2030 Agenda for Sustainable Development is to fulfil its commitment to "leave no one behind," the need to identify routes to supporting adaptation and climate resilient development in these areas is especially pressing.

This chapter further narrows its focus to give consideration to the ways in which female-owned MSMEs experience and respond to climate risks in SALs. Literature emphasizes that climate change impacts the activities of social groups differently, based on social relations that define roles and access to resources for adaptation and which shape overall adaptive capacity (Nyukuri 2016; Omolo 2010). These differences have emerged and manifest in the context of established gender roles that often shape what women and men, in different social contexts, are allowed, able, or expected to do, or access.

Female-owned MSMEs are increasingly recognized as key in promoting resilience at micro- (e.g., household level) and macroscales (Spath 2016). In literature that considers female contributions to entrepreneurship (Omolo 2010; Terry 2009), a number of studies indicate that, due to their strategic position in families, women often more efficiently allocate returns from MSMEs to the most critical household assets, including health, education, and food security, which themselves shape and enable household and business adaptive capacity (Nichter and Goldmark 2009; Omolo 2010; Terry 2009). Another study suggests that women involved in entrepreneurship activities make relatively higher contributions to family and social welfare than men (Minniti, 2010). In recognition of this strategic role, promotion of female entrepreneurship, through advocacy and microcredit facilities, is often promoted as a tool through which to not only empower women but also to address broader poverty reduction and development agendas as well (Kevane and Wydick 2001). Nevertheless, while a significant body work has already accumulated on women's contribution to entrepreneurship (Omolo 2010; Terry 2009) and vulnerability to climate change (Bardasi et al. 2007; Kelly et al. 2011), little research or policy attention has been given to exploring how female-led MSMEs experience climate risks or to identifying options for adaptation.

This chapter contributes to addressing this knowledge gap by exploring how female-owned MSMEs in Kenya's SALs respond to climate risks, as well as the roles, resources, and entitlements that female-headed MSMEs draw on to enable these responses. It does this with the aim of identifying factors within MSMEs and within MSME business environments that both enable and constrain female experience of MSME adaptive capacity. Given that adaptation is contextually bound and situated, this chapter explores these questions through a case study of the semiarid county of Narok. Situated in the Kenya's Rift Valley, Narok is less than 150 kms from Nairobi.

The rest of this chapter is organized in a further six sections. The next section provides an overview of methods. The chapter then provides a brief overview of the fieldsite, reflecting on general challenges for private sector adaptation in Kenya's SALs. Drawing on empirical research in Narok, it then considers experience of climate risk among female-led MSMEs in Kenya's SALs and the way in which female-led MSMEs respond to climate risk. We then offer insights on some of the challenges to adaptation and climate resilient business development in female-led MSMEs in the hope of identifying priority areas for addressing the broader structural deficits that condition female vulnerability to climate change and barriers that limit adaptive capacity among female entrepreneurs in SALs. Ultimately this chapter is going to argue that strong sociocultural orientations around gender roles and resource use and access not only confine female-led MSMEs to sectors – most notably agriculture – that experience higher exposure to climate risk. But it is also going to provide evidence to suggest that women in Kenya's SALs face additional, and more pronounced, barriers to building resilience within their business activities.

Data Collection Methods

The first stage of this research involved review of academic and policy literature on female enterprise and private sector adaptation in SALs. The documents were identified through key word searches in databases including Google Scholar and Science Direct. The contents of the documents were coded into various thematic areas around gender roles, representations, and entitlements, among others. Relevant policy documents were also scoped from government archives, with specific attention given to documents outlining climate change response strategies enshrined in climate change papers such as the National Climate Change Action Plan, Kenya's Nationally Determined Contributions, and the Climate Change Act, as well as to MSME-related interventions, such as those relating to environmental and social safeguards underpinning the operations of MSMEs. These documents were also analyzed and incorporated into the coding frame. In line with the research questions, the review focused on the role of women in MSMEs, including within Kenya and within SALs specifically, and sought to identify some of the formal and informal structural challenges female-led MSMEs face, as well as options to address these structural challenges and to enhance adaptive capacity.

Literature review was then supplemented by fieldwork involving focus group discussions and key informant interviews in Narok county. The focus group discussions took the form of a 1-day, in-depth workshop with 17 female entrepreneurs drawn from the 3 constituencies of Narok county. About two thirds of the participants were involved in agricultural enterprises, including both livestock and crops, while the one third were engaged in nonagricultural MSMEs, such as running a tailoring shop and small construction companies and restaurants within Narok town. The full-day meeting was held at the Maralink Hotel in Narok. The discussions were structured around three main components.

Firstly, to gain insight into the broader conditions in which female entrepreneurship operates in Narok, in plenary, open lines of questioning were pursued (Johnson and Weller 2001; Krueger and Casey 2000). Participants were asked to outline their business journeys and reveal moments of change when their MSMEs experienced particular growth, stagnation, or even diminishing returns, either due to climatic or non-climatic drivers. They were then asked to reflect specifically on how this business journey may have been shaped by their identity as a female-led business. Since researchers sought to gain an understanding of participant's engagement with climate risk within their own frame of reference, participants were then engaged more directly, but in as neutral a format as possible, in discussions about how climate impacts their MSMEs.

Participants were then divided into three groups of six women, to each discuss specific issues that had arisen in plenary, pertaining to how their businesses respond to climate pressures – as well as to discuss some of the challenges (including financial, technological, institutional, cultural, etc.) they face to managing risk within their business. In these discussions, participants were also encouraged to explore opportunities for building the adaptive capacity of MSMEs, including ideas for potential policy options. These smaller, breakout groups offered participants the opportunity to provide more in-depth accounts of their experiences than the plenary setting had permitted. The final session of the day involved presentations and discussions where each small group presented their findings and allowed the wider audience to verify, elucidate, and interrogate their insights. This combination of breakout and plenary approaches proved a useful means of stimulating discussions and triangulating various insights.

Five purposively sampled key informants, with different positions of experience and oversight with respect to female-led MSME activities and management, were also engaged in semi-structured interviews. Three informants were women involved in running both agricultural and nonagricultural MSMEs, while one was a male respondent involved in managing beekeeping cooperatives in Narok that incorporate female-led beekeeping businesses. The fifth informant was a policy maker. More open lines of questioning were pursued first, with the interview protocol then prompting for insights on enablers, barriers, and opportunities for female-led MSMEs.

A 1-day stakeholder workshop in Narok was also held, in which 30 stakeholders including county government, development NGO, microfinance, and various microenterprises representatives participated. One third of the workshop participants were female, while the others were male. This workshop had a broad remit, including to raise awareness of the Pathways to Resilience in Semi-Arid Economies (PRISE) project and to identify potential partnerships. However, this workshop also generated further insights around how MSMEs in Narok experience climate risk and around enabling adaptation for female-led MSMEs. The workshop also served as a useful forum for triangulation, as it involved discussion of preliminary findings from the focus group discussions and interviews. It thus served as a useful tool to explore broader implications of our findings for a wider range of actors in Narok county, as well as elsewhere.

Private Sector Adaptation in Kenya's SALs

Eighty percent of Kenya's terrestrial land is classified as arid or semiarid, receiving less than 500 mm of rainfall per year (Atela et al. 2014). Kenya's arid and semiarid lands are home to over ten million people whose livelihoods are widely dependent on natural resources and associated enterprises. Agriculture (both livestock and crops) and tourism are key sectors in the national economy (Republic of Kenya 2010a), and arid and semiarid economies make important contributions to both of these sectors. Pastoralists in Kenya's arid and semiarid areas contribute about 70% of Kenya's livestock (Ole Riamit 2010) and host most of the country's wildlife conservancies, such as the Tsavo National Park and Maasai Game Reserve, which are important sources of tourist revenue to both local and national economies. As is characteristic of semiarid regions (Wade et al. 2015), Kenya's SALs experience high rainfall variability. Thus, this high social and economic dependence on natural resources means SALs in Kenya face great vulnerability to climate risk. In Kenya, this manifests in cycles of drought and hunger, occurring with seemingly increasing frequency (Republic of Kenya 2010a).

MSME adaptive capacity – the ability of businesses to adapt to this exposure, to minimize its adverse impacts, and to take advantage of new opportunities (c.f. Adger et al. 2005) – is shaped and constrained by diverse social, political, and economic processes (Smit and Wandel 2006), including local sociocultural circumstances, policies, and market arrangements. Recent literature has sought to identify routes to enabling MSMEs to cope with the heightened exposure to climate risk that they will face in areas such as SALs. This literature has highlighted, for example, the need for enabling policies and an institutional environment that supports individual participants in the private sector (Crick et al. 2018a; Pauw 2015; Stenek et al. 2013; Trabacchi and Stadelmann 2013). It has also highlighted the need for climate information, sensible regulatory structures, and economic incentives (Biagini and Miller 2013; IFC 2010). Also focusing on MSMEs, Crick et al. (2018a) seek to synthesize literatures on private sector adaptation and private sector development into a coherent framework to identify "building blocks" of an enabling environment for private sector adaptation and climate resilient development and to assist in the identification of gaps in enabling conditions for policy intervention. The resulting framework is structured around four interlinked pillars: policies and institutions, infrastructure and markets, financial environment, and information and capabilities (see Fig. 1).

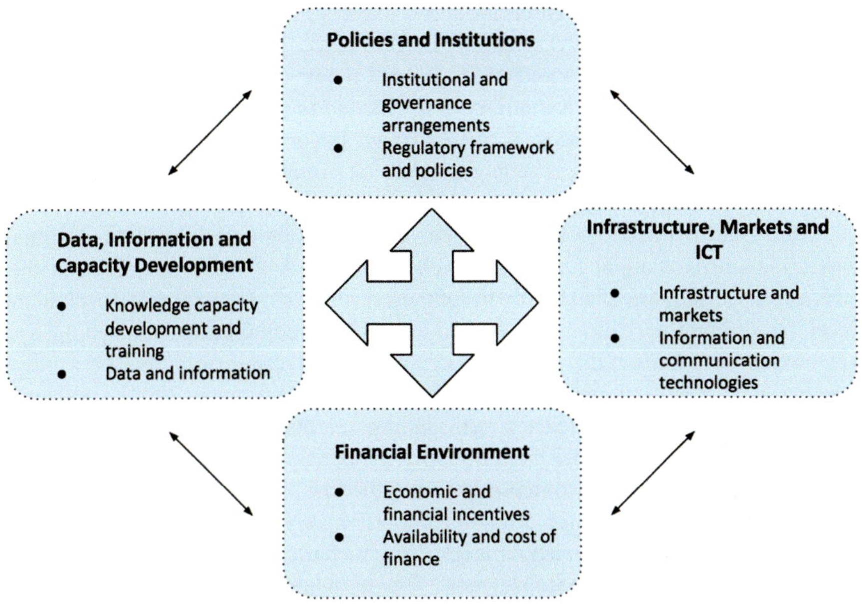

Fig. 1 Core elements of an enabling environment for private sector adaptation and interlinkages across elements. (Source: Crick et al. 2018a)

The four pillars of Crick et al.'s framework for enabling private sector adaptation can ultimately work to promote or constrain the adaptive capacity of enterprises, and when qualitatively evaluated against Crick et al.'s framework, Kenya as a whole is found to lack many of the conditions that support private sector adaptation (Crick et al. 2016). Key gaps surround the availability and use of data and information, capacity development and training activities, and the financial environment. MSMEs in Kenya's SALs are further constrained by limited infrastructure and market development, leading to weak interfirm linkages and restricted access to new technologies. MSMEs in SALs experience these constraints to their business development and adaptive capacity, regardless of gender. But there is evidence to suggest that some constraints, including access to finance and training opportunities, affect female-owned enterprises more severely (Brindley 2005; Singh and Belwal 2008; Welsh et al. 2013).

Exposure to Climate Risk among Female-Led MSMEs in Narok, Kenya

Literature suggests that the majority of female entrepreneurs in SSA are confined to microenterprises, in the informal sector, with limited growth potential (Bardasi et al. 2007). It also suggests that female-owned enterprise tends to be concentrated in economic sectors such as agriculture and agricultural processing, which typically require less capital for participation (Bardasi et al. 2007; International Labour

Organisation 2015). Our research in Narok reflects these findings, suggesting that most women in Kenya's SALs are confined to agriculture-based MSMEs – often involving small-scale trade in cereal, milk, and poultry products. Through extreme temperatures, high rainfall variability, and fluctuations in natural capital, such as grazing and water resources – which may become scarce, unavailable, or reprioritized for immediate livelihood needs at times of climate stress – these industries are highly exposed to the effects of climate change.

In the face of this exposure, our focus group participants often linked climate impacts, such as drought and climate-related livestock disease, to losses to their business assets and to reduced growth for their businesses. For example, respondents described drought conditions resulting in loss of livestock products emanating from low or zero birth rates, reduced production of milk, and diminishing market value of animals as a result of weight loss (c.f. McPeak and Barrett 2001). Reflecting interactions and co-dependencies surrounding resilience within households and businesses in SALs, livestock is not simply a business asset to MSMEs in Narok. Rather it also represents a source of protein, income, savings, security, and social status. To lose animals through droughts, therefore, may well represent an economic and social disaster at the household level. Similar challenges occur in relation to lost crops. And, furthermore, because most female-owned businesses around Narok town rely on customers and raw materials from the surrounding areas, participants emphasized that times of reduced productivity, supply chain disruption, and agricultural income, coupled with the subsequent reduction in the purchasing power of farmers, also strongly impact nonagricultural MSMES.

Respondents additionally highlighted that conflict around scarce water and grazing resources at times of drought has resulted in further insecurity in agricultural MSMEs in Narok and placed stress on livelihoods in SALs more broadly. Respondents also reflected that at times of drought, due to vulnerability and hunger, men in Narok – who generally control land resources – often resort to selling off land at "crisis" values, significantly below the actual value of the lands. MSMEs led by women can be significantly affected by such coping strategies, since the money received from land sales is rarely invested into female enterprise and land disposal means that land can no longer be used for farming or as security to access credit from microfinance institutions. Extreme flooding events have also been witnessed in Narok in the recent past, which participants linked to extensive loss and damage through destruction of transport links – limiting the timely transportation of often perishable goods – and of premises within which business assets, stock, and produce were stored.

Ways in which Female-Led MSMEs in Narok Respond to Climate Risks

Literature on private sector adaptation distinguishes between *sustainable* forms of adaptation, such as taking out insurance, which seek to maintain business operations at existing levels, and *unsustainable* coping strategies, such as the distress sale of assets, as part of which business activity is scaled back (Crick et al. 2018b). Despite being generally aware of and experiencing climate risk, female MSME owners interviewed in Narok broadly perceived limited adaptation options to be available for their businesses and often described dependency on unsustainable coping strategies at times of climate stress. Respondents indicated that one main way of responding to climate pressures involved stock reduction at times of water scarcity, to preserve resources and reduce the risk of losses. In response to reduced water availability, tomato traders from Narok South, for example, described reducing their risk of loss and damage by reducing the amount of land under production, so as to limit the area of land in need of irrigation. Yet while such strategies may help businesses to cope in the short term, they do nothing to help build longer-term MSME adaptive capacity. Indeed, over time reducing stocks and production may prove counterproductive, as the resulting loss in profits and business that such downscaling presents may limit the resources that women have available to help them to cope with future climate change impacts – presenting potential longer-term implications for broader household resilience.

At other times, female entrepreneurs we spoke to noted that reducing stocks resulted in reduced profits to an extent that sometimes women may close, or diversify, their businesses, to pursue alternative livelihood options. One major diversification option reported by participants is the charcoal business, in which women engage in charcoal burning, or buy charcoal from charcoal burners, for sale in Narok town. Indeed, our research suggested that women involved in agricultural MSMEs often diversify or shift toward charcoal-based enterprise, believing this to be a livelihood strategy that is less strongly affected by climatic variables. Yet, while providing a short-term coping strategy, charcoal production may also weaken longer-term resilience by destroying vegetation, exposing agricultural land to erosion, and further diminishing productivity, leading to loss of biodiversity. Reflecting this trend, studies indicate that Kenya's arid and semiarid areas are now frontiers of deforestation and charcoal production, as demand for charcoal has increased and humid forests have become restricted as "water towers" for national economic interests (e.g., Republic of Kenya 2013).

> I am part of a women's group in Narok West involved in tomato production for sale to various institutions and customers in Narok. However, due to drought that happened in the last two years, it has become very difficult to maintain production of tomatoes because of water problems. Most of us have therefore abandoned tomato production and businesses and engage in charcoal burning and trade which has ready markets from hotels and restaurants serving tourists visiting Maasai Mara Game Park. Through the sale of charcoal, I have managed to buy a motorbike to ease the distribution of charcoal and tomato products to the market.
>
> – Female entrepreneur, Narok May 2017

Another coping strategy employed in Narok, pertains to land sales, as referenced above. Sale of land is a thriving business in Narok town, with immigrants into the Narok area providing a ready market for land. Participants argued that proceeds from these sales could usefully support their businesses against climatic risk, providing a boost of funds to female-led businesses and even helping them to secure access to other adaptation options and safety nets, such as insurance schemes. However, participants highlighted that the ability of women to benefit from the proceeds of land sales typically depends on the willingness of the male members of their family – who are generally the custodians of the land – to share the proceeds to support female-led businesses. In cases when the participants in our research had benefited from these proceeds, they described having had to negotiate with men to secure this access, at times committing to sharing business proceeds with their husbands. But these women were the exception as, in most cases, female-led MSMEs were said to not receive any direct impact from the sale of land. Sometimes women felt they benefitted indirectly from these sales if the men in their families used some of the proceeds from land sales to purchase food and other basic household items. In such circumstances, female entrepreneurs described a reduced emotional burden, due to the temporarily reduced dependency on their own enterprise to support household needs.

Although family land is predominantly held by men, land disposal remains salient to female entrepreneur's adaptive capacity more broadly, since participants highlighted that their businesses may benefit informally from land and livestock assets. Most participants, for example, reported using land and livestock as informal security for their MSMEs, especially when seeking to access credit from family networks, or from women's group initiatives, such as table banking groups. A number of participants indicated that following recent flooding events, which resulted in destruction of business premises and assets in Narok, female entrepreneurs, including those running small restaurants and tailoring stalls, were able to access capital support from family networks and table banking groups based on the availability of family land as security. But land disposal as an adaptation strategy depletes the resources available for enterprises to protect themselves against future climate risk; thus when land has been sold, the ability of female entrepreneurs to access these informal benefits of land ownership may be lost.

Overall, adaptation behaviors highlighted by participants were largely localized, with participants identifying very limited opportunities for adaptation beyond their immediate environments. Reflecting uptake trends in SSA more broadly (Crick et al. 2018b; Mahul and Stutley 2010), when pressed, the female entrepreneurs we spoke to perceived very limited capability currently for their businesses to access sustainable adaptation strategies, such as business and climate insurance products and other savings safety nets. Typically facing structural exclusion from more formal savings mechanisms, respondents from both agricultural and nonagricultural MSMEs highlighted that they sometimes hoard maize during high harvest seasons for sale during extreme climatic events such as drought. This hoarding of cereals – which is the main staple food in Narok – respondents suggested allows them to fetch higher prices and returns during climatic extremes, which they characterized as instrumental in cushioning the impacts of adverse climatic conditions.

Participants also highlighted the importance of formal and informal support networks, such as women's groups, in supporting female entrepreneurs in the face of the multiple and concurrent social and economic challenges they experience.

Wider literature has suggested that social networking remains a major adaptation tool for women both in rural and urban areas, as women often come together in groups to share and support each other (Atela 2012). A key example of these networks in Narok manifests in "merry-go-round" and table banking initiatives, which bring groups of women together to save and lend money among themselves. In table banking, women make collective savings through which they are able to earn interest and access credit to invest in their MSMEs. Participants suggested these networks were an important means of accessing cash for business and household needs. However, most participants felt that money from table banking was not adequate to protect businesses from the impacts of climate extremes, not least because, given the temporally bound nature of climate shocks, resources may be required by businesses concurrently.

Table banking and other women's groups may also undertake a range of other activities with potential to increase the resilience of MSMEs, such as the group purchase of inputs (e.g., drought-resilient seeds), and initiating cooperatives and other common pool resource management initiatives, such as reforestation/afforestation, and greenhouse farming. They may also provide a forum for knowledge sharing and technology demonstration. Women's groups were an important forum for initiating diversification into beekeeping enterprise and the development of beekeeping cooperatives in Narok. The Kenya Hives Limited partnership, for example, develops market linkages and provides capacity building services around beehive management and harvest quality to beekeeping entrepreneurs in the Narok region. Reflecting on this partnership, respondents suggested that implementation of new beekeeping technologies often requires approval from men who control land resources and suggested women therefore typically find it difficult to access new beekeeping technologies. Nevertheless, they offered examples where women in the partnership have broken through these cultural barriers and have been able to access and utilize modern hives, which, since they can be shifted from one geographical region to another, are more responsive to climate hazards. Participants detailed that these women have been able to increase their honey production and sales in the face of climate change and are now major suppliers of honey in Narok and its environs. The women that have achieved this business success in honey production participants suggested had typically received better educational opportunities and were thought to have benefitted from greater exposure to training seminars thorough the partnership, allowing them to learn, share experiences, and establish networks.

Challenges that can Impede Adaptation of Female-Led MSMEs

As discussed earlier, the adaptive capacity of female-led MSMEs in Kenya's SALs is constrained by many of the same factors that limit male-led MSMEs, such as deficient infrastructure, including power, transportation, water, and telecommunications and a gap in access to and use of finance (Beck and Cull 2014; Crick et al. 2018a; Stein et al. 2013). Yet participants in our research identified several factors that they suggested constituted additional barriers to adaptation for female-led MSMEs in Kenya's SALs.

In general terms, respondents emphasized that socially institutionalized gender roles are a major challenge to female entrepreneurship, with the burden of caring for children and running internal family affairs, as well as meeting household livelihood needs, most often falling on women in Narok (c.f. Omolo 2010). Participants emphasized that these responsibilities present a high opportunity cost, in terms of reducing the time available for women to invest in business development, including in identifying and executing strategies to build resilience to climate risk within their MSMEs. They also emphasized that the additional emotional pressures that arise from facing these responsibilities under conditions of uncertain food and financial security form an additional barrier to adaptation and cited physical impacts on their health, such as headaches and exhaustion, as a consequence of this pressure. Capturing the idea of a "double vulnerability," female entrepreneurs in our research emphasized that at times of climate stress, their experience of vulnerability occurs simultaneously at the business and household level and is often interconnected. Participants suggested that resilience within female-led MSMEs is therefore intimately linked to resilience within households, and thus participants identified that a key route to enabling private sector adaptation among female-led MSMEs is to build resilience at the household level. Yet they felt this relationship is often overlooked and poorly understood by policy makers.

Participants reported that women's access to entrepreneurship and adaptation resources, such as land, capital, and new technologies, is widely restricted by traditional governance systems in Narok. This reflects broader trends in social barriers faced by women entrepreneurs in SALs (Omolo 2010). As discussed, most women in Kenya's SALs lack ownership rights to land or legally secured entitlements, as these resources are still largely under traditional customary tenure, where men have exclusive right to land inheritance and subsequent ownership. As such, women generally depend on their husbands' land entitlements. This has several further notable implications for the adaptive capacity of female-led MSMEs. Firstly, as exemplified in the example of the Kenya Hives Limited partnership, without land tenure, women in Narok are often restricted in their ability to make decisions about the management and development of their business assets, which can prevent them from adopting more climate-resilient business strategies.

Further, MSMEs without land tenure may find it challenging to form market linkages and partnerships with other larger companies that operate in formal settings. Land tenure offers legislative protection and resource security; thus larger companies sometimes choose to only work with other businesses with secure entitlements. In Narok, participants identified this dynamic, suggesting that the preference of large companies to work with actors that hold secure tenure results in a vulnerability feedback loop, where male entrepreneurs and private investors become more powerful and female-led enterprise becomes further marginalized. These ideas also emerged in another Kenyan case study of Maasai and Turkanas, inhabiting arid and semiarid areas such as Kajiado and Turkana, where analysis revealed that most land in these areas has been sold off to private estate investors or farm owners, who have since restricted access to water sources and livelihood resources especially for women (Orindi et al. 2007).

An additional consequence of women having limited access to land tenure is that without land tenure to serve as a surety, female entrepreneurs may face further barriers to securing access to credit through formal or informal channels. World Bank Enterprise Survey data found that more than 25% of firms in Africa consider finance their most important constraint (Beck and Cull 2014) and participants in our research also highlighted lack of capital as a major challenge for female-led MSMEs. The challenge of accessing finance is further exacerbated by female-led MSMEs being more likely to be unregistered and operating in the informal sector: Indeed, there are currently multiple other disadvantages to operating informally, which also include restricted access to new market opportunities, as well as to public sector services (Fjose et al. 2010; USAID 2005).

Female-led MSMEs also face barriers to accessing new climate-resilient technologies, finance, and other forms of business safety nets, such as climate insurance products. Female educational attainment is low in Narok, and, as such, female entrepreneurs can struggle, for example, to keep records and track their business growth in a manner that would enable them to convince creditors to support them to access new technologies and inputs. These challenges are reflected in the wider private sector adaptation literature, which recognizes that internal factors in companies, such as the education of the work force, influence firm's willingness or ability to adapt (Crick et al. 2018a).

The factors described above – low education levels and structural exclusion from land, capital, markets, and new technologies – help confine most female entrepreneurs in SALs to more limited livelihood and income generating activities which, as discussed, are often those exposed to greatest climate risk. As discussed, these same factors can limit the ability of female entrepreneurs to identify and implement strategies to build resilience to climate risk. But, importantly, they also impede women's agency in decision-making and consultative fora, meaning the voices, aspirations, and capabilities of female entrepreneurs are not yet being clearly articulated in public policy, legislative, and investment domains (Orindi et al. 2007).

Reflecting literature that argues that such exclusion and underrepresentation of women in key policy making processes impedes gender-responsive climate-related planning, policymaking, and implementation (Atela et al. 2017), participants in our research described this gap in consultation as representing a significant barrier to realizing inclusive climate-resilient development in SALs. Through their, often concurrent, responsibilities within households, female entrepreneurs were understood to have incomparably close awareness of key livelihood needs, assets, opportunities, and stressors, affording them unique and important potential to design and operate MSMEs that are more responsive to the livelihood – and adaptation – needs of SAL households (c.f. Horrell and Krishnan 2007). This situated insight and experience was deemed important for autonomous adaptation within female-led MSMEs, but, emphasizing opportunities for instrumental and substantive benefits to be realized from greater consultation, participants also believed female entrepreneurs are currently underutilized at a societal level in the design of strategies to build adaptation and resilience in SALs. In a similar vein, many of the female entrepreneurs we spoke to believed they suffered from a lack of role models, to demonstrate techniques to identify opportunities arising from climate change, and to build resilience into their business operations.

Despite these barriers to consultation, Kenya has a sophisticated business and climate policy landscape, and there are provisions aimed at supporting women in business in national policies and strategies. The revised Kenya Procurement Act of 2010 requires that all public agencies set aside one third of tenders for women and youth entrepreneurs. Similarly, Kenya's constitution now advocates for one third representation of women in various decision-making platforms, including those related to climate change (Republic of Kenya 2010b). Kenya's climate change strategies, such as the National Climate Change Action Plan and the Climate Change Act, are also accompanied by national recognition of the importance of enhancing gender mainstreaming in climate change actions and of supporting female-led businesses. Nevertheless, our research suggests that there are opportunities to enhance the effectiveness of the implementation of these policies.

Firstly, participants identified aspects of the policy environment which they felt were poorly adapted to suit their needs. Some participants involved in non-agricultural MSMEs, for example, highlighted that opportunities, such as the Women Enterprise Fund, established by the Kenyan government, mainly target the poorest women in society but provide limited opportunity for slightly larger and more established – but still vulnerable – businesses (see also Fjose et al. 2010). Since large enterprises within the formal sector find it easier to access more formal bank loans, this creates a "missing middle" when it comes to accessing finance for businesses in Africa (c.f. Crick et al. 2018a; Intellecap 2015). Secondly, a small number of respondents expressed hope that policies such as the "one third procurement rule" may hold particular potential to help break down cultural barriers and to spur robust and resilient financial, technological, and knowledge opportunities for women entrepreneurs in Narok and elsewhere. However, these broader policy incentives and platforms were suggested to be inadequately implemented, monitored, and enforced. More national policies on gender may therefore not be the answer to dismantle sociocultural barriers to inclusive private sector adaptation. Instead, the way forward may require a move away from simple policy statements to designing and reinforcing clearer and more inclusive implementation platforms, which ensure that the broader enabling frameworks, established in national policy, are mainstreamed and implemented at ground level.

Conclusion

The chapter has presented empirical evidence on ways in which female-led MSMEs in Narok experience and respond to climate risk, and it has highlighted some of the challenges that they face to climate change adaptation. This research has shown that climate risk is strongly felt by female-led MSMEs in Narok, and participants often link climate impacts such as drought, flooding, and climate-related disasters to losses to their business assets and to reduced business growth. Exposure is heightened by the concentration of female-led MSMEs in agricultural sectors that are highly vulnerable to climate change, although nonagricultural MSMEs also experience spillover impacts of climate extremes, such as conflict, supply chain disruption, and depressed local markets.

Socially institutionalized gender roles limit female entrepreneurs' access to land, capital, markets, new technology, and educational opportunities and typically confine female-led MSMEs to smaller, informal enterprise, which furthers their structural exclusion. Despite participants recognizing that women have high awareness of livelihood needs, assets, opportunities, and stressors – and thus that they hold unique potential to build climate resilience into the livelihoods of communities in SALs – these factors also limit the ability of female entrepreneurs to advocate for their interests at household and policy levels. This impedes the opportunity for female entrepreneurs to shape more gender-responsive policy making and a more supportive enabling environment, which participants suggested can serve to perpetuate, and further enhance, female vulnerability.

Reflecting the many barriers to adaptation that they face, female entrepreneurs in Narok often depend upon unsustainable forms of coping at times of climate stress, through which business activity is scaled back. Such strategies, such as reducing production to limit the risk of loss and damage, may help enterprises to cope in the short term, but they do nothing to help build longer-term MSME adaptive capacity. Indeed, through reduced profits, loss of business, and the sale of valuable business assets such as land, some of the coping strategies female entrepreneurs in Narok adopt may undermine the longer-term resilience of livelihoods in SALs.

Collective action and collaboration through formal and informal social networks is currently an important private sector adaptation tool for women in Narok. Women's groups and lending networks offer a means of sharing knowledge, of identifying new opportunities to diversify enterprise, and of pooling resources to facilitate the introduction of new climate resilient technologies and to build market linkages. Social networking is likely to remain an important adaptation tool for female entrepreneurs; however our research shows that those who are best able to maximize the opportunities of these networks are still those that benefit from a wider suite of broader enabling factors, such as higher educational attainment. Kenya recognizes the need to support female entrepreneurs in various national policies, including in national climate change legislation. However, these policies are currently poorly implemented and mainstreamed at ground level, with more work needed for the benefits of these policies to be realized in real terms. Future action seeking to build adaptive capacity among female-led MSMEs in SALs should recognize the links between climate resilience in female-led enterprise and resilience at the level of households, as building resilience at the household level is likely to serve as a key route to enabling private sector adaptation among female-led MSMEs.

Acknowledgments The research for this chapter was carried out, in collaboration with Kenya Markets Trust, as part of the PRISE project, under the Collaborative Adaptation Research Initiative in Africa and Asia (CARIAA) program, with financial support from the UK Government's Department for International Development (DfID) and the International Development Research Centre (IDRC), Canada. Joanes Atela was also supported by funding from the Community-Based Adaptation project under the African Centre for Technology Studies. Kate Gannon and Florence Crick were also supported by funding from the UK Economic and Social Research Council (ESRC) through the Centre for Climate Change Economics and Policy (CCCEP) and the Grantham Foundation for the Protection of the Environment.

References

Ackerman F, Kozul-Wright R, Vos R (2012) Climate protection and development. Bloomsbury Academic, London

Adger WN, Arnell NW, Tompkins EL (2005) Successful adaptation to climate change across scales. Glob Environ Chang 15:77–86

Atela JO, (2012) The Politics of Agricultural Carbon Finance: The Case of the Kenya Agricultural Carbon Project, STEPS Working Paper 49, STEPS Centre, Institute of Development Studies and University of Sussex, Brighton, UK. https://core.ac.uk/download/pdf/19917944.pdf

Atela JO, Quinn CH, Minang PA (2014) Are REDD+ projects pro-poor in their spatial targeting? Evidence from Kenya. Appl Geogr 52:14–24

Atela J, Huq S, Ochieng C, Owiyo T, Orindi V (eds) (2017) Enhancing effectiveness of community based adaptation to climate change: unlocking some nuts and bolts'. ACTS Press, Nairobi, Kenya, p 250

Averchenkova A, Crick F, Kocornik-Mina A, Leck H, Surminski S (2016) Multinational and large national corporations and climate adaptation: are we asking the right questions? A review of current knowledge and a new research perspective. WIRES Clim Chang 7:517–536

Bardasi E, Blackden CM, Guzman JC (2007) Chapter 1.4 gender, entrepreneurship, and competitiveness in Africa: Africa competitiveness. Report: 69–85

Beck T, Cull R (2014) Small- and medium-sized enterprise finance in Africa. Africa growth initiative working paper 16

Biagini B, Miller A (2013) Engaging the private sector in adaptation to climate change in developing countries: importance, status, and challenges. Clim Dev 5:242–252

Brindley C (2005) Barriers to women achieving their entrepreneurial potential. Int J Entrep Behav Res 11:144–161

Crick F, Diop M, Sow M, Diouf B, Diouf B, Muhwanga J (2016) Enabling private sector adaptation in developing countries and their semi-arid regions – case studies of Senegal and Kenya. Grantham research institute on climate change and the environment working paper no 258. http://www.lse.ac.uk/GranthamInstitute/wp-content/uploads/2016/12/Working-Paper-258-Crick-et-al.pdf

Crick F, Gannon KE, Diop M, Sow M (2018a) Enabling private sector adaptation to climate change in sub-Saharan Africa. WIRES Clim Chang 9(2):e505

Crick F, Eskander S, Fankhauser S, Diop M (2018b) How do African SMEs respond to climate risks? Evidence from Kenya and Senegal. World Dev 108:157–168

De Souza K, Kituyi E, Harvey B, Leone M, Murali KS, Ford JD (2015) Vulnerability to climate change in three hot spots in Africa and Asia: key issues for policy-relevant adaptation and resilience-building research. Reg Environ Chang 15:747–753

Dougherty-Choux L, Terpstra P, Kammila S, Kurukulasuriya P (2015) Adapting from the ground up. Enabling small businesses in developing countries to adapt to climate change. World Resources Institute and United Nations Development Programme, Washington DC, USA

Fjose S, Grünfeld Leo A, Green C (2010) SMEs and growth in Sub-Saharan Africa: identifying SME roles and obstacles to SME growth. MENON Business Economics Publication no. 14/2010

Food and Agriculture Organisation (FAO) (2016) FAO study provides the most detailed snapshot to date on trees, forests and land use in the world's drylands. Food and Agriculture Organisation of the United Nations. http://www.fao.org/news/story/en/item/425931/icode/

Gannon KE, Crick F, Rouhaud, E, Conway D, Fankhauser S (2018) Supporting private adaptation to climate change in semi-arid lands in developing countries, Pathways to Resilience in Semi Arid Economies (PRISE) Briefing. http://www.lse.ac.uk/GranthamInstitute/publication/supporting-private-adaptation-to-climate-change-in-semi-arid-lands-in-developing-countries/

Horrell S, Krishnan P (2007) Poverty and productivity in female-headed households in Zimbabwe. J Dev Stud 43(8):1351–1380

Intellecap (2015) Closing the gap Kenya: update on key challenges for the "Missing middle" in Kenya. www.intellecap.com/sites/default/files/publications/26.10.15%20(1)_0.pdf

International Finance Corporation (IFC) (2010) A strategy to engage the private sector in climate change adaptation in Bangladesh

International Labour Organisation (2015) Small and medium-sized enterprises and decent and productive employment creation. In: Report IV. International labour conference, 104th session

IPCC (2014) Climate change 2014: synthesis report. Contribution of Working Groups I, II and III to the fifth assessment report of the intergovernmental panel on climate change [Core Writing Team, RK Pachauri, LA Meyer (eds)]. IPCC, Geneva

Jobbins G, Conway D, Fankhauser S, Gueye B, Liwenga E, Ludi E, Mitchell T, Mountfort H, Suleri A (2016) Resilience, equity and growth in semi-arid economies: a research agenda. Overseas Development Institute, London

Johnson JC, Weller SC (2001) Elicitation techniques for interviewing. In: Gubrium JF, Holstein JA (eds) Handbook of interview research. Sage, London

Kelly DJ, CG Brush, Greene PG, Y Litovsky (2011) Global Entrepreneurship Monitor 2010 women's report. www.gemconsortium.org/docs/download/768

Kevane M, Wydick B (2001) Microenterprise lending to female entrepreneurs: sacrificing economic growth for poverty alleviation? World Dev 29(7):1225–1236

Krueger RA, Casey M (2000) Focus groups 3rd edition: a practical guide for applied research. Sage, London

Mahul O, Stutley C (2010) Government support to agricultural insurance: challenges and options for developing countries. World Bank, Washington, DC

McPeak JG, Barrett CB (2001) Differential risk exposure and stochastic poverty traps among East African pastoralists. Am J Agric Econ 83(3):674–679

Minniti M (2010) Female Entrepreneurship and Economic Activity, The European Journal of Development Research, 22(3):294–312

Mountfort H (2015) COMMENT: a triple whammy – avoiding a bleak future for semi-arid regions. Pathways to Resilience in Semi-Arid Economies (PRISE) Blog. August 18th

Ngare M (2013) Female enterprise development: a case study of graduate women in Nairobi. Lahti University of Applied Sciences. Master's Thesis 2013. https://publications.theseus.fi/bitstream/handle/10024/65831/Ngare_Mercy.pdf?

Nichter S, Goldmark L (2009) Small firm growth in developing countries. World Dev 37(9): 1453–1464

Nyukuri E (2016) Gender approaches in climate compatible development: lessons from Kenya. A publication of CDKN. www.practicalaction.org/cdkn-gender-and-climate-change-initiative

Ole Riamit K (2010) REDD+ in Kenya:An indigenous peoples' perspective. In: Alangui WV, Subido G, Tinda-an R (eds.) Indigenous peoples, forests & REDD plus: state of forests, policy environment & ways forward. Baguio, Philippines: Tebtebba Foundation, pp. 367–458.

Omolo NA (2010) Gender and climate change-induced conflict in pastoral communities: case study of Turkana in northwestern Kenya. African J Confl Resolut 10:2

Orindi VA, Nyong A, Herrero M (2007) Pastoral livelihood adaptation to drought and institutional interventions in Kenya. In: United Nations Development Programme. Human development report 2007–2008: fighting climate change: human solidarity in a divided world. Human Development Report Office, Occasional Paper 2007/54. http://citeseerx.ist.psu.edu/viewdoc/download?doi=10.1.1.497.4003&rep=rep1&type=pdf

Pauw P (2015) Not a panacea: private-sector engagement in adaptation and adaptation finance in developing countries. Clim Pol 15:583–603

Pauw P, Pegels A (2013) Private sector engagement in climate change adaptation in least developed countries: an exploration. Clim Dev 5:257–267

Pricewaterhouse Coopers (PWC) (2010) Business leadership on climate change adaptation: encouraging engagement and action

Republic of Kenya (2010a) Revised Public Procurement and Disposal Act, 2010. Government Printer, Nairobi

Republic of Kenya (2010b) The constitution of Kenya. Government Printer, Nairobi

Republic of Kenya (2013) Analysis of drivers and underlying causes of forest cover change in the various forest types of Kenya. Government Printer, Nairobi

Singh G, Belwal R (2008) Entrepreneurship and SMEs in Ethiopia: evaluating the role, prospects and problems faced by women in this emergent sector Gurmeet. Gender Manag 23:120–136

Smit B, Wandel J (2006) Adaptation, adaptive capacity and vulnerability. Glob Environ Chang 16:282–292

Spath B (2016) Women's economic empowerment in technical assistance programmes, examples of good practice in private sector development. GIZ, Frankfurt

Stein P, Hommes M, Pinar AO (2013) Closing the credit gap for formal and informal micro, small and medium enterprises. IFC advisory services. Access to finance. International Finance Corporation, Washington DC, USA

Stenek V, Amado JC, Greenall D (2013) Enabling environment for private sector adaptation – an index assessment framework, International Finance Corporation. International Finance Corporation, Washington DC, USA

Terry G (2009) No climate justice without gender justice: an overview of the issues. Gend Dev 17(1):5–18

Tompkins EL, Eakin H (2012) Managing private and public adaptation to climate change. Glob Environ Chang 22:3–11

Trabacchi C, Stadelmann M (2013) Making adaptation a private sector business: insights from the pilot program for climate resilience in Nepal. Climate Policy Initiative. https://climatepolicyinitiative.org/wp-content/uploads/2013/12/SGG-Case-Study-Pilot-Program-for-Climate-Resilience-in-Nepal.pdf

Tucker J, Daoud M, Oates N, Few R, Conway D, Mtisi S, Matheson S (2015) Social vulnerability in three high-poverty climate change hot spots: what does the climate change literature tell us? Reg Environ Chang 15:783–800

USAID (2005) Removing barriers to formalization: the case for reform and emerging best practice

Wade CT, Touré O, Diop M (2015) Gestion des risques climatiques. Rapport d'étude, pathways to resilience in semi-arid economies (PRISE) working paper, IED Afrique, Dakar, Senegal

Wedawatta G, Ingirige B, Amaratunga D (2010) Building up resilience of construction sector SMEs and their supply chains to extreme weather events. Int J Strateg Prop Manag 14:362–375

Welsh DHB, Melimi E, Kaciak E, Ahmed S (2013) Sudanese Women Entrepreneurs. J Dev Entrep 18:1350013

Assessment of Barriers to Food Security in North-Eastern Nigeria

51

Hubert Fudjumdjum, Walter Leal Filho, and Desalegn Yayeh Ayal

Contents

H. Fudjumdjum (✉)
Research and Transfer Centre Sustainable Development and Climate Change Management, Hamburg University of Applied Sciences, Hamburg, Germany
e-mail: fudjumh@yahoo.fr

W. Leal Filho
International Climate Change Information and Research Programme (ICCIRP), Faculty of Life Sciences, Hamburg University of Applied Sciences, Hamburg, Germany

Research and Transfer Centre Sustainable Development and Climate Change Management, Hamburg University of Applied Sciences, Hamburg, Germany
e-mail: walter.leal2@haw-hamburg.de; info@iccip.net

D. Y. Ayal
Center for Food Security Studies, College of Development Studies, Addis Ababa University, Addis Ababa, Ethiopia
e-mail: desalula@gmail.com

W. Leal Filho (ed.), *Handbook of Climate Change Resilience*,
https://doi.org/10.1007/978-3-319-93336-8_99

Abstract

Scholars have long pointed out the relationship between climate change and food security. The impact of climate change on food security tends to be of grave concern in places where rainfed agriculture is at the ground source of food and income. That is the case among the majority of rural people in Nigeria who are disproportionately poor and malnourished. However, the impact of climate change on food insecurity in the context of conflict, displacement, and waning policy response has not been an essential focus of scholars. Based on a reassessment of available studies and latest reports about the nature of climate change, weak adaptation capacity of farmers, poverty, inconsistent policy response, and the increasing occurrence of conflict and displacement, this chapter draws attention to the overlooked unique vulnerability of people prone to conflict and displacement to food insecurity. Accordingly, this chapter highlights existing barriers to food security in conflict-ridden areas in Nigeria from a holistic perspective. It argues that, to date, policies, funding, and infrastructure development meant to ensure food security never approached a scale of priority the problem deserves. It posits that the question of food security in Nigeria reduced to a mere climate change issue and that critical policy, budget, and political issues must tackle as core components of food security response. It concludes by suggesting possible solutions to the multifaceted origins of food insecurity in northern Nigeria.

Keywords

Climate change · Conflict · Displacement · Food security · Policy · Vulnerability

Introduction

The burning of fossil fuels (coal, oil, and gas) to meet increasing energy demand and the spread of intensive agriculture to increase food production (IPCC 2014) increased accumulation of greenhouse gasses in the atmosphere, thereby increasing the mean global temperature over the past decades (IPCC 2007). Ecological exhaustion and poor socioeconomic conditions made sub-Saharan Africa more vulnerable to the ill effects of climate change on livelihoods and food security (Ayal and Muluneh 2014; IPCC 2007). As a result, access to water, energy, and transport jeopardized (see Fig. 1). Given that about 239 million people are food insecure in sub-Saharan Africa, the threat of climate change on livelihoods could increase shortly (IPCC 2007).

Around 235,000 people in West Africa are facing emergency food insecurity, with a further 6.4 million facing crisis conditions, particularly in the conflict-affected Lake Chad Basin areas of Nigeria, Chad, Niger, and Cameroon (Sasson 2012). The most popular country in Africa with a population of 182 million and the seventh in the world (WB 2016), Nigeria, accounts for 47% of the West African population (WB 2012). Nigeria's economy is heavily dependent on sectors that are either climate sensitive or contribute to climate change such as agriculture, forestry, fishery (which together employ up to 70% of the workforce), oil, and gas (Nigeria – 2013–2017 – Country Strategy Paper 2016). Although influenced by climate

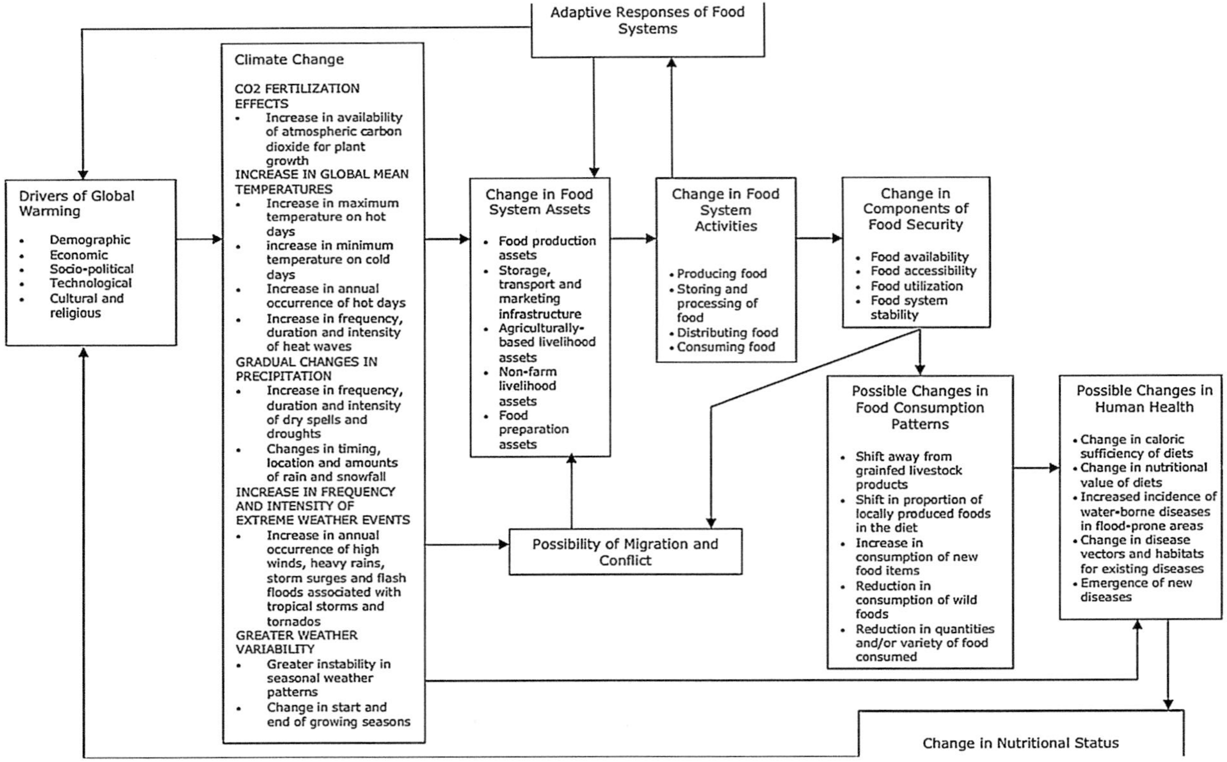

Fig. 1 Climate change and food security. (Source: FAO/NRCB. Figure produced for Food and Agriculture Organization of the United Nations Rome, 2008 Report)

change, agriculture remains the bedrock of the Nigerian economy, while the oil sector is essential (Mohammed and Charles 2016; Ijirshar 2015; Liverpool-Taste et al. 2011; Gbadebo 2008). Agriculture is the primary supplier of food for the ever-growing population (Mohammed and Charles 2016; Mogues and Dillon 2008). The agricultural production growth rate of 3.7% is lagging behind the 6.5% food demand growth rate (Onwuka 2014). The food production and demand imbalance necessitate food import which hardly contributes to reduce poverty and achieve sustainable development (Astou 2015).

People considered food secure if safe, sufficient, and nutritious food is available and accessible all times to keep them healthy and active (WFP 2015, 2016; FAO 2004). However, Nigeria's food insecurity is widespread (Adebayo 2012; Babatunde et al. 2008) to the extent that 62% of the population live in extreme poverty (Joe Brock 2013, cited in Astou 2015). Rising food prices and food riots are among the many symptoms of the prevailing food crisis and insecurity. The particular food insecurity situation of women and children is grave (Sasson 2012). One reason is that food production could not keep pace with the rising population and volume of the agricultural output (Metu et al. 2016; Liverpool-Taste et al. 2011). In Nigeria, the food insecurity situation is severe (Ojo and Adebayo 2012) mainly in urban areas (Omuemu et al. 2012).

Clover (2003) reported that availability, access, and affordability are all elements of food security that encompass a wide range of interrelated economic, social, and political factors which challenge Africa's ability to address food security. Ultimately, hunger is a political creation which must be ended by political means. The government has reacted through promoting of low-cost technologies to improve agricultural productivity and food security. However, the effort is far from alleviating the unique predicaments of food insecure and malnourished people displaced from their villages because of conflict. By discussing the need to customize government response to the impact of climate change on the food insecurity of conflict-prone regions, this chapter attempts to cast a new approach to food insecurity challenges. The model deemed appropriate for the new plan is adopted from climate change and adaptive response mechanisms. The FAO/NRCB 2008 framework depicts causes and multifaceted impacts of climate change and response in the food system (see Fig. 1). The structure allows discussing the reciprocal effects of climate change and food security and policy responses.

Figure 1 illustrated the cause and effect relationship of climate change and socioeconomic and cultural factors. Anthropogenic activities have increased the frequency and severity of climate change and extremes and its impact on food system and food security. Climate change and extreme events affect all components of food system (food production, food processing, food distribution, and food consumption) and food security (availability, accessibility, utilization, and stability). This in turn forces people to devise coping and adaptation responses which encompass change in production system, technology, and food consumption patterns. Increasing frequency and severity of climate change and extremes create favorable conditions for climate change-sensitive food-, water-, and air-borne diseases. In the developing nations like Nigeria, climate change and extremes impact on food system and food security is a cause of change in nutritional status and induces people to look for innovative adaptive strategies.

The range of stated coping and adaptive responses to climate extremes, wittingly or unwittingly, has induced massive cultural shift. This cultural shift is evident from new nutrition sources to mode of production.

Aims and Objectives

This chapter aims to highlight the unique food insecurity of people in conflict-ridden areas of Nigeria. Specifically, it attempts to suggest a framework by which the barriers of food security in North-Eastern Nigeria could be addressed.

Materials and Methods

The information needed to assess existing barriers to food security in Nigeria was obtained from electronic databases such as Google Scholar, BioMed, Pubmed, Earth, and Environment Science websites. The Medical Subject Heading keywords, used in isolation or combination, were food security, climate change, North-West Africa, food security barriers, food security improvement and conflict, and displacement. Published journal articles on related topics in the English language were used for this chapter. The searched items were from published journals on the subject with no year limit. The search generated 575 articles and abstracts. These published sources were further filtered based on relevance so that only the most appropriate were selected and used for this chapter. The current food insecurity situation reports of North-Eastern Nigeria of World Food Program Nigeria, UNICEF Nigeria, WHO Nigeria, and others were used for the chapter. Hence, the chapter was based on a reassessment of existing claims and reports rather than collection and analysis of a fresh field data.

Barriers to Improving Food Security in Nigeria

Policy Inconsistency

Designing context-specific and practical policy and strategy helps to adapt and build resilience to climate change and improve food security. On the contrary, the governance regimes, policies, legislation, and financial inconsistency worsen direct and indirect impact of climate change on food security (FMARD 2016). As Adebayo (2012) reported, the Nigerian agricultural policy suffers from wide gaps between intent and actual practices.

However, recently the Nigerian government had attempted to foster food security through various policies and programs (Metu et al. 2016). Among them are the National Plan of Action for Food and Nutrition in Nigeria launched in 2004 and the National Strategic Plan of Action for Nutrition (2014–2019). The National Special Program for Food Security (NSPFS) is an initiative of the Federal Government of Nigeria and the Food and Agricultural Organization (FAO) for poverty reduction in line with the thrust of the National Economic Empowerment Strategy.

Table 1 Items under the utilization component of food security

Utilization	Old policy (1990–2000)	New policy (2001–2010)
Access to improved water sources (%)	50	59.3
Access to improved sanitation facilities (%)	35	30.3
Percentage of children under 5 years of age affected by wasting	16.7	13
Percentage of children under 5 years of age who are stunted	44.7	42.3
Percentage of children under 5 years of age who are underweight	32.5	26.5
Prevalence of anemia among pregnant women (%)	61.7	60
Prevalence of anemia among children under 5 years of age (%)	76.8	74

Source: FAOSTAT, FAO of the UN (2014)

However, the needs of farmers were no more a matter of priority. Research and development in the sector slowed down thereby causing a stagnant food production. Policies on agricultural productivity had little popular participation in implementation, and there was no determined effort in customizing the food production improvement plans to local conditions (Metu et al. 2016; Liverpool-Taste et al. 2011; Enete and Amusa 2010).

Although new policy launched in 2001 had an initial positive impact on access to improved water sources, the stability of food security and access to improved sanitation facilities declined. Besides, the percentage of children under 5 years affected by underweight, stunting, and anemia increased. Similarly, the new policy after 2001 didn't decrease the rate of women affected by anemia (see Table 1).

According to the journal *The Sun* voice of the Nation in his article entitled "The green alternative and food security" (2016), agricultural policy implementation approach, lack of coherence, and program discontinuity are highlighted as barriers to food and nutrition security in Nigeria. Food security policy mainstreaming and implementation gap accompanied with administrative reshuffling of the federal departments in charge of the multiple programs were also attributed for food insecurity in Nigeria (Liverpool-Taste et al. 2011).

The situation of ecological degradation, conflict, and food insecurity is further exacerbated by poor road infrastructure (Abur et al. 2015). Farmers maintained existing fragile road networks by themselves, although the local government offers negligible assistance. According to ADB (2013), Nigeria is an "infrastructure deficit" country that has nearly the worst rank in Africa. In effect, low output and income among farmers have to do with poor road infrastructure which precludes affordable and timely acquisition of agricultural inputs, such as fertilizers and pesticides, and convenient and remunerative access to the market. It could, therefore, be discerned that climate change, poor infrastructure, and conflict destabilize rural livelihoods through increasing production cost (Abur et al. 2015) and reducing agricultural productivity.

Besides, decline in soil fertility, poor credit services, fragmented and decreasing farmland size, poor grain storage facility, and conflict have worsened food insecurity in Nigeria. Consequently, more than 70% of households do not have the amount and quality of food needed for a healthy and productive life (Abu and Soom 2016). Hence, Nigeria spent billions of dollar to import food grain annually. The magnitude of food insecurity could be summarized from the assessment of the Nigerian Ministry of Health and information which dubbed the situation in Borno State as "nutrition emergency." That being the cause, access to food has become increasingly precarious so that a prospect of famine is predicted to be inevitable in some parts of Borno State. Still worse is that the devaluation of *Naira's* exchange rate against the US dollar would raise the price of local and imported foods, especially staple foods such as rice, millet, maize, and sorghum. Depreciation of the national currency destabilizes the entire price structure, thereby rendering many households with minimal or acute food insecurity (ADB 2013).

Problems of Funding, Credit, Training, and Poor Agricultural Practices

Figure 2 shows that the average value of food production from 1989 onward had an increasing trend. However, since around 2004, the average cost of food production in Nigeria significantly decreased for various reasons. Underfunding is a crucial factor for low food production and severe food insecurity in Nigeria (Vintagesam 2014). The performance of agriculture was constrained by inadequate funding and poor budgetary execution (Adofu et al. 2012; Mogues and Dillon 2008). The level of public spending on agriculture and the share of agriculture in the national GDP have been decreasing since about 2009.

As illustrated in Fig. 3, downward trends in agricultural investment and intensity of spending are attributable to neglect by the state to the sector.

Although Vision 2020 was launched in 2010 to make Nigeria among the top 20 economies by 2020, agriculture suffered from underfunding.

Having studied farming households in Nigeria, Godwin and Aondonenge (2016) concluded that lack of access to credits, low availability of land, and poverty are the leading factors which undermined food security (Dekunle et al. 2012). Traditional farming, based on dull tools and small farmland, restricts farmer's access to credit facilities (Metu et al. 2016). Soil infertility, lack of off-farm income-generating activities, poor grain storage, and processing problems are other causes of food insecurity causes. Due to poor socioeconomic status and illiteracy, farmers do not have adequate access to new technology transfer and market infrastructure to tackle the poor output and distribution of agricultural produces (Nnamani 2015). Moreover, lack of training needed to adapt to climate change and variability has negatively influenced farm production (GAFSP 2016). The precarious nature of funding, farming inputs, and institutional access in conflict areas are worse than elsewhere.

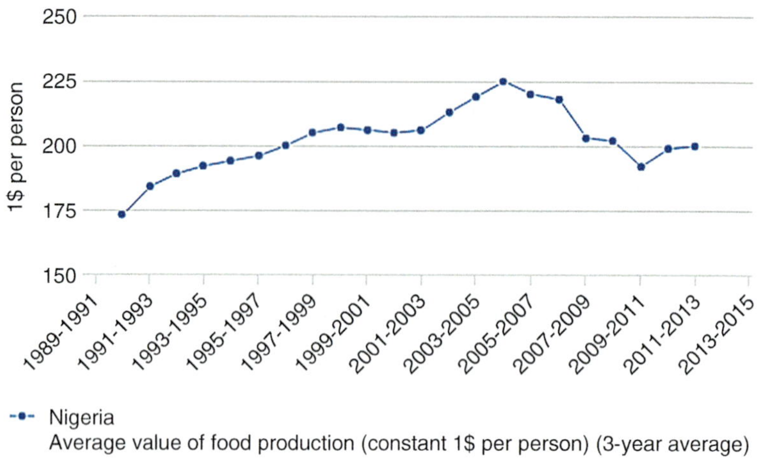

Fig. 2 Average value of food production (constant I$ per person) (3-year average) 1990–2015. (Source: FAOSTAT 2016)

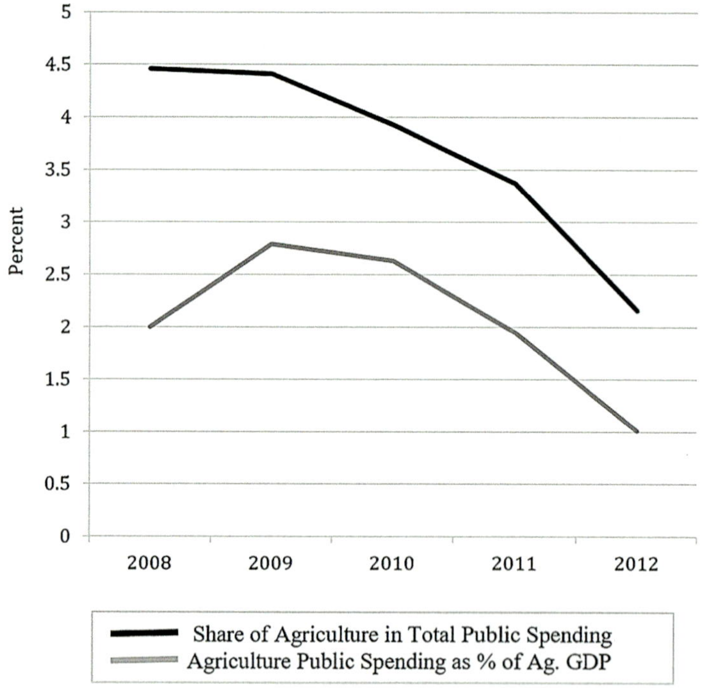

Fig. 3 Comparison of the degree of government commitment to investment in agriculture and intensity of spending in Nigeria, 2008–2012. (Source: Authors' graph based on data from)

Barriers to Food Security in North-Eastern Nigeria

Climate Change, Conflict, and Food Insecurity Nexus in North-Eastern Nigeria

Climate change and extremes increase the incidence of conflict and the vulnerability of rural livelihoods to food insecurity (Kayode et al. 2016; Ukpe et al. 2016; Arora-Jonson 2011; Raleigh and Urdal 2009; IPCC 2007; Moore et al. 2005; Konczacki 1978). Conflict disrupts livelihood activities, erodes asset background, and worsens food insecurity (Orhero 2015; Ajodo-Adebanjoko and Walter 2014; Hendrix and Brinkman 2013). Hendrix and Brinkman (2013) attested that food insecurity and conflict have a reciprocal relationship.

Climate change has decreased the availability of fresh water in northern Nigeria, thereby instigating conflict between farmers and pastoralists. Given the small amount of rainfall, rivers and streams are drying, underground water levels are depleted, and the water volume of lakes is remarkably declined. Consequently, competition for freshwater and grazing lands caused conflict shattered the old social fabric of peaceful coexistence and interdependence between farmers and pastoralists. The mutually reinforcing factors of conflict and resources depletion are exacerbated by climate change which is beyond the control of conflicting parties. As indicated by Ukpe et al. (2016), it is no wonder, therefore, that rural livelihood in general and food security are actively undermined by the effects of climate change.

Although all people are affected by food insecurity and conflict, the level of harm and propensity of exposure differs across gender (Kayode et al. 2016), age, income, educational, occupational, and religious groups whereby pastoralists are the greatest losers. Desperate for fresh water and forage, pastoralists could not but let their cattle to feed on crops of sedentary farmers. Desperate for more farmland, sedentary farmers encroach into the traditionally grazing fields of pastoralists and impede the flow of rivers, thereby jeopardizing the rights of pastoralists for pasture and fodder supply.

The cumulative effect of such trajectories is food insecurity, instability, and joblessness such that people become easy preys of membership to extremist armed dissidents such as Boko Haram. In return, it fuels conflict between religious, ethnic and vocational groups. There is evidence that resources competition and mismanagement significantly increase farmers and pastoralists climate change vulnerability, food insecurity and further ecological degradation and conflict (Ajodo-Adebanjoko and Walter 2014; Sunday 2012; Girard et al. 2012).

Climate change and conflict adversely affect food availability, food access, the stability of access, as well as food utilization. Both reduce the amount and quality of food available, disrupt people's access to food, and limit families' access to food preparation facilities and health (IFPRI 2015; Orhero 2015; Girard et al. 2012; Das Gupta et al. 2011; IPCC 2007). The conflict also impaired adaptation enabling factors such as infrastructure expansion, credit service provision, and improvement of the agricultural sector (Dekunle et al. 2012; Okojie 2009). The ongoing conflict in

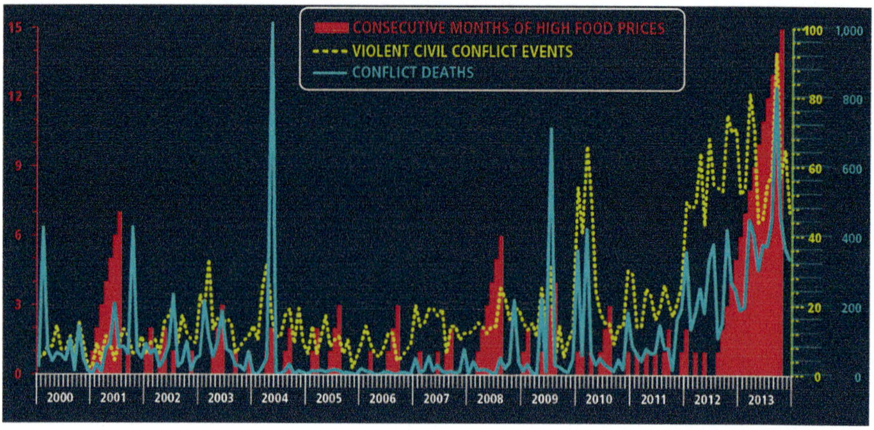

Fig. 4 Food price hikes and intensity of civil conflict in Nigeria. (Source: Food consumer price index (CPI) and conflict data from the Food and Agriculture Organization of the United Nation's Statistical Division (FAOSTAT) and Armed Conflict Location and Event Data Project (2014))

the North-Eastern part of Nigeria has strongly affected food security as people suffer from severe restrictions on their livelihood (Ajodo-Adebanjoko and Walter 2014). Famine Early Warning Systems Network (2016) underscored that conflict in North-Eastern Nigeria had left a significant portion of the population without access to adequate food, water, and health services. The looming threat of attack from insurgents in rural northeast Nigeria has disrupted agricultural activities (Aluko et al. 2016; IFPRI 2015).

Some farmers are afraid of planting their crops, while others have entirely abandoned their land in the course of fleeing violence. Violence, death, and food prices in the conflict-affected areas increased (see Fig. 4). According to Aluko et al. (2016), Boko Haram insurgency has drastically affected the supply and distribution of agricultural production between North-Eastern and Southern part of Nigeria due to the unwillingness of traders to embark on a journey to the troubled zone for fear of attack by insurgents.

Below-average harvests and flooding undermined access to food in many parts of northern Nigeria where humanitarian assistance is difficult because of poor infrastructure and conflicts between Boko Haram and government forces, besides disputes between pastoralists and farmers. Although the government and NGOs attempt to extend food and non-food support to internally displaced people, the amount and quality of food assistance in aid camps are inadequate, leaving most people with only two meals a day. There are also people displaced by conflict who are outside the reach of humanitarian support. That is the case in Bama, Damboa, Dikwa, Monguno, Gamboru, Banki, and Baga (UNICEF 2016). Poor shelter and healthcare facilities further worsen the effect of malnutrition as washing is a severe problem. All of these suggest that adaption to climate extreme events should go beyond production increment and embrace sociopolitical bottlenecks.

Ineffective Agricultural Policy and Food Insecurity

Although the Nigerian government has given a renewed focus to boost the agricultural sector through its Agricultural Transformation Agenda (ATA), the emphasis is on lending to farmers to reduce dependence on food imports (FMARD 2016). ATA's 2011–2016 plan hardly aims at improving the plights of farmers and pastoralists in conflict-ridden zones. Even measured against ATA's aim, the actual performance is such that Nigeria remains heavily reliant on food imports so that the agricultural sector could not be a source of foreign currency (Metu et al. 2016; FMARD 2016; Adebiyi 2012; Obadan and Edo 2004). That was the reason for the birth of The Agricultural Promotion Policy (APP) which aimed at achieving food self-sufficiency and generating foreign currency from quality food export. Thus, even the latest policy focuses on creating jobs through agribusiness and reducing Nigerians reliance on food imports, rather than addressing the unique vulnerability of farmers and pastoralists in conflict regions (Abu and Soom 2016; FMARD 2016; Ijirshar 2015). In general, the Nigerian government effort to build farmers and pastoralists adaptive capacity against climate change and improve food security is not yet successful.

Conclusions

Food and nutrition insecurity is becoming acute especially in conflict-ridden areas of Nigeria, affecting millions of people. The federal spending has not been evenly distributed in all vulnerable areas. Robust expenditure on agricultural development and research remains vital to build resilience to climate change-induced shocks. Well-funded and well-managed food security measures would limit the breadth and depth of conflict and violence as well as strengthen national-level governance systems and institutions. The major challenge facing Nigeria is creating a sound, integrated, and enduring policy to ensure sound management of available resources to end the conflict-climate change-food insecurity nexus. Agriculture should assume a lucrative and climate-resilient business model with linkages to agro-based industries so as to propel economic growth, generate employment, reduce poverty, and ensure the nation's food security.

Farmers and pastoralists are slow in changing their farming practices due to lack of adequate training to adapt to climate change and variability. Conflict on livelihood bases further compounds food insecurity, malfunctioning market, and poor humanitarian response to improve the predicaments of both resident and displaced people. To overcome such challenges, agricultural research and funding schemes should be decentralized and participatory. Agricultural policy should allow space for continuous capacity building to farmers and pastoralists for the dissemination of viable climate change adaptation schemes. Food insecurity in conflict-prone regions of Nigeria is burdened with meager government budget, ineffective policy response, malfunctioning marketing, constrained access to technology, and other challenges. There is no unique mechanism put in place to address the unique predicaments posed on the food security of conflict-prone regions. That being the case, tackling the food insecurity in those areas requires approaching the solution beyond the prism of climate change and agricultural adaptation.

Recommendations

- The government should improve the financing of agriculture through the provision of additional funds for state-owned rural banks to enable them to grant more soft loans and encouraging the conventional banks to give low-interest loans to all classes of farmers and pastoralists.
- Improvement of irrigation to tackle heavy reliance on rainfed agriculture which requires proper drainage and irrigation materials.
- Empower the international center for soil fertility and agriculture development to engage in research that help improve soil fertility.
- Formulate and implement sound agricultural financing policy specifically targeting poor farmers and pastoralists to build their adaptive capacity.
- Mainstream smart climate agricultural practices in all layers and branches of government and public institutions. The emphasis should be given to locally effective and affordable climate change adaptation technologies and capacity building of the vulnerable groups to increase their agricultural production and food security.

References

Abu G, Soom A (2016) Analysis of factors affecting food security in rural and urban farming households of Benue State, Nigeria. Int J Food Agric Econ 4(1):55–68

Abur C, Ademoyewa GR, Damkor M (2015) Impact of rural roads infrastructure on the income and productivity of households farmers in North Central Nigeria. J Agric Environ Manag 4(10):451–458. Retrieved from https://en.wikipedia.org/wiki/

ADB (2013) African Development Bank Group Federal Republic of Nigeria, Country Strategy Paper 2013–2017, pp 17–20. Retrieved May 2017, from https://www.afdb.org/en/documents/document/2013-2017-nigeria-country-strategy-paper-31073/

Adebayo O (2012) Effects of family size on household food security in Osun State, Nigeria. Asian J Agric Rural Dev 2(2):136–141

Adebiyi B (2012) Jonathan's Agricultural Business Plan. Nigeria Tribune, Wednesday 13 June 2012. Retrieved May 2017, from https://issuu.com/nigeriantribuneonline/docs/210416

Adofu I, Abula M, Agama J (2012) The effects of government budgetary allocation to agricultural output in Nigeria. Sky J Agric Res 1(1):1–5

Ajodo-Adebanjoko A, Walter U (eds) (2014) Poverty and the challenges of the insecurity in Nigeria to development. Eur Sci J 10(14):361–372

Aluko O, Osikabor B, Adejumo A, Sumade S (2016) Perceived effect of Boko-Haram insurgency on means of accessing Cowpea from North-East Nigeria to Bodija Market, Ibadan, Oyo State, Nigeria. Open Access Library J 3:e2723. Retrieved 29 Jan 2017, from https://doi.org/10.4236/oalib.1102723

Arora-Jonson S (2011) Virtue and vulnerability: discourses on women, gender, and climatic change. Glob Environ Chang 21:744–751

Astou D (2015) Food imports as a hindrance to food security and sustainable development: the cases of Nigeria and Senegal. CUNY Academic Works 1–20. https://academicworks.cuny.edu/cc_etds_theses/554

Ayal D, Muluneh A (2014) Smallholder farmers' vulnerability to climate variability in the highland and lowland of Ethiopia: implications for adaptation strategies. Doctoral thesis, University of South Africa, Geography Department, pp 136–156

Babatunde R, Olorusanya E, Adejola A (2008) Assessment of rural household poverty; evidence from South Western Nigeria. Am Eurasian J Agric Environ Sci 3(6):900–905

Clover J (2003) Food security in sub-Saharan Africa. Afr Secur Rev 12(1):5–15. https://doi.org/10.1080/10246029.2003.9627566

Das Gupta M, Bongaarts J, Cleland J (2011) Population, poverty, and sustainable development: a review of the evidence. Policy research working paper series 5719. The World Bank, Washington, DC

Dekunle A, Adegbite D, Kayode S (2012) Rural financial services and poverty alleviation: the roles of the Nigerian agricultural cooperative and rural development bank (Nacrdb) in Oyo State, Nigeria. Nig J Agric Food Environ 8(1):1–9

Enete AA, Amusa TA (2010) Challenges of agricultural adaptation to climate change in Nigeria: a synthesis of the literature. Field Actions Science Reports. The Journal of Field Actions 4. Retrieved from https://factsreports.revues.org/678

Famine and Early Warning Network (2016) NIGERIA food security outlook June 2016 to January 2017. pp 3–10. Retrieved Sept 2016, from https://fews.net/sites/default/files/documents/reports/Nigeria_OL_06_2016_Full%20OL_07_18.pdf

FAO (Food and Agriculture Organization) (2004) FAO statistics on food and agriculture organization of the United Nations, Statistic Division, Socio-Economic Statistics, and Analysis Service. pp 12–36. Retrieved 17 Sept 2016, from http://www.fao.org/employment/irecruitment-access/en/

FAOSTAT (Food and Agriculture Organisation Statistics) (2016) Retried on 17 September 2016, from http://www.fao.org/faostat/en/#country/159

FMARD (Federal Ministry of Agriculture and Rural Development) (2016) The Agriculture Promotion Policy (2016–2020): the building on the successes of the ATA, closing key gaps policy, and strategy document. Abuja FCT, Nigeria. pp 26–47. Retrieved Mai 2018, from http://fscluster.org/sites/default/files/documents/2016-nigeria-agric-sector-policy-roadmap_june-15-2016_final1.pdf

Gbadebo O (2008) Crude oil and the Nigerian economic performance. J Oil Gas Bus 1–9. Retrieved 09 Feb 2017, from http://www.ogbus.ru/eng/

Girard AW, Dzingina C, Akogun O, Mason JB, McFarland DA (2012) Public health interventions, barriers, and opportunities for improving maternal nutrition in Northeast Nigeria. Food Nutr Bull 33(2 suppl1):S51–S70. https://doi.org/10.1177/15648265120332S104

Global Agriculture and Food Security Program (GAFSP) (2016) Retrieved May 2018, from http://www.gafspfund.org/content/about-gafsp

Godwin AA, Aondonenge S (2016) Analysis of factors affecting food security in rural and urban farming households of Benue State, Nigeria. J Food Agric Econ 4(1):55–68

Hendrix C, Brinkman H (2013) Food insecurity and conflict dynamics: causal linkages and complex feedbacks. Stab: Int J Secur Dev 2(2, 26):1–18

IFPRI (International Food Policy Research Institute) (2015) 2014–2015 Global Food Policy Report. Washington, DC: International Food Policy Research Institute. Retrieved on 18 September from https://doi.org/10.2499/9780896295759

Ijirshar VU (2015) The empirical analysis of agricultural exports and economic growth in Nigeria. J Dev Agric Econ 7(3):113–122

IPCC (2007) In: Pachauri RK, Reisinger A (eds) Climate Change 2007: synthesis report. The contribution of working groups I, II and III to the fourth assessment report of the Intergovernmental Panel on Climate Change. IPCC, Geneva

IPCC (2014) In: Barros VR, Field CB, Dokken DJ et al (eds) Climate Change 2014: impacts, adaptation, and vulnerability. Part B: regional aspects. The contribution of working group II to the fifth assessment report of the Intergovernmental Panel on Climate Change. Cambridge University Press, Cambridge, UK/New York. Accessed on 19 June 2017. https://books.google.de/books?hl=en&lr=&id=aJTBQAAQBAJ&oi=fnd&pg=PA1142&dq=extreme+events+climate+change&ots=v0TvNP85CH&sig=QnUMMk1NNVbS9jeFVsxCYYkc34#v=onepage&q=extreme%20events%20climate%20change&f=true

Kayode J, Jeje A, Faramade A (2016) Assessment of the impact of climate change and variability on rural women in sampled communities of Ondo and Ekiti States of Nigeria. Int J Agric Pap 1:1–9

Konczacki Z (1978) The economics of pastoralism: a case study of sub-Saharan Africa. Frank Cass, London

Liverpool-Taste LS, Kuku O, Ajibola A (2011) Review of literature on agricultural productivity, social capital and food security in Nigeria. NSSP working paper no. 21. International Food Policy Research Institute (IFPRI), Washington, DC. Retrieved 2 Feb 2017, from https://ideas.repec.org/p/fpr/nsspwp/21.html

Metu A, Okeyika K, Maduka O (2016) Achieving sustainable food security in Nigeria: challenges and way forward. In: 3rd international conference on African Development Issues (CU-ICADI 2016)

Mogues T, Dillon A (2008) Nigeria agriculture public expenditure review (NAGPUR), and planned research on agricultural subsidies and the health productivity nexus. IFPRI, 2008. Supported by DFID, CIDA, and World Bank

Mohammed N, Charles P (2016) Political economy analysis of small scale farmers and food security in Nigeria. Eur J Bus Soc Sci 5(04):35–49

Moore KM, Bertelsen M, Cisse S, Kodio A (2005) Conflict and agro-pastoral development in the Sahel. In: Moore KM (ed) Conflict, social capital and managing natural resources: a West African case study. CABI Publishing, Oxfordshire

Nnamani C (2015) Innovative stride to zero hunger beyond 2015 in Nigeria. Recent Pat Food Nutr Agric 7(1):14–21

Obadan M, Edo S (2004) Overall economic direction, strategy, and performance. In: Bello-Imam IB, Obadan MI (eds) Anthology of democratic governance and development management in Nigeria's Fourth Republic, 1999–2003. HEB, Plc, Ibadan

Ojo E, Adebayo P (2012) Food security in Nigeria: an overview. Eur J Sustain Dev 1(2):199–222

Okojie C (2009) Decentralization and public service delivery. Nigeria Strategy Support Program (NSSP) background paper no. NSSP 004

Omuemu V, Onyiriuka U, Otasowie E (2012) Prevalence of food insecurity in Egor local government area of Edo State, Nigeria. Ann Afr Med 11(3):139

Onwuka E (2014) Another perspective of look at the impact of Nigeria's growing population on country's development. Afr Popul Stud 21(1):1–18

Orhero A (2015) The Nigerian's economic and security implications of Boko Haram terrorism. Int J Afr Asian Stud 10:6–11

Raleigh C, Urdal H (2009) Climate change, demography, environmental degradation, and armed conflict. Environmental change and security program (ECSP) report, Issue 13. pp 14–20. Responsibilities. Retrieved 17 Sept 2016, from http://www.unwater.org/downloads/unw_transboundary

Sasson A (2012) Food security for Africa: an urgent global challenge. Agric Food Secur 1(2):1–16

Sunday DA (2012) Conflicts between farmers and pastoralists in Northern Nigeria induced by freshwater scarcity. Dev Ctry Stud 3(12):337–352

The Sun voice of the Nation (2016) The green alternative of food security. pp 2–8. Retrieved 17 Oct 2016, from http://sunnewsonline.com/the-green-alternative-and-food-security/

Ukpe O, Nweze N, Arena C (2016) Food insecurity vulnerability status of farm households in Niger- Delta, Nigeria. Int J Food Agric Econ 4(1):109–123

UNICEF (2016, Sept 1–7) Nigeria Humanitarian Situation Report. Retrieved 17 Sept 2016, from http://reliefweb.int/report/nigeria/nigeria-humanitarian-situation-report-september-2016

Vintagesam M (2014) 2014 National Budget. The wane in agricultural sector budget allocation. INTEL, Nigeria, p 8

WFP (2015) What is food security? | WFP | United Nations World Food Programme – Fighting Hunger Worldwide. Retrieved 13 Sept 2016, from https://www.wfp.org/node/359289

WFP (World Food Program) (2016, Sept) Nigeria situation report no. 01, 13 September 2016. Retrieved 17 Jan 2016, from http://reliefweb.int/report/nigeria/wfp-nigeria-situation-report-01-13-september-2016

World Bank (2012) Can small farmers protect themselves against adverse weather? From evidence to policy, Note 71392. Washington, DC

World Bank (2016) The World Bank annual report 2016. Washington, DC. © World Bank. https://openknowledge.worldbank.org/handle/10986/24985. License: CC BY-NC-ND 3.0 IGO

Rural Farmers' Adaptation Decision to Climate Change in Niger Delta Region, Nigeria

52

Nsikak-Abasi A. Etim and NseAbasi N. Etim

Contents

Abstract

The agricultural sector remains the major source of livelihood for rural communities in Nigeria. But the sector is faced with challenges of changes in climate. These continuous changes have adversely affected rural farmers who are dependent on rainfall for farming and livelihood as evident in low precipitation and drought. The continuous changes in pattern of climate have necessitated adaptation to its impact. A study was therefore conducted to estimate the factors influencing rural farmers' adaptation decision to the effects of climate change. Multistage sampling technique was employed to select the representative farmers

N.-A. A. Etim (✉)
Department of Agricultural Economics and Extension, University of Uyo, Uyo, Akwa Ibom State, Nigeria
e-mail: nsikakabasietim@uniuyo.edu.ng

N. N. Etim
Department of Animal Science, Akwa Ibom State University, Obio Akpa Campus, Abak, Akwa Ibom State, Nigeria
e-mail: etimbobo@yahoo.com

W. Leal Filho (ed.), *Handbook of Climate Change Resilience*,
https://doi.org/10.1007/978-3-319-93336-8_100

for the study. Primary data were obtained from a cross section of 300 rural farmers using questionnaire. Multinomial logit model was used to estimate the determinants of rural farmers' adaptation decision to climate change in the Niger Delta region of Nigeria. Adaptation strategies identified were crop diversification, conservation agriculture, adjustment of planting date, and crop and livestock integration. Findings indicate that educational level, access to weather information through extension contact, family size, and income were significant factors influencing farmers' adaptation decisions to climate change. The most used adaptation strategy was adjustment in planting date (28.67%), followed by crop diversification (27.33%), conservation agriculture (22.67%), and crop and livestock integration (16%). Results further revealed that the major source of information on climate change was from extension personnel. Policies to improve the literacy level of rural farmers and increase in the awareness of information on weather and climate would be sensible decisions.

Keywords
Adaptation · Climate change · Farmers · Nigeria

Introduction

Climate change is one of the most critical challenges in the world and has been a subject of debate and discussion among researchers, development stakeholders, and policy makers. The continuous change in climate has posed severe risk to livelihood particularly for the poorest of the poor and the most vulnerable populations due to its adverse consequences on human health, food security, economic activities, natural resources, and physical infrastructure (IPCC 2007). Many farmers in sub-Saharan Africa are adversely affected by the current changes in climate. As reported by Deressa et al. (2009) and Kurukulasuriya and Mendelssohn (2008), agriculture in Africa is negatively affected by climate change, and care has to be taken to cushion the effect; however, climate change in sub-Saharan Africa is already impacting negatively on rain-fed agriculture and livestock systems (ATPS 2013). Rural farmers are (and will be) likely to be most affected by the effects of changes in climate due to their lack of resources and capacity to adapt (Fosu-Mensah et al. 2012). Smith and Skinner (2002) reported that without adaptation to changes in climate, agricultural production will be severely affected making rural farmers more vulnerable, since their adaptation to climate change is based on their conviction of the future benefits. Climate change is a global environmental threat to all economic sectors particularly the agricultural sector (Abid et al. 2015). But the sector which remains the major source of livelihood for rural communities is seriously and adversely challenged by changes in climate. Empirical evidence suggests that the global climate has continued to change due to increase in greenhouse gas (GHG) emissions. The concentration of the emissions has raised the average temperature and altered the amount and distribution of rainfall globally. There is high level of confidence that temperatures will increase in the West African region and rainfall is likely to decrease in the region. These continuous changes have adversely affected rural farmers who are dependent on rainfall for farming and livelihood. This is evident in low precipitation, frequent drought, and low yield. These continuous changes have necessitated the adaptation to its impacts. The economic costs of climate change are principally dependent on the extent of adaptation to the changing climate, but the decision by a farmer to a use an adaptation strategy to mitigate the effects of climate change is heavily dependent on a number of factors such as education, income, access to weather information, and age. According to FAO (2009), climate change adaptation strategies for agriculture are now a matter of urgency and are not separated from other priorities but are integrated into development planning, programs, and projects (World Bank 2008). Farmers in developing nations are developing resilience to climate change-related risks like droughts and floods through practicing diverse adaptation strategies (Belay et al. 2017). Many studies have been done on farm-level adaptation to climate change across different disciplines in various countries which explored farmers' adaptive behavior and its determinants (Bryan et al. 2013; Deressa et al. 2009; Hassan and Nhemachena 2008; Thomas et al. 2007, Abid et al. 2015). In spite of the global extensive research on adaptation of the agriculture sector to climate change such as Amusa et al. (2015) and Enete and Amusa (2010), among others, limited studies have been done regarding the adaptation decisions to climate change by rural farmers in Akwa Ibom State. This study was therefore informed on this premise to fill the lacuna by investigating rural farmers' adaptation decisions to climate change. The objective of the study was to empirically estimate the factors influencing rural farmers' adaptation decisions to climate change in Akwa Ibom State, Niger Delta region of Nigeria.

Methodology

Study Area

The study was conducted in Akwa Ibom State, Niger Delta region of Nigeria. Akwa Ibom State is one of the nine states that make up the Niger Delta region of Nigeria. It lies between latitude 4°33′ and 5°53′ North and longitude 7°25′ and 8°25′ East. The state has an population of 3.9 million and a total land area of 7249 square kilometers (National Population Commission 2006). For administrative convenience, there are 31 local government areas in the state. It has six Agricultural Development Project (ADP) zones, namely, Abak, Eket, Etinan, Ikot Ekpene, Oron, and Uyo. The state is typically agrarian, and most inhabitants depend on rainfall for farming. The state has long period of rainfall, and annual precipitation ranges between 2000 mm and 3000 mm. It is characterized by two seasons – the long rainy (wet) season (March to October) and short dry season (November to February). It has tropical rainforest climate.

Sampling and Data Collection Technique

Three hundred (300) rural farmers were selected for the study using the multistage sampling procedure. First, three out of the six ADP zones were randomly selected. The second stage involved the selection of 20 villages to make 60. Lastly, 5 rural farmers were selected per village to make a total of 300 farmers. Primary data were obtained from the rural farmers with the aid of questionnaire and focus group discussion. Both structured and unstructured questionnaires were used to obtain information from the rural farmers. During the focus group discussions which were held on different days, information was elicited from informed persons in the communities including adult young men and women (Fig. 1).

Analytical Technique of Data

The determinants of rural farmers' adaptation decisions to changes in climate were analyzed using the multinomial logit (MNL). In this study, the model was used to analyze the choices the rural farmers make with respect to adaptation strategies and the factors that determine their choices. Studies by Deressa et al. (2009), Abid et al. (2015), and Belay et al. (2017) on determinants of farmers' adaptation to climate change have used this model. This model suits such analyses since it allows the analysis of decisions across more than two categories, permitting the determination of choice probabilities for different categories (Maddala 1986; Wooldridge 2002). Nonetheless, the model requires that households are associated with only their most preferred option from a given set of adaptation strategies. Unbiased and consistent parameter estimates using the MNL model need to assume independence of irrelevant alternatives that requires that the probability of using a certain adaptation method by a given household is independent of the probability of choosing another adaptation method. Collecting and using only the most preferred adaptation option for each household risk underemphasizing the known relevance to rural farmers of using multiple adaptation strategies (Dasgupta et al. 2014).

Model Specification

The model is specified as follows:

Let y denote a random variable with values (1,2 ..... j) for a positive integer and x set of variables (Deressa et al. 2011). In this study, y is a dependent variable and represents the adaptation alternatives (strategies) from the set of adaptation measures, whereas the x represents the factors that influence choice of the adaptation strategies which contains household attributes, $p_1, p_2 \ldots p_j$, which are associated probabilities such that $p_1 + p_2 + \ldots + p_j = 1$. This tells us how a certain change in x affects the response probabilities $p(y = 1/x)$, $j = 1, 2 \ldots j$. Since the probabilities must sum to unity, $p(y = j/x)$ is determined once the probabilities for $j = 2 \ldots j$ are known.

$$p\,(y = 1/x) = 1 - \left(p_2 + p_3 + \ldots p_j\right) \qquad (1)$$

It is common to designate one as the reference category in the MNL model. The probability of inclusion in other categories is then compared to the probability of inclusion in the reference category. Consequently, for a dependent variable with j

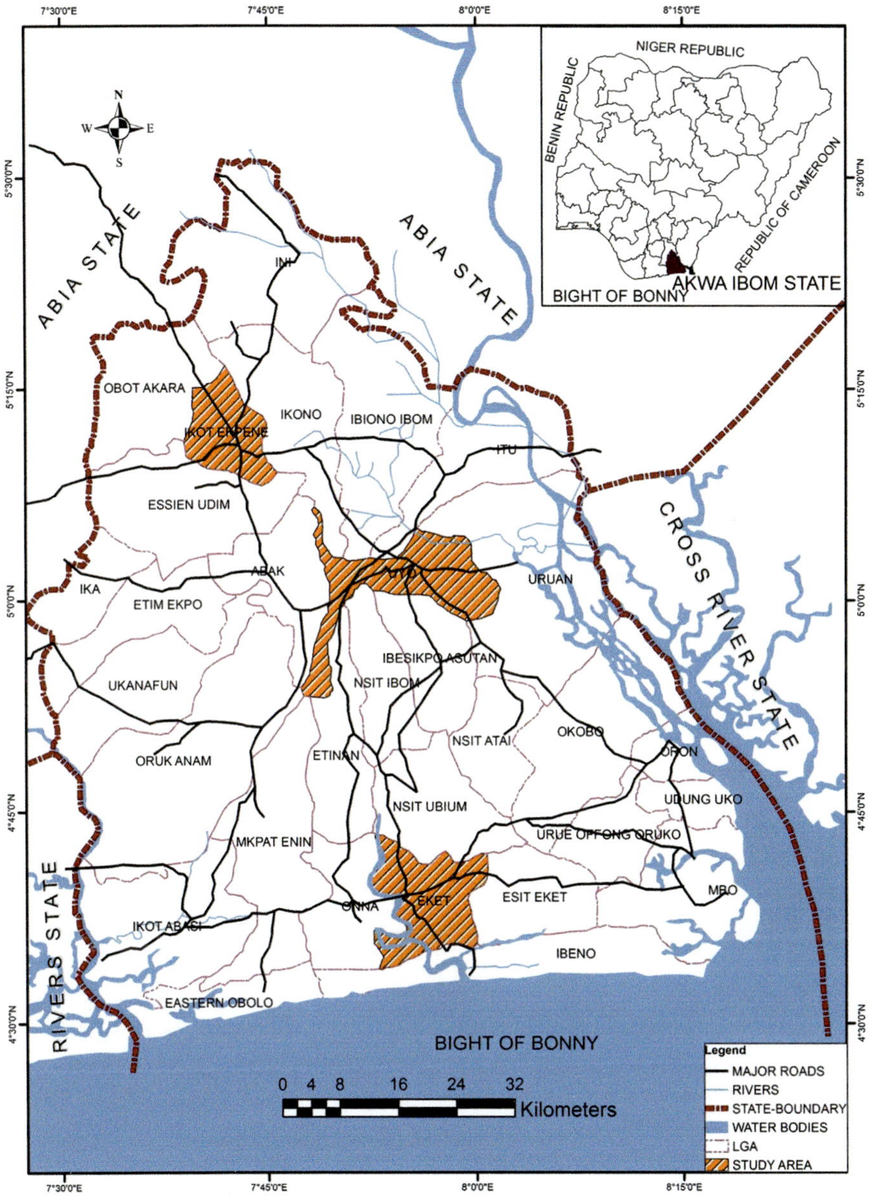

Fig. 1 Map of Akwa Ibom State, Nigeria, showing location of study (Source: Akwa Ibom State Ministry of Lands and Town Planning 2014)

categories, this requires the computation of j-1 equations, one for each category relative to the reference category, to describe the relationship between the dependent variable and the independent variables.

The choice of the reference category is arbitrary but should be theoretically motivated. The estimation of MNL model for this study was conducted by normalizing "base category" or "reference estate." The adaptation measures were grouped into four, and the base category was "no adaptation strategy." The theoretical explanation of the model implies that in all situations, the estimated coefficient should be compared with the base category (Gujarati 2004). Therefore, the choice of the reference category is based on empirical literature and theoretically motivated. The generalized form of probabilities for an outcome variable with j categories is n:

$$\Pr(\mathrm{y_i} = \mathrm{j}/x) = \mathrm{prij} = \frac{\exp(\mathrm{x}^1\beta j)}{1 + \sum_j^j = 2\exp} \qquad (2)$$

$$\text{for } \mathrm{j} > 1$$

The parameter estimates of the MNL model only provide the direction of the effect of the independent variables on the dependent (response) variables; estimates represent neither the actual magnitude of change nor the probabilities. Differentiating eq. 2 with respect to the explanatory variable provides the marginal effect of the independent variables given as

$$\frac{\delta pi}{\delta xk} = \mathrm{pj}\left(\beta jk - \sum_{\mathrm{j=1}}^{\mathrm{j=1}} \mathrm{pj\beta jk}\right) \qquad (3)$$

The marginal effect of marginal probabilities is a function of probabilities and measures the expected change in probabilities where particular adaptation decision choice is being made by a unit change of the independent variable from the mean (Greene 2000). The empirical specification for examining the influence of explanatory variables which are described in Table 1 on the adaptation decisions choice is given as follows:

Table 1 Variable used in the MNL model and their expected significance

Variables	Definition and measurement	Expected sign
Adaptation strategy	Choice set of adaptation strategies	
Sex	Sex of the chief decision maker (1, male; 0, female)	±
Age	Age of the chief decision maker in years (continuous)	±
Education	Number of years of formal education of household head (continuous)	±
Farming experience	Years of experience in farming (continuous)	±
Farm size	Size of farmland in hectares (continuous)	±
Household size	Number of household members (continuous)	±
Labor	Labor in man-days employed in farm operations (continuous)	
Access to weather information	Access to weather information (1, yes; 0, no)	±
Farm income	Income from farming activities in naira (continuous)	±

Table 2 The variance inflation factors (VIF) test result for multicollinearity of variables used in the analysis

Variable	VIF estimates
Age	2.102
Education	3.056
Farm size	1.522
Household size	4.031
Farming size	2.588
Farming experience	1.381
Labor	2.003

Source: Computed from field survey 2016

$$\begin{aligned} y = 1 - j/\beta_o + \beta_1\,(\text{sex}) + \beta_2\,(\text{edu}) + \beta_3\,(\text{farming exp}) + \beta_5\,(\text{farm size})\,) \\ + \beta_6\,(\text{household size}) + \beta_7\,(\text{labor}) + \beta_8\,(\text{access to weather information}) \\ + \beta_9\,(\text{farm income}) + \mu. \end{aligned}$$

Results and Discussion

Test result for collinearity among specific variables in the model. Table 2 presents the VIF result for multicollinearity between the dependent variable and the explanatory variables used in the multinomial logit equation. No multicollinearity problem was detected. Result implies that the estimates of the multinomial logit model have minimum variance, were consistent, and were probably unbiased.

Socioeconomic Characteristics of Rural Farmers

Most (60%) of the farmers were women, while only 40% were men (Fig. 2).

Most of the rural farmers (54.67) were within the age of 21–40 years suggesting that they were within the economically active and productive age. Result is synonymous with earlier empirical study by Edet and Etim (2014) and Etim et al. (2017) in their study of urban farming and its potentials for waste recycling and adoption choice of soil enhancing materials (Fig. 3).

Figure 4 below revealed the marital status of the rural farmers. About 26.67% by farmers were single, whereas 73.33% were married.

The educational background of the rural farmers is shown in Fig. 5. The result revealed that most farmers (70%) had primary and secondary education. This is an indication that the rural farmers were literate implying that they were faster in the adoption of innovations. Similar empirical findings were reported by Etim et al. (2017).

The experience of rural farmers in farming is shown in Fig. 6. Majority of the farmers (53.33%) had between 11 and 20 years experience in farming, whereas 26.67% had 6–10 years experience in farming. About 20% had between 1 and 5 years experience in farming.

Fig. 2 Sex of farmers (Source: Field survey 2016)

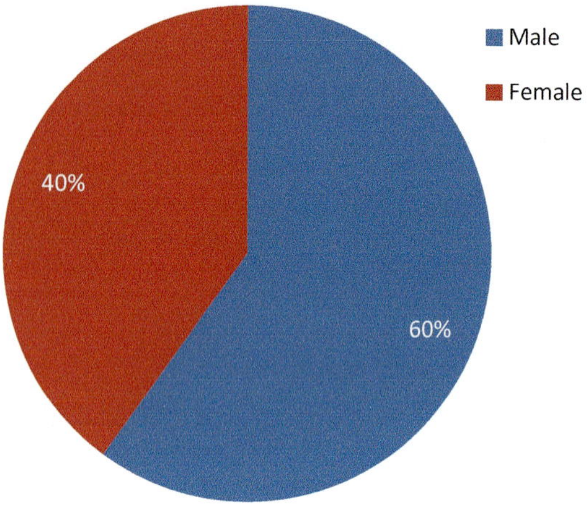

Fig. 3 Age of rural farmers (Source: Field Survey 2016)

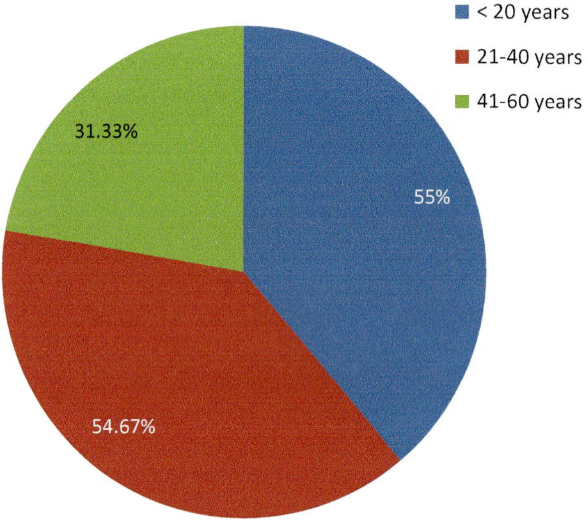

Factors Influencing Adaptation Decision of Rural Farmers to Changes in Climate

Results on Table 3 revealed that age of the rural farmers significantly influenced the probability of diversification and adjustment in planting date by 5%. An increase in age of the farmer by one year increased the likelihood to use conservation agriculture as an adaptation strategy by 1.1%. Result suggests that younger farmers are more interested in trying out innovations due to their risk-taking habit.

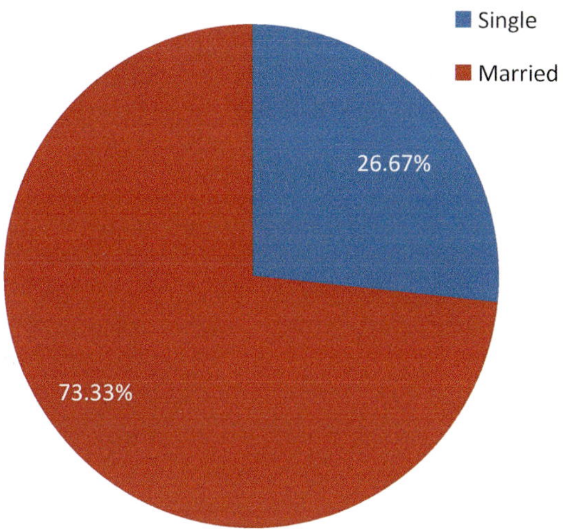

Fig. 4 Marital status of rural farmers (Source: Field Survey 2016)

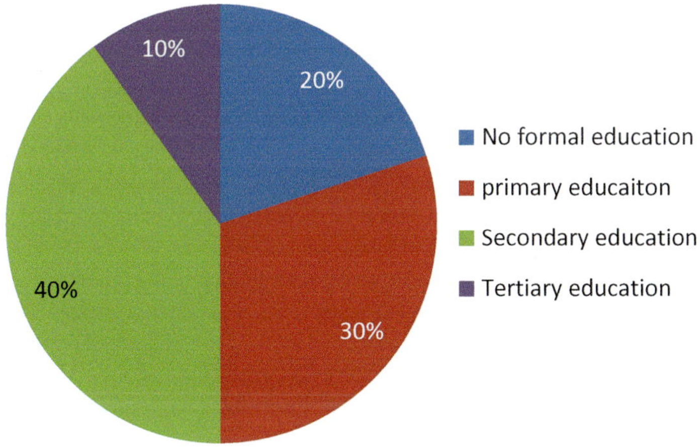

Fig. 5 Educational status of rural farmers (Source: Field Survey 2016)

The educational level of the rural farmers was positively significant with a direct influence on crop diversification and conservation agriculture by 1%. Increasing the level of education by 10 years increased the probability of integrating crops and livestock as an adaptation strategy by 11.30%. This may not be unconnected to the fact that educated farmers are expected to adopt new technologies because of their awareness of the potential benefits from the proposed adaptation measures to climate change (Hassan and Nhemachena 2008). Similar empirical results were obtained by Belay et al. (2017) and Etim et al. (2017). Findings are also synonymous with earlier empirical results by Feder et al. (1985), Udoh and Etim (2006, 2008), Etim and Okon (2013), Etim and Edet (2013), and Etim (2015) who found that higher education empowers farmers to interpret and respond to new information and ideas faster than

Fig. 6 Farming experience of rural farmers (Source; Field Survey 2016)

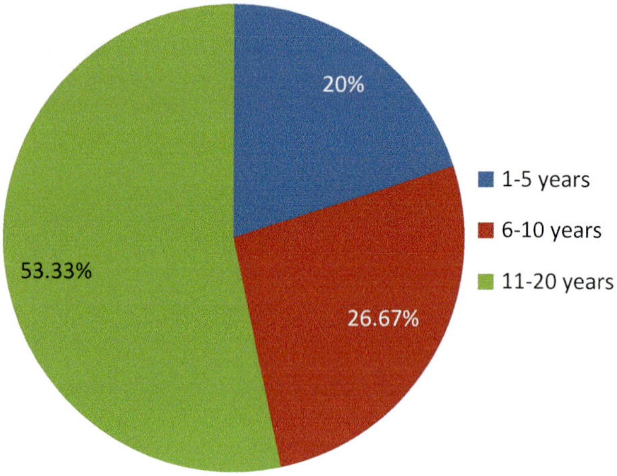

their counterpart with lower education. This result also conforms with the hypothesis that human capital plays a positive role in the acquisition and evaluation of innovations (Nkamleu and Adesina 2000; in Cameroon; Bacha et al. 2001 and Zegeye et al. 2001 in Ethiopia; Chirwa 2005 in Malawi; Chianu and Tsujii (2004), and Etim 2015; Etim et al. 2017 in Nigeria) obtained similar empirical results.

The adaptation options to climate change were also affected by access to weather and climate information as expected. A rural farmer with better access to weather information made better-informed adaptation decisions. Result revealed that rural farmers with access to weather information had a higher probability and were more likely to implement climate change adaptation strategies like adjustment in planting date and conservation agriculture at 5% and 1% significance level, respectively. Information on rainfall and temperature variability increased the likelihood of adjustment in planting date by 23.03%. Result is synonymous with earlier empirical finding by Hassan and Nhemachena (2008), Maddison (2006), Melka et al. (2015), Negash (2011), and ACCCA (2010).

The variable household size is significant and has positive impact on climate change adaptation. Result on Table 3 reveals that a unit increase in productive household member increases the probability of using crop diversification, conservation agriculture, and crop and livestock integration by 8.7%, 2.09%, and 9.31%, respectively. This may not be unconnected to the fact that larger productive household members tend to raise output due to its association with labor-intensive agricultural practices (Kurukulasuriya and Mendelssohn 2008; Gbetibouo 2009 and Belay et al. 2017).

Farm size was positively related to three out of the four adaptation strategies. Increasing farm size by 1 hectare increases the probability of adapting conservation agriculture, crop diversification, and adjustment in planting date by 7.11%, 2.24%, and 1.81%, respectively. Maddison (2006) also reported that the likelihood of adaptation to climate change is positively influenced by farm size. Result implies that larger farmers give room for the experimentation of technology on a small plot

Table 3 Marginal effects from the multinomial logit on the choice of adaptation practice

	No adaptation		Crop diversification		Conservation agriculture		Adjustment of planting date		Crop and livestock integration	
Explanatory variables	$\frac{\delta y}{\delta x}$	p-level	$\frac{\delta y}{\delta x}$	p-level	$\frac{\delta y}{\delta x}$	p-level	$\frac{\delta y}{\delta x}$	p-level	$\frac{\delta y}{\delta x}$	p-level
Sex	0.0561	0.0012	0.0631*	0.0021	0.1800	0.0016	0.6510**	0.0100	0.0130	0.000
Age	0.0266	0.0081	0.1457**	0.0133	0.61080	0.011	0.4110**	0.0661	0.0210	0.0200
Education	0.0711	0.0054	0.3810*	0.0008	0.8012*	0.0701	0.0812	0.0077	0.1130*	0.0110
Farming experience	0.0126	0.0132	0.0716	0.0120	0.0289	0.0088	0.2110	0.0031	0.6334	0.0027
Farm size	0.0551	0.0202	0.0224*	0.0300	0.0711**	0.0120	0.0181*	0.0021	0.0712	0.0100
Household size	0.1481	0.5100	0.0871*	0.0410	0.0209*	0.0240	0.0071	0.0027	0.0931**	0.0066
Labor	0.2088	0.0044	0.0521	0.0082	0.0312	0.0460	0.0613	0.0900	0.5208	0.0100
Access to weather information	0.3807	0.0069	0.0232	0.0120	0.5006*	0.023	0.2303**	0.110	0.0034	0.306
Farm income	0.2005	0.0013	0.1109*	0.0410	0.3011*	0.0331	0.8401	0.500	0.3066	0.212

Source: Computed from multinomial logit results 2016

of larger lands without fear of compromising the household food security. This practice is called small plot adoption technique (SPAT). Earlier empirical findings by Zepeda (1994), Abara and Singh (1993), Fernandez Cornejo (1996), Adesina (1996), Onyenweaku et al. (2010), Etim (2015), and Etim et al. (2017) illustrate the benefits derivable from large-scale adoption of innovation or new technologies are higher for larger farms as there was a positive effect of farm size on technology adoption.

Farm income is positive and impacts significantly on crop diversification and conservation agriculture. Result implies that a unit increase in household income increases the probability of adapting crop diversification and conservation tillage by 11.09% and 30.11% at 1% level of significance. Result conforms to earlier empirical findings by Negash (2011) and Belay et al. (2017) who reported the positive association of income with crop diversification and conservation agriculture.

The rate of adaptation strategies to climate change is shown in Fig. 7. The most used strategy to reduce the impact of climate change was adjustment of planting time. In order to evade planting crops during periods of drought and to ensure that there was adequate rains, about 28.67% of farmers adjusted their dates of planting to April when rains are adequate and stable. In order to counteract the effect of climate change, 27.33% of the farmers diversified their crops and planted cassava, pumpkin, maize, and a mixed variety of vegetables. Sixteen percent integrated crops and livestock. To preserve moisture, 22.67% of the rural farmers tilled the soil to the minimum and maintained soil cover in the form of organic matter and also practiced crop rotation. Rural farmers affirmed that the practice helped not only in soil and water conservation but also added fertility to the soil and improved it. Whereas only 5.33% of the rural farmers used no adaptation strategy.

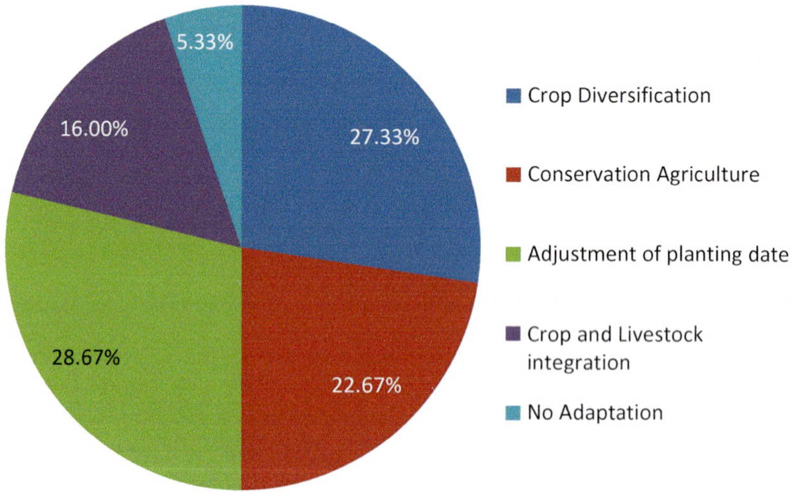

Fig. 7 Rate of adaptation to climate change (Source: Field Survey 2016)

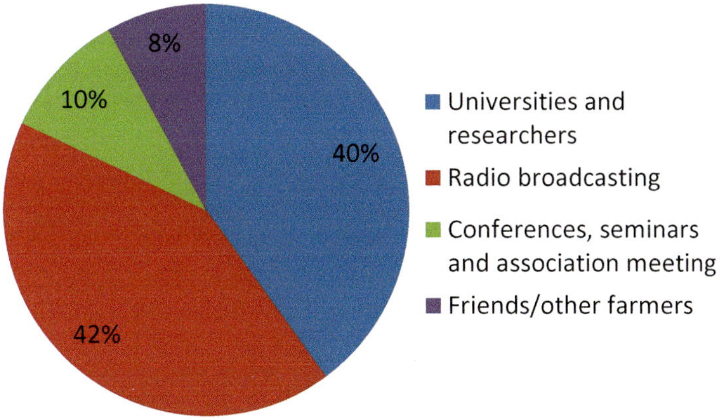

Fig. 8 Source of information about climate change (Source: Field Survey 2016)

Figure 8 shows that the about 42% of rural farmers obtained information on climate change from radio programs, whereas 40% of farmers received climate change information from extension personnel. Eight (8) percent of the farmers were aware of climate change through conference, attendance, seminars, and association meetings and 10% from friends/relatives.

Conclusion

There is no doubt that changes in climate have adversely affected farming and threatened the survival of many farm families. But adaptation to the changing patterns of temperature and rainfall is a panacea to this environmental challenge. This study was conducted to empirically analyze the adaptation decisions of rural farmers to changes in climate in the Niger Delta region of Nigeria. Findings of the study revealed that age, educational level, farm size, household size, access to weather information, and farm income significantly influenced rural farmers' choice of adaptation strategies. In order to cope with the changes in climate, majority of the farmers resorted to adjustment in planting dates as a coping strategy in the drought period and to ensure stability of rainfall before planting. Results further show that most rural farmers obtained information climate change through radio programs.

References

Abara IOC, Singh S (1993) Ethics and biases in technology adoption: the small farm argument. Technol Forecast Social Change 43:289–300

Abid M, Scheffran J, Schneider UA, Ashfaq M (2015) Farmers perceptions of and adaptation strategies to climate change and their determinants: the case of Punjab Province. Pakistan Earth Syst Dyn 6(1):225–243

ACCCA (Advancing Capacity to Support Climate Change Adaptation) (2010) Improving decision-making capacity of small holder farmers in response to climate risk adaptation in the drought prone districts of Tigray, Northern Ethiopia. In Farm level climate-change perception and adaptation in drought prone areas of Tigray, Northern Ethiopia. Project No. 093

Adesina AA (1996) Factors affecting the adoption of fertilizer by Rice farmers in cote d'Ivoire. Nutr Cycl Agroecosyst 46:29–39

Akwa Ibom State Ministry of Lands and Town Planning (2014) Map of Akwa Ibom State of Nigeria

Amusa TA, Okoye CU, Enete AA (2015) Gender-based vulnerability and contributions to climate change adaptation decisions among farm households in Southwest Nigeria. American –Eurasian J Agric Environ Sci 15(9):1779–1789

ATPS (African Technology Policy Studies) (2013) Farmers' response and their adaptation strategies to climate change in Mafeteng District, Lesotho [Sekaleli, T.s.T, Sebusik] ATPS. Research Paper No. 26

Bacha D, Aboma G, Gemeda A, De Groote H (2001) The determinants of fertilizer and manure use in maize production in Western Oromiya, Ethiopia. Seventh Eastern and Southern Africa Regional Maize Conference, 11–15 February, Pretoria

Belay A, Recha JW, Woldeamanuel T, Morton JF (2017) Smallholder farmers' adaptation to climate change and determinants of their adaptation decisions in the central rift valley of Ethiopia. Agric Food Security 6(24):1–13

Bryan E, Ringler C, Okoba B, Roncoli C, Silvestri S, Herrero M (2013) Adapting agriculture to climate change in Kenya: household strategies and determinants. J Environ Manag 114:26–35

Chianu JN, Tsujii H (2004) Determinants of farmers' decision to adopt or not adopt inorganic fertilizer in the savannas of northern Nigeria. Nutr Cycl Agro Ecosyst 70(3):293–301

Chirwa EW (2005) Adoption for fertilizer and hybrid seeds by smallholder maize farmers in southern Malawi. Development Southern Africa 22(1):1–12

Dasgupta P, Morton JF, Dodman D, Karapinar BMF, Rivera-Ferre MG, Toure Sarr A, Vincent KE (2014) Rural areas. In: Field CB, Barros VR, Dokken DJ, Mach KJ, Mastrandrea MD, Bilir TE, Chatterjee M, Ebi KL, Estrada YO, Genova RC, Girma B, Kissel ES, Levy AN, MacCracken S (eds) Climate change 2014: impacts, adaptation and vulnerability. Part A: global and sectoral aspects. Contribution of working group II to the fifth assessment report of the intergovernmental panel on climate change. Cambridge University Press, Cambridge, pp 613–657

Deressa TT, Hassan RM, Ringler C (2011) Perception of and adaptation to climate change by farmers in the Nile basin of Ethiopia. J Agric Sci 149(1):23–31

Deressa TT, Hassan RM, Ringler C, Alemu T, Yusuf M (2009) Determinants of farmers' choice of adaptation methods to climate change in the Nile Basin of Ethiopia. Glob Environ Change 19(2):248–255

Edet GE, Etim NA (2014) Urban farming and its potentials for waste recycling. Am J Social Sci 2(1):16–20

Enete AA, Amusa TA (2010)Challenges of Agricultural Adaptation to Climate Change in Nigeria: a synthesis from the Literature. The Journal of Field Actions. Field Actions Science Report Vol. 4 (online). http://journal.openedition.org/factsreport/678

Etim NA, Okon S (2013) Sources of technical efficiency among subsistence maize farmers in Uyo, Nigeria. J Agric Food Sci 1(4):48–53

Etim NA (2015) Adoption of inorganic fertilizer by urban crop farmers in Akwa Ibom State, Nigeria. Am J Exp Agric 5(5):466–474

Etim NA, Edet GE (2013) Adoption of inorganic fertilizer by resource poor cassava farmers in Niger Delta region, Nigeria. Int J Agric Innovat Res 2(1):94–98

Etim NA, Okon SE, Ebenezer M (2017) Adoption choice of soil enhancing materials by resource poor farmers. Russian J Agric Socio-Econ 3(63):160–169

FAO (2009) Coping with a changing climate: considerations for adaptation and mitigation in agriculture. Rome

Feder G, Just R, E. and Zilberman D. (1985) The adoption of agricultural innovations in developing countries a survey. Econ Dev Cult Chang 32(2):255–298

Fernandez-Cornejo J (1996) The microeconomic impact of IPM adoption: theory and application. Agri Res Econ Rev:149–160

Fosu-Mensah BY, Vlek PLG, MacCarthy DS (2012) Farmers' perceptions and adaptation to climate change: a case study of Sekyere dumase district in Ghana. Environ Dev Sustain 14:495–505

Gbetibouo GA (2009) Understanding farmers' perceptions and adaptations to climate change and variability: the case of the Limpopo Basin, South Africa, vol 849. International Food Policy Research Institute, Washington, DC

Greene WH (2000) Econometric analysis, 4th edn. Prentice Hall International Inc., New York

Gujarati DN (2004) Basic econometricsm, 2nd edn

Hassan R, Nhemachena C (2008) Determinants of African farmers' strategies for adapting to climate change: multinomial choice analysis. Afri J Agric Resour Econ 2:83–104

IPCC (Intergovernmental panel on climate change) (2007) Climate change 2007. Impacts, adaptation and vulnerability. Contribution of working group II to the fourth assessment report. Cambridge University Press, Cambridge, UK

Kurukulasuriya P, Mendelssohn R (2008) Crop switching as a strategy for adapting to climate change. Afr J Res Econ 2(1):105–126

Maddala GS (1986) Limited-dependent and qualitative variables in econometrics (No. 3). Cambridge University Press, Cambridge

Maddison D (2006) The perception and adaptation to climate change in Africa. Centre for Environmental Economics and Policy in Africa (CEEPA). Discussion paper No. 10. Centre for Environmental Economics and Policy in Africa, University of Pretoria, South Africa

Melka Y, Kassa H, Ketema M, Abebaw D, Schmiedel U (2015) The effect of drought risk perception on local people coping decision in the central. Rift Valley of Ethiopia 7(9):292–302

NPC (National Population Commission) (2006) Population Census of the Federal Republic of Nigeria. Analytical Report at the National Population Commission, Abuja

Negash M (2011) Analyzing the determinants of farmers preference for adaptation strategies to climate change: Evidence from north Shoa zone of Amhaha region, Ethiopia

Nkamleu GB, Adesina AA (2000) Determinants of chemical input use in peri-urban lowland systems. Bivariate probit analysis in Cameroon. Agri Syst 63(111):21

Onyenweaku CE, Okoye BC, Okorie KC (2010) Determinants of fertiliser adoption by Rice farmers in Bende local government area of Abia State, Nigeria. Nigerian Agri J 41(2):1–6

Smith B, Skinner MW (2002) Adaptation options in agricultural and to climate change: a topology. Mitigation and Adaptation Strategies for Global Change 7(1):85–114

Thomas DS, Twyman C, Osbahr H, Hewitson B (2007) Adaptation to climate change and variability: farmers responses to intra seasonal precipitation trends in South Africa. Climate Change 83:301–322

Udoh EJ, Etim NA (2006) Cocoyam farms in Akwa lbom State. Nigeria. A stochastic production frontier approach. J Sustain Dev Agric Environ 2:41–48

Udoh EJ, Etim NA (2008) Measurement of farm-level efficiency of waterleaf (*Talinumtriangulare*) production among city farmers in Akwa Ibom State. J Sustain Dev Agric Environ 3(2):47–54

Wooldridge JM (2002) Inverse probability weighted M-estimators for sample selection, attrition and stratification port econ. J 1(2):117–139

World Bank (2008) Climate change adaptation and mitigation in development programmes: a practical guide

Zegeye T, Tadesse B, Tesfaye S (2001) Determinants of adoption of improved maize technologies in maize growing regions in Ethiopia. Second national maize workshop of Ethiopia 12–16 November, Addis Ababa, Ethiopia

Zepeda L (1994) Simultaneity of technology adoption and productivity. J Agric Res Econ 19:46–57

Economic Viability of Alternative Small-Scale Irrigation Systems Used in Vegetables Production in Koulikoro and Mopti Regions, Mali

53

A. M. Kane, J. K. Lagat, T. Fane, J. K. Langat, and B. Teme

Contents

A. M. Kane (✉)
Egerton University, Njoro, Kenya

Institut d'Economie Rurale (IER), Bamako, Mali
e-mail: kanemali2003@gmail.com

J. K. Lagat · J. K. Langat
Egerton University, Njoro, Kenya

T. Fane
Faculty of Economic Sciences and Management, University of Bamako, Bamako, Mali

B. Teme
Institut d'Economie Rurale (IER), Bamako, Mali

W. Leal Filho (ed.), *Handbook of Climate Change Resilience*,
https://doi.org/10.1007/978-3-319-93336-8_101

Abstract

Majority of the households in Mali depend on rain-fed agriculture for their food production. However, overreliance on rain-fed agriculture limits the production output due to unreliable rainfall and drought. To mitigate this, the government of Mali has invested in rehabilitation of irrigation schemes to reduce dependence on rainfall. Through adoption of appropriate irrigation technologies and improved agronomic management practices, increased agricultural productivity is envisaged. This study therefore determined the contribution of different irrigation systems to produce vegetables on household welfare in rural communities. The objective of the study was to investigate the changes in livelihoods of smallholder farmers using the irrigation systems in vegetable production. This research was carried out in three localities corresponding to two specific climatic regions favorable to vegetable crop production in Mali (Baguineda, Kati, Koulikoro Region and Mopti Region) in the year 2016. This study was guided by the production theory. The primary data was collected from 273 farmers selected proportionately from four wards (Fanafiecoura and Tieman in Koulikoro Region and Mopti and Dialango in Mopti Region) using face-to-face interviews. Secondary data was obtained through literature reviews. Descriptive statistics and benefit-cost analysis (BC ratio) is used for analysis. The Statistical Product and Service Solutions (SPSS) and Excel software are used to process and summarize the data. This study concludes that use of manual irrigation system is not economically viable in the production of potatoes, shallots, and tomatoes as confirmed from a low BCR of less than unity. The average BCR of California, Drip, Sprinkler, and Canal IP irrigation systems in the production of potatoes, shallots, and tomatoes are all greater than unity (>1) implying that they lead to greater benefits as compared to costs. The government and other agricultural development stakeholders should support smallholder farmers of Mali in adopting non-manual irrigation methods of crop production, which can minimize the constraint of irrigation systems used in vegetable production for food security in Mali.

Keywords

Drought · Irrigation systems · Vegetables production · Water management · Smallholders · Economic evaluation · Mali

Introduction

Climate change is adversely affecting crop production in the world. These adverse effects are expected to continue until proper adaptive measures are implemented. The effects of climate change on yields are mainly more negative than positive. The negative effects of climate change on crops occurs more in high-latitude regions. Without adaptation, gradual increase in temperature of 2 °C or more that will result to decreases in the yields of major crops in tropical and temperate regions is likely in the near future (IPCC 2014).

In the future, a sizeable degree of further climate change is certain due to the lag between greenhouse gas emissions and global warming (IPCC 2014; USDA 2013). Whether greenhouse gas emissions cease or not, continued warming of the earth's land surfaces and its oceans is expected for decades and centuries, respectively, due to past emissions (USDA 2013). Global temperatures will increase at an increasing rate due to committed warming and future emissions (IPCC 2014). Therefore, in order to avoid future yield declines, agriculture will need to adapt to the projected changes in climate.

Mali is a large, land-locked country in the Sahelian zone. Mali has a unique climate that exhibits large interannual variability, particularly with regard to rainfall. The maximum temperatures in Mali are often 45 °C, with little interannual and only small seasonal variations. A major issue in the country in relation to climate change is the southward encroachment of the Sahelian and Saharan climatic and vegetation zones over the past few decades. In these regions, rainfall has decreased. The interannual variability has recently increased with the rainy season becoming extremely unpredictable. There has been intense increase in temperatures in the country with constant drought conditions that has resulted to significant migration as a coping strategy. The visibly now clear climate change situation is obviously expected to increase interannual variability and the occurrence of extreme climatic events. Intra-seasonal variability is also likely to increase. This includes increase in the number of dry spells during the rainy season and sometimes rains even during the non-rainy seasons.

Mali economy is highly supported by its primary sector (agriculture), which employs 83% of the population. Agriculture contributes about 50% of the GDP. Agricultural activities in Mali are highly vulnerable to the impacts of climate change. Mali ranks extremely low on the Human Development Index (168 out of 179). This implies the country's poor adaptive capacity and inability to respond to climate change. The projections of changes in Mali with respect to precipitation are unclear. However, it is certain that increasing temperatures will cause greater evapotranspiration, which will inevitably result to drier soil conditions and increase in demand of water availability, a factor that agricultural activities have to adapt and cope with. Improper adaptation will definitely cripple the backbone of the economy (agriculture). The decrease in water availability may result to conflict unless irrigation strategies are implemented.

Climate change is also certain that it would increase variability and the incidence of extreme weather events, like droughts, floods, and intense rainfall. With lack of proper planning and management, the occurrences of disasters that are to increase may harm the economy. The frequency of floods in the country may harm crops and property, increase soil erosion, and destroy some weakly constructed dams. Increased maximum temperatures have a bad effect on health. Increase in cases of disease such as diarrhea is very certain with increased incidences of floods. It is also expected that longer-term conditions related to malnutrition as a result of food security will persist.

One of the recommended adaptation measures with respect to climate change in Mali is use of irrigation. Without adaptation measures there are likely to be adverse effects on agriculture associated with changes in climate. Adaptation measures need to be put in place to minimize the impact of climate change in Mali. Malians have dealt with a variable and unpredictable climate through coping strategies such as diversification and soil conservation and venturing into nonfarm activities. It is advisable that adaptation strategies must be based on traditional coping mechanisms for enhanced sustainability. Adaptations must also originate from the ideas and needs of the local communities in order to be successful.

Other popular adaptation options include agricultural improvements (adoption of crops with shorter growing seasons), agroforestry systems and seasonal forecasts, water harvesting and management techniques, and the strengthening of relevant government and nongovernmental institutions (agricultural extension). With the uncertainties associated with the impacts of climate change, efforts on strengthening the capacity of key institutions in Mali are key. Government agencies, research organizations, and NGOs must integrate climate change into their activities as well as respond to the best available information on climate change. Increased harmonization and sharing of information and best practice between groups should be encouraged.

In the Fourth Assessment Report of IPCC, resilience was defined as "ability of a social or ecological system to absorb disturbances while retaining the same basic structure and ways of functioning, the capacity for self-organisation, and the capacity to adapt to stress and change." Vulnerability and resilience are interlinked and cannot be viewed differently. The human knowledge and their practices within their communities can reduce their vulnerabilities to adverse impact of climate change which are their resilient qualities (IPCC 2014).

According to the Resilience Alliance (2002), resilience has three distinct characteristics: system capacity to undergo change and still be in the same state, self-organization capability, and capacity to build and increase learning and adaptation. Resilience can be analyzed at various levels ranging from individual to household, community, ethnic group, and global level (Jordan 2009). Resilience is a complex, context-specific, and highly dynamic (Schwarz et al. 2011). The resilience of households increases their adaptation potential and helps lessen the impact from climate change. Resilient households are less vulnerable to the climate change and are therefore less prone to suffer the same magnitude as the non-resilient group from the climate change (Speranza 2010).

Most sub-Saharan African countries are characterized by low agricultural productivity. One of the reasons for poor production is that African agriculture is predominantly rain-fed, which is in most cases unreliable resulting in poor yields. The changing weather conditions would further exacerbate the situation, exposing smallholders to negative impact of climate change (Todaro and Smith 2012). It is becoming increasingly evident that the required food supplies cannot be met by rain-fed conditions alone (Shiferaw et al. 2004). In Mali, the economy is dependent on agriculture contributing to 36% of income derived from cereal, vegetable, cotton, and sugarcane (Office Statistique des Communautés Européennes 2013).

Agricultural production in Mali is beset by a number of challenges arising from climate change. Some of the most visible consequences are the declining groundwater from the months of January to February which comes at a crucial time for horticultural crops, and this limits the growth of fruits and vegetables in several localities. The immediate impact is a reduction in crop yield by up to 50% in rain-fed agriculture (Todaro and Smith 2012). In a study that aimed at modeling a watershed in the semiarid Southern Great Plains area of North America, Kalaugher (2012) observed that groundwater pumping and irrigation are already causing effects on water budgets similar to those that climate change is likely to bring. Water management has a big impact on the hydrology of basins around the world. Groundwater pumping decreases the flow of streams with widespread irrigation increasing the groundwater recharge.

Due to sinking water levels, the use of manual pumps in deep wells in some areas is no longer sustainable. It has been argued that one strategy which would be used to mitigate water scarcity and dependence on rain fall is irrigation. Indeed, Pinstrup and Watson (2011) and Hussain (2004) revealed that investing in small-scale irrigation schemes is one of the strategies to improve production levels especially for smallholder farmers. The general belief is that irrigated agriculture limits crops failure and external shocks and increases yield, thus leading to better food security (FAO 2010).

According to IPA (2017), Malian government has over the years endeavored to expand the country's irrigation infrastructure in order to improve agricultural production and enhance food security. The aim was to bring about a large expansion of the total irrigated area. Bagayoko et al. (2007) noted that before the year 2008, Dougabougou and Siribala were the only irrigated large-scale land in sugarcane plantation. Currently, the government of Mali is implementing food sustainability program for smallholder farmers through a long-term national program for food security by 2025. This may be attained by targeting small-scale irrigation systems to increase production.

The types of irrigation technologies that are practiced in Mali are drip irrigation, California system, sprinkler system, and gravity system. The drip irrigation technology consists of bringing water under pressure in a system of pipelines. This water is then distributed in drops in the field by a large number of gutters distributed all along rows of plants. This irrigation system is used to grow tomato, onion, shallot (*Allium fistulosum*), banana, papaya, and oranges.

The general objective of the study is to investigate the improved livelihood of smallholder farmers in rural areas by using irrigation systems in vegetable production in Koulikoro and Mopti regions, Mali.

Literature Review

Irrigation Potential in Sub-Saharan African (SSA) Countries

According to Hillel (2000) and Reuben et al. (2012), the key to maximizing crop yields per unit of supplied water in dry lands irrigated areas is by ensuring that the available moisture is used through plant transpiration and little is lost through soil evaporation, deep filtration, and transpiration from weeds. In recent years there has been increasing concern at the performance of conventional irrigation systems in sub-Saharan Africa.

Poor development in irrigation technologies has resulted in underutilization of the agricultural potential in reducing food insecurity. This has been one of the reasons irrigation has lost its appeal as an investment strategy. Good performance in irrigation systems is pegged on its ability to result in high output and efficient use of available water resources. For irrigation systems to be effective and efficient, it must ensure that maintenance of soil fertility, drainage, and salinity-control measures are employed (Hillel 2000).

Small-Scale Irrigation Technologies

There are many technologies used in small-scale irrigation in Mali. These technologies comprise manual drilling and mechanical drilling including pedal pumps (popularly known as *Nafasoro*), Aeolian pump, electric pump, and motor pump. For the manual drilling, water is lifted from wells, rivers, or other water source of the surface by human effort using rope and a container (Keita et al. 2009). These devices are mainly used by resource-poor farmers. In the mechanical drilling, water is shifted by manual or motor pumping systems. This method is used for small and big irrigation by average and wealthy farmers. With both methods, water is distributed to plots through the gravity irrigation, Californian system, sprinkling system, drip irrigation, and manual watering.

Economic Efficiency of Water Use in the Small-Scale Irrigation Systems in Mali

Water use efficiency is a term commonly used to describe the relationship between water (input) and agriculture product (output). When used in this way, the term is, strictly speaking, a water use index. Water use efficiency is also often used to express the effectiveness of irrigation water delivery and use. Barrett Purcell and Associates (1999) correctly pointed out that efficiency is in fact a dimensionless term obtained by dividing figures with the same units such as volume of water used (output) divided by a volume of water supplied (input). Consequently, the ton of produce per megaliter of water used is an index, not efficiency. This common misuse of the term "water use efficiency" has created great confusion. Water use efficiency (WUE) concepts have evolved over a century of irrigation development (Bos and Wolters 2000; Clemmens and Solomon 2007; Solomon 2000).

Sub-Saharan Africa has vast untapped water resources. Expansion of the irrigated area has the potential to make a substantial contribution to agricultural development and address the problem of food insecurity. Many irrigation schemes in the past failed due to a combination of factors, including high investment costs, poor planning, and a lack of maintenance. It is recommended that new irrigation schemes are initiated in response to demand from farmers, increasing the chances of local management and maintenance of the schemes. Irrigation cannot be treated in isolation and must be considered alongside other elements of agricultural development, including improved markets (proximity, information), institutional and legal transparency, clarity of land rights, efficient use of inputs (seeds, fertilizer, pesticides), extension services for farmers, research and development, and environmental management (Reuben et al. 2012).

In sub-Saharan Africa the available groundwater resources are 100 times those of renewable surface water. But farmers often hold back from investing in groundwater irrigation because of the high drilling costs of tube wells and lack of information about groundwater availability. Furthermore the evidence suggests that the region has significant groundwater resources but that in most cases the hydrology is suited for low yielding boreholes that can only be operated by hand pumps. Newer technologies such as drip and sprinkle systems have the potential to increase productivity but are really only accessible to those farmers who can afford to buy them and who are growing cash crops such as vegetables, fruits, and flowers (Reuben et al. 2012).

Economic Viability of Small-Scale Irrigation Systems in Mali

According to the World Bank (2011), most research studies carried out on performance of irrigation schemes have targeted to monitor the performance over time, for example, to evaluate the impact of change in management, or to examine the performance of similar projects. Mostly these evaluations focus on analysis of inputs and outputs of irrigation projects that is water, land, labor, value of production, cost of operation, and maintenance. These indicators are often referred to as external indicators. These indicators do not provide significant information when comparing projects. Obviously projects producing fruit and vegetables have a better productivity than single-crop rice projects. The use of internal indicators as a tool for the diagnosis of irrigation projects need to be defined and discussed.

The concept of viability can be defined at different levels and in various contexts. In a general context, it includes the ability of the scheme to generate sufficient income to satisfy the household income expectations of the irrigators and to cover basic operational and maintenance (O + M) costs of the irrigation infrastructure while not mining the natural resources (soil and water). Though income expectations may differ widely across cultures and among individuals, it is much related to the relative role irrigation plays in the income functions of individual irrigators (Kamara et al. 2001). Further considerations include the ability of the scheme to maintain cash flows and consistency of income generation over time and management of risks and shocks associated with small-scale farming.

Technical Efficiency of Small-Scale Vegetables Production Under Different Irrigation Systems

The justification of this study is drawn from a case study from ICBA-SSA (2015) in Mali which suggests that improving irrigation technologies can significantly contribute toward improving the rural economy and livelihoods. A well-defined scientific framework is needed for successful expansion of this technology. This framework can be developed based on studies of the efficient utilization and management of available water resources. The extent and scope of the challenges facing small farmers can be identified through studies to develop and test suitable technologies for increasing agricultural productivity. Such efforts should lead to more effective utilization of soil and water resources, improved cropping systems, and strengthening the institutions serving farmers (including the private sector) as well as improvements in agricultural productivity and farm income.

Types of Irrigation Systems

Drip Irrigation

Drip irrigation has many advantages over sprinkler or flood irrigation, including application uniformity, the ability to apply water exactly where it is needed, and the potential reduction of disease and weed incidence in irrigated systems. Drip irrigation refers to both rigid ½ in. poly tubing with inline emitters and the thin wall tubing commonly referred to as "drip tape." Drip tape is available in an assortment of wall thicknesses and emitter spacing and is relatively low cost but also much less durable compared to the rigid poly tubing. Drip tape is commonly used in small-scale vegetable production systems as a means of conserving water and minimizing weed and disease pressure. Depending on the water source, drip tape and tubing often require filtration to limit clogging of emitters. Drip tape and poly tubing with inline emitters require pressure regulation to optimize application uniformity. Drip tape and poly tubing with inline emitters require a grade of 2% or less and run no more than 300 ft for optimum distribution uniformity. Careful consideration must be given to design when setting up a drip irrigation system to optimize distribution uniformity and system function.

Sprinkling Irrigation Systems

Sprinkler irrigation involves applying irrigation water which is similar to natural rainfall. Using a system of pipes through pumping, water is distributed to the target crops. Through the sprinklers, water is sprayed into the air so that it breaks up into small water drops which fall to the ground. This method is best for ensuring uniform application of water. Sprinkler irrigation is good for most row, field, and tree crops, and water can be sprayed over or under the crop canopy. Not all sprinklers are good for all crop irrigation. Some produce large water drops that may damage the crop. Unlike some types of irrigation systems, sprinkler irrigation can be applied to any form of land in relation to its slope, whether flat or otherwise. It is advisable that the lateral pipes supplying water to the sprinklers be laid out along the land contour whenever possible. This minimizes the pressure changes at the sprinklers, thereby providing a uniform irrigation.

Although sprinkler irrigation is best suited to sandy soils with high infiltration rates, it is also adaptable to most other types of soils. It is however not suitable for soils which easily form a crust. If sprinkler irrigation is the only method available, then light fine sprays should be used. Large sprinklers that produce larger water droplets should be avoided. To avoid problems of sprinkler nozzle blockage, good clean supply of water, free of suspended sediments, is required. This also avoids spoiling of crop through sediment coating.

Sprinkle irrigation is used on approximately 5% of irrigated land throughout the world, the majority of which is in developed countries. It is unlikely to replace the large areas under surface irrigation, (essentially the remaining 95%, except for a small amount of trickle). Sprinkle irrigation has a distinct advantage, because good water management practices are built into the technology. Sprinkle irrigation technology can provide the flexibility and simplicity required for successful operation, independent of the variable soil and topographic conditions. Pumps, pipes, and on-farm equipment can all be carefully selected to produce uniform irrigation at a controlled water application rate, and, provided simple operating procedures are followed, the irrigation management skills required of the operator are minimal. This puts the responsibility for successful irrigation in the hands of the designer rather than leaving it entirely to the farmer. Sprinkle can be much simpler to operate and requires fewer water management skills. However, it requires sophisticated design skills and on-farm support in terms of maintenance and the supply of spare parts.

Sprinkle is potentially less wasteful of water and uses less labor than surface irrigation. It can be adapted more easily to sandy soils subject to erosion on undulating ground, which may be costly to regrade for surface methods. There are many types of sprinkle systems available to suit a wide variety of operating conditions. The most common for smallholders is a system using portable pipes (aluminum or plastic) supplying small rotary impact sprinklers. Because of the portability of sprinkle systems, they are ideal for supplementary as well as total irrigation.

Manual Irrigation

Manual irrigation systems are not only cheap but simple and require no technical equipment. They however suffer a major disadvantage of high labor requirement. A common and very simple technique for manual irrigation is, for instance, the use of watering cans as it can be found in peri-urban agriculture around large cities in some African countries. There are many methods for manual irrigation, which are easy to install and simple to use. In general, all of these methods have high self-help compatibility and a relatively high performance (FAO 1997).

Most smallholder farmers in developing countries irrigate their field with watering cans. The rose on the top of the outlet creates a sprinkler effect (FAO 2011). Irrigation by watering cans is a very basic way but is still widely used. This creates a lot of work for the labors especially if this technique is used for large fields. A common way to make this work easier is a carrying pole across the shoulders. With watering cans, the field worker is able to irrigate very specific and only where it is necessary.

Gravity Irrigation Systems (Canal IP)

A gravity-fed irrigation system is a cheap effective way to provide water for a smaller-sized crop area. It would be especially cost-effective if the climate of the area can provide enough precipitation to consistently keep a reservoir filled using rainwater harvesting techniques. The basic system is very simple consisting of an elevated reservoir with a pipe coming out the bottom that feeds water into a basic drip irrigation system that is all controlled either by hand or with a very efficient battery-powered timer that controls the rate at which the crop is watered.

This involves diverting water into a farm field. There are two primary ways of diverting surface water and groundwater: gravity diversions and pumping plants. When water surface elevations or heads at the water source are sufficient, gravity diversions are used. A pumping plant is used to lift and/or offer pressure for conveying and/or applying irrigation water.

A gravity diversion is the most common type of gravity diversion. It uses a turnout to admit water from an open water source into farm canals and pipelines. A turnout consists of an inlet, a conduit or other means of conveying water through the bank of the supply canal, and, where required, an outlet transition. Turnouts normally include a means of regulating and measuring inflow to the farm such as weirs, sluice gates, or valves.

On farms that obtain water from pressurized pipelines, a valve is used in lieu of a turnout to admit water into the farm pipeline. A pumping plant is necessary only when the delivery pressure (from the off-farm pipeline) is not sufficient to provide the head needed to operate the farm irrigation system. The inflow rate to the farm is controlled by regulating the delivery pressure and valve opening.

Pumping plants are used when water must be lifted from the water source and/or when sufficient head (pressure) is not available to operate the farm irrigation system. Pumping plants normally have one or more horizontal or vertical centrifugal pumps powered by either electric motors or internal combustion engines.

California Irrigation Systems

Karlinsky (2005) explains that California irrigation system is a network of PVC pipelines put underneath the soil to permit to reduce water loss by infiltration. The term "California irrigation method" was derived to mean the irrigation system used in the citrus industry in California. It implies the underground installation of water tubes meant for distribution by means of spigots. This irrigation system is suitable for use in tree irrigation where spacing is about 6–7 m. This provides adequate room for burying of tubes as well as further space for cultivation between the rows by a variety of means such as animals or tractors. A furrow system was used to irrigate the roots themselves, instead of ditches. Spigots that are attached to the underlying water tubes are used to bring water to the surface while discharging it into dirt furrows, plowed in the orchards by mechanical means. The roots are served by water through percolating from the furrows, precluding the need for basins.

Theoretical Framework

Production Theory

The study will be based on production theory. In this theory, farmers will be maximizing their revenues by trying to attain the highest profits possible given certain constraints. This can be expressed as

$$\text{Max } \pi = p_a q_a - p_x x - wl, \tag{1}$$

where:

π is the profit.
q_a is the quantity of product which in this case it is vegetable that the farmer gets from the farm.
p_a is product price in this case the price of vegetables.
p_x is the price of variable factors.
wl is the amount of labor multiplied by the wage rate.

The inputs p_x are a vector of a number of inputs like seeds, maintenance costs, costs of transporting vegetables, and binding costs in a contract. The cost of labor (l) will be obtained by multiplying the wage rate (w) by the quantity of labor. The farmers' revenue is income derived from the sale of vegetables at the given market price. The maximization equation is set up as follows:

$$\text{Max } \pi = (p_a q_a - p_x x - wl)$$

Subject to: $g(q_a, x, l; z^q) = 0$,

Supply function:

$$q_a = q_a(p_a, p_x, w, z^q) \tag{2}$$

Factor demands:

$$x = x(p_a, p_x, w, z^q) \tag{3}$$

$$l = l(p_a, p_x, w, z^q) \tag{4}$$

z^q (fixed capital, farm size) = fixed factors and farm characteristics

Thus the farmers will be maximizing profits from sale of the farm products subject to the constraints being faced which may be management, institutional, and financial constraints (McIntosh et al. 2005). This can be represented as

$$\text{Max.Profit} : \pi^* = \pi^*(p_a, x, y, z) \tag{5}$$

p_a – price of vegetables and its products

x – institutional constraints which include information availability, customer search costs, length of supply chain, cost of contracts, groups, opportunity cost of time, and standards of measurement
y – financial constraints which include debt, debt asset ratio, asset base, and financial records
z – managerial constraints which include farm size, farmer characteristics, production system, and vegetable type

The profit equation is maximized through as follows:

$$\pi = \beta_i x_i + \beta_j x_j + \beta_k x_k + \varepsilon \quad (6)$$

Where:

π – profitability
x_i – institutional constraints for the ith farmer
x_j – financial constraints for the jth farmer
x_k – managerial constraints for the kth farmer

Methodology

The Study Area

The study area was Koulikoro and Mopti regions of Mali. Mali's population is estimated at 18.54 million with an average density of 15 inhabitants/km^2 (Worldometer 2017), which varies from 20 in the south (regions Koulikoro, Kayes, Segou, Sikasso, Mopti, and Gao) and less than 1 inhabitant/km^2 in the north (Tombouctou and Kidal regions) (Fig. 1).

In Mali, the bulk of agricultural production is dependent on rainfall, while the country has land and water resources adequate for the development of agricultural production from irrigation.

The population in this study comprised of smallholder farmers in Koulikoro and Mopti regions in Mali. These would be farmers who would be holding at most 1 ha of land for vegetable farming using small-scale irrigation systems such as drip, sprinkler, and California.

Research Design

In research design and sampling, a cross-sectional survey research design was used. The design was useful in describing the economic viability of alternative small-scale irrigation systems used in vegetables production in Koulikoro and Mopti regions without having to manipulate variables as in experimental research.

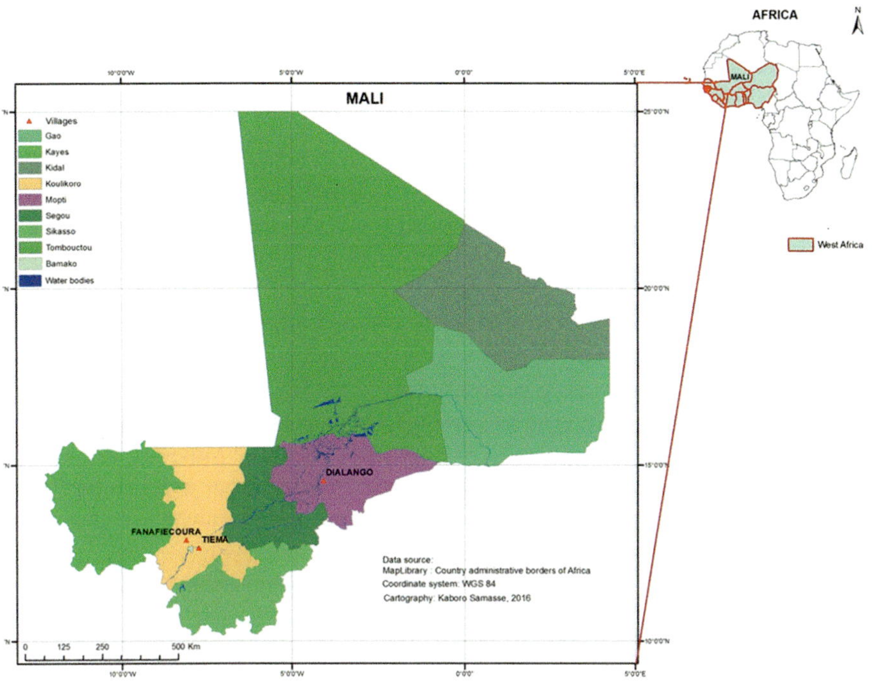

Fig. 1 Maps showing location of study area. (Source: PIB 2011)

The sampling technique employed was multistage sampling. Data was collected using a structured questionnaire. A total of 273 respondents were interviewed as derived from the sample size determination formula proposed by Kothari (2004):

$$n = \frac{K^2 P(1 - P)}{D^2}$$

Confidence level (K) (Z-value) 95% (2-tail) = 1.96; expected proportion in population (P) (50% most conservative); acceptable margin of error in percent (D) which is 9.5%; hence n = 273.

A multistage sampling procedure was used to select farming households of local irrigation scheme in the three regions. The first stage was to purposely select three regions, Fanafiecoura, Tieman, and Dialango, on the basis of the fact that they are within the area mapped for vegetable crops and the incidence of irrigation systems that is already taking place. The second stage involved the use of cluster sampling where the sampling frame was separated into two clusters with one cluster being the vegetable crops integrated farm households using small-scale irrigation system, and the other cluster was composing of vegetable farmers not engaged in small-scale irrigation system. Random sampling was used to select a representative sample from each cluster shown in Table 1.

Table 1 Sample for the survey

Enterprises	Different irrigation technologies T_1	T_2	T_3	Population	Sample size	Proportion (%)	Location
Tomato	31	31	31	3923	93	34.0	Tieman
Potato	30	30	30	2727	90	33.0	Fanafiecoura
Shallot	30	30	30	3549	90	33.0	Dialango
Total	91	91	91	10,200	273	100.0	

Source: DNSI, Report RGA 2005; Kelly 2008
T_1 = (drip irrigation system), T_2 = (sprinkling), T_3 = (Californian irrigation system)

A questionnaire was used to collect primary data which varied from household characteristics to irrigation farming. The questionnaire consisted of both closed-ended and open-ended questions. Closed-ended questions provided a basis for quantifying the data obtained. The open-ended ones provided useful information that was used in explaining observation. All responses were entered into the SPSS program.

Methods of Data Analysis

Descriptive statistics (means, frequencies, and percentages) were used to describe the irrigation systems and the household. The benefit-cost ratio (BCR) analysis was used for judging the economic soundness of irrigation projects. From an economic point of view, an irrigation project with BCR greater than 1.5 is generally accepted. BCR is obtained by dividing the annual total benefits by the annual total cost. The net annual benefit is computed as a difference between the value of produce and total running costs incurred in production, while running cost included expenditure on seeds, manure/fertilizers, pesticides, hired labor/equipment, and government taxes/levies, among others.

The interest on capital cost of the project at the prevailing rate is included in the denominator. BC ratio is thus a division of the total discounted benefits by the total discounted costs.

In calculating the simple benefit-cost ratio, the following formula is used (Boardman 2011):

$$\Sigma\left(B_i/(1+d)_i\right)/\Sigma\left(C_i/(1+d)_i\right),$$

where:

B_i = the benefits of the project in year i, i = 0 to n
C_i = the costs of the project in year i
n = the total number of years for the project duration
d = the discount rate

Results and Discussions

Descriptive Results

Descriptive statistics were used to describe the basic features of the data in this study. Summaries about the sample and the measures are provided. Together with simple graphics analysis, they formed the basis of every quantitative analysis of data.

The household characteristics considered were sex, age, level of education, occupation, household size, income from nonfarming activities, housing, use of irrigation in crop production, period that respondents used irrigation on their farms, difficulties in using irrigation, proposed solutions, sources of irrigation water, mode of access of irrigation water, membership to Water Users Association (WUA), functionality status of the CGB (Water User Association), major type of irrigation systems used practiced, size of irrigated, area per household, access to institutional support, information, education and training, farmers training programs, farmers' visit of extension demonstration site, road conditions in the region, and farmers' extension providers.

Sex of the Household Heads

The results of sex of the household heads are presented in Table 2.

Majority of the household heads were male (90.7%) and only a small proportion being female (9.3%). This implies that majority of the farming decisions such as use of irrigation systems may be dominated by male gender. This may also imply that the measures put to ensure gender balance in economic activities often place more emphasis to supporting men than women (World Bank 2009). Involvement of both men and women in economic activities is the key to successful gender mainstreaming (Guslits and Phartiyal 2010). Women are increasingly being seen as active agents of change and the dynamic promoters of social transformations that can alter the life of all members in society. According to Nyanjom (2011) the exclusion of women in decision-making not only delays delivery of benefits but also affects equity and institutional efficiency. Gender sensitivity is therefore important.

Age of the Head of Household

According to Table 3 the mean age of the household heads was 38.49 years with a standard deviation of 11.37 years. A cumulative percentage of 64.4% of the household heads was aged between 25 and 50 years.

Table 2 Sex of the household heads

	Frequency	Percent (%)
Female	25	9.3
Male	245	90.7
Total	**270**	**100.0**

Table 3 Age of the head of household (years)

Age bracket	Frequency	Percent	Cum. percent
25–30	7	2.6	2.6
31–35	23	8.5	11.1
36–40	49	18.1	29.2
41–45	43	15.9	45.1
46–50	52	19.3	64.4
51–55	25	9.3	73.7
56–60	61	22.6	96.3
Above 60	10	3.7	100
Total	**270**	**100.0**	

Note. Range = 25–65 years, mean age = 38.49, std. deviation = 11.37, n = 270

Different from other developing countries where farming is reserved for middle aged and older generation, this implies that farming is also popular among the young persons in Mali. Age has an influence on crop farming productivity and production due to the effect of technology adoption. According to Truong and Ryuichi (2002) young and middle-aged farmers are generally receptive to adoption of new technology in farming.

The youth who fall in the age bracket of 18–35 years of age were not well represented (cumulative of 11.1%), since most of them were majorly engaged in full-time education.

Age have an influence on the household head experience in crop farming. Alonso and Lewis (2001) assert that age is highly correlated with farming experience. According to Mosca and Pastore (2008), young and middle-aged household heads are more productive and generally receptive to new technology.

Household Heads' Level of Education

Level of formal education plays a major influence in farming. In this study, majority (47.6%) of the farmers were illiterate with no formal education (Fig. 2). About 21.1% had primary, 17.5% had secondary, while 9.7% had local language proficiency (alphabetization) level of education.

Majority of the vegetable farmers lack adequate formal education which is necessary for better modern farming. In addition to this, the level of education of the household head can influence the kind of decision that may be made on behalf of the entire household with regard to use of technology like irrigation on crop farming. More educated farmers are likely to make better decisions as well as quickly adopt new technologies in farming as compared to their less-educated counterparts. Education is of key importance as a means of empowering farmers with the necessary skills and knowledge to perform in their respective farming activities. The higher level of education increases farming success. High literacy levels also reduce gender parity in economic activities such as farming. In addition to this, education of a farmer is correlated to their agricultural skills such as production and marketing (Perry and Johnson 1999).

A study conducted by World Bank (2017) revealed that illiteracy is a major constraint magnifying the state of poor performance in agricultural activities.

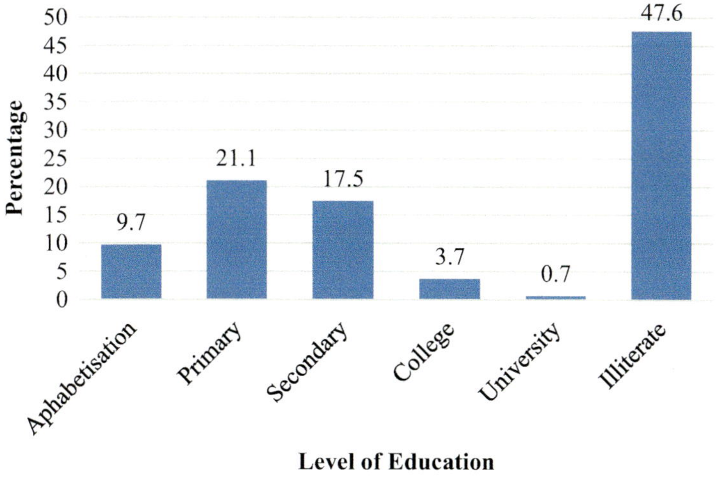

Fig. 2 Level of education of the head of the household

Uneducated persons are unable to understand and utilize technical information and therefore unable to make informed decisions in the light of increasing research findings in agriculture.

Main Occupation of the Farmers

The most popular occupation practiced was agriculture (crop and livestock) as depicted in Fig. 3.

Majority of the households were engaged in crop and livestock agriculture occupation as represented by 37.9% of the households. About 34.1% of the households were engaged in crop (only) agriculture. Livestock-only farming was practiced by only 7.6% of the households. About 12.1% of the households were engaged in paid employment as their major occupation (Fig. 3).

Household Size

The size of households in adult equivalent varied as shown in Table 4.

Majority of the households had 4–6 adults as represented by 57.2%. About 25.0% of the households had 7–9 adult members, while 15.0% had 1–3 adult members. It is just about 2.8% of the households who had ten adult members or more.

The mean household size (in adult equivalent) in the study area was 5.51 with a standard deviation of 2.184. This mean household size is relatively similar to the national average for agricultural households which is about 5.1 (IER, Annual Report 2017).

Income Earning from Nonfarming Activities

The distribution of household members who earned incomes from nonfarming activities is shown in Table 5.

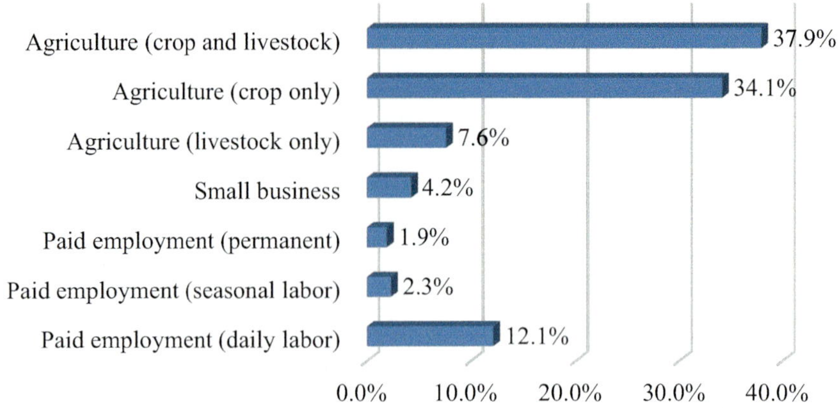

Fig. 3 The main occupation of the farmers

Table 4 Household sizes in adult equivalent

Household sizes	Frequency	Percent
1–3	55	15.0
4–6	211	57.2
7–9	93	25.0
10+	11	2.8
Total	**370**	**100.0**

Minimum = 1, maximum = 13, mean = 5.51, standard deviation = 2.184

Majority of the households (53%) had 1–2 members who were earning income from nonfarming activities. About 24.8% had no member earning income from any nonfarming activities. About 14.4% and 3.7% of the households had 3–4 and 5–6 members who earned incomes from nonfarming activities, respectively. It was only 4.1% of the households that had seven members and more earning incomes from nonfarming activities.

Type of Residential House Owned by Households

Majority of the households owned houses that were iron-roofed with either cemented floor (56.7%) or mud floor (33.7%). This is as depicted in Table 6.

There were few households (2.2%) that owned houses with roof made of grass/other materials whether with cemented floor or mud floor. The type of housing by respondents can be taken as a good representation of the level of standard of living.

Size of Land

Households' land size is taken to mean the size of land owned by the household plus that which is rented. Table 7 shows the amount of land that was available to households.

Table 5 Number of household members who earn income from nonfarming activities

Household members	Frequency	Percent	Cum. percent
None	67	24.8	24.8
1–2 members	143	53.0	77.8
3–4 members	39	14.4	92.2
5–6 members	10	3.7	95.9
7 members and more	11	4.1	100.0
Total	**270**	**100.0**	

Table 6 Type of residential house owned by households

Type of residential house	Frequency	Percent
Iron-roofed and mud floor	91	33.7
Iron-roofed and cemented floor	153	56.7
Roof made of grass/other materials and cemented floor	19	7.4
Roof made of grass/other materials and mud floor	6	2.2
Total	**270**	**100.0**

Table 7 Land size

Land size (ha)	Frequency	Percent (%)
<5	183	67.8
5–10	68	25.2
>10	19	7.0
Total	**270**	**100.0**

The results in Table 7 show that majority (67.8%) of the households had access to land that was less than 5 ha. About 25.2% accessed land 5–10 ha, while only 7.0% accessed more than 10 ha of land. These results suggest farmers are basically smallholders. Majority of the households therefore may not have enough land available for irrigation.

Crop Irrigation by Households

All (100.0%) of the households practiced irrigation farming. Water is not a scarce resource, and every household could practice irrigation farming. The main water sources to the households irrigating their crops include dams/water ponds, boreholes/shallow well, water pumps, dams, rivers, and tanks for water harvesting.

Majority of the households (48.8%) indicated to be irrigating their crop in the months of November to January. About 33.5% irrigated between the months of August and October (Fig. 4). A few (14%) irrigated their fields between the months of February and April (14.0%) and May and July (3.7%).

The greater practice of irrigation in the period between August and January is mainly associated with the fact that it coincides with the dry season, while the period between February and July is mainly a wet season.

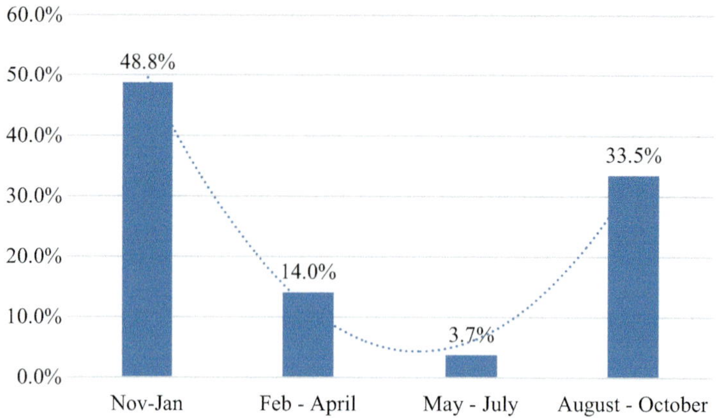

Fig. 4 Period that respondents used irrigation on their farms

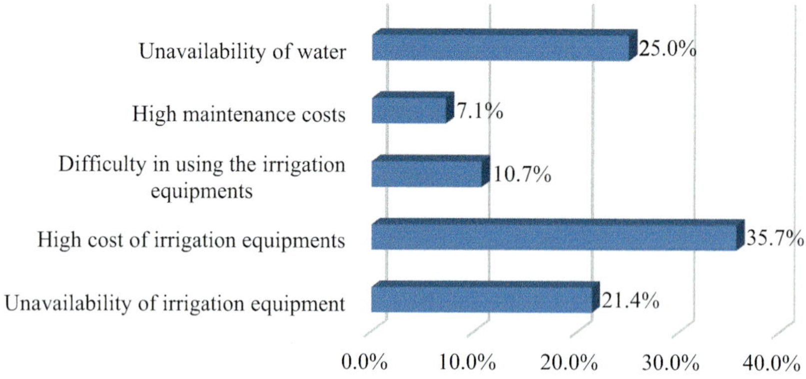

Fig. 5 Reasons for non-practice of irrigation

Respondents were requested to indicate the main challenges faced in the practice of irrigation and the results summarized in Fig. 5.

Majority of the respondents cited the high cost of irrigation equipment (35.7%) as the main challenge in irrigation. Other respondents indicated unavailability of water (25.0%), unavailability of irrigation equipment (21.4%), difficulty in using irrigation equipment (10.7%), and high maintenance cost (7.1%) as the major challenges in their implementation of irrigation farming (Fig. 6).

Majority of the respondents suggested that the best solution for the challenges facing the adoption of irrigation farming in the study area is government subsidy on equipment costs as represented by 31.1% of the total responses (Fig. 6).

Other suggestions that were presented included capacity building/trainings (22.6%), rational choice of equipment (20.0%), high reliance on ground-dug water

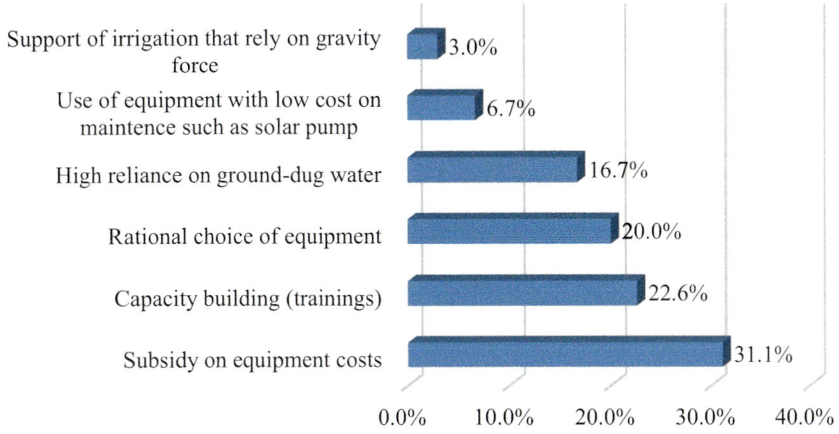

Fig. 6 Proposed solutions for challenges in irrigation farming

Table 8 Source(s) of irrigation water

Sources of water	Frequency	Percent
River	68	25.2
Dug well	127	47.0
Dam	31	11.5
Forage	44	16.3
Total	**270**	**100.0**

(16.7%), use of equipment with low cost on maintenance such as solar pump (6.7%), and support of irrigation that rely on gravity force (3.0%).

Respondent's main source of irrigation water is summarized in Table 8.

Majority of the respondents (47.0%) sourced their irrigation water from dug well. Other sources of irrigation water included river (25.2%), harvested water (16.3%), and dams (16.3%).

This study sought to know the different modes of access of irrigation water used by respondents. Majority of the respondents indicated to be accessing their irrigation water through private means as represented by 62.6% of the total responses (Table 9).

About 30.7% of the respondents accessed their irrigation water through free means (communal access), while 4.8% and 1.9% accessed the irrigation water resource through payment of money to others and membership to particular groups, respectively.

This study noted that majority of the respondents were not members of water management association (CGB Association) as represented by 60.0% of the responses. It was only 40.0% of the respondents who were members of water management association (Table 10).

Respondents were requested to indicate whether the CGB (Water User Association) or water user association that operated in the study area was functional or not and the results summarized in Table 11.

Table 9 Mode of access of irrigation water

Mode of access of irrigation water	Frequency	Percent
Private means	169	62.6
Free/communal access	83	30.7
Through payment of money to others	13	4.8
Membership in a group	5	1.9
Total	**270**	**100.0**

Table 10 Membership to water management association

Membership status	Frequency	Percent
Yes	108	40.0
No	162	60.0
Total	**270**	**100.0**

Table 11 Functionality of the water management association (CGB)

Response	Frequency	Percent
Yes	45	16.7
No	13	4.8
Do not know	212	78.5
Total	**270**	**100.0**

Majority of the respondents (78.5%) indicated that they did not know about the functionality of CGB association. It was only 16.7% of the respondents who indicated that the water user association was functional. A few respondents indicated that the water management association was not functional.

This study sought to determine the major type of irrigation systems practiced in the study area (Table 12). With respect to the last three cropping seasons (2014–2016), this study observed a major practice of drip, sprinkling, Californian, canal-gravity/IP, and manual irrigation systems.

The highest practiced irrigation system was Californian (32.2%), manual irrigation (26.3%), sprinkling irrigation (20.4%), drip irrigation (17.0%), and IP and gravity canal irrigation (4.1%).

This study found that the amount of land under various irrigation systems dedicated for potatoes, tomatoes, and shallots production varied as indicated in Table 13.

Drip irrigation was practiced on an average of 0.49 ha (with a standard deviation of 0.70 ha). Other irrigation systems that were practiced by respondents included sprinkling (0.41 ha with a standard deviation of 0.69 ha), Californian (0.32 ha with a standard deviation of 0.16 ha), canal IP (0.34 ha with a standard deviation of 0.19 ha), and manual (0.42 ha with a standard deviation of 0.45 ha).

Table 12 Major type of irrigation systems used for crop production

Irrigation system	Frequency	Percent
Drip	46	17.0
Sprinkling	55	20.4
Californian	87	32.2
Canal-gravity/IP	11	4.1
Manual	71	26.3
Total	**270**	**100.0**

Table 13 Area of different irrigation systems used on potatoes, shallots, and tomatoes

		Irrigated Area			
Irrigation system	Crop name	Min	Max	Mean	Std. dev.
Drip	Potatoes	0.02	1.00	0.27	0.34
	Shallots	0.10	7.00	0.68	0.92
	Tomatoes	0.10	7.00	0.52	0.83
	Aggregate	**0.02**	**7.00**	**0.49**	**0.70**
Sprinkling	Potatoes	0.10	8.00	0.80	1.31
	Shallots	0.10	3.00	0.54	0.74
	Tomatoes	0.20	7.00	1.88	0.04
	Aggregate	**0.10**	**8.00**	**0.41**	**0.69**
Californian	Potatoes	0.10	0.50	0.29	0.15
	Shallots	0.10	0.50	0.30	0.18
	Tomatoes	0.10	0.50	0.37	0.16
	Aggregate	**0.10**	**0.50**	**0.32**	**0.16**
Canal IP	Potatoes	0.10	1.00	0.29	0.20
	Shallots	0.10	0.50	0.39	0.17
	Tomatoes	0.10	0.50	0.34	0.19
	Aggregate	**0.10**	**1.00**	**0.34**	**0.19**
Manual	Potatoes	0.05	1.00	0.40	0.31
	Shallots	0.05	3.00	0.51	0.69
	Tomatoes	0.05	1.00	0.35	0.34
	Aggregate	**0.05**	**3.00**	**0.42**	**0.45**

Access to Extension Services

Only 31.9% of the farmers in the study area have access to agricultural extension services. This is a depicted in Fig. 7.

Due to their limited number in the study area, most extension service officers prefer the use of field days in disseminating agricultural information. Majority of the extension officers use analogue rather than digital methods of access to and use of agricultural information. Effective extension services can equip farmers with knowledge and skills necessary for efficient production.

This study sought to understand the different providers of farmers' extension services. Therefore, respondents who accessed extension services were requested to

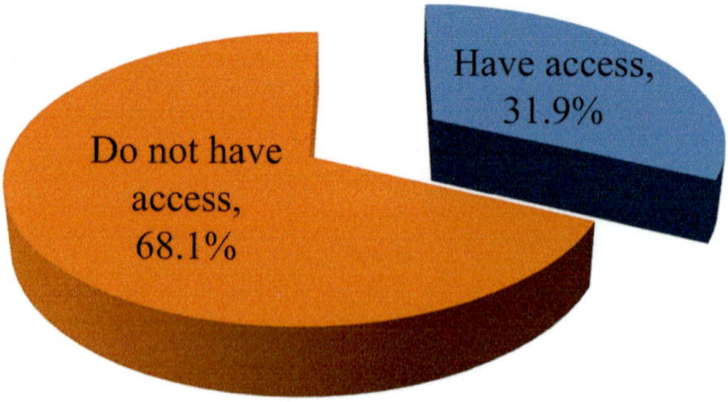

Fig. 7 Access to extension services

indicate the extension service provider that had served them and the results summarized in Fig. 8.

Majority of the respondents indicated that they had received extension services from NGOs and development agencies as represented by 60.5% of the total responses. About 34.9% of the respondents had accessed extension services from government extension workers, while only 4.7% of the respondents had accessed extension services from other farmers.

In terms of frequency of extension provider visits (number of times visited in the last year) to the respondents, this study observed that majority of the farmers were never visited by the extension service providers as represented by 68.1% of the total responses. An average farmer was visited about two times (2.423) by extension providers during the previous year (Fig. 9).

About 12.6% of the farmers had received 1–3 visits from extension providers. About 9.6%, 7.4%, and 2.2% had received 4–6 times, 7–9 times, and 10 times and more, respectively.

This study sought to know about farmers' participation in various irrigation training programs during their previous cropping year. The results are summarized in Fig. 10.

Majority of the respondents indicated to have participated in various irrigation training programs during their last cropping season as represented by 78% of the total responses. About 22% of the respondents did not participate in any training program.

Majority of the trainings that were received by farmers that received capacity building extension indicated to have related to crop production as represented by 84.8% of the total responses (Fig. 11). About 37.0% of the trainings related to soil and water conservation, while 13.0% related to agricultural marketing. Other trainings included irrigation and water management (17.4%), livestock production (10.4%), postharvest and storage practices (4.1%), and crop protection (3.3%).

When asked about the usefulness of the training topics that farmers received during the trainings received during the previous cropping season, majority of the respondents were indifferent as represented by 57.1% of the total responses (Fig. 12).

A few respondents (23.8%) were of the opinion that the training received was very useful, while 9.5% considered the trainings as somehow useful. Some

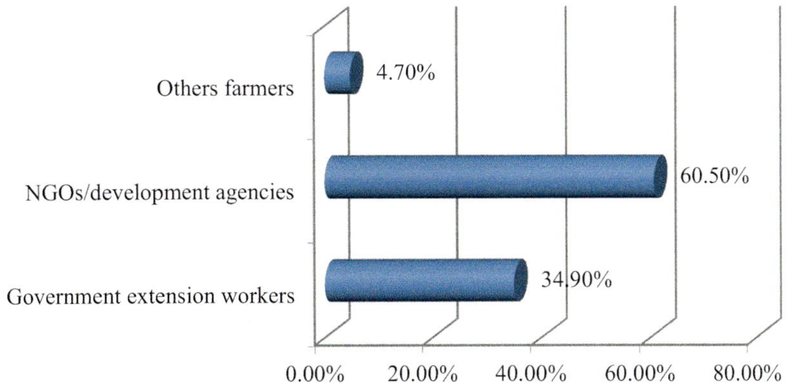

Fig. 8 Farmers' extension providers for different services

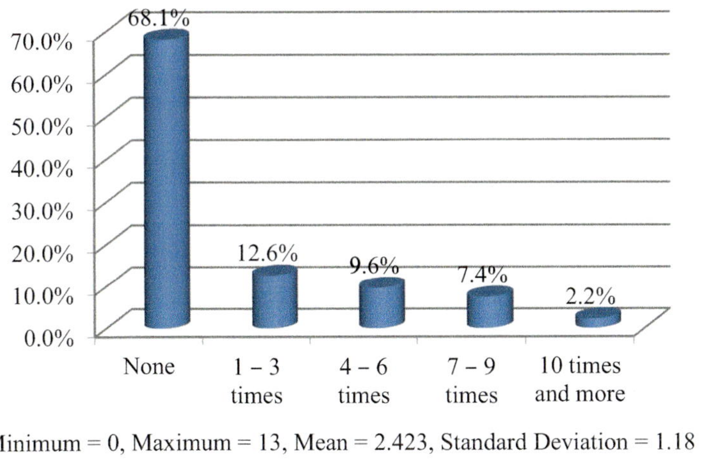

Fig. 9 Frequency of extension providers visits to the respondents

respondents (6.8%) considered the trainings to have been less useful, while 2.7% considered the trainings to have been not useful at all.

Majority of the respondents indicated to have visited extension demonstration sites within the last cropping season as represented by 68% of the total responses. On the contrary 32% of the respondents did not visit any extension demonstration site (Fig. 13).

Farmers' access to useful information related to agricultural production and marketing from various sources formed an important focal point in this study (Fig. 14).

About 60.7% of the respondents had accessed agricultural production and marketing from television. Other sources where respondents had accessed agricultural information included newspapers (65.6%), radio (53.3%), other farmers (44.1%), and marketplaces (41.1%).

Fig. 10 Farmers' participation in irrigation training programs during the last cropping year

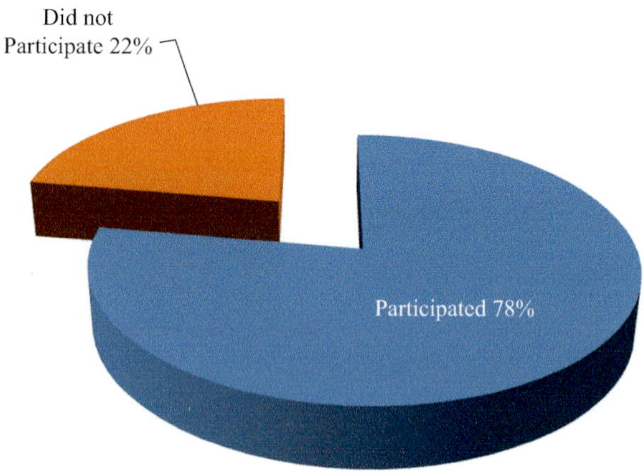

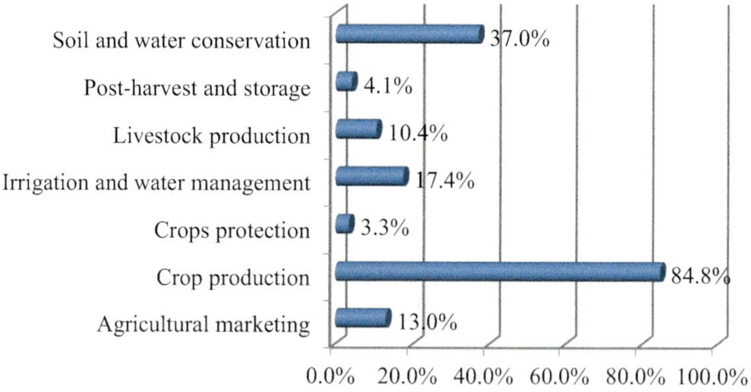

Fig. 11 Topics accessed during farmers training

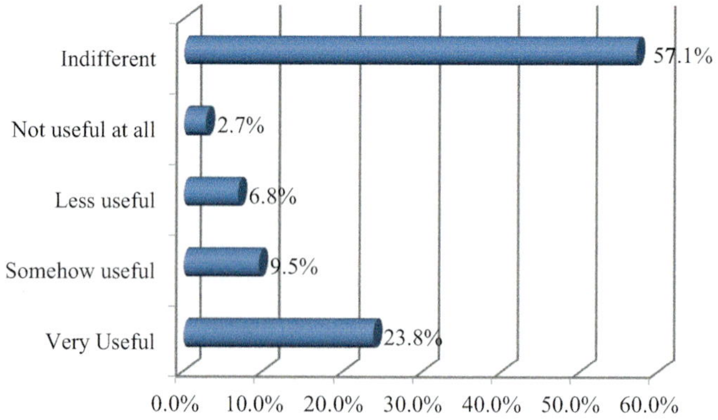

Fig. 12 Usefulness of the training topics that farmers received

Fig. 13 Farmers' visit of extension demonstration site

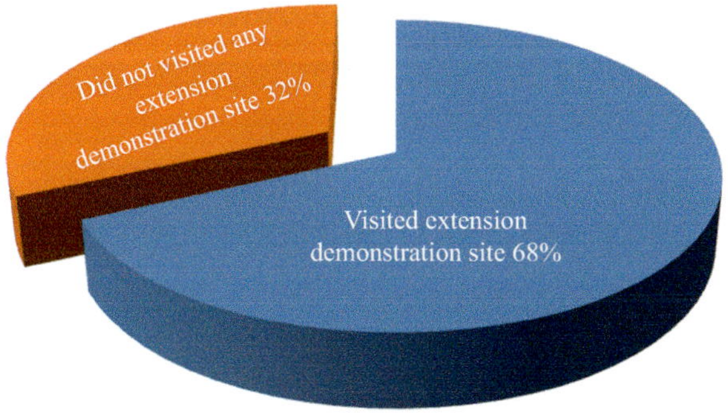

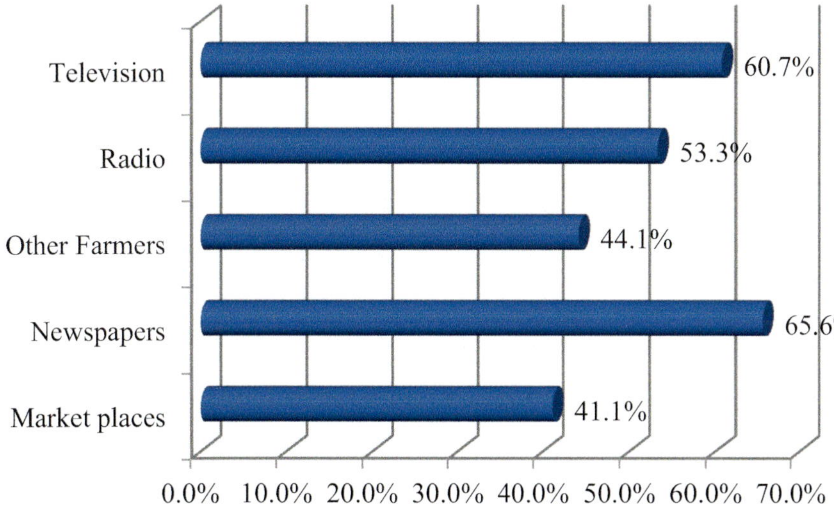

Fig. 14 Farmers' sources of agricultural information

Road Conditions in the Region

Respondents were requested to indicate the distance that they travel to the nearest shopping and centers through all-weather and tarmacked roads. The results are shown in Table 14.

The results indicated in Table 14 show that an average farmer in the study area was positioned a distance of 6.23 km (with a standard deviation of 6.56 km) away from the nearest shopping center through all-weather roads. An average farmer was 8.06 km (with a standard deviation of 10.32 km) away from the nearest urban center through all-weather roads.

There are few tarmacked roads in the study area. Therefore, majority of farmers are several kilometers away from tarmacked roads. An average farmer in the study area had to travel 15.47 km (with a standard deviation of 16.10 km) and 19.97 km (with a standard deviation of 9.98 km) to the nearest shopping and urban centers, respectively.

Table 14 Farmers' perception on road conditions in the region

Center	All-weather portion	Tarmacked portion	Fare (Kshs)
Nearest shopping center	6.23 (6.56)	15.47 (16.10)	800.57 (93.35)
Nearest urban center	8.06 (10.32)	19.97 (9.98)	505.56 (47.88)

An average farmer was spending a fare of 800.57 Fcfa (with a standard deviation of 93.35) and 505.56 Fcfa (with a standard deviation of 47.88) to the nearest shopping and urban centers, respectively.

Benefit-Cost Ratio (BCR) Analysis

The objective in this study sought to determine the economic viability of the alternative small-scale irrigation systems. The benefit-cost ratio (BCR) analysis was used also for judging the economic soundness of irrigation projects. From an economic point of view, an irrigation project with BCR greater than 1.5 is generally accepted. BCR is obtained by dividing the annual total benefits by the annual total cost. The net annual benefit is computed as a difference between the value of produce and total running costs incurred in production. Running cost included expenditure on seeds, manure/fertilizers, pesticides, hired labor/equipment, and government taxes/levies, among others.

The interest on capital cost of the project at the prevailing rate is included in the denominator. BC ratio is thus a division of the total discounted benefits by the total discounted costs.

In calculating the simple benefit-cost ratio, the following formula is used (Boardman 2011):

$$\Sigma\left(B_i/(1+d)_i\right)/\Sigma\left(C_i/(1+d)_i\right),$$

where:

B_i = the benefits of the project in year i, i = 0 to n
C_i = the costs of the project in year i
n = the total number of years for the project duration
d = the discount rate

The economic viability of the alternative small-scale irrigation systems was analyzed. Hereafter, the decision-making criteria are applied to determine the economic viability/desirability of the alternative small-scale irrigation systems in the study area.

The benefit-cost analysis in this study made use of market prices, not shadow prices. This is because there is no evidence of distortions of market prices of factors and products. Furthermore, while it is possible to calculate values for intangible impacts, such will be discussed in qualitative terms. Although sophisticated

Table 15 Running cost (Fcfa), duration to maturity, and yield and market information in the production of tomatoes, potatoes, and shallots per hectare

Parameters	Crops	Tomato	Potato	Shallot
Running costs	Seeds	400,000	750,000	150,000
	Manure	100,000	125,000	120,000
	Fertilizers	150,000	210,000	200,000
	Pesticides	95,000	175,000	70,000
	Labor	250,000	415,625	350,000
	Transportation	125,000	125,000	115,000
	Other costs	2,500	5,000	3,000
Duration to maturity and yield	Vegetative cycle (month)	2–3	3–4	3–4
	Yield (t/ha)	10–40	25	70
Marketing/commercialization	Producers (Fcfa/kg)	150	125–275	100
	Market (Fcfa/kg)	500	156–533	400
Return	Gross (Fcfa/ha)	37,500,000.00	34,450,000.00	112,000,000.00
	Net	36,377,500	32,644,375	110,992,000

equipment is not used in manual irrigation system, such an arrangement was not considered as costless.

This study determined the running cost, duration to maturity, and yield and market information in the production of potatoes, tomatoes, and shallots in the various irrigation systems (Table 15). These costs included expenditure on seeds, manure, fertilizers, pesticides, hired labor, and others.

To verify the feasibility of alternative small-scale irrigation systems, the benefit-cost ratio was calculated. The interest on capital cost of the irrigation systems was calculated at the prevailing rate. BCR is thus a division of the total discounted benefits by the total discounted costs.

The main reason for discounting is because a dollar available some years in the future is not as good as a dollar available now. This is because a dollar available now can be invested and earn interest and would be worth more than a dollar in future. With an interest rate of r, a dollar invested for t years will increase to $(1 + r)^t$. Therefore the amount of money that would have to be deposited now so that it would grow to be one dollar t years in the future is $(1 + r)^{-t}$. This is called the discounted value or present value of a dollar available t years in the future. The benefits and costs are in constant value dollars, and therefore the discount rate used must be the real interest rate. In this study, the present value of the streams of benefits and costs is discounted at a 13% back to time zero since the interest rate on long-term bonds is 18% and the rate of inflation is 5%.

In calculating the simple benefit-cost ratio, the following table (Table 16) was to compare BCR of the four irrigation systems in the production of potatoes, shallots, and tomatoes.

Table 16 A comparison of BCR of the selected irrigation systems in producing potatoes, shallots, and tomatoes in the year 2017

	Benefit-cost ratio (BCR)		
Irrigation systems	Potatoes	Shallots	Tomatoes
California	1.499	2.024	2.147
Drip	1.964	2.825	2.948
Sprinkler	1.727	2.252	2.375
Canal IP	1.395	1.920	2.043
Manual	0.165	0.690	0.813

NB: Discount rate = 18%, inflation rate = 5%

The BCR of 0.165, 0.690, and 0.813 when producing potatoes, shallots, and tomatoes, respectively, using manual irrigation system confirms the economic undesirability involved in system. An average BCR of 0.556 on using manual irrigation system for the production of all the three major crops also is far less than 1 confirming that it is not economically viable to use the system in the production of selected major crops under investigation. A lower than 1 BCR implies that the costs incurred during production are greater than the benefits. The BCR of other irrigation systems (California, drip, sprinkler, and canal IP) are all greater than unity (1) implying that they lead to greater benefits as compared to costs. The excess benefits (compared to costs) are realized more with drip irrigation system (BCR = 2.579) with the second best being sprinkler (BCR = 2.118) and the third being California (1.890). The fourth best irrigation system is canal IP (BCR of 1.786).

Conclusions

This study concludes that use of manual irrigation system is not economically viable in the production of potatoes, shallots, and tomatoes as confirmed from a low BCR of less than unity. The BCR of California, drip, sprinkler, and canal IP irrigation systems in the production of potatoes, shallots, and tomatoes are all greater than unity (1) implying that they lead to greater benefits as compared to costs. This study recommends that the government and other agricultural development stakeholders should embark on supporting smallholder farmers in drought-prone areas of Mali in adopting nonmanual irrigation methods of crop production as opposed to reliance on rain-fed agriculture. This investment is capable of offering smallholder farmers an opportunity to reap greater benefit as compared to cost.

References

Alonso, P., & Lewis, G. B. (2001). Public service motivation and job performance: Evidence from the federal sector. American Review of Public Administration, 31, 363–380.

Bagayoko M, N'Diaye MK, Dicko M, Tangara B (2007) Characterisation of soil degradation under intensive rice production in Office du Niger zone of Mali. In: Bationo A, Waswa B, Kihara J,

Kimetu J (eds) Advances in integrated soil fertility management in sub-Saharan Africa: challenges and opportunities. Springer, Dordrecht
Barrett, P. & Associates Pty Ltd. (1999). Determining a framework, terms and definitions for water use efficiency in irrigation. Land and Water Resources Research and Development Corporation. Retrieved from http://www.insidecotton.com/xmlui/handle/1/2130
FAO (1997) Small-scale irrigation for arid zones. Principles and options. Food and Agriculture Organisation of the United Nations (FAO), Rome. Retrieved from https://www.sswm.info/glossary/2/lettero#term3403
FAO (2010) How design, management and policy affect the performance of irrigation. FAO, Rome
FAO (2011). Global Action on Climate Change in Agriculture: Lnkages to food security, Market and Trade Policies In Developing Countries. Rome: FAO.
Guslits, B., & Phartiyal, J. (2010). Women & water in the age of globalization: Protecting our most vital resource. Women & Environments International Magazine (82/83), 12–14.
Hillel, D. (2000). Salinity management for sustainable irrigation: integrating science, environment, and economics. Washington, DC: World Bank.
Hussain I (2004) Assessing impacts of irrigation on poverty: approaches, methods, case studies and lessons. International Water Management Institute (IWMI). http://www.iwmi.cgiar.org/propoor/files/ADB_Project/Research_Studys/poverty%20study%20-ethiopia.pdf
IPA (2017) Irrigation and property rights for farmers in Mali. Retrieved from https://www.poverty-action.org/study/irrigation-and-property-rights-farmers-mali
IPCC (2014) Climate change 2014: synthesis report. Contribution of Working Groups I, II and III to the Fifth Assessment Report of the Intergovernmental Panel on Climate Change. IPCC, Geneva
Jordan J (2009) Rethinking community resilience to climate change: does a social capital lens help? Current crises and new opportunities. Development Studies Association, Coleraine. Retrieved from http://www.devstud.org.uk/aqadmin/media/uploads/4ab779da220b7_1-jordan-dsa 09.pdf
Kalaugher L (2012) Groundwater pumping and irrigation have similar effect to climate change. Retrieved from http://environmentalresearchweb.org/cws/article/news/51754
Kamara, A., Van Koppen, B. and Magingxa, L. (2001). Economic viability of small-scale irrigation systems in the context of state withdrawal: the Arabie Scheme in the Northern Province of South Africa. International Water Management Institute (IWMI), Pretoria, South Africa, 2nd WARSFA/Water Net Symposium: Integrated Water Resources Management: Theory, Practice, Cases; 1(1), 30–31.
Karlinsky N (2005) California dreaming: ideology, society and technology in the citrus industry of Palestine, 1890–1939. State University of New York Press, Albany
Keita, D.M., Bedingar, T., Lubunga, R., Mousseau, L.P. and Matila, M.P.S. (2009). African development fund, Republic of Ghana; Northern rural growth programme, Osan 2. Appraisal Report, Osan 2.
Kelly, V. (2008). Agricultural Statistics in Mali: Institutional Organization and Performance. Background paper for the World Bank (WB) on agricultural. Washington D.C: World Bank.
Kothari (2004) Research methodology, 2nd rev ed. New Delhi, India: The New Age International (P) Ltd.
McIntosh, C., A. De Janvry, and E. Sadoulet (2005) How rising competition among microfinance lenders affect incumbent village banks, Economic Journal, 115, 987–1004.
Mosca, M. & Pastore, F. (2008). Wage Effects of Recruitment Methods: The Case of the Italian Social Service Sector. Retrieved from https://doi.org/10.1111/j.0042-7092.2007.00700.x.
Nyanjom, O. (2011). Devolution in Kenyans' new constitution. Constitution Working Paper Series no. 4. Nairobi: Society for International Development.
Office Statistique des Communautés Européennes (2013) Statistiques de Base: Agriculture Elevage. Office Statistique des Communautés Européennes, Bamako
Perry, J., & Johnson, J. (1999). What makes a small farm successful? Economic Research Service-USDA. Agricultural Outlook, November.
Pinstrup A, Watson D (2011) The role of government in global, national and local food systems: food policy for developing countries. Cornell University. Driving forces of food system

Resilience Alliance (2002) The Resilience Alliance. Retrieved from http://www.resalliance.org/index.php/resilience

Reuben, M J., Kadigi, G.T., Bizoza, A., Zinabou, G., (2012). Irrigation and water use efficiency in Sub-Saharan Africa, Briefing Paper Number 4 | 2012, GDN Agriculture Policy Series.

Schwarz A-M, Béné C, Bennett G, Boso D, Hilly Z, Paul C, Andrew N (2011) Vulnerability and resilience of remote rural communities to shocks and global changes: empirical analysis from Solomon Islands. Glob Environ Chang 21:1128–1140. https://doi.org/10.1016/j.gloenvcha.2011.04.011

Shiferaw B, Freeman HA, Swinton S (2004) Natural resource management in agriculture: methods for assessing economic and environmental impacts. CABI Publishing, Wallingford

Speranza CI (2010) Resilient adaptation to climate change in African agriculture. Deutsches Institut für Entwicklungspolitik gGmbH, Bonn. Retrieved from http://www.adaptationlearning.net/sites/default/files/Studies%2054.pdf

Todaro MP, Smith SC (2012) Economic development policy, 11th edn. New York University. Developing Countries Economic Policy. Adison-Wesley Publishers USA, Boston

Truong, T.N.C & Ryuichi, Y. (2002). Factors affecting farmers' adoption of technologies in farming system: A case study in OMon district, Can Tho province, Mekong Delta. Omonrice 10:94–100.

USDA (2013) Climate change and agriculture in the United States: effects and adaptation. United States Department of Agriculture technical bulletin 1935. Washington, DC, USA

Cultivar Selection and Management Strategies for *Lablab purpureus* (L.) Sweet in Africa

54

Neil R. Miller, Wilfred Mariki, Alison Nord, and Sieglinde Snapp

Contents

Abstract

The single greatest impact of climate change for farmers throughout the tropics is rainfall uncertainty. This crisis calls for popularization of resilient crop species which can withstand periods of drought and respond with vigorous growth and food production when rainfall resumes. *Lablab purpureus* (L.) Sweet, a crop native to Africa, demonstrates just such resilience and yet was nearly lost to

N. R. Miller (✉)
Canadian Foodgrains Bank, Arusha, Tanzania
e-mail: NRMiller@FoodgrainsBank.ca

W. Mariki
Selian Agriculture Research Institute, Arusha, Tanzania

A. Nord · S. Snapp
Department of Plant, Soil and Microbial Sciences, Michigan State University, East Lansing, MI, USA

W. Leal Filho (ed.), *Handbook of Climate Change Resilience*,
https://doi.org/10.1007/978-3-319-93336-8_102

African farmers during the colonial era when faster-producing *Phaseolus* sp. beans replaced it in all but a few geographic areas. Unfortunately, the decline in lablab production was accompanied by a loss of genetic diversity and a loss of knowledge of appropriate agronomic practices for the crop.

This chapter presents initial results of field trials of 28 lablab cultivars at three locations in northern Tanzania. Cultivars selected from the core collection developed by Pengelly and Maass (Genet Resour Crop Evol 48:261–272, 2001) as well as farmer landraces varied widely in relative maturity, growth habit, insect colonization, and grain yield. Intercropping with maize reduced productivity of most cultivars but suppressed the population of pod-sucking bugs. Lablab cultivars were more drought resilient than cowpea (*Vigna unguiculata*) grown for comparison. Strategies for the improvement and repopularization of this "lost" crop are discussed.

Keywords

Lablab purpureus · Cover crops · Legumes · Drought resilience

Introduction

Background

Lablab purpureus beans and leaves have fed African families for millennia. The crop is thought to have originated in wild form on the African continent but was domesticated independently by African and Indian farmers (Maass et al. 2005). Subsequently it has spread as a minor crop throughout the tropics. In many Africa cultures, lablab is valued for its nutritional and healing properties, especially for women following childbirth. The Kikuyu, in central Kenya, further assign religious significance to the crop, which plays a key role in cultural rites (Robertson 1997).

Despite its historical and cultural significance, lablab cultivation diminished drastically through the colonial era (Maass et al. 2010; Robertson 1997), and today, lablab is grown mostly in isolated pockets scattered throughout the continent. The main exception to this trend is central Kenya and northern Tanzania where demand from Kikuyu consumers has sustained widespread production.

Lablab production in northern Tanzania is largely undocumented but has expanded over the past 15 years to an estimated 20,000 ha producing 8000 tons of lablab for export to Kenya. Lablab-producing areas are often dry environments which do not have reliable enough rainfall for consistent *Phaseolus* bean production. The crop is sometimes planted in association with maize, but in the driest regions, it is planted as a sole crop. In addition to the grain, which is largely grown for sale, the green foliage which continues to grow after grain harvest is highly valued as a dry season forage for livestock.

The ability of lablab to survive extended periods of dry weather positions it as a key tool for adaptation to climate change. Increasingly erratic and unpredictable rainfall is the single greatest impact of climate change on small farmers in the tropics (FAO 2009). In this context, a crop which can survive months of drought stress and then respond to capricious rainfall events with new growth and production is a valuable tool in maintaining food security. Drought *resilience* exists in other tropical legumes, notably pigeon pea (*Cajanus cajan*) and cowpea, but has been relatively unexplored by the scientific community which has, instead, focused on species and cultivars which produce quickly and are thereby drought *avoidant.*

Although interest in lablab has grown in recent years, significant challenges exist in developing this crop as a significant component in food security. First, the genetic identity of lablab cultivars being grown by African farmers is largely unknown. Kenya is the only country on the continent where varieties have been formally identified and released, and despite its importance in the Kikuyu diet, the only widely planted certified seed is that of a cultivar released in 1978 (KEPHIS 2017). Thus, the need for further agronomic screening and selection is evident.

A second challenge for lablab producers throughout the tropics is the crop's extreme attractiveness to insect pests, especially at flowering and podding stages. Chief among these pests in northern Tanzania are lepidopteran larvae (*Maruca vitrata, Helicoverpa armigera, Etiella zinckenella*, etc.) commonly referred to as pod borers, but often more damaging at flowering stage, and pod-sucking bugs (*Riptortus pedestris, Clavigralla tomentosicollis*, etc.). Of 39 known hosts of *M. vitrata*, lablab is considered to be the most suitable (Sharma 1998). Attack by these pests can cause near-complete losses in grain yield, and commercial producers typically spray insecticides at least twice. Smaller-scale producers, who have less access to insecticides, often suffer substantial losses. While some cultivar resistance to these pests has been found on cowpea and pigeon pea (Kaoneka et al. 2016; Sharma 1998), results on lablab have not been particularly encouraging, and virtually no research has been done among the genetically diverse collection of African accessions.

Finally, while one significant market exists for black-seeded lablab in Africa (in Kenya) elsewhere in post-colonial Africa, consumer preference has swung largely toward *Phaseolus* beans. Furthermore, while the Kikuyu palate insists on black-seeded lablab, other cultures prefer milder-flavored cream and brown cultivars. Traditional lablab recipes have largely been lost, and because the crop takes longer than *Phaseolus* beans to cook, successful promotion of lablab in African diets will also need to take into account consumer preference issues.

In the face of these challenges, there is an enormous reserve of genetic diversity waiting to be tapped. The world collections of lablab include over 3000 accessions with a broad range of geographic, morphological, and genetic characteristics (Maass et al. 2010). The genetic diversity of lablab cultivars being grown on Kenyan farms was studied by Kimani et al. (2012) and found to be surprisingly uniform. If this pattern holds true elsewhere, it suggests that selection and breeding could bring significant benefits to African lablab producers.

The present study aims to investigate a diverse set of 29 lablab cultivars, evaluating their performance as a sole crop and intercrop with maize in the context of northern Tanzania.

Description of the Study Areas and Methodology

Study Area

The study was conducted at three sites in the Northern Zone of Tanzania. The Selian Agricultural Research Institute (SARI) is located in Arusha and is typical of mid-land locations at an altitude of 1387 MASL, with a mean annual temperature of 19.5 °C and a mean annual rainfall of 1052 mm. The second site was located at the Tropical Pesticides Research Institute (TPRI) research farm in Miwaleni, representing lowland conditions at 719 MASL with mean temperature of 23.5 °C and a mean annual rainfall of 600 mm. The third site was located at the Tanzania Coffee Research Institute (TACRI) in Lyamungo at an elevation of 1287 MASL. Situated on the lower slopes of Mt. Kilimanjaro, the weather at this site is typical of the Tanzanian highlands despite its moderate altitude. The mean annual temperature is 15.6 °C and mean annual rainfall is 1546 mm.

Rainfall at these sites is weakly bimodal with highly variable short rains and more consistent long rains. Field crops are generally planted in the long rains beginning in March to May. Crops are rarely planted in the unreliable short rainy season of November–December.

Research Methods

Study Site Selection and Sampling Methods

Experiments at the SARI and TPRI sites were designed as a modified split plot with rows of 8 (SARI) and 7 (TPRI) plots to which maize-lablab intercropping or sole lablab cropping were randomly assigned. Lablab accessions were randomized within these strips with a total of three replications at all sites. This structure was allowed for in the analysis by including the strips of plots as a random effect. Pannar 15 maize was planted in six rows per plot, 90 cm apart and 50 cm within rows with two seeds per hill for a total seeding rate of 44,444 seeds/ha. Plot length was 4.5 m with a 1.5 m alley between plot ends. 18-46-0 fertilizer was applied in maize planting holes at a rate of 77 kg/ha and 46-0-0 was side dressed at 110 kg/ha, for a rate of 65 kg N/ha and 35 kg phosphate/ha. Lablab cultivars were seeded 12 and 8 days after maize planting at SARI and TPRI, respectively, in rows 90 cm apart and 50 cm within rows with two seeds per hill with rows located equidistant between maize rows in intercropped plots. One maize row was planted at the side borders of all sole-cropped lablab plots so that maize shading was uniform whether they were adjacent to maize intercropped plots or not.

The TACRI site did not include maize intercropping and was laid out in three randomized complete blocks with rows 70 cm apart and 60 cm within rows with two seeds per hill. Plots were six rows wide with 1.4 m between plots, and plot length was 5.0 m with 2.0 m alley between plot ends.

Insect pests were monitored closely beginning at flowering stage, and insecticides were applied when populations reached presumed damaging levels. The SARI, TPRI, and TACRI sites were sprayed 4, 3, and 3 times, respectively.

Data Sources and Collection Methods

Twenty plants in each plot were counted for flowering on a 10–14-day schedule, and these counts were used to estimate 50% flowering dates. Early and late season soil cover, lablab plant height at maize harvest, and plant survival at the end of the dry season were measured in all plots, while biomass and nitrogen content were measured on a subset of 14 lablab accessions and 1 cowpea accession at the SARI and TPRI sites.

Intensive insect pest sampling was conducted twice at SARI in order to measure differences between cultivars and intercropping versus sole cropping. Ten inflorescences were selected at random from each plot and tapped eight times on the inside of a 10 l plastic bucket. Care was taken to prevent flying insects from escaping before recording the type and number of insects collected. This method has proved both reproducible and highly efficient, allowing for comparisons of insect infestations on large numbers of plots as utilized in these experiments.

Maize yields were recorded from 3 m of 2 rows at the center of each intercropped plots. Plant number and ear number were recorded, ears were sun dried and threshed, and the weight and moisture content of the grain was recorded. Lablab harvest began as soon as dried pods were present on the earliest-maturing plots and was repeated on a roughly monthly schedule with five harvest dates at SARI and two each at TPRI and TACRI. Dry pod weight was recorded at each harvest date, but pods from all harvest dates were combined for threshing and determination of overall plot yield.

Late-season plant cover was estimated at SARI and TPRI using digital photographs taken from a 3 m height above each plot. A standardized rating scale was developed using photographs of plots with a range of 0–90% soil cover (as measured along 4, 2 m transects). This standardized rating scale was then compared visually to rate % soil cover of other plots. Photographs were taken at SARI on 5 October and at TPRI on 8 October, both after more than 5 months with no significant rainfall.

Data Analysis

Data were compiled and subjected to analysis of variance (ANOVA) for hypothesis testing. Lablab grain yields were log transformed in order to stabilize variance. In order to reduce the effect of a relatively high maize stand variability, maize plant density was included as a covariate in the ANOVA model for maize yield. For parameters with a significant ($P < 0.05$) F-statistic for the accession main effect, individual accession performance was further separated using the Bonferroni Least Squares Means method (SAS v9.4).

Results and Discussion

Weather

Rainfall at the SARI site was abundant prior to planting. After lablab seeding, however, only 124 mm of rain fell in the following 4 months (Fig. 1). The TPRI site received 231 mm rainfall in the 11 days immediately following planting, but thereafter, rain fell only once over the following 8 months. TACRI, in contrast, was wet through most of the early growth period with a total of 376 mm of rainfall during the 4 months following planting.

Relative Maturity

Lablab and cowpea accessions differed significantly in relative maturity at all three sites (Tables 1, 2, and 3). At SARI the 50% flowering date ranged from 52 to 130 days after planting. Relative maturity at TACRI averaged 20 days later than at SARI due to cooler temperatures. Flowering of early-maturing lablab and cowpea accessions at this site was delayed more than flowering of later-maturing accessions. At TPRI, flowering began considerably earlier than at either of the other sites due to relatively hot conditions, and the earliest accessions passed 50% flowering before counts began, while several late-maturing accessions succumbed to drought stress before reaching 50% flowering. Thus, flowering data for this site is incomplete. Cream-seeded lablab accessions flowered later overall than the darker-seeded accessions.

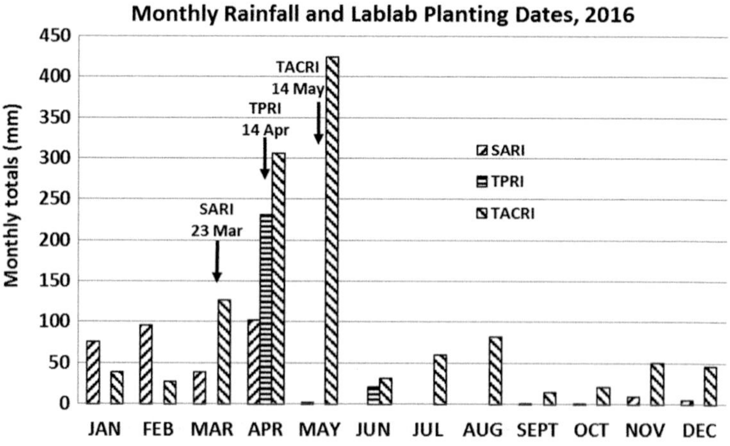

Fig. 1 Monthly rainfall totals at three lablab trial locations in 2016. Arrows indicate lablab planting dates

Table 1 Relative maturity and grain yield of lablab accessions grown at SARI in 2016

	50% Flower[a]		Grain yield (kg/ha)[b]		
Accession	Sole crop	Inter-crop	Sole crop	Inter-crop	Mean separation[c]
Karamoja Red	62.6	66.3	1059	944	a
Highworth	59.1	62.5	844	412	ab
Q 6880B	55.5	52.1	1031	300	abc
DL1002	62.2	64.7	728	312	abcd
SARI Chungu	64.5	66.7	598	333	abcde
ILRI 14437	66.6	71.6	472	377	abcde
SARI Rongai	63.0	71.1	401	393	abcde
Echo Cream	69.3	76.7	525	257	abcdef
Fadhari cowpea	61.9	61.1	499	291	bcdefg
CIAT 22759	64.4	69.0	397	200	bcdefg
Dodoma white	82.2	107.8	365	335	bcdefg
Eldoret G1 Black 2	61.2	64.3	359	216	bcdefg
Karatu Black	72.9	72.7	522	204	bcdefg
Eldoret M5 Cream	63.4	68.3	375	257	bcdefgh
Vuli 2 cowpea	61.4	61.4	237	183	cdefghi
ILRI 6930	66.3	67.8	318	149	cdefghi
Eldoret B1 Maridadi	65.9	71.3	317	172	defghi
Rongai	82.4	92.0	328	96	defghi
ILRI 13700	72.1	71.9	135	264	defghi
DL1001	90.5	92.2	289	94	defghi
SARI Nyeupe	86.5	95.1	433	54	defghi
ILRI 6536	79.3	88.8	152	122	efghi
ILRI 14411	64.5	75.2	80	237	fghi
CPI 81364	60.6	67.1	115	122	ghij
ILRI 10979	61.3	67.2	181	140	ghij
Kola Black	118.6	130.3	92	111	hij
ILRI 11630	80.9	90.0	76	70	ijk
Karatu White	97.1	119.1	91	46	jkl
PI 195851	118.5	124.0	132	53	kl
ILRI 10953	63.6	67.2	75	27	l
Means	**71.9**	**77.8**	**371**	**220**	

[a]Number of days after planting when 50% of plants produced flowers. Flowering dates of accessions were significantly different ($P < 0.0001$) as was the effect of intercropping ($P = 0.02$); however no accession by intercropping interaction was found

[b]Grain yields are presented as simple means, but were log transformed for ANOVA

[c]Means within each accession not followed by a common letter differ according to Bonferroni's test ($P < 0.05$)

Intercropping delayed lablab and cowpea flowering at an average of 5.9 days at SARI and 6.8 days at TPRI. This intercropping effect appeared more pronounced on late-maturing lablab accessions than on early-maturing accessions.

Table 2 Relative maturity and grain yield of lablab accessions grown at TPRI in 2016

	50% Flower[a]		Grain yield (kg/ha)[b]		
Accession	Sole crop	Inter-crop	Sole crop	Inter-crop	Mean separation[c]
Fadhari cowpea	68.0	69.2	145	304	a
Vuli 2 cowpea		68.5	250	293	ab
Kitui cowpea	68.4	69.6	164	23	abc
CIAT 22759	67.2	70.6	154	77	abcd
Eldoret G1 Black 2	59.2	62.7	109	72	abcdef
Eldoret W7 Black 1	62.9	65.3	181	40	bcdefg
DL1002	59.2	67.3	22	92	bcdefgh
ILRI 6930	68.0	71.6	79	11	bcdefgh
Q 6880B			26	65	bcdefghi
Karamoja Red	64.2	71.0	8	49	cdefghi
SARI Rongai	66.7	74.4	105	40	cdefghi
Eldoret M5 Cream	58.4	64.4	7	84	cdefghij
ILRI 14411	65.9	71.2	14	53	cdefghijk
SARI Chungu	65.9	71.9	30	13	cdefghijkl
Highworth	53.1	63.1	12	8	defghijklm
ILRI 6536	86.1		28	26	defghijklm
Rongai	95.3		19	14	efghijklm
ILRI 6930 Cream	76.7		83	5	efghijklm
CPI 81364	49.0	58.8	2	19	fghijklmn
Karatu Black	66.1	72.4	3	15	ghijklmn
ILRI 14437	68.7		5	12	ghijklmn
ILRI 11630	102.5		45	8	ghijklmn
Eldoret B1 Maridadi	56.5	71.3	3	5	ghijklmn
ILRI 10953	55.6	64.3	4	22	ghijklmn
ILRI 10979	56.4	64.1	0	33	ghijklmn
ILRI 13700	69.5	85.9	20	3	hijklmn
Echo Cream	108.3		5	1	ijklmn
DL1001	98.1		4	2	jklmn
SARI Nyeupe			9	0	klmn
Dodoma white			2	0	lmn
Kola Black			2	0	mn
PI 195851			0	0	n
Karatu White			0	0	n
Means	**69.8**	**68.2**	**50**	**42**	

[a]Number of days after planting when 50% of plants produced flowers. Flowering dates of accessions were significantly different ($P < 0.0001$) as was the effect of intercropping ($P = 0.01$) as well as the accession by intercropping interaction ($P = 0.001$)

[b]Means within each accession not followed by a common letter differ according to Bonferroni's test ($P < 0.05$)

[c]Grain yields are presented as simple means but were log transformed for ANOVA

Table 3 Relative maturity and grain yield of lablab accessions grown at TACRI in 2016

Accession	50% Flower[a]	Grain yield (kg/ha)[b]	Yield mean separation[c]
Echo Cream	97.3	1438	a
Eldoret W7 Black 1	90.2	1314	ab
SARI Rongai	91.7	1262	abc
ILRI 6536	93.0	1189	abc
SARI Nyeupe	105.9	1003	abcd
DL1002	78.0	993	abcd
SARI Chungu	92.3	983	abcd
Dodoma white	100.1	926	abcd
CIAT 22759	88.6	864	abcde
ILRI 6930 Cream	92.0	609	abcdef
ILRI 11630	100.4	509	abcdefg
ILRI 14437	89.4	479	abcdefg
Q 6880B	92.3	421	abcdefg
ILRI 6930	95.9	421	abcdefg
Eldoret G1 Black 2	81.3	408	abcdefg
Karatu White	106.0	370	abcdefg
Kitui cowpea	103.7	366	abcdefg
ILRI 10953	89.3	216	bcdefgh
Fadhari cowpea	90.7	195	cdefgh
Rongai	105.6	177	defgh
DL1001	93.2	177	defgh
CPI 81364	81.9	168	defgh
Eldoret M5 Cream	95.8	160	defgh
Karatu Black	102.1	137	efgh
Highworth	89.4	121	fgh
Eldoret B1 Maridadi	90.1	106	fghi
PI 195851	112.7	84	ghi
ILRI 14411	94.8	47	hij
ILRI 13700	98.2	18	ij
Vuli 2 cowpea	102.1	13	j
Kola Black	115.1	13	j
ILRI 10979	63.6	6	j
Mean	**97.3**	**464**	

[a]Number of days after planting when 50% of plants produced flowers. Flowering dates of accessions were significantly different ($P < 0.0001$)
[b]Grain yields were log transformed for analysis and back transformed for presentation
[c]Means within each column not followed by a common letter differ according to Bonferroni's test ($P < 0.05$)

Grain Yield

Grain yields varied significantly by accession at all sites. Yields at SARI and TACRI were significantly higher than at TPRI, where extreme drought stress and insect damage limited production. Mean yields at SARI varied from over 1002 kg/ha to only 51 kg/ha. Relative maturity of lablab accessions did not appear to have a strong effect on grain yield at SARI, with early-, medium-, and late-maturing accessions all appearing toward the top of the list. Medium- and early-maturity cowpea controls yielded mid-way and at the lower end, respectively, relative to lablab accessions.

Mean lablab yields at TPRI varied from 116 kg/ha to 0 kg/ha, while the two cowpea varieties included as controls outyielded all lablab accessions. Many of the late-maturing lablab accessions never flowered fully and thus produced no seed.

Lablab yields at TACRI varied from 1438 kg/ha to 6 kg/ha. Several later-maturing accessions appeared to yield relatively higher compared with their performance in the hotter locations, as did cream-seeded accessions. Cowpea control varieties yielded relatively poorly compared to their performance in the hotter locations.

Intercropping significantly reduced lablab and cowpea grain yields at both sites. Intercropped lablab at SARI yielded an average of 59% of sole-crop grain yields, while intercropped yields at TPRI were 83% of sole-crop yields. There was no significant interaction between accessions and intercropping at either site. Neither was there any measurable effect of lablab accessions or intercropping on maize yield at either site.

Insect Damage

Reproductive-stage insects caused significant damage at all locations, with pod-sucking bugs predominating at SARI and flower-feeding caterpillars causing extreme damage at TPRI. Individual plot ratings were conducted only at the SARI site, and while caterpillar counts were too low to distinguish between accessions, pod-sucking bug counts varied significantly between accessions at both 118 days and at 139 days after planting ($P < 0.01$). Early-maturing accessions tended to have higher sucking-bug counts at both dates (Fig. 2), making it difficult to distinguish whether observed differences were due solely to the presence of an earlier food source or whether insect preference was affected by other characteristics of these accessions.

Intercropping, on the other hand, resulted in a significantly lower infestation of plant-sucking bugs at SARI at 118 days after planting (Fig. 3).

Drought Resilience

Drought resilience was measured at SARI and TPRI by recording the percentage of soil covered by live plant growth 5 months after the last rainfall (see section "Data Sources and Collection Methods"). Lablab accessions differed significantly

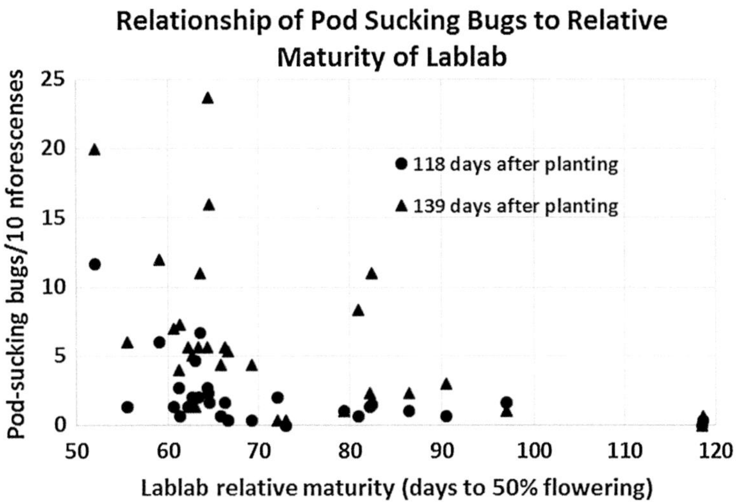

Fig. 2 Pod-sucking bug populations on lablab accessions with different maturities at two sampling dates, SARI 2016. Points represent mean values for each accession (sole-cropped plots only)

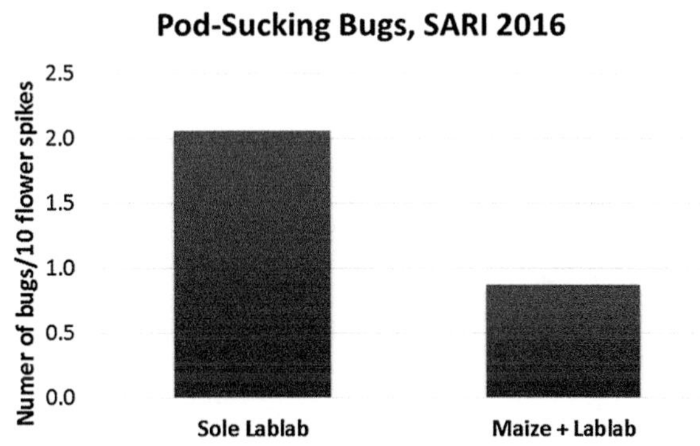

Fig. 3 Population of pod-sucking bugs in sole lablab and lablab intercropped with maize at SARI, 2016. Means are significantly different ($P = 0.03$)

in late-season soil cover at both sites (Table 4). At SARI, a strong association was identified between relative maturity and late-season soil cover with late-flowering accessions remaining greener late into the dry season (Fig. 4). Many accessions remained alive well beyond the 5 October rating date, and following a 9 mm rainfall in November, plants produced a final harvest. At TPRI, in contrast, plants never regrew, though many plots retained green, living plants through December.

At both sites, lablab accessions remained green significantly longer than did the cowpea controls (Table 4). At SARI, lablab accessions averaged 41.3% and 29.0% soil cover for sole and intercropped treatments, respectively, while the cowpea controls averaged only 5.0% and 7.5%. At TPRI, lablab soil cover was much lower than SARI (6.1% and 3.2%), but cowpea cultivars were completely dead

Table 4 Late-season soil cover of lablab and cowpea accessions at SARI and TPRI in 2016[a]

	SARI		TPRI	
Accession	Sole crop (%)	Inter-crop (%)	Sole crop (%)	Inter-crop (%)
CIAT 22759	31.7	15.0	8.3	5.0
CPI 81364	60.0	33.3	1.7	0.0
DL1001	36.7	40.0	15.0	6.7
DL1002	25.0	18.3	3.3	3.3
Dodoma white	26.7	25.0	10.0	5.0
Echo Cream	38.3	20.0	8.3	6.7
Eldoret B1 Maridadi	38.3	28.3	8.3	1.7
Eldoret G1 Black 2	38.3	25.0	3.3	1.7
Eldoret M5 Cream	31.7	23.3	8.3	3.3
Eldoret W7 Black 1			3.3	1.7
Highworth	33.3	21.7	1.7	1.7
ILRI 10953	35.0	36.7	5.0	0.0
ILRI 10979	43.3	15.0	5.0	3.3
ILRI 11630	40.0	25.0	8.3	5.0
ILRI 13700	46.7	33.3	6.7	3.3
ILRI 14411	33.3	31.7	6.7	0.0
ILRI 14437	38.3	26.7	5.0	1.7
ILRI 6536	61.7	31.7	8.3	3.3
ILRI 6930	58.3	36.7	6.7	3.3
Karamoja Red	36.7	30.0	6.7	3.3
Karatu Black	40.0	40.0	6.7	3.3
Karatu White	58.3	26.7	5.0	5.0
Kola Black	53.3	50.0	5.0	5.0
PI 195851	80.0	43.3	3.3	3.3
Q 6880B	25.0	31.7	3.3	1.7
Rongai	60.0	26.7	10.0	3.3
SARI Chungu	33.3	33.3	6.7	1.7
SARI Nyeupe	46.7	35.0	5.0	5.0
SARI Rongai	18.3	16.7	5.0	5.0
Fadhari cowpea	8.3	11.7	0.0	0.0
Vuli 2 cowpea	1.7	3.3	0.0	0.0
LSD ($P = 0.05$)	14.8	14.8	3.3	3.3
Mean: Lablab	**41.3**	**29.0**	**6.1**	**3.2**
Mean: Cowpea	**5.0**	**7.5**	**0.0**	**0.0**

[a]Estimates from aerial photographs taken on 5 and 8 October, 2016, at SARI and TPRI, respectively

by the time of late-season soil cover rating. Intercropping reduced lablab ability to survive drought. Late-season soil cover of intercropped lablab relative to sole cropping dropped from 41.3% to 29.0% at SARI ($P = 0.05$) and from 6.1% to 3.2% at TPRI ($P = 0.02$).

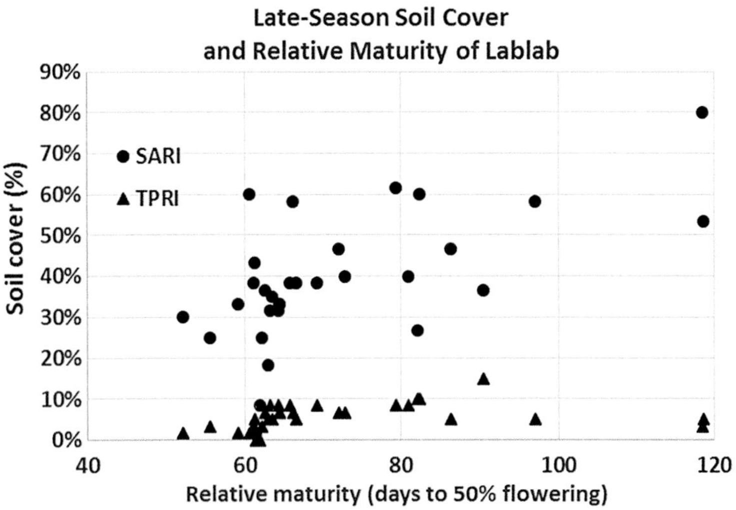

Fig. 4 Late-season soil cover of lablab accessions with different maturities at SARI and TPRI in October, 2016. Data from sole-cropped plots only

Conclusion

Lablab accessions screened by this study demonstrated a high degree of variability in phenotypes, including relative maturity, yield, proneness to insect attack, and drought resilience. This broad diversity is supported by other reports in the literature and suggests that there is great potential for advancement of this species through selection and breeding.

The importance of effective reproductive-stage insect management was illustrated by the severe caterpillar damage experienced at flowering stage at TPRI and the high population of pod-sucking bugs at SARI. While the SARI experiment documented varied cultivar colonization by pod-sucking bugs, this effect was strongly correlated with the relative maturity of the accessions, and thus using this finding as evidence of genetic resistance could be disingenuous. Nevertheless, in addition to the late-maturing accessions, several medium-maturity accessions also rated low for sucking-bug colonization. If these same accessions continue to experience low levels of bug infestation in subsequent years, they may provide sources of resistance for future crossing.

Intercropping also reduced sucking-bug populations. Similar phenomena have been reported for sucking insect pests of other legumes in the literature (Kennedy et al. 1990; Soundararajan and Chitra 2012), though data for lablab intercropping is scarce. Maize-lablab intercropping is common in northern Tanzania, though in drier areas lablab is often sole-cropped. Where rainfall is adequate to support the growth of both crops, intercropping should be encouraged as one means to manage insect pests. While intercropping and plant resistance are integral parts of an integrated pest management system, they will likely *not* provide adequate control of reproductive-stage insects to assure economic lablab production. In this context, there is a need for development of treatment thresholds to guide prudent use of insecticides.

The remarkable drought resilience of lablab was illustrated clearly by these experiments, particularly at the SARI location where many plots remained productive after 6 months with minimal rainfall. Once established, lablab plants were able to survive extreme drought, clearly superior to the drought tolerance of cowpea. In situations of extreme early drought, such as that experienced at TPRI, lablab plants do not have a chance to become established, and their ability to produce is compromised. In these conditions, the drought-avoidant cowpea cultivars out-produced all the lablab accessions tested. Lablab plants out-survived their cowpea controls, but in the absence of any late-season precipitation, they did not have the ability to produce grain.

Superior agronomic characteristics alone will not lead to a resurgence of interest in lablab consumption in Africa. Consumer preference must also be considered, and the project of which this study is part also plans to screen accessions for cooking time and palatability. Preliminary screening suggests that cultivars for consumption within Tanzania, where consumers are attracted to lighter-seeded cultivars, will differ from those demanded by the Kikuyu export market. Fortunately, the broad variability in seed types represented by this set of accessions provides the means with which to address varied consumer preferences.

References

Food and Agriculture Organization (2009) Climate change in Africa: the threat to agriculture. FAO Regional Office for Africa, Accra. 7 pp

Kaoneka SR, Saxena RK, Silim SN, Odeny DA, Rao NVPRG, Shimelis HA, Shiambi M, Varshney RK (2016) Pigeonpea breeding in eastern and southern Africa: challenges and opportunities. Plant Breed 135:148–154

Kennedy FJS, Rajamanickam K, Raveendran TS (1990) Effect of intercropping on insect pests of groundnut and their natural enemies. J Biol Control 4:63–64

KEPHIS (2017) National crop variety list – Kenya. Kenya Plant Health Inspectorate Service, Nairobi http://www.kephis.org/images/VarietyList/april20170525.pdf

Kimani EN, Wachira FN, Kinyua MG (2012) Molecular diversity of Kenyan Lablab Bean (*Lablab purpureus* (L.) Sweet) accessions using amplified fragment length polymorphism markers. Am J Plant Sci 3:313–321

Maass BL, Jamnadass RH, Hanson J, Pengelly BC (2005) Determining sources of diversity in cultivated and wild *Lablab purpureus* related to provenance of germplasm by using amplified fragment length polymorphism. Genet Resour Crop Evol 52(6):683–695

Maass BL, Knox MR, Venkatesha SC, Angessa TT, Ramme S, Pengelly BC (2010) *Lablab purpureus* – a crop lost for Africa? Trop Plant Biol 3:123–135

Robertson CC (1997) Black, white, and red all over: beans, women, and agricultural imperialism in twentieth century Kenya. Agric Hist 71(3):259–299

Sharma HC (1998) Bionomics, host plant resistance and management of the legume pod borer *Maruca vitrata*, a review. Crop Prot 17:378–382

Soundararajan RP, Chitra N (2012) Impact of intercrops on insect pests of blackgram, *Vigna mungo* L. J Entomol 9:208–219

Introduction of a Biodiversity Farming Concept in Amuru District, Uganda

55

Nyeko Pen-Mogi and Martine Nyeko

Contents

Abstract

It is difficult to abandon farming systems that a community has developed for years in favor of a new system. In northern Uganda, the major crops grown in the area are annual food and cash crops. These crops are millet, rice, sorghum, simsim, cotton, tobacco, and many others. These crops are highly susceptible to tree shades. This means for every acre of land opened for these crops, an equal acre of land must be cleared of trees. With the advent of climate change vulnerability, rain-fed agricultural practices, especially in annual seasonal agroecological regions like those in northern Uganda, are no longer sustainable.

N. Pen-Mogi (✉)
Gulu University, Gulu, Uganda
e-mail: penmogi@gmail.com

M. Nyeko
Faculty of Agriculture and Environment, Gulu University, Gulu, Uganda
e-mail: mnyeko@gu.ac.ug

W. Leal Filho (ed.), *Handbook of Climate Change Resilience*,
https://doi.org/10.1007/978-3-319-93336-8_103

It is against this background that there is urgent need to introduce a new biodiversity farming concepts to the communities in the region. This work describes how instructors from regions where perennial food and cash crops such as banana, oil palm, coffee, and fruit trees including the high-value horticultural crops help introduce these crops in northern Uganda. The paper describes the adaptation and adoption of the new environmentally friendly farming systems of growing high-value horticultural crops such as tomatoes, cabbages, bananas, onions, Irish potatoes, and watermelon. The next strategy is to introduce perennial food and cash crops such as banana, coffee, cocoa, palm oil, and other fruit trees under agroforestry farming systems aimed at preserving existing forests and planting and establishing new forests.

Keywords
Adaptation · Adoption · Climate change mitigation · Biodiversity farming system · High-value crops · Perennial crops

Introduction

River Nile plays a central role in the livelihoods of millions of people in eastern, central, and northern African countries. Egypt in particular solely depends on the Nile for all its water needs in agriculture and domestic uses including recreation and transportation. The former Egyptian President Anwar Sadat was quoted to have said "the only matter that could take Egypt to war is water," and in 1988, the former Egyptian Foreign Affairs Minister Boutros Boutros-Ghali who later became the Secretary-General of the United Nations was also quoted to have predicted that "the next war in the Middle East would be fought over the waters of the Nile" (Kameri-Mbote 2007). These two statements clearly show the importance of River Nile in the livelihood of the Egyptians.

River Nile starts from Lake Victoria commonly known as the source of the Nile. The longest part of the Nile in Uganda is in Nwoya, Amuru, and Adjumani districts running from Karuma Falls down to the border with Southern Sudan at Nimule. From Karuma Falls, the Nile flows through Murchison Falls National Park up to Pakwach Bridge where the park ends. The environment in this part of the Nile is a National Game Park. From Pakwach to Zoka Forest in Adjumani, there used to be a game reserve on the eastern bank of the Nile. Over the last 30 years, the game reserve had been de-gazetted and people have started occupying the area. There are still thick forests covering a distance of about 10–20 km especially along the eastern side of the Nile (Fig. 1). With increasing human settlement now, there is likely going to be a very extensive environmental degradation with serious consequences on River Nile.

Agriculture and forest production depends on a number of resources including, among others, land and farm resources, water resources, crop and genetic resources, and knowledge resources, to mention a few. Land and water form the most important agriculture and forest resources. In the Albert Nile segment, declining soil fertility

Fig. 1 There are still some very thick forests in the region. There is great danger if such forests will soon be degraded. Planting such as coffee, banana, and cocoa will protect the forestsperennial crops

due to land degradation and the erratic rainfall pattern linked to climate change remain major challenges to agricultural production and food security. In order to guarantee an effective sustainable agricultural development in the long term, holistic approach that takes into consideration economic, social, and environmental dimensions is required in managing the complex processes that affect land productivity and provide environmental security.

Soil erosion presents a major problem that relates not only to environmental quality but also to food security in the Albert Nile Basin. In addition, the potential for disastrous flooding is great due to widespread deforestation and sedimentation in reservoirs that reduces the storage capacity. The uncontrolled conversion of forests into agricultural land undermines the capacity of the basin to feed the population and improve the livelihoods of the rural subsistence farmers.

Consequently, despite the fact that many people are engaged in production, food insecurity is still a major issue especially among the rural subsistent farmers. FAO in 2006 estimated that the growth rate in food production in Uganda stands at 1.5% against population growth rate of 3.2% (National Census Report 2001; Leliveld et al. 2013). In this regard, transforming the agricultural practices and expanding the production capacity are prerequisites for alleviating rural poverty and household food deficits and reducing exploitation of the environmental capital.

According to the Ministry of Energy (2014), 90% of Ugandans use biomass energy and the country loses about 90,000 ha of forest cover annually through charcoal burning and wood fuel and forest clearing for agriculture (Mafabi 2014). According to MEMA (2008), deforestation in Uganda has seen 5 million hectares of forest in 1990 dwindle down to 3.5 million hectares by 2005 (Figs. 2, 3, and 4). Poor agricultural farming practices have affected a number of rivers and streams that

Fig. 2 Massive destruction of trees for timber, charcoal, and firewood

Fig. 3 Fire is set to burn the stumps of big trees, and they dry out to pave way for growing crops such as maize and simsim

supply River Nile whereby many of them are silting upstream. This study aims at introducing new sustainable farming system in the Albertine Region of River Nile with an overall objective of protecting the mighty waters of the Nile.

Fig. 4 This shows part of the forest that has been cut down to open land for growing crops. In the background, one can still see thick forest waiting to be cut down

Increasing the productivity of the limited available land through water use innovation in rain-fed agriculture, which includes agroforestry, water harvesting, and other conservation farming technologies, requires concerted efforts based on actions that are well coordinated. In 2000, the Uganda Government launched the Plan for Modernization of Agriculture (PMA) program, which is an operational framework for poverty eradication. The PMA (2000), Forest Policy (2001), and the National Forest Plan (2002) all these recognized agroforestry as one of the options for improving farm productivity, and thereby eliminating poverty through increased household incomes. Through these policies, the Uganda Government was committed to providing extension and advisory services that support farmers, communities, organizations, and entrepreneurs in the development of agroforestry.

Nair (2003) argued that agroforestry farming systems provide a better alternative to traditional agricultural systems in land management. Numerous benefits include, among others, reducing runoff and erosion, maintaining soil organic matter and physical properties, increasing nutrient inputs through nitrogen fixation and uptake from deep soil horizon, and promoting more closed nutrient cycling (Young 1997). There are still food production challenges, but the bigger challenges facing humanity are not increasing yields but maintaining productivity in the face of an increasingly hostile climate and a need to stabilize the climate before it deteriorates further.

Protected forests in northern Uganda are under strain as people clear trees for crop production and for making charcoal. In Gulu District, forest reserves have been significantly degraded over the past decade. Reported massive degradation of forest covers in Gulu District. Over the years, the rise in agricultural production was through expanding land use and not by raising yields. The yields of major crops have been stagnant or declining since the early 1990s (Deininger and Okidi 2001).

Through agroforestry practices, there are still some thick forests that can be protected (Fig. 1). This will require sensitizing the communities to adapt and adopt new biodiversity farming systems. There are perennial crops such as coffee, cocoa, palm oil, and various fruit trees that are available for inter-planting with trees. These crops earn good income to the rural farmers. Once the farmers see the economic benefits of such crops integrated with forests, they will start protecting and valuing the forest rather than turning them into charcoal.

Problem Statement

Many permanent rivers and streams in northern Uganda flow into Albert Nile Basin from both sides of the river. The farming practices on both eastern and western sides of the Albert Nile Basin are seasonal annual cropping systems and rain-fed. All the crops grown in this region are invariably intolerant to heavy tree shades. Therefore, the farmers cut down all the trees to eliminate shades. Fire is set on the lower stumps of the big trees, and they eventually dry off (Figs. 3 and 4). It is therefore right to say that for every one acre of land cultivated, there is one acre of trees destroyed (Nyeko 2009).

The farmers are experiencing poor agricultural yields due to climate change, coupled with poor planting materials (seeds), and they have resorted to exploit the environmental capital through cultivating along the rivers, streams, and wetlands. Charcoal burning has become a big industry in rural areas, and the product is sold to residents in cities and towns who cannot afford to use electricity. Due to over-exploitation of the environmental capital, leading to environmental degradation and climate change, many of these rivers and streams are drying upstream. With increasing population and poor agricultural production per environmental unit (Nyeko 2009), efforts to increase food production now rely heavily on opening new land including forests and wetlands.

The need to provide food security to the rapidly growing population in Uganda should not be at the expense of environmental insecurity. In the Albert Nile Basin, extensive agriculture is responsible for deforestation, loss of soil nutrient, and persistent soil erosion that has caused siltation of streams and in some cases complete loss of streams. Coupled with land degradation, the erratic nature of rainfall associated to changes in global temperatures, the impact on food security poses serious threat to the livelihood of many rural subsistent farmers. Effort to reduce this risk in the near future should emphasize improving information base and devising innovative strategies for ensuring successful adaptation. Adaptive management is particularly essential for devising means and strategies to respond to the risk of climate variability and natural resources degradation, which undermine the development of the country.

It is our considered observation and we fear that if this trend is to go on unchecked, it will negatively affect the rainfall patterns in the Albert Nile Basin with serious consequences on food production. It is against this background that we introduce agroforestry farming practices. This practice will guarantee food security and economic well-being of the people living along the Albert Nile Basin. We strongly believe that by ensuring good income and food security through sustainable farming practices, it will be easy to sensitize the people to protect the environment.

Northern Uganda suffered civil strife for over 20 years. During the civil conflict due to Lord's Resistance Army (LRA), thousands of people moved to live in camps. The youths who were under 10 years old and those born in camps grew eating food supplied by World Food Programme (WFP). They never had the opportunity to engage even in the basic cultural or traditional agricultural practices. The displaced peoples' camps were all located in trading centers, and the people generally got used to living in a semi-urban kind of environment. Many of these youths missed the opportunities to attend formal education. That group of youths are now 15 years and above. They spend time in trading centers playing cards and betting European football leagues. We established the Institute of Agricultural Knowledge for Practice (AIKAP) to teach agricultural skills to such youths who lost the opportunity to attend formal education so that they can earn a living on the land and be able to send their children to school.

There is a general statement that education in all its forms has the potential to empower people, by increasing their self-confidence, their capacity to improve their livelihoods, and their participation in wider processes of social and economic change (Franco Flattini 2009). The great majority of the hard-to-reach poor people are in the rural areas. The G8 identified that giving them a wider access to education was one of the crucial issues in promoting development. The concentration of hunger in rural areas suggests that no sustained reduction in hunger is possible without special emphasis on agriculture and rural development (FAO 2006). This project aims to provide such youths in rural communities with better tools, skills, and capacities in agricultural production through informal education.

About 36% of all African labor force is engaged in subsistence agriculture. If that percentage of the population could have access to methods of improving their agricultural techniques, increasing production, and gaining ability to transform agriculture into an income-earning enterprise, African nations would benefit in terms of increased GDP, standard of living, infrastructure development, and political and economic stability. One way to accomplish this is to develop agricultural training systems that are both formal and informal to improve the rural farmers' skills and abilities to create livelihoods out of agriculture rather than simply subsistence (Juma 2011).

Over 80% of Ugandans live in rural areas. They continue to grow food and cash crops in the traditional ways with very low productivity. Because they have no formal training in basic agricultural practices and the extension workers are not able to reach them, they carry out cultivation in the most destructive manner to the environment. The IAKP will address the need for sustainable farming, energy conservation, climate change, biodiversity conservation and mitigation, and regeneration of the degraded environment.

Implementation Strategies

1. To promote biodiversity farming concepts.
2. The training module will be on-farm and practical.

3. We shall identify skilled farmers who are progressive and recruit them as instructors.
4. We shall provide accommodation for such skilled farmers (instructors).
5. We shall provide a variety of planting materials (horticultural and perennial crops, including tree fruits and other multipurpose trees seedlings).

The instructors will use on-farm demonstration to beneficiaries or trainees.

Food insecurity has negative economic impacts, exacerbates poverty, and poses today a problem to hundreds of millions of people in the African continent, especially in rural communities of semiarid regions (UN 2015). There will be competition for energy, land, and water with growing demand for food. Much of the food production will continue to come from the rural smallholder production systems, and this is at the heart of sustainable development agenda (Nicol et al. 2015). According to UNDP (2016) report, the expansion of the economy and the eradication of poverty in particular are critically dependent on agriculture, yet rain-fed agriculture continues to bear the largest burden of food production in the sub-Saharan Africa (Falkenmark and Rockstrom 2004). Climate change affects rainfall variability, water scarcity, soil degradation, and food insecurity, among other things (Nicol et al. 2015).

Adaptation and Adoption Problems

Technology adoption and adaptation is a function of cultural orientation. It is very difficult to introduce new crops to communities that have practiced particular agro-production systems for years. There is urgent need to introduce new farming systems that can provide sustainable food production, reduce poverty, and conserve the environment. With the impending large-scale investments in exotic tree plantations, it is important to carry out a scientific study to evaluate the biodiversity ecosystem potential in natural forest against the artificial forest plantation. This study will attempt to assess the role of local tropical trees in carbon capture potential and climate change mitigation including the introduction of a biodiversity farming systems that take care of food security and poverty reduction to communities.

Adoption of agroforestry practices by the rural farmers as the primary means of production and livelihoods will not advance significantly without quantifying the basic economic information to guide financial decisions. Several authors described the difficulties associated with adaptation and adoption of new farming systems (IPCC 2007). There is a need to develop the cost-benefit information and business case studies for agroforestry production in the Albert Nile Basin. A logical starting point for developing this knowledge base will be the demonstration or pilot projects.

In the 1980s and 1990s when Prof. Wangari Maathai was advocating for tree planting in Kenya, many people in her hometown Muranga did not take her seriously. If you visit Muranga County, now, you would be amazed with the high level of afforestation. According to Mwangi and Mbugua (2017), people in the region started experiencing extreme weather changes with unpredictable rain pattern and that was the time they realized the importance of tree planting. This story proves patience and determination are important in adaptation and adoption of new technologies by communities.

In Uganda, three major constraints exist in adapting agroforestry into the farming systems. The first limitation is inadequate scientific research findings needed to divulgate appropriate agroforestry technologies through demonstration projects and actions. The second limitation is inadequate coordination of the different agroforestry initiatives. The third and most important one is the cultural and social acceptance to adopt and adapt new farming systems by the communities.

Broad Objectives

The broad aims of this study are (a) to introduce agroforestry farming to the communities, (b) to conduct scientific basic research through well-structured analysis of the agroforestry systems, and (c) to assess the photosynthetic potential of exotic tree used in afforestation programs against the local native trees and forests.

Specific Objectives

This project aims at:

(a) Introducing new high-value food and cash crops under agroforestry system (biodiversity farming)
(b) Assessing the carbon uptake potential of local trees against the exotic trees to help reduce any ecological imbalance in the environment

The current study is an innovative idea attempting to adopt and adapt agroforestry to a community that has practiced annual crop production system for years. The study will focus on the following issues:

(a) Assessing the carbon dioxide sequestration potential of a number of local and exotic trees (which will form the basis for policy on tree plantation in the country)
(b) Conserving the watershed in the Albertine Region of River Nile
(c) Introducing a biodiversity farming system to a community for sustainable agricultural production systems
(d) Reducing deforestation, increasing soil fertility, ensuring food security, and reducing poverty

Activity Plans

(a) Introduce a biodiversity farming systems to the rural communities so that they become less dependent on the environmental capital. We shall create social awareness and acceptability of biodiversity farming system for sustainable management of agroforestry resources and realization of food and environmental security. The enhancement of CO_2 uptake due to extensive biomass growth in agroforestry systems is an extended important role of agroforestry in climate change mitigation.
(b) Carry out studies to analyze the carbon uptake potentials of natural forests trees and the exotic forests in the region (we shall get a PhD student to undertake this studies). This study will help to formulate policies as to what extent exotic forests should be encouraged in the region.

The Project Area

The study will involve the communities living along the Albert Nile Basin extending from Lake Albert in Uganda down to the border with Southern Sudan at Nimule. The need to increase food production in the study area to meet the growth in food demand and to provide energy (biofuel) are the most contentious issues affecting land use in the area. The main option for increased food production is the expansion of land. There is a need to exploit the increase in food productivity through innovative agricultural practices such as agroforestry. The massive depletion of the natural forest resources in the long run affects the quality and quantity of water supply with eventual reduction in food production. Oil palm plantation (Fig. 9) is good for protecting the environment in the Albertine Nile, and we plan to encourage its production.

The surveys will focus on the following issues: profile of the current agroforestry production and marketing activities, participation and interest in various agroforestry practices, and motivation for adopting agroforestry practices. Five focus group sessions to be considered are as follows: Ministry of Agriculture (MoA), National Forestry Authority (NFA), local government, Department of Agriculture, and community groups and the NGOs. The discussion shall be geared toward understanding the production interests and issues, markets and marketing, and policy and legislative concerns, prioritizing the agroforestry support topics and preferred support methods (logistics), and assessing the existing agroforestry initiative program parameters and gathering input on the desired delivery options for potential future programs.

Preliminary Results

We know the sociocultural and economic difficulties in accepting perennial cash and food crops by communities that produce annual crops. Perennial crops take years before they start yielding fruits. This fact must be in the minds of everyone who wants to introduce perennial crops. In this project, we introduced the growing of high-value horticultural crops such as tomatoes, cabbages, onions, and watermelon that take only 3 months to mature. The main crops for the northern region are millet, rice, sorghum, sweet potatoes, and cassava. The introduction of Irish potatoes and bananas (Figs. 5 and 6) in the region was readily accepted and has become popular

Fig. 5 Banana plantation, a perennial crop being introduced under the biodiversity farming system

Fig. 6 Irish potatoes not grown in northern Uganda. It is being introduced under the biodiversity farming system

with over 150 households involved. The people named the banana Pen-Mogi (after my name) on the basis that I introduced the crop to the region. The demands for these crops are high with good market prices. One does not need a lot of land to grow these crops, but one gets very high income from these crops. We believe the adaptation and adoption of these high-value horticultural crops will reduce demand for opening more land as commonly required for maize, rice, simsim, and many other cereal crops grown in the region. This will reduce agricultural actions that accelerate climate change.

Over 130 youths (Fig. 7) have attended basic training in growing and managing tomatoes, onions, and watermelon. All those youths are now actively engaged in the production of these high-value crops. One farmer specialized in growing watermelon reported to have earned 18,000,000 Uganda shillings which is about $6,000 in a year on 2 acres of land. Under the present cultural cropping system, one would need 20–30 acres of maize plantation to earn this kind of money, equivalent to 20–30 acres of trees destroyed for crop production (Figs. 3 and 4) (Table 1).

Currently, the market demands for produce such as tomatoes, watermelon, cabbages, and onions in South Sudan are very high. These crops come from southern and western parts of Uganda and are transported through northern Uganda to South Sudan. The people in northern Uganda do not grow these crops. It is our intention to first introduce the production of these high-value early-maturing crops in the region (Table 2). When we establish the growing of horticultural crops, we shall then start introducing perennial crops such as palm oil, coffee, and cocoa. There are thick forests that still exist in the region (Fig. 1). Our plan is to introduce the integration of such forests with coffee, cocoa, and banana. The introduction, adaptation, and

Fig. 7 A group of youth attending a biodiversity on farm skills training. (Courtesy of Mr. Ocen)

Table 1 Comparing crops grown in southern and western Uganda and those in northern Uganda and the acreage required to earn same income

Southern and western crops		East and northern crops		Extra land required	
Crops	Income/acre	Crops	Income/acre	Difference	Extra acres
Coffee	16 (4000)	Cotton	1–5 (400)	14.5 (3400)	10.67
Coffee	16 (4000)	Maize	1–2 (300)	14.8 (3450)	13.3
Tomatoes	6 (1500)	Beans	0.8 (210)	5.2 (1400)	7.5
Irish potatoes	9 (2250)	S. potatoes	0.75 (200)	8.25 (2200)	12
Banana	10 (2500)	Rice	4 (1050)	6 (1500)	2.5
Onions	30 (8000)	Maize	1.2 (300)	28.2 (7700)	25
Onions	30 (8000)	Sunflower	0.8 (210)	29/2 (7710)	37

Table 2 The yield and income of some of the high-value horticultural crops we are introducing in our demonstration gardens (brackets USD)

Demonstration crops	Yield/harvest	Income	Income per acre
Banana (1 acre)	100 bunches every 2 months at 15,000=	1,500,000=	9,000,000 = ($2500)
Tomatoes half acre	1050 kg × 2,500	2,625,000=	5,350,000 = ($10,000)
Irish half acre	8 bags × 150,000	1,200,000=	2,400,000 = ($750)
Onions half acre	750 kg × 2,500	1,875,000=	3,750,000 = ($1000)

adoption of this new farming concept will improve crop production, increase household income, and reduce poverty and demands on the environmental capital.

We have not started the livestock component, and we do expect difficulties introducing livestock since people in this region have been traditional cattle keepers for years. Our emphasis will be to introduce fast-maturing animal breeds that produce more milk in the case of cows and goats. The bio-wastes from these animals will be used for bioenergy and bio-fertilizer production. The other aspects of planting exotic trees to replace the local trees are for future research.

Discussion

The introduction of new farming concepts to communities that have for years practiced a particular farming system is extremely difficult, and this has been described by Franzel (1999) and Place and Dewees (1999). The rural population in Africa will depend on agriculture for many years to come. With increasing population, agricultural land is getting smaller and smaller. There is urgent need to introduce farming systems that will take care of the biodiversity ecosystem components in temporal and spatial scales (Nyeko 2009). The impact of biodiversity farming system on development will be noticeable within 5–10 years as farmers start harvesting bananas, coffee, palm oil, and other fruits to boost their income. There will be reduced demand on the environmental capital, noticeable improvement in soil fertility, and environmental regeneration and conservation of water shades. The impact of this farming system on reducing climate change through carbon capture mechanism and the regeneration of the degraded environment will be overwhelming.

Various economic research documents from the Economic Policy Research Centre based at Makerere University talk of chronic poverty in northern Uganda (Sarah Ssewanyana 2010; Ssewanyana and Kasirye 2013; Ssewanyana and Batageka 2007; Okidi and Mugambe 2002). Despite several programs rolled out by the government to combat poverty in the region over the years such as Poverty Eradication Action Plan (PEAP 2000, 2001), Vision 2025 (1999), and Plan for Modernization of Agriculture (PMA) (2000), they have not helped to reduce poverty in the region. The government has instituted several interventions to reduce poverty in northern Uganda during and after civil conflicts such as Northern Uganda Social Action Fund (NUSAF), Northern Uganda Rehabilitation Programme (NURP), and recently Peace Recovery and Development Plan (PRDP). The World Bank and several agencies funded these programs.

Unfortunately, all these projects and programs have not resulted into any visible reduction in poverty level and improved welfare of the population. All these programs emphasized the encouragement and improvement of the traditional crops that are low value and destructive to the environment. These low-value crops have made the region chronically poor. Table 1 shows a coffee farmer earning 16 M from one acre of land. Table 1 also shows a cotton farmer earning 1.5 M from one acre of land. The cotton farmer must plant 10.7 acres of cotton to earn 16 M. Whereas one acre of coffee is manageable by family labor, 10.7 acres of cotton is far beyond family labor alone. This work is an attempt to introduce new cropping systems that integrate environmentally friendly farming practices, bringing in new high-value crops that require few acres of land with high financial returns to the farmers and conserve the biodiversity.

Poverty has remained elusive in Africa (Kaduru 2011). It is not possible to simply sensitize and appeal to the population to stop cutting trees for timber, firewood, and charcoal production because this is their main source of economic livelihood. They will continue opening more forest land for agricultural production to increase their agricultural yields. There must be alternative income generating activities that are readily accessible to such communities before one can convince the people to stop trading in wood products.

The introduction of perennial crops to communities that have been depending on short-maturing annual crops is almost impossible (Nyeko 2009). The approach adopted in this work is to introduce some high-value but fast-maturing (3–4 months) crops to such communities. The communities find it easy to adapt and adopt the cultivation of such crops. When the new crops are adapted and adopted, the next stage will be the introduction of perennial crops under agroforestry practices. We plan to introduce coffee and palm oil plantation (Figs. 8 and 9) which are environmentally friendly in Albert Nile Basin.

Fig. 8 Coffee plantation in southern region of Uganda. Coffee requires tree shade and earns high income. We shall introduce coffee growing in northern Uganda to reduce poverty and protect the environment

Fig. 9 Palm oil plantation in Kalangala Islands in Lake Victoria. This is the perennial type of plantation we want to introduce along River Nile to protect the environment

We have presented here that some farmers earned as high as 18 million shillings ($6,000) from growing watermelon on just 2 acres of land. None of the traditional crops such as maize, rice, millet, or simsim could earn this kind of money on 2 acres of land. It would require over 10–15 acres for any of the above crops to earn such money. The introduction of these high-value crops has been readily adapted and adopted. The environmental and biodiversity conservation of this new farming system alone has created tremendous impact, as it would have required opening vast acres of land (Table 1) to obtain the same earning from the traditional annual crops in the northern region.

The popularization of these high-value and early-maturing crops will now lead the project to introduce perennial crops such as banana, palm oil, cocoa, and other fruit trees. We believe these measures will ensure drastic reduction in the demands and over-exploitation of the environmental capital. It will be at this stage that one can sensitize the community to cut one tree and plant five being advocated by many political actors.

The other section of this research will involve the introduction of wood-saving cooking stoves. Many communities in the rural areas in African countries use firewood for cooking. The present open cooking places are very wasteful with wood fuel. Many people living in cities and towns use firewood or charcoal as they cannot afford the cost of domestic electrical appliances including the high cost of electricity. The high domestic consumption of firewood, charcoal, and timber is of great concern. They are partly responsible for the rapid forest cover depletion in Africa, and it requires urgent intervention.

Conclusion

This work is still at a very early stage. However, within the short time we have operated, we have witnessed significant progress in mobilizing the community to adopt the new farming system. We have trained over 130 youths with skills to grow high-value horticultural crops. These youths will not need a lot of land to grow these crops as they would if they continue growing maize or cotton. We believe that the traditional crops such as rice, maize, millet, etc., will mainly be for domestic consumption. It is clear from this work that the long-term strategy of introducing a biodiversity farming system in northern Uganda is achievable. We are looking for funds to set up a few agricultural skills training programs to roll out this idea to a wider youth groups in the region in order to create impact.

References

Deininger K, Okidi J (2001) Growth and poverty reduction in Uganda, 1992–2000. Panel data evidence. World Bank, Economic Policy Research Centre, Washington, DC

Falkenmark M, Rockstrom I (2004) Balancing water for man and nature: the new approach to eco-hydrology. EarthScan, London

FAO (2006) Global food security, biodiversity conservation and the future of agricultural intensification. FAO report. FAO, Rome
Flattini F (2009) In: Acker D, Gasperini L (eds) Education for rural people. FAO, Rome
Forestry Policy (2001) Uganda forestry policy. Ministry of Water, Lands and Environment
Franzel AR (1999) Socioeconomic factors affecting the adoption of potential of improved tree fallows in Africa. Agrofor Syst 47(1–3):305–320
IPCC (2007) Climate change 2007: adaptation and vulnerability, contribution of Working Group II to the 4th Assessment Report of the IPCC. Cambridge University Press, Cambridge
Juma C (2011) The new harvest, agricultural innovation in Africa. Oxford University Press, New York
Kaduru R (2011) Is anyone listening? Strategies for eradicating poverty in LDCs. The Fourth UN Conference on the Least Developed Countries, 2011: Civil Society Forum session
Kameri-Mbote P (2007) Navigating peace: water, conflict and cooperation: lessons from the Nile Basin. www.nilewater.org/water
Leliveld A, Diez T, Foeken D, Klaver W (2013) Agriculture dynamics and food security tends in Uganda: developmental regimes in Africa (DRA) Project. ASC-AFCA collaborative research group: Agro-food clusters in Africa (AFCA). Research report 2013-ASC-2. Agro-Food Clusters in Africa (AFCA) report, London/Leiden, December 2013
Mafabi P (2014) Daily Monitor News Paper, March 19th 2014
Ministry of Energy (2014) Tree resources: looking at the future of biomass sector. Daily Monitor Publications, Sunday 14th December 2014
Mwangi HM, Mbugua SK (2017) Personal communication in Muranga, Kenya
Nair PKR (2003) Carbon sequestration potential of agroforestry systems. Opportunities and challenges. Advances in agroforestry no. 8. Springer, Dordrecht
National Census Report (2001) Organic agriculture, environment and food security. In: Nadia ES, Hattam C (eds) Environment and natural resources services sustainable development department. Uganda Printing and Publishing Corporation, Entebbe
National Forestry Plan (2002) Republic of Uganda, National Forestry Plan: Ministry of Water, Lands and Environment, October 2002
NEMA (2008) State of the environment report. National Environment Management Authority, Ministry of Water and Environment, Kampala
Nicol A, Langan S, Victor M, Gonsalves J (2015) Water –smart agriculture in East Africa, Colombo, Sri Lanka: International Water Management Institute (IWMI). CGIAR research program on water, land and ecosystems (WLE). Global Water Initiative East Africa (GWI), Kampala, 352p
Nyeko JHP (2009) Environmental mitigation and regeneration through sustainable farming and food security. Sci Res Assay 4(8):773–779
Okidi J, Mugambe GK (2002) An overview of chronic poverty and development policy in Uganda. EPRC working paper 11. Chronic Poverty Research Centre, Manchester
Place F, Dewees P (1999) Policies and incentives for the adoption of improved fallows. Agrofor Syst 47(1–3):323–343
Plan for Modernization of Agriculture (2000) Eradicating poverty in Uganda. Ministry of Agriculture and Ministry of Finance, Planning and Economic Development, Kampala
Poverty Eradication Action Plan (2000) Poverty Eradication Action Plan is a comprehensive development framework for economic growth in Uganda, Ministry of Agriculture, Animal Industry and Fisheries
Poverty Eradication Programme (2001) Poverty Eradication Action Plan is a comprehensive development framework for economic growth in Uganda, Miistry of Agriculture, Animal Industry and Fisheries
Ssewanyana S (2010) Combating chronic poverty in Uganda: towards a new strategy: EPRC. Research series 67 MIO2-10. Economic Policy Research Centre, Kampala
Ssewanyana S, Bategeka L (2007) Chronic poverty and economic growth in Uganda: the role of markets: EPRC. Makerere University

Ssewanyana S, Kasirye I (2013) The dynamics of poverty income in Uganda, Economic Policy Research Centre, occasional paper No 35, Makerere University
UN (2015) Millennium development goals report 2015. United Nations, New York
UNDP (2016) United Nations Development Programme; Uganda Annual Report 2016
Vision 2025 (1999) Strategic framework for national development by Uganda. National Long Term Perspective Studies Project (Uganda), Kampala
Young A (1997) Agroforestry for soil management. International Centre for Research in Agroforestry, Wallingford

Scaling Up Climate-Smart Agricultural (CSA) Solutions for Smallholder Cereals and Livestock Farmers in Zambia

56

Idowu Kolawole Odubote and Oluyede Clifford Ajayi

Contents

I. K. Odubote (✉)
School of Agricultural Sciences, Zambian Open University, Lusaka, Zambia
e-mail: kola.odubote@zaou.ac.zm

O. C. Ajayi
Technical Centre for Agriculture and Rural Co-operation, Wageningen, The Netherlands
e-mail: Ajayi@cta.int

W. Leal Filho (ed.), *Handbook of Climate Change Resilience*,
https://doi.org/10.1007/978-3-319-93336-8_109

Abstract

The project to scale up climate-smart agriculture (CSA) in Zambia is an example of a partnership between public, private, and research institutions to address impact of climate change. It aims to contribute to improve smallholder farmers' resilience to climatic shocks through a number of innovative multilayer CSA interventions. These include increased access and use of stress tolerant maize seeds by linking farmers and agro-dealers to stress tolerant seed suppliers, increased access and use of ICT-enabled weather information services by farmers, increased and sustained private sector investment in weather index insurance (WII) and farmer's uptake, and promotion of integrated crop-livestock farming systems. This chapter presents some successes that have been recorded to scale up the interventions through the project: a total of 331 camp extension officers and 132 agro-dealers were trained; 3591 smallholder farmers attended seed fairs held; and 25,632 farmers have been reached with ICT-enabled weather and agronomic practices information services in the 12 districts of Zambia. It also presents challenges, opportunities, and synergies of the partnership in CSA solutions offerings. It emphasizes that private sector investments in CSA and smallholder farmer uptake are hinged on market facilitation approach. Preliminary results show that farmers are willing to embrace CSA solutions, and this is useful in designing CSA solution platforms.

Keywords

Climate-smart agriculture · Public-private partnership · Extension services · Weather indexed insurance · Agro-dealers · Seed fairs · Climate solutions

Introduction

In Zambia, poverty and vulnerability are largely a rural phenomenon and are associated primarily with rain-fed smallholder agricultural system. The smallholder rural farmer is constrained in a number of ways which hampers productivity and profitability, thus subjecting the household to a vicious poverty cycle. The Second National Agriculture Policy (Ministry of Agriculture and Ministry of Fisheries and Livestock 2016) noted that the agricultural sector generates approximately 10% of the gross domestic product (GDP) and provides livelihoods for more than 70% of Zambia population. It is, therefore, evident that the sector contributes very marginally to substantive poverty reduction, especially in rural areas.

The above scenario is likely to get worse with the rising level of climate change which has resulted in flooding in some areas and drought in other areas leading to severe food shortages as crops are either washed away or stunted and dried off. Animals are also known to have died due to lack of drinking water and grasses/forages. Extremes of temperatures have also been observed with obvious negative effects on plant and animal physiology, growth, and production. Human life is also likely to be affected due to reduced food production and threatened food security and food nutrition. Climate change-related losses in agriculture are expected to amount to US$2.2–3.13 billion over the next 10–20 years in Zambia (CIAT and World Bank 2017).

This means that the poor and vulnerable smallholder farmers, especially those in rural, low rainfall, and drought-prone areas, are subjected to perennial economic and social stress which can erode their adaptive abilities. Droughts and floods can erode incomes, severely threaten food security, and weaken the very foundation upon which smallholder farmers build assets and capabilities to reduce risk and increase resilience to climate change (Ministry of National Development Planning 2016).

It is important to note that technologies and practices have been developed over time and are available to reduce or eliminate the negative impact of climate change. These technologies and practices which will be described later on are classified as adaptation, mitigation, or resilience measures. Adaptation refers to actions aimed at reducing the vulnerability of the people by managing the known and unknown impacts of climate change. Mitigation on the other hand refers to efforts that seek to prevent or slow down the increase of atmospheric greenhouse gases (GHG) concentrations by limiting current and future emissions and enhancing potential sinks for greenhouse gases. Resilience, however, is the ability of a system and its component parts to anticipate, absorb, accommodate, or recover from the events of the hazardous event in a timely and efficient manner.

Notwithstanding the availability of these technologies and practices which comprised climate-smart agriculture (CSA), they are not within the reach of the vulnerable smallholder farmers who needed them most and who are the worst hit by the negative effects of climate change. Some of the limiting factors for the smallholder rural farmers include:

Finance and insurance: Commercial banks and insurance service providers perceive poor and vulnerable farmers, especially women and youth as high-risk borrowers. Smallholder farmers are unable to afford the type of premiums that can make private sector investment in property and life insurance products commercially viable and sustainable.

Extension services: High illiteracy rates in rural areas amidst an overstretched agricultural extension and support system magnify smallholder farmer vulnerability to climate change. Evidence based on the study by UNDP (2015) shows that very low extension officer to farmer ratios exists in Zambia and this further constrains the ability of smallholder farmers to respond to and adequately invest in CSA adaptation and mitigation strategies. For example, currently, the agriculture extension officer to cereal farmer ratio is 1:2500 farmers as compared to the planned 1:200. On the other hand, the livestock and fisheries sub-sector has an almost nonexistent farmer extensions system to support good husbandry practices, integrated crop-livestock practices, and diversified CSA systems and livelihoods.

Early warning system: At the core, smallholder farmers lack access to an early warning agricultural information system, markets, transport and communication infrastructure to help them make timely informed decisions about the types of crops and seeds to cultivate and purchase, management requirements, and expected yields based on observed and forecasted rainfall patterns. Specifically, farmers do not have recourse to any stress tolerant maize seeds nor diversified means of livelihood in times of crop failure due to drought and/or flood. This is worsened by the smallness or absence of private investment in the sub-sector by formal financial providers, including insurance companies to mitigate against economic and social risk due to poor rainfall, drought, or floods. Consequently, during times of economic and social stress, smallholder farming households often look up to the Disaster Management and Mitigation Unit (DMMU), the Ministry of Community Development and Social Welfare (MCDSW), and other civil society organizations (CSO)-supported social protection interventions as a way of coping.

The above is a typical pathway to chronic smallholder poverty, hunger, and food insecurity due to increasingly erratic weather and climate shifts. Lack of capacity of farmers to invest in CSA practices at the farm level, climate, and weather-induced shocks like droughts and perennial floods often lead to prolonged episodes of food insecurity, indebtedness, and loss of assets and reinforce adverse social cultural practices with severe gender equity implications. For example, when hit by economic and social stresses, rural agricultural households tend to explore a mix of coping strategies. These coping strategies might involve withdrawing children, especially the girl child from school to support income-earning activities, and girls may forcefully or willfully enter into early or teenage marriages at the expense of their education for economic benefits of self or the parents (World Vision 2015; UNICEF 2015).

Breaking this cycle of poverty, food insecurity and hunger will require concerted and coordinated efforts on many fronts. Primarily, it calls for increased and integrated investments in scaling up piloted innovations in support of CSA values and practices at farm level. This is likely to contribute to reducing the long-term risk in responsive, manageable, and commercially viable ways aimed at enhancing the resilience of these communities to the effects of global warming.

Grasso (2004) noted that climate change is the exemplary global public good, because each country's emissions of greenhouse gases contribute cumulatively to the increase of the overall concentration, and each country's abatements entail higher cost than benefit, unless effective concerted collective actions take place. Conversely, private sector (profit making) decision-makers will both bear the cost and enjoy the benefits of their own decisions. The bottom line is that it has long been recognized that most of these CSA technologies and practices are best advanced by either the government as a public good or the private sector for profit, if a business case can be made.

Climate-Smart Agriculture Practices (CSA)

Climate-smart agriculture has been described as agriculture that sustainably increases productivity and resilience (adaptation), reduces/removes GHGs

Table 1 Climate-smart practices useful in smallholder agricultural production

Land and water management	Crop management	Livestock management	Agroforestry	Integrated food energy systems
Participatory land use planning and management Conservation agriculture (e.g., minimum tillage) Contour planting Terraces and bunds Planting pits Water storage (e.g., water pans) Alternate wetting and drying (rice) Dams, pits, ridges Improved irrigation (e.g., drip)	Intercropping with legumes Crop rotations New crop varieties (e.g., drought resistant) Improved storage and processing techniques Greater crop diversity	Rotational grazing Fodder crops Grassland restoration and conservation Manure treatment Improved livestock health Animal husbandry improvements Improved feeding strategies (e.g., cut and carry in small ruminants and dairy animals)	Boundary trees and hedgerows Nitrogen fixing trees on farms Multipurpose trees Improved fallow with fertilizer shrubs Woodlots Fruit orchards	Biogas Production of energy plants Improved stoves

Source: Ministry of Agriculture and Livestock (2017)

(mitigation), and enhances achievements of national food security and development goals (Ministry of Agriculture and Livestock 2017). FAO (2012) observed that CSA is an approach that helps to guide actions needed to transform and reorient agricultural systems to effectively support development and ensure food security in a changing climate. It also noted that CSA is not a set of practices that can be universally applied but rather an approach that involves different elements embedded on farm and beyond the farm and incorporates technologies, policies, institutions, and investment.

Table 1 presents a suite of CSA practices that encompass land and water management, crop management, livestock management, agroforestry, and integrated food energy systems. It is impossible for smallholder farmers to engage in all the abovementioned practices. However, through careful selection, the project decided to focus on promoting the three pillars of conservation agriculture (minimum tillage or soil disturbance, permanent soil cover and diverse crop rotation); greater crop diversity (new crop varieties and drought-resistant seeds) and animal husbandry improvements and including improved feeding strategies.

Business Case for Climate-Smart Agriculture Project

As the Food and Agriculture Organization (FAO 2012) observed, innovations that enable farmers to make more climate-smart choices must be supported. Innovative, promising, and scalable interventions are required to reduce smallholder farmer's risk of exposure to climate change, rural poverty, food insecurity, hunger, and malnutrition and to increase their resilience. Investing in agricultural innovations and technologies, targeting smallholder producers and especially women, is clearly a key priority for achieving sustainable agricultural growth and poverty reduction among these key groups. Technologies to increase resilience, such as drought- and heat-resistant crop varieties, are clearly important in the context of climate change adaptation and risk management.

The principal strategy for the new project was to adapt and scale up technologies which sought to catalyze long-term behavioral change among players in the agricultural sector toward more sustainable CSA-oriented practices. Among these is the pilot Vuna project which sought to reduce the risk and vulnerability of smallholder livestock and cereal farmers and increase their resilience by promoting diversified, adaptive, and climate-resilient production systems (Vuna 2017). Alongside, it sought to drive adoption of weather index insurance (WII) by smallholder farmers in order to mitigate risks of climatic extremes to a rain-fed production system.

Evidence based on the evaluation, learning, and change functions of the pilot Vuna project demonstrate that the intervention exposed smallholder farmers to early warning information around CSA and facilitated a suite of inputs and technical knowledge in support of CSA practices at farm level. It also led into increased access to and use of CSA information at the farm level and, therefore, laid a strategic entry point and foundation to building long-term resilience of the target groups to climate variability and food insecurity. The WII did open up private sector participation in the smallholder farming sector and holds the promise for increasing access to formal finance and financial inclusion of the smallholder farmers.

This new project is being promoted in order to contribute toward finding a lasting solution to the problem of food insecurity and hunger especially in the drought-prone agricultural districts of Central, Eastern, and Southern provinces of Zambia. Overall, it seeks to promote the adoption of effective CSA strategies that will reduce present and future risks associated with climate change faced smallholder in order to build their resilience. This project will also contribute to addressing the complications faced in promoting synergies and connecting the dots across mitigation, adaptation, and food security measures in order to minimize the potential negative trade-offs of climate change.

The new project intends to increase the willingness and ability of the agricultural inputs sector and particularly the rural agro-dealer networks, to influence the behavior of smallholder farmers toward more diversified and climate-resilient production systems by offering a suite of CSA-relevant inputs, technologies, and information. It leverages and builds on the Ministry of Agriculture's Farmers Input Support Programme (FISP) e-voucher initiative that dramatically increases the purchasing power of the smallholder community and allows for a wide spectrum of choice of agricultural inputs.

The 2-year project is initiated and supported by the ACP-EU Technical Centre for Agricultural and Rural Cooperation (CTA) implemented in the field by a consortium of national partners including Zambian Open University (ZAOU), Musika Development Initiatives (MUSIKA), and Professional Insurance Corporation Zambia Limited (PICZ) with the endorsement of the Ministry of Agriculture (MOA) in a public-private research institution partnership. The partnership aims to scale up the piloted and innovative Vuna Programme on CSA, which was implemented by Musika during the 2016/2017 farming season.

This project as part of its approach seeks to integrate CSA messaging and awareness into the information dissemination process around the FISP e-voucher initiative in the targeted districts but not restricted to farmers participating in the initiative. Complimentary to the awareness campaigns, the project will make available appropriate CSA-relevant information to participating farmers and the wider farming community through training of authorized agro-dealers and their networks.

Alongside, the project plans to scale up private sector-driven models of promoting affordable access and use of weather index insurance (WII) by smallholder farmers and integrated crop-livestock farming systems in order to mitigate risks of climatic extremes to a rain-fed production system. It will invest in the promotion of integrated CSA agronomic and information systems by riding on the FISP e-voucher initiative in the targeted districts but will further open up participation to farmers who are not targets of this initiative.

In order to embed risk adaptation and mitigation measures within the smallholder agriculture farming systems, the project will promote the multiplication and use of stress tolerant seeds including their management requirements amidst climate change variability. The strategy is to adopt and ride on the existing agro-dealers networks and farmer outreach infrastructure to make inputs available at last mile agro-dealer distribution points. Agribusiness multinationals will also be engaged to support aggregation of agriculture produce and products as a way of increasing market access of the target farmers.

The scaling-up climate-smart agricultural solutions for smallholder cereals and livestock farmers in Zambia, therefore, intend to tap into these opportunities to contribute to transforming the Zambian agricultural input and output market systems to effectively support socioeconomic development and ensure food security in the advent of climate change. The intervention has four component activities, namely:

(i) To improve smallholder farmer access to and use of stress tolerant maize seeds
(ii) To increase smallholder farmer access to and use of weather and agronomic information
(iii) To increase farmer uptake of weather index insurance through a private sector investment
(iv) To promote diversified livelihood opportunities for farmers through an integrated crop-livestock farming systems

The Public-Private Research Institution Partnership

The choice of partners and their participation were based on several factors which include interest, core business, competence, and willingness/determination to contribute to the success of the partnership. Table 2 gives a summary of the participating partners, their roles, and the rationale.

Project Theory of Change

This theory of change is based on a firm conviction that once there is a compelling business case, private sector insurance companies, seed suppliers, agribusiness aggregators, and authorized agro-dealers will be more willing to leverage donor-funded initiatives to support the development and efficiency of the smallholder input and output markets. If the CSA solutions project can demonstrate that smallholder input and output markets are lucrative and viable to private sector, then private sector will be more willing to risk their investment. On the other hand, smallholder farmers will be more willing and able to increase the uptake of stress tolerant seed varieties, invest in WII, and practice integrated crop-livestock system to enhance their resilience against climate change. This will in turn guarantee future investments by private sector and smallholder farmers in WII and CSA practices.

Project Intervention Rationale

FAO (2012) outlines six commitments that governments need to implement if they are to contribute to a world without hunger and with sustainable development. Among these, governments are called to establish and protect the rights to resources, especially for the most vulnerable; to promote fair and well-functioning agricultural and food markets; and to reduce risk and increase resilience of the most vulnerable.

The livestock sub-sector, which has great potential to contribute to employment and wealth creation, only contributes about 0.58% of the total agriculture employment. Meanwhile, the livestock is very important given that it contributes to food supply, food and nutritional security, improvement of human health, income generation, means of finance, employment, source of manure as organic fertilizer, manure as source of energy, transportation of agricultural inputs, and produce. It is also a store of assets. Livestock thus underpins the smallholder agricultural production in the rural areas. Therefore, strategies to improve production and productivity and its integration into farming systems will be beneficial to the smallholder farmer and lead to higher and regular incomes throughout the year.

Due to smallholder farmers' dependence on rain-fed agriculture, erratic weather, and climate patterns, smallholder farming households, especially women and youth, can no longer guarantee their livelihoods. The most feasible and more sustainable

Table 2 Roles and rationale for participating partners

Partners	Roles and participation	Rationale
Ministry of Agriculture (HQ) and field staff	Endorse the intervention as a key value- adding component of its existing agriculture programs, especially FISP and e-vouchers Support farmer agro-dealer mobilization and provide locally imbedded extension services Support dissemination of project CSA information through National Agricultural Information Services (NAIS) Collect weather data from the meteorological department	The Ministry of Agriculture is the custodian of agriculture policy, policy strategy and overall implementation plan. It has the outreach infrastructure, which the project will leverage, for mobilizing lead farmers and other smallholder farmers to provide agriculture extension services The Ministry of Agriculture has an existing relationship with the meteorological department through NAIS
Professional Insurance Corporation of Zambia	Invest in weather index insurance Provide weather index insurance cover for the target smallholder farmers Collect project M&E data on the level of uptake of WII Support market days and farmer awareness campaigns Verify and pay out WII claims when triggered Will host the ICT platform that will be used to send CSA information to farmers	PICZ is the biggest general insurance company in Zambia with an estimated market share of 21% with a very large agricultural portfolio. It has an annual gross written premium of €35.3 million. It is also the best-rated insurance company with an A+ rating. This investment is supported by SWISS Re, which is a renowned global reinsurer. The company is further committed to promoting financial inclusiveness of smallholder agriculture Being the insurer, it is in their interest that insured farmers exhibit good agricultural practices as it reduces disputes at claims stage. This also ensures sustainability of the intervention
Musika Development Initiatives	Musika will take a lead role in facilitating market linkages across the key project stakeholders. Specifically, they will undertake the following: Support training of agricultural extension officers Reach out to seed companies and agro-dealers to ensure availability of stress tolerant maize seed Reach out to livestock and crop aggregators to create market for farmers Support farmer CSA training Support the roll-out of WII	Musika's core mandate is to stimulate and support private sector investment in smallholder rural input and output markets. It has the requisite outreach infrastructure and networks, which are national in scope on which this project will leverage

(*continued*)

Table 2 (continued)

Partners	Roles and participation	Rationale
Zambian Open University (ZAOU)	Project manager, in charge of overall planning, implementation, coordination, client liaison, M&E, and grant management and reporting Conduct training in animal husbandry and nutrition Conduct research in dry season feeding strategies	ZAOU through its School of Agriculture has been involved in transformative community program initiatives to scale up CSA training, to undertake research, gather and document evidence, and promote innovative market facilitation model for rural agriculture input and output markets. ZAOU is also the lead applicant and agency of CSA project

way of protecting the rights to resources of the most vulnerable smallholder farming households is to ensure that their exposure to risk due to climate change is minimized through the promotion of CSA practices.

On the demand side, two primary measures are critical. The first is to ensure that farmers have access and are willing to use stress tolerant maize seeds that can withstand adverse weather and climate patterns and support adequate yields. The second is to invest in farmer early warning CSA information systems that combine agronomic training on the selection and management of stress tolerant seeds, using more innovative and ICT-based infrastructure. On the other hand, supply-side interventions are vital. For example, smallholder farmers are among the poorly integrated into the financial sector, and smallholders severely lack access to formal credit and insurance products. Insufficient rainfall, prolonged drought, and floods can, therefore, spell doom, which is manifest through food insecurity, hunger, rising indebtedness, and erosion of income-earning opportunities and asset base.

Increasing insurance penetration using private sector-driven models that respond to the service needs of the smallholder sector is, therefore, vital. In this respect, protecting the rights to resources of the vulnerable requires private sector investments aimed at expanding access and use of weather index insurance products by the poor and vulnerable farmers in order to reduce their exposure to risk and increase their resilience to weather- and climate-based economic and social shocks. At the same time, smallholder farmers select crops and livestock based on availability and ease of access to markets. Studies have shown that many smallholder farmers are confined to the production of maize not because it is necessarily a commercially viable crop but because it has a ready market. Supporting horizontal diversification in the smallholder farming sector will therefore require a deliberate strategy that links them to markets of both crops and livestock through aggregation.

The project is aligned to the Zambian government's developmental goals as outlined in the Second National Agricultural Policy. Similarly, through its research, monitoring, and evaluation activities, the project will build an evidence base for the adaptation and mitigation benefits that can be generated through the adoption of sustainable CSA practices. It will also establish low transaction cost mechanisms to channel CSA information and insurance to small-scale agriculture producers.

Methodology

Project Target Groups

The project targets smallholder farmers in Central, Eastern, and Southern provinces of Zambia. Specifically, 12 districts have been targeted for this intervention, and these include Mazabuka, Monze, Choma, and Kalomo in Southern Province; Mumbwa, Chibombo, Kapiri Mposhi, and Serenje in Central province; and Nyimba, Petauke, Chipata, and Lundazi in Eastern Province. Primarily these districts have been selected because of their comparatively higher levels of poverty beyond the national average, their low rainfall patterns and increased exposure to climate change, and opportunity to support both crop and livestock production. Figure 1 below provides a visual justification of the geographical targeting.

The project leverages on already existing farmer producer cooperatives, farmer associations, women clubs, or any other recognized farmer groups. The project will benefit 80,000 smallholder farmers and will live up to the objective of promoting gender mainstreaming, by having at least 40% of the total beneficiaries being women.

The choice of the three provinces is hinged on the fact that they fall in the agroecological regions I and II a, which receive on the average less than 800 mm and between 800 and 1000 mm of rainfall, respectively. These regions are suitable for crops and livestock production and can easily support integrated crop-livestock farming systems as the people are mostly dependent on agricultural activities. However, the regions are also generally prone to poor rainfall patterns. Thus, smallholder farmers in these provinces are very vulnerable to the impact of climate change.

Some of the selected districts do share boundaries with Zimbabwe and Malawi, which may give room for possible regional evaluation. It must be borne in mind that the project is the CTA Southern Africa regional flagship covering the aforementioned countries.

Project Activities and Key Outputs

Table 3 above provides a detailed breakdown of the planned activities and key outputs.

Intervention Activities

The project is an innovative public-private research institution partnership forged with government, donor projects, seed suppliers, agro-dealers, commodity aggregators, lead farmers, and private insurance company. A field scoping exercise prior to the implementation revealed the need for a consortium to drive the process, and the following members were selected: Musika Development Initiatives a market

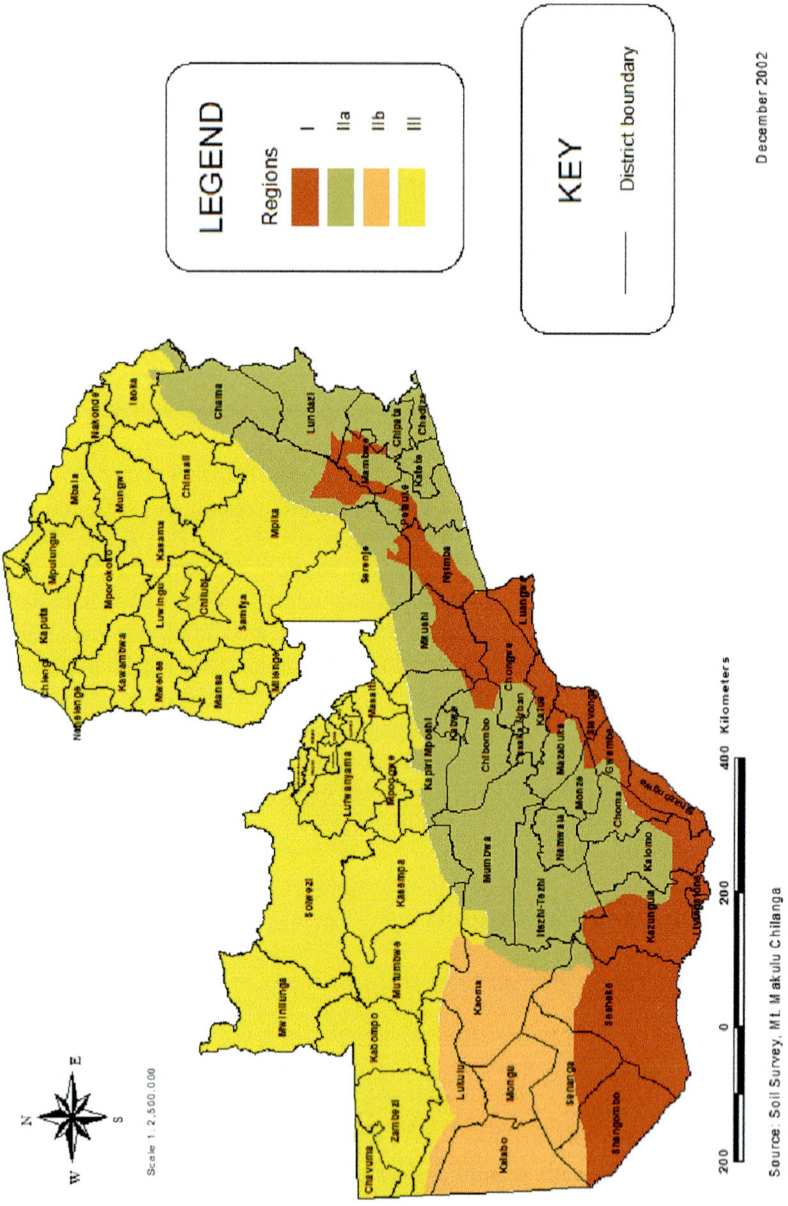

Fig. 1 Agroecological zones of Zambia

Table 3 Project components, activities, and key outputs

Project component	Activities	Key outputs
Increased uptake/investment in weather index insurance (WII)	Develop appropriate weather index insurance (WII) products Conduct farmer sensitization and market linkage activities on WII Train and support agriculture and livestock extension officers, lead farmers, and agro-dealers on WII Conduct digital registration of farmers for WII	New WII products developed Farmer sensitization and needs assessment conducted Farmer training on WII undertaken Digital registration and database of farmers for the project carried out Uptake and use of WII by farmers documented and shared Activities relating to development of WII products, sensitization, and training of project stakeholders documented and shared
Improved access to ICT-based CSA weather and agronomic information by farmers	Conduct farmer sensitization on ICT-based climate information services Disseminate ICT-based CSA information Conduct digital registration and develop a database of smallholder farmers for mobile-enabled CSA information services Train and support farmers to access and use CSA information	Farmer sensitization and market campaign on ICT-based climate information services conducted Partnerships with ICT service providers to support dissemination of CSA information established Digital database of farmers for CSA project created Training of farmers on the use of ICT- based CSA information conducted Activities related to farmer sensitization, project partnership building with ICT service providers, creation of farmer digital database, and use of CSA information documented and shared
Increased access to and adoption of CSA-relevant inputs and practices	Link farmers and agro-dealers to stress tolerant seed suppliers Train public and private partners including agro-dealers on basic knowledge on relevant stress tolerant seed Conduct farmer awareness of CSA-related inputs, technologies, and information	Stress tolerant seeds (legume and cereal) are made available to smallholder farmers through local agro-dealer outlets Public, private partners, and local agro-dealers trained on stress tolerant seeds Smallholder farmers access CSA-relevant inputs Activities related to training, awareness, distribution, and use of stress tolerance seed documented and shared

(*continued*)

Table 3 (continued)

Project component	Activities	Key outputs
Enhanced integrated crop-livestock farming systems	Train lead farmers on livestock husbandry, nutrition, and relevance of integrated crop-livestock farming Conduct smallholder farmer awareness on the relevance of integrated crop-livestock farming practices Create market linkages and provide market information for crops and livestock farmers	Lead farmers trained on livestock husbandry, nutrition and relevance of integrated crop-livestock farming Target smallholder farmer are aware of the relevance of integrated crop-livestock farming practices Activities relating to sensitization and training of project stakeholders documented and shared

linkage nongovernmental organization (NGO); Professional Insurance Corporation Limited, an insurance company; Ministry of Agriculture for extension services delivery; and the Zambian Open University to coordinate research and knowledge documentation.

Deliveries will be through trainings of extension officers and agro-dealers, holding seed fairs and meetings with other stakeholders. The project activities will be implemented by each of the project partners under the overall coordination of the Zambian Open University. The activities were assigned based on the core business, project mandate, and expertise of each consortium member.

Intervention Methodology Overview

The implementation of this intervention will take a "market facilitation" approach in which focus is to empower market actors to make commercially viable and sustainable investment decisions. This is a guaranteed way to building their adaptive capacity and resilience to climate change and its impacts. The project will forge partnerships with government, donor projects, seed suppliers, rural agro-dealer networks, crop and livestock extension system, commodity aggregators, lead farmers, and private insurance company. The strategy of this partnership approach is to catalyze private investment and bolster producers' ability to actively integrate into the input and output market as an adaptive and coping strategy to climate change. The implementation philosophy hinges on the basic sustainability principle that places the incentives for change, and the abilities to make change, into the "market" as opposed to retaining them as the preserve of the intervention itself.

The project will embark on trainings for all camp extension officers (CEOs) in the 12 selected districts. Training will include having an overview of the project, climate change concept, best agronomic practices, crop-livestock integrated farming system, and use of Android phones for digital farmers registration using Open Data Kit (ODK) software.

The project will create an ICT-based information platform, which will be an output of the digital farmer registration process. This platform will consist of weather and climate information, agronomic information linked to weather and rainfall patterns, agro-dealer outlets for stress tolerant seeds, crop and livestock market information, and good animal husbandry practices and feeding especially during the long dry season.

Firstly, to support increased productivity and adaptive capacity, the project will link primary seed producers to the smallholder farmer market through last mile agro-dealer outlets. The project will invest in training agro-dealers in CSA information in order for them to become a locally imbedded extension support system to farmers. This is a strategy for enhancing customer loyalty to the agro-dealer network, which in turn will increase agro-dealer sales and their profitability and enhance availability of seed and efficiency of the input market. Creating an informal outreach infrastructure for locally imbedded crop and livestock extension services delivery is an additional strategy for addressing the challenge of low extension staff to farmer ratios and glaring skills and knowledge gaps among crop and livestock farmers that severely constrain their productivity.

Secondly, the project will promote the development of stress tolerant seed market by deliberately establishing partnerships with primary seed suppliers to provide information on types of seed varieties available, their management requirements, level of yields, and associated risks that have to be proactively managed. To ensure that the smallholder market has the relevant information, about the input and output market, the project will conduct joint market days and seed fairs with primary seed producers or suppliers, agro-dealers, insurance companies, commodity aggregators, and livestock marketing and processing companies. This activity will support the creation of demand but also will directly link production to existing output market.

Thirdly, the project will work with private aggregators in the agriculture value chain in soya, cotton, groundnuts, maize, and livestock to support output bulking to reduce transaction costs and guarantee a captive output market for farmers. To increase their adaptive capacity and incomes, their produce must reach lucrative and reliable markets.

Fourthly, to enhance the adaptive capacity of the farmers to climate extremes and build their medium to long-term resilience, the project will deliberately support expansion of integrated crop-livestock livelihoods practices. This will build on locally existing market days for crops and livestock and will establish holding pens for small and large livestock and where applicable use the existing livestock aggregation centers in order to provide a locally available market for livestock. To ensure sustainably of the initiative, smallholder livestock producers will be trained in good husbandry practices and nutrition especially during dry periods and linked to the market aggregators and processors. Zambian Open University will provide training to lead farmers and extension officers on livestock nutrition and relevance of integrated crop-livestock farming.

Lastly, an equally important marketing and adaptive strategy is that the project will support market penetration and market development of private micro-insurance and financial markets in the smallholder crop and livestock sector, generally seen as largely traditional and not commercial lucrative and viable. The project has collaborated with Professional Insurance Corporation of Zambia (PICZ), which is the largest local insurance company on the Zambian market. PICZ with the support of a global reinsurer will engage an actuarial consultant to develop and roll out an appropriately priced weather index insurance cover to smallholder farmers on the project list, but this will also extend to other farmers.

The actuarial consultant to be engaged by PICZ will regularly monitor weather and climate data using satellite imaging and feed this information back into the CSA information platform, through PICZ and ZAOU monitoring and evaluation (M&E) functions. The complementary approach to collecting meteorology and satellite data is to increase the accuracy of data to inform CSA practices. To facilitate uptake of weather index insurance by smallholder farmers and adoption of stress tolerance seed, the price of WII will be built into the seed price, and the project will discuss and reach consensus with the project partners on how to competitively price the products in a way that does not distort the market but also proves viable and sustainable.

Weather index insurance is a new type of insurance that can pay out when there are crop losses caused by unfavorable weather. The unfavorable weather can be too little rain, too much rain, extremely cold days, extremely hot days, or extremely high humidity. It is not based on the actual damages on a farmer's field but rather automatic payout when specific weather events are recorded by the reference weather station as the monitoring instrument. Measuring risk in this way allows the insurance to be affordable but still cover crop losses the farmer may experience. Therefore, WII products pay out based on the value of a "formula," not on losses measured in the field. The index is a variable that is highly correlated with losses and that cannot be influenced by the insured farmer (Recha 2016).

According to Risk Shield Consultants (2017), these WII products have given insurance access for the first time to 75,000 farmers over 2013–2016 period. These products are low-cost (e.g., premiums between US$6 and US$23 per farmer per season, depending on options selected). The insurance products reduce the livelihood risk for farmers and give them better access to farming inputs and credit, including access to inputs early in the season in case of failed germination due to late onset of rains. The products are reinsured in the international reinsurance markets. Due to the severe droughts in recent years, there have been weather index triggered payouts every year over the last four seasons.

The product design is based on satellite rainfall data and, in addition, feedback from various preselected farmers, camp extension officers, yield data, and agronomical information from the Ministry of Agriculture. Using these data sources, a typical WII product developed should answer as much as possible the following questions:

- Products: What type of product? What is covered? When does coverage start and end?
- Monitoring: How is weather monitored? When does coverage start? How is dry spell measured? How is excessive rainfall measured? How is weather data shared?

- Payouts: How is early dry spell payout made? Are partial payouts applicable?
- Marketing: How are farmers aggregated? When will product be marketed? How will farmers be enrolled?
- Insurer: Who are the insurers? Who are the reinsurers? Sum insured? Premium? Deductible or Excess?

Results and Discussions

The project in line with the core activities embarked on a training campaign that was attended by 331 Ministry of Agriculture camp extension officers and **132** agro-dealers from the 12 districts. The participants were trained on the concept of weather index insurance (WII), access and use of stress tolerant seeds, the need to follow best agronomic practices, and also crop-livestock integration.

The CEOs were later taken through digital registration of farmers and survey administration using the procured Android phones and Open Data Kit (ODK) installed on the Android devices. The project also conducted 12 seed fairs with a total 3591 participants, aimed at promoting the use of stress tolerant maize seed varieties. The seed fairs also served as an opportunity for farmers, agro-dealers, and seed production companies to interact and share knowledge about the various stress tolerant maize seed varieties that exist on the market. Figure 2 below gives a breakdown of the camp extension officers that were trained in each of the 12 districts.

Engagement of Actuarial Consultant and Designing of Weather Index Products

These activities could not be carried out as the Government of the Republic of Zambia made a policy pronouncement and immediately embarked on a nationwide

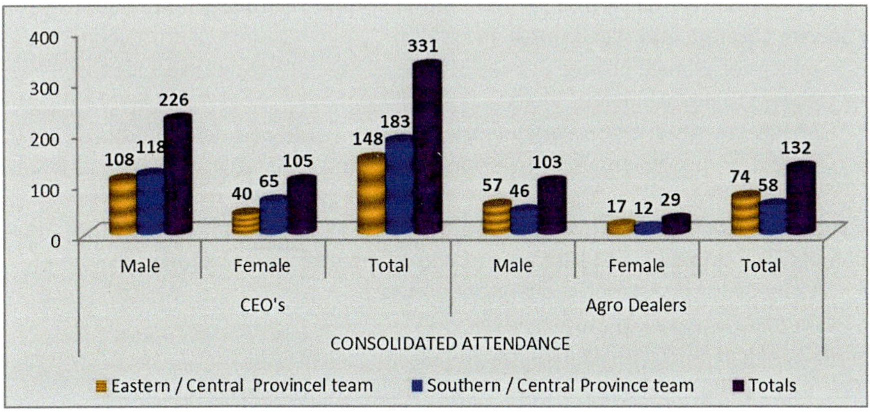

Fig. 2 Consolidated attendance at trainings of CEOs and agro-dealers

drive for uptake of WII products by farmers under the e-voucher system of the FISP. Thus, the engagement of an actuarial consultant to develop a WII product was shelved as it would lead to unnecessary duplication. We therefore lost control of the WII design and deployment. Nevertheless, the project team carried out the review of the WII manual developed under the Vuna Pilot, and this was used for our trainings and also for the nationwide campaign. PICZ still coordinated the trainings despite not being the developer of the product but have entered into discussions to coinsure "facultative inwards" with the insurance company selected by the government.

Training of Extension Officers

The primary target of the trainings that took place in the 12 project districts were the camp extension officers of the Ministry of Agriculture and local agro-dealers. This was so because they work in close contact with farmers and are tasked with the responsibility of providing training and technical support to farmers on various issues concerning agriculture.

An agricultural extension officer directly works with farmers and companies related to agriculture. Their primary role is to aid these groups to make better decisions to increase agricultural production. The extension officers are constantly armed with the latest techniques and information related to agriculture, and they relay this information to farmers and agricultural business. Agricultural extension officers often provide consultation to farmers and agricultural businesses. In these consultations, they give talks, guidance, and actual demonstration on the latest technologies related to agriculture and on how farmers can take advantage of such technologies. They also attend seminars and work with other experts in agriculture to learn more or even develop new methods that could advance production. It is for this reason that all the camp extension officers, including the District Agricultural Coordinators (DACO) and the Senior Agricultural Officers (SAO), were trained.

The CEOs were taken through the following topics during the trainings:

Climate Change and Agronomic Practices:

- Concept of climate change
- Concept of stress tolerant seeds
- Concept of conservation agriculture
- Concept of climate-smart agricultural practices

Crop- Livestock Integrated Farming Systems:

- Improved animal husbandry practices: management, housing, and feeding
- Strategies for dry season feeding of animals
- Competitive usage of crop residues
- Advantages of crop-livestock integration

The strategy is that the CEOs will incorporate the knowledge they received from the training in their day-to-day extension activities that they conduct and will now be in a position to explain to the farmers what weather index insurance is and how it benefits them.

Seed Fairs

In each of the 12 districts, the trainings were followed up with a seed fair so as to accord seed producers, agro-dealers, and smallholder farmers an opportunity to interact and share knowledge on stress tolerant maize seeds and where farmers can access these seeds.

Prior to the trainings of CEOs, all seed companies in Zambia were invited in writing to participate in the seed fairs by displaying and informing the smallholder farmers about the different stress tolerant seed varieties they stock. During training of CEOs, the DACO and his team were requested to suggest both the date and venue for the seed fairs. Once this was agreed, the information was then passed on to the seed companies, and a steering committee was established comprising SAO, CEOs, and District Agricultural Information Officer (DAIO), representatives of seed companies, agro-dealers, and farmer organizations, in consultation with the project team to plan and organize the seed fairs. Resources for communication and other logistics were provided by the project in order to provide a level playground for all participants.

On the actual day of the seed fair, each participating seed company was given about 10–15 min to address the farmers before farmers were allowed to visit the booths of various seed companies and agro-dealers exhibiting. In total, 3591 smallholder farmers attended and 8 seed companies exhibited (Zamseed, Seedco, MRI, Dekalb, Pannar, Pioneer, K2, and good nature seed). Other exhibitors (non-seed producers) were Stanbic Bank, Vitalite (solar energy company), ATS Agrichem, Zambia Meteorological Department, Airtel, agro-dealers, and farmers from various farming camps.

Given the above successfully conducted activities, it is expected that there will be increased investment in and adoption of stress tolerant maize seeds and increased number of farmers practicing mixed farming as well as increased awareness among farmers on weather index insurance and how it can benefit them.

Digital Farmers' Registration

The CEOs were provided with Android phones and trained on usage of ODK for data collection and farmers registration. Most of the CEOs were using mobile phones to conduct survey for the first time. As of the 28th of February 2018, the project had recorded 27,642 registered smallholder farmers of the targeted 50,000 farmers in the first year.

Short Messaging Services (SMS)

The mobile phone SMS platform was set up to provide farmers with information on various weather data, CSA agronomic practices, animal husbandry, and nutrition strategies. Weekly weather information are received from the Zambia Meteorological Department. Messages to farmers were then drafted by the agronomic officer in the Ministry of Agriculture together with expert at the Zambian Open University. The National Agricultural Information Services (NAIS) then packaged the message in a ready-to-use format for the farmers. Maintenance of the database and server for delivery of the SMS is by the Zambian Open University. Four text messages have since been sent to 25,632 farmers who were registered on the SMS platform.

The project is still ongoing hence assessment of the impact of the interventions has not yet been determined. It is pertinent to mention that Indaba Agricultural Policy Research Institute of Zambia (IAPRI) an independent research and consultancy firm was contracted to collect reliable and sufficient baseline data that can be drawn upon to undertake impact studies.

Opportunities and Challenges of the Public-Private Research Institution Partnership

The partnership provided a wide range of opportunities and challenges in the implementation. The CEOs and agro-dealers applauded their increased knowledge and understanding of climate change in general, the concepts of weather index insurance, use of stress tolerant maize seeds, and crop-livestock integrated farming systems as tools in combating the negative effect of climate change.

Sustainability Measures

The seed companies were very appreciative of the seed fairs as it syncs with their marketing and sales strategy. The agro-dealers also see the seed fairs as a morale booster and avenue to know more about the seeds they are selling. It has since been agreed that all seed fairs for subsequent years should be held in the camps and, if possible with the support of the seed companies and the project partners, hold more than one seed fair per district.

Synergies

The partnership resulted in signed memorandum of understanding among the partners which made it possible to utilize organizational structures and facilities for mutual benefit of the project. It led to several meetings with the DACOs and SAOs in the various districts for both the training of CEOs and hosting of seed fairs. The Ministry of Agriculture facilitated the training of the CEOs. The project team also had several meetings with National Agricultural Information Services under the Ministry of Agriculture which led to an agreement to support the project by helping to package the weather and agronomic practices information in a form that it could be disseminated via the SMS platform to the farmers.

PICZ released members of staff to carry out the WII trainings. Musika through their network on the other hand helped to facilitate the seed fairs hosting and agro-dealers trainings. Musika also facilitated the revision of the earlier version of the weather index insurance manual and production of the revised weather index insurance manual which is being used throughout the country for WII training (Musika Development Initiatives 2017).

Our interactions with the Zambia Meteorological Department led to an agreement to provide the weather data and forecasts that were sent to farmers. The National Agricultural Information Services helped to package the messages in a format easily understood by an average smallholder rural farmer.

Outcome/Impact Stories

Most CEOs expressed optimism that the knowledge gained will go a long way in helping them conduct their official duties and encourage farmers to practice CSA, while some farmers expressed gratitude for the knowledge gained from the seed fairs and that they will be more cautious with the choice of seeds they plant. The DACOs expressed gratitude to the project team for putting together the seed fairs which were being held for the first time in most of the districts, and they have since requested that the event be held annually to help spread awareness on the importance of using stress tolerant seed varieties and combat fake seeds distribution.

The seed companies expressed their appreciation for the opportunity provided to have direct access to farmers and eliminate or reduce use of uncertified seeds.

Challenges

The main challenge to the project implementation is the numerous meetings held among various partners and stakeholders and exchange of several correspondences. It was time-consuming and demanding at the same time. However, this was necessary due to location, distance, and activities to be conducted; it helped to boost confidence in the partnerships.

Another challenge faced was the configuration of the Android devices which required the use of Internet connectivity services. This was solved by carrying out the configuration before venturing out into the field. The phones also failed to pick the GPS coordinates. Power supply failure was also encountered in some districts because of electricity load shedding.

Seed fair attendance in the districts ranged from as low as 59 in Choma to 958 in Lundazi partly because of location. Districts that suffered from low turnout of farmers had their seed fairs held at the central business district (CBD). This resulted in low turnout because events organized and held at district level are usually thought of as political and farmers always shun such events. Furthermore, though the CBDs are central in location, they are usually far from farmer's settlements.

Districts that had their seed fairs held early in the morning experienced low turnout of women compared to those that had their seed fairs held much later in the day primarily because of household chores commitments by women.

Conclusion

This chapter confirmed that climate change requires concerted, coordinated, and multilayered efforts as evidenced in the synergy of the players. The intervention complements existing Zambia Agricultural Policy and National Agricultural Extension and Advisory Services Strategy and builds capacity of technical staff. It is anticipated that the Ministry will continue to integrate CSA into wider agricultural policy beyond the life of the interventions. This chapter also has implications for crop diversification and livestock options for small-scale farmers. The implementation of the interventions was hinged on a "market facilitation" approach in which focus is to empower market actors to make increased and integrated investments in support of CSA private sector investments at farm level.

This chapter showed that assigning predefined roles based on core business, expertise, and project activities allowed for sustainability and synergies in implementation. Partners were seen as equal and contributing to jointly shared goals. Competence, relevance, objectivity, transparency, and mutual benefit as criteria for choice of partners proved useful. Partners were able to leverage on other partners in the delivery of their mandate such as access to weather data, messaging platform, trainings, and hosting of seed fairs. Traditional and political leaders are equally important in information dissemination. Preliminary results showed that farmers are willing to embrace CSA solutions.

References

CIAT and World Bank (2017) Climate – smart agriculture in Zambia. CSA country profiles for Africa series. International Centre for Tropical Agriculture (CIAT), Washington, DC, 25p

FAO (2012) The State of food and agriculture. FAO, Rome

Grasso M (2004) Climate change: the global public good. Working papers 75. University of Milano-Bicocca, Department of Economics, Milano, revised May 2004

Ministry of Agriculture and Livestock (2017) Climate smart agriculture (CSA) training manual. Lusaka

Ministry of Agriculture and Ministry of Fisheries and Livestock (2016) Second national agriculture policy. Lusaka

Ministry of National Development Planning (2016) National policy on climate change. Lusaka

Musika Development Initiatives (2017) Revised weather index insurance manual: supporting climate smart agriculture in Zambia

Recha JW (2016) Weather index insurance manual: supporting climate smart agriculture in Zambia. Vuna manual. Pretoria. Vuna online http://vuna-africa.com/

Risk Shield Consultants (2017) Product design for e-voucher linked weather index insurance for 2017–2018. Lusaka

UNDP (2015) Zambia human development report. Lusaka

UNICEF (2015) Qualitative study of child marriage in six districts of Zambia. Lusaka

Vuna (2017) Zambia evoucher scale up. Project Ref Number 6924-A005

World Vison (2015) A situation report on child marriages in Zambia. Lusaka

Multidimensional Framework for Achieving Sustainable and Resilient Food Systems in Nigeria

57

Kyle Frankel Davis and Olawale Emmanuel Olayide

Contents

Abstract

Africa faces the grand challenge of feeding a growing, more affluent population in the coming decades while reducing the environmental burden of agriculture. Approaches that integrate food security and environmental goals offer promise for achieving a more sustainable and resilient food system for the continent and for adapting to a changing climate. Here we outline a multidimensional

K. F. Davis (✉)
The Earth Institute, Columbia University, New York, NY, USA

The Nature Conservancy, New York, NY, USA

Data Science Institute, Columbia University, New York, NY, USA
e-mail: kd2620@columbia.edu

O. E. Olayide
Centre for Sustainable Development, University of Ibadan, Ibadan, Nigeria

W. Leal Filho (ed.), *Handbook of Climate Change Resilience*,
https://doi.org/10.1007/978-3-319-93336-8_115

framework to be applied to the case of staple crop production in Nigeria with the eventual intended purpose being to inform sustainable and resilient food security pathways for the country. This chapter begins by presenting a broad overview of global food security challenges and distant and local drivers of food production decision-making, with the focus of then turning toward the food security and sustainability issues facing sub-Saharan Africa and Nigeria. The chapter then introduces a novel international collaboration focusing on food security and sustainability in Nigeria – the Assessing Climate Resilient and Nutritious Crops (CRENUT) project – the primary aim of which is to link stakeholder involvement, technical expertise, and multiple types of information to achieve policy-relevant research outputs that can enhance science-based agricultural decision-making in Nigeria. This project will follow a set of five key steps to move from basic scientific discovery to its application through agricultural policy: (1) in-depth understanding of stakeholder priorities and goals; (2) an assessment of data needs, data availability, and options for data generation; (3) analysis to develop strategies to align multiple goals and to fully understand the potential co-benefits or trade-offs associated with a particular agricultural policy; (4) identification of policy mechanisms by which the desired food security and sustainability outcomes can be achieved; and (5) the transfer of this collective knowledge to inform science-based decision-making related to agriculture and food security. In doing so, this chapter present a universal approach that can be used to identify a host of desirable policy pathways which, if pursued, can promote the sustainable development of Nigeria's agricultural sector and can realize co-benefits for nutrition, rural livelihoods, and climate change adaptation.

Keywords

Sustainability · Climate change · Food security · Water resources · Greenhouse gas emissions · Rural livelihoods · Agricultural policy · Food prices · Diets · Nutrition

Introduction: Achieving Sustainable Food Security in a Changing Climate: A Global Challenge

Global food supply has nearly tripled over the past half-century (FAO 2017) driven by demands resulting from demographic growth and improvements in income (Alexandratos and Bruinsma 2012; Tilman et al. 2011). Since World War II, global population has more than doubled from 3.1 billion (1961) to 7.6 billion people (2017), with South and East Asia experiencing the most substantial increases (UN-DESA 2015). Rising incomes have also meant that households can afford richer diets with higher calorie and protein intake per capita (Tilman et al. 2011). This general improvement in household economic status has meant that 216 million people in developing countries were able to escape undernourishment between 1991 and 2015 alone (FAO/IFAD/WFP 2015). Yet substantial nutritional deficiencies persist in many parts of the developing world with roughly one in nine people receiving inadequate protein and calories and still more lacking access to important micronutrients (FAO/IFAD/UNICEF/WFP/WHO 2017). Food security challenges related to ensuring nutritious food supply and equitable food access continue to disproportionately affect nations in the developing world, while habits of overconsumption have become commonplace elsewhere (Alexandratos and Bruinsma 2012).

Rising incomes and associated shifts in diets have also led to a growing percentage of agricultural production not being used for direct human consumption. Due in large part to the usage of energy-rich oil cakes as feed, 51% of the world's crop calories are currently devoted to animal production (Davis and D'Odorico 2015; FAO 2017). The growth in demand for animal products – combined with a shift toward a more crop-dependent livestock sector – has substantially increased competition for crop use between direct human consumption and feed to support livestock (Thornton 2010). Due in large part to the efficient feed conversion ratios of monogastrics as well as the inherent variability in rangeland biomass production (Steinfeld et al. 2006), the world's livestock systems have been transitioning from an extensive, beef-dominated system toward a focus on concentrated, feed-reliant pig and chicken production (Davis and D'Odorico 2015).

In addition to demographic and dietary changes, there has also been a rapid increase in demand for crop-based biofuels since the start of the century (OECD/FAO 2017; D'Odorico et al. 2018). This has led to the growing diversion of crop supply – and the agricultural resources required to support it (Rulli et al. 2016) – toward the production of bioethanol and biodiesel. The increasing proportions of crop supply used for feed or biofuel has also been accompanied by substantial food waste. Currently, one-quarter of food production is lost or wasted at various steps along the food supply chain – from losses during production and storage to uneaten food on a person's plate – with distinct regional patterns (Gustavsson et al. 2011). The production of this lost and wasted food is supported by large environmental footprints (Kummu et al. 2012; West et al., 2014). This ultimately means that a large fraction of the natural resources dedicated to food supply in the places of production is not efficiently exploited while also made unavailable for natural systems and other uses.

The recent boom in global population was made possible by massive increases in crop production over the past 50 years, with global crop supply more than tripling and animal production increasing by 2.5-fold during this time (FAO 2017). For certain, the wide diffusion of high-yielding crop varieties has led to much of the tripling in food supply and avoided substantial cropland expansion. At the same time, these changes have rendered global food production highly dependent on external inputs of fertilizer, pesticide, and irrigation. Hence, the use of these crops to feed billions more people and to prevent the massive conversion of natural systems has come with important trade-offs, promoting cultivation practices with extensive environmental consequences (e.g., water resource depletion, overapplication of fertilizers, use of marginal or sloped lands) where policies and subsidies inadvertently allowed (Pingali 2016). As a result, the global food system has become one of the most extensive ways by which humanity has modified natural systems.

Croplands and rangelands now cover 34% of the planet's ice-free surface (Ramankutty et al. 2008). More than half of accessible runoff is withdrawn for human use (Postel et al. 1996), with the vast majority of consumptive water use devoted to agriculture (Hoekstra and Mekonnen 2012). Fertilizer production has more than doubled the amount of reactive nitrogen in the environment (Schlesinger 2009), while greenhouse gas emissions from food production (e.g., ruminant digestion, fertilizer denitrification) and land-use change contribute 19–29% of humanity's greenhouse gas emissions (Vermeulen et al. 2012). Shifts in diets toward animal-based products have enhanced the environmental impact of food production, as the livestock sector contributes disproportionately to the environmental burden of food production (Kastner et al. 2012; Mekonnen and Hoekstra 2012; Herrero et al. 2013; Eshel et al. 2014; West et al. 2014; Davis et al. 2015a). Trends within animal production toward more feed-reliant systems have also led to important environmental trade-offs, where improvements in land-use efficiency and GHG emissions per unit of animal production have been offset by the mounting water and nitrogen requirements of feed production (Davis et al. 2015a).

In addition to these mounting environmental impacts, the nutritional quality of global cereal production has also declined steadily with time, as nutrient-rich cereals have been supplanted by high-yielding rice, wheat, and maize varieties (DeFries et al. 2015). This has been in part driven by the increasing prevalence of large farms which generally produce a less nutritionally diverse set of crops (Herrero et al. 2017) and has resulted in dwindling amounts of key nutrients such as protein, iron, and zinc being produced per ton of cereal crop (DeFries et al. 2015). In addition, climate change is already impacting crop yields in key agricultural regions (Lobell et al. 2011). Enhancements of atmospheric CO_2 concentrations are also expected to exacerbate these declines by adversely affecting crop nutrient content in plant tissue, especially in C3 crops (e.g., rice, wheat) (Myers et al. 2014). Though food supply remains largely nutritionally adequate at the global scale, the persisting challenges of food access and widespread malnourishment and nutrient deficiencies amplify these trends in nutritional quality.

Massive levels of production, pressures on crop supply for multiple uses, and expansive environmental burden all characterize the current state of the global food system. This system has also become increasingly globalized, with 23% of food calories currently traded internationally and three out of four countries reliant on food imports to meet domestic demand (D'Odorico et al. 2014; Davis et al. 2014a). The steady rise in connectedness has ultimately been fueled by rising incomes, both directly and indirectly. For example, increasing demand for animal products – and the feed required to support their production – has meant that countries with emerging economies and a rising middle class (e.g., China) have had to depend more heavily on feed imports in order to support domestic animal production (Davis et al. 2015a). International food trade can act as either a buffer or a source of vulnerability for nations relying on food imports. On one hand, trade can allow countries to maintain populations above what locally available water resources would support (e.g., Suweis et al. 2013) and can act as a stabilizer when local production conditions are dependent on variable weather conditions, including precipitation patterns. On the other hand, this can leave importing countries more exposed to economic or environmental shocks that occur beyond their borders and beyond their direct control (e.g., D'Odorico et al. 2010; Puma et al. 2015; Suweis et al. 2015). The growing reliance on international food trade has also had differential effects on natural systems within importing and exporting countries. Because exporting countries can typically produce food with relatively high resource-use efficiencies, international food trade has led to important resource savings globally (Chapagain et al. 2006). However, for individual trade connections, a country may decide that the economic benefits of producing a particular crop may outweigh the environmental costs. The increasing importance of international food trade makes this kind of unsustainable production decision more likely, as trade has served to separate consumer choices from the environmental impacts of production (DeFries et al. 2010; Lenzen et al. 2012; Dalin et al. 2017) and allowed many high-income countries to displace environmental pressure into the developing world (Weinzettel et al. 2013).

Many countries have begun actively addressing their potential vulnerabilities within the global food system through various food security mechanisms, such as food self-sufficiency policies for key staple crops, fertilizer and irrigation subsidies, strategic reserves, and trade agreements. One such phenomenon that has recently received attention is the global land rush (Deininger 2013; Dell'Angelo et al. 2017). Since the start of the century, countries lacking in investment capital or access to agricultural technologies began encouraging direct investments in their agricultural lands in an effort to enhance local agricultural productivity, to support rural development and job creation, and to promote technology transfers. In response to the flexibility of these policies as well as to the growing perception of unreliable food trade linkages (e.g., 2008 food crisis), investors from import-dependent countries began acquiring large tracts of agricultural land in the developing world in an effort to expand the land and water resources under their control and to thereby enhance their capacity for absorbing shocks within the global food system (Davis et al. 2015b). While the initial intent of these land deals may have been noble, there is a growing body of scientific and anecdotal evidence showing that the development and food security goals of these investments are often not achieved (e.g., De Schutter 2011; von Braun and Meinzen-Dick 2009) and instead frequently bring substantial social and environmental consequences (see, e.g., Davis et al. 2015c; D'Odorico et al. 2017).

These realized and potential impacts are especially concerning for rural livelihoods, food security, and access to agricultural resources in the countries where the investments take place (including in Nigeria). One estimate found that the incomes of more than 12 million people could be lost globally if communities living within the boundaries of newly leased lands were displaced by large-scale land acquisitions (Davis et al. 2014b). Other studies have shown that such deals can place substantial additional pressure on the freshwater resources on which smallholder farmers traditionally depend (Rulli and D'Odorico 2013; Rulli et al. 2013) and can break cycles of local food production and consumption by facilitating exports (Rulli and D'Odorico 2014). For Nigeria and other developing countries, identifying mechanisms that support yield enhancements, technology transfers, and secure land tenure for these critical stakeholders is therefore essential for advancing global and local food security, promoting poverty alleviation, enhancing food system resilience, and ensuring the sustainable and equitable use of available (and often limited) natural resources.

As a result of these complex and interacting phenomena, the enhanced crop demands for food, feed, and biofuel – both locally and internationally – have increased the pressure on land and water resources. At the same time, global food production must increase substantially in the coming decades to meet the demands of a growing, more affluent world population. There is also universal recognition that strategies for achieving this additional food supply must enhance the nutritional quality of production while also avoiding additional environmental degradation and reducing the vulnerability of agriculture to climate change impacts (Lobell et al. 2008; Challinor et al. 2014; DeFries et al. 2015). In many agriculturally important regions, yield trends of major crops are already experiencing effects related to climate change (Lobell et al. 2011; Ray et al. 2015) and air pollution (e.g., Van Dingenen et al. 2009) and are expected to be further impacted if appropriate mitigation and adaptation strategies are not promptly adopted (Wheeler and von Braun 2013). For developing countries in particular, these potential declines in crop productivity combined with persistent undernourishment and steady demographic growth present formidable food security and sustainability challenges. It is thus essential for these nations to identify solutions that take into account multiple food security, environmental, and economic dimensions (e.g., MacDonald et al. 2015; Davis et al. 2017a). Sustainability and food security policies in countries with agriculture-based economies must determine whether additional natural resources are and will be adequately available to enhance production, especially in low-yielding areas (e.g., Davis et al. 2017b), and, if so, how to avoid their injudicious use (e.g., Pradhan et al. 2015). In this way developing nations can ensure that increases in production occur in places where and when natural resources can support it (see e.g., Hoekstra et al. 2012; Brauman et al. 2016; Mekonnen and Hoekstra 2016; Davis et al. 2017a). Only by accounting for the local economic, social, and environmental aspects of crop production can nations ensure that their food security goals are achieved sustainably and equitably (Pingali 2012; Garnett et al. 2013).

Local–Global Linkages, Drivers, and Impacts across the Global Food System

Global demands for food and feed are largely defined by population, dietary preferences, resource availability, and markets. Through their influence on the price that a farmer can ultimately obtain, all of these factors largely determine local choices regarding crops and management, thus linking global demands to local resource use. This local dimension may be further impacted by the local land tenure policy and property structure. With smallholder agriculture contributing about 70% of the world's food production (Samburg et al. 2016), how all of these local and global factors may benefit or impact rural livelihoods has important implications for the sustainable development of developing nations. Indeed, the inverse relationship between field size and yield (e.g., Chand et al. 2011; Larson et al. 2012) – after controlling for inputs – points to the central role that smallholder agriculture can play in ensuring future food security in the face of increasing climatic variability. By growing a greater variety of crop species, smallholders also help to retain resilience and diversity within croplands and agricultural systems. However, the fact that the majority of cultivated areas are rainfed also means that these farmers must cope with high and ever-increasing interannual variability in precipitation as well as more frequent extreme climate events and their negative effects on yield stability. To help minimize the exposure of these farmers to crop stress and potential failure, governments will need to help in expanding their adaptive capacity through improved access to irrigation, drought-resistant crop varieties, crop insurance, and alternative income sources. For irrigation expansion in particular, this will need to be implemented selectively so as to avoid the expansion of water-stressed areas. Overall, how governments and farmers choose to utilize these coping strategies will have significant implications for the local balance of water availability and demand and for the long-term sustainability of food systems.

The ongoing emergence of water scarcity – and stress on other agriculturally important resources – provides strong evidence that growing global demands are not well matched with sustainable food production sources (Fig. 1). In addition, the potential impacts of climate change on natural resource availability and the expected increases in natural resource demand from population growth will have important consequences for the magnitude and distribution of resource use and scarcity in the coming decades (Veldkamp et al. 2016). Exporting countries do not reap the food security benefits of a portion of their croplands but undergo the environmental costs (Suweis et al. 2013). Hence, they may need to reevaluate whether the economic benefits of doing so outweigh the impacts. The extent to which food production in both export- and import-dependent nations can cope with these challenges will therefore likely propagate to other parts of the world via food trade linkages. As such, how countries choose to meet domestic food demand – on the spectrum from complete food self-sufficiency to total import reliance – and how they respond to global market influences will determine the balance between their internal and external environmental footprints of consumption and their dependence on distant natural resources. For a resource-scarce country, this reliance on distant resources may prevent unsustainable resource use that might have occurred had the country produced the good domestically. However, food trade linkages can also expose that same country to economic or environmental shocks that occur beyond their control (e.g., Puma et al. 2015; Suweis et al. 2015). Based on the expected frequency and intensity of both remote and local shocks as well as the economic and environmental trade-offs they may be willing to accept, countries in sub-Saharan African and other developing regions will therefore need to decide how they focus resources – either on the improvement of local yields and their stability or on diversifying their trade

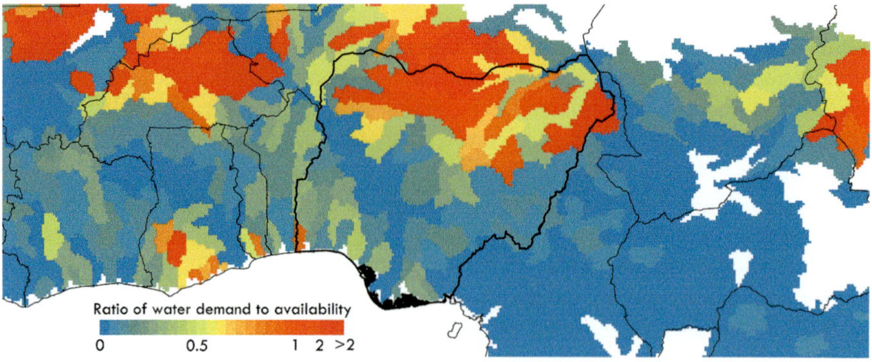

Fig. 1 Annual average water stress across croplands of West Africa. Water stress is calculated as the ratio of human consumptive water demand for crop production to annual renewable water availability (as defined in Brauman et al. (2016)). Water-stressed areas are defined as those areas where this ratio is greater than or equal to 1, i.e., where consumptive water demand exceeds renewable water availability. In these places, unsustainable water use is likely to occur, leading to the depletion of groundwater resources and/or of environmental flows. (Data on water stress came from Davis et al. (2017a). Nigeria is outlined with a bold black border)

linkages – in order to minimize their exposure to risk from climate change and variability.

While pressure on natural resources is a direct result of food production choices, these crop selections are ultimately dictated by global and local dietary preferences, economic mechanisms, and infrastructure that link the farmer to the consumer. The teleconnections of production and consumption decisions are essential to remember when developing solutions aimed at ensuring food system sustainability, rural livelihood development, and responsible resource use within agroecosystems. A radical shift in perspective is required in order to determine whether, where, and how food production should occur in the face of growing demand and a changing climate and to integrate food security goals with other sustainability objectives. All sustainable solutions for enhancing food security must begin by assessing whether sufficient resources are locally available – particularly in low-yielding areas (e.g., Davis et al. 2017b) – and to what extent these places can reliably support additional crop production. Only in this way can decision-makers ensure that increases in food production occur in places where and when natural resources can regularly support it (e.g., Wackernagel et al. 2002; Brauman et al. 2016), in order to avoid the local emergence of more severe stressed conditions. For many rainfed agricultural systems, this means not only accounting for the average amounts of food production and resource demand but also their potentially high interannual variability and their inherent trade-offs. By developing a baseline knowledge on resource availability and demand and how they are locally affected by management strategies, planners and policy-makers can make science-based decisions about which social, economic, and environmental outcomes and trade-offs will be acceptable to them and the people they represent.

The Challenges of Climate-Smart Agriculture in Nigeria and Sub-Saharan Africa

Achieving sustainable and climate-smart agricultural systems, addressing widespread undernourishment and nutrient deficiencies, and reducing dependence on food imports are a formidable set of challenges for sub-Saharan Africa. The region's population is expected to more than double by the year 2050, with major crop demand predicted to triple (van Ittersum et al. 2016). Further, much of the population of sub-Saharan Africa continues to rely heavily on staple crops (rice, maize, sorghum, millet, cassava, yams, and wheat) to meet nutrient requirements and – despite efforts at economic growth and human development – continues to experience substantial undernutrition. With the largest population on the continent (190 million people) and an expected population doubling before mid-century (410 million people by 2050), how and to what extent Nigeria will be able to sustainably meet future domestic food demand will have serious implications for global and regional food security in the coming decades. Compounding this demographic pressure, Nigeria already faces challenges related to food security and malnutrition (25% of the population is severely food insecure (FAO/IFAD/UNICEF/WFP/WHO 2017)), employment and poverty (the unemployment rate has tripled within this decade (NBS 2018)), and infrastructure (Foster and Pushnak 2011). In addition, a large portion of the Nigerian population relies directly on agriculture as their main source of income, a livelihood that leaves them especially vulnerable to climate variability and climate change. There is therefore a need not only to increase the production of and access to nutritious foods but also to buffer against anticipated climate impacts, minimize water use and greenhouse gas emissions, and promote sustainable and diverse rural livelihoods. While the country's large yield gaps and substantial potential for increased irrigation offer much promise for meeting its future food demand locally (van Ittersum et al. 2016), assessments of the nutrition quality, water use, emissions, climate resilience, and economic value of each staple crop in Nigeria are still missing but of great importance. Such work will provide an invaluable information base for the long-term sustainability of Nigeria's agricultural system and serve as an important next step in advancing the country's food and nutrition policies (MBNP 2016).

Indeed, recent work has shown that – in many areas of the world (including Nigeria) – the current distribution of crops still produces marked trade-offs between nutrition, water use, and economic value (Davis et al. 2017a). Other studies (e.g., DeFries et al. 2016) have also incorporated climate sensitivity into these considerations, showing that it is possible to make crop selections that realize co-benefits in terms of nutritional value, land-use efficiency, and climate resilience relative to current cropping patterns and practices. In Nigeria (and across sub-Saharan Africa), farmers that adopt appropriate crop diversification strategies will be able to better protect themselves from the adverse effects of climate change (i.e., climate change crucially affects farmers' crop choice decisions). This tactic of diversifying crop portfolios has already proven effective for mitigating the effects of climate variability experienced by farmers in developing countries (Barrett and Carter 2013; Mitter et al. 2015). Importantly, the extent to which this strategy can be adopted depends strongly on a household's capacity to perceive risk and to absorb shocks (e.g., price volatility, low rainfall) (Dercon and Christiaensen, 2011; Stuart et al. 2014). However, the potential advantages of selecting crops and cropping systems that better align outcomes in terms of nutrition, rural livelihoods, climate resilience, and environmental stewardship remain unclear for Nigeria.

The importance of agricultural transformation for economic growth, rural development, and poverty reduction is well documented (Babu et al. 2018), and this sector has been a priority in Nigeria's recent efforts to advance the country's economic transformation agenda. With the introduction of the Agricultural Transformation Agenda in 2011, Nigeria has made major strides toward defining and executing its policy objectives in the food and agricultural sector. Building on this progress, the new Nigerian government launched the Agriculture Promotion Policy in 2016. While these policies have improved the organizational and implementation strategies of the agricultural policy process (Babu et al. 2014), transformation of the sector remains both a challenging task and a critical lever for achieving sustainable development in the country. It is therefore essential to understand whether the current priorities of Nigerian agricultural policy also align with sustainable development objectives of the country (Babu and Blom 2014; Babu et al. 2014, 2018; FMARD 2016) and how and where agricultural strategies and crop choices can be improved to realize multiple economic, social, environmental, and food security benefits. To do this requires an assessment of the state of agricultural and sustainability information available in the country as well as the formulation of a multidimensional approach that can determine the various synergies and trade-offs associated with a particular policy pathway.

Toward Science-Based Agricultural Decision-Making: Steps for Achieving Policy-Relevant Research Outputs

The pathway in moving from basic scientific discovery to its application through agricultural policy involves multiple key steps, each of which can alone be difficult to achieve. This section provides a brief overview of the information, stakeholder involvement, and technical expertise required to achieve policy-relevant research outputs that can promote science-based agricultural decision-making. The five steps identified here are (1) in-depth understanding of stakeholder priorities and goals; (2) an assessment of data needs, data availability, and options for data generation; (3) analysis to develop strategies to align multiple goals and to fully understand the potential co-benefits or trade-offs associated with a particular agricultural policy; (4) identification of policy mechanisms by which the desired food security and sustainability outcomes can be achieved; and (5) the transfer of this collective knowledge to inform science-based decision-making related to agriculture and food security (Fig. 2). To complete this progression requires the involvement of and feedback from a diverse set of stakeholders and experts, as linking sustainable agricultural solutions with policy is a deeply cross-disciplinary challenge.

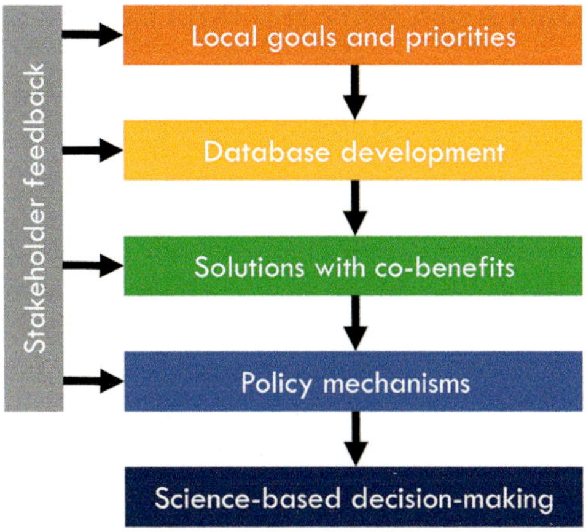

Fig. 2 Steps for informing sustainable agricultural decision-making

Understanding Stakeholder Priorities and Goals

Nigeria and the region of West Africa face well-known barriers to sustainability including pressing needs for enhancements of rural development, food security, environmental stewardship, and climate change adaptation. However, understanding how the country is choosing to address the various facets of sustainability in relation to agriculture – and which aspects they are prioritizing – demands close interaction with local experts, policy experts, and other stakeholders. These key persons can be intimately involved in informing all of the steps leading from scientific research to policy options, and these interactions between investigators and various stakeholders can serve as critical checks regarding the feasibility of particular strategies that could in principle help to move toward a sustainable agriculture system but lack real-world applicability or practicality. In addition, these collaborations can enhance the collective pool of expertise areas and resources available to a group of researchers, qualities which are essential in attempting to address inherently cross-disciplinary issues at the nexus of sustainability, agriculture, food security, and climate change. Input from local experts can also help to highlight linkages, trade-offs, or limitations that may not be apparent to investigators. For instance, if a researcher finds that a particular crop is drought-resistant and nutritious and has a low water footprint, they may recommend that this crop replace what is currently being cultivated in a particular area. However, it may be the case that the soil composition is unsuitable for the cultivation of any other crops, and any policy aimed at replacing the existing crop in these places would be doomed to fail. Access to the wealth of knowledge on these nuances of agriculture, food security, and agricultural policy therefore can only be achieved through sustained engagement with stakeholders, the relationships with whom require considerable effort to foster. Only in this way can academic researchers ensure that their work and insights reach those in a position to effect long-term sustainable changes within agriculture.

Data Availability and Data Generation

A formidable challenge facing agriculture and food security in Nigeria is a dearth of basic data availability and a lack of regular data collection. Depending on the specific institution or agency responsible for curating and updating particular information, these data issues may be manifested in a variety of ways. It may be the case that the data – though perhaps important for addressing certain key knowledge gaps – are simply not collected as there was never a program in place for their acquisition. Alternatively, the structure and financial support may be in place to support data collection – as with the case of Nigeria's agricultural census – but data collection simply does not occur. Even when the data are collected, the sampling methodology, data structure and representativeness, and other associated documentations can lack transparency and leave potential users unable to accurately assess the reliability of the information or to perform independent validation. This is not to say that the data collection schemes put in place are inadequate or deficient, simply that the supporting data description can make it difficult for independent investigators to fully understand the limitations of the data and are therefore often prevented from using it. It is also often the case that government data are either not publicly available, are not digitized or centrally stored, or face other bureaucratic or organizational challenges at the institutions in which they are housed. While many of these issues are beyond the control of the various bodies responsible for the management, oversight, and advancement of agriculture in Nigeria, they are ubiquitous – to a greater or lesser extent – across these agencies. Ultimately, these challenges significantly hinder the ability of the nation's experts and representatives to make informed recommendations and decisions related to agriculture, food security, rural livelihoods, and options for climate change adaptation of the sector.

In particular, there is little information on the environmental impacts of crop production in Nigeria. Yet this information is essential for informing food security and sustainability efforts and for examining the multidimensional outcomes of policy pathways for climate change adaptation in the country. To this end, the authors have established the Assessing Climate Resilient and Nutritious Crops (CRENUT) project with two main objectives. The first objective is to combine agricultural census data, crop water models, climate datasets and projections, remote sensing techniques, and economic information to quantify the greenhouse gas emissions, water use, nutritional quality, climate resilience, input costs, and prices of the staple crops grown throughout Nigeria. Information on the total energy use of the agricultural sector, the number and types of irrigation pumps, irrigation use by crop, shares of surface and groundwater use, groundwater depth, fertilizer application, and machinery will be combined to estimate crop-specific energy requirements and, as an extension thereof, greenhouse gas footprints (Rao et al. in review). Gridded crop-specific water footprints will be calculated as the crop water demand (i.e., the water required to prevent crop water stress) using a process-based crop water model. Crop water demand will be disaggregated between the crop water demand met by precipitation (green) and irrigation (blue) and will be calculated under current and future climate conditions (Chiarelli et al. 2016; Olayide et al. 2016; Davis et al. 2017a). Data on the nutrient content of staple crops (e.g., calories, protein, iron, vitamin A, etc.) will come from Nigeria's national nutrient composition report (Oguntoya and Akinyele 1995). These values will then be used to convert tons of production into nutritional supply to better target specific macro- and micronutrient outcomes. The climate sensitivity of each staple crop will be assessed based on the historical variation of each crop's yields in relation to interannual variability in rainfall and temperature using a mixed-effects modeling approach following DeFries et al. (2016). Model coefficients will be calculated separately for individual crops based on the geographic distribution of their cultivation and analyzed at the scale of agroecological zones. Output coefficients for precipitation and temperature will then be used as proxies for climate sensitivity. Input costs will be assessed utilizing available production cost data combined with much of the information used in the quantification of crop-specific greenhouse gas emissions. Lastly, information on consumer price and producer price will come from the FAO (2017).

Identifying Strategies for Aligning Multiple Sustainability Goals

The second objective of the CRENUT project is to assess the potential benefits and shortcomings of alternative, high-nutrient staple crops in Nigeria agricultural production. The above analyses will provide essential information to enable comparisons across the multiple aspects of nutrition, economic and cultural value, climate resilience, emissions, and water use (Fig. 3). These data be used to compare the climate resilience of different staple crops and assess where and to what extent particular staple crops can potentially contribute to Nigeria's nutrition security, income security, emissions reductions, and water sustainability under climate change. It will also be possible to consider

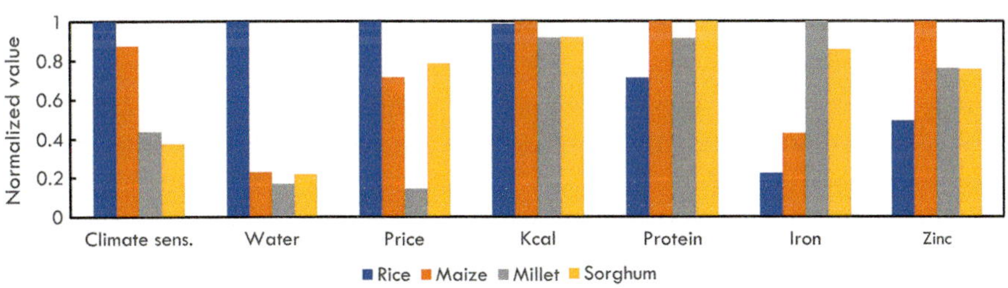

Fig. 3 Example comparison of crop dimensions for India. For each dimension, values are reported relative to the highest value in that category among the crops considered. For crop prices, for instance, rice had the highest price per ton; the per ton price for each crop was then divided by that of rice. Centered on the year 2006, data on price, climate sensitivity, calories, protein, and iron are from DeFries et al. (2016). Zinc data are from the USDA (2018). Irrigation water requirement data are from Mekonnen and Hoekstra (2010) for all of India. All data were originally reported on a per weight basis – except for climate sensitivity (unitless). A high value for climate sensitivity indicates greater vulnerability to rainfall and temperature variability. A high value for water indicates a high irrigation water requirement. The same multidimensional approach can be applied to Nigeria

scenarios in which each dimension is prioritized and examine the effect of such a choice on the other dimensions. For instance, by maximizing nutrient supply, it will be possible to determine its effect on economic value, overall climate resilience, and water use. The suite of possible outcomes will be examined for both current and future climatic conditions to understand whether incentives for alternative crops would be economically, socially, and environmentally meaningful in the long-term for Nigeria. In a more policy-relevant vein, this combined information will allow for an informed examination of the nutritional, economic, environmental, and climatic implications of a variety of possible food security and sustainability policy pathways and will thus allow decision-makers to determine whether the resultant co-benefits or trade-offs represent acceptable outcomes and, if so, to then explore policy levers for the achievement of those changes. It may be the case that specific policy mechanisms (e.g., government agricultural support programs) may already provide the necessary levers for achieving a particular outcome and that a modification of the specific included crops or subsidies may be all that is needed. Pursuing existing policy avenues can also help to ensure that a desired change is more rapidly enacted and that the necessary budgetary allocations are already in place. Overall, the planned data generation component of the CRENUT project will not only provide valuable agricultural information that is lacking for Africa's most populous country but can also (more importantly) be used to inform science-based agricultural decision-making in Nigeria and to identify optimum pathways for enhancing food security and sustainability within the country.

Policy Mechanisms for Implementation

A deep understanding of agricultural policy is essential for developing realistic and readily implementable solutions to the sustainable development of Nigeria's agricultural sector. It is possible to develop a multitude of hypothetical scenarios for enhancing food system sustainability within the country, but if a clear pathway to their realization does not exist, then their real-world applicability comes into question. Close engagement with local experts and stakeholders is therefore vital for eliminating glaring erroneous assumptions and for identifying feasible scenarios that account for multiple considerations related to agricultural policy, nutrition, environmental impacts, rural livelihoods, cultural preference, climate resilience, food prices, and input costs. The multidimensional approach introduced here offers promise for determining if and how trade-offs may occur, but it is ultimately the responsibility of policy-makers to decide whether such trade-offs would be acceptable to the people that they represent.

Science-Policy Interface and Science-Based Decision-Making

All of the previous steps are ultimately aimed at providing a collection of evidence that can be used to inform sustainable and resilient agricultural decision-making. By quantifying the various characteristics of the production of a variety of staple crops and by examining feasible scenarios of potential future crop production patterns, this can provide policy-makers with a spectrum of outcomes and allow them to readily examine the synergies and trade-offs associated with a particular future trajectory or policy pathway (Fig. 3). As such, this approach attempts to address a major challenge with which decision-makers must frequently contend – formulating sustainability policies with incomplete or imperfect information. In examining these multiple dimensions related to nutrition, livelihoods, environmental stewardship, and climate resilience together, policy-makers can then make more informed determinations as to whether any potential trade-offs would be acceptable for them and the people that they represent and ideally identify strategies that achieve a host of desirable outcomes simultaneously.

Feedback from and interactions with stakeholders is an essential component for all of the steps outlined here. This feedback is vital for better understanding local goals and priorities, for identifying dimensions or outcomes that are of greatest importance, for determining the data availability and needs to assess those outcomes comprehensively, and for developing solutions and strategies that conform to the nuances of local policy, culture, and society. In general, feedback between the first three steps – all of which occur before a policy change is instituted – can occur relative rapidly. By comparison, the feedback from monitoring and evaluation of an enacted policy change aimed at sustainability and climate change adaptation occurs on much slower iterations and within the limited scope and responsiveness allowable under that particular policy. This information is critical in determining if an intervention is not only having the intended outcomes but also avoiding any unintended consequences or negative externalities.

The ability of Nigeria to make informed science-based decisions related to agriculture and food security depends on a host of factors that extend beyond sustainability science alone (e.g., governance, research funding). Nonetheless such decisions cannot take into account considerations of sustainability unless sufficient information is available on which to base those decisions. The aim of the CRENUT project will be to fill important information gaps related to agriculture and the environment and to support decision-making related to food security, agriculture, and sustainability at all levels of government. Based on this knowledge, it will then be possible for the country to examine the long-term viability of the Nigerian Agricultural Policy and other guiding food security policies, to determine whether current practices will allow Nigeria to meet longer term sustainability goals (e.g., Sustainability Development Goals; Paris Accord), and to identify additional or alternative strategies that can enhance climate resilience and food system sustainability.

Sustainable Pathways for Meeting Future Food Demand in the Face of Climate Change

There is wide agreement that humanity's rate of resource use exceeds what can be sustainably generated and absorbed by Earth's systems (Wackernagel et al. 2002; Rockström et al. 2009; Galli et al. 2014; Hoekstra and Wiedmann 2014; Steffen et al. 2015). It is also clear that a continuation of current agricultural practices will enhance the vulnerability of the global food system to economic and environmental shocks (Suweis et al. 2015). A radical transformation of food systems across the planet is therefore required in order to increase nutritious food production while minimizing its environmental impacts and facing uncertainties related to demand and climate impacts (FAO 2009; Tilman et al. 2011; Alexandratos and Bruinsma 2012; Rosenzweig et al. 2014; DeFries et al. 2015). In addition, there is a growing need for agricultural solutions that better accommodate future uncertainties in natural variability in agricultural systems. Large crop yield gaps still exist in Nigeria and across sub-Saharan Africa (Mueller et al. 2012; van Ittersum et al. 2016) and offer farmers and the country the potential to substantially increase crop production, the local supply of food, and its stability. In addition, many places with the potential to transition to double-cropping systems have yet to do so (Ray and Foley 2013). By enhancing crop yields in those places where water resources can sustainably support it, Nigeria can improve its ability to cope during future climatic stress and can increase the strategic food reserves at its disposal. Doing so can also prevent additional agricultural expansion, the consequences of which would be profound and undesirable for the functioning of natural systems (Foley et al. 2011; Tilman et al. 2011). On the one hand, this agricultural intensification offers great promise for increasing the food self-sufficiency of regions of Nigeria that are subject to substantial interannual variability in rainfall (van Ittersum et al. 2016; FAO 2017). On the other hand, these remaining yield gaps raise questions about how best to promote the diffusion of high-yielding crop varieties and agricultural technologies – given that these agricultural advancements have yet to reach many places even 50 years on from the start of the Green Revolution (Pingali 2012) – and whether the primary focus of agroecosystems in northern Nigeria (where rainfall is inherently highly variable) should be to maximize yields or stabilize them. How the nation deals with these regional trade-offs between system redundancy and efficiency will be central in determining its long-term sustainability in the coming decades (e.g., Suweis et al. 2015).

Other recent work has shown the potential to maintain or reduce current levels of resource use while increasing crop production – thereby eliminating large inefficiencies in production systems. One such study demonstrated the possibility of using different irrigation and soil management strategies to close the crop yield gap by half without increasing cropland area or irrigation use (Jägermeyr et al. 2016). Other work showed that by redistributing crops based on their suitability, it is possible to increase food production while also realizing substantial water savings (Davis et al. 2017a, c). Still further work is needed to better understand how economic and environmental factors influence farmers' planting decisions and how the sum of individual farmer decisions ultimately dictates food availability, large-scale resource demand, and the overall climate resilience of the agricultural sector. These and other pursuits point to the need to integrate considerations of food, livelihoods, and the environment in developing more effective strategies for achieving sustainable food systems – ones that enhance the incomes and nutritional status of rural communities and that have a greater capacity to accommodate substantial interannual variability in precipitation. In identifying these food security strategies, Nigerian decision-makers will need to determine how best to effectively distribute the nation's risk to climatic variability and uncertainty in order to ensure that no single economic or environmental shock overwhelms its ability to buffer against it. Building resilience into all steps of these food supply chains – through the adoption of location-specific solutions and local knowledge – will be essential for achieving the long-term sustainability of agricultural systems in Nigeria.

More generally, the dilemma between increasing food supply and the environmental burden of production is a feature of historical agricultural approaches, and a large body of literature focuses on identifying strategies for its resolution (e.g., Godfray et al. 2010; Foley et al. 2011; Tilman et al. 2011). These potential solutions include enhancing crop yields on current croplands, improving resource-use efficiency, reducing food waste, and shifting to less resource-intensive diets – all of which can offer multiple benefits for the environment and for human health. Indeed, recent work has shown that employing a suite of these solutions can achieve food security and environmental co-benefits in the coming decades (Tilman and Clark 2014; Davis et al. 2016; Jagermeyr et al. 2016). Other work has shown that – by better aligning food security and environmental goals – it is possible to substantially reduce resource demand while also producing more food (e.g., Mueller et al. 2014; Davis et al. 2017a). While all of this research is a cause for optimism, the major challenge that remains in actualizing food system sustainability is in adapting these solutions to specific locations in ways that incorporate local priorities, that engender local buy-in, and that allow sufficient flexibility in adapting to climate change.

One key to addressing these challenges in Nigeria is to better understand the "wedges" – be they policy-based, cultural, infrastructural, or economic – that have produced discrepancies between current patterns of food production and those that optimize outcomes along the dimensions of nutrition, environmental sustainability, climate resilience, and livelihoods (e.g., Wada et al. 2014, NatGeo). Identifying and prioritizing certain "wedges" for intervention will require interdisciplinary approaches that link a global perspective – which accounts for the sometimes distant drivers of environmental change in Nigeria – with direct stakeholder engagement. In addition, understanding how local and distant policy and consumption decisions influence specific production choices – and acknowledging the growing role of international food trade in redistributing food (Wood et al. 2018) and its impacts of production – will be essential for effectively addressing sustainability challenges within food systems. Nigeria can complement these efforts through awareness initiatives and educational programs that inform the public of the environmental consequences of their dietary choices, including through the use of footprint tools (e.g., Leach et al. 2012), sustainability food labels (e.g., Leach et al. 2016), and guidelines for healthy eating (e.g., Tilman and Clark 2014), among other strategies. It is clear that any solution aimed at achieving sustainable food systems must be tailored to a specific place, taking into account nutritional, cultural, political, economic, and environmental factors. The approaches that will offer the most promise for achieving sustainable and climate-smart food systems will be those that best identify and minimize potential trade-offs and that seek ways to realize multiple co-benefits.

Conclusion

This chapter has provided a brief overview of the food security and sustainability challenges facing Nigeria and much of sub-Saharan Africa. Following from this, this chapter has outlined a five-step approach by which Nigeria (and other data-limited countries) may begin to address sustainability challenges related to agriculture and food security: (1) in-depth understanding of stakeholder priorities and goals; (2) an assessment of data needs, data availability, and options for data generation; (3) analysis to develop strategies to align multiple goals and to fully understand the potential co-benefits or trade-offs associated with a particular agricultural policy; (4) identification of policy mechanisms by which the desired food security and sustainability outcomes can be achieved; and (5) the transfer of this collective knowledge to inform science-based decision-making related to agriculture and food security. A nation can examine the co-benefits and trade-offs of a particular policy decision only if the information on dimensions of interest is available. The newly established CRENUT project introduced here will attempt to address these critical knowledge needs of the country by combining local expertise and technical capacity with in-depth and frequent input from government officials, decision-makers, and a variety of other stakeholders. These collaborations offer promise for best identifying the priorities of local communities and for aligning research objectives with national food security and sustainability goals. In doing so, this chapter has presented a universal approach that can readily incorporate the economic, nutritional, and environmental dimensions of greatest interest to a country. This approach can then be used to identify a host of desirable policy pathways which, if pursued, can promote the sustainable development of Nigeria's agricultural sector and can realize co-benefits for nutrition, rural livelihoods, and climate change adaptation.

References

Alexandratos N, Bruinsma J (2012) World agriculture towards 2030/2050: the 2012 revision. FAO, Rome

Babu SC, Blom S (2014) Capacity development for resilient food systems: issues, approaches, and knowledge gaps. IFPRI, Washington, DC

Babu SC, Gyimah-Brempong K, Nwafor M, Edeh H (2014) Capacity assessment for achieving the agricultural transformation agenda in Nigeria. NSSP working paper 26. IFPRI, Washington, DC/Abuja

Babu SC, Mavrotas G, Prasai N (2018) Integrating environmental considerations into the agricultural policy process: evidence from Nigeria. Environ Develop 25:111–125

Barrett CB, Carter MR (2013) The economics of poverty traps and persistent poverty: policy and empirical implications. J Develop Stud 49:976–990

Brauman K, Richter B, Postel S, Malsy M, Flörke M (2016) Water depletion: an improved metric for incorporating seasonal and dry-year water scarcity into water risk assessments. Elementa 4:83

Challinor AJ, Watson J, Lobell DB, Howden SM, Smith DR, Chhetri N (2014) A meta-analysis of crop yield under climate change and adaptation. Nature Clim Change 4:287–291

Chand R, Prasanna PAL, Singh A (2011) Farm size and productivity: understanding the strengths of smallholders and improving their livelihoods. Econom Polit Weekly 46(Suppl rev agric):5–11

Chapagain AK, Hoekstra AY, Savenije HHG (2006) Water saving through international trade of agricultural products. Hydrol Earth Syst Sci 10:455–468
Chiarelli DD, Davis KF, Rulli MC, D'Odorico P (2016) Climate change and large-scale land acquisitions in Africa: quantifying the future impact on acquired water resources. Adv Wat Resour 94:231–237
D'Odorico P, Rulli MC, Dell'Angelo J, Davis KF (2017) New frontiers of land and water commodification: socio-environmental controversies of large-scale land acquisitions. Land Develop Degrad 28:2234–2244
Dalin C, Wada Y, Kastner T, Puma MJ (2017) Groundwater depletion embedded in international food trade. Nature 543:700–704
Davis KF, D'Odorico P (2015) Livestock intensification and the influence of dietary change: a calorie-based assessment of competition for crop production. Sci Tot Environ 538:817–823
Davis KF, D'Odorico P, Rulli MC (2014a) Moderating diets to feed the future. Earth's Future 2:559–565
Davis KF, D'Odorico P, Rulli MC (2014b) Land grabbing: a preliminary quantification of economic impacts on rural livelihoods. Popul Environ 36:180–192
Davis KF, Yu KL, Herrero M, Havlik P, Carr JA, D'Odorico P (2015a) Historical trade-offs of livestock's environmental impacts. Environ Res Lett 10:125013
Davis KF, Rulli MC, D'Odorico P (2015b) The global land rush and climate change. Earth's Future 3:298–311
Davis KF, Yu K, Rulli MC, Pichdara L, D'Odorico P (2015c) Accelerated deforestation driven by large-scale land acquisitions in Cambodia. Nat Geosci 8:772–775
Davis KF, Gephart JA, Emery K, Leach A, Galloway JN, D'Odorico P (2016) Meeting future food demand with current agricultural resources. Glob Environ Change 39:125–132
Davis KF, Rulli MC, Seveso A, D'Odorico P (2017a) Increased food production and reduced water use through optimized crop distribution. Nat Geosci 10:919–924
Davis KF, Rulli MC, Garrassino F, Chiarelli D, Seveso A, D'Odorico P (2017b) Water limits to closing yield gaps. Adv Water Resour 99:67–75
Davis K, Seveso A, Rulli M, D'Odorico P (2017c) Water savings of crop redistribution in the United States. Water 9:83
De Schutter O (2011) The green rush: the global race for farmland and the rights of land users. Harvard Intl Law J 52:503–559
DeFries RS, Rudel T, Uriarte M, Hansen M (2010) Deforestation driven by urban population growth and agricultural trade in the twenty-first century. Nat Geosci 3:178–181
DeFries R, Fanzo J, Remans R, Palm C, Wood S, Anderman TL (2015) Metrics for land-scarce agriculture. Science 349:238–240
DeFries R, Mondal P, Singh D, Agrawal I, Fanzo J, Remans R, Wood S (2016) Synergies and trade-offs for sustainable agriculture: nutritional yields and climate-resilience for cereal crops in Central India. Glob Food Secur 11:44–53
Deininger K (2013) The global land rush. In: Barrett CB (ed) Food security and sociopolitical stability. Oxford University Press, Oxford, pp 95–119
Dell'Angelo J, D'Odorico P, Rulli MC (2017) Threats to sustainable development posed by land and water grabbing. Curr Opin Environ Sust 26-27:120–128
Dercon S, Christiaensen L (2011) Consumption risk, technology adoption and poverty traps: evidence from Ethiopia. J Develop Econom 96:159–173
D'Odorico P, Laio F, Ridolfi L (2010) Does globalization of water reduce societal resilience to drought? Geophys Res Lett 37:L13403
D'Odorico P, Carr JA, Laio F, Ridolfi L, Vandoni S (2014) Feeding humanity through global food trade. Earth's Future 2:458–469
D'Odorico P, Davis KF, Rosa L, Carr JA, Chiarelli DD, Dell'Angelo J, Gephart J, MacDonald GK, Seekell DA, Suweis S, Rulli MC (2018) The global food-energy-water nexus. Rev Geophys (in press)
Eshel G, Shepon A, Makov T, Milo R (2014) Land, irrigation water, greenhouse gas, and reactive nitrogen burdens of meat, eggs, and dairy production in the United States. Proc Natl Acad Sci USA 111:11996–12001
Federal Ministry of Agriculture and Rural Development (2016) The green alternative: the agricultural promotion policy 2016–2020. Federal Ministry of Agriculture and Rural Development, Federal Republic of Nigeria, Abuja

Foley JA, Ramankutty N, Brauman KA, Cassidy ES, Gerber JS, Johnston M, Mueller ND, O'Connell C, Ray DK, West PC, Balzer C, Bennett EM, Carpenter SR, Hill J, Monfreda C, Polasky S, Rockström J, Sheehan J, Siebert S, Tilman D, Zaks DPM (2011) Solutions for a cultivated planet. Nature 478:337–342

Food and Agriculture Organization of the United Nations (2009) Global agriculture towards 2050. FAO, Rome

Food and Agriculture Organization of the United Nations (2017) FAOSTAT database. FAO, Rome http://www.fao.org/faostat/en/#home

Food and Agriculture Organization of the United Nations, International Fund for Agricultural Development, United Nations Children's Fund, World Food Programme, World Health Organization (2017) The state of food security and nutrition in the World 2017. Building resilience for peace and food security. FAO, Rome

Food and Agriculture Organization of the United Nations, International Fund for Agricultural Development, World Food Programme (2015) The state of food insecurity in the world 2015. Meeting the 2015 international hunger targets: taking stock of uneven progress. FAO, Rome

Foster V, Pushak N (2011) Nigeria's infrastructure: a continental perspective. Policy research working paper 5686. The World Bank, Washington, DC

Galli A, Wackernagel M, Iha K, Lazarus E (2014) Ecological footprint: implications for biodiversity. Biol Conserv 173:121–132

Garnett T, Appleby MC, Balmford A, Bateman IJ, Benton TG, Bloomer P, Burlingame B, Dawkins M, Dolan L, Fraser D, Herrero M, Hoffmann I, Smith P, Thornton PK, Toulmin C, Vermeulen SJ, HCJ G (2013) Sustainable intensification in agriculture: premises and policies. Science 341:33–34

Godfray HCJ, Beddington JR, Crute IR, Haddad L, Lawrence D, Muir JF, Pretty J, Robinson S, Thomas SM, Toulmin C (2010) Food security: the challenge of feeding 9 billion people. Science 327:812–818

Gustavsson J, Cederberg C, Sonesson U, van Otterdijk R, Meybeck A (2011) Global food losses and food waste: extent, causes and prevention. FAO, Gotherburg/Rome

Herrero M, Havlik P, Valin H, Notenbaert A, Rufino MC, Thornton PK, Blummel M, Weiss F, Grace D, Obersteiner M (2013) Biomass use, production, feed efficiencies, and greenhouse gas emissions from global livestock systems. Proc Natl Acad Sci USA 110:20888–20893

Herrero M, Thornton PK, Power B, Bogard JR, Remans R, Fritz S, Gerber JS, Nelson G, See L, Waha K, Watson RA, West PC, Samberg LH, van de Steeg J, Stephenson E, van Wijk M, Havlík P (2017) Farming and the geography of nutrient production for human use: a transdisciplinary analysis. Lancet Plan Health 1:e33–e44

Hoekstra AY, Mekonnen MM (2012) The water footprint of humanity. Proc Natl Acad Sci USA 109:3232–3237

Hoekstra AY, Wiedmann TO (2014) Humanity's unsustainable environmental footprint. Science 344:1114–1117

Hoekstra AY, Mekonnen MM, Chapagain AK, Mathews RE, Richter BD (2012) Global monthly water scarcity: blue water footprints versus blue water availability. PLoS One 7:e32688

Jägermeyr J, Gerten D, Schaphoff S, Heinke J, Lucht W, Rockstrom J (2016) Integrated crop water management might sustainably halve the global food gap. Environ Res Lett 11:025002

Kastner T, Rivas MJI, Koch W, Nonhebel S (2012) Global changes in diets and the consequences for land requirements for food. Proc Natl Acad Sci USA 109:6868–6872

Kummu M, de Moel H, Porkka M, Siebert S, Varis O, Ward PJ (2012) Lost food, wasted resources: global food supply chain losses and their impacts on freshwater, cropland, and fertiliser use. Sci Tot Environ 438:477–489

Larson DF, Otsuka K, Matsumoto T, Kilic T (2012) Should African rural development strategies depend on smallholder farms? An exploration of the inverse productivity hypothesis. The World Bank, Washington, DC

Leach AM, Galloway JN, Bleeker A, Erisman JW, Kohn R, Kitzes J (2012) A nitrogen footprint model to help consumers understand their role in nitrogen losses to the environment. Environ Develop 1:40–66

Leach AM, Emery KA, Davis KF, Gephart JA, Carr JA, Pace ML, D'Odorico P, Galloway JN (2016) Environmental impact food labels combining carbon, nitrogen, and water footprints. Food Policy 61:213–223

Lenzen M, Moran D, Kanemoto K, Foran B, Lobefaro L, Geschke A (2012) International trade drives biodiversity threats in developing nations. Nature 486:109–112

Lobell DB, Burke MB, Tebaldi C, Mastrandrea MD, Falcon WP, Naylor RL (2008) Prioritizing climate change adaptation needs for food security in 2030. Science 319:607–610

Lobell DB, Schlenker W, Costa-Roberts J (2011) Climate trends and global crop production since 1980. Science 333:616–620

MacDonald GK, Brauman KA, Sun S, Carlson KM, Cassidy ES, Gerber JS, West PC (2015) Rethinking agricultural trade relationships in an era of globalization. Bioscience 65:275–289

Mekonnen MM, Hoekstra AY (2010) The green, blue and grey water footprint of crops and derived crop products, Value of water research report series no. 47. UNESCO-IHE, Delft

Mekonnen MM, Hoekstra AY (2012) A global assessment of the water footprint of farm animal products. Ecosystems 15:401–415

Mekonnen MM, Hoekstra AY (2016) Four billion people facing severe water scarcity. Sci Adv 2:e1500323

Ministry of Budget and National Planning (2016) National policy on food and nutrition in Nigeria. Ministry of Budget and National Planning. Federal Republic of Nigeria, Abuja

Mitter H, Heumesser C, Schmid E (2015) Spatial modeling of robust crop production portfolios to assess agricultural vulnerability and adaptation to climate change. Land Use Policy 46:75–90

Mueller ND, Gerber JS, Johnston M, Ray DK, Ramankutty N, Foley JA (2012) Closing yield gaps through nutrient and water management. Nature 490:254–257

Mueller ND, West PC, Gerber JS, MacDonald GK, Polasky S, Foley JA (2014) A tradeoff frontier for global nitrogen use and cereal production. Environ Res Lett 9:054002

Myers SS, Zanobetti A, Kloog I, Huybers P, Leakey ADB, Bloom AJ, Carlisle E, Dietterich LH, Fitzgerald G, Hasegawa T, Holbrook NM, Nelson RL, Ottman MJ, Raboy V, Sakai H, Sartor KA, Schwartz J, Seneweera S, Tausz M, Usui Y (2014) Increasing CO_2 threatens human nutrition. Nature 510:139–142

National Bureau of Statistics (2018) Unemployment rate. National Bureau of Statistics, Federal Republic of Nigeria, Abuja

Oguntoya EB, Akinyele IO (eds) (1995) Nutrient composition of commonly eaten foods in Nigeria – raw, processed, and prepared. Food Basket Foundation International, Ibadan

Olayide OE, Tetteh IK, Popoola L (2016) Differential impacts of rainfall and irrigation on agricultural production in Nigeria: any lessons for climate-smart agriculture? Agric Wat Manag 178:30–36

Organisation for Economic Co-operation and Development, FAO (2017) Agricultural outlook 2017–2026. FAO, Rome

Pingali PL (2012) Green revolution: impacts, limits, and the path ahead. Proc Natl Acad Sci USA 109:12302–12308

Pingali PL (2016) The hunger metrics mirage: There's been less progress on hunger reduction than it appears. Proc Natl Acad Sci USA 113:4880–4883

Postel SL, Daily GC, Ehrlich PR (1996) Human appropriation of renewable fresh water. Science 271:785–788

Pradhan P, Fischer G, van Velthuizen H, Reusser DE, Kropp JP (2015) Closing yield gaps: how sustainable can we be? PLoS One 10:e0129487

Puma MJ, Bose S, Chon SY, Cook BI (2015) Assessing the evolving fragility of the global food system. Environ Res Lett 10:024007

Ramankutty N, Evan AT, Monfreda C, Foley JA (2008) Farming the planet: 1. Geographic distribution of global agricultural lands in the year 2000. Glob Biogeochem Cyc 22:GB1003

Rao N, Poblete-Cazenave M, Bhalerao R, Davis KF, Parkinson S (n.d.) Food grains, energy demand and GHG emissions in India (in review)
Ray DK, Foley JA (2013) Increasing global crop harvest frequency: recent trends and future directions. Environ Res Lett 8:044041
Ray DK, Gerber JS, MacDonald GK, West PC (2015) Climate variation explains a third of global crop yield variability. Nat Commun 6:5989
Rockström J, Steffen W, Noone K, Persson A, Chapin FS, Lambin EF, Lenton TM, Scheffer M, Folke C, Schellnhuber HJ, Nykvist B, de Wit CA, Hughes T, van der Leeuw S, Rodhe H, Sorlin S, Snyder PK, Costanza R, Svedin U, Falkenmark M, Karlberg L, Corell RW, Fabry VJ, Hansen J, Walker B, Liverman D, Richardson K, Crutzen P, Foley JA (2009) A safe operating space for humanity. Nature 461:472–475
Rosenzweig C, Elliott J, Deryng D, Ruane AC, Müller C, Arneth A, Boote KJ, Folberth C, Glotter M, Khabarov N, Neumann K, Piontek F, Pugh TAM, Schmid E, Stehfest E, Yang H, Jones JW (2014) Assessing agricultural risks of climate change in the 21st century in a global gridded crop model intercomparison. Proc Natl Acad Sci USA 111:3268–3273
Rulli MC, D'Odorico P (2013) The water footprint of land grabbing. Geophys Res Lett 40:6130–6135
Rulli MC, D'Odorico P (2014) Food appropriation through large scale land acquisitions. Environ Res Lett 9:064030
Rulli MC, Saviori A, D'Odorico P (2013) Global land and water grabbing. Proc Natl Acad Sci USA 110:892–897
Rulli MC, Bellomi D, Cazzoli A, De Carolis G, D'Odorico P (2016) The water-land-food nexus of first-generation biofuels. Sci Rep 6:22521
Samberg LH, Gerber JS, Ramankutty N, Herrero M, West PC (2016) Subnational distribution of average farm size and smallholder contributions to global food production. Environ Res Lett 11:124010
Schlesinger WH (2009) On the fate of anthropogenic nitrogen. Proc Natl Acad Sci USA 106:203–208
Steffen W, Richardson K, Rockstrom J, Cornell SE, Fetzer I, Bennett EM, Biggs R, Carpenter SR, de Vries W, de Wit CA, Folke C, Gerten D, Heinke J, Mace GM, Persson LM, Ramanathan V, Reyers B, Sorlin S (2015) Planetary boundaries: guiding human development on a changing planet. Science 347:1259855
Steinfeld H, Gerber P, Wassenaar T, Castel V, Rosales M, de Hann C (2006) Livestock's long shadow: environmental issues and options. FAO, Rome
Stuart D, Schewe RL, McDermott M (2014) Reducing nitrogen fertilizer application as a climate change mitigation strategy: understanding farmer decision-making and potential barriers to change in the US. Land Use Pol 36:210–218
Suweis S, Rinaldo A, Maritan A, D'Odorico P (2013) Water-controlled wealth of nations. Proc Natl Acad Sci USA 110:4230–4233
Suweis S, Carr JA, Maritan A, Rinaldo A, D'Odorico P (2015) Resilience and reactivity of global food security. Proc Natl Acad Sci U S A 112:6902–6907
Thornton PK (2010) Livestock production: recent trends, future prospects. Phil Trans Roy Soc B 365:2853–2867
Tilman D, Clark M (2014) Global diets link environmental sustainability and human health. Nature 505:518–522
Tilman D, Balzer C, Hill J, Befort BL (2011) Global food demand and the sustainable intensification of agriculture. Proc Natl Acad Sci U S A 108:20260–20264
UN-DESA (2015) World population prospects: the 2015 revision. UN, New York
United States Department of Agriculture (2018) USDA food composition database. USDA, Washington, DC https://ndb.nal.usda.gov/ndb/
Van Dingenen R, Dentener FJ, Raes F, Krol MC, Emberson L, Cofala J (2009) The global impact of ozone on agricultural crop yields under current and future air quality legislation. Atmos Environ 43:604–618
van Ittersum MK, van Bussel LGJ, Wolf J, Grassini P, van Wart J, Guilpart N, Claessens L, de Groot H, Wiebe K, Mason-D'Croz D, Yang H, Boogaard H, van Oort PAJ, van Loon MP,

Saito K, Adimo O, Adjei-Nsiah S, Agali A, Bala A, Chikowo R, Kaizzi K, Kouressy M, Makoi JHJR, Ouattara K, Tesfaye K, Cassman KG (2016) Can sub-Saharan Africa feed itself? Proc Natl Acad Sci USA 113:14964–14969

Veldkamp TIE, Wada Y, Aerts JCJH, Ward PJ (2016) Towards a global water scarcity risk assessment framework: incorporation of probability distributions and hydro-climatic variability. Environ Res Lett 11:024006

Vermeulen SJ, Campbell BM, Ingram JSI (2012) Climate change and food systems. Ann Rev Environ Res 37:195–222

von Braun J, Meinzen-Dick R (2009) "Land grabbing" by foreign investors in developing countries: risks and opportunities. IFPRI, Washington, DC

Wackernagel M, Schulz NB, Deumling D, Linares AC, Jenkins M, Kapos V, Monfreda C, Loh J, Myers N, Norgaard R, Randers J (2002) Tracking the ecological overshoot of the human economy. Proc Natl Acad Sci USA 99:9266–9271

Wada Y, Gleeson T, Esnault L (2014) Wedge approach to water stress. Nat Geosci 7:615–617

Weinzettel J, Hertwich EG, Peters GP, Steen-Olsen K, Galli A (2013) Affluence drives the global displacement of land use. Glob Environ Change 23:433–438

West PC, Gerber JS, Engstrom PM, Mueller ND, Brauman KA, Carlson KM, Cassidy ES, Johnston M, MacDonald GK, Ray DK, Siebert S (2014) Leverage points for improving global food security and the environment. Science 345:325–328

Wheeler T, von Braun J (2013) Climate change impacts on global food security. Science 341:508–513

Wood SA, Smith MR, Fanzo J, Remans R, DeFries RS (2018) Trade and the equitability of global food nutrient distribution. Nature Sust 1:34–37

Potential of Smart Aquaculture Technologies on Improving Fish Production in Malawi

58

M. Mulumpwa

Contents

Abstract

The effects of climate change have not only affected crop production in Malawi but also aquaculture (fish farming). Aquaculture in Malawi begun in as early as 1906 with the culture of rainbow trout for angling. In 1956/1957 pond culture of indigenous species (*Oreochromis shiranus* and *Tilapia rendalli*) begun. However, most of wetlands, rivers, and streams that were perennial and could be diverted for the purpose of fish farming are becoming annual. This scenario has made fish farming increasingly becoming a challenge in areas that were once suitable for the enterprise, as fish ponds are drying up annually due to lack of adequate water supply. This is defeating the potential of fish farming in successfully supplementing the declining fish catches from capture fisheries on the market.

Rainwater harvest can be adopted for fish farming to do away with a problem of water availability throughout the year. Integrated aquaculture agriculture systems are also known to increase resilience of fish farming to climate change. Cage culture is another technology that can improve fish farming in the face of

M. Mulumpwa (✉)
Sengabay Fisheries Research Centre, Salima, Malawi
e-mail: mulumpwa.mexford@gmail.com

W. Leal Filho (ed.), *Handbook of Climate Change Resilience*,
https://doi.org/10.1007/978-3-319-93336-8_116

climate change. Cage culture enables fish farmers to overcome the problem of ponds drying up by culturing the fish in perennial water bodies such as lakes, dams, and rivers in confinements.

Keywords
Cage culture · Rainwater harvest · Aquaculture · Climate change · Food security · Malawi

Introduction

Studies have shown that fishery resources in Malawi are declining (Bulirani et al. 1999; Kanyerere 2003 and Department of Fisheries 2012), while artisanal fishers are on increase posing a worry on sustainable exploitation of the fisheries resource. This scenario is in turn posing a doubt on continued provision of fish as an ecosystem services in the future as fish stocks are being heavily exploited. The government and nongovernmental organizations are tirelessly ensuring that the fishery resources are sustainably exploited to avoid collapse of part/whole resource (Department of Fisheries 2012). This is very important because in Malawi fish constitutes about 70% of animal protein intake. Currently fisheries provide the cheapest source of proteins for the majority of poor Malawians (Russel and Dobson 2008) more especially sun-dried small fish species such as *E. sardella* (*bonya*). However, increasing human population of Malawi, overfishing, dwindling in catches, as well as rising of real fish prices have reduced per capita fish consumption from 12 to 18 kg/person/year in the 1970s to 9.4 kg/person/year in 1990, to 5.4 kg/person/year in 2008, to 5.6 kg/person/year in 2011 (Yaron et al. 2010; Government of Malawi (GoM) 2012), and to 8.12 kg/person/year in 2014 (National Fisheries and Aquaculture Policy (NFAP) 2016). The decline from 14 kg/person/year in the 1970s to 8.12 kg in 2014 represents a 58% decline of per capita consumption of fish in Malawi (GoM 2012). This decline in the fish per capita consumption is a serious indicator of a threat of national food security and may have serious nutritional implications for the nation more especially the poor, under-five children, expectant women, the elderly, and those affected or infected by HIV/AIDS.

The increased demand for fish on market is making more people to opt for fishing, thereby increasing population of small-scale or artisanal fishers on Lake Malawi and other water bodies. This increased number of artisanal fishers is making management of fisheries in Malawi so complex and challenging as most of them are employing more destructive gears and fishing activities to catch more fish and realize more returns. This is because it is requiring more fishing effort to catch the same unit of fish that was caught with less fishing effort when fish was abundant in the lakes. It is for this reason that the Department of Fisheries in Malawi is promoting fish farming among others to supplement dwindling fish catches from the wild so as to maintain the provision of animal protein on the market. Unfortunately, fish farming is being halted by climate change as well as lack of adequate extension services to local communities. This is negatively affecting adoption rate of the fish farming technologies.

Aquaculture Technologies that Can Overcome Climate Change Effects

Cage Culture

The abundant freshwater resource has made cage culture development suitable in Malawi. Water bodies in the country cover almost 20% of total surface area which is unfortunately underutilized for aquaculture production. The environmental conditions, for instance, in Lake Malawi is very suitable for cage culture more especially in bays which are mostly not heavily affected by prevailing winds; hence, the surface waves are relatively low. This reduces incidences of cages washed away to the shore by strong waves. The bays usually have a sandy, gentle, or flat bottom surface with high dissolved oxygen as a result of self-aeration from the waves. All these make bays conducive for cage culture as most of the environmental conditions fall within the limits required for the technology on both biotic and abiotic factors.

Cage culture is relatively new aquaculture technology in Malawi and becoming increasingly promising to aquaculture investors (FAO 2005). The largest intensive cage culture in Malawi is being operated by Maldeco Aquaculture Limited which was established in 2003 with a targeted production of 3000 tonnes per year. Cage culture has the potential of producing large quantities of fish for export and domestic market thereby supplementing capture fisheries.

The Department of Fisheries has carried out exploratory surveys to identify better sites on Lake Malawi for cage culture, and several sites were identified in Salima, Mangochi, and Nkhatabay districts. The Department of Fisheries through Sengabay Fisheries Research Centre carried out cage culture trials with funding from the Presidential Initiative on Aquaculture Development (PIAD) in 2007 and 2009 at Michesi (Fig. 1) in Mangochi district and the Food and Agriculture Organization (FAO) in 2012 at Chigumukile (Fig. 2) in Salima district both on Lake Malawi. The studies identified a better species candidate for cage culture as *Tilapia rendalli* and stocking density being 100 fish/m^3. However, there is more to be done to adequately understand the technology so as to make it work in Malawi in the face of climate change.

Large-scale cage culture is very expensive; as such its adoption rate in Malawi could be relatively low. Medium- and small-scale cage culture enterprises constructed from locally available materials could be relatively cheaper and affordable to many investors and hence may have high adoption rate. The small- and medium-scale commercial enterprises are the most efficacious engines of economic growth (Moehl et al. 2006) and have a potential of increasing fish production far above the current production from ponds in the country. Other factors influencing low rate of adoption of the technology among others include lack of information, extension services, and guidelines for cage culture, high cost of feed, lack of quality fingerlings, and theft.

Lack of Information

There is a need to conduct more research to understand the technology better and eventually making it work in Malawi in the face of climate change. This is

Fig. 1 Fish cages ready for stocking then at Michesi in Mangochi district

Fig. 2 Fish cage installation then in progress at Chigumukile in Salima district

challenging as Sengabay Fisheries Research Centre that is mandated to conduct cage culture trails lacks resources to generate more information for our policy-makers on this technology. There is a need to collaborate with other research centers and institutions of high learning such as Lilongwe University of Agriculture and Natural Resources (LUANAR) even those abroad to conduct thorough studies on the technology and its socioeconomic aspects in Malawian context.

Lack of Extension Services

The technology although ready for adoption by farmers, cooperatives, and investors among others is still not adequately known to the general public. Sengabay Fisheries Research Centre is ready to provide technical assistance, installation, and site exploratory survey to potential investors. However, not many realize that assistance to venture into cage culture is available. Majority of extension workers are not aware of the technology. It is therefore of paramount importance to sensitize both the extension workers and general public of the available technology.

Lack of Guidelines for Cage Culture

There is a need to generate guidelines for cage culture to assist anyone to follow and venture into the enterprise with little or no assistance. This is mostly important for husbandry practices. The stakeholders should spearhead the formulation of such a material to influence the adoption rate of the technology.

High Cost of Feed

It is hypothesized that feeding in aquaculture enterprises accounts to over 60% of total production cost. However, just like most of agricultural enterprises, the relative cost of production in cage culture decreases with an increase in the number of cages. The cost of production can be significantly reduced by training investors to make their own feed using locally available cheap ingredients. Investors can as well buy in bulk the feed ingredients while in season and store them in their warehouses for future use. However, it should be considered that since the fish will be confined, it requires complete diets with relatively good acceptability and digestibility, among others. High-quality feed for fish are being sold by Maldeco among others. On the other hand, Malawi lacks producer of floating feed which has relatively a high feed conversion ration unlike the sinking feed being produced. There is currently an increase of importation of floating feed from Zambia which is as well increasing feeding costs hence increasing operation costs even further. There is a need to have a feed manufacturing plant that manufactures floating feed in Malawi to significantly reduce feeding cost hence reducing operating costs.

Lack of Quality Fingerings

Quality of fingerings or seed is very important. As it is with crops that poor seed beget poor crop yields, the same is true with fish production. Fish farmers whether in ponds or cages among others should take issue of quality of fingerings/seeds very seriously. Quality fingerings can be sourced from certified fingering producers which include government owned National Aquaculture Centre at Domasi in Zomba district. However, most of these certified fingering producers do not meet the market demand.

Theft

The cage culture is sometimes hit by problem of theft although this is a challenge in some areas, in others it is not. However, measures should be taken to adequately control the problem in the areas it exists. A guard can be employed to guard the cages from thieves.

Use of Water Harvest in Fish Pond Production

Of late, water harvest has been advocated for household use and irrigation of crops to overcome droughts and ensure food security in the face of climate change. However, very few realize that the same principle can be used in fish production to deal with drying up of fishing ponds annually. Water can be harvested in a dam and used throughout the year to provide water adequately to fish ponds. The climate change is increasingly making more areas that were once conducive for aquaculture no longer so, as evaporation rate is very high coupled with seepage and low rainfalls. At the moment, many ponds that could hold water throughout the year are becoming annual making fish farming relatively becoming a challenge. The water harvest technologies can assist in making aquaculture more resilient to climate change in Malawi.

Integrated Aquaculture Agriculture Systems (IAAS)

Integrated aquaculture agriculture systems promote sharing of resources among farming systems more especially where by-products of a particular farming system are used as inputs of another. The shared resources can be manure, management, feed ingredients, crop residues, and water among others. IAAS ensures that there is efficient utilization of limited resources, thereby significantly reducing wastage. Chopin and Robinson (2004) reported that there is also another form of aquaculture integrated system that is slowly gaining popularity known as integrated multitrophic aquaculture which involves culturing of different fish species that occupy different trophic niches, for instance, a cage culture with both pelagic and demersal fish species.

The Role of Extension Services in Smart Aquaculture in the Face of Climate Change

The aquaculture sub-sector has a potential to increase fish production in the country. Enhanced aquaculture production especially at commercial levels would improve supply of fish protein in rural areas, creation of wealth, and employment in such areas. The aquaculture sub-sector can also be one of the major sources of fish exports, thereby contributing to the country's economic growth. There are 6000 fish farmers with varying sizes of ponds in the aquaculture sub-sector. Fish production in the sub-sector has been increasing from 800 tonnes per annum in 2006 to 3600 tonnes per annum by 2015. This is still low if compared to the demand of fish on the market. The production can significantly increase in this sector if the technologies discussed above are fully adopted in the country. It is for this reason that the role of extension is very important if the sub-sector of aquaculture is to develop in Malawi, more especially in ensuring that the Department of Fisheries realizes its aim of increasing aquaculture production from the current 3600 metric tonnes to 10,000 metric tonnes by 2021 (NFAP 2016). The extension workers should fully understand these technologies and carry them to the farmers and other investors who can adopt them. If this process of adoption of the smart aquaculture is adequately facilitated by extension workers, fish production in Malawi will significantly increase. This will contribute positively to food security in the country.

Conclusion

It can be concluded that aquaculture has a potential of supplementing the declining fish on the market from capture fisheries. However, this is possible if aquaculture is made more resilient to climate change. Climate smart aquaculture is a good candidate to this course. Some of these smart aquaculture are using water harvesting in pond production, integrated aquaculture agriculture systems, and cage culture. Extension services can assist in increased adoption of these technologies.

Cross-References

- ▶ Agriculture in the Face of Climate-Mediated Flooding in Tropical Africa: Technical Innovations of Fish Farmers in Southwestern Nigeria
- ▶ Climate Resilience in African Coastal Areas: Scaling up Institutional Capabilities in the Niger Delta Region
- ▶ Impacts of Climate Change on Food Security: An Appraisal
- ▶ Scaling Up Climate-Smart Agricultural (CSA) Solutions for Smallholder Cereals and Livestock Farmers in Zambia
- ▶ Water Management as a Means for Climate Change Adaptation and Sustainable Development

References

Bulirani A, Pålsson OK, Banda MC, Weyl OLF, Kanyerere GZ, Manase MM, Sipawe RD (1999) Fish stocks and fisheries of Malawian waters: resource report. Government of Malawi, Fisheries Department, Fisheries Resource Unit, Lilongwe. 54pp

Chopin T, Robinson S (2004) Defining the appropriate regulatory and policy framework for the development of integrated multi-trophic aquaculture practices: introduction to the workshop and positioning of the issues. Bull Aquac Assoc Can 104(3):4–10

Department of Fisheries (2012) National fisheries policy 2012–2017. Fisheries Department, Lilongwe
FAO (2005) Aquaculture production, 2004. Year book of Fishery Statistics - Vol.96/2. Food and Agriculture organization of the United Nations, Rome, Italy
Government of Malawi (GoM) (2012) Annual frame survey 2012. Fisheries Department, Lilongwe
Kanyerere GZ (2003) Age, growth and yield-per-recruit analysis of ndunduma, *Diplotaxodon limnothrissa* (Teleostei: Cichlidae), in the Southeast Arm of Lake Malawi. MSc thesis, Rhodes University, Grahamstown
Moehl J, Brummett R, Boniface Mk, Cohre A (2006) Guiding principles for promoting aquaculture in Africa: Benchmarks for sustainable development, Rome
National Fisheries and Aquaculture Policy (NFAP) (2016) 2nd edn. Department of Fisheries, Lilongwe
Russel A, Dobson T (2008) Adaptive organizational learning framework for resilience in fisheries co-management: based on an analysis of fisheries regime in Malawi. World Fish Centre, Cairo
Yaron G, Mangani R, Mlava J, Kambewa P, Makungwa P, Mtethiwa A, Munthali S, Mgoola W, Kazembe J (2010) Economic valuation of sustainable natural resource use in Malawi. Ministry of Finance and Development Planning, Lilongwe

Climate Change and Variability in the Mixed Crop/Livestock Production Systems of Central Ethiopian Highland

59

Nigatu Alemayehu, Mary Masafu, Abule Ebro, Azage Tegegne, and Getachew Gebru

Contents

Abstract

Teff and wheat are the two major food crops grown in the central Ethiopian highlands. Teff is staple food for over 60 million Ethiopians. Wheat is one of the five top crops grown in Ethiopia, and the country is the second largest wheat

N. Alemayehu (✉)
Consultant, Addis Ababa, Ethiopia
e-mail: nigatualemayehu16@gmail.com

M. Masafu
University of South Africa, Pretoria, South Africa

A. Ebro · A. Tegegne · G. Gebru
International Livestock Research Institute, Addis Ababa, Ethiopia
e-mail: abuleebro@gmail.com; ggebru09@gmail.com

W. Leal Filho (ed.), *Handbook of Climate Change Resilience*,
https://doi.org/10.1007/978-3-319-93336-8_120

grower in sub-Saharan Africa. Crop residues mainly from teff and wheat remained main sources for livestock feed. The objective of this paper is to assess the impact of climate change and explore adaptive responses of these two crops. The observed climate data for six decades (1951–2009) revealed increases in the rainfall, increases in both maximum and minimum temperatures, and no significant change in relative humidity. Increased sunshine hours and decreased wind speed were observed too. The most recent GCM (global climate model) climate projections that are used in the Fifth Assessment IPCC Report were used to predict the impact of the climate change on teff and wheat. DIVA-GIS version 1.5 was used for crop suitability analysis. It was found that teff will expand in terms of area due to its adaptation to wider agroecological zone. Teff is C_4 crop that allows more efficient use of water and faster photosynthesis under high heat and light condition than C_3 crops like wheat.

Keywords
Climate change · Ethiopia · Wheat · Teff · Highland

Introduction

Climate change has already influenced agricultural ecosystem in several regions of the world (IPCC 2007) in which there is ample scientific and historical proof of such effect. The world's history showed that human being was able to and had a capacity to adapt to changing climate for chronicles. However, in the last few decades, unexpected disaster on natural ecosystems has occurred very frequently such as drought and flooding in many parts of the world (Easterling et al. 2007). This has directly or indirectly affected existing agricultural production systems and hence food security particularly in less developed countries. In addition to this disaster, global human population is increasing at the faster speed than agricultural production, societies are expected to face changes in lifestyle through growing urbanization and modernization, and the increasing number of middle-income population is reported in many parts of the world that demands high-quality protein and energy food stuff (World Bank 2011). Thus agricultural production systems will require effective adaptive strategies to overcome gaps of demand and supply of food and other societal needs (Easterling et al. 2007).

The central Ethiopian Highland constitutes one of the major food-producing agroecology of Ethiopia. The Ethiopian Highland farmers have been struggling to maintain their livelihood by continuously trying to produce food for subsistence. But, the production system has often become non-resilient to shocks caused by climate variability and change. As a result, famine and hunger have killed many millions of people, and occasionally appearing flooding has displaced thousands from their home (Meze-Hausken 2004).

Temperature and precipitation are the two important weather parameters consistently sensitive to crop production; particularly rainfall amount and distribution over crop and pasture growing period significantly affect the livelihood of Ethiopian smallholder farmers and pastoralists (Tefera 2012). Historical weather station data evidenced that the seasonality of rainfall varies in different areas of Ethiopia (Li et al. 2015). In the eastern part of the country in Somali region, rains come twice a year – during the March–June Belg season and during the October–December Deyr season. In the north, western, central, and south parts of the country, most areas receive both Belg (March–June) and summer (June–September) Kiremt rains (NMSA 2007).

Many farmers plant slowly maturing but relatively high-yielding "long cycle" crops that grow during both the Belg and Kiremt seasons like maize, sorghum, and millet that account for half of national food production (Tefera 2012). However, during the past 20 years, the areas receiving sufficient Belg rains have contracted by 16%, exposing densely populated areas in the rift Valley and southern and eastern Ethiopia at risk (USAID 2015). The decline in Belg rain means areas growing long-season growing crops will be affected and more inhabitants will be exposed to chronic food shortage. Therefore, farmers in these areas already have started adaptive measures in shifting their choice of crop from long-growing season to short-growing crops like teff, wheat, and pulses (Tefera 2012). The challenge here is that the productivity of short-growing crops is lower than long-growing crops, which should be compensated by the use of high-yielding short-seasoned improved varieties and cultivars.

If recent warming trends continue, most of Ethiopia will experience more than a 1.0 Celsius (°C) increase in air temperature at the end of the century, with the warming tendency projected to be greatest in the south and eastern parts of the country (Li et al. 2015). This warming will intensify the impacts of droughts and crop failure and could particularly reduce the area of productive crop and pasture land. The current low precipitation and increased warming in the eastern and southern part of Ethiopia coincided with already heavily populated, where all available fertile land has been cultivated and the size of landholdings is diminishing due to population growth (USAID 2015). It appears likely that the combination of population growth, land degradation, and more frequent droughts will result in more frequent food-related crises and social unrest.

Fortunately different climate and crop grow models forecasted that the northern, central, and western parts of Ethiopia continue to receive adequate rainfall for crop growth (NMSA 2007; Funk et al. 2012; Li et al. 2015). The current study area is located in the central Ethiopian Highland that will help to justify the opportunities for better agricultural production under global climate change and variability scenario. The positive trend in the future rainfall in parts of the country will help to offset a catastrophic national failure to food security and drought observed in the rift valley, south, and east parts of Ethiopia (Funk et al. 2012; USAID 2015). Improved crop varieties suitable to climatically secure areas could help mitigate the impact of climate change in the rift valley, south, and east part of Ethiopia (Tefera 2012).

Finally, it is worth important to understand the direct effect of elevated CO_2 on crop yield too. Atmospheric CO_2 concentrations for C_3 crops (i.e., wheat, rice, and soybean) have shown an increase in average yield, while C_4 crops' photosynthetic pathway is unresponsive to elevated CO_2 (Stallknecht 1997). Teff, maize, and soybean are among C_4 crops that would not be positively affected by carbon fertilization.

The crop-livestock mixed production systems of the current study area are characterized by mild climatic conditions which suite for growing a wide range of annual short-seasoned crops by rotating cereals with pulses (IPMS 2004). The crop residues of these crops are the major sources of animal feeds. The agroecology characteristic provides farmers with very high flexibility for modifying management practices to better adapt to climate variability and change. These would include expansion of lowland crop into cooler highland regions, switches to new varieties, or shifts toward earlier planting dates. However, availability of improved crop varieties and other inputs, capital, land access, and policies limit to quickly respond to adaptive measure to climate change and variability. On the other hand, increasing food price can be considered as an incentive to innovate in agricultural production and productivity.

This manuscript is aimed to assess the impact of climate change and variability on agriculture in the central Ethiopian highlands based on historical weather data and to predict future geographically suitable areas for two predominant crops, namely, teff and wheat. The paper also highlights adaptive management practices to improve the agricultural production systems and decision-making in relation to climate change and variability.

Research Methodology

Description of the Study Area

The current study area, Ada'a district, ecologically belongs to high potential livestock and crop production system in Ethiopia (IPMS 2004). Administratively it is 1 of the 12 districts in East Shoa Zone, Oromia Regional State of Ethiopia, and is located about 45–60 km southeast of the capital, Addis Ababa. The study site falls within the western margin of the great East African Rift Valley.

The study area has a bimodal rainfall pattern, with the small rains occurring from March to May (Beleg rain) and the main rains from June to September (Mehere rain). Based on the long-term data, more than 75% of the annual rainfall was received during the main rains, when cropping normally takes place (Kahsay 2004). The highest annual rainfall was recorded in the year 1966, where the annual rainfall reached 1287 mm, much higher than the average 890 mm. The small rainy season (Beleg rain) could sometimes be erratic, but there is no cropping during this period, except it is important for softening the land which will facilitate land preparation in addition to growth of some grasses for livestock.

Sixty-year data from 1940 to 2000 (Kahsay 2004) indicated that the average minimum and maximum air temperature ranged from 7.9 °C to 28.2 °C, respectively, with mean annual temperature 18.5 °C. The highest average temperature is recorded for the months of March, April, May, and June. While the months of October, November, and December had the lowest average temperatures.

The agroecology in the district is best suited for diverse agricultural productions. The district is nationally popular for its excellent quality teff (Teff is an important food grain in Ethiopia where it is used to make Injera/bread which is staple food for 60% of Ethiopians). Wheat is the next abundant crop grown in the district. Pulses, especially chickpea, grow in the bottomlands on residual moisture and in rotation with cereals.

Altitude of the district ranges from 1500 m to greater than 2000 m above sea level. Two major agroclimatic zones identified in the district are (IPMS 2004) (see Fig. 1 for topography of Ethiopia and the study site in parallelogram shape):

- The mountain zone located over 2000 m, covering 150 km^2, (9%) of the area.
- The highland zone at an elevation range of 1500–2000 m extending over 1600 km^2 covers 91% of the area.

Meteorological Data

Sixty-year meteorological data from 1950 to 2009 was obtained from the Ethiopian National Meteorological Service Agency (NMSA) from weather station located in the study area. The meteorological data are on temperature in Celsius degree centigrade, rainfall in annual mm averaged for 365.25 days, wind speed measured in meter per second, sunshine in hours, and relative humidity in percentage. The meteorological data were compared by decades. The weather data reported here is based on information from Debrezeit Air Force weather station which is one of the oldest weather stations in the country. The data used was six-decade (1951–2009) daily record for rainfall, temperature, and relative humidity, three-decade daily record for sunshine hours, and two-decade daily record for wind speed.

Socioeconomic Survey

As part of the study, 110 farmers were selected to review their perception about climate change and the variability of their adaptation measures. These farmers were those who had participated in extension system provided by the Ethiopian Ministry of Agriculture. The reason for the selection was that these farmers were most likely able to provide information on climate change indicators and adaptation around their location. The farmers were asked about their perception of climate change, indicators for climate change, and the adaptation methods they have followed to mitigate climate change. Structured questionnaires written in English were used as research tools to collect data, and interview was done in local language, Oromifa. Questionnaires were made simple by the use of open-ended questions. Interviews were conducted orally with household heads (either male heads or female heads) in their homes. The collected data was transferred into a spreadsheet using the Statistical Package for Social Sciences (SPSS 2008) 17.0.1 version.

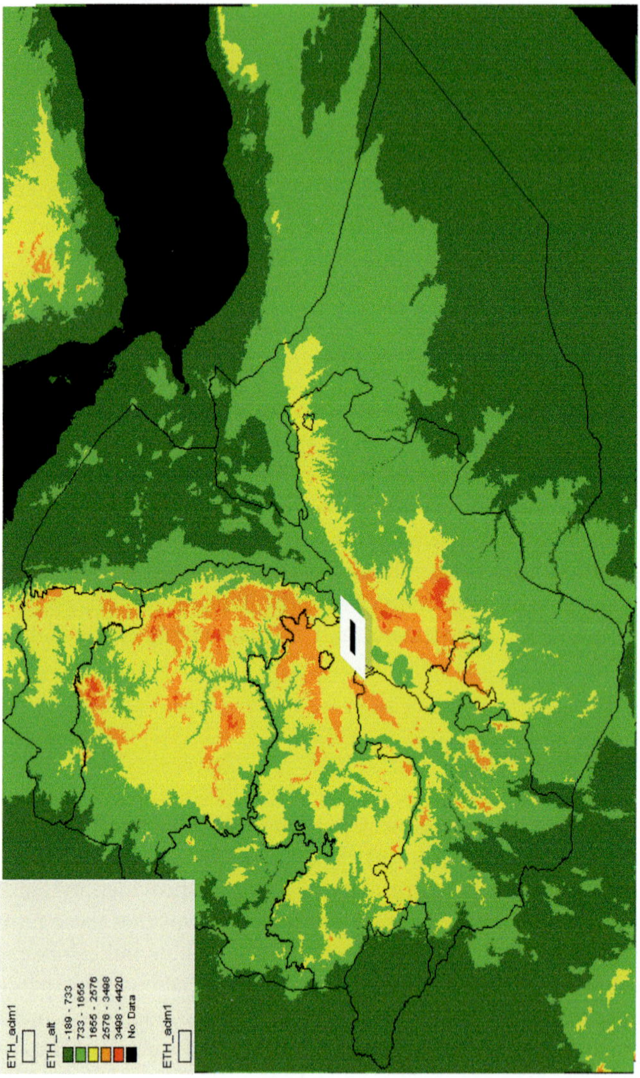

Fig. 1 Topography of Ethiopia. (Study site designated in parallelogram)

Suitability Analysis Using DIVA-GIS

DIVA-GIS is a geographical information system to analyze plant biodiversity data. This model used FAO EcoCrop database of the environmental requirements of a long list of plant species (Hijmans et al. 2005). This can be used to identify possible crop to grow in particular environment. DIVA-GIS implemented EcoCrop to predict the adaptation of crop over geographical areas (White et al. 2001) that could also be used to estimate the crop residue potential for future livestock feed. Currently, only temperature and precipitation data are used to make this prediction. To determine the suitability of a growing season for a certain crop, the following temperature parameters were used:

KTMP: absolute temperature that will kill the plant
TMIN: minimum average temperature at which the plant grows
TOPMN: minimum average temperature at which the plant will grow optimally
TOPMX: maximum average temperature at which the plant will grow optimally
TMAX: maximum average temperature at which the plant will cease to grow

To determine the suitability of the rainfall in an area for particular crop, the following rainfall parameters are used:

Rmin: minimum rainfall (mm) during growing season
Ropmin: optimal minimum rainfall (mm) during the growing season
Ropmax: optimal maximum rainfall (mm) during the growing season
Rmax: maximum rainfall (mm) during the growing season

EcoCrop model runs for rainfall, for temperature, or for both (the boxes to tick are in the predict table). To run the model, we selected a species in the select tab, searching for species using options in Filter by. Then, we selected the crop of interest by clicking on it. In the parameters tab, we inspected the temperature and rainfall parameters from the EcoCrop database and changed them as desired. The most recent GCM climate projections that are used in the Fifth Assessment IPCC Report were applied. Teff and common wheat were the two selected crop types to run DIV GIS version 1.5 (Fick and Hijmans 2017); these two crops are selected for their dominance in area coverage and production as well as relevance as crop residue for livestock feeding in the highlands of Ethiopia (Berhe 2010). For future conditions, downscaled global climate model (GCM) data from CMIP5 IPCC Fifth Assessment was used (IPCC 2014). The change of suitability in the present and future was calculated by the difference of future-current use by grid overlay and subtraction. The data available here are the IPPC5 climate projections from global climate models (GCMs) for four representative concentration pathways (RCPs). The GCM output was downscaled and calibrated (bias corrected) using WorldClim 1.4 as baseline "current" climate (Fick and Hijmans 2017).

Methods of Data Analysis

Detailed statistical analysis of the data was carried out by using SPSS (2008) version 17.0.1. Weather data were analyzed for decade mean and tested for their significant difference among decades. Variables that were the most representative of the study area extracted from the socioeconomic survey were computed for univariate analysis (effects of climate change on agricultural production, perceptions on climate change, and factors that influenced decisions to adapt to climate change).

Results

Historical Weather Data for the Past 60 Years (1951–2009)

Climate is said to be stable if the long-term average does not change significantly over time. However, climate in Ethiopia and worldwide is changing. Climate change refers to a statistically significant trend in climate over many decades. It is superimposed on natural yearly-to-decadal climate variability. Sixty-year climatic data for rainfall, temperature, wind speed, sunshine hours, and relative humidity in the weather station in the study area is reviewed in this section of the paper.

Rainfall Variability and Trend

This baseline climate data was developed for precipitation from 1951 to 2009 at Debrezeit Air Force stations in the central Ethiopian highland. The data was obtained from National Meteorological Agency of Ethiopia archives. Mean annual rainfall shows large spatial and temporal variation. The Ethiopian National Meteorological Services Agency (NMSA 2007) defines 50 to 75% of a 30-year average as "below-normal" rainfall and designated as dry whereas 0 to 50% of average rainfall as "much-belownormal" rainfall and indicator of drought (Meze-Hausken 2004, Wilhite 2000). An annual rainfall above 30-year average is considered wet year (2007). The study area has experienced both dry and wet years over the last 60 years. (NMSA 2007) reported that years like 1965, 1984, and 2001 were extremely dry and drought years, while 1961, 1964, 1967, 1977, 1996, and 2007 were very wet years.

In the current study area, which is located in a climatically secured area, decade-based trend analysis of annual rainfall shows that rainfall remained more or less constant when averaged over the whole decades. The least significant difference (LSD) analysis shows that beginning from decade three to six, there is no significant difference in rainfall amount among decades. However, there is a significant difference between decade one and two and the other decades of three, four, five, and six. Generally speaking, the rainfall has increased by 7.5% in the past three decades (1990–2009) with an increase of rainfall amount by 2% for each decade (Fig. 2). This is in agreement with

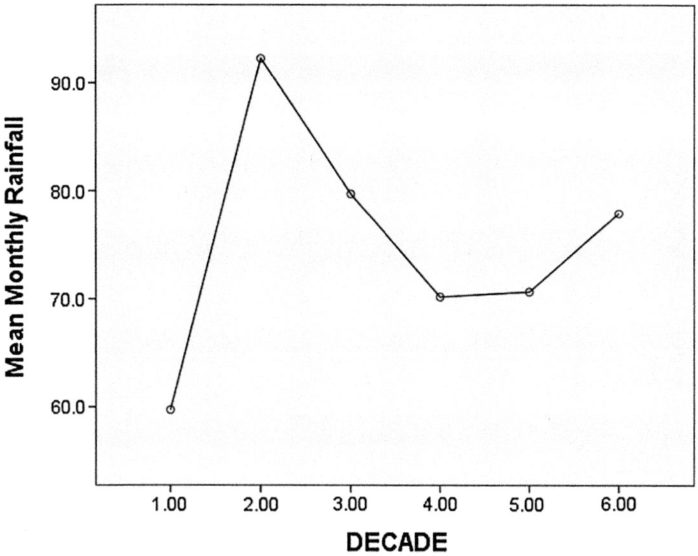

Fig. 2 Estimated monthly means of rainfall in mm by decade. (Decade 1 = 1951–1959, Decade 2 = 1960–1969, Decade 3 = 1970–1979, Decade 4 = 1980–1989, Decade 5 = 1990–1999, Decade 6 = 2000–2009)

NMSA (2001) and that of Funk (2012) who reported an increasing trend in annual rainfall from 1990 to 2009 along central and northern Ethiopian highlands and western lowlands.

Temperature Variability and Trend

Temperature variability is high in Ethiopia; for instance, years like 1957, 1958, 1973, 1987, 1995, and 2002 were very warm, while 1964, 1967, 1968, 1975, 1977, 1989, and 2003 were very cool years (NMSA 2007). From the current weather data, the average annual maximum temperature of the study area has increased on average by about 0.08 °C per decade totaling 0.5 °C over six decades (Fig. 3); similarly average annual minimum temperature has increased by about 0.08 °C every decade (Fig. 4), which is close to the estimate of NMSA (2001) of 0.10 °C per decade at country level. However, Funk et al. (2012) reported rapid increases in later part of *Belg* (March–May) (0.6 °C/decade) – but significant increases in the *Kiremt* (June–September) (+0.4–0.6 °C/decade) for highlands of Ethiopia based on meteorological data extrapolated from five selected stations for the years 1951–2009.

In the present study, it is interesting to note that the average annual minimum temperature and the average annual maximum temperature increased at the same pace (Figs. 3 and 4). However, in both maximum and minimum temperatures, the LSD analysis shows that there are no significant temperature differences across most decades except decade six in case of maximum temperature, which indicated that

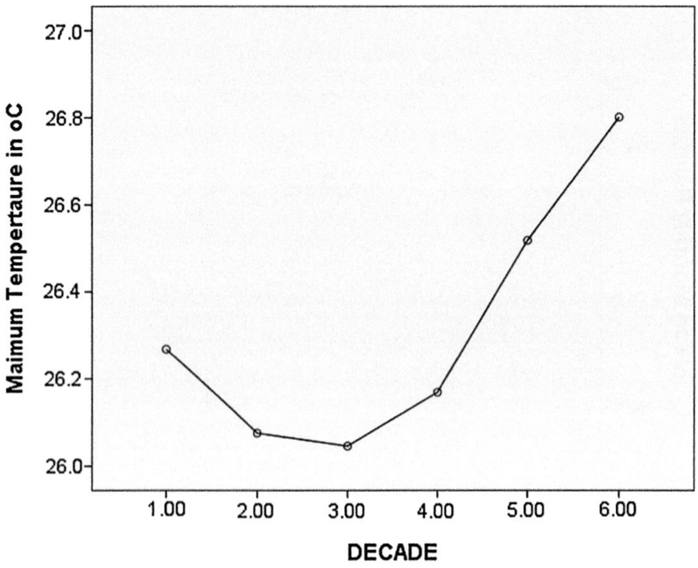

Fig. 3 Estimated monthly means of maximum temperature by decades. (Decade 1 = 1951–1959, Decade 2 = 1960–1969, Decade 3 = 1970–1979, Decade 4 = 1980–1989, Decade 5 = 1990–1999, Decade 6 = 2000–2009)

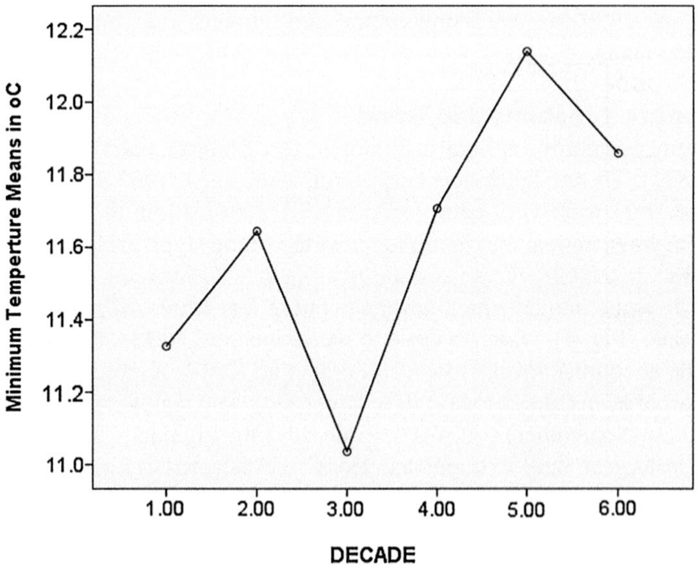

Fig. 4 Estimated monthly means of minimum temperature by decades. (Decade 1 = 1951–1959, Decade 2 = 1960–1969, Decade 3 = 1970–1979, Decade 4 = 1980–1989, Decade 5 = 1990–1999, Decade 6 = 2000–2009)

maximum temperature increased significantly in the sixth decade (2000–2009). For minimum temperature, the significance level started since fourth decade (1980s) and was highest in fifth decade (1990s) and declined back in the sixth decade.

Relative Humidity

Humidity together with temperature and pressure is one of the fundamental variables that characterize the local state of the atmosphere (ASA 2005). Within the realm of agriculture, humidity measurements have a wide variety of applications, including prediction of plant disease and prediction and maintenance of livestock feed requirement (ASA 2005). It is critical to note that for crop production, it is actually soil moisture that is the key variable, not rainfall as soil moisture is affected by evaporation (Funk et al. 2012). In many locations, the primary threat to soil moisture is increased temperature as this increases evaporative losses both directly and through transpiration. Uncertainty in future rainfall amounts is therefore often of less relevance than sometimes thought, as increasing temperatures can result in soil moisture deficits even under increased rainfall conditions (USAID 2015). In the present study, there is no significant difference for relative humidity across the decades from 1951 to 2009, as shown in Fig. 5 below. This likely shows central Ethiopian Highland humidity remained constant for crop growth.

Sunshine Hours

Solar radiation is the largest energy source and is able to change large quantities of liquid water into water vapor (ASA 2005). The potential amount of radiation that can reach the evaporating surface is determined by its location and time of the year. Due to differences in the position of the sun, the potential radiation differs at various

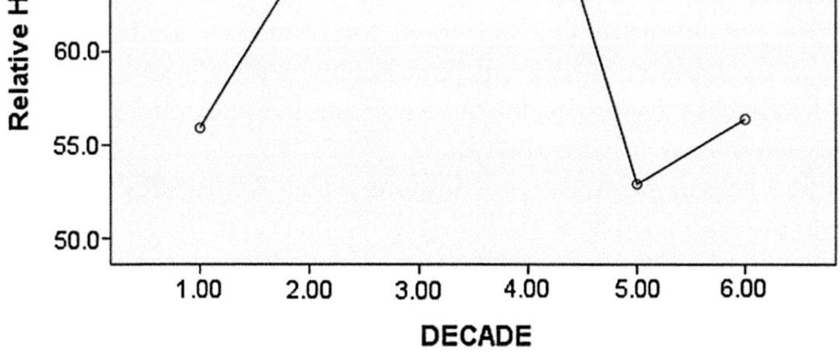

Fig. 5 Estimated monthly means of relative humidity by decades. (Decade 1 = 1951–1959, Decade 2 = 1960–1969, Decade 3 = 1970–1979, Decade 4 = 1980–1989, Decade 5 = 1990–1999, Decade 6 = 2000–2009)

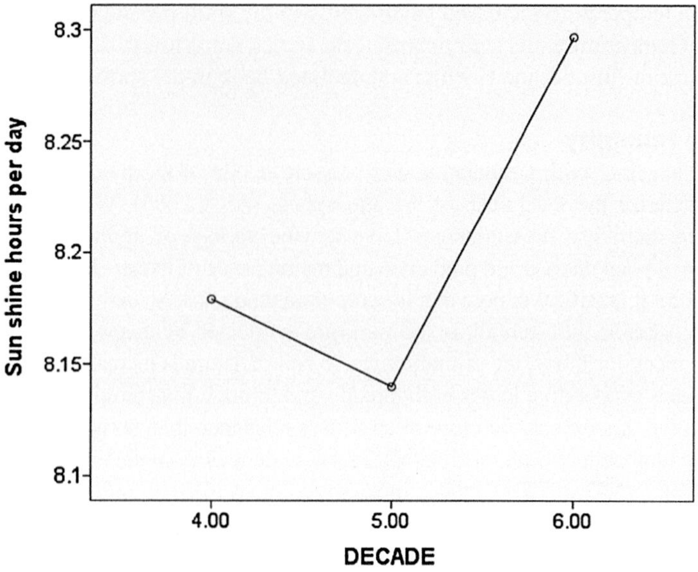

Fig. 6 Estimated monthly means of sunshine hours by decade. (Decade 4 = 1980–1989, Decade 5 = 1990–1999, Decade 6 = 2000–2009)

latitudes and in different seasons. The actual solar radiation reaching the evaporating surface depends on the turbidity of the atmosphere and the presence of clouds which reflect and absorb major parts of the radiation (ASA 2005).

In the current study area, statistically there was no significant difference for the sunshine hours in the last three decades. But, sunshine hours increased from average 8.15 h per day to 8.3 h per day (Fig. 6). The relative humidity remained constant in Fig. 5 probably due to the less likelihood of sunshine hours to affect evapotranspiration and hence crop production.

Wind Speed

Agriculturists are interested in wind speed and direction, as these data are used to determine the location of different farm buildings relative to each other and the design of windbreaks that protect the farmstead and livestock during the winter (ASA 2005). Real-time wind speed and direction information is critical to the spraying of crops and pest control. Wind current transports pollen that ensures the fertilization of crops. Wind provides ventilation for livestock buildings, cools animals in the summer, and disperses odor. Light wind speeds tend to move odors along the surface with little mixing, resulting in a concentration of the odors and complaints. However, strong wind can damage crops and farm structures. Wind also transports insects and disease that affect crops. Wind also affects evapotranspiration. The wind speed in the study area showed significant difference between decades and showed declining trend in the last two decades estimated at −1.6% as shown on Fig. 7 below.

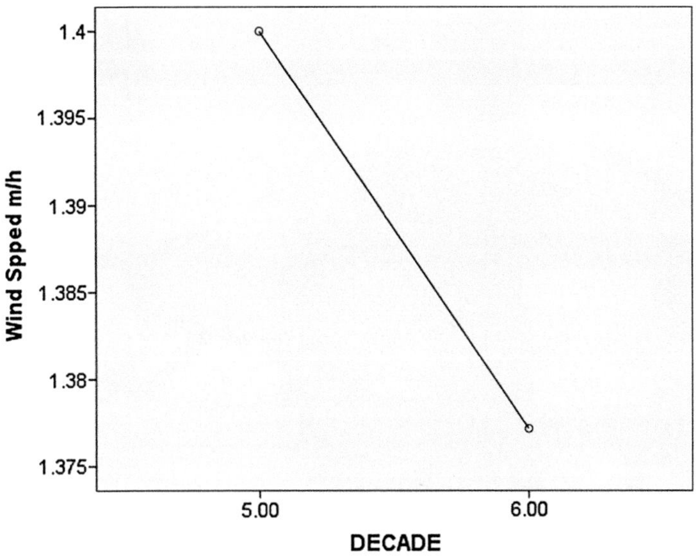

Fig. 7 Estimated monthly means of wind speed by decades. (Decade 5 = 1990–1999, Decade 6 = 2000–2009)

The decline in wind speed may have negative consequence for open-pollinating crops with long growth cycle like maize but might have less yield impact for closed- and self-pollinating crops like teff and wheat.

Present Teff and Wheat Production Conditions

Higher temperatures resulted in reduced soil moisture availability, water availability, and water quality. It also induced changes to timing and distribution of agricultural pests (USAID 2015). Higher temperature could also decrease reliability on groundwater sources and surface water during droughts or a prolonged dry season. The negative changes in rainfall pattern result in reduced crop productivity or failure, create new ecological niches for plant pests and diseases, and lead to the loss of pasture and animals. Similarly excessive rainfall damage to crop and infrastructure contributes to erosion and farmland degradation.

In this study, suitability assessment for crop sector has been made taking wheat and teff as the two major crops growing in highlands of Ethiopia. For example, wheat is one of the five top-level crops grown in Ethiopia, and the country is the second largest grower of wheat in sub-Saharan Africa (White et al. 2001). Precipitation and minimum temperature are the key determinants of potential wheat production. Rainfall with a minimum of at least 350 mm and a minimum temperature between 6 °C and 11 °C during the wettest quarter

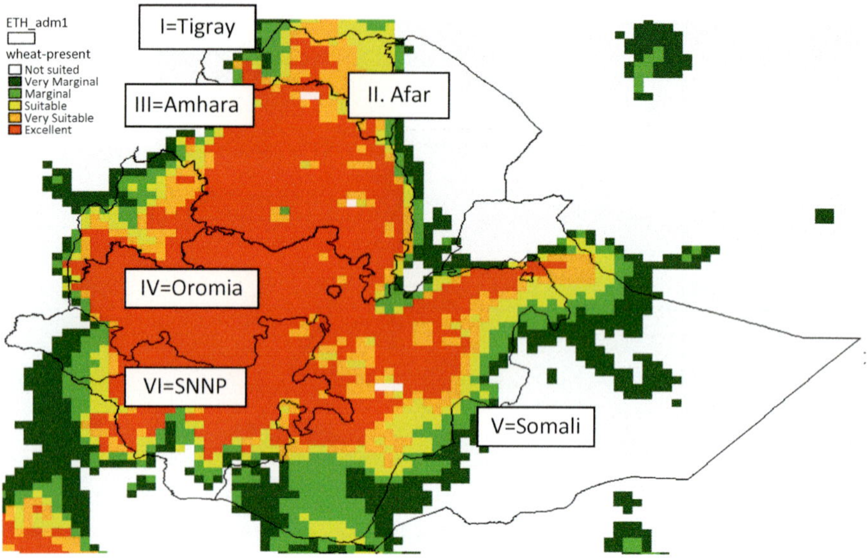

Fig. 8 Current suitability area for common wheat in Ethiopia

of 3 months of growth period are important for wheat growing in Ethiopia (White et al. 2001). Figure 8 shows the current potential wheat-growing areas in Ethiopia mainly located on the highlands over 1500 m above sea level in Tigray, Amhara, Oromia, and northern SNNPR regional states. From these figures, it is possible to argue that wheat is more adaptive to highland cooler agroecology than teff that is adapted to all agroecologies including the lowlands of western Ethiopia (Ketema 1997).

Teff is a C_4 plant and is intermediate between a tropical and temperate grass (Stallknecht 1997). Teff is adapted to environments ranging from drought stress to waterlogged soil conditions and diverse soil types (Ketema 1997). Teff production has been reported at altitudes from 0 to 3000 m above sea level. Under current condition, teff is suitable in most area of the country covering different agroecology (Fig. 9). In Ethiopia, teff annually occupies over 2.8 million hectares which is 25–30% of the total area covered by cereals (Wondimu and Tekabe 2001). It is a daily food staple for about 60 million Ethiopians, 60% of the 100 million Ethiopian population. The quality of crop residue from teff is equivalent to grass species like Rhodes grass (Berhe 2010); hence the crop residue from teff is an important animal feed. At present, teff grows in all administrative regions of Ethiopia predominantly on the highland of Tigray, Amahra, Oromia, and SNNPR regions (Fig. 7). From DIVA-GIS suitability analysis, teff is excellent to grow in lowlands regions of the country in Gambella, Benishangule, and south SNNPR, areas which aren't growing teff at present (Fig. 7). This result shows the future potential area for expansion of teff production in the western lowlands of Ethiopia.

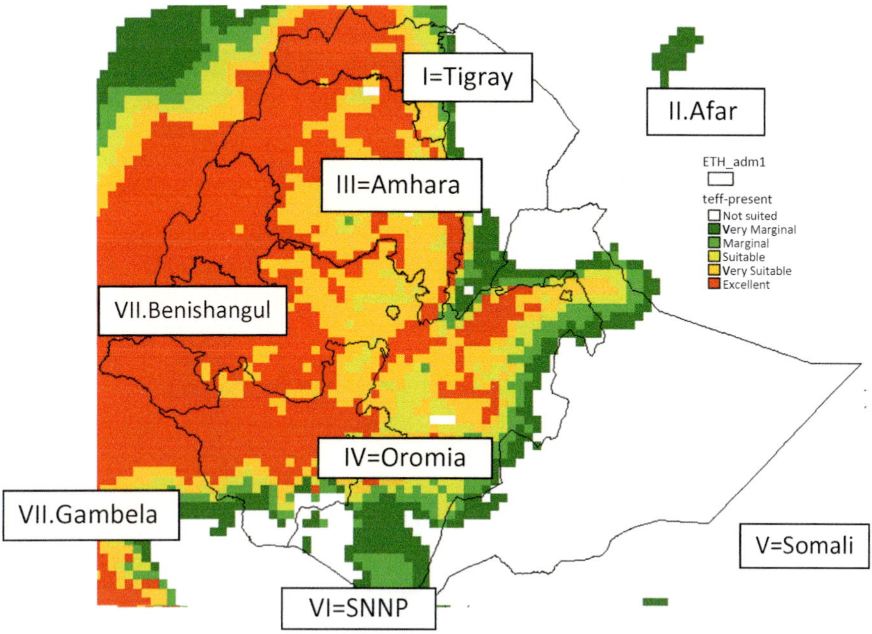

Fig. 9 Current suitability area for teff in Ethiopia

Future Geographic Coverage of Teff and Wheat

Potential Wheat Production Areas

Water deficits and warm night temperatures seemed to be key factors delimiting bread wheat production areas in Ethiopia (White et al. 2001). Potential new areas for bread wheat production were identified by assuming that technologies could be developed to allow wheat to be grown in drier or warmer environments that lead to an increase in yield (White et al. 2001). In the climate change crop suitability assessment, wheat growth expands to the east and south part of the country particularly in south Oromia, south of SNNP, and north of Somali regions (see Fig. 10). The unsuitable areas for wheat that are currently located in south of Oromia and north of Somali regions will turn into suitable area; this areas' suitability increases by 13–40% (see Fig. 11). On the other hand, there is shrinkage in wheat suitability area in western side of the Ethiopian highlands, namely, in the regions of Tigray, Amhara, Oromia, and SNNPR that suitability is estimated to reduce by 24–40% (see Fig. 11).

Teff

Teff and other C_4 species, such as maize, sorghum, and millet, have an evolutionary advantage of a photosynthetic pathway over C_3 species such as wheat and cool season grasses (Stallknecht 1997; Wondimu and Tekabe 2001). This C_4 pathway results in efficient water use and allows plants to photosynthesize faster under high

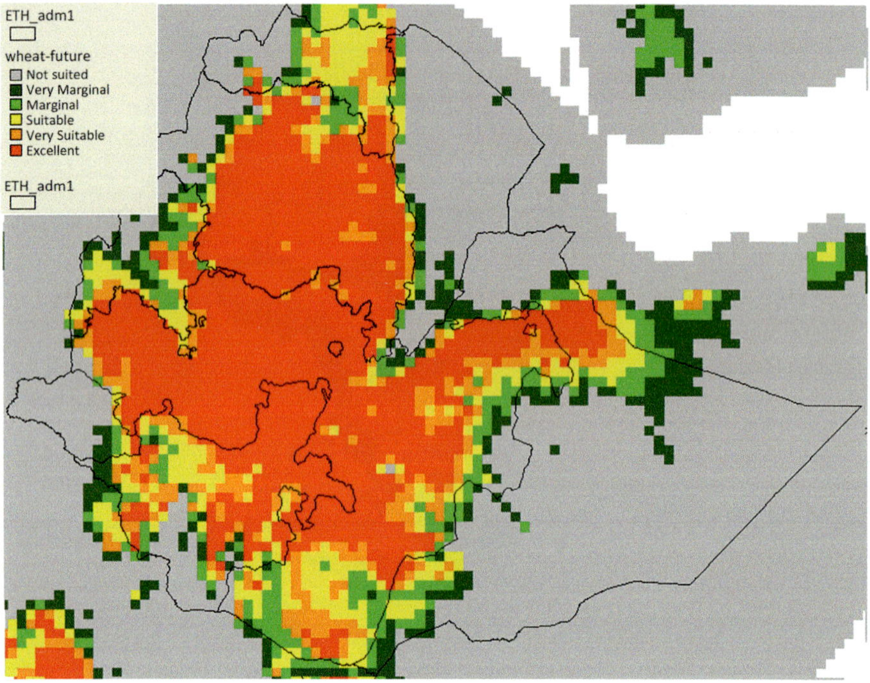

Fig. 10 Wheat production potential areas in the future

heat and light conditions than C_3 species (Stallknecht 1997). DIVA GIS analysis on the agroecological suitability of teff in relation to climate change from 2050 to 2070 indicated that generally areas suitable to teff under climate change scenarios will increase in almost all administrative regions of Ethiopia (Fig. 12). Marginalized currently unsuitable areas to the crop get bigger in suitability, which indicates the creation of appropriate agroecological niche to teff under climate change scenarios that predict increased precipitation and higher temperature. The most benefited regions are the eastern lowland like Afar region, south Oromia, South SNNPR, and Somali regionals (see Fig. 13). The highland areas remain suitable, even some agroecological zones become excellent under climate change scenario. Since teff is a C_4 grass-like corn and sorghums, it produces best during periods of maximum heat unit accumulation provided that moisture is at optimum.

Farmers' Perception of Climate Change

Farmers in the study area were asked about the communal resources accessibility to all community members. Accordingly they have listed and ranked the resources indicated in the Table 1. The highly accessed resources were water, wood, and grasses. They are mainly used to sustain life of rural community and their livestock.

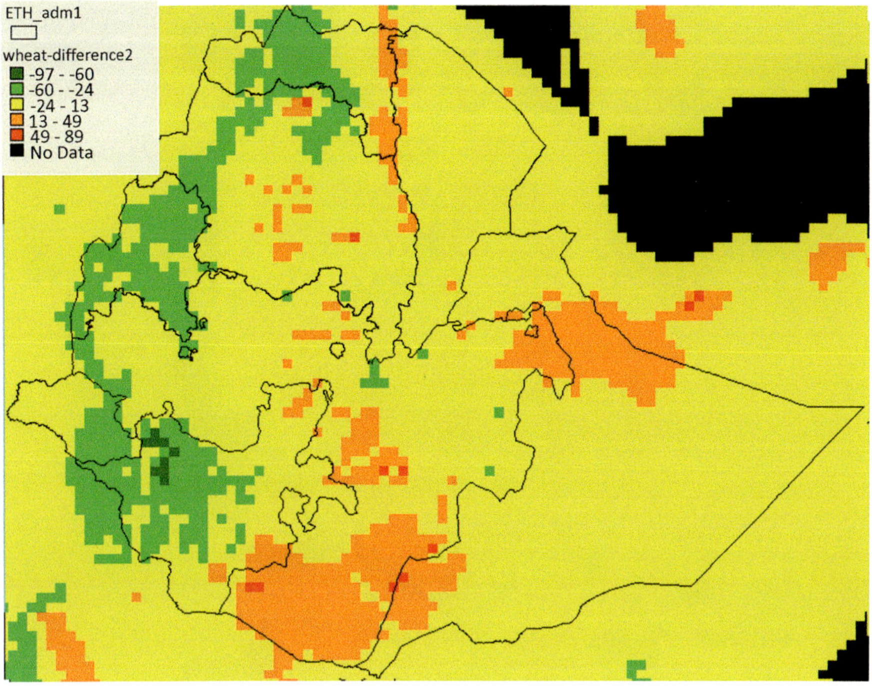

Fig. 11 Wheat suitability difference between the future and the present time

These resources were reported highly depleted for lack of management and ownership. Some of the resources like grass for house roofing, wild fruits, and medicinal plants were exhaustively depleted so that their accessibility is limited due to unavailability. As human and livestock number expands, still more pressure was put on natural resources, including the dwindling forest, grazing land, as well as freshwater.

Farmers listed detritus and decomposers which were existing during their forefathers and had important role in decomposition and humus formation. These detritus and decomposers were listed as earthworm, millipede, army worm, fiddler crabs, termites, ants, wood battles, snail, and birds. The risk status of detritus and decomposers as described by 39% of the respondents were decreasing listed as under threat by 20% of who said at risk, endangered by 50%, and disappeared by 5%. Farmers believed that the presence of decomposers and detritus was a sign of good harvest for 15% of the respondents, a sign of good agroecological balance for 10%, and an indication of soil fertility for 50%. To farmers, weather indicators for climate change were mentioned as change in patterns of rainfall, temperature, and sunshine intensity. For rainfall, 40% of the respondents mentioned erratic rain, nonseasonal (coming late from usual schedule and stopping early), and torrential that cause floods. In case of temperature, 35% of respondents reported increasing temperature caused difficulty for long-hour labor workout in the field, inconvenience for

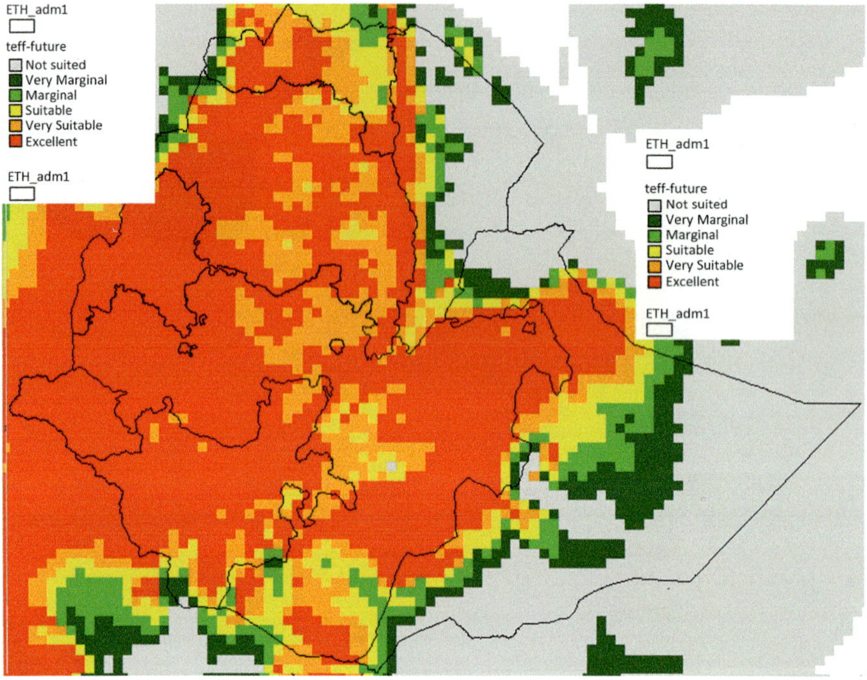

Fig. 12 Teff production potential area in the future

physiological breathing, and increased water consumption. Crop growth also is reported to become faster than ever. Temperature increase also changed the lifestyle such as wearing light cloth throughout the year without seasonal difference. Light intensity was increased as reported by 25% of the respondents who described symptoms like human facial discoloration into dusty dark and early wilting of plants. All respondents unanimously reported status of indigenous trees and wild animals as endangered or in disappearance stage. Cultivation area and livestock population have relatively increased.

Farmers' Adaptation Measures to Climate Change

Some farmers (25%) consider climate change as God's punishment for his anger toward human sin. These farmers believed that climate will come to normal upon God's forbearance and forgiveness to human sins. About 75% believed that anthropogenic factors contributed to climate change. In the current study, it was found that lack of land for cultivation was the major obstacle to farmers' adaptation to climate change. To practice different land rehabilitation practices such as conservation tillage, fallowing, intercropping, and area enclosure, land has to be available. Other major issues identified by farmers for adaptation to

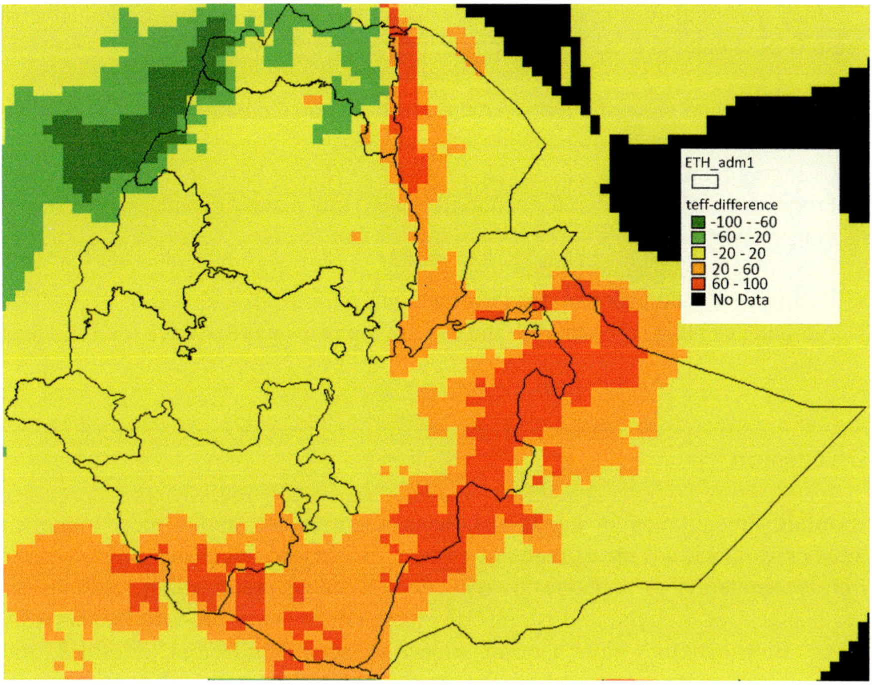

Fig. 13 Teff suitability difference between the future and the present time

Table 1 Communal resources accessed by community

	Communal resource type	Rank
1	Water for human use	1
2	Wood for household energy use	2
3	Wood for construction	3
4	Water for livestock use	4
5	Communal grazing	5
6	Charcoal making for income generation	6
7	Medicinal plants	7
8	Ritual trees and plants	8
9	Wild fruit	9
10	Grass for roofing	10

climate change were lack of information on adaptation strategies, the absence of credit, shortage of labor and availability of water, as well as poor soil fertility. In the past three decades, farmers have adopted crops like chickpea and vetch in response to climate change by expanding cultivation area by threefold from base year 2005 to 2009. Chickpea and vetch can stand moisture stress and could survive on residual moisture and could also grow in crop rotation with cereals to increase soil fertility.

Other potential adaptation measures listed by farmers to cope with adverse impacts of climate change on crop production and crop residue were:

- Improving and changing management practices and techniques such as planting date, seeding rate, fertilizer application rate, etc.
- Changing crop regions
- Proper use of climate information for land use planning and early warning systems, etc.
- Promoting irrigation agriculture
- Enhancing erosion control and reforestation
- Adopting suitable crop varieties and developing new ones that have food and feed value

Discussion

Rainfall: Precipitation in liquid and frozen form is one of the most important weather variables for an agricultural system. To survive, all plants rely on water from precipitation to recharge the soil. Periods of no precipitation result in the development of drought, and periods of excess precipitation produce floods (ASA 2005). Both extremes have a major impact on agricultural crops, livestock, and operations. However, many other natural and human factors can affect water availability (Wilhite and Glantz 1985) whether for crop growth, pastures, or household use. The interconnectedness of changes in temperature, sunshine hours, wind condition, evapotranspiration, or other physical factors such as soil fertility, soil depth, soil type, and vegetation cover also affect water availability (Meze-Hausken 2004).

The rainfall pattern of Ethiopia is affected by the interactions among seasonal intertropical convergence zone (ITCZ) movement, land surface heating, and topography that provide favorable conditions for the generation and development of precipitation in Ethiopia (Funk et al. 2012). El Niño-Southern Oscillation (ENSO) significantly affects rainfall intensity in Ethiopia, strong El Niño years lead to decreased Kiremt-season precipitation, and strong La Niña years result to increased precipitation (Segele et al. 2009). Likewise, the Indian Ocean Dipole (IOD) and Mediterranean Ocean sea surface temperatures affect wind dynamics and moisture supply, respectively (Li et al. 2015). In addition, sea surface temperature anomalies (SSTAs) over the North Atlantic can mainly influence the low-frequency variability of precipitation (Slingo et al. 2005).

According to USAID (2015) between the mid-1970s and late 2000s, Belg and Kiremt rainfall decreased by 15–20% across parts of southern, southwestern, and southeastern Ethiopia. Similarly during the years 1990–2009, the areas receiving sufficient Belg rains have contracted by 16%, exposing densely populated areas in the rift valley, south, and east part of Ethiopia to climate change vulnerability (Funk et al. 2012).

However, areas of Ethiopia at 8°N and areas west of 38°E appear likely to continue to receive more than 800 mm of precipitation during the Belg and Kiremt season (Funk et al. 2012). Therefore, Ethiopia does not face a catastrophic national failure of rainfall but rather is regional hot spots with a tendency toward more frequent droughts. Although these droughts will tend to affect most pastoralists, agropastoralists, and some of the highlanders living in climatically marginal areas, improved yields in climatically secure areas could help mitigate the impact of those effects (USAID 2015). The crop suitability analysis of this study also confirms the marginalized east and west parts of Ethiopia will be very suitable for teff growing due to climate change scenarios that predict increased precipitation in coming century in Ethiopia. The highland areas remain suitable for both teff and wheat, even some agroecological zones become excellent under climate change scenario.

Temperature: For each plant species, a defined range of maximum and minimum temperatures are set to form the boundaries of observable crop growth. For example, common wheat Triticum aestivum grows best at an optimum temperature between 15 °C and 23 °C, and the maximum temperature to cease growth is 27 °C (White et al. 2001). Teff Eragrostis tef grows comfortably at an optimum temperature between 22 °C and 28 °C, and the maximum temperature to cease growth is 30 °C (Stallknecht 1997). Extreme exposure of plants to extreme temperature affects yield which is determinant during reproductive stage. Hatfield and Prueger (2015) reported increases in minimum air temperature may be more significant in their effect on growth and phenology of crop than maximum temperature. Minimum air temperatures affect nighttime plant respiration rates and can potentially reduce biomass accumulation and crop yield (Hatfield and Prueger 2015). Rasul et al. (2009) found that higher minimum temperatures reduced grain yield in rice, while higher maximum temperature raised yields because the maximum temperature seldom reached the critical optimum temperature of particular crop. Funk et al. (2012) reported that if recent warming trends in Ethiopia continue, in the end of twenty-first century, most of Ethiopia will experience more than a 1.0 Celsius (°C) increase in air temperature, with the warming tendency projected to be greatest in the south and east part of the country. In Pakistan, rice grain yield declined by 10% for each 1 °C increase in minimum temperature (Rasul et al. 2009). Funk et al. (2012) observation based on 1951–2009 weather station data indicated that central Ethiopian Highland temperature turns out to became hotter – with more rapid increases from +0.4 to 0.6 °C/decade. This warming will intensify the impacts of droughts and could particularly reduce the amount of productive crop land for highland crops like wheat, since wheat plants typically prefer temperatures cooler than 23 °C. One potential strategy to mitigate the impact of increasing temperature is to select varieties which open their pollen in the early morning when temperatures are cooler (Rasul et al. 2009). Teff is one of the crops that open their pollen early in the morning between 4 AM and 6 AM (Ketema 1997). Therefore, teff is a candidate crop to adapt to extreme hotter climate condition in the future Ethiopian agriculture. Since teff is a C_4 grass-like corn and sorghums, it produces best during periods of maximum heat unit accumulation provided that moisture is at optimum. However, wheat crop suitability analysis shows that due to increase in temperature, shrinkage in agroecological suitability is observed in the southwestern part of Ethiopia (Fig. 11).

When assessing the effect of solar radiation on evapotranspiration, it shows that not all available energy is used to vaporize water. Part of the solar energy is used to heat up the atmosphere and the soil profile (ASA 2005). In the current study, no significant change is observed in evapotranspiration due to increased solar radiation, which calls for further investigation for no change in the result. Jhajharia et al. (2018) reported a decrease in sunshine duration in Northeast India that modified the evapotranspiration process and affects the crop water requirements, which, in turn, negatively affected the agricultural production. A similar research topic should be conducted in Ethiopia to assess the impact of climate change on crop performance at the laboratory and field level. While there is a very strong link between soil moisture and rainfall, other variables such as soil depth, slope, temperature, and winds strongly influence soil moisture conditions and hence crop performance (Rasul et al. 2009). Therefore, the future climate and agriculture research should encompass other physical characteristics like soil type, soil depth, land slope, soil moisture retention capacity, etc. The decline in wind speed may have negative consequence for open-pollinating crops with long growth cycle like maize but might have less yield impact for closed-pollinating crops like teff and wheat.

Climate change in Ethiopia is related with agricultural production and hence directly linked to food security. In addition to climate change, the food security situation in Ethiopia is also affected by rapid population growth and land and water resource degradation. Current estimated human population of Ethiopia is over 100 million; at a population growth rate of 3.2%, the Ethiopian population doubles every 22 years (Worldometers 2017). Crop statistics from Food and Agriculture Organization of United Nations indicate that per capita cereal production in Ethiopia in 2017 is relatively low estimated at 150 kg per person per year (FAO and STAT 2017). Although agricultural yield is expected to improve, it is difficult to double agricultural yield every 22 years in order to keep the pace with human population growth. If these trends of gaps in population growth and food production continue, millions more Ethiopians will be exposed to hunger and malnutrition. Therefore, in addition to climate change adaptation, Ethiopia has multitudes of tasks to cope up climate change with social and economic development such as improving and revisiting country's population policy, rehabilitation of degraded land and water resources, and finding alternative livelihood option to its population. If the country doesn't mitigate with existing climate change adaptation measures and continued with population explosion, social unrest, emigration, terrorism, hunger, and malnutrition will follow.

Conclusion

Based on 60-year weather data, climate change in central Ethiopian highlands is real. Climate adaptation has to be accepted, and that uncertainty in climate change should be recognized as unavoidable. Effective communication will benefit all the direct vulnerable groups and decisionmakers for adaptation to climate change. In 2018, Ethiopia, with a population of 100 million and a population growth rate of 3.2%

(a doubling time of 22 years), faces increased levels of food insecurity. For Ethiopia to cope with climate change-related shocks and population pressure, there are four mitigation options: (1) improved water and agricultural management practices, (2) family planning and limiting child birth, (3) changing or mixing the current staple food items into a high calorie per hectare crops, and (4) livelihood diversification to generate off-farm income. These four options need decision at policy and household level. Raising yields in the more agriculturally viable areas may be a better option than extending agriculture into ever more marginal areas. Agricultural development in the central and western highland areas is much less threatened and, with better agricultural practices and water management, could dramatically enhance food security at the national level. Under climate change scenarios, there are food and feed crops and cultivars that could have best adapt to the changing environment. The future agriculture research should be geared toward identifying the type of crops adapting to each agroecological niche, improving their productivity and nutritional availability. Raising motivation is very important to direct vulnerable groups to change farming practices and adapt so as to use the opportunity to their own advantage.

References

American Society of Agronomy (ASA) (2005) Micrometeorology in agricultural system. Agronomy monograph no. 47. American Society of Agronomy Inc., Madison

Berhe T (2010) Increasing productivity of teff: new approaches with dramatic results. Sasakawa Africa Association, Regional Office, Addis Ababa. Report, January 2010

Easterling WE, Aggarwal PK, Batima P, Brander KM, Erda L, Howden SM, Kirilenko A, Morton J, Soussana JF, Schmidhuber J, Tubiello FN (2007) Food, fibre, and forest products. In: Parry ML, Canziani OF, Palutik JP, van der Linden PJ, Hanson CE (eds) Climate change 2007: impacts, adaptation and vulnerability. Contribution of Working Group II to the Fourth Assessment Report of the Intergovernmental Panel on Climate Change. Cambridge University Press, Cambridge

FAO, STAT (2017) http://www.fao.org/faostat/en/#country

Fick SE, Hijmans RJ (2017) Worldclim 2: new 1-km spatial resolution climate surfaces for global land areas. Int J Climatol 37:4302–4315

Funk C, Rowland J, Eilerts G, Kebebe E, Biru N, White E, Galu G (2012) A climate trend analysis of Ethiopia. US Geological Survey, Fact Sheet 3053 USGS. https://www.usaid.gov/sites/default/files/documents/1860/A%20Climate%20Trend%20Analysis%20of%20Ethiopia.pdf

Hatfield JL, Prueger JH (2015) Temperature extremes: effect on plant growth and development. Weather Clim Extremes 10:4–10. ELSEVIER

Hijmans RJ, Cameron SE, Parra JL, Jones PG, Jarvis A (2005) Very high resolution interpolated climate surfaces for global land areas. Int J Climatol 25:1965–1978

IPCC (2007) In: Core Writing Team, Pachauri RK, Reisinger A (eds) Climate change 2007: synthesis report, Contribution of Working Groups I, II and III to the Fourth Assessment Report of the Intergovernmental Panel on Climate Change. IPCC, Geneva, p 104

IPCC (2014) Climate change 2014: impacts, adaptation, and vulnerability. The fifth assessment report of the intergovernmental panel on climate change. Cambridge University Press, Cambridge/New York

IPMS (2007) Improving Productivity and Market Success for Ethiopian Farmers (2004) Ada Liben Woreda diagnostic survey. ILRI (International Livestock Research Institute), Addis Ababa, p 74

Jhajharia D, Pandey PK, Pandey V, Dabral PP, Choudhary RR, Kumar R, Singh VP (2018) Changes in sunshine duration in humid environments of Agartala (Tripura), India. In: Singh VP et al. (eds), Climate change impacts. Water science and technology library, vol 82. Springer Nature Singapore. https://doi.org/10.1007/978-981-10-5714-4_2

Kahsay B (2004) Land use and land cover changes in the central highlands of Ethiopia: the case of Yere Mountains and its surroundings. MSc thesis, Addis Ababa University, p 147

Ketema S (1997) Tef. *Eragrostis tef* (Zucc.) Trotter. Promoting the Conservation and Use of Underutilized and Neglected Crops. 12. Institute of plant genetics and crop plant research, Gatersleben/Int. plant genetic resources institute, (IPGRI), Rome

Li L, Li W, Ballard T, Sun G, Jeuland M (2015) CMIP5 model simulations of Ethiopian Kiremt-season precipitation: current climate and future changes. Clim Dyn 1–13. https://doi.org/10.1007/s00382-015-2737-4

Meze-Hausken E (2004) Contrasting climate variability with perceived climate change response of a catchment to climate and land use changes in Tropical Africa: case study South Central Ethiopia. J Hydrol 275:67–85

NMSA (2001) National Meteorological Services Agency under the GEF supported climate change enabling activities project of Ethiopia. United Nations Educational, Scientific, and Cultural Organization. Federal Democratic Republic of Ethiopia Ministry of Water Resources National Meteorological Services Agency. Initial National Communication of Ethiopia to the United Nations Framework Convention on Climate Change (UNFCCC) June 2001, Addis Ababa

National Meteorological Services Agency (NMSA) (2007) Climate change national adaptation program of action (NAPA) of Ethiopia. NMSA, Addis Ababa

Rasul G, Chaudhry QZ, Mahmood A, Hyder KW(2009) Effect of temperature rise on crop growth & productivity. Pak J Meteorol 8(15):53–62

Segele ZT, Lamb PJ, Leslie LM (2009) Large-scale atmospheric circulation and global sea surface temperature associations with Horn of Africa June–September rainfall. Int J Climatol 29:1075–1100

Slingo J, Spencer H, Hoskins B, Berrisford P, Black E (2005) The meteorology of the Western Indian Ocean, and the influence of the East African Highlands. Phil Trans R Soc A 363:25–42

Stallknecht GF (1997) New crop fact sheet "Teff". Purdue University, Centre for New Crops & Plant Products. https://www.hort.purdue.edu/newcrop/cropfactsheets/teff.html#Scientific%20Names. Accessed Jan 2011

Statistical Package for Social Sciences (2008) SPSS version 17.0.1 for windows. SPSS Inc., Chicago

Tefera A (2012) Ethiopia: grain and feed annual report. Global Agricultural Information Network. USDA Foreign Agriculture Service, Report number ET 1201

USAID (2015) Climate variability and change in Ethiopia, Summary of findings. Technical report. United States Agency for International Development Climate Change Adaptation Thought Leadership and Assessments (ATLAS). Chemonics International Inc., Washington, DC

White JW, Taneer DG, Corbett JD (2001) An agro-climatological characterization of bread wheat production areas in Ethiopia. NRG-GIS series 01-01. CIMMYT, Mexico

Wilhite DA (2000) Drought: a global assessment. Routledge, London

Wilhite DA, Glantz MH (1985) Understanding the drought phenomenon: the role of definitions. Water Int 10:111–120

Wondimu A, Tekabe F (2001) Utilization of Teff in the Ethiopian diet. In: Tefera H, Belay G, Sorrells M (eds) Narrowing the rift. Teff research and development. Proceedings of the international workshop on teff genetics and improvement, Debrezeit, 16–19 October 2000, pp 239–244

World Bank (2011) Vulnerability, risk reduction and adaptation to climate change country profile: Ethiopia. The World Bank Group, Washington, DC

Worldometers (2017) www.Worldometers.info

Impact of Smart Crop-Livestock Diversification as Climate Change Adaptation Strategies on Farmers' Living Conditions, Tahoua State, Niger Republic

60

Zakou Amadou and Zakari Saley Bana

Contents

Abstract

Farming practices are becoming increasingly affected by the changing climate. Many studies have focused on farmers' adaptation strategies capable of enhancing vulnerable households' resilience building capacity. However, few studies have focused on how smart crop-livestock diversification as climate adaptation decisions influence farmers' living conditions. Based on synthesis from previous research and direct interview with farmers, six climate change adaptation options,

Z. Amadou (✉)
Faculty of Agricultural Sciences, Department of Rural Economics and Sociology, Tahoua University, Tahoua, Niger
e-mail: zakouamadou77@gmail.com

Z. Saley Bana
Disaster Risks Reduction Program Officer (DRR/O), Catholic Agency For Overseas Development (CAFOD), Niamey, Niger
e-mail: zakbana@gmail.com

W. Leal Filho (ed.), *Handbook of Climate Change Resilience*,
https://doi.org/10.1007/978-3-319-93336-8_122

each varying at five levels, were compiled and included in this study. In this study, the authors use fractional factorial design to construct five versions of the questionnaire with each five questions, and data were collected from 464 farmers randomly assigned to one version. For each question, respondents were asked to make repeated choices about his best adaptation strategies, and this process is consistent with random utility maximization. Mixed logit model was used to model crop-livestock diversification adaptation actions on farmers' living conditions. Results indicate that cereal crops such as millet (32%), followed by livestock such as cattle and goat (30%), perennial crops such as *Adansonia digitata* (17%), and tuber and root crops such as cassava (11%), are the most preferred crop-livestock diversification as climate change adaptation strategies. Results also reveal that farmers' living conditions would be more improved when this combination is implemented in their production systems. The aggregate annual benefit of introducing smart crop-livestock as climate change adaptation was estimated to be 9,722,763.08 FCFA ($19,445.53). These findings may be useful to build resilience capacity and to address future challenges related to climate change, thereby ensuring sustainable development in the study area.

Keywords
Climate smart agriculture · Crop-livestock diversification · Adaptation actions · Farmers' living conditions

Introduction

Farming practices and livestock husbandry are becoming increasingly affected by the changing climate. Climate change affects crop growth and quality and livestock and human health, limits access to drinking water, negatively affects the health of people, and thereby poses a real threat to food security (Mugambiwa and Tirivangasi 2017). Therefore, there is an urge need to develop and disseminate adaptation strategies capable of improving and sustaining agricultural farming systems and thereby ensuring livelihoods for the present and future generations. Adaptation strategies are also essential to reduce damages and take initiatives globally, nationally, regionally, and locally to reinforce farmers' resilience capacity building because a complex adaptation is a better candidate to mitigate climate change.

At the global level, several studies have well-documented that climate change is projected to have a negative impact on agricultural production when adaptative measures are not undertaken to minimize its negative impacts. The changing climate is a global challenge that consistently affects sustainable livelihoods and economic development especially in developing countries where its impacts are being felt and threatened sustainable human development. Studies have also reported that climate change is not only a global externality that negatively impacts households and communities, but also its potential to disorganize economies is real, and its challenge at global, national, and local levels can no longer be denied (Kurjia et al. 2011; Sultan and Gaetani 2016). The negative externality is being felt to the extent that during the Paris summit (2015), wealthy nations have pledged to mobilize \$100 billion to help developing nations to have a smooth ecological transition. Climate smart crop production systems have been proposed to ensure sustainability, address climate hazards, and reduce disaster risk. Resilience building by practicing sustainable natural resource management and conservation is also an excellent candidate to respond and adapt to climate dynamic (Scherr et al. 2012).

At the national level, the Niger Government financially assisted by the United Nations Development Programme (UNDP) has financed a project related to climate change. The project called PANA-resilience (2010), operating in areas most vulnerable to climate change, has carried out actions to help farmers across eight municipalities distributed in eight regions of the country to developing resilience building capacity. After having identified farmers' needs, the project has developed technologies such as improved seed variety, vegetable production, creation of sewing centers, cereal bank construction, fertilizer and pesticide bank creation, and rehabilitation of degraded land. These farmers have also benefited from training on market gardening and community life to manage the committee of the site. Finally, farmers have been able to diversify their food and generate profits from the sale of products helping them to get out of poverty. Additionally, they were also able to save money which is invested in other activities aim at not only reducing their vulnerability in the face of climate change and food insecurity but also stopping the rural migration.

At the Sahel or regional level, synthesis of previous studies (Epule et al. 2013, 2017) related to the most commonly used adaptation action in the entire Sahel was income diversification (12%), followed by water harnessing (11.5%). These studies also reported that the technically related adaptation actions (water harnessing, soil conservation, climate monitoring early harvesting, and mixed farming) were turned to be dominant (49%). Drought-resistant species, rainwater harvesting, and high yielding varieties are the technical adaptation actions developed by the research community and disseminated by technical field experts. A meta-analysis on climate change adaptation in the Sahel highlighted that climate change adaptation strategies can be broadly categorized in four options, namely, technically related adaptation, indigenous problem-solving adaptation, social safety net adaptation, and economics safety net adaptation. Additionally, each adaptation option has been further divided into five adaptation actions. Thus, mixed farming also known as crop-livestock diversification actions which are included in the technically related adaptation options is the focus of this study (Epule et al. 2017). Finally, the Sahel ecosystem is one of the most fragile on the African continent facing recurrent droughts and floods, high winds, and decreasing precipitation and thereby reducing access to food and water resources.

At the local level, research has also determined that a single adaptation action cannot completely enhance resilience building capacity. Therefore, indigenous people adaptation actions should be considered as the saying goes thinking locally and acting globally. The Niger Government and its partners have understood that climate change adaptation strategies designed without farmers' participation are less likely to succeed. The new vision of development focuses on a better understanding of the people's aspirations and their direct involvement in the development process.

Therefore, people make the project and the project ensures the development of the community is a well-known saying largely shared by development partners and keeps producing successful stories. Despite several climate change adaptation actions proposed to vulnerable households, relatively little is known how these climate change adaptation actions influence farmers' living conditions.

Few studies have considered how mixed farming as adaptation actions impacts farmers' living conditions. Research shows that communities have indigenous knowledge but lack modern scientific knowledge and technology. However, disaster risk reduction practitioner has modern scientific knowledge and lack indigenous knowledge. Therefore, tremendous gain can be achieved if these actors work hand in hand (Dube and Munsaka 2018).

This chapter addresses the key research question related to determining how farmers' preferences for crop-livestock diversification actions as climate change adaptation options and to evaluate its impact on farmers' living conditions. Furthermore, farmers will be able to make use of these results via extension agent and make more efficient and strategetic production decisions in what concerns climate change.

Literature Review

Conditional and multinomial logit have been increasingly used to evaluate the value of product characteristics and predict consumers' behavior. A variety of literature has been developed and applied in conditional, multinomial, and mixed logit models to study consumers' behavior but no previous study investigating farmers' behavior related to smart crop-diversification action using mixed logit. The literature review focuses on historical and empirical studies reported on climate change adaptation strategies.

Historical Studies on Climate Change Adaptation in the Sahel from 1975 to 2016

Research on climate change adaptation dates as back as 1970, but only studies from 1975 to 2016 were considered in this research. A systematic review study conducted on climate change actions provided the status by which climate change can be inventoried as well as a baseline from which changes can be spatially and temporarily monitored (Epule et al. 2017). Table 1 reports systematic review study from 1975 to 2016. Table 1 shows the evolution of climate change adaptation overtime. Results indicate that income diversification is the most common reported adaptation strategies, revealing that income diversification is very important to enhance climate change resilience building capacity. They have concluded that this chapter offers a proxy or indicator against which the status of climate change adaptations in the Sahel can be inventoried as well as providing a baseline from which changes can be monitored over time and space.

Table 1 Evolution of climate change adaptation strategies from 1975 to 2016

Climate change adaptation	Decades	Frequency
Income diversification	2008–2016	21
Water harnessing		15
Migration		8
High-yield variety		13
Soil conservation		8
Income diversification	1997–2007	8
Farm inputs		6
Migration		3
Income diversification	1985–1996	18
Water harnessing		27
Soil conservation		19
Income diversification	1985–1975	6
Migration		6
Help from kin and network		5

Source: Adapted from Epule et al. (2017)

Fadina and Barjolle (2018) have recently studied "Farmers' Adaptation Strategies to Climate Change and Their Implications in the Zou Department of South Benin." Their results show that Benin farmers have not only a well-developed perception of climate change but also adopt many climate change adaptation strategies to build climate change resilience. These include crop-livestock diversification, agroforestry, and diversification of income-generating activities. Results from the binary logit model also reveal that farming experience and educational level of household head are determinants of climate change adaptation decisions. Similarly, results from the multinomial logit analysis show that farming experience, educational level, farm size, and gender have a significant impact on climate change adaptation strategies. Finally, a deep analysis of each strategy suggests that crop diversification and agroforestry have been identified as the most promising and sustainable adaptation strategies in the study zone (Fadina and Barjolle 2018).

Empirical Climate Change Adaptation Studies

Several studies have investigated the impact of climate change adaptation strategies on rural farmers' resilience capacity building. The evaluation of farmers' adaptation strategies to climate change has been well-documented in the academic literature (Tabbo and Amadou 2017; Burnham and Ma 2015; Epule and Bryant 2016; Traoré et al. 2015; Snorek et al. 2014; Nyong et al. 2007). Similarly, the evaluation of technically related climate change adaptation package among rural households has been well researched in the academic literature (Tabbo et al. 2016; Labbe et al. 2017; Lwasa 2015; Recha et al. 2016; Sietz and Van Dijk 2015).

Fewer academic studies have determined the impact of climate change adaptation on farmers' living conditions. The study investigating "Aid relief in Haiti after the earthquake: Haitians' preferences for food and other basic commodities" provides insight toward smart crop-livestock diversification actions and an overview of a methodology that should be a good candidate in analyzing change in farmers' living conditions as a result of climate change adaptation. Before the work of Morris and Wodon (2003) and Lusk and Rock (2017), there are not many studies investigating people' needs following a natural disaster. Lusk and Rock (2017) have studied surveyed respondents' preferences for different aids as well as the change in their quality of life as a result from changing in house, food medical care, income, and education. The results indicate that education is the most important attribute in the Haitian's life.

Epule et al. (2017) have studied climate change adaptation in the Sahel using systematic review study approach. They have concluded that the technically related adaptation options are the most important, followed by indigenous adaptation options, social adaptation options, and economic adaptation options. Tabbo and Amadou (2017) have also investigated climate change adaptation packages offered by development partners using the best-worst scaling (BWS). Results revealed that herd rebuilding, human capacity building, introduction of fishing, soil and water conservation, introduction of the *Moringa oleifera*, financial credit, forage seed marketing, and farm inputs were the most desirable strategies for local farmers.

In Africa, farmers' perceptions of climate change adaptation strategies have been studied and well-documented (Simelton et al. 2013; Comoé 2013; Assoumana et al. 2016; Gebreeyesus 2017). These studies have concluded that all respondents have observed changes in climate pattern and that variation in the duration and the amount of rainfall have been identified as the main perceptions. Previous studies have also highlighted that climate change adaptation strategies depend largely on context and locality (Fadina and Barjolle 2018).

Methodology

This section presents information on how questionnaire was designed to collect data and how data were analyzed and interpreted.

Data Collection Method

Following Lusk and Rock (2017), the authors describe different types of climate change adaptation strategies that vary systematically with factors such as the type of crops and livestock and the annual household income and ask respondent which crop-livestock diversification action they would prefer (as shown in Table 2).

The choice experiment also known as choice-based conjoint was used to determine crop-livestock diversification actions farmers prefer. Based on this method, each attribute or variable is represented at different levels which systematically vary

Table 2 Smart crop-livestock diversification action attributes with their different levels

Cereal crops
-Cereal 1: Sorghum
-Cereal 2: Millet
-Cereal 3: Maize
-Cereal 4: Rice
-Cereal 5: Fonio
Cash crops
-Cash 1: Cowpea
-Cash 2: Peanut/groundnut
-Cash 3: Bambarra nut
-Cash 4: *Hibiscus sabdariffa*
-Cash 5: Sesame
Tuber and root crops
-Tuber and root crops 1: Onion
-Tuber and root crops 2: Potatoes
-Tuber and root crops 3: Sweet potatoes
-Tuber and root crops 4: Cassava
-Tuber and root crops 5: Carrot
Livestock
-Livestock 1: Camel
-Livestock 2: Cattle
-Livestock 3: Goat
-Livestock 4: Sheep
-Livestock 5: Poultry
Perennial crops
-Perennial crops 1: *Moringa oleifera*
-Perennial crops 2: Sahelian apple
-Perennial crops 3: *Faidherbia albida*
-Perennial crops 4: *Adansonia digitata*
-Perennial crops 5: Acacia species
Annual household income
-Income 1: Less than 150,000 FCFA
-Income 2: Between 150,000 FCFA and 200,000 FCFA
-Income 3: Between 200,001 FCFA and 250,000 FCFA
-Income 4: Between 250,001 FCFA and 300,000 FCFA
-Income 5: Between 300,001 FCFA and 350,000 FCFA

Source: Field work, 2018; Field Survey, 2018

to create a set of choice questions. The usefulness of this approach is that it can be used to determine the extent to which, for example, smart crop-livestock diversification would benefit farmers.

Respondents made repeated choices between two different crop-livestock diversification actions according to six attributes presented in Table 1: cereal crop, cash

Which smart crop-livestock diversification action would you prefer?

Attributes	Option A	Option B	None
Cereal crops	Millet	Sorghum	
Cash crops	Peanut	Peanuts	
Tuber and root crops	Potatoes	Sweet potatoes	
Livestock	Cattle	Sheep	
Perennial crops	Sahel apple	Acacia species	
Annual income	150,000 FCFA	250,000 FCFA	
I would choose.......	[ ]	[ ]	[ ]

Fig. 1 A sample of choice question about smart crop-livestock actions

crop, tuber and root crops, livestock, perennial crops, and annual income. Each attribute is varied at five levels. For example, for each cereal crops, respondents had to choose between sorghum, millet, maize, rice, and fonio.

For each question, respondents were asked to choose between "Option A" and "Option B." Therefore, there were $6^5 = 7{,}776$ options that could be defined. Thus, the full factorial design with two options is given as follows: $6^5 * 6^5 = 7{,}776 * 7{,}776 = 60{,}466{,}176$ possible choices. From this full factorial design, a fractional factorial design that is balanced and orthogonal was created. The resulting design consisted of 25 choice questions. In order to make the task easier for respondents to answer, the authors have created five blocks or survey versions each containing five questions. Each respondent received one block of five questions to answer. An example of one of the choice question is presented in Fig. 1.

The responses to choice questions were analyzed using random utility framework proposed by McFadden (1974). The utility function is divided in two parts: a systematic part which is assumed to depend on attributes and stochastic part which is assumed to capture unobserved preferences. The random utility function can be mathematically expressed as follows:

$$U_{ij} = V_{ij} + \varepsilon_{ij} \tag{1}$$

where U_{ij} is the utility for a person i facing option j, V_{ij} is the deterministic component, and ε_{ij} is the stochastic part (Train 2009).

Choice-Based Conjoint Modeling Method

The authors used choice-based conjoint analysis also known as a choice experiment to determine the quality of life after climate change hit. Most choice experiments have focused on the variation of product characteristics and price, while the authors describe the impact of crop-livestock diversification as climate change adaptative strategy on quality of life that vary systematically with factors such as income and crop-livestock diversification adaptation strategies.

The utility function is assumed to be a function of type of income, cereal, cash, tubers and root crops, livestock, and perennial crops. The authors estimated the model using income as a continuous variable. The utility function can be presented as following:

$$\begin{aligned} Vi = {} & \alpha_1(\text{Income})_i + \mu_{21}(\text{Cereal})_{1i} + \mu_{22}(\text{Cereal})_{2i} + \mu_{23}(\text{Cereal})_{3i} \\ & + \mu_{24}(\text{Cereal})_{4i} + \mu_{25}(\text{Cereal})_{5i}\, \mu_{31}(\text{Cash})_{1i} + \mu_{32}(\text{Cash})_{2i} \\ & + \mu_{33}(\text{Cash})_{3i} + \mu_{34}(\text{Cash})_{4i} + \mu_{35}(\text{Cash})_{5i} + \mu_{41}(\text{Tuberroot})_{1i} \\ & + \mu_{42}(\text{Tuberroot})_{2i} + \mu_{43}(\text{Tuberroot})_{3i} + \mu_{44}(\text{Tuberroot})_{4i} \\ & + \mu_{45}(\text{Tuberroot})_{5i} + \mu_{51}(\text{Livestock})_{1i} + \mu_{52}(\text{Livestock})_{2i} \\ & + \mu_{53}(\text{Livestock})_{3i} + \mu_{54}(\text{Livestock})_{4i} + \mu_{55}(\text{Livestock})_{5i} \\ & + \mu_{61}(\text{Perennial})_{1i} + \mu_{62}(\text{Perennial})_{2i} + \mu_{63}(\text{Perennial})_{3i} \\ & + \mu_{64}(\text{Perennial})_{4i} + \mu_{65}(\text{Perennial})_{5i} \end{aligned} \tag{2}$$

where the index i represents individual i and μ_{kj} the coefficient for each variable (Lusk and Rock 2017).

Results from likelihood ratio test indicate that the mixed multinomial logit model is better in determining the relative importance of crop-livestock diversification as a climate change adaptative measure on the quality of life than multinomial logit. Previous studies showed how to determine relative importance score (Orme 2010; Lusk and Rock 2017). By following Lusk and Rock (2017), the impact each attribute can make in the utility function by taking the difference between the level with the highest utility and the one with the lowest utility was initially considered. Secondly, all those differences are summed up to come up with a total utility range. Thirdly, the attribute importance is then determined as the ratio between the difference in utility for one attribute and the total utility range, as shown below:

$$\text{Attribute importance (AI)} = \begin{aligned}[t] & (\text{attribute utility range (AUR)}) \\ & /(\text{utility range total (URT)}) \end{aligned}$$

For instance, attribute importance for cereal crops can be presented as following:

$$\begin{aligned} AI_1 & = AUR_1/(URT) \text{or } AI_1 \\ & = AUR_1/(AUR_1 + AUR_2 + AUR_3 + AUR_4 + AUR_5 + AUR_6) \\ & = \mu_{4j}/\left(\mu_{1j} + \mu_{2j} + \mu_{3j} + \mu_{4j} + \mu_{5j} + \mu_{6j}\right) \end{aligned} \tag{3}$$

where:

$AUR_1 = \mu_{1j} =$ attribute utility range for income
$AUR_2 = \mu_{2j} =$ attribute utility range for cereal crops
$AUR_3 = \mu_{3j} =$ attribute utility range for cash crops
$AUR_4 = \mu_{4j} =$ attribute utility range for tuber and root crops
$AUR_5 = \mu_{5j} =$ attribute utility range for livestock
$AUR_6 = \mu_{6j} =$ attribute utility range for perennial crops

The authors hypothesized that perennial crops' production as climate change adaptative actions would greatly improve farmers' living conditions than cereal crops' production as climate change adaptation actions attributes. The authors also hypothesized that the mixed logit model outperforms the multinomial logit. Finally, the authors also hypothesized that adaptation strategies based on crop production are more welfare enhancing than adaptation strategies based on livestock production.

The parameters estimated from the model where income is considered as a continuous variable were used to calculate willingness to pay (WTP). The WTP is defined as the amount of money when taking from farmer's income making him indifferent between two situations. Considering the kth attribute, WTP for option j relative to option k is the income level one would give up to have j instead of k. Based on the previous statement and following Eq. (2), WTP for millet relative to sorghum is $WTP_{cereal2-cereal1} = -(\mu_{22} - \mu_{21})/\alpha_1$. Similarly, WTP for *Faidherbia albida* relative to *Adansonia digitata* is $WTP_{albida-digitata} = -(\mu_{62} - \mu_{63})/\alpha_1$. In general, μ_{ij} is the coefficient estimated for the levels of each attribute, while α_1 is the estimated coefficient for income.

The authors also sought to determine farmers' quality of life as a result of climate change. The authors assumed that major changes have occurred particularly in farmers' food security and livelihoods situation as compared to their situation after climate change decision was made. The authors asked them questions about crop-livestock diversification strategies they were using before climate change and types of crop-livestock diversification they are using after climate change. The authors then computed the change in utility resulting from change in crop-livestock diversification situation for each farmer. Using the Eq. (2) and assuming ceteris paribus, the total change in cereal crops can be mathematically represented as follows:

$$\sum (1/\alpha_1) \left[\mu_{21}\left(\text{Cereal}_{1j}\right)^{\text{after}} + \mu_{22}\left(\text{Cereal}_{2j}\right)^{\text{after}} + \mu_{23}\left(\text{Cereal}_{3j}\right)^{\text{after}} + \mu_{24}\left(\text{Cerreal}_{4j}\right)^{\text{after}} - \mu_{21}\left(\text{Cereal}_{1j}\right)^{\text{before}} - \mu_{22}\left(\text{Cereal}_{2j}\right)^{\text{before}} - \mu_{23}\left(\text{Cereal}_{3j}\right)^{\text{before}} - \mu_{24}\left(\text{Cereal}_{4j}\right)^{\text{before}}\right] \quad (4)$$

Average change can be calculated by dividing Eq. (4) by the total respondents (N).

Generally, $(\text{Cereal}_{ij})^{\text{before}}$ and $(\text{Cereal}_{ij})^{\text{after}}$ are dummy variables taking the values of 1 or 0. For example, a farmer who used to grow local millet varieties and decided to adopt improved millet varieties as adaptation strategies, ceteris paribus, his utility can then be represented as follows:

$$V_i^{\text{After}} - V_i^{\text{Before}} = \mu_{24}\left(\text{Cereal}_{4j}\right)^{\text{after}} - \mu_{21}\left(\text{Cereal}_{1j}\right)^{\text{before}} = \mu_{24} - \mu_{21} \quad (5)$$

Therefore, this following relationship generally holds: $(\text{Cereal}_{4j})^{\text{after}} = (\text{Cereal}_{1j})^{\text{before}} = 1$.

Results and Discussions

This section summarizes characteristics of surveyed respondents, mixed logit and total change in farmers living conditions in monetary terms, and ratio scores and attribute utility range across two estimation models. Table 3 presents summary statistics of respondents included in the analysis. Results indicate that most respondents have an age ranging from 36 to 40 (22%). Most of the respondents were male (70%), married (80%), and uneducated (88%). They had an average family size between one and four persons (54%) and with monthly average income of 55,142 FCFA (currency for French-speaking countries of West Africa), which is equivalent to 110 US dollar. Results presented in Table 3 also revealed that most of the respondents were farmers (64%) and most of them received information on climate

Table 3 Characteristics of surveyed respondents (%)

Age	**Percentage (%)**
20–25	9%
26–30	16.5%
31–35	18%
36–40	22%
41–45	9%
46–50	6%
Above 55	19.5%
Gender	
Male	70%
Female	30%
Education	
Educated	12%
Uneducated	88%
Marital status	
Married	80%
Not married	20%
Monthly income ($)	$110.28
Family size	
1–4	54%
5–6	22.5%
7–8	10.5%
9–10	13%
Profession	
Farmers	64%
Herdsmen	36%
Information on climate change	
Yes	82%
No	18%

Note that $1 is equal to 500 FCFA

change (82%). This implies that young male farmer that is married and having formal education should be targeted by extension agent to make a significant change in the study area.

Table 4 presents results of the mixed logit based on farmers' utilities for smart crop-livestock diversification action attributes as climate change strategies. The model was estimated with income included as continuous variable, while all other five attributes were considered at five levels. A base level is also set for each attribute, and the other levels were estimated relative to the base. A positive coefficient for an estimate indicates that this level is more important to the respondents as compared to the base, while a negative coefficient indicates that level is less important.

Table 4 also reveals that coefficients of the standard deviation for millet, rice, fonio, *Hibiscus sabdariffa*, potatoes, sweet potatoes, cassava, carrot, cattle, and poultry are highly significant, indicating that these coefficients do indeed vary in the population. Additionally, these estimated means and standard deviations of these coefficients provide useful information on the share of the population placing a positive value on climate change adaptation attributes and the share placing a negative value.

Results suggest that farmers prefer to have millet instead of sorghum. Results from estimation lead to a coefficient of 8.993 for millet and a coefficient of 3.797 for rice. In terms of cash crop, results show that *Hibiscus sabdariffa* is the most preferred with a coefficient of 1.798 relative to cowpea. This result is not surprising because hibiscus species have been used by local farmers to control weeds and to bring land colonized by *Sida cordifolia* to active production. Our research hypothesis stating that perennial crops' production has a higher impact on farmers' living conditions than cereal crops' production is rejected, revealing that cereal crops' production is more welfare enhancing than perennial crops' production.

Results reveal that for tuber and root crops, farmers preferred to have cassava (3.017) as adaptation strategies to onion. Our results are consistent with the study conducted by Mupakati and Tanyanyiwa (2017), revealing that cassava has an extensive root system which is well-adapted to poor soils, which may not support crops like maize. They have also concluded that increase cassava production as its tubers can be value added to produce a range of products that include livestock feed and porridge.

In terms of livestock, farmers surveyed preferred cattle (8.178) relative to camel, followed by goat (7.728), sheep (7.423), and poultry (5.440). This result is not surprising due to the fact that a recent study on climate change (Tabbo and Amadou 2017) reported that herd rebuilding is the most important for farmers. This is particularly illustrated by the increasing introduction of goats in the northern part of Niger, because goats are easy to maintain, they are very prolific, and they are capable of producing high-quality meat and milk. Goats are also known to impact environmental regeneration through their waste, which involves the return of the vegetation, and this idea is confirmed by research findings stating that the Sahel is getting more and more greener (Tabbo and Amadou 2017).

Table 4 Mixed logit estimation with income as continuous variable

Variables	Estimate	Standard error	t-value	Pr(>|t|)
Income	0.013	0.001	9.647	<0.0001
Cereal crops				
Base: Sorghum	0.000	0.000	0.000	0.000
Millet	8.933	3.462	2.581	0.010
Maize	−2.397	1.660	−1.444	0.149
Rice	3.797	1.941	1.956	0.050
Fonio	−0.731	1.399	−0.522	0.601
Cash crops				
Base: Cowpea	0.000	0.000	0.000	0.000
Peanut	−1.610	1.417	−1.136	0.256
Bambarra nut	−1.495	1.397	−1.070	0.285
Hibiscus sabdariffa	1.798	1.642	1.095	0.274
Sesame	−1.928	1.203	−1.603	0.109
Tuber and root crops				
Base: Onion	0.000	0.000	0.000	0.000
Potatoes	−3.107	1.482	−2.097	0.036
Sweet potatoes	−2.463	1.444	−1.706	0.088
Cassava	3.017	1.834	1.645	0.100
Carrot	−3.131	1.700	−1.842	0.065
Livestock				
Base: Camel	0.000	0.000	0.000	0.000
Cattle	8.178	3.054	2.678	0.007
Goat	7.728	2.846	2.715	0.007
Sheep	7.423	2.764	2.686	0.007
Poultry	5.440	2.632	2.067	0.039
Perennial crops				
Base: *Moringa oleifera*	0.000	0.000	0.000	0.000
Sahelian apple	−3.144	1.506	−2.088	0.037
Faidherbia albida	4.004	1.916	2.090	0.037
Adansonia digitata	4.763	2.244	2.123	0.034
Acacia species	−7.174	2.483	−2.889	0.004
Standard deviation estimates				
sd.millet	0.314	5.521	2.840	0.005
sd.wheat	0.445	2.928	1.825	0.068
sd.rice	0.330	4.921	2.655	0.008
sd.fonio	0.352	4.060	2.614	0.009
sd.peanut	0.994	2.551	−0.507	0.612
sd.bambaranut	0.951	2.800	0.486	0.627
sd.hibiscus	0.420	3.028	2.064	0.039
sd.sesame	3.256	1.870	0.310	0.756
sd.potatoes	0.384	3.217	−2.499	0.012
sd.sweet potatoes	0.336	4.293	2.864	0.004

(*continued*)

Table 4 (continued)

Variables	Estimate	Standard error	t-value	Pr(>|t|)
sd.cassava	0.363	3.792	2.516	0.012
sd.carrot	0.294	6.722	2.986	0.003
sd.cattle	0.365	3.729	2.515	0.012
sd.goat	0.512	2.249	1.700	0.089
sd.sheep	0.571	2.031	−1.495	0.135
sd.poultry	0.274	9.270	2.887	0.004
sd.sahel_apple	0.527	2.004	−1.792	0.073
sd.faidherbia	1.795	2.105	0.103	0.918
sd.adansonia	0.857	2.069	0.750	0.453
sd.acacia	0.492	2.805	1.491	0.136
Number of observations	464			
Number of choices	6960			
Loglikelihood	−2083.9			

Finally, when it comes to perennial crops, *Adansonia digitata* (4.763) was the highest ranked attribute, followed by *Faidherbia albida* (4.004) by farmers interviewed. These perennial crops have been selected as the most desirable because they are leguminous crops which are capable of fixing nitrogen, improving soil fertility, and thereby increasing crop yield. They are used as windbreak and excellent forage for livestock. Similar studies have reported that crop diversification and agroforestry are the most promising strategies and capable of providing several economic and environmental benefits and thereby ensuring the sustainable development in the study area (Fadina and Barjolle 2018; Zoysa and Inoue 2014).

Table 7 reports the multinomial logit model. Likelihood ratio test shows that the mixed logit model outperformed the multinomial logit model. Therefore, coefficients of multinomial logit model though being presented (Appendix 1) are not interpreted.

For easier interpretation and comparison among the importance relative of different attributes, Fig. 2 shown below was used. Figure 2 presents relative importance comparison across mixed and multinomial logit. However, only results from mixed logit model were reported and discussed.

Figure 2 shows that cereal production ($AI_2 = 32\%$) is the most important attribute to change farmers' living conditions, followed by livestock ($AI_5 = 30\%$), perennial crops ($AI_6 = 17\%$), and tuber and root crops ($AI3 = 11\%$). Income (AI_1) and cash crops (AI_3) have each close score, falling between the range 4% and 6%. Results from Table 4 were used to generate Table 5 as shown below.

In this study, levels of each attribute were estimated and compared to a base for mixed logit models. The attribute with the highest utility will then be compared to attribute with zero utility. The results indicate that utility total range (UTR) for mixed logit is higher than that for multinomial logit. Therefore, only results from mixed logit were reported. As shown in Table 5, attribute utility range for each attribute was computed. Thus, the attribute utility range is $AUR_1 = 0.970$ for income ($0.013*80-0.013*5 = 0.970$).

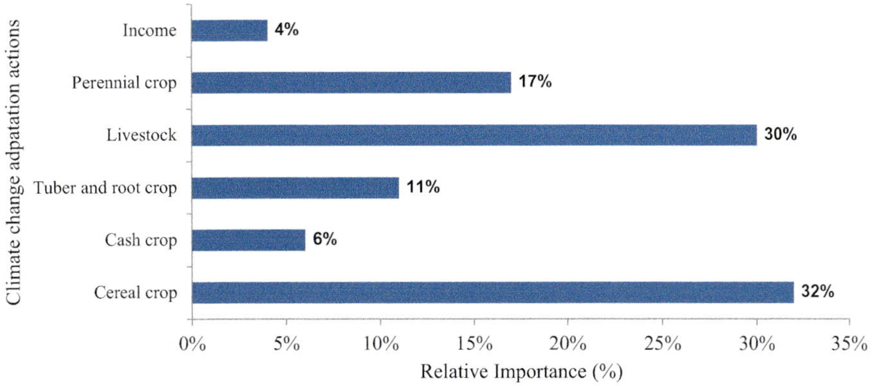

Fig. 2 Relative importance of crop-livestock diversification actions based on mixed logit

Table 5 Ratio scores and attribute utility range across two estimation models

Multinomial logit			Mixed logit		
Variables	AUR	Ratio scores	Variables	AUR	Ratio scores
Cereal crop	0.540	23%	Cereal crop	8.933	32%
Cash crop	0.299	12%	Cash crop	1.798	6%
Tuber and root crop	0.062	3%	Tuber and root crop	3.017	11%
Livestock	0.898	37%	Livestock	8.178	30%
Perennial crop	0.416	17%	Perennial crop	4.763	17%
Income	0.186	8%	Income	0.975	4%
UTR	2.400	100	UTR	27.659	100

AUR attribute utility range, *UTR* utility total range

Similarly, $AUR_2 = 8.933$ for cereal crop, $AUR_3 = 1.798$ for cash crop, $AUR_4 = 3.017$ for tuber and root crop, $AUR_5 = 8.178$ for livestock, and $AUR_6 = 4.763$ for perennial crop, and the utility total range (UTR) obtained by adding up those attribute utilities becomes 27.66. The relative importance score of each attribute as reported in Fig. 2 is obtained as ratios of attribute utility to utility total range.

For instance, Eq. (3) can be used to compute attribute utility range for cereal as follows:

$$AI_2 = 8.933/(0.970 + 8.933 + 1.798 + 3.017 + 8.178 + 4.763) = 8.933/27.66 \\ = 32\%$$

Similarly, the attribute utility range for livestock is:

$$AI_5 = 8.178/(0.970 + 8.933 + 1.798 + 3.017 + 8.178 + 4.763) = 8.178/27.66 \\ = 30\%$$

Results indicate that total utility range for all crops (26.682) is lower than all total utility range for all livestock species (29.739), showing that livestock production is better than that of crops. Thus, the attribute utility total range for all crops is 26.682/56.421 = 47.29%, while the attribute utility total range for animals is 29.739/56.421 = 52.71%.

Willingness-to-Pay (WTP) Estimates

Results reveal that farmers are WTP 430,897.65 FCFA ($861.8) annually to have millet as cereal crops relative to sorghum. The annual WTP for wheat is 226,940.70 FCFA ($453.88) lower relative to sorghum. The annual marginal WTP for millet relative to wheat is 657,838.40 ($1315.67), indicating millet is the most desirable attribute.

In terms of cash crop, the annual WTP for farmers surveyed to have peanut, bambaranut, hibiscus, and sesame were lower relative to cowpea, implying that cowpea is the most preferred cash crop in the study area. Also, in terms of tubers and root crops, respondents are WTP 31,862.63 FCFA ($63.72) annually to have carrot relative to onion.

Regarding livestock, respondents in the study were WTP 383,276.53 FCFA ($766.55) annually to have goat relative to camel. Similarly, the annual WTP for respondents to have sheep is 192,371.46 FCFA ($384.74). The annual marginal WTP for respondents to have goat relative to sheep is 190,905.07 FCFA ($381.81).

In terms of perennial crops, results show that respondents interviewed were WTP 157,711.20 FCFA ($315.42) annually to have *Adansonia digitata* relative to *Moringa oleifera*. Similarly, respondents surveyed were WTP 79,478.70 FCFA ($158.95) to have *Faidherbia albida* relative to *Moringa oleifera*. Results also indicate that the annually marginal WTP for respondents to have *Adansonia digitata* relative to *Faidherbia albida* is 78,232.50 FCFA ($156.46).

Change in Farmers' Living Conditions

Table 6 presents total utility and monetary changes in farmers' living conditions as a result of climate change adaptation options. Table 6 shows that utility total range for crops (25.712) is lower than utility total range for livestock (28.769), revealing that farmers derive more utility with livestock as climate change adaptation strategies. However, utility total range for cereal crops such as millet (8.933) is higher than utility total range for livestock such as cattle (8.173), implying that farmers derive more utility with growing millet as climate change adaptation strategies. The tubers and root crops such as cassava (3.017) have the lowest utility range followed by the cash crops such as *Hibiscus sabdariffa* (1.798). The total aggregate utility range for crop-livestock is 55.956, while the total aggregate monetary impact is 9,722,763.08 FCFA ($19,445.53).

Table 6 Total utility and monetary changes on farmers' living conditions

Crops	Utility range	Total change in FCFA
Millet	8.933	1,594,196.92
Rice	3.197	570,541.54
Hibiscus	1.798	320,873.85
Cassava	3.017	538,418.46
Faidherbia albida	4.004	714,560.00
Adansonia digitata	4.763	850,012.31
Utility total range for crops	25.712	4,588,603.08($9,177.206)
Livestock	Utility range	
Cattle	8.178	1,459,458.46
Goat	7.728	1,379,150.77
Sheep	7.423	1,324,720.00
Poultry	5.44	970,830.77
Utility total range for livestock	28.769	5,134,160($10,268.32)
Income share	0.975	
Grand utility total range	**55.956**	
N	464	
Number of choice per person	5	
Income coefficient	0.013	
Total change in FCFA		9,722,763.08 ($19,445.53)

The utility for a person growing millet is given by $V_i^{before} = 8.933 + 0.013*\text{income}$

If this person decides to grow sorghum, then his utility becomes $V_i^{after} = 0.000 + 0.013 * \text{income}$

The change in utility is given by $V_i^{after} - V_i^{before} = -8.933$. This value can be put in monetary term as: $-8.933/0.013 = -68{,}153.85$ FCFA ($1371.31).

The total change in utility caused by the change in cereal crops growing is calculated by adding up change across surveyed respondents $\sum (V_i^{after} - V_i^{before})$. The resulted average change is given:

$\sum (V_i^{after} - V_i^{before})/N = \sum (V_i^{after} - V_i^{before})/464$. Thus, Table 6 presented above reports total change in monetary terms for each attribute level. Results from mixed logit reveal that cereal crops such as millet (1,594,196.92 FCFA) have the highest impact followed by livestock such as cattle (1,459,458.46 FCFA), followed by goat (1,379,150.77FCFA), by sheep (1,324,720.00 FCFA), by poultry (970,830.77 FCFA), by perennial crops such *Adansonia digitata* (850,012.31 FCFA) and *Faidherbia albida* (714,560.00FCFA), while cash crops such as *Hibiscus sabdariffa* (320,873.85 FCFA) have the lowest impact. Results also show that the aggregate impact for livestock production (5,134,160 FCFA) as climate change adaptation strategies has the highest impact than the aggregate impact for crop production (4,588,603.08 FCFA) as climate change adaptation strategies. The research hypothesis stating that crop production is more welfare enhancing than livestock production is rejected, implying that the aggregate impact for livestock production has a higher impact on farmers living conditions.

Conclusions and Implications

Previous research pointed out that climate change is one of the most global challenges that the world is facing and its effects significantly contribute to the development challenges of ensuring food security, poverty reduction, and access to drinking water and as well as providing affordable healthcare in the developing world. Although local, regional, and global efforts have been undertaken to enhance vulnerable households' resilience building capacity, little is relatively known how smart crop-livestock diversification actions impact farmers' living conditions. This research sought to determine farmers' preferences for smart crop-livestock diversification actions, to assess its impact on farmers living conditions, and to propose a sustainable resilience-building tool.

Results suggest that a young male farmer that is married and having Quranic education should be targeted for effective climate change adaptation program. Results reveal that millet, rice, *Hibiscus sabdariffa*, cassava, cattle, goat, sheep, poultry, *Adansonia digitata*, and *Faidherbia albida* are the optimal crop-livestock combination that should be promoted in the study area for effective climate change resilience building capacity. Results indicate that cereal crops are the most important attribute in farmers' living conditions. The second most important attribute is livestock, while the third most important attribute on farmers' living conditions is the perennial crops.

The application of survey-based methods is used to determine the economic consequences of climate change adaptation options resulting from change in farmers' living conditions.

The authors have focused on change on cereal crops and livestock production situation on farmers' living conditions. Farmers' utility increased annually by 4,588,603.08 FCFA ($9,177.206) and 5,134,160 FCFA ($10,268.32) because of change in crop production and livestock production, respectively, after climate change adaptation options were made. The total aggregate change as a result of crop-livestock diversification on farmers' living conditions is 9,722,763.08 FCFA ($19,445.53).

This research represents an attempt related to how farmers value smart crop-livestock diversification options after climate change had occurred as well as to propose a resilience-building tool capable of ensuring sustainable agricultural development in the study area. The findings of this study should be included in the development program, thereby assisting policymakers to better plan and guide agricultural policies by focusing on climate change adaptation dimension as suggested by farmers. Future research direction is to evaluate technical-based options, socially based options, indigenous knowledge-based options, and economic-based options on farmers' living conditions. It might be also important to track a panel of people in order to study the stability of farmers' preferences for climate change adaptation options over time.

Appendix 1

Table 7 Multinomial logit estimation as income considered as continuous variable

Variables	Estimate	Standard error	t-value	Pr (>|t|)
Income	0.0025	0.0004	6.8790	0.0000
Cereal crop				
Base: Sorghum	0.0000	0.0000	0.0000	0.0000
Millet	0.5405	0.1263	4.2809	0.0000
Maize	−0.2189	0.1351	−1.6204	0.1051
Rice	0.2321	0.1128	2.0585	0.0395
Fonio	0.1124	0.1108	1.0144	0.3104
Cash crop				
Base: cowpea	0.0000	0.0000	0.0000	0.0000
Peanut	0.0454	0.1217	0.3728	0.7093
Bambarra nut	0.0637	0.1238	0.5144	0.6070
Hibiscus sabdariffa	0.2989	0.1026	2.9134	0.0036
Sesame	0.1522	0.1062	1.4325	0.1520
Tuber and root crops				
Base: Onion	0.0000	0.0000	0.0000	0.0000
Potatoes	−0.2695	0.1023	−2.6349	0.0084
Sweet potatoes	−0.2007	0.1048	−1.9159	0.0554
Cassava	0.0618	0.1076	0.5742	0.5658
Carrot	−0.1030	0.1110	−0.9277	0.3535
Livestock				
Base: Camel	0.0000	0.0000	0.0000	0.0000
Cattle	0.8979	0.1385	6.4806	0.0000
Goat	0.7347	0.1099	6.6867	0.0000
Sheep	0.5836	0.0925	6.3108	0.0000
Poultry	0.6000	0.1177	5.0963	0.0000
Perennial crop				
Base: Moringa	0.0000	0.0000	0.0000	0.0000
Sahel apple	−0.1367	0.1106	−1.2352	0.2167
Faidherbia albida	0.4158	0.1251	3.3235	0.0009
Adansonia digitata	0.3637	0.1092	3.3310	0.0009
Acacia species	−0.4123	0.1223	−3.3719	0.0007
Number of observations	464			
Number of choices	6960			
Loglikelihood	−2143			

References

Assoumana BT, Ndiaye M, Puje G, Diourte M, Graiser T (2016) Comparative assessment of local farmers' perceptions of meteorological events and adaptations strategies: two case studies in Niger Republic. J Sustain Dev 9:118–135

Burnham M, Ma Z (2015) Linking smallholder farmer climate change adaptation decisions to development. Clim Dev 8(4):289–311

Comoé H (2013) Contribution to food security by improving farmers' responses to climate change in northern and central areas of Côte D'ivoire. ETH, Zurich

Dube E, Munsaka E (2018) The contribution of indigenous knowledge to disaster risk reduction activities in Zimbabwe: a big all to practitioners. Jàmbá 10(1):a493. https://doi.org/10.4102/jamba.v10i1.49

Epule T, Bryant C (2016) Small scale farmers' indigenous agricultural adaptation options in the face of declining or stagnant crop yields in the Fako and Meme divisions of Cameroon. Agriculture 6(2):22

Epule TE, Peng C, Lepage L, Chen Z (2013) The causes, effects and challenges of Sahelian droughts: a critical review. Reg Environ Chang. https://doi.org/10.1007/s10113-013-0473-z

Epule TE, Ford JB, Lwasa S, Lepage L (2017) Climate change adaptation in the Sahel. Environ Sci Policy. https://doi.org/10.1016/j.envsci.2017.05.018

Fadina AMR, Barjolle D (2018) Farmers' adaptation strategies to climate change and their implications in the Zou Department of South Benin. Environments 5:15. https://doi.org/10.3390/environments5010015. www.mdpi.com/journal/environments

Gebreeyesus KA (2017) Impact of climate change on the agro-ecological innovation of coffee agroforestry systems in Central Kenya. PhD thesis, SupAgro, Montpellier, 28 Apr

Kurjia P, Nanja D, Stern R (2011) Exploring daily rainfall data to investigate evidence of climate change in Southern Zambia and its implication for farmers in the area

Labbe J, Ford JD, Araos M, Flynn M (2017) The government-led climate change adaptation landscape in Nunavut, Canada. Environ Rev 25(1):1–14. https://doi.org/10.1139/er-2016-0032

Lusk JL, Rock A (2017) Aid relief in Haiti after the earthquake: Haitians' preferences for food and other basic commodities. Dev Policy Rev 35:O303–O321. https://doi.org/10.1111/dpr.12321

Lwasa S (2015) A systematic review of research on climate change adaptation policy and practice in Africa and South Asia deltas. Reg Environ Chang 15:815–825. https://doi.org/10.1007/s10113-014-0715-8

McFadden D (1974) Conditional logit analysis of qualitative choice behavior. Frontiers in econometrics. Zarembka, New York

Morris SS, Wodon Q (2003) The allocation of natural disaster relief funds: Hurricane Mitch in Honduras. J World Dev 31(7):1279–1289

Mugambiwa SS, Tirivangasi HM (2017) Climate change: a threat towards achieving Sustainable Development Goal number two (end hunger, achieve food security and improved nutrition and promote sustainable agriculture) in South Africa. Jàmbá 9(1):a350. https://doi.org/10.4102/jamba.v9i1.350

Mupakati T, Tanyanyiwa VI (2017) Cassava production as a climate change adaptation strategy in Chilonga Ward, Chiredzi District, Zimbabwe. Jàmbá 9(1):a348. https://doi.org/10.4102/jamba.v9i1.348

Nyong A, Adesina F, Osman Elasha B (2007) The value of indigenous knowledge in climate change mitigation and adaptation strategies in the African Sahel. Mitig Adapt Strateg Glob Chang 12(5):787–797

Orme B (2010) Getting started with conjoint analysis: strategies for product design and pricing research, 2nd edn. Research Publishers LLC, Madison

Recha JW, Mati BM, Nyasimi M, Kimeli PK, Kinyangi JM, Radeny M (2016) Changing rainfall patterns and farmers' adaptation through soil water management practices in semi-arid eastern Kenya. Arid Land Res Manag 30(3):229–238. https://doi.org/10.1080/15324982.2015.1091398

Scherr SJ, Shames S, Friedman R (2012) From climate-smart agriculture to climate-smart landscapes. Agric Food Secur 1(12):1–15. https://doi.org/10.1186/2048-7010-1-12

Sietz D, Van Dijk H (2015) Land-based adaptation to global change: what drives soil and water conservation in western Africa? Glob Environ Chang 33:131–141

Simelton E, Quinn CH, Batisani N, Dougill AJ, Dyer JC, Fraser EDG, Mkwambisi D, Sallu S, Stringer LC (2013) Is rainfall really changing? Farmers' perceptions, meteorological data, and policy implications. Clim Dev 5:123–138

Snorek J, Renaud FG, Kloos J (2014) Divergent adaptation to climate variability: a case study of pastoral and agricultural societies in Niger. Glob Environ Chang 29:371–386

Sultan B, Gaetani M (2016) Agriculture in West Africa in the twenty-first century: climate change and impacts scenarios, and potential for adaptation. Front Plant Sci 7:1262. https://doi.org/10.3389/fpls.2016.01262

Tabbo AM, Amadou Z (2017) Assessing newly introduced climate change adaptation strategy packages among rural households: evidence from Kaou local government area, Tahoua State, Niger Republic. Jàmbá 9(1):a383. https://doi.org/10.4102/jamba.v9i1.383

Tabbo AM, Amadou Z, Danbaky AB (2016) Evaluating farmers' adaptation strategies to climate change: A case study of Kaou local government area, Tahoua State, Niger Republic, Jàmbá: Journal of Disaster Risk Studies 8(3), a241. https://doi.org/10.4102/jamba.v8i3.241

Train KE (2009) Discrete choice methods with simulation. Cambridge University Press, New York

Traoré B, Van Wijk MT, Descheemaeker K, Corbeels M, Rufino MC, Giller KE (2015) Climate variability and change in Southern Mali: learning from farmer perceptions and on-farm trials. Exp Agric 51(04):615–634

Zoysa MD, Inoue M (2014) Climate change impacts, agroforestry adaptation and policy environment in Sri Lanka. J For 4:439–456

Climate Change Implication on Airborne Infections: Roadmap for Nigeria's Health Sector Development

61

Tosin K. W. Gbadegesin

Contents

Abstract

Airborne diseases also termed as droplets infection are those diseases that are caused by pathogenic microbial agents which get discharged through coughing, sneezing, laughing, or through close personal contact. And it is well-known that the health of a population, if it is to be sustained, requires clean air, safe water, adequate food, tolerable temperature, stable climate, and high levels of biodiversity. This chapter states that climate change has become a great challenge to our generation and its impact is felt in almost every society in the world. It alters temperature, air movement, and precipitation in various ways and to varying degrees across Nigeria with consequences on human health. Nigeria, as a developing country, is vulnerable and has low coping capability. This vulnerability is due in part to existing problems of poverty, weak institutions, and armed conflict, which limit a population's capacity to deal with the additional health challenges posed by climate change. It therefore posit that addressing the health aspects of climate change need not depend on the federal government; it can happen at the level of cities, states, local governments, businesses or institutions, as well as through collaborations.

T. K. W. Gbadegesin (✉)
Centre for Sustainable Development (CESDEV), University of Ibadan, Ibadan, Nigeria
e-mail: cymbalawesley@gmail.com

W. Leal Filho (ed.), *Handbook of Climate Change Resilience*,
https://doi.org/10.1007/978-3-319-93336-8_131

Keywords
Airborne · Diseases · Vulnerable · Pathogenic · Population · Droplet · Infection

Introduction

Over the last 50 years, human activities – particularly the burning of fossil fuels – have released sufficient quantities of carbon dioxide and other greenhouse gases to trap additional heat in the lower atmosphere and affect the global climate. In the last 100 years, the world has warmed by approximately 0.75 °C. Over the last 25 years, the rate of global warming has accelerated, at over 0.18 °C per decade (UK-Met Office 2008). Sea levels are rising, glaciers are melting, and precipitation patterns are changing. Extreme weather events are becoming more intense and frequent and climate change has continued to affect social and environmental determinants of health: clean air, safe drinking water, sufficient food, and secure shelter (Arnell 2004).

Between 2030 and 2050, climate change is expected to cause approximately 250,000 additional deaths per year, from malnutrition, malaria, diarrhea, heat stress, etc. The direct damage costs to health (i.e., excluding costs in health-determining sectors such as agriculture, water, and sanitation) is estimated to be between US$ 2–4 billion/year by 2030 (Ekpoh 2014). Areas with weak health infrastructure as in the case of developing countries will be the least able to cope without assistance to prepare and respond. These areas are also dealing with a growing epidemic of chronic diseases such as diabetes and cancer (Africa Adaptation Programme 2016).

The United Nations Framework Convention on Climate Change (UNFCCC) sees it as a change of climate that is attributed directly or indirectly to human activity, altering the composition of the global atmosphere. Human activity includes the pollution that arises from industrial activity and other sources that produce greenhouse gases (UNFCCC 2015). These gases, such as carbon dioxide, have the ability to absorb the spectrum of infrared light and contribute to the warming of our atmosphere. Once produced, these gases can remain trapped in the atmosphere for tens or hundreds of years (IPCC 2007a).

The term was coined in 1827 by Joseph Fourier, a French mathematician and physicist, who envisioned that the warming process of the Earth acted in the same way as a greenhouse traps heat – a process of visible light and invisible radiation, with Earth's atmosphere acting as the glass barrier. The visible radiation from the Sun that passes through the Earth's atmosphere can be absorbed by land, water, and vegetation, but some of this infrared radiation remains trapped in the atmosphere. While this is a natural process, it remains in a state of fine balance (UNFCCC 2015).

Greenhouse gases act like a blanket around the Earth. They trap warmth from the sun and make life on Earth possible. Without them, too much heat would escape and the surface of the planet would freeze. However, increasing the concentration of greenhouse gases in the atmosphere causes the Earth to heat more and the climate to change (Stern 2006). The greenhouse effect is the cycle by which these gases become trapped in the atmosphere and heat the planet (Stern 2006).

Studies have shown that the burning of fossil fuels, such as oil and coal, which emits greenhouse gases (primarily carbon dioxide) into the atmosphere as a primary cause of climate change. Other human activities, such as agriculture and deforestation, also contribute to the proliferation of greenhouse gases. While some quantities of these gases are naturally occurring and are critical part of Earth's temperature control system, the atmospheric concentration of carbon monoxide did not rise above 300 ppm between the advent of human civilization roughly 10,000 years ago and 1900. Today, it is at about 400 ppm, a level not reached in more than 400,000 years (Zhou et al. 2008; Gitay et al. 2002).

Nigeria is experiencing adverse climate conditions with negative impacts on the welfare of millions of people. Persistent droughts and flooding, off-season rains, and dry spells have sent growing seasons out of orbit in a country dependent on rain-fed agriculture. Alarm bells are ringing with lakes drying up and a reduction in river flow in the arid and semi-arid regions (McMichael 2002). The result is fewer water supplies for use in agriculture, hydropower generation, etc. Several environmental air pollutants, which include carbon monoxide, oxides of nitrogen, sulfur dioxide, ozone, lead, and particulate matter, have been associated with increased hospitalization and mortality (BNRCC 2011).

Researchers have therefore called climate change the biggest health threat the world has ever faced as it is now an unavoidable truth that the health of the planet is linked to the health of its people. Reducing emissions of greenhouse gases through better transport, food, and energy-use choices can result in improved health, particularly through reduced air pollution (Barrett et al. 2015). Therefore, this chapter theoretically explores the relationship between climate change and airborne diseases, looks at climate change and air quality, as well as trends of climate change patterns in Nigeria and profer recommendation for Nigeria's health sector development.

Effect of Climate Change on Air Quality

Greenhouse gases have an effect on climate change which in turn has an effect on air quality. Weather elements like temperature, humidity, wind, precipitation, and extreme weather events (such as drought-driven wildfires and dust storms) can affect air pollutant emission, chemistry, deposition, and transport (Dennekamp and Carey 2010). For example, higher atmospheric concentration of carbon dioxide and a warmer climate prolong the pollen season and increase pollen levels. High pollen concentrations will contribute to asthma morbidity (IPCC 2014).

The relationship between climate change and air quality can also be described as complex with many interacting mechanisms. For example, air quality influences can be of climatic/meteorologic (temperature, humidity, wind speed, wind direction, and mixing height), natural (ground-level ozone and light-catalyzed air chemistry reactions, aeroallergens, forest fires, and dust from dry soils), or anthropogenic (using carbon-based fuels for local energy use, transportation, and agriculture) origin, resulting in eventual deposition of air pollutants (Sapkota et al. 2005). Particulate matter is a pollutant of concern as it is a complex mixture of extremely small particles and liquid droplets. When inhaled, these particles can reach the deepest regions of the lungs. Exposure to particle pollution is linked to a variety of significant health problems (Prüss-Üstün and Corvalán 2006).

Sunlight and high temperatures, combined with other pollutants such as nitrogen oxides and volatile organic compounds, can cause ground-level ozone to increase. Ozone forms in the troposphere by the action of sunlight on ozone precursors (through photochemistry) from the by-products of burning carbon-based fuels (D'Amato et al. 2013). At the surface, an increase in temperature accelerates photochemical reaction rates (strong correlation between higher ozone levels and warmer days – but not always). Ground-level ozone can damage lung tissue and is especially harmful for those with asthma and other chronic lung diseases. Even those with moderate disease may be at risk from temperature rises above 16.5 °C (Cohen et al. 2004; Hileman 2006).

Climate Change Implication on Health

It is well known that the health of a population, if it is to be sustained, requires clean air, safe water, adequate food, tolerable temperature, stable climate, and high levels of biodiversity (WHO 2004; IPCC 2007b). Globally, climate change is expected to alter temperature, air movement, and precipitation in various ways and to varying degrees across Africa with consequences on human health. With the strong connection between a population's health and economic and environmental health, the impact of climate change on each is one of the major ways in which climate change may impede the development of the African continent (IPCC 2007a).

For example, climate change may affect health through increased frequency and intensity of extreme weather events (EWEs) (such as hurricanes, heatwaves, floods, and droughts), each of which are drivers of malnutrition and changes in the distribution of diseases (Stern 2006). Variations in health problems for each country is due to many factors such as geographic differences in temperature and precipitation, socioeconomic conditions, the quality of existing health infrastructure, communication capacity, and underlying epidemiology (Menactra 2011).

The extent and magnitude of effects is expected to increase with further climatic changes. Changes in the mean and variance of climate variables (such as temperature and precipitation) alter the incidence and geographic range of many climate-sensitive infectious diseases, with impacts varying across geographic and temporal scales (WHO 2002). Many African countries suffer serious health consequences due to impacts of climate change as they have populations that are among the most vulnerable to climatic changes in the world. This vulnerability is due in part to existing problems of poverty, weak institutions, and armed conflict, which limit their capacity to deal with the additional health challenges posed by climate change (WHO 2004).

Meningitis – an airborne meningococcal disease – caused by the bacteria *N. meningitidis* exists all over the world. However, in no region of the world is it as great a problem as in the Sahel region of Africa. For Africa as a whole, meningitis is one of the continent's top three climate sensitive diseases, with roughly 350 million people living in endemic zones for this disease (Stern 2006). Humans are the only natural reservoir for this disease, and it is often spread between humans via respiratory droplets or saliva (i.e., through coughing, sneezing, kissing or other forms of close and direct contact), with the symptoms apparent on some individuals and nonapparent on others (Menactra 2011).

Airborne Diseases and Climate Change

Climate change affects air quality through several pathways, including production and allergenicity of allergens and increase regional concentrations of ozone, fine particles, and dust. Some of these pollutants can directly cause respiratory disease or exacerbate existing conditions in susceptible populations, such as children or the elderly (Arnell 2004).

Airborne diseases are those diseases that are caused by pathogenic microbial agents which get discharged through coughing, sneezing, laughing, or through close personal contact. These pathogens ride on either dust particles or small respiratory droplets and can stay suspended in air and or are travelling distances on air currents. Infectious diseases spread by "droplet infection" have long been one of the most deadly branches of disease. It has been calculated that about half of all episodes of human illness are caused by respiratory viruses (Patz et al. 2005).

Most of these are of course quite trivial infections like the most frequent of them all, the common cold. Airborne diseases are being about twice as significant as water and food-borne diseases. "Droplet infection" continues as by far the commonest and most important in both developed and developing countries (Ashrae 2012).

Airborne transmission occurs when bacteria, fungi, or viruses travel on dust particles or on small respiratory droplets that may become aerosolized when people sneeze, cough, laugh, or exhale. They hang in the air much like invisible smoke; they can travel on air currents over considerable distances. These droplets are loaded with infectious particles (McMichael et al. 2003).

With airborne transmission, direct contact with someone who is infected is not necessary to become ill. The amount of exposure necessary varies from disease to disease. With chickenpox, a child could easily catch it from another aisle in a supermarket. With tuberculosis, closer contact and less air circulation are often needed (Charron et al. 2005).

Diseases such as the common cold, flu, measles, mumps, rubella, fifth disease, and chickenpox are airborne diseases caused by virus. Measles, mumps, and chicken pox are usually considered childhood diseases, although they can affect adults. Cold and flu virus are transmitted in a similar fashion, although influenza viruses are stronger pathogens and can cause potentially deadly global epidemics (Arnell 2004).

Tuberculosis, pneumonia, whooping cough, diphtheria, meningitis, and anthrax are examples of bacterial airborne diseases. Tuberculosis and bacterial pneumonia, despite the availability of vaccines, are still considered strong threats to mortality in Africa and the rest of the world (Wolf et al. 2015).

Fungal airborne diseases are primarily spread through the inhalation of spores. Unlike viral and bacterial airborne diseases, fungal airborne diseases are almost entirely noncommunicable. Carpet and floor dusts, latex paint, moist building materials, molds, and mildews are all sources of fungal spores. Outdoor spores vary between climates and seasons and can also be inhaled in agricultural areas such as barns or work sheds resulting in a disease called Farmer's Lung (Charron et al. 2005).

Climate Change Trend in Nigeria

About four decades ago, the Lake Chad covered an area of over 40,000 square kilometers, whereas it now encompasses a mere 1300 square kilometers. While the negative trend continues unabated, the land is laid to waste by rising temperature, leading to the rapid southward expansion of the Sahara Desert. Farmlands and surrounding villages have become barren because they were swallowed up by advancing desertification, which led to massive migration of people in search of more fertile terrain from the north east towards the greener plateau and middle belt region (Ekpoh 2014).

Growing desertification forced thousands of Fulani herdsmen to move to the south and middle belt, leading to clashes with crop farmers that culminated in death of hundreds. Nigeria's Guinea Savannah region is not spared either. Logging and over dependence on firewood for cooking have stripped a greater part of this area of its vegetation cover. The situation is similarly replicated in the south, where the forest around Oyo has long been reduced to grassland. The south-eastern part of the country has been struck by a different problem. There, gulley erosion has devastated many settlements and farmlands, leading to poverty among local populations (New Climate Institute 2015).

Just as desertification is devastating vast areas of the north, rising sea levels are threatening Nigeria's coastal regions. Although a source of oil wealth, the Niger Delta's low-lying terrain and waterways makes it extremely vulnerable to flooding. In southern Nigeria, climate change also resulted in massive flood experienced in 2012 and 2017 where houses, farms, farm products, properties, and even human beings were swept away (Nigeria Emissions Reduction Policy 2016).

Large Nigerian cities are currently developing resilience strategies with Lagos having a climate change policy and action plan in place. Enugu also appointed a chief resilience officer to oversee initiatives, and Kano, Kaduna, and Abuja are all developing holistic urban resilience strategies. The Clean Technology Fund is also supporting the development of transformative public transport schemes in Lagos, Kano, and Abuja. The Department of Climate Change has established a pioneering Climate Finance Unit that will enhance knowledge and information on climate finance opportunities and help develop robust project proposals (BNRCC 2011).

As it stands, Nigeria is yet to have any climate change specific law enacted by the legislative arm of government and assented to by the President (London School of Economics and Political Science 2013). The country has however adopted several environmental and sectoral policies, strategies, and plans where climate change adaptation could apply, though at present their use in enabling and supporting climate change adaptation is limited (UNFCCC 2015). In 2012, the Federal Executive Council adopted a comprehensive strategy policy on climate change: the Nigeria Climate Change Policy Response and Strategy (New Climate Institute 2015). The overarching objective of the policy is to promote low-carbon, high-growth economic development and build a climate-resilient society through the achievement of the following targets (London School of Economics and Political Science 2013; UNFCCC 2015):

- Implement mitigation measures that will promote low carbon as well as sustainable and high economic growth
- Enhance national capacity to adapt to climate change
- Raise climate change-related science, technology, and R&D to a new level that will enable the country to better participate in international scientific and technological cooperation on climate change
- Significantly increase public awareness and involve private sector participation in addressing the challenges of climate change
- Strengthen national institutions and mechanisms (policy, legislative, and economic) to establish a suitable and functional framework for climate change governance

Allied to this, the Great Green Wall of the Sahara and the Sahel Initiative is a planned project to plant a wall of trees across Africa at the southern edge of the Sahara desert as a means to prevent desertification. It is to be implemented in Nigeria in 11 frontline states of Adamawa, Bauchi, Gombe, Kebbi, Sokoto, Zamfara, Katsina, Kano, Jigawa, Yobe, and Borno. It will cover 43 LGAs in the frontline states to be covered to rehabilitate 225,000 Ha of lands. It involves establishment of green wall or shelterbelt from Kebbi State to Borno State, a distance of 1500 km and 15 km wide (Africa Adaptation Programme 2016).

Conclusion

The effects of climate change have reached a point where the emphasis has moved from proof of its existence to policies of mitigation. Action against climate change such as recycling initiatives, cleaner source of energy, etc. as well as how to maintain economies that throughout the twentieth century been reliant on fossil fuels and experienced unrestricted energy consumption for their development has become a focal point of this generation.

Addressing the health aspects of climate change need not depend on the federal government; it can happen at the level of cities, states, local governments, businesses, or institutions, as well as through regional collaborations. Hospitals and health care systems have some specific challenges to tackle and can be a kind of beacon in a community, a place people can turn to. Therefore, policies and measures need to ensure that actions taken to reduce climate-related health risks address health burdens and explicitly consider key uncertainties (2015 Annual Global Climate Report 2016).

Infectious diseases, especially those transmitted via air, insect vectors, water, etc. are sensitive to climatic conditions. Disease incidence data is needed to provide a baseline for epidemiological studies. The lack of precise knowledge of current disease incidence rates makes it difficult to comment about whether incidence is changing as a result of climatic conditions (Kuhn 2005). Monitoring to assess climate-change impacts on health therefore requires data gathering coupled with analytical methods able to quantify climate-attributable portion of such diseases.

A successful policy-focused assessment of climate change impact on diseases should include a multidisciplinary assessment team, responses to questions asked by all stakeholders, evaluation of risk management adaptation options, identification and prioritization of key research gaps, characterization of uncertainties and their implications for decision-making, and tools that support decision-making processes.

Finally, in 2009, the World Health Assembly endorsed a new WHO work plan on climate change and health (WHO 2009). This includes:

- Advocacy: To raise awareness that climate change is a fundamental threat to human health.
- Partnerships: To coordinate with partner agencies within the UN system and ensure that health is properly represented in the climate change agenda.
- Science and evidence: To coordinate reviews of the scientific evidence on the links between climate change and health and develop a global research agenda.
- Health system strengthening: To assist countries to assess their health vulnerabilities and build capacity to reduce health vulnerability to climate change.

References

2015 Annual Global Climate Report. Asheville, NC National Oceanic and Atmospheric Administration, U.S. National Centers for Environmental Information. Available at http://www.ncdc.noaa.gov/sotc/summary-info/global/201512. Accessed on 21 Jan 2016

Africa Adaptation Programme (2016) Nigeria, AAP Nigeria. Available at: https://www.undp-aap.org/countries/nigeria. Accessed 26 Sept 2016

Arnell NW (2004) Climate change and global water resources: SRES emissions and socio-economic scenarios. Glob Environ Chang 14:31–52

Ashrae (2012) Ashrae position document on airborne infectious diseases. Joshua David Kneifel. Twenty three agenda to ANSI/ASHRAE/IES Standard, Atlanta, 90:1–2012

Barrett B, Charles JW, Temte JL (2015) Climate change, human health, and epidemiological transition. Prev Med 70:69–75

Building Nigeria's Response to Climate Change Project (2011) National Adaptation Strategy and Plan of Action on Climate Change for Nigeria (NASPA-CCN), BNRCC. Available at: http://citeseerx.ist.psu.edu/viewdoc/download;jsessionid=27C7F6A7FD5FCAAD495C69457F2C5F7D?doi=10.1.1.367.6707&rep=rep1&type=pdf

Charron DF, Edge T, Fleury MD, Galatianos W, Gillis D et al (2005) Links between climate, water and waterborne illness, and projected impacts of climate change. HPRP 6795-15-2001/4400016c

Cohen AJ, Anderson HR, Ostro B, Pandey KD, Krzyzanowski M, Künzli N, Gutschmidt K, Pope CA III, Romieu I, Samet JM, Smith KR (2004) Urban air pollution. In: Ezzati M, Lopez AD, Rodgers A(eds), Health Effect Institute, Boston, Massachusett

D'Amato G, Baena-Cagnani CE, Cecchi L et al (2013) Climate change, air pollution and extreme events leading to increasing prevalence of allergic respiratory diseases. Multidiscip Respir Med 8(1):12

Dennekamp M, Carey M (2010) Air quality and chronic disease: why action on climate change is also good for health. NSW Public Health Bull 21(5/6):115–121

Ekpoh IJ (2014) Slow response to climate change in Nigeria: need for urgent and comprehensive action. Stud Soc Sci Hum 1(1):19–29. Available at: http://eprints.covenantuniversity.edu.ng/3787/1/corporate%20machiavellism-ijm.pdf

Gitay H, Suarez A, Watson RT, Dokken DJ (2002) Climate change and biodiversity. IPCC technical paper V. Intergovern-mental Panel on Climate Change, Geneva

Hileman B (2006) Malaria control, resurging use of the banned pesticide DDT to prevent malaria poses dilemma for health, environment. Chem Eng News 84:30–31

IPCC (2007a) Impacts, adaptation and vulnerability. (Contribution of working group II to the fourth assessment report of the Intergovernmental Panel on Climate Change), Geneva

IPCC (2007b) Fourth assessment report. Intergovernmental Panel on Climate Change Secretariat, Geneva. http://www.ipcc.ch/

IPCC (2014) 2014: summary for policymakers. In: Climate change 2014: mitigation of climate change. Contribution of working group III to the fifth assessment report of the Intergovernmental Panel on Climate Change [Edenhofer O, Pichs-Madruga R, Sokona Y, Farahani E, Kadner S, Seyboth K, Adler A, Baum I, Brunner S, Eickemeier P, Kriemann B, Savolainen J, Schlömer S, von Stechow C, Zwickel T, Minx JC (eds)]. Cambridge University Press, Cambridge/New York

Kuhn Katrin (2005) Using climate to predict infectious disease epidemics. Communicable diseases surveillance and response protection of human environment roll back malaria. WHO library cataloguing-in-publication data

London School of Economics and Political Science (2013) Uncertainties and decision in climate change economics. Centre for climate change economics and policy, working paper 128

McMichael AJ (2002) Population, environment, disease, and survival: past patterns, uncertain futures. Lancet 359:1145–1148

McMichael A, Campbell-Lendrum D, Corvalán C, Ebi K, Githeko A, Scheraga J (2003) Climate change and human health: risks and responses. World Health Organisation, Geneva

Menactra (2011) Menactra: causes and presentations of invasive meningococcal disease (IMD). Menactra: meningococcal (Groups A, C, Y and W-135) polysaccharide diphtheria toxoid conjugate vaccine. Retrieved 27 Aug 2011 from http://menactra.ca/index.cfm?fa=view.show&cname=Causes

New Climate Institute (2015) National policy on climate change Nigeria 2013. Available at: http://www.climatepolicydatabase.org/index.php?title=National_Policy_on_Climate_Change_Nigeria_2013

Nigeria Emissions Reduction Policy (2016, November 20) Ministry of Environment, Federal Government of Nigeria

Patz JA, Campbell-Lendrum D, Holloway T, Foley JA (2005) Impact of regional climate change on human health. Nature 438:310–317

Prüss-Üstün A, Corvalán C (2006) Preventing disease through healthy enviroments. World Health Organisation, Geneva

Sapkota A, Symons JM, Kleissl J (2005) Impact of the 2002 Canadian forest fires on particulate matter air quality in Baltimore city. Environ Sci Technol 39:24–32

Stern N (2006) The Stern review on the economics of climate change. HM Treasury. Cambridge University Press. http://www.hm treasury.gov.uk/independent_reviews/stern_review_economics_climate_change/sternreview_index.cfm

UK-Met Office (2008) The United Kingdom Government Met Office: HadCRUT3 annual time series, Hadley Research Centre

United Nations Framework Convention on Climate Change (2015) Nigeria's intended nationally determined contribution, UNFCC. Available at: http://www4.unfccc.int/submissions/INDC/Published%20Documents/Nigeria/1/Approved%20Nigeria's%20INDC_271115.pdf
WHO (2002) The World Health Report 2002. Reducing risks, promoting healthy life. World Health Organization, Geneva. Last accessed 14 May 2002
WHO (2004) The World Health Report 2003 – shaping the future. World Health Organisation, Geneva. http://www.who.int/whr/2003/en/index.html
WHO (2009) Protecting health from climate change: connecting science, policy and people. World Health Organization, Geneva
Wolf T, Lyne K, Martinez GS, Kendrovski V (2015) The health effects of climate change in the WHO European Region. Climate 3:901–936
Zhou XN, Yang GJ, Yang K, Wang XH, Hong QB et al (2008) Potential impact of climate change on schistosomiasis transmission in China. Am Trop Med Hyg 78:188–194

Herdsmen on the Move: The Burdens of Climate Change and Environmental Migration in Nigeria

62

Toyib Aremu and Praise Abraham

Contents

Abstract

The changes occurring in the climate, such as rainfall pattern change, can induce human (and animal) migration as a form of adaptive strategy especially by families that are agricultural or natural-resources dependent. Climate change is a global concern now that different countries including Nigeria are living in the reality of its impacts. These impacts are felt more in northern Nigeria (the arid region) where there is an evident ecological decline resulting in drought and desertification. Herders from this region, who are predominantly into cattle rearing as a source of wealth and livelihood, are forced to embark on a

T. Aremu (✉)
Department of Environmental Sustainability, Centre for Sustainable Development, University of Ibadan, Ibadan, Oyo, Nigeria
e-mail: aremutoyib@gmail.com

P. Abraham
Department of Social Sustainability, Centre for Sustainable Development, University of Ibadan, Ibadan, Oyo, Nigeria
e-mail: praisemark15@gmail.com

W. Leal Filho (ed.), *Handbook of Climate Change Resilience*,
https://doi.org/10.1007/978-3-319-93336-8_146

north–south migration in response to drought and unavailability of forage and water to feed their cows. Often, however, this change of environment is characterized by conflicts between migrating herdsmen and receiving communities for various reasons including competition emanating from scarcity of natural resources. This review is relevant at this time of a national debate in Nigeria on finding the right policies to abate the continuous clashes between herdsmen and farmers. The paper assesses migration in the context of climate change, the challenges associated with migration of herdsmen, the mutual cooperation that existed between herdsmen and farmers, and the conflict that now characterize their relationship.

Keywords
Environmental migration · Climate change · Natural resources · Conflict · Herdsmen · Farmers · Nigeria

Introduction

Climate change are irreversible observable variations in the state of the climate over time. It is caused by an increased concentration of greenhouse gases (GHGs) in the atmosphere. Greenhouse gases which includes methane (CH_4), carbon dioxide (CO_2), water vapor (H_20), and nitrous oxide (N_20) are emitted through various human activities like burning of fossil fuels, wood burning, agricultural practices, gas flaring (Ogbo et al. 2013), and overgrazing. The wanton release of greenhouse gases into the atmosphere as a result of burning of fossil fuels, burning of wood, gas flaring, agriculture, and other economic activities has continued to cause an increase in the average world's temperature. Recently, climate change has gained global attention as never before (Odjugo 2010) due to the problems it has triggered around the world. Nigeria particularly is experiencing the adverse effects of climate change evident in the occurrence of drought and desertification especially in the northern parts of the country (which are the arid region). The irregular rainfall has affected the agriculture system, the reason being that it is very much rain-fed (Idowu et al. 2011; Ogbo et al. 2013; Olaniyi et al. 2014). The subsistence farmers (and pastoralists) who are unable to access better technologies and improved agricultural practices are at the receiving end of these climate risks (Hendrix and Salehyan 2012).

Climate change affects the availability of pasture and water for animal consumption and is responsible for droughts, desertification, extreme cold, and storms. Lakes are drying up and there is a decline in river flow in the northern regions. There is fewer supply of water for agricultural purposes like crop production and maintenance of pasture for cattle rearing. When herdsmen are short-changed by these basic needs for their cattle due to climate change, it is common to find them moving from northern Nigeria down south in search of better environmental and weather conditions. This migration usually for adaptation purposes creates problems for the receiving communities. This kind of encroachment, according to Odoh and Chigozie (2012), could engender conflict given the pressure on land, food shortage, conflicts of interest, cultural differences, overpopulation, social disorganization, and religious, social, and cultural intolerance that follows. This paper has been divided into five sections "Introduction," "Conceptual Clarifications," "Conceptual Framework," "Climate Change, Herdsmen, and Migration in Nigeria," "Climate Change and Environmental Conflicts" and "Why the Farmers-Herdsmen Crises have become Intensified" before the conclusion is made.

Conceptual Clarifications

Climate Change

An important component of this paper is the concept of climate change. Climate itself encompasses the statistics of temperature, humidity, atmospheric pressure, wind, rainfall, atmospheric particle count, and other meteorological elemental measurements in a given region over long periods. The climate of a region functions by its climate system which has five components: atmosphere, hydrosphere, cryosphere, land surface, and biosphere. The more the quantity of greenhouse gases in the atmosphere, the more the amount of solar energy trapped by the planet which results in global warming or cooling (IPCC 2007; Odoh and Chigozie 2012). Climate changes are irreversible observable variations in the state of the climate over time; it is caused by two major factors: the natural process also known as bio-geographical and human activities also known as anthropogenic (IPCC 2007).

Environmental Migration

Migration is the movement by people from one place to another with the intention of settling permanently or for a period of time. It could be within or outside a country. People can migrate to get a better life and seek refuge or asylum or for tourism (Oiarzabal and Reips 2012). The definition of environmental migrants adopted in this study is that of El-Hinnawi which was modified by the International Organization for Migration. According to this definition, they are "persons or groups of persons who, for compelling reasons of sudden or progressive changes in the environment that adversely affect their lives or living conditions, are obliged to leave their habitual homes, or choose to do so, either temporarily or permanently, and who move either within their country or abroad" (Brown 2007). The decision to migrate as a result of environmental degradation can be voluntary or made out of compulsion. Voluntary migrants migrate for a number of reasons, particularly to seek better economic opportunities. On the other hand, those who are compelled to migrate do so for external factors and reasons and take the status of refugees. In between voluntary migrants and environmental refugees however are anticipatory refugees who have prior knowledge of impeding degradation and have the ability to move before they are forced to (Bates 2002; Kunz 1973). Poor people are more vulnerable to a deteriorating environment and the migration resulting from it. Households in this situation usually send a member, usually young men, in what is referred to as "dispatch migration" to take advantage of distant opportunities. The movement of the entire household is a last resort, and when it does happen, it is within the country (Bates 2002).

Herdsmen

Herdsmen are owners or keepers of herds of animals like cattle. They usually move their herds between places to search for pasture and water. In Nigeria, it is predominantly the occupation of the Fulani ethnic group which means they are the majority who own and rear cattle for commercial purposes (Oli et al. 2017). They are concentrated principally in Nigeria, Mali, Guinea, Cameroon, Senegal, and Niger. Their indigenous language is known as Fula and it is classified within the Atlantic branch of the Niger-Congo language family. They constitute the fourth-largest ethnic group in Nigeria with an estimated population of over seven million (Burton 2016). Pastoralism is a lifestyle that "obtains livelihood from mobile livestock rearing on unimproved, communal pastures. It is essentially a subsistence system where the people involved derive all or large part of their food and other living needs using the herds (cattle, sheep, goats, camels, yaks, llamas, reindeer) they tend" (RASS 2006 cited in Msuya 2015).

Conceptual Framework

The conceptual framework (Fig. 1) adopted for this study illustrates the complex path to climate conflict originating from climate change. Climate change, from this framework, leads to the loss of means of livelihoods through increase in extreme events such as droughts and desertification. Loss of means of sustenance, in this case grazing land and water source for the cattle, influences migration decision by herdsmen. Migration in turn leads to pressure and competition on limited resources among other things and finally leads to conflicts. Different from the common explanation that the clashes between herdsmen and farmers are ethno-religious, this pathway offers a new perspective on how resource scarcity can engender conflict.

Climate Change, Herdsmen, and Migration in Nigeria

The Fulanis in Nigeria are known for their nomadic and pastoral nature and have a relatively large and wealthy population. According to Odoh and Chigozie (2012), "Nigeria has a restricted inventory of pastoral peoples, the Fulani, the Kanuri-related groups, the Shuwa, the Yedina and the Uled Suleiman." Fulanis are the most widely spread around Africa and around the Volta region of Ghana, Gambia River, and Chad and most likely entered Nigeria before the fifteenth century. The raising of concerns

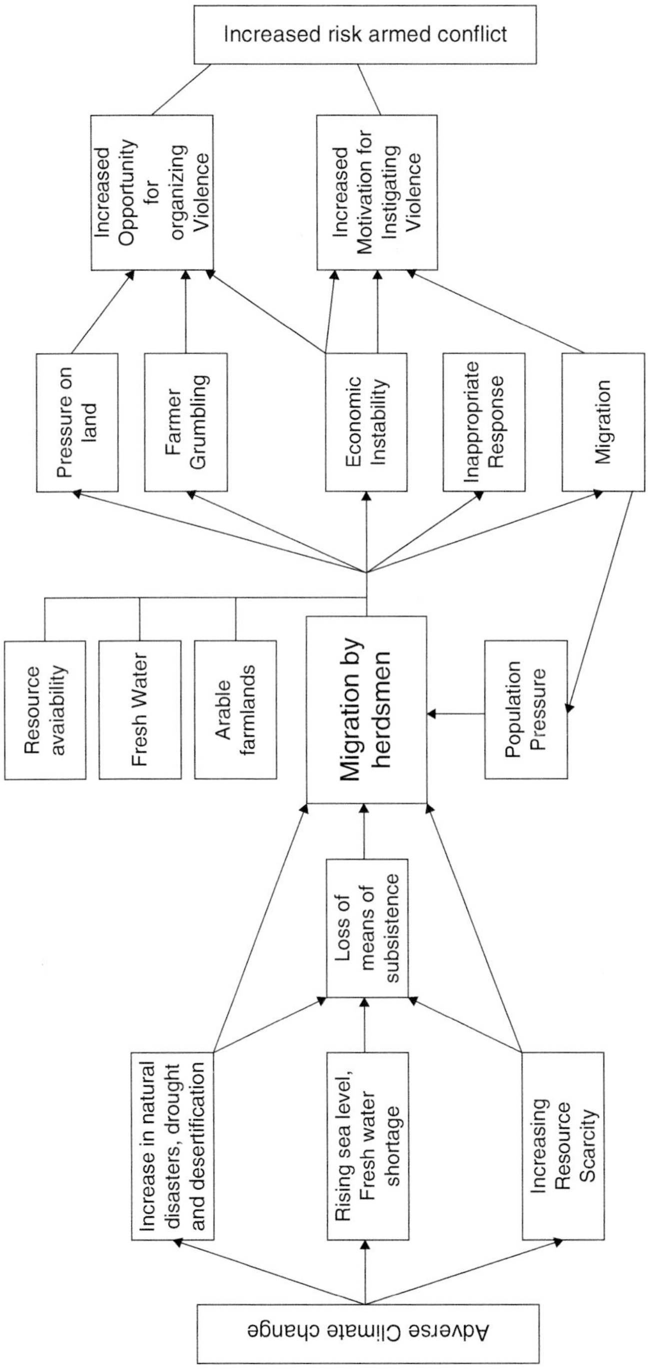

Fig. 1 Conceptual Framework showing pathway to climate conflicts. (Source: Odoh and Chigozie 2012)

about the rise of environmental migrants began about three decades ago when the then director of the United Nations Environment Programme projected a figure of about 50 million environmental refugees at the time (Bates 2002; Tolba 1989). In time past, the herders were restricted to the edge of the desert, but during the twentieth century they began to migrate and settle in different zones which was formerly impossible for them to do. The West African Sahelian drought and famine of 1968–1974, consequent upon climate change in the region, led to migration of Fulani herdsmen and their cattle and resulted in pressure on and competition in resource use and conflict between farmers and herdsmen (Shettima and Tar 2008). The incidence of severe heat and less rain in northern Nigeria predisposes the region to climate impacts such as desertification and water and land scarcity. Pasture and water for cattle become scarce and found at different times of the year in different places thereby necessitating what is referred to as "seasonal migration" of the herdsmen and their herds (Emmanuel 2017). With almost two-thirds of the north western and eastern states at the risk of turning into desert or semi-desert, the risk of climate conflicts cannot be ruled out.

When pastoralists move out of their climate-affected communities down south in search of water and forage for their cattle, they do so into communities of farmers who fear that their crops will be destroyed by the herders and their herds. Since climate change does not only affect pastoralists, it has also pushed farmers to expand their cultivation area which limits the area of land available for pastoralists to graze their animals. This leads to competition over the use of resources between the sedentary farmers and migrating herdsmen often resulting in fights. Unfortunately, many of the communal clashes are misinterpreted as ethnic and religious clashes between Muslim Fulani herdsmen and Christian farmers (e.g., see Adamu and Ben 2017) when actually the control of natural resources is a significant cause of the conflicts between the two groups. Apart from environmental degradation, resource scarcity, and climate change which are responsible for these conflicts, there are conflict entrepreneurs, who, according to Frerks (2007) and cited in Shettima and Tar (2008), use political and socioeconomic factors often as tools to fuel distrust and hate among herdsmen and farmers.

Climate Change and Environmental Conflicts

The connection that exists between climate change and conflict has been simplified by Odoh and Chigozie (2012) when they pointed that poor responses to climatic shifts (such as changes in temperature, rainfall pattern, sea levels, storm, etc.) in a region, population, or sector lead to shortages in resources (such as land, water, vegetation/forages, etc.) and poor responses to these shortages engender conditions like sickness, hunger, job losses, and poor economic growth which could heighten conflict risks. Poor governance and societal inequalities in the country has also promoted political and economic instability, social fragmentation, and migration. Climate change impacts the social and economic life of the herders because the well-being of the people is affected as there is unbearable heat and limited water supply

Table 1 Yearly death statistics resulting from herdsmen-farmer clashes (2014–2018)

S/No	Period	Number of deaths	Source
1.	January–April 2018	937	Bloomberg (https://www.bloomberg.com/news/articles/2018-05-22/qaddafi-regime-s-legacy-fuels-bloody-conflicts-in-west-africa)
2.	2017	549 (14 states)	Amnesty International (http://www.dailypost.ng/2018/01/30/amnesty-international-reveals-many-nigerians-fulani-herdsmen-killed-2018)
3.	2016	2500	Voice of Africa news (https://www.voanews.com/a/farmer-herder-conflict-rises-across-nigeria/3326151.html)
4.	2015	4780	The Cable News (https://www.thecable.ng/path-of-a-ruthless-killer-all-boko-haram-deaths-in-2015)
5.	2014	6347	Armed Conflict Location and Data Project (http://www.theguardian.com/global-development/2015/jan/23/boko-haram-nigeria-civilian-death-death-toll-highest-acled-african-war-zones)

for household consumption. Environmental degradation and desert encroachment in the northern region have caused herdsmen to migrate southwards to search for forage and water to cater for their cattle and this has eventually led to increased motivation to instigate violence due to resource scarcity (Olakunle and Adejoke 2013; Theisen et al. 2013).

Farmer-herder conflicts have for years affected economic livelihood in West Africa with consequences on human and animal lives, properties, peaceful coexistence, and orderliness. This conflict is not a new phenomenon although recent conflicts seem more pronounced than in the past. Farmers and herders enjoyed low-intensity conflicts in the past because scarcity of resources, land degradation, and increase in population did not pose as much a challenge as they do now. But recently in the country there have been incessant massacre and continual crisis (see Table 1) which has consequently caused host communities to lose peace and lives in the heat of the crisis (Ubelejit 2016). In spite of the cooperation that existed between farmers and herdsmen, competition for resources and environmental degradation have now threatened such relationships and resulted to different kinds of communal clashes, destruction of agricultural produces, and death of those involved in the conflicts.

For example, there was a violent conflict in 2009 when Udeni Gida village was attacked by herders. This was as a result of a previous clash that had occurred when some herdsmen led their cattle into a rice plantation, destroying crops and even leading to the death of a farmer. The local authorities immediately reacted expelling about 700 herders from Borno state and 2,000 from Plateau state in the same year. Recently, armed conflict between the Fulani herdsmen and the sedentary farmers had been reported to take place in over 20 villages in Nigeria, including, Ago Nla, Ago Kekere, Idi OpeKekere, Olukore, Olokuta, Apena, Jaramosan, Akeeran, Alagigo, Aaaro, Temidire, Iya Ibadan, Ideto, Apapa, Akele, Agwo, Bodija, Olufayo,

OkoTeku, Moniya, OkeOgun, Shaki, Egbedore. Ilobu, EfonAlaye, Alu, and Iyamoye (Odoh and Chigozie 2012). A study carried out by Sulaiman and Ja'afar (2010) and cited in Odoh and Chigozie (2012) in Bauchi state (Northern Nigeria) from 2003 to 2007 showed that livestock lost in the course of the farmers-herders clash included 34 cattle and 11 sheep/goats, worth about 1.8 million Naira and 66, 000 Naira respectively for cattle and sheep/goat. Six persons lost their lives during the period with anticipated economic contribution valued at 2.8 million Naira. The number of persons injured in the farmer-pastoralists conflict as revealed by the study was 45 among whom 15 persons were injured and were not able to perform their economic activities during the period of receiving treatment.

Why Has the Farmer-Herdsmen Crises Become Intensified?

The relationship between farmers and pastoralists is supposed to be mutual rather than parasitic. They have enjoyed a lot of mutualism in the past both in Nigeria and other African countries before the turning of the tide and launching of attacks on one other. According to Shettima and Tar (2008), pastoralists needed the farmlands of crop farmers to feed their cattle and the local farmers require the products from the pastoralists' cattle. Different kinds of relationships have existed between nomadic herders and sedentary crop farmers such as the provision of clothes and cooking utensils among other things by farmers in exchange for salts (Webb 1995 cited in Hussein 1998); trading of manure by herders for grains (Blench 2000) and the soliciting of animals by farmers to graze farms after harvesting in exchange for money, sugar, and tea (Horrowitz and Little 1987 in Shettima and Tar 2008); and farmers putting animals in the care of herders in exchange for grazing and exchanging agricultural produce for milk, cattle, and manure (Shettima and Tar 2008). Conflicts between herders and crop farmers, however, begin when the activity of one group starts affecting that of the other. The coexistence and complementary relationship between farmers and nomads are tried when planting of crops prevents the movement of cattle and their herders to a point where they cannot access pastures and water or when the herders and their cattle get on farms and destroy the crops during planting season.

The reason for the prolonged crisis may be a result of lack of a representative voice by the herdsmen or lack of power as pointed by Shettima and Tar (2008). It is common to hear in the news how many lives and crops were destroyed by "Fulani herdsmen," but it is difficult to read how much of the brunt is worn by the herdsmen. Shettima and Tar (2008) cited the arguments of Toulmin (1983) and Winter and Rowley (1998). Toulmin (1983) argues that when conflicts between farmers and pastoralists get in the hands of state arbitration structures, they listened to the farmers which means the pastoralists are left "defenseless and powerless." In Winter and Rowley's (1983) similar views, pastoralists are often criminalized as being the cause of the conflicts because, according to the judges, "It is the cattle that move to meet the farm and not the farm that move to meet the cattle." But Traore (1996) and Shettima and Tar (2008) think that while "straying animals" are used as an excuse to put all the blame on the pastoralists, it also makes sense to talk about "straying fields" into grazing lands which would mean that farmers also play a part in causing the conflicts. This was the case in northern Benin where encroachment of farmers into grazing land and the disappearance of paths to grazing and water areas were some of the reasons that led to conflicts with the pastoralists, according to De Haan (1998).

Conclusion

This study examined the causes and effects of the Fulani herdsmen and farmer crisis. Farmer-herder conflicts, being a common feature of economic livelihood in West Africa, has been identified as a threat to peaceful coexistence, security, and orderliness. From the points argued in this paper, it is clear that environmental scarcity which is a result of migration (and migration results from climate change) is a cause of environmental conflicts. The crises have, however, been heightened owing to the way it has been handled by the government or the people. While government's body language seems like it approves of the killing of herdsmen, media reportage, too, has made the situation look like only the farmers suffer in this climate-induced conflict. To holistically and sustainably address environmental conflict issues in Nigeria, therefore, a lot of trust needs to be developed: trust between farmers and herders, trust between government and parties involved in conflicts, and trust between arbitrators and victims of the conflicts. There should be no indication that one party is being favored at the expense of another. Traditional and indigenous peoples also have valuable lessons to offer about successful and unsuccessful adaptations to climate change, lessons which could be vital in the context of global climate change (Egeru 2012; Macchi 2008). Incorporating indigenous knowledge into climate change policies can lead to the development of effective adaptation strategies that are cost-effective, participatory, and sustainable (Egeru 2012; Robinson and Herbert 2001). The media houses, too, have a big role to play in managing the overarching crises by maintaining professionalism in reporting. The government should provide climate-smart interventions in the Sahelian areas which are more prone to climate change impacts for adaptation and mitigation; diversify crops, adopt improved silage and forage, and improve range management; and implement other improved agricultural systems as suggested in Nigeria's National Adaptation Strategies and Plan of Action towards climate change (BNRCC 2011).

References

Aaron S (2011) Climate change adaptation and conflict in Nigeria: special report. United States Institute of Peace, Washington, DC

Adamu A, Ben A (2017) Nigeria: Benue State under the shadow of "herdsmen terrorism" (2014–2016). World Watch Research and Voice of Martyrs Canada

Awogbade MO (1983) Fulani pastoralism: Jos case study. Ahmadu Bello Press, Zaria

Bates DC (2002) Environmental refugees? Classifying human migrations caused by environmental changes. Popul Environ 23(5):465–477. Available online at https://www.jstor.org/stable/27503806

Blench R (2000) Extensive pastoral livestock systems: Issues and options for the future. ODI, London

BNRCC (Building Nigeria's Response to Climate Change) (2011) National Adaptation Strategy and Plan of Action on Climate Change for Nigeria (NASPA-CCN). https://csdevnet.org/wp-content/uploads/NATIONAL-ADAPTATION-STRATEGY-AND-PLAN-OF-ACTION.pdf

Brown O (2007) Climate change and forced migration: observations, projections and implications. UNDP, Winnipeg

Burton G (2016) Background report: the Fulani Herdsmen. Project Cyma Publication. pp 1–18

De Haan L J (1998) Gestion de terrroir at the frontier: village land management of peasants and pastoralists. In Beninin Bruins HJ and Lithwick H (eds) The Arid Frontier: Interactive Management of Environment and Development. Dordrecht and Boston: Kluwer Academic Publisher

Egeru A (2012) Role of indigenous knowledge in climate change adaptation: a case study of the Teso Sub-region, Eastern Ugandan. Int J Trad Knowl 11(2):217–224

Emmanuel A (2017) Farmer-Herder conflict in Africa: an assessment of the causes and effects of the sedentary Farmers-Fulani Herdsmen conflict. A case study of the Agogo traditional area, Ashanti region of Ghana

Frerks G (2007) Linking environment and conflict: building block for a knowledge, innovation and research strategy. The hague: prepare for the Buza-NOW Workshop on Conflict and Natural Resources

Hendrix CS, Salehyan I (2012) Climate change, rainfall and social conflict in Africa. J Peace Res 49(1):35–50

Holdren JP (2010) Climate-change science and policy: what do we know? What should we do? Keynote Address, Kalvi Prize Science Forum, International Cooperation in Science, Oslo, September 6

Hussein K (1998) Conflict between farmers and herders in the semi-arid Sahel and East Africa: A review. Pastoral Land Tenure Series No. 10. Drylands Programme. International Institute for Environment and Development & Overseas Development Group. London

Idowu AA, Ayoola SO, Opele AI, Ikenweiwe NB (2011) Impact of climate change in Nigeria. Iran J Energy Environ 2(2):145–152. ISSN 2079-2115

IPCC (Intergovernmental Panel on Climate Change) (2003) Climate change 2003: synthesis report. A contribution of Working Groups I, II, and III of the Intergovernmental Panel on Climate Change. Cambridge University Press, Cambridge

Intergovernmental Panel on Climate Change (IPCC) (2007) Climate change 2007. The fourth assessment Report (AR4). Synthesis report for Policymakers

John M, Ronald A, Abdul-Rahim A (2016) Governing interests of Fulani herdsmen and peasant farmers in natural resources in the Asante Akim North District of Ghana. Adv Soc Sci Res J 3(10). https://doi.org/10.14738/assrj.310.2238

Kunz EF (1973) The refugee in flight: Kinetic models and forms of displacement. International Migration Review 7. Summer

Macchi M, Oviedo G (2008) Indigenous and traditional peoples and climate change: Issues Paper. International Union for Conservation of Nature

Magnus TO, Gleditsch NP, Buhaug H (2013) Is climate change a driver of armed conflict. Clim Chang 117(3):613–625

Msuya DG (2015) Pastoralism beyond ranching: a farming system in severe stress in semi-arid Tropics especially in Africa. J Agric Ecol Res Int 4(3):128–139. https://doi.org/10.9734/JAERI/2015/19208

Odjugo PAO (2010) General overview of climate change impacts in Nigeria. J Hum Ecol 29(1):47–55. https://www.krepublishers.com/02-Journals/JHE/JHE-29-0-000-10-Web/JHE-29-1-000-10-Abst-PDF/JHE-29-1-047-10-1993-Odjugo-P-A-O/JHE-29-1-047-10-1993-Odjugo-P-A-O-Tt.pdf

Odoh SI, Chigozie CF (2012) Climate change and conflict in Nigeria: a theoretical and empirical examination of the worsening incidence of conflict between Fulani herdsmen and farmers in Northern Nigeria. Arabian J Bus Manag Rev 2(1):110–124

Ogbo A, Lauretta NE, Ukpere W (2013) Risk management and challenges of climate change in Nigeria. J Hum Ecol 41(3):221–235. https://www.krepublishers.com/02-Journals/JHE/JHE-41-0-000-13-Web/JHE-41-3-000-13-Abst-PDF/JHE-41-3-221-13-2390-Ukpee-W/JHE-41-3-221-13-2390-Ukpee-W-Tx%5B5%5D.pmd.pdf

Oiarzabal PJ, Reips UD (2012) Migration and diaspora in the age of information and communication technologies. J Ethn Migr Stud 38(9):1333–1338

Olakunle MF, Adejoke OF (2013) Climate change and inter-ethnic conflict in Nigeria. Peace Rev: J Soc Just 25:104–110. https://doi.org/10.1080/10402659.2013.759783

Olaniyi OA, Funmilayo OA, Olutimehin IO (2014) Review of climate change and its effect on Nigeria ecosystem. Int J Environ Pollut Res 2(3):70–81

Oli NP, Ibekwe CC, Nwankwo IU (2017) Prevalence of herdsmen and farmers conflicts in Nigeria. Int J Innov Stud Sociol Humanit 3(2):30–39

Peter AO (2010) General overview of climate change impacts in Nigeria. J Hum Ecol 29(1):47–55

Robinson JB, Herbert D (2001) Integrating climate change and sustainable development. International Journal of Global Environmental Issues 1(2):130–149. https://www.researchgate.net/profile/John_Robinson11/publication/5172137_Integrating_Climate_Change_and_Sustainable_Development/links/02bfe50eefb41847c8000000.pdf

Shettima AG, Tar UA (2008) Farmer-pastoralist conflict in West Africa: exploring the causes and consequences. Inform Soc Justice 1(2):163–184. https://doi.org/10.3734/isj.2008.1205

Tolba MK (1989) Our biological heritage under siege. Bioscience 39:725–728

Toulmin C (1983) Herders and Farmers or Farmer-herders and Herder-farmers? ODI pastoral development network paper 15d. ODI, London

Traore (1996) Problems in Pastoral land management related to tenure: policies versus basic practice. In: GRET and IIED (eds) Managing Land Tenure and resource access in West Africa. Proceedings of a Workshop held in Goree, Senegal

Ubelejit NT (2016) Fulani herdsmen and communal conflicts: climate change as precipitator. J Polit Sci Leadersh Res 2(1)

Ute TU (2016) Fulani herdsmen and communal conflicts: climate change as precipitator. J Polit Sci Leadersh Res 2(1):26–32

Assessing Socioeconomic Factors Influencing Production and Commercialization of Bambara Groundnut as an Indigenous Climate-Resilient Crop in Nigeria

63

Olawale Emmanuel Olayide, Samuel A. Donkoh, Isaac Gershon Kodwo Ansah, William Adzawla, Patrick J. O'Reilly, Sean Mayes, Aryo Feldman, Razlin Azman Halimi, George Nyarko, Christopher O. Ilori, and Tunrayo Alabi

Contents

O. E. Olayide (✉)
Centre for Sustainable Development, University of Ibadan, Ibadan, Nigeria
e-mail: waleolayide@yahoo.com; oe.olayide@ui.edu.ng

S. A. Donkoh · I. G. K. Ansah · W. Adzawla
Faculty of Agribusiness and Communication Sciences, University for Development Studies, Tamale, Ghana

P. J. O'Reilly · S. Mayes · A. Feldman
Crops for the Future (CFF), Jalan Broga, Selangor Darul Ehsan, Malaysia

R. A. Halimi
Crops for the Future (CFF), Jalan Broga, Selangor Darul Ehsan, Malaysia

Southern Cross Plant Science, Southern Cross University, East Lismore, NSW, Australia

G. Nyarko
Faculty of Agriculture, University for Development Studies, Tamale, Northern Region, Ghana

C. O. Ilori
Crop Protection and Environmental Biology, University of Ibadan, Ibadan, Nigeria

T. Alabi
Geographic Information Systems Unit, International Institute of Tropical Agriculture, Ibadan, Nigeria

W. Leal Filho (ed.), *Handbook of Climate Change Resilience*,
https://doi.org/10.1007/978-3-319-93336-8_158

Abstract

Climate change is impacting the cropping system, landscape, livelihoods, and nutrition diversity of farming households and communities in Africa. Climate change adaptability and resilience are emerging as important criteria for setting national priorities for promoting indigenous crops to enhance food and nutrition security, especially of resource-poor smallholders. However, many climate resilient indigenous crops have been lost due to inappropriate policies that fail to prioritize climate resilience and nutritional diversity. Bambara groundnut (*Vigna subterranea*) is an indigenous crop in Africa. It is tolerant to drought, poor soils, and short spells of elevated temperatures. It, therefore, offers several advantages over other legumes as a source of nutrition, food security and improved welfare in the face of climate change. The research investigated farmers' perceptions and socioeconomic factors that influenced the cultivation and commercialization of bambara groundnut and the effect of commercialization on smallholder farmers' welfare in two local government areas (LGAs) of Benue State, Nigeria. In all, 300 smallholder farmers were sampled through a multistage sampling technique. The method of analysis involved the estimation of a fractional regression and treatment effect models. We found that older farmers who perceived that bambara groundnut is a climate-resilient and food security crop allocated more of their total farmland to its production. The perception that bambara groundnut is a climate-resilient crop also impacted positively on the commercialization of bambara groundnut. Formal education coupled with the commercialization of bambara groundnut led to increased farmers' welfare. We recommend that more sensitization and education should be given to farmers on the good characteristics of bambara groundnut as a climate-resilient and food security crop while they are also supported to upscale its production for commercialization purposes.

Keywords

Indigenous crop · Bambara groundnut · Commercialization · Climate resilience · Food security and nutrition · Household welfare

Introduction

Bambara groundnut (BG) is an indigenous drought-tolerant African legume (Chai et al. 2016; Mwale et al. 2007). Nigeria is the biggest producer of BG in Africa (Hillocks et al. 2012). Bambara is mainly cultivated and consumed in the northern parts of Nigeria by smallholders. The crop is important for its climate resilience and nutritive value (Mabhaudhi et al. 2013; Stadler 2009; Brough and Azam-Ali 1992). Its gross energy exceeds that of other common pulses such as cowpea, lentils, and pigeon pea (FAO 1982), and it also contains significant quantities of important dietary nutritional components such as protein, carbohydrate, and fats (Goli 1997). In this respect, it offers several advantages over other legumes as a source of nutrition, including protein content as high as 30% for some landraces (Adebowale et al. 2011). Its protein is a good source of essential amino acids, specifically methionine, cysteine, and lysine (Adebowale et al. 2011; Yao et al. 2015) comparable to other legumes such as cowpea and chickpea, and its carbohydrate has up to 53% starch with high amylose content (Yao et al. 2015; Oyeyinka et al. 2015). BG is also a good source of important dietary fatty acids such as linoleic (18:2, omega-6), linolenic (18:3, omega-3) and oleic acids (18:1, omega-9) (Yao et al. 2015; Minka and Bruneteau 2000; Adeleke et al. 2018; Aremu et al. 2013). In addition to its suitability as a food security crop for subsistence farmers, BG is also capable of fixing nitrogen into the soil, thereby contributing to soil fertility improvement. BG is reported to be tolerant to drought, poor soils, and extreme heat, hence making it a suitable crop to the low-input production systems (Karikari 1996). BG is a significant indigenous legume crop for promoting resilience to climate change, food security, and nutrition (Hillocks et al. 2012; National Research Council 1996).

Bambara groundnut is categorized as a neglected and underutilized crop species (Hillocks et al. 2012; Mkandawire 2007), and it is one of the indigenous legumes that require urgent rescue and inclusion in priority crops for Africa (National Research Council 1996). The search for food system resilience in response to climate change is leading to the re-discovery of the crop. However, the cultivation and scalability of production of BG face some challenges due to the nature of the crop, production potentials, utilization, and market access. In particular, its cropping system appears to be at least partly responsible for BG's underutilization. While, BG is believed to yield well in a mono-cropping system, such systems are not always advantageous to the farmer because local, resource-poor farmers can utilize only a small portion of land for intensive mono-cropping. Another factor limiting the production of BG is attributed to problems with processing and poor knowledge of its utilization for a range of purposes. The utilization of BG, as with other legumes like cowpeas, faces other challenges. It is mainly consumed directly or with little processing but requires a long cooking time to attain a desirable soft texture and palatability (Azam-Ali et al. 2001; Taiwo et al. 1998).

Despite the exclusion of the BG in the priority crop list of most African nations, the significance of BG lies in its potentials for promoting food security and nutrition, climate resilience, ease of production, multiple uses, market potentials, and potential for genetic improvement (Chai et al. 2016; Hillocks et al. 2012; Mwale et al. 2007; Azam-Ali et al. 2001). Further, there is a dearth of information on the socioeconomic factors that influence BG as an indigenous climate-resilient crop in Nigeria. More importantly, sustainable adoption and scalability of BG depends on how farmers perceive it. Therefore, this chapter assessed the influence of smallholder farmers' perceptions and the socioeconomic characteristics relating to the allocation of land to, and commercialization of BG in Nigeria. It further examined the effect of commercialization of BG on the welfare of farmers who are engaged in growing it.

The rest of the paper is organized as follows. Section "Materials and Methods" discusses the materials and methods, while section "Results and Discussion" presents and discusses the results from the data analysis. In section "Conclusion and Recommendations," the paper concludes with some policy implications.

Materials and Methods

Study Area

The study site is Benue State, Nigeria. Benue State was selected for the study because it is the "Food Basket of the Nation," and lies in the savannah and northern area of Nigeria. In addition, BG is known to be widely cultivated, traded, and consumed in Benue State. Gboko and Kwande local government areas (LGAs) in Benue State correspond with the development domains (Olayide et al. 2009a; Thornton et al. 2002) which were implemented for the study. The development domains involve the combination of population density/market access and agricultural potentials.

Sampling and Spatial Analysis Using GIS Methods

The study utilized a geographic information system (GIS) for spatial analysis (Olayide et al. 2008, 2009a; Omamo et al. 2006; Thornton et al. 2002) to gain a better appreciation of the patterns of agriculture, agricultural development challenges, and opportunities in the study areas. Many types of spatial analysis and mapping are feasible. The current analysis focuses on agricultural development domains representing particular realizations of agricultural potential and population density/access to markets, to visualize differences and similarities in agricultural development priorities and options.

Agricultural Potential and Population Density

Agricultural potential of any location is a strong indicator of agricultural productivity and human welfare. Crop productivity (measured in terms of output per hectare) in BG will hinge on important drivers like length of growing period. Agricultural potential largely influences the absolute advantage (productivity) of a location in production of particular agricultural commodities, while access to markets and infrastructure and population pressure help to determine the comparative advantage (profitability) of particular livelihoods, given the absolute advantages (Pender et al. 1999, 2004). For example, an area with suitable climate and soils may have an absolute advantage in producing high-value perishable vegetables, but little comparative advantage if it is remote from markets and roads.

Improvements in markets or access roads are thus expected to favor production of higher value commodities as well as non-farm activities and should contribute to higher incomes and welfare (Pender et al. 2004). Improved access to markets and infrastructure has more ambiguous theoretical impacts on land use, land management practices, and resource conditions, depending upon the relative impacts on costs of productive factors (Pender et al. 2004) and because of ambiguous effects of output prices on incentives to conserve land. Population density is expected to influence the labor intensity of agricultural production, including the choice of commodities as well as production technologies and land management practices, by affecting the land/labor ratio (Pender 2004).

Population growth may drive expansion of agricultural production into forest or grazing areas, reduce the duration of fallow periods, or induce adoption of land-saving commodities or technologies through crop-livestock intensification, investments in land improvement, and adoption of labor-intensive land management practices through mechanization, among other changes (Olayide et al. 2009a). Without improvements in technologies, markets, or infrastructure, crop-livestock intensification is unlikely to improve welfare, though it may improve resource conditions by inducing land conservation (Olayide et al. 2009b).

Agricultural Potentials

The key factor that endows any location with a comparative advantage to support agriculture-based livelihoods is its biophysical potential to nurture the growth of economically important plants and animals. Location-specific factors strongly influence the types of crops and livestock that will perform well, the risk of exposure to harmful pests, diseases, floods, droughts, and erosion hazards (Omamo et al. 2006). Within Africa (particularly Nigeria), dominated by a smallholder, subsistence-oriented agricultural production base, some of the most binding constraints to improved agricultural production potential have been recognized as the amount and variability of water supply, the biotic pressure of pests and diseases, and soil fertility.

The length of growing period (LGP), defined as the number of days in a year when sufficient water is available in the soil profile to support plant growth, is a shorthand way of capturing agricultural potential in a single layer of attribute. The final number of classes in the development domain surface is the product of the classes in each component layer (agricultural potentials and population density/market access). The length of growing season (LGP) approach (Olayide et al. 2009a) was adopted. Two patterns of LGP were identified for the study area: short (210–239 days) and long (240–269 days). Kwande local government area is classified as long LGP while Gboko local government area has a short LGP (Fig. 1).

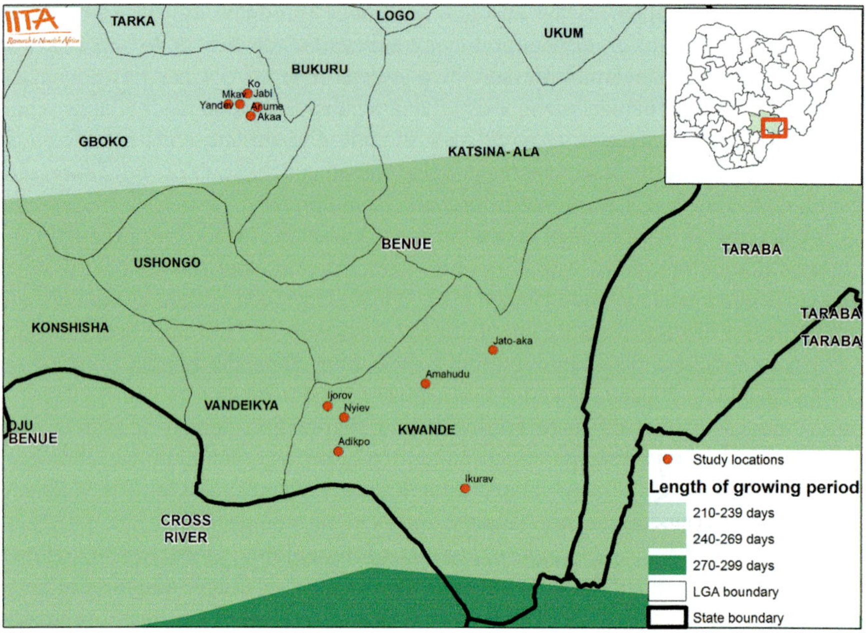

Fig. 1 Agricultural potential map of the study area

Population Density and Market Access

The land-to-labor ratio has been theorized to have consequences for land management and other production technology strategies. This means, holding other factors constant, farmers in areas of high population density are more likely to adopt labor-intensive production strategies than are those in areas of low density. As such, population density is a useful organizing frame for examining land management decisions. Because the development domains will be used to help gauge the potential effects of innovations in rural production and natural resource management strategies, it makes sense to examine the patterns of rural population densities only, to the extent that they can be isolated spatially from urban areas (Omamo et al. 2006).

Another important metric in geospatial analysis of development domains is market access. Market access and population density are often correlated and can be used interchangeably (Olayide et al. 2008). This involves measurement of physical distance or travel time between production locations and predefined market locations. Such market locations are usually determined using settlement population size criteria. The study area delineated by population density was defined as follows: High population density was more than 150 persons per square kilometer while low population density was 75–150 persons per square kilometer. Areas with population density less than 75 persons per square kilometers were not considered (Fig. 2).

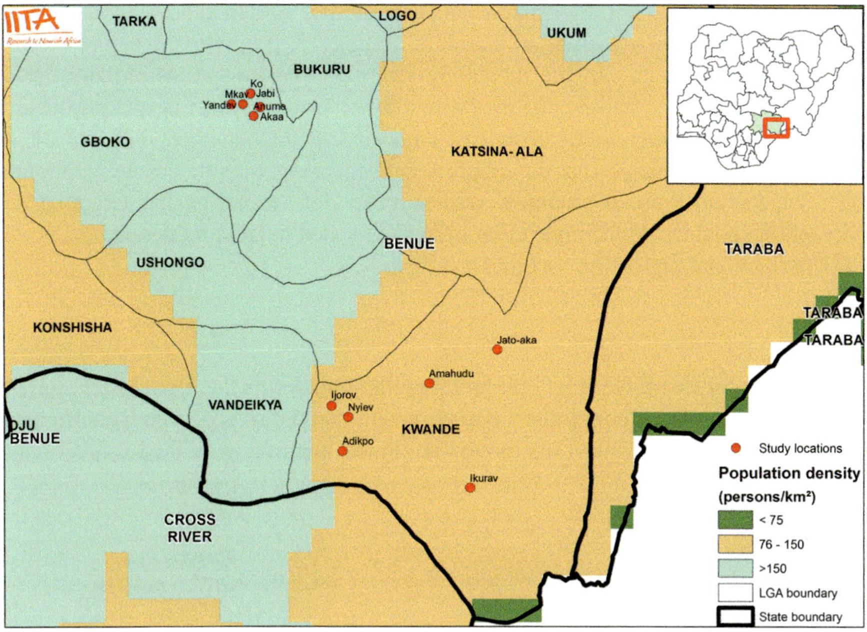

Fig. 2 Population density map of the study area

Data Collection

Farm-household level data were collected using structured questionnaires. Field observations and market surveys were also conducted. A total of 300 farmers were sampled at random from a list of farming households in the study area. The sample comprised 25 farmers from each of the 12 communities in the development domains in Gboko and Kwande local government areas of Benue State, Nigeria. Data were analyzed using descriptive statistics and the estimation of a fractional regression and a treatment effect models. The analytical models are explained in sections "Econometric Model of the Effect of Farmers' Perception of Bambara Groundnut as a Climate Resilient Crop on Land Allocation to Production" and "Modelling the Effect of Commercialization on Welfare of BG Farmers."

Econometric Model of the Effect of Farmers' Perception of Bambara Groundnut as a Climate Resilient Crop on Land Allocation to Production

Consider farmer *i*, who allocates a proportion y, of total arable crop land to BG cultivation. We argue that, the farmer's perception of BG's characteristics as a climate-resilient crop will influence what proportion of land to allocate to its production. Our dependent variable, y, therefore assumes a proportional nature, which requires a proportional variable modelling technique (Papke and Wooldridge 1996, 2008). The summary statistics of the proportion of land allocated to BG production indicates that all farmers cultivate at least some area of BG, but others cultivate BG as the sole crop. This means that the proportional dependent variable has no zero, but a one. We therefore use a fractional regression model to analyse the effect of the perception variables on the proportion of land allocated to BG production.

The fractional regression assumes the form

$$g(y) = x_i'b \tag{1}$$

g is a nonlinear distribution or link function that transforms the model such that all predictions from the model fulfil the requirement that $0 \leq \bar{y} \leq 1$. The b coefficients are estimated using the quasi-maximum likelihood function, with the log-likelihood given by

$$\ln L = \sum_{i=1}^{N} w_i y_i \ln\{G(x_i'b)\} + w_i(1 - y_i)\ln\{1 - G(x_i'b)\} \tag{2}$$

where $N =$ sample size, y is the proportion of land allocated to BG production by farmer i, w is an optional weight, G is the probit or logit link function, and x is the vector of explanatory variables, with specific emphasis on the perception of BG as a climate resilient crop. Empirically, therefore, the land allocation to BG function is expressed as follows:

$$\begin{aligned} g(y|x) = {} & b_0 + b_1x_1 + b_2x_2 + b_3x_3 + b_4x_4 + b_5x_5 + b_6x_6 + b_7x_7 + b_8x_8 \\ & + b_9x_9 + e \end{aligned} \tag{3}$$

The definition, measurement, and summary statistics of these variables are presented in Table 1.

Modelling the Effect of Commercialization on Welfare of BG Farmers

We examine how farmers decision to commercialize (produce BG beyond subsistence needs) influence their welfare outcomes in a linear framework as follows:

$$w_i = b_0 + x_i'b + \delta_i c_i + e_i \tag{4}$$

where w_i is the welfare measure for farmer i; b is a coefficient vector associated with control variables x; c is the decision to commercialize (1 if farmer commercializes BG production, 0 otherwise), and e is an unobserved stochastic variable. The effect of commercialization on farmers' welfare, measured by δ, is the primary objective in this model.

Table 1 Definition and measurement of variables

Variable	Definition and measurement	Mean	Std. dev.	Min	Max
y	Proportion of land allocated to BG production	0.72[a]	0.3850	0.0002	1
c	Commercialization dummy; 1 if farmer produces beyond subsistence, 0 otherwise	0.65[a]	0.4778	0	1
w	Welfare (in naira), proxied by natural log of household expenditure	10.521	0.7088	9.190	12.504
x_1	Age, in years	42.00	13.259	19	83
x_2	Primary education dummy; 1 if yes, 0 otherwise	0.18[a]	0.3876	0	1
x_3	Secondary education dummy; 1 if yes, 0 otherwise	0.38[a]	0.4878	0	1
x_4	Tertiary education dummy; 1 if yes, 0 otherwise	0.26[a]	0.4394	0	1
x_5	Vocational education dummy; 1 if yes, 0 otherwise	0.04[a]	0.1963	0	1
x_6	Perception that BG requires little rainfall (drought-tolerant) compared to other legumes, index	2.13	1.8444	1	4
x_7	Perception that BG is a food security crop and that it is readily available; index	2.34	1.8915	1	4
x_8	Perception that BG is a food security crop and that it is accessible; index	2.37	1.8304	1	4
x_9	Perception that BG is a food security crop and that it is affordable; index	2.36	1.8633	1	4
x_{10}	Perception that BG requires little or no manure or fertilizer; index	2.14	1.8999	1	4
x_{11}	Perception that BG is less tedious to cultivate; index	2.51	1.9865	1	4
x_{12}	Community; 1 if Kwande, 0 if Gboko	0.5[a]	0.5008	0	1
x_{14}	Household size, number of persons	6.89	3.0707	1	18
x_{15}	Formal grain market: dummied; 1 if present in community, 0 absent	0.26[a]	0.4394	0	1
x_{16}	Nature of road: dummied; 1 if tarred, 0 otherwise	0.72[a]	0.4505	0	1
x_{17}	Fertilizer: dummied; 1 manual, 0 if otherwise	0.75[a]	0.4356	0	1
x_{18}	Weeding: dummied; 1 if manual, 0 otherwise	0.68[a]	0.4659	0	1
x_{19}	Ploughing: dummied; 1 if manual, 0 otherwise	0.92[a]	0.2665	0	1
x_{20}	Sex: dummied; 1 if male, 0 if female	0.39[a]	0.4893	0	1

[a]indicates that the value is a proportion

But, the decision variable (commercialization) is endogenous in the model because the decision to commercialize or not is nonrandom. Farmers could self-select into being commercial based on inherent unmeasured attributes such as innovativeness, risk-aversion, comparative advantage, among others which could also influence welfare (Cerulli 2014; Maddala 1983). Failure to account for this self-selection and going ahead to estimate the model with ordinary least squares (OLS) regression could lead to biased estimates (Greene 2003; Wooldridge 2009).

Therefore, to address this self-selection bias we estimate the welfare model augmented with an endogenous treatment effect commercialization variable. This process generates the latent continuous commercialization function:

$$c_i^* = z_i' \gamma + v_i, \tag{5}$$

where $c_i = \begin{Bmatrix} 1 \ if \ c_i^* > 0 \\ 0 \ if \ c_i^* \leq 0 \end{Bmatrix}$

where γ is a vector of unobserved parameters to be estimated, z is a vector of observed explanatory variables influencing the decision to commercialize, and v is an independently and identically distributed error term. The latent variable model describes the process that generates observed farmers' decisions towards commercialization of BG. The two error terms, e_i and v_i, follow a bivariate normal distribution with mean vector zero and covariance matrix:

$$\begin{bmatrix} \sigma_1^2 & \rho\sigma_1 \\ \rho\sigma_1 & 1 \end{bmatrix} \tag{6}$$

where ρ is the covariance estimate of the correlation between the two error terms, indicating how unobservable variables that influence welfare correlate with unobservable variables that influence decision to commercialize. A statistically significant test of $\rho = 0$ indicates evidence of selectivity bias, giving credence to the endogenous treatment effect as opposed to the simple OLS regression. A positive value of ρ indicates that unobserved variables that increase welfare tend to be associated with unobserved variables that increase the likelihood of commercializing. A negative value shows that unobserved variables that increase welfare occur with unobserved factors that decrease the likelihood of commercializing (Cameron and Trivedi 2010).

Empirical Model

We regress the natural log of welfare (measured in terms of household expenditure) on commercialization and other variables as follows:

$$lnw_i = b_0 + \sum_{k=1}^{7} b_k x_{k,i} + \gamma_i \lambda_i + e_i \tag{7}$$

where λ is the Inverse of Mills Ratio (IMR), which measures the ratio of the probability density function of the standard normal distribution to its cumulative distribution function. The IMR is generated from the treatment equation before being included in the substantive model to estimate the effect of commercialization on welfare.

The treatment equation is empirically specified as

$$c_i = \gamma_0 + \sum_{j=1}^{10} \gamma_j z_{j,i} + v_i \quad (8)$$

Results and Discussion

Descriptive Statistics of Variables Used in the Estimation Model

Table 1 shows the descriptive statistics of the variables used in the estimation of the models. The sample involved 39% male and 61% female farmers. On average, farmers allocated 72% of their total farmland to the production of BG. In fact, some farmers allocated all their plots to the cultivation of the legume. These are perhaps the mono-croppers. Also 65% of the farmers produced BG beyond home requirements for consumption. We categorized these farmers as commercially oriented while the subsistence farmers were those who produced just enough for home consumption. This is the dependent variable for the selection equation of the treatment effect model. The average welfare (in Naira) was 37,049.12, ranging from 9,798.65 and 268,337.29. The average age of the farmers was 42 years also ranging from 19 to 83 years. This shows that the BG farmers were young adult. This is good for the long-term production of the crop. The educational background of the farmers is quite encouraging, considering the fact that the highest percentage of them had secondary education (38%), followed by tertiary education (26%), primary education (18%) and vocational education (4%). The percentage of farmers with no formal education was 14%. A high level of formal education augur well for quick technology uptake and other modern forms of farming and marketing of agricultural produce (Sheikh et al. 2003).

With respect to the perception variables, as indicated earlier, the Likert scale was given as 1 for strongly agreed SA and 4 strongly disagreed (SD). The mean scores for almost the perception variables are approximately 2.00. This means that on a whole the respondents agreed with the statement they were given about BG. For instance, with the mean scores of 2.13 and 2.14, respectively, the respondents generally agreed with the statements that "Bambara groundnut is drought tolerant" and "Bambara groundnut requires little or no fertilizer or manure." Similarly, the respondents agreed with the statements that Bambara groundnut as a food security crop is readily available (2.34); it is affordable (2.36) and it is accessible (2.37). However, with the mean score of 2.51, there was less agreement among respondents with the statement that BG is less tedious to cultivate.

The average household size was 6.89 ranging from 1 to 18 members. Only 26% of the respondents indicated that there was a legume market in their community. Also, 72 farmers had tarred roads leading to their communities. Similarly, technology used on the farms as a whole (BG farms and other farms) was traditional (manual). For instance, 75%, 68%, and 92% used traditional means (manual means) to apply fertilizer, weed, and plough their farms respectively. We took equal samples of farmers from Kwande (50%) and Gboko (50%) LGAs. The total sample consisted of 39% males and 61% females confirming the general notion that BG is women's crop (Adzawla et al. 2016b).

Farmers' Perceptions of BG as a Climate-Resilient and Food Security Crop and its Effect on Land Allocation to BG Production

The first objective was to investigate the extent to which respondents' perceptions about BG as a climate-resilient/drought tolerant and food security crop influence the proportion of land allocated to BG production. BG was perceived as a climate-resilient crop/drought tolerant, suggesting that the greater the farmer knowledge in regards to the reality of climate change and the crop's resilience to climate change, the greater the proportion of land she/he would allocate to BG. A large proportion of land allocated to the production of a crop does not necessarily mean higher output. Intensification of production may sometimes even lead to higher output per unit area of land than just increasing the area cultivated (extensification) with less proportional increase in output per land area.

From the results in Table 2, eight of the variables are significant in explaining the dependent variable (i.e., proportion of land allocated to BG production). While three variables are negatively related to the dependent variable, the other five significant variables are positively related to the dependent variable. The Wald chi-square test is significant at 1% indicating that the explanatory variables jointly determine the dependent variable. Also, the pseudo R-squared is almost 50% implying that the model is good in that the estimator was able to predict 50% of the variation in the dependent variable. Normally, in the estimation of a fractional regression model, the marginal effects are relevant. However, discussion on this is not considered due to the very low values of 0.001 or less.

It should be recalled that the Likert scale used to measure the dependent variable was such that one (1) represents strongly agreed (SA) while four (4) represents strongly disagreed (SD). This means that the negative coefficient of the climate change-related perception variable "bambara groundnut requires little rainfall" and "bambara groundnut requires little fertilizer or manure" implies that farmers who perceived that bambara groundnut is a resilient crop that does well with less rainfall and poor soils, allocated a greater proportion of their total farmland (72%) to its production. Negative signs for coefficients indicate the agreement on the independent variables. In Adzawla et al. (2016b) also, a majority of the farmers shared the opinion that BG requires little rainfall and does not require the use of fertilizers. The policy implication of this finding is that if more farmers were to expand the area of their farms under BG, then there would be the need for awareness of the reality of climate change and the resilience of BG to thrive in drought condition or harsh climatic environments. Wasula et al. (2014) also found that there was limited knowledge about the importance of BG in the Kakamega county of Kenya.

Table 2 Determinants of bambara groundnut land share

Variable	Coefficient	Standard error	Marginal effect	Standard error
Age	0.0276**	0.0107	0.000002***	0.0000009
Primary education	0.5078	0.5093	0.000057	0.000055
Secondary education	0.1678	0.4546	0.000016	0.000041
Tertiary education	0.5606	0.4749	0.000037	0.000036
Vocational education	1.3409*	0.7670	0.000064*	0.000037
Perception rainfall	−0.419***	0.1468	−0.000035***	0.000012
Food availability perception	−0.3953*	0.2154	−0.000033*	0.000018
Food accessibility perception	0.1571	0.3002	0.000013	0.000025
Food affordability perception	0.4768*	0.2439	0.000040*	0.000020
Fertilizer perception	−0.3926**	0.1571	−0.000033**	0.000013
Perception on BG is tedious	0.4627***	0.1183	0.000038***	0.000010
Domain (LGA)	19.2889***	0.2598	0.001601***	0.000165
Constant	−1.4424	0.9632		

$N = 300$
Wald chi2(12) = 11002.92
Prob>chi2 = 0.0000
Pseudo $R^2 = 0.5018$

Note: *** and ** and *represent 1%, 5% and 10% significant levels respectively

However, in the case of the perception that BG cultivation is tedious, the positive sign means that farmers who perceived this allocated a smaller proportion of their farmland to cultivating the crop. This is also plausible and instructive.

Similarly, farmers who perceived that BG was a food security crop, in terms of it being available, allocated a greater percentage of their farmland to its cultivation. However, farmers who perceived that the crop was affordable cultivated less. Both findings are plausible because if the crop is affordable, then one can decide to cultivate less and concentrate on major cash crops so that incomes from the sale of such cash crops may be used to buy BG. The negative coefficient of the "availability" variable is also as expected because availability does not necessarily mean that one can have it for free. In other words, if the crop is available but there is no money to buy it, then you have to cultivate (more of) it. In the Wasula et al. (2014) study, 92% of the respondents perceived that the production of BG ensures food security because it is drought and pest resistant.

Furthermore, the positive sign of the coefficient of the education variable shows that the smallholder farmers with vocational education (mostly full-time farmers without formal education) cultivated BG more than those with formal education. Such farmers are likely to know more about the indigenous crop and climate-resilient nature of BG and therefore allocate more of their farmlands to cultivation of BG. This finding was consistent with Adzawla et al. (2016a) where the probability of going into BG cultivation was lower for famers who had formal education. In the studies by Awunyo-Vitor et al. (2013), Donkoh et al. (2013) and Shehu et al. (2010) farmers with formal education were more technically efficient than those with little or no formal education (including vocational education). Hence, any policy intended to upscale the production of BG could therefore target smallholder farmers that have both formal education and vocational skills.

The positive coefficient of the age variable also indicates that the older farmers have a greater probability of allocating a greater proportion of their farmlands to BG cultivation than the relatively young farmers. This is also plausible, considering that BG is an indigenous crop which has been with the farmers for a long time. The danger, however, is that if the young farmers are not aware and educated about the good characteristics of the crop, BG may remain a crop for the older generation without formal education and prone to extinction. Therefore, there is an urgent need to bring young farmers on board the cultivation of BG.

Further, Kwande farmers tend to allocate more of their farmlands to BG than Gboko farmers. This is also plausible because Kwande has high agricultural potentials, including longer growing season and market access.

Effect of Bambara Groundnut Commercial Orientation on Farmers' Welfare

The second objective of the study was to determine the effect of commercial orientation of BG on the welfare (consumption expenditure) of farmers' household. This was done by estimating a treatment effect model with the commercialization equation being the selection equation and the welfare being the substantive equation. The negative coefficients are indications that the responses tend to zero while positive signs means otherwise (please refer to results in Table 1).

From Table 3, the factors that are significant in influencing the market orientation of farmers are the climate change (rainfall) variable, the availability of a legume market in their community, and the technology variables. The results indicate that the probability of a farmer going commercial in the production of BG was greater for farmers who perceived that BG required less rainfall. This is logical because with this knowledge, the farmer would expand the BG farm so as to realize greater output for sale. Similarly, with the farmers' orientation to go commercial, they would want to be oriented towards modern/improved technology adoption rather than maintaining traditional production methods. The results indicate that the probability of going commercial was lower for farmers who did fertilizer application and weeding manually as opposed to those who used modern or improved technologies on their farms. Similarly, the negative sign of the legume market variable is inconsistent with the a priori expectation. It was expected that the availability of legume markets in a community would motivate farmers to go into commercial production of BG. But theoretically (and in practice), availability of a market in a location of production may not necessarily encourage production since markets serve to move commodities from a point of surplus to a point of scarcity (Olayide et al. 2009b). It

Table 3 Factors influencing the commercialization of bambara groundnut

Variable	Coefficient	Standard error
Age	0.0002	0.0097
Household size	0.0027	0.0389
Food availability perception	0.1984	0.1463
Rainfall perception	−0.2729***	0.0926
Fertilizer perception	0.1204	0.1402
Legume market	−0.6761***	0.2399
Nature of road	0.0723	0.2238
Fertilizer application technology	−0.5698**	0.2577
Weeding technology	−0.5022***	0.1868
Ploughing technology	−0.6048**	0.2934
Constant	1.5980	0.9844

N = 125
Wald chi2 (8) = 78.88
Prob>chi2 = 0.0000
Wald test of independent equations (rho): chi2(1) = 22.28
Prob>chi2 = 0.0000

Note: ***, ** and *represent 1%, 5% and 10% significant levels respectively

implies that BG may be found in higher concentration outside its primary area of production or ex situ markets.

Table 4 shows the results of the determinants of welfare of farmers. The positive and significant sign of the commercialization variable confirms the a priori expectation that producing BG on a commercial basis would lead to increased welfare of farmers. It should be noted that producing in commercial quantities (marketable surplus) does not only increase the welfare of the producers but it goes to bring development to the community and indeed the nation as a whole. Of course, this is based on the assumption that BG does not crowd out the production of other staple crops. Farmers with formal primary and secondary education also have a greater welfare than those without formal education. This is also consistent with our a priori expectation because farmers with formal education have access to information and efficient farm management practices. The negative sign on household size variable is also consistent with other findings. This is, arguably, because large household size is an indication that the household has more persons to feed (assuming that the income level is fixed), which means that the household would be poor or have relatively low welfare index. Adzawla et al. (2016a) and Donkoh et al. (2014) also found negative correlation between household size and welfare.

The non-significance of the output variable is surprising because the expectation was that farmers who increased output would have increase welfare. However, the significance of the commercialization variable is instructive as it implies that increased output may not necessarily lead to increased welfare, but it is when the output is sold that it would necessarily lead to increased welfare. The results also

Table 4 The Determinants of farmers' welfare

Variable	Coefficient	Standard error
Age	−0.0082	0.0056
Primary education	0.3528*	0.1815
Secondary education	0.2755*	0.1558
Tertiary education	0.1261	0.1665
Vocational education	0.2397	0.2329
Household size	−0.1129***	0.0219
Output	0.0602	0.0593
Commercialization	0.7872***	0.1845
Constant	9.1551***	0.5734

Note: *** and *represent 1%, and 10% significant levels respectively

revealed that the increase in output may be used to satisfy domestic food and nutrition needs of the households, and not necessarily for generating income for the households.

Conclusion and Recommendations

This chapter investigated the extent to which farmers' perceptions about BG as a drought-tolerant, climate-resilient, and food security crop influence the allocation of their farmlands to the production of the indigenous legume. It also estimated the effect of commercialization of BG on the welfare of smallholder farmers. A total of 300 smallholder farmers were sampled from Kwande and Gboko local government areas of Benue State, Nigeria, through a multistage sampling technique. A fractional regression model was estimated to analyze the extent to which farmers' perceptions about BG as a drought-tolerant, climate-resilient,food security, and nutrition crop influence the allocation of their farmlands to the production of the indigenous legume while a treatment effect model was also estimated to analyze the effect of commercialization of BG on the welfare of smallholder farmers.

It was found that the percentage of farmers' total farmlands allocated to BG production was high for the following: older farmers; farmers who had vocational/ extension service training; farmers who perceived that BG production requires less rainfall (that is, drought tolerant); farmers who perceived that BG production requires less manure/fertilizer; farmers who perceived that BG is a food security crop and is available; and farmers who lived in area of high agricultural potential as opposed to those who lived in area of low agricultural potential. The percentage of farmers' total farmland allocated to BG production was also high for the following: farmers who perceived that the cultivation of BG was less tedious; and farmers who perceived that BG as a food security and nutrition crop. Similarly, the following categories of farmers were more commercially oriented: those who perceived that BG production requires less rainfall (climate resilient); farmers who had a legume market outside their community; farmers who adopted modern/improved methods of applying fertilizers and herbicides.

Household welfare was also relatively high for commercially oriented farmers, farmers with small household size, and those who had primary education. The following conclusions can be made from the key findings: BG is cultivated because it is perceived as a climate resilient, food security, and nutrition crop. This is a good motivation and incentive to scale up the production of the crop. Secondly, commercializing BG can help to increase farmers' welfare. It is hereby recommended for promotion as a welfare-improving crop. Thirdly, awareness and education should be given to farmers about the crop's characteristics as a drought resilient, food security, and nutrition crop.

References

Adebowale YA, Schwarzenbolz U, Henle T (2011) Protein isolates from Bambara groundnut (*Voandzeia subterranean* L.): chemical characterization and functional properties. Int J Food Prop 14(4):758–775. https://doi.org/10.1080/10942910903420743

Adeleke OR, Adiamo OQ, Fawale OS (2018) Nutritional, physicochemical, and functional properties of protein concentrate and isolate of newly-developed Bambara groundnut (*Vigna subterrenea* L.) cultivars. Food Sci Nutr 6(1):229–242. https://doi.org/10.1002/fsn3.552

Adzawla W, Donkoh SA, Nyarko G, O'Reilly P, Mayes S (2016a) Use patterns and perceptions about the attributes of Bambara groundnut (*Vigna subterranea* (L.) Verdc.) in Northern Ghana. Ghana J Sci Technol Dev (GJSTD) 4(2):56–71

Adzawla W, Donkoh SA, Nyarko G, O'Reilly P, Olayide Olawale E, Mayes S, Feldman A, Azman Halimi R (2016b) Adoption of Bambara groundnut production and its effects on farmers' welfare in northern Ghana. Afr J Agric Res 11(7):583–594

Aremu MO, Mamman S, Olonisakin A (2013) Evaluation of fatty acids and physicochemical characteristics of six varieties of bambara groundnut (*Vigna subterranea* L. Verde) seed oils. Riv Ital Sostanze Grasse 90(2):107–113

Awunyo-Vitor D, Bakang J, Cofie S (2013) Estimation of farm level technical efficiency of small-scale cowpea production in Ghana. Am Eurasian J Agric Environ Sci 13(8):1080–1087

Azam-Ali SN, Karikari SK, Massawe FJ (2001) Assessing the potential of an underutilized crop – a case study using bambara groundnut. Exp Agric 37(4):433–472. https://doi.org/10.1017/s0014479701000412

Brough SH, Azam-Ali SN (1992) The effect of soil moisture on the proximate composition of Bambara groundnut (*Vigna subterranea* (L.) Verdc). J Sci Food Agric 60:197–203

Cameron AC, Trivedi PK (2010) Microeconometrics using stata. Stata Press, College Station

Cerulli G (2014) Ivtreatreg: a command for fitting binary treatment models with heterogeneous response to treatment and unobservable selection. Stata J 14:453–480

Chai HH, Massawe F, Mayes S (2016) Physiological responses of a Bambara groundnut segregating population to mild drought stress In: Hall RA, Rudebjer P (eds) 3rd international conference on neglected and underutilized species (NUS): for a food-secure Africa. Accra, 25–27 Sept 2013. Proceedings, Bioversity International, Rome, Italy and International Foundation for Science, Stockholm, Sweden

Donkoh SA, Tachega M, Amowine N (2013) Estimating Technical Efficiency of Tomato Production in Northern Ghana. Am J Exp Agric 3(1):56–75

Donkoh SA, Alhassan H, Nkegbe PK (2014) Food expenditure and household welfare in Ghana. Afr J Food Sci 8(3):164–175. FAO

Food and Agricultural Organisation (FAO) of the United Nations. 1982. Legumes in human nutrition. FAO Food and Nutrition Paper No. 20. FAO, Rome

Goli AE (1997) Bambara groundnuts: introduction. In: Promoting the conservation and use of underutilized and neglected crops. Bambara groundnut, *Vigna subterranean* (L). Verdic. International Plant Genetic Resources Institute, Rome, Italy

Greene WH (2003) Econometric analysis, 5th edn. Upper Saddle River. Prentice Hall, Upper Saddle River, New Jersey, USA

Hillocks RJ, Bennett C, Mponda OM (2012) Bambara nut: a review of utilisation, market potential and crop improvement. Afr Crop Sci J 20(1):1–16

Karikari SK (1996) The status of Bambara groundnut genetic resources in Botswana SACCAR Newsletter No. 34, June, 1996

Mabhaudhi T, Modi AT, Beletse YG (2013) Growth, phenological and yield responses of a Bambara groundnut (*Vigna subterranea* L. Verdc) landrace to imposed water stress: II. Rain shelter conditions. Water SA 39:191–198

Maddala GS (1983) Limited-dependent and qualitative variables in econometrics. Cambridge University Press, Cambridge, UK

Minka SR, Bruneteau M (2000) Partial chemical composition of bambara pea [*Vigna subterranea* (L.) Verde]. Food Chem 68(3):273–276. https://doi.org/10.1016/S0308-8146(99)00186-7

Mkandawire CH (2007) Review of Bambara groundnut production in sub-saharan Africa. Agric J 2:464–470

Mwale SS, Azam-Ali SN, Massawe FJ (2007) Growth and development of Bambara groundnut (*Vigna subterranea*) in response to soil moisture 1. Dry matter and yield. Eur J Agron 26:345–353

National Research Council (1996) Lost crops of Africa: volume I: grains. The National Academies Press, Washington, DC. https://doi.org/10.17226/2305

Olayide O, Alene A, Ikpi A, Nziguheba G (2008) Population density and distance to market does not influence the farmers' use of organic manure. Available at http://orgprints.org/12310/

Olayide OE, Alene AD, Ikpi A, Nziguheba G (2009a) Manure marketing in the Savannas of Nigeria: implications for sustainable food security. J Food Agric Environ 7(2):540–545

Olayide OE, Alene AD, Ikpi A (2009b) Determinants fertilizer use in northern Nigeria. Pak J Soc Sci 6(2):91–98

Omamo SW, Diao X, Wood S, Chamberlin J, You L, Benin S, Wood-Sichra U, Tatwangire A (2006) Strategic priorities for agricultural development in eastern and Central Africa. Research report 150. International Food Policy Research Institute

Oyeyinka SA, Singh S, Adebola PO, Gerrano AS, Amonsou EO (2015) Physicochemical properties of starches with variable amylose contents extracted from bambara groundnut genotypes. Carbohydr Polym 133:171–178. https://doi.org/10.1016/j.carbpol.2015.06.100

Papke LE, Wooldridge JM (1996) Econometric methods for fractional response variables with an application to 401 (k) plan participation rates. J Appl Econ 11:619–632

Papke LE, Wooldridge JM (2008) Panel data methods for fractional response variables with an application to test pass rates. J Econ 145:121–133

Pender J (2004) Development pathways for hillsides and highlands: some lessons from Central America and East Africa. Food Policy 29(4):339–467

Pender J, Place F, Ehui S (1999) Strategies for sustainable agricultural development in the East African highlands. EPTD discussion paper 41. International Food Policy Research Institute, Washington, DC

Pender J, Jagger P, Nkonya E, Sserunkuma D (2004) Development pathways and land management in Uganda. World Dev 32(5):767–792

Shehu JF, Iyortyer JT, Mshelia SI, Jongur AAU (2010) Determinants of yam production and technical efficiency among yam farmers in Benue state, Nigeria. J Soc Sci 24(2):143–148

Sheikh AD, Rehman T, Yates CM (2003) Logit models for identifying the factors that influence the uptake of new 'no-tillage'technologies by farmers in the rice–wheat and the cotton–wheat farming systems of Pakistan's Punjab. Agric Syst 75(1):79–95

Stadler F (2009) Analysis of differential gene expression under water-deficit stress and genetic diversity in Bambara groundnut (*Vigna subterranea* (L.) Verdc.) using novel highthroughput technologies. PhD thesis. Technical University of Munich, Munich

Taiwo KA, Akanbi CT, Ajibola OO (1998) Regression relationships for the soaking, and cooking properties of two cowpea varieties. J Food Eng 37:331–344

Thornton PK, Kruska RL, Henninger N, Kristjanson PM, Reid RS, Atieno F, Odero AN, Ndegwa T (2002) Mapping poverty and livestock in the developing world. International Livestock Research Institute, Nairobi

Wasula SL, Wakhungu J, Palapala V (2014) Farmers' perceptions on adoption of Bambara nut production as a food security crop in Kakamega County, Kenya. Int J Disaster Manage Risk Reduct 6(1)

Wooldridge JM (2009) Introductory econometrics: a modern approach, 4th edn. South-Western Cengage Learning, Mason 50–62

Yao DN, Kouassi KN, Erba D, Scazzina F, Pellegrini N, Casiraghi MC (2015) Nutritive evaluation of the Bambara groundnut Ci12 landrace [*Vigna subterranea* (L.) Verdc. (Fabaceae)] produced in cote d'Ivoire. Int J Mol Sci 16(9):21428–21441. https://doi.org/10.3390/ijms160921428

Perceptions of Challenges Facing Pastoral Small Ruminant Production in a Changing Climate in Kenya

64

Rosemary N. Ngotho-Esilaba, J. O. Onono, J. N. Ombui, J. F. Lindahl, and H. O. Wesonga

Contents

R. N. Ngotho-Esilaba (✉) · H. O. Wesonga
Veterinary Sciences Research Institute, Veterinary Research Centre, Kenya Agricultural and Livestock Research Organisation, Nairobi, Kenya
e-mail: r.esilaba@gmail.com; hezron.wesonga@gmail.com

J. O. Onono · J. N. Ombui
Department of Public Health, Pharmacology and Toxicology, College of Agriculture and Veterinary Sciences, University of Nairobi, Nairobi, Kenya
e-mail: joshua.orungo@uonbi.ac.ke; jombui@uonbi.ac.ke

J. F. Lindahl
International Livestock Research Institute, Uppsala University and Swedish University of Agricultural Sciences, Nairobi, Kenya
e-mail: J.Lindahl@cgiar.org

W. Leal Filho (ed.), *Handbook of Climate Change Resilience*,
https://doi.org/10.1007/978-3-319-93336-8_179

Abstract
The impacts of climate change on domestic animal health in pastoral areas have not been well studied or documented. Many challenges facing and compromising efficiency of pastoralism remain amorphous, with the result that as trialed climate adaptation initiatives continue to be ineffective in some sub-Saharan arid and semiarid areas. At the heart of faced challenges is the impact of heightened and accelerated climate variability on the epidemiology of livestock diseases. This paper presents a description of general and disease challenges facing small ruminants' producers in a pastoral county in Kenya. A disease lens is used in deconstructing the challenges using participatory multi-stakeholder approaches. In conclusion, the paper outlines pathways for building resilience in small ruminant health and production in pastoral areas. The recommendations may be used to inform policymakers and organizations dealing with disease risk, mitigation, and control, as well as those concerned with sustainable pastoral development.

Keywords
Climate-change impacts · Pastoralism · Sheep and goats · Health and disease · Resilience

Introduction

There is evidence that the most prominent perceived threats to life on Earth are climate change, terrorism, and inappropriate government policies. The Intergovernmental Panel on Climate Change (IPCC 2001) indicates that Africa is the most vulnerable continent to projected climate change impacts because of widespread poverty and low levels of technical development which constrain adaptation capability despite the fact that African agriculture has a 97% role in food security and livelihoods of the poor. Synergistic options between climate change mitigation and adaptation exist (IPCC 2014).

Agriculture is the main source of income for the Kenyan economy, contributing 26% to the GDP (KNBS 2013). Climate predictions for Kenya indicate wetter humid zones, drier arid zones, and an upward altitudinal shift in agro-ecological zoning (Seo et al. 2008; Jaetzold et al. 2010). Viability of traditional pastoralism has been questioned (Eriksen and Lind 2009), and it has been suggested that the future of pastoralists in Kenya is dismal due to overwhelmed coping capacity to climate change impacts. On the flipside of this argument is livestock's role as the best agricultural option in arid and semiarid areas (ASALs); in particular, small ruminants.

Environmental challenges are in favor of small versus large ruminant production to improve food and nutrition security and livelihoods in Kenya (CCFAS 2015). According to studies by the CGIAR's research program on Climate Change and Food Security, CCFAS, small ruminants are being viewed as a panacea for vulnerability alleviation, especially for nomadic communities because of their smaller ecological footprint in a climate threatened world.

There are many challenges facing the livestock subsector in Kenya, putting to doubt its ability to sustain a rapidly growing and urbanized population or attain animal source food and nutrition security as guided by the first Millennium Development Goal (MDG) and the Sustainable Development Goals (SDGs). In 2014, the Kenyan agricultural sector recorded decreased crop and livestock production due to erratic and depressed rainfall resulting in a decelerated increase in agricultural value addition to the economy of 3.5% (KNBS 2015).

Specific threats to livestock agriculture include adverse weather, as well as localized, emerging and migratory pests and diseases. Amongst these challenges, weather stands out since it is intimately intertwined with productivity, health, or disease outcomes, albeit covertly. Shortage of water, fodder and other livestock nutritional inputs in pastoral set ups leads to conflicts while decimation of livestock by diseases and famine impacts negatively on livelihoods, household food, and nutritional security and incomes.

The foregoing indicates the need for a community-based threats and risk analysis which is the foundation for creating a sustainable future for pastoralism. Understanding perceptions of pastoral vulnerability and experiences, climate-related disease risks, drought coping mechanisms and devising mitigation strategies at the grassroots is vital for sustainable pastoralism since effects are felt locally and therefore need a local framework. The United Nations' Framework Convention on Climate Change (UNFCCC) driven Nationally Appropriate Mitigation Actions (NAMAs) have aimed to enhance livestock productivity and production efficiency whilst reducing environmental impacts. However, NAMAs have neither prevented recurrent famines in livestock and human populations, nor disease morbidity and mortality as evidenced by perpetual livestock losses, endemic diseases and recurrent epidemics with dire socioeconomic impacts on livestock systems and pastoral livelihoods. Gaps in information on challenges facing pastoral herders at the grass roots, including disease risk profiles, deficient animal disease management strategies and inadequate disease control decision support frameworks (WHO-OIE-WBG 2014) need to be flagged up and addressed. This, in an endeavor to fortify pastoral resilience, support compliance with the World Animal Health standards (OIE 2010, 2013) and World Trade Organization's (WTO) Sanitary and Phytosanitary Standards (SPS) agreement and enhance sustainable smart animal agriculture.

Methodology

As part of a larger study, pastoralist communities, the key stakeholders from the sheep and goat value chain were engaged using focus group discussions in order to identify and describe challenges facing pastoral small ruminant production in a changing climate. The research was conducted using a checklist of a few domains' questions in four randomly selected study sites in Laikipia County, which is the gateway to the northern pastoral frontier of Kenya.

Key Stakeholders

Entry points into each area were identified during reconnaissance surveys. The Laikipia county deputy director of veterinary services provided connections to an active central government *Agricultural Sector Development Support Programme,* the *ASDSP* project. Using the area chiefs' assistance, contacts were used to recruit representative participants from each sub-county and to choose suitable venues.

In one site, Kimanjo, 20 participants from 13 group ranches found in Laikipia North were identified and recruited. The female and male participants were mainly smallholder producers and traders, with a few processors and community health providers. Discussions were held in a hall belonging to the *Laikipia Livestock Marketing Cooperative Society Limited,* a facility constructed using funding from the African Wildlife Foundation, as a corporate social responsibility project.

Meetings in Laikipia West were organized and facilitated by a subcounty livestock health officer, assisted by an area chief in one site and an assistant chief in a second site. Two meetings were held in the pastoralist and agro-pastoral-forested areas. For venues, the former was held in a church while the later was held in a community hall constructed using the Danish International Development Agency, DANIDA aid. The pastoralists' meeting had 15 smallholder livestock producers and traders, mainly males, while the small-holder agro-silvo-pastoralists were 13 people of mixed gender.

The contact from Laikipia North East organized a meeting with 10 pastoralist elders in the Mukogodo Forest reserve. These were mainly medium-scale producers and traders, who gave us great insights into this authentic and unadulterated community's ways of life and livestock rearing livelihoods.

Constraints Identification Methods

To identify challenges to small ruminants' production, prioritize and rank diseases, participatory epidemiological tools and methods as described by Oppenheim (1997), Pretty et al. (1995), Catley (1997, 1999, 2000, 2005), Mariner (2000) and Catley and Mariner (2002) and social survey techniques as described by Horton (1991), Dvorak and Izac (1992), Rhoades (1992), and De Vaus (1996) were used.

Site Description

The first site *Kimanjo* is a semi-desert in the North, juxtaposed between wildlife reserves and conservancies like the *Ol Jogi,* while the second *Nadungoru* in the North East is a plains zone surrounded by the *Mukogodo* Hills and Forest which borders the bespoke wildlife conservancy *Borana* and the *Il Ngwesi* Community Trust ranch. The third site *Maundu ni Meri* in the West is a riverine zone adjacent to a major livestock market, along an animal trading route, as is the fourth site known as *Lorien,* which is also a riverine zone near the *Nyahururu* Forest (Figs. 1 and 2).

Fig. 1 Impacts of drought in Kimanjo

Fig. 2 Grazing sheep and goats in Laikipia East

Typology of Small Ruminant Herders in the Study Sites

Using currently known production zoning, herders in *Kimanjo* are pastoralists, those in Nadungoru can be described as silvo-pastoralists, while those in *Maundu ni Meri* and *Lorie*n are agro-pastoralists.

These classifications, as described by FAO (1978), KSS (1982), Jaetzold et al. (2010), and Yu and Guo (2016), may need modification for the sheep and goat value chain, since this study discovered that small ruminants' movements differ from those of cattle. Four types of movements were identified: there are those sheep and goats that rarely move far from homesteads and these include the young ones and the sick. Depending on the season, some animals move farther away from homesteads for

Fig. 3 A sheep and goat trader in Rumuruti

distances of up to 10 km daily. In extremely dry seasons, sheep and goats are translocated to a different ecological zone altogether, where pastures and water are easily available. Conflicts over natural resources, raids and rustling are other reasons for relocation. The last class of movement is to and from distant markets using motorbikes and motorcars. All these dynamic movements buttress disease epidemiological profiles for different production zones. The fourth classification in particular has negative implications for the mitigation of greenhouse gases emissions resulting from transportation and crowding around markets (Fig. 3).

Perceptions of the Challenges Faced by Sheep and Goat Producers

General Constraints to Small Ruminant Production

The first factor contributing to what challenges a site faces, is its unique ecological character as described earlier. The second is the dynamisms of daily and seasonal movements. A third factor that was observed and discussed is the limited access or lack of extension and service delivery systems. Overall, it was found that challenges were perceived and scored or ranked differently in all sites as listed below (Table 1).

Diseases were ranked highest in three of the sites. This can be attributed to effects of communal grazing and mixing of flocks at watering points and markets. Absence of a formal animal health infrastructure in most of the study sites seemed to aggravate the disease challenges. The research was carried out in a drought season, which is why lack of pastures and water were pressing needs in most sites (Fig. 1).

Nadungoru is a quaint little village surrounded by hills and forests (Fig. 4), whose challenge can only be lack of water and pasture, before any other. There are no rivers nearby, the only borehole is barely functional, and the watering points are far away in the hills. When the pastoralists here exhaust the forest undergrowth, they move the

Table 1 Constraints to small ruminant production in Laikipia

Constraint	Kimanjo	Maundu ni Meri	Lorien	Nadungoru
1	Diseases	Diseases	Diseases	Lack of pasture
2	Lack of water	Wildlife conflicts	Marketing constraints	Lack of water
3	Lack of pasture	Marketing constraints	Lack of information	Diseases
4	Wildlife conflicts	Lack of pasture	Lack of pastures	Rustling
5	Rustling	–	Wildlife conflicts	Wildlife conflicts

Fig. 4 Mukogodo Forest near Nadungoru

sheep and goats to the forests on the slopes of Mt. Kenya, which in itself is very challenging both to the herder and to the animals due to cold weather, which enhances the risk of cold-related diseases in both.

Laikipia is home to various and numerous species of wildlife (Fig. 5). These roam freely into homesteads, leading to lots of conflicts including human-wildlife, wildlife-livestock and wildlife-crops. This is the reason why all sites prioritized wildlife conflicts as important. An extension of these conflicts is those between pastoralists and owners and managers of wildlife conservancies, when the former overrun boundaries and invade private properties, in search of pasture and water. This last conflict normally escalates during droughts, as was the case during the study period in 2016–2017. At this time, the worst ever conflicts were experienced and seen by the world, especially in the *Maundu ni Meri* research site, which was affected by clashes.

Fig. 5 Giraffes in the savanna plains of Laikipia West

Conflicts in the Study Sites

The northern sites of *Kimanjo* and *Nadungoru* border other predominantly pastoralist counties whose residents make sporadic "visits" especially during drought seasons, resulting in animal rustling with losses of up to 100% for an individual herder. Some of the research participants in *Nadungoru* lost animals in the course of the surveys and most moved away to safety in neighboring communities.

In the *Nadungoru* research site, there were sporadic localized conflicts between neighboring communities due to drought induced competition over natural resources. Animals belonging to participants in the study had to be moved away to far off sites where they could access safe pastures and water sources. Ensuing studies were relocated to these new sites.

In the *Maundu ni Meri* site, there were dynamic external invasions into individual small-holder and private pasturelands which escalated into unfortunate livestock, wildlife and a few human deaths. These were reported in national electronic and print media. One report quoted an anthropological study conducted by an unnamed author (Nameless 2017), whose conclusion was that the invasions into the County emerged out of two coinciding crises: a decades-old pastoralists' crisis and a political crisis that gained pace in the post devolution era, in 2013. Stern actions undertaken by the government did weather down the crises. A grave outcome for the study was that it was practically impossible to follow up participants and their flocks because they relocated permanently to far off undisclosed safe sites, after losing many animals in the clashes. A nearby site was selected for ensuing studies.

Marketing Constraints in Rumuruti

Marketing was described as a constraint to production in West Laikipia. This was surprising considering that the largest livestock market in Kenya is found nearby in *Rumuruti*. The pastoralists in *Maundu ni Meri* complained about levies and low sale prices at the *Rumuruti* market. They expressed the desire to have their own market.

Impacts of Climate Change on Small Ruminant Rearing

The pastoralists barely dwelt on the issue of climate change and its impacts on their livestock rearing; they intimated that their most pressing issues were disease prevention and control and lack of veterinary staff on site. In lieu of this, it was decided that the issue of climate change challenges as perceived by herders; that is climate change, impacts are important and constitute a central link to the other challenges they face, should be investigated in further studies.

Disease Constraints to Small Ruminant Production

The fact that the research team came from a veterinary background may have had something to do with the respondents' perception of disease as the most important constraint; probably, they hoped that the team would solve their immediate animal health problems. Some efforts were made to do this, in order to build a lasting partnership with the community. It would be interesting to see the constraint scoring with a water management team or any other discipline.

The fact that diseases were found to be most important in most sites says a lot about the pastoralist communities' love for their animals; they do not like to see them suffering. The apt descriptions of signs and symptoms for all diseases by the participants were exceptional in detail and accuracy. They had excellent indigenous knowledge about the diseases' causes, symptoms, prophylaxis, and treatments through use of modern drugs, homeopathic, or alternative medicine. However, they did admit to having limited knowledge about combatting new and emerging diseases like *peste des petits ruminants* (PPR). As stated earlier, there is a lot of variability in disease dynamics in the different sites following patterns of flock movements and migrations, which are dictated by lack of water and pastures as a result of prolonged droughts and location of markets. The prioritization of disease constraints for the four sites was as listed below (Table 2).

The wide spectrum of diseases for all sites read like texts in a small ruminants' disease book. The list being so variable for all sites, it is hard to prioritize one or two diseases for the Laikipia county disease management team to address. This notwithstanding, questions have to be asked why preventable and treatable diseases and conditions are and continue to be a challenge. One would ask, "what are the contributing factors"? The factors contributing to high prioritization of infectious diseases by participants in the research have been addressed in a separate study.

Conclusions and Recommendations

The pathway towards building resilience in pastoral small ruminants' health and production begins with an analysis of constraints and challenges that the producers face. Then comes the next step of managing critically important challenges. In the case presented here, thorough disease risk analyses are indicated. Such analyses will

Table 2 Small ruminant diseases in Laikipia

Priority	Kimanjo	Maundu ni Meri	Lorien	Nadungoru
1	CCPP[a]	Sheep & goat pox	Helminthiases	Helminthiases[b]
2	Heart-water	PPR	CCPP	CCPP
3	Pulpy kidney[c]	Helminthiases	Mange	Pulpy kidney[c]
4	Sheep & goat pox	CCPP	Enterotoxaemia	Heart-water
5	Foot and mouth disease	Jaundice[d]	Foot rot	Hydatidosis
6	Jaundice[d]	Hydatidosis	Sheep and goat pox	Foot and mouth disease
7	Helminthiases	Enterotoxaemia	Orf	Jaundice[d]

[a]CCPP is *Contagious caprine pleuropneumonia.*
[b]The Nadungoru participants particularly mentioned the terrestrial annelid worm or leech (*Hirudinea*).
[c]Flock owners insisted that this was due to worms. Further investigations revealed that this could have been due to *Clostridium perfringens* infection.
[d]Flock owners called it "yellow fever." The experts viewed this as jaundice likely due to liver fluke *Fasciola hepatica* infestation.

identify factors contributing to exceeded thresholds for the diseases to become a problem to a community; only then, can these critical factors be managed, through investments of time, money, and other resources including people.

Just as the communities were involved in identifying challenges constraining production of small ruminants, so must they be part of the solutions to their problems. They must be part and parcel of all analyses and be incorporated into strategic planning and intervention teams for effective strategies and interventions' implementation. The whole process calls for wide collaboration for the sheep and goat value chain and industry actors, as well as coordination by the county government.

One of the sites already has in operation, a producers' group; the *Laikipia Livestock Marketing Cooperative Society LTD (LLMCS).* The group has six partners and five agendas.

The *LLMCS* Agenda

1. Livestock and pasture improvement.
2. Rangeland rehabilitation.
3. Bee-keeping conservation.
4. Water conservation and tree planting.
5. Capacity building: awareness creation and conflict resolution amongst pastoralist communities and sensitization on forest encroachment.

LLMCS is a good entry point for animal health and production stakeholders to address the disease and other related pressing challenges. The group needs to get a further reach into many communities and enlist more members in all of Laikipia. This will certainly ensure success on the road to sustainable pastoral small ruminant development in Laikipia.

Cross-References

- ▶ Building Community Capacity in Fragile Environments: Case study of the Mara Serengeti Ecosystem
- ▶ Building Pastoral and Agro-pastoral Community Resilience Against Drought in the Context of the Paris Agreement: The Case of Isiolo County, Kenya
- ▶ Climate Change Vulnerability among Pastoralists and Farmers in Ethiopia
- ▶ Herdsmen on the Move: The Burdens of Climate Change and Environmental Migration in Nigeria
- ▶ Livestock Market Improvement with Anthropological Approach in Drought Resilience Project in Northern Kenya
- ▶ Planning and Climate Change in Serbia: Integrated Responses to the Causes and Impacts
- ▶ Understanding Kenyan Farmers' Perceptions of and Responses to Climatic variability to Build their Resilience

References

Catley A (1997) Adapting Participatory Appraisal (PA) for the veterinary epidemiologist: PA tools for use in livestock disease data collection. Proceedings of the Society for Veterinary Epidemiology and Preventive Medicine, April 1997, Chester

Catley A (1999) Methods on the move: a review of veterinary uses of participatory approaches and methods focusing on experiences in Dryland Africa. Sustainable Agriculture and Rural Livelihoods Programme, International Institute for Environment and Development (IIED), Department for International Development-UK (DFID) and Organization of African Unity/Inter-African Bureau for Animal Resources (AU/IBAR), Nairobi. 2005

Catley A (2000) The use of participatory appraisal by veterinarians in Africa. Office International des Epizooties Paris Sci Techn Rev 19(3):702–714

Catley A (2005) Participatory epidemiology: a guide for trainers. African Union/Inter-African Bureau for Animal Resources (AU/IBAR), Nairobi. 2005. ISBN 9966-7044-1-8 (electronic)

Catley A, Mariner J (2002) Where there is no data: participatory approaches to veterinary epidemiology in pastoral areas of the Horn of Africa. Drylands Programme, vol 110. International Institute for Environment and Development (IIED), London

CCFAS (2015) Galla goats: http://bit.ly/1HdVoh8. The SmartAG Partner. CCFAS, East Africa. Quarterly Newsletter. January–March 2015. CGIAR Research Program on Climate Change, Agriculture and Food Security

De Vaus DA (1996) Surveys in social research, 4th edn. UCL Press Ltd., London. 411pp. © DA de Vaus 1985, 1990, 1991, 1993, 1996

Dvorak K, Izac AM (1992) Exploratory surveys in research planning. IITA research guide, vol 28. Training Program, International Institute of Tropical Agriculture (IITA), Ibadan. 21p

Eriksen S, Lind J (2009) Adaptation as a political process: adjusting to drought and conflict in Kenyan Drylands. Environ Manage May 2000 43(5):817–835. E Pub 2008 Aug 23

FAO (1978) Report on agro ecological zones project. Methodology and results for Africa: world resources report, 48/1. Food and Agricultural Organization, Rome

Horton D (1991) Tips for planning formal farm surveys in developing countries: IITA research guide, vol 31. Training Program, International Institute of Tropical Agriculture (IITA), Ibadan. 34p

IPCC (2001) Draft (19 February 2001), Summary for policymakers; Climate Change 2001: impacts, adaptation and vulnerability. Approved by Working Group II, Geneva, 13–16 Feb 2001

IPCC (2014) Summary for policymakers In: Barros VR, Field CB, Dokken DJ, Mach KJ, Mastrandrea MD, Bilir TE, … White LL (eds) Climate change 2014: impacts, adaptation and vulnerability. Part A: global and sectoral aspects. Contribution of Working Group II to the Fifth assessment report of the intergovernmental panel on climate change. Cambridge, UK/New York, Cambridge University Press, pp 1–32

Jaetzold R, Schmidt H, Hornetz B, Shisanya C (2010) Farm management handbook of Kenya: vol. II. Natural conditions and farm management information: 2nd edn, Part II/B, Rift Valley

KNBS (2013) Economic report 2013. Kenya National Bureau of Statistics. Herufi House, Nairobi. Available at http://www.knbs.or.ke

KNBS (2015) Economic survey 2015. Kenya National Bureau of Statistics. Herufi House, Nairobi. Available at http://www.knbs.or.ke

KSS (1982) Exploratory soil map and agro-climatic zones of Kenya, scale 1:1000000, Rep. E1. Kenya Soil Survey, Nairobi

Mariner JC (2000) Manual on participatory epidemiology. FAO Animal Health Manual, vol 10. Food and Agricultural Organization (FAO), Rome

Nameless (2017) Cattle Barons: political violence, land invasions and forced displacement in Kenya's Laikipia County

OIE (World Organisation for Animal Health) (2010) OIE PVS tool. Tool for the evaluation of performance of veterinary services

OIE (World Organisation for Animal Health) (2013) Terrestrial Animal Health Code. Available at: www.oie.int/en/international-standard-setting/terrestrial-code/accessonline/

Oppenheim AN (1997) Questionnaire design, Interviewing and attitude measurement, New Edition. Pinter Publishers Ltd., London. 303p. © A. N. Oppenheim, 1966, 1992, 1996, 1997

Pretty J, Guijt I, Thompson J, Scoones I (1995) Participatory Learning and Action: A Trainers Guide. Sustainable Agriculture Programme, International Institute for Environment and Development (IIED), London

Rhoades RE (1992) The art of the informal agricultural survey: IITA research guide, vol 36. Training Program, International Institute of Tropical Agriculture (IITA), Ibadan. 57p

Seo SN, Mendelsohn R, Dinar A, Hassan R, Kurukulasuriya P (2008) A Ricardian analysis of the distribution of climate change impacts on agriculture across agro-ecological zones in Africa. Policy research working paper 4599. World Bank, Washington, DC

WHO-OIE-WBG (2014) Operational framework for good governance at the human-animal Interface: bridging WHO and OIE tools for the assessment of National Capacities. World Health Organisation. World Organization for Animal Health. World Bank Group EURO and SEARO WHO. Available at http://www.who.int/ihr/publications/WHO-OIEOperationalFramework/en/

Yu B, Guo Z (2016) Typology of agricultural productivity zones. In: Benin S (ed) Agricultural productivity in Africa: trends, patterns and determinants. International Food Policy Research Institute (2016) IFPRI, Washington, DC

Assessment of Climatic Variability Effect on Millet Production and Yield

65

Olumuyiwa Idowu Ojo, Abdulafeez Olalekan Olaniyan,
Adeniyi Suleiman Gbadegesin, and Masengo Francois Ilunga

Contents

Abstract

The information on climatic variability with its impacts on crop yields are needed more than ever especially on grains that serve as staple food in most part of the world. Hence, the objectives of the authors of this chapter were to collect and analyze rainfall, temperature, and millet yield data of the six states in North Central Nigeria, to generate geo-spatial variation maps for the chapter area using GIS techniques. All the six states in North Central of Nigeria (Benue, Kogi, Kwara, Nasarawa, Niger, and Plateau) including Federal Capital Territory (FCT), Abuja, were selected, and the rainfall and temperature data for 22 years

O. I. Ojo (✉)
Department of Civil Engineering, University of South Africa, Pretoria, South Africa

Ladoke Akintola University of Technology, Ogbomoso, Nigeria
e-mail: olumuyiwaojo@gmail.com

A. O. Olaniyan · A. S. Gbadegesin
Ladoke Akintola University of Technology, Ogbomoso, Nigeria

M. F. Ilunga
Department of Civil Engineering, University of South Africa, Pretoria, South Africa

W. Leal Filho (ed.), *Handbook of Climate Change Resilience*,
https://doi.org/10.1007/978-3-319-93336-8_188

(1991–2012) were used for the analysis. The relationship between millet yield and rainfall gave a negative relationship, while the relationship of both rainfall and temperature on millet yield showed a strong negative relationship deducing that there is variation in the climatic requirement. There is no doubt that farmers increasingly need detailed maps of this kind to plan crop planting schemes and monitor rate of yield (smart agriculture).

Keywords
Agriculture · Climatic variability · Millet production yield · North Central

Introduction

The agricultural sector has a multiplier effect on any nation's socioeconomic and industrial fabric because of the multifunctional nature of the sector (Adejuwon and Odekunle 2004). In Nigeria, agriculture is the main source of food and employer of labor employing about 60–70% of the population (Mayong et al. 2005). Agriculture in Nigeria is mostly rain-fed; it follows therefore that any change in climate is bound to influence its productivity in particular and other socioeconomic activities in the country. The impact could, however, be measured in terms of effects on crop growth, availability of soil water, soil erosion, incident of pest and diseases, sea level rises, and decrease in soil fertility (Adejuwon 2004). According to Woodward et al. (2010), climate is a measure of the average pattern of variation in temperature, precipitation, humidity, atmospheric pressure, atmospheric particle count, wind, and other meteorological variables in a given region over long periods. Climatic change is one of the environmental hazards that are life threatening to economic development and sustainability of humanity worldwide. Natural climate cycle and human activities have contributed to an increase in the accumulation of heat-trapping greenhouse gases in the atmosphere thereby contributing to increase in temperature in the global climate (UNFCCC 2007). As further explained by UNFCCC, the effect of climate change implies that the local climate variability, which people have previously experienced and adapted to, is changing and this change is observed in a relatively great speed. Temperature and precipitation are the two most important aspects of climatic variation in Nigeria. Temperature has been an important climatic factor determining the cropping pattern and quality of products in the country. The seasonal and spatial variation of temperature and rainfall largely regulates the agricultural activities in the country. As most of the farmers have to depend on suitable weather conditions to commence their farm activity, seasonal variation in climate has remarkable implication in the sustainability in Nigeria's agriculture.

Millets are a group of highly variable small-seeded grasses, widely grown around the world as cereal crops or grains for fodder and human food. Millets are important crops in the semiarid tropics of Asia and Africa (especially in India, Nigeria, and Niger), with 97% of millet production in developing countries. The crop is favored due to its productivity and short growing season under dry, high-temperature conditions (FAO 2007). In Nigeria, millet is sown annually on about five million hectares between latitude 70°N and 140°N with a yield of about four million metric tons of grains (Ikwelle 1998). Ong and Monteth (1994) have also demonstrated that the potential millet yield is strongly influenced by temperature, with optimum temperature for the process being 30–35 °C. Above this optimum, germination and shoot elongation rates decline early and rapidly. The crop is well adapted to some extent to growing areas characterized by drought, low fertility, and high temperatures. It could relatively perform better on soils with high salinity or low pH values. Because of its tolerance to difficult growing conditions, it can be grown where other cereals such as maize or wheat would not survive. This grain is used in Nigeria primarily for human consumption (Anyawu et al. 2003).

According to Tim (2005), geographical information system (GIS) can be defined as a computer system that stores and links non-geographical attributes or geographically referenced data with graphic map features to allow for a wide range of information processing and display operations, as well as map production, analysis, and modeling. GIS is a computer-based technology that helps in management, manipulation, analysis, display (mapped) data, and decision-making process. The aim was to determine the climatic variability effect on millet production and yield in North Central of Nigeria (comprising of Benue, Kogi, Kwara, Nasarawa, Niger, Plateau states, and FCT, Abuja) from 1991–2012 using GIS technique. This was achieved by collection and analysis of rainfall, temperature, and millet yield data (1991–2012) of the six states in North Central Nigeria. In addition, geo-spatial variation maps showing rainfall, temperature, and millet yields in the chapter area were generated in order to assess the correlation of rainfall and temperature to millet yield in North Central region of Nigeria.

Materials and Methods

Description of the Chapter Area

The North Central Nigeria with a land area of 296, 898 Km^2 representing nearly 32% of the country's total land area (NBS 2008) was covered. There are six states in the zone and the FCT, Abuja. The states include Benue, Kogi, Kwara, Nasarawa, Niger, and Plateau. It is located in the central part of Nigeria and in the subhumid region of the country and bounded to Bauchi, Kaduna, Zamfara, and Kebbi States to the north; Cross-River, Ebonyi, Enugu, Edo, Ondo, Ekiti, Osun, and Oyo States to the south; Taraba State and Republic of Cameroon to the east; and the Republic of Benin to the west. The zone has about 20.36 million people with the rural population constituting 77% situated between latitudes 6° 30″ and 11° 20″N and longitude 7° and 10°E, (NPC 2006). The position of the North Central Nigeria area in the map of Nigeria is

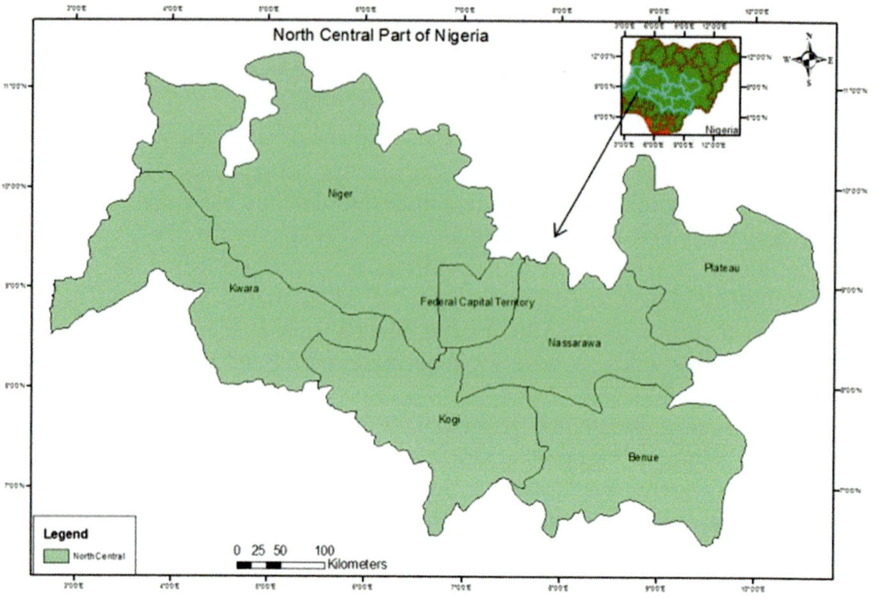

Fig. 1 Map showing the location of North Central Nigeria in the map of Nigeria

Table 1 Coordinates of North Central states. (Source: Adejuwon 2004)

S/N	States	Longitude (Degree, minute, seconds)	Latitude (Degree, minute, seconds)
1	Benue	10°0′0″ E	8°8′0″ N
2	Kogi	6°45′0″ E	7°45′0″ N
3	Kwara	4°32′32″ E	8°29′48″ N
4	Nasarawa	8°31′0″ E	8°28′60″ N
5	Niger	3°30′0″ E	8°20′0″ N
6	Plateau	100°38′0″ E	80°24′0″ N
7	FCT, Abuja	8°0′0″ E	9°0′0″ N

shown in Fig. 1, and the coordinates of the six states including FCT, Abuja (Benue, Kogi, Kwara, Nasarawa, Niger, and Plateau), of the North Central are shown in Table 1.

Collection of Rainfall, Temperature, and Millet Yield Data

The data of rainfall, temperature, and yield for North Central (Benue, Kogi, Kwara, Nasarawa, Niger, and Plateau) from 1991 to 2012 were collected from the archives of the Nigerian Meteorological Agency (NIMET), Ilorin, Kwara state. The annual millet yield data was obtained based on the ratio of expected annual production and exact area of farming. The production rate and farming area for the six states of North Central based on annual basis were collected from the Federal Ministry of Agriculture, Nigeria.

Application of Geo-spatial Analysis

With the intention of producing a highly reliable result, ArcGIS (9.2 version) was used to analyze annual rainfall, temperature, and millet yield spatial variation using Kriging method of interpolation. The map of Nigeria in JP2 format is imported into ArcGIS software by making use of adding data icon at the ArcGIS interface. The geographic information systems (GIS) database creation and mapping techniques were used to create interactive maps of the millet production areas, areas harvested, and the yield. GIS is a computer-based technology that helps in management, manipulation, analysis, display (mapped) data, and decision-making process. GIS offers a relatively efficient way for rainfall, temperature, and millet yield modeling. GIS techniques have the ability to integrate spatially explicit information into different formats from multiple sources. Through interpolation, maps of sensitivity were drawn using ArcGIS. Thus, the results are presented geographically for better understanding. GIS makes it easy to work with spatial data, to see patterns, which could not be seen originally, revealing hidden trends and distributions, thus gaining new insights. Kriging analysis was done by computed weighted average of the known values from a surrounding set of sampled points and thereafter used for the estimation of the rainfall and crop yield point distribution map.

Results and Discussion

The area covered by average rainfall, average annual temperature, and average annual millet yield range as shown in Table 2 and Figs. 2, 3, and 4.

Generation of Geo-spatial Variation Maps

Annual Rainfall and Annual Millet Yield Maps

Annual rainfall and millet yield maps were produced by taking into consideration a period of 6 years at interval of 4 years, respectively (1991, 1995, 1999, 2003, 2007, and 2011), as depicted in Figs. 5, 6, 7, 8, 9, and 10. For annual rainfall variation map, the area in deep blue shows the area with highest seasonal rainfall, while the area in light blue shows the area with lowest seasonal rainfall. For annual millet yield variation map, the area in deep green shows the area with heavy annual millet yield, while the area in tinted green represents area with lowest annual millet yield. Mean rainfall had inverse relation on the mean millet yield, which mean areas with increased mean rainfall turned out to have decreased millet yield output. With the present rainfall, by year 2012, mean annual millet yield declined.

Table 2 Averages of land area, annual millet yield, annual rainfall, and annual temperature (1991–2012)

S/N	States	Area (ha)	Yield (t/ha)	Rainfall (mm)	Temperature (°C)
1	Benue	36.149	1.497	943.7	7.9
2	Kogi	17.700	0.873	865.9	23.7
3	Kwara	14.896	1.521	907.2	21.2
4	Nasarawa	29.693	1.241	802.8	23.9
5	Niger	265.782	0.958	1000.9	18.8
6	Plateau	98.560	1.225	934.6	0.2
7	FCT, Abuja	11.200	0.994	796.6	20.4

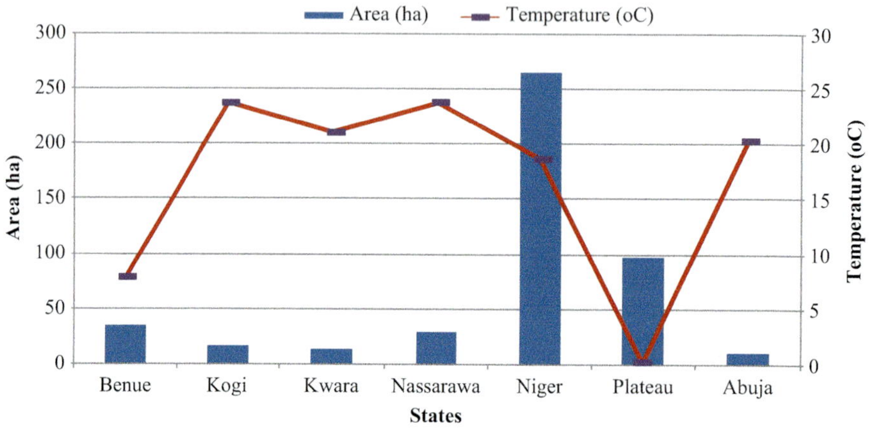

Fig. 2 Area of North Central states and average rainfall (1991–2012)

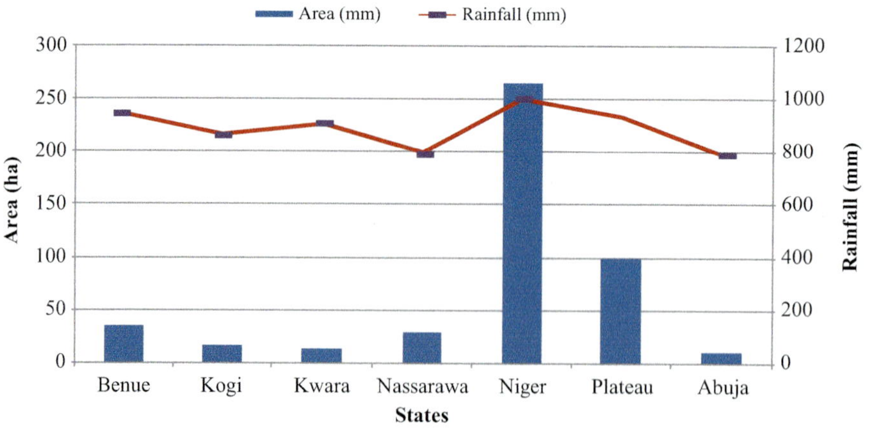

Fig. 3 Area of North Central states and average temperature (1991–2012)

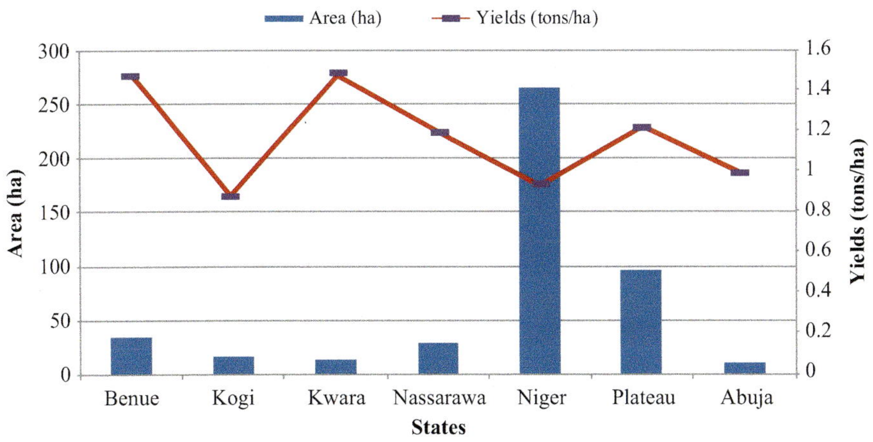

Fig. 4 Area of North Central states and average yield (1991–2012)

Annual Temperature and Annual Millet Yield Maps

Annual temperature and annual millet map was produced by taking into consideration a period of 5 years at interval of 4 years, respectively (1991, 1995, 1999, 2003, 2007, and 2011), as shown in Figs. 11, 12, 13, 14, 15, and 16. For annual temperature variation map, the area in deep red shows the area with highest annual temperature, while the area in yellow shows the area with lowest annual temperature. For annual millet yield variation map, the area in deep green shows the area with high annual millet yield, while the area in faded green represents area with lowest annual millet yield.

Mean annual millet yield however varied from state to state based on temperature variation in the area. Mean temperature was seen to have an inverse relation with the mean millet yield, which means that areas with increased mean temperature had declined mean millet yield. Mean annual millet yield for the year 2012 is highest at Benue State (1.497 t/ha) and lowest at Kogi (0.873 t/ha).

Discussion

The study of the effect of climate change on the millet production and yield showed that there were losses induced by increasing temperature, which may have led to increased evapotranspiration and possibly to reduction of the crop development period. From the results, it can be said that millet yield decreases because of the climate change, irrespective of whether rainfall increases or decreases. The results show their negative impact (trends) on crop yield, which is consistent with findings by Schlenker and Lobell (2010). There were changes in mean temperature and rainfall, as changes in their variability also affect crop yield (Wheeler et al. 2000). In addition, the role of temperature increase is a major factor affecting yield change

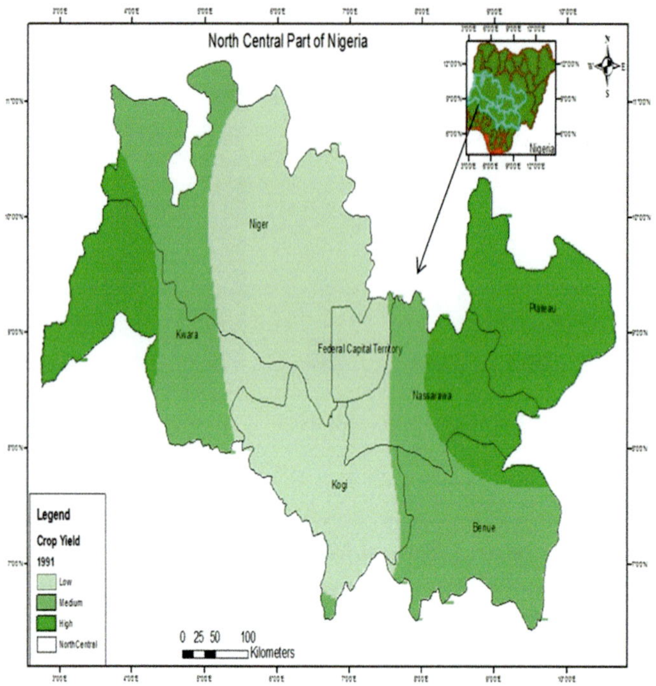

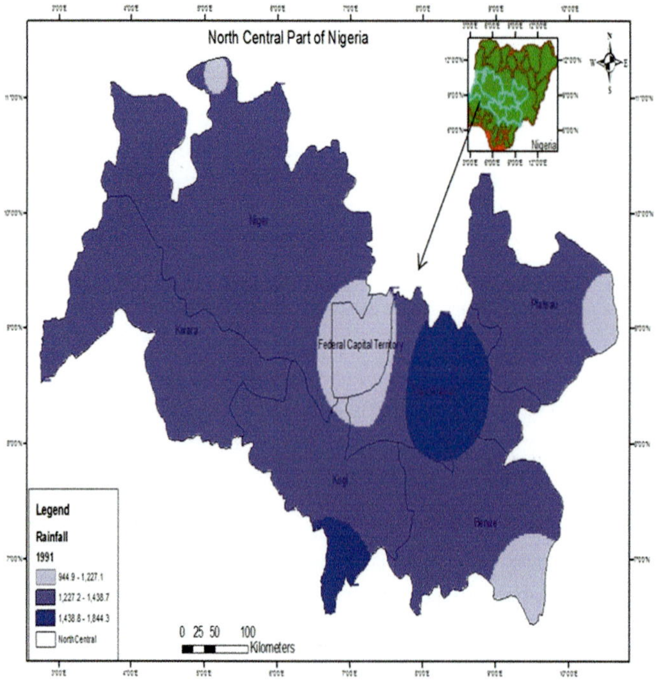

Fig. 5 Geo-spatial variation map of annual millet yield and annual rainfall of the area for the year 1991

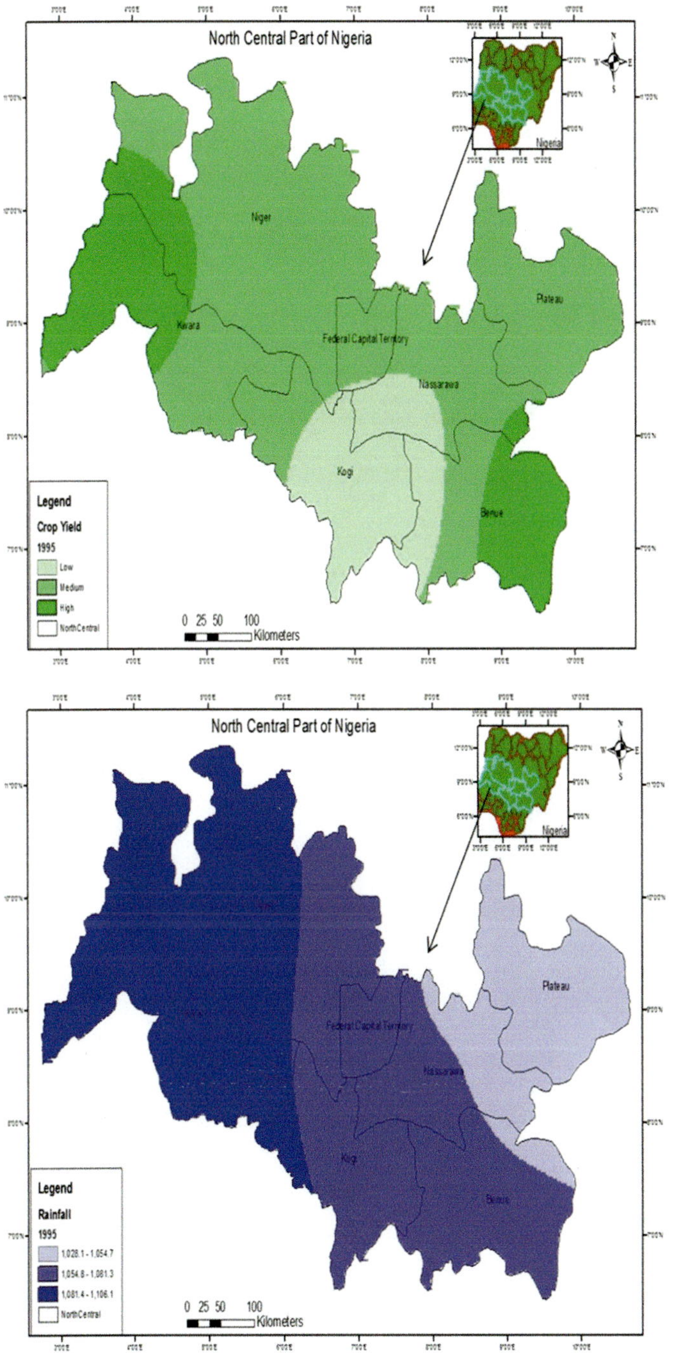

Fig. 6 Geo-spatial variation map of annual millet yield and annual rainfall of the area for the year 1995

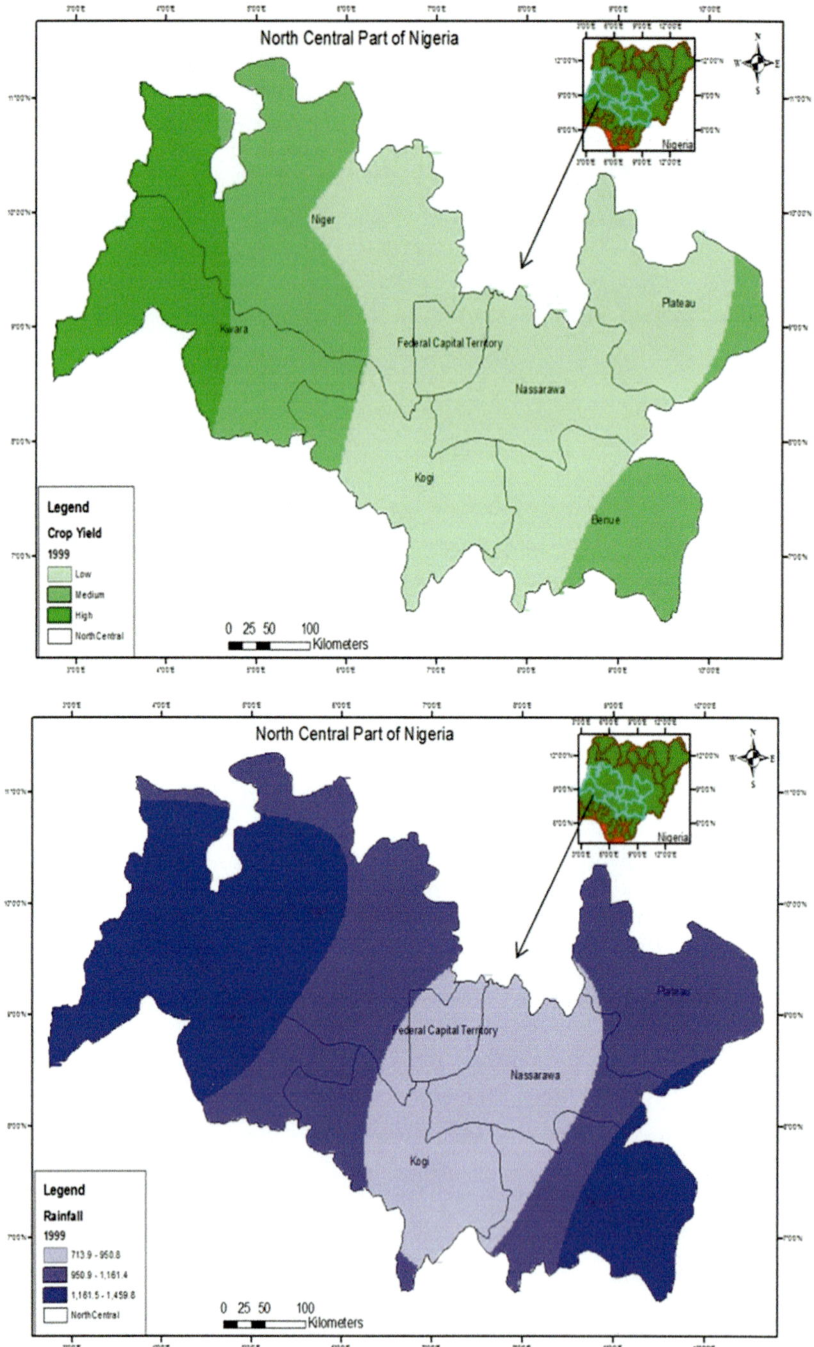

Fig. 7 Geo-spatial variation map of annual millet yield and annual rainfall of the area for the year 1999

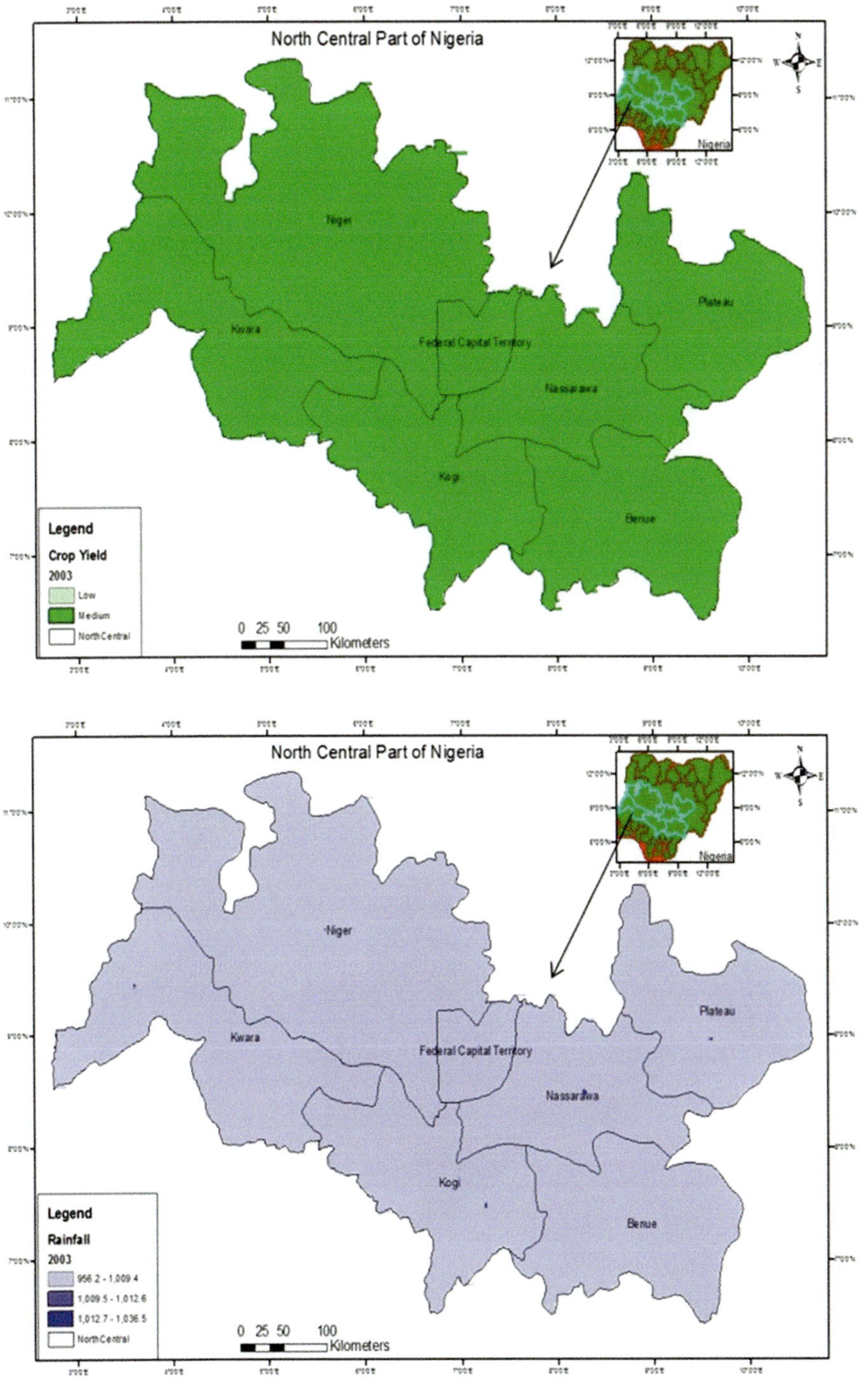

Fig. 8 Geo-spatial variation map of annual millet yield and annual rainfall of the area for the year 2003

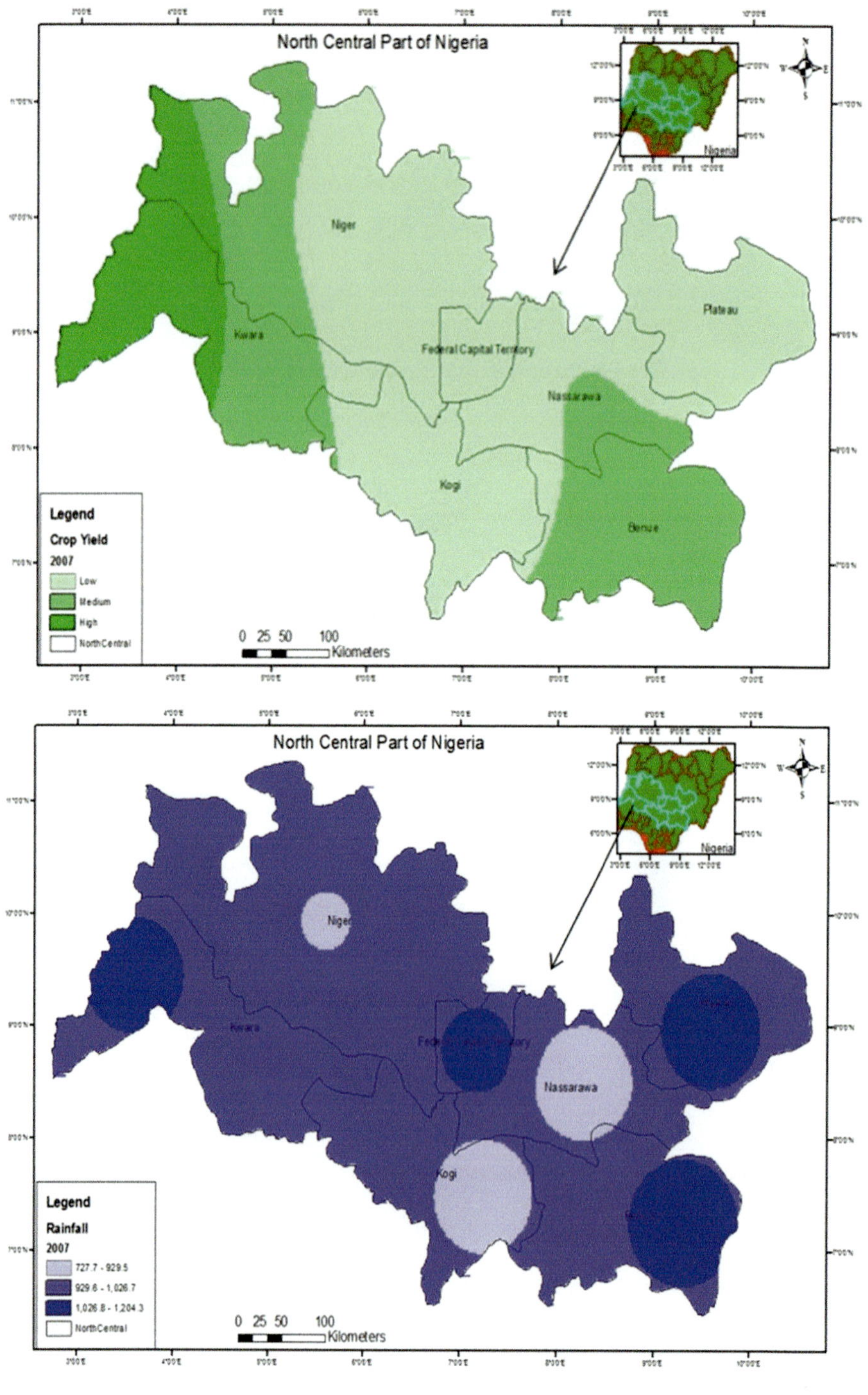

Fig. 9 Geo-spatial variation map of annual millet yield and annual rainfall of the area for the year 2007

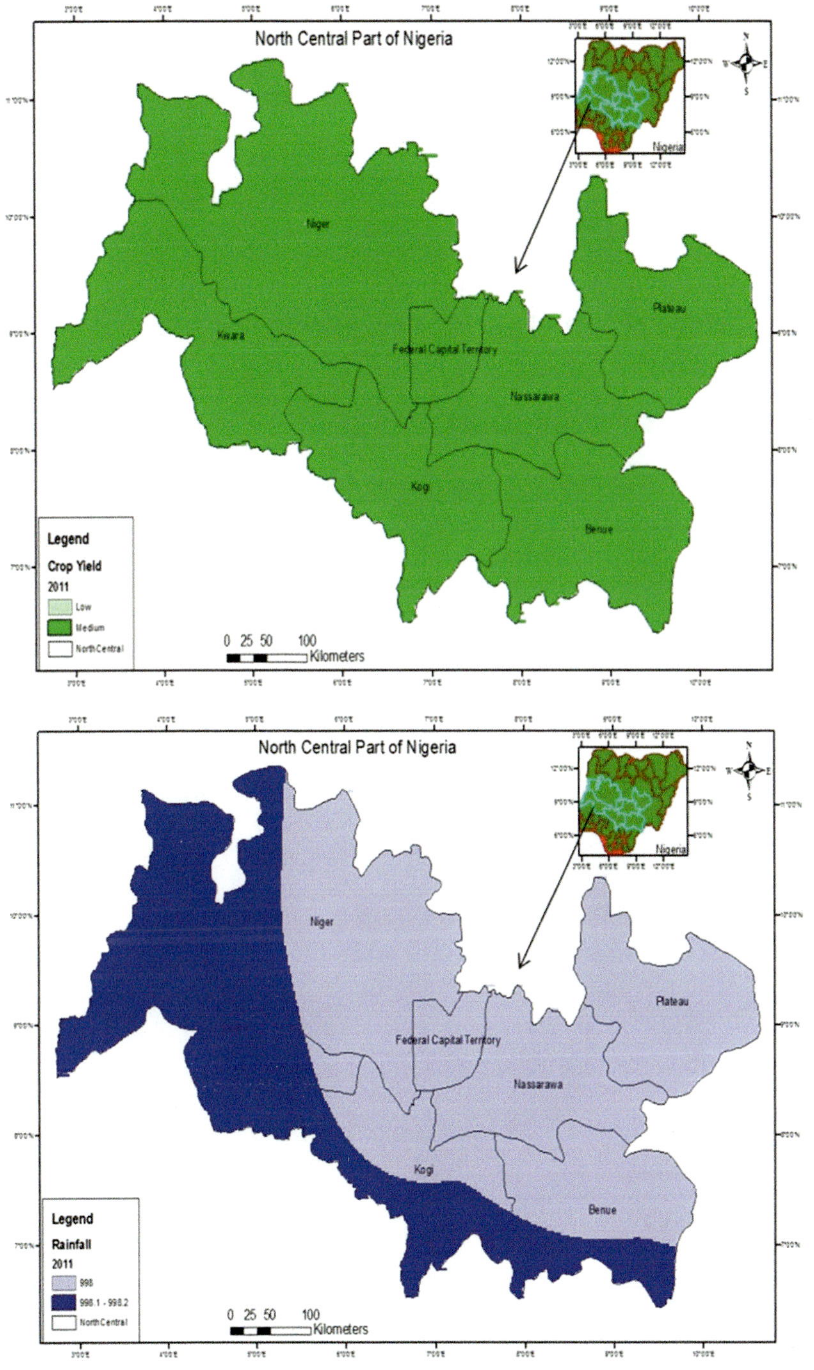

Fig. 10 Geo-spatial variation of map of annual millet yield and annual rainfall of the area for the year 2011

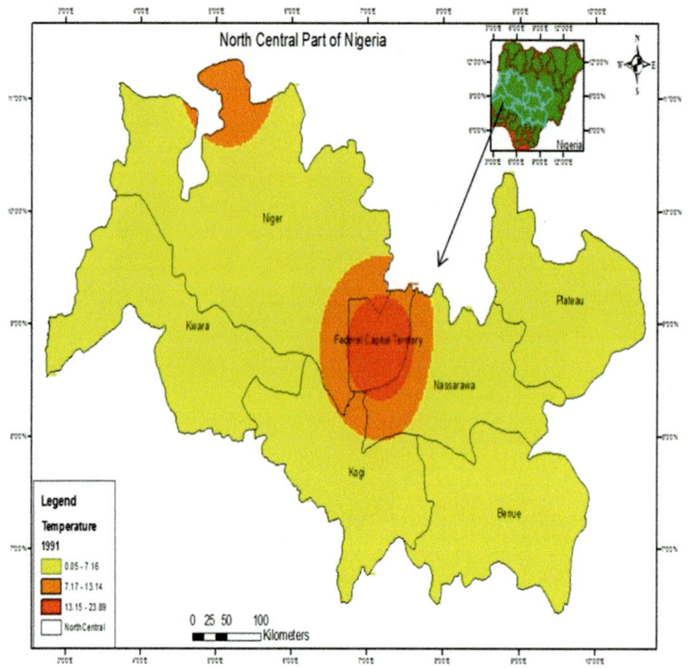

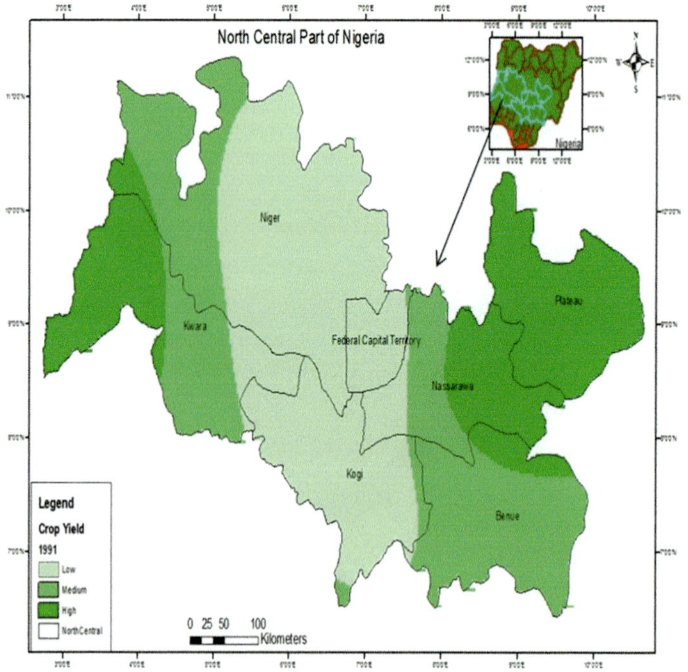

Fig. 11 Geo-spatial variation of map of annual temperature and annual millet yield of the area for the year 1991

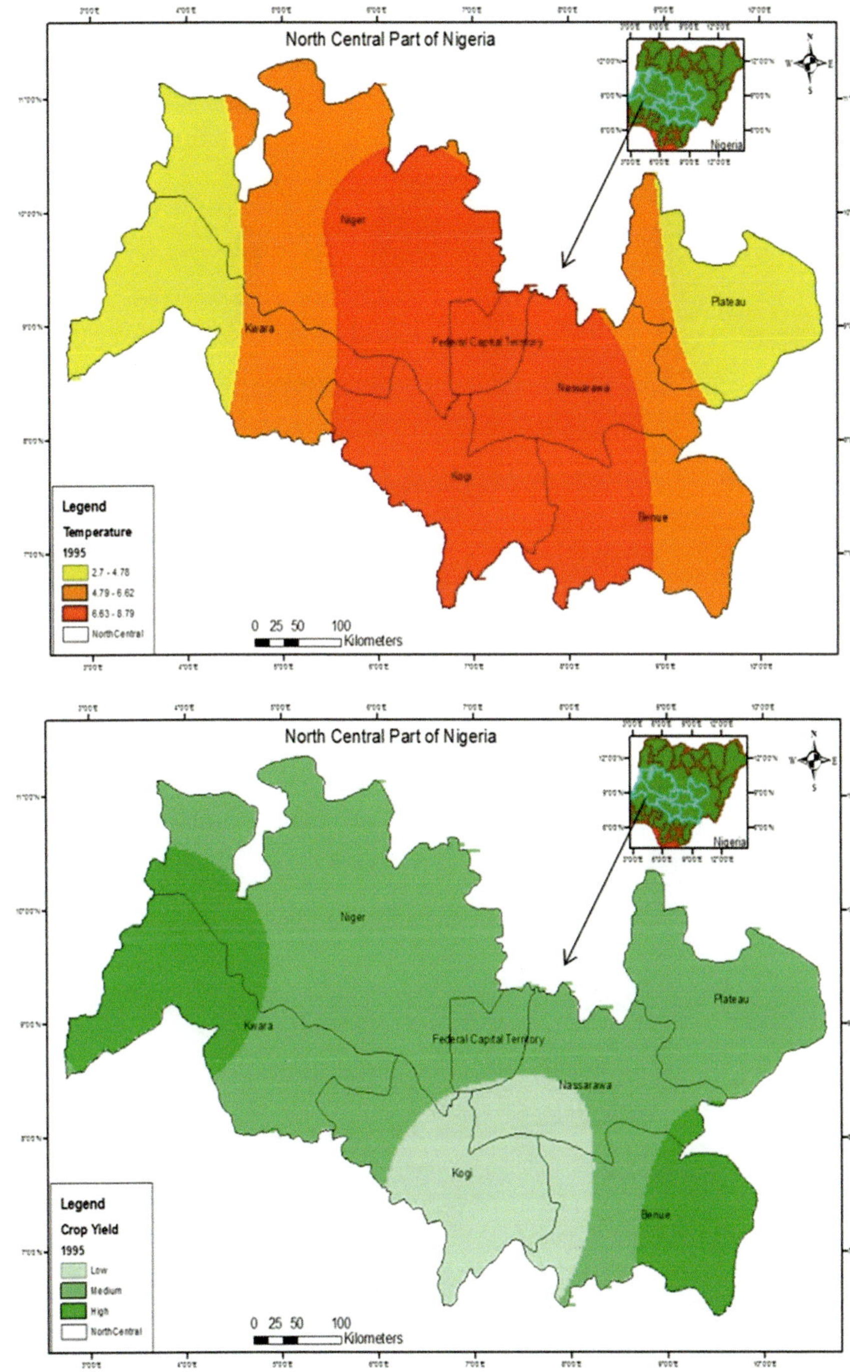

Fig. 12 Geo-spatial variation of map of annual temperature and annual millet yield of the area for the year 1995

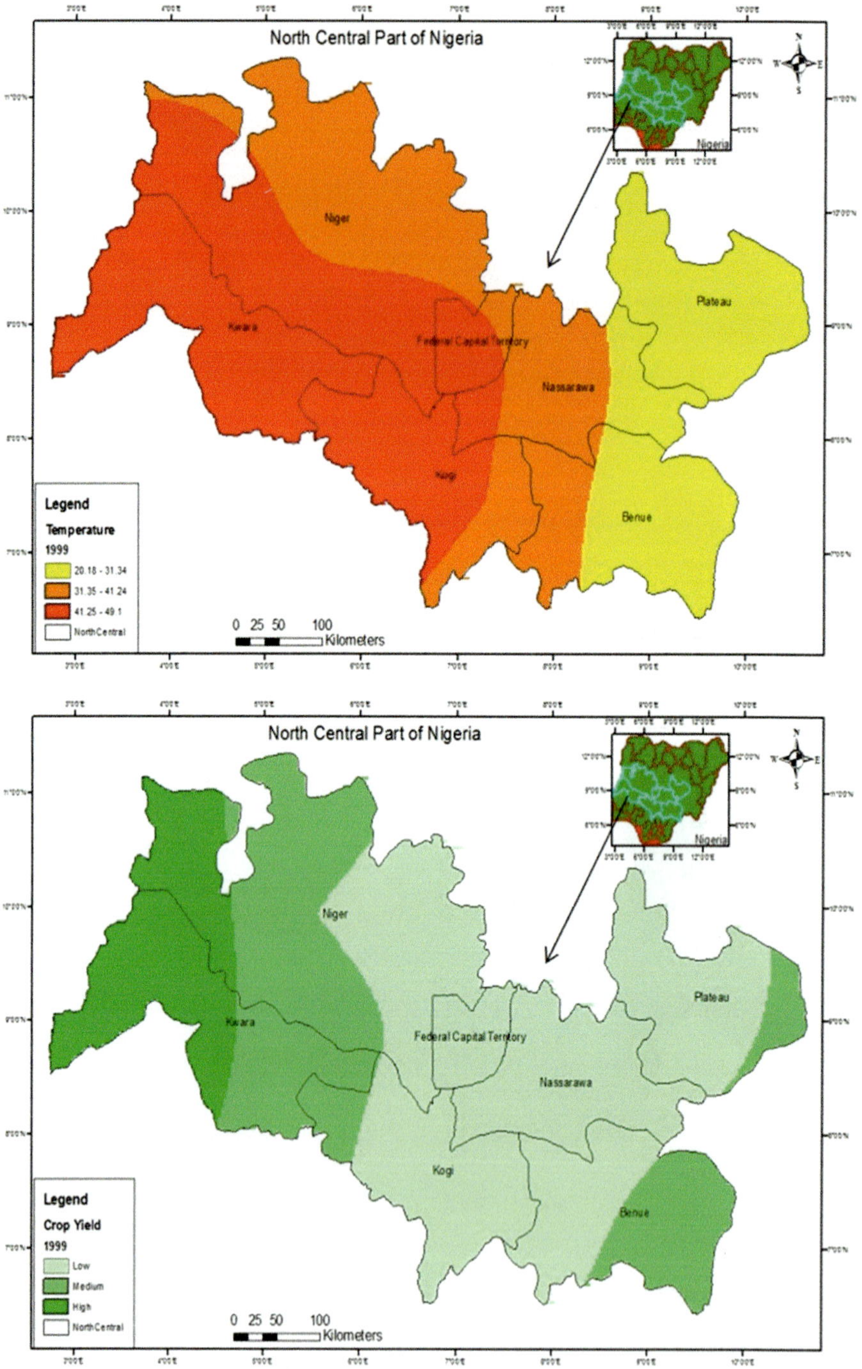

Fig. 13 Geo-spatial variation of map of annual temperature and annual millet yield of the rea for the year 1999

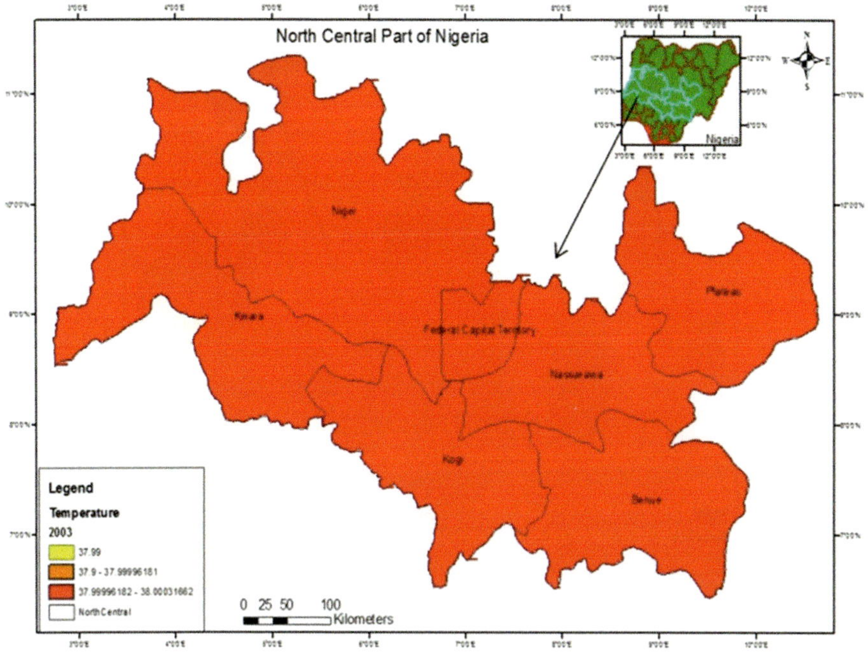

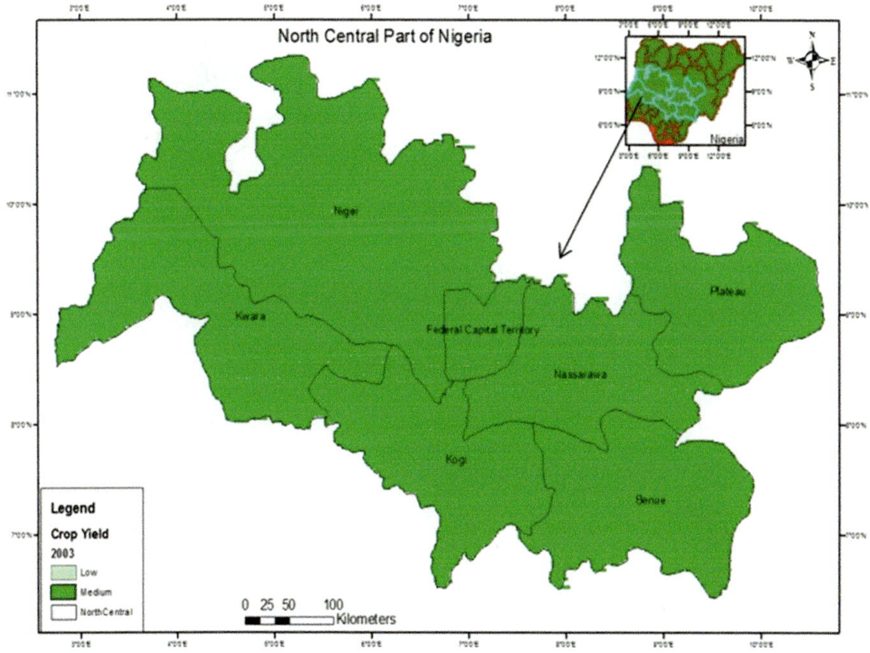

Fig. 14 Geo-spatial variation of map of annual temperature and annual millet yield of the area for the year 2003

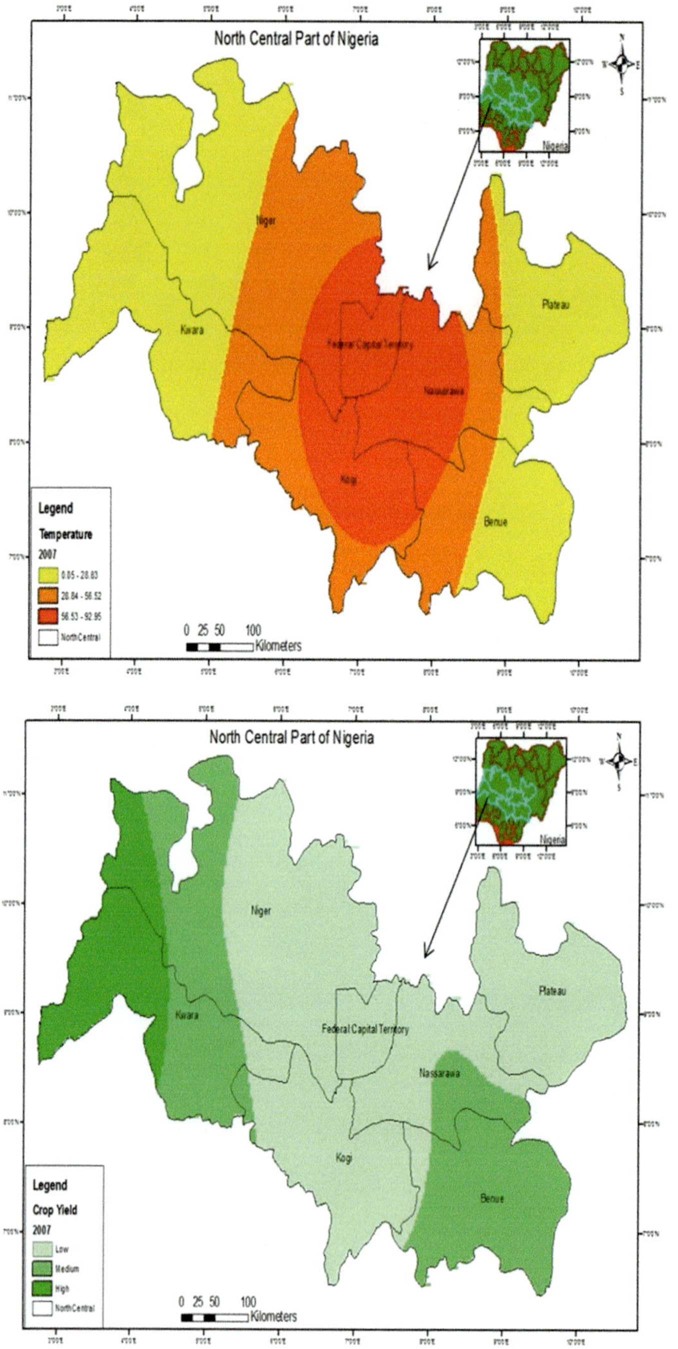

Fig. 15 Geo-spatial variation of map of annual temperature and annual millet yield of the area for the year 2007

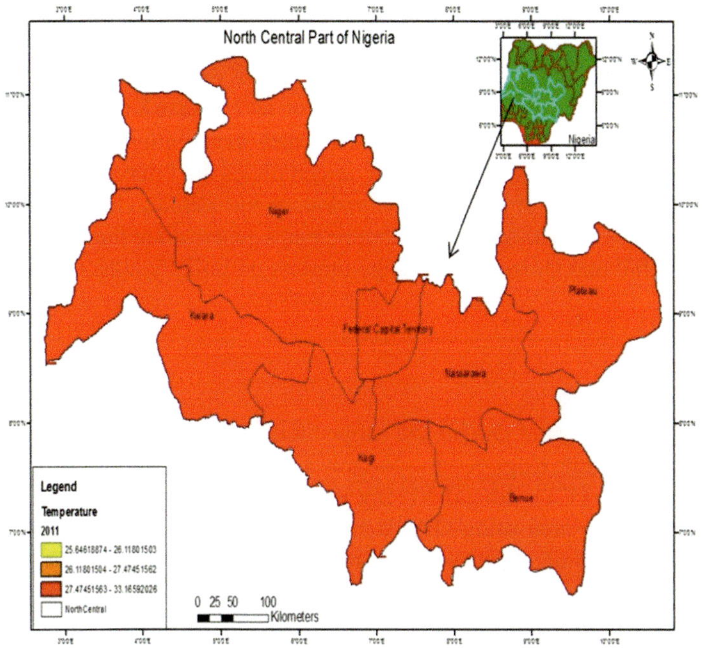

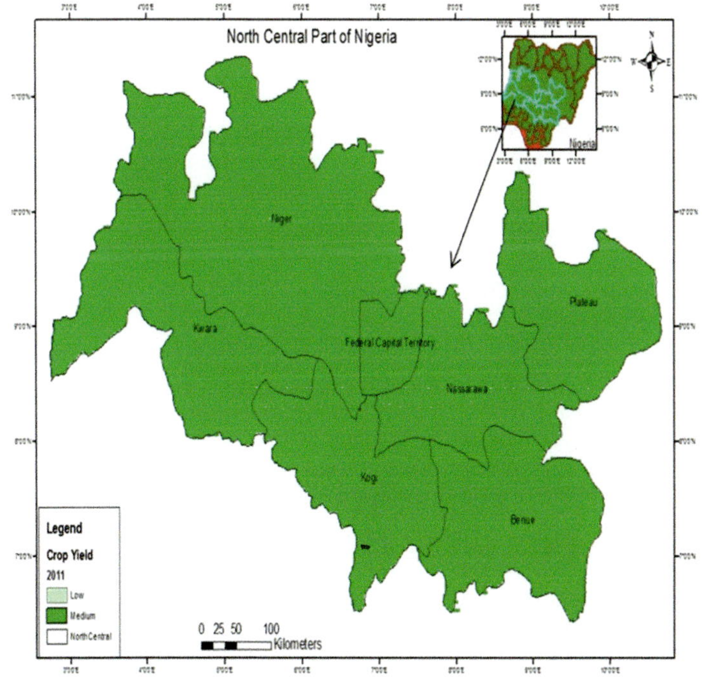

Fig. 16 Geo-spatial variation of map of annual temperature and annual millet yield of the area for the year 2011

(Thornton et al. 2009), and those crop yield losses are expected to be greater with the overall impacts of the climate change. According to Nortes et al. (2009) that decreased amounts of rainfall can cause an increase or decrease in developmental rates, depending on the stage of development as increased temperature leads to increased evapotranspiration which affect water availability for photosynthesis. The analysis may help to account for uncertainties in future climate scenarios, where the adverse role of increasing temperature and rainfall changes may be seen on crop yields, as the geo-spatial distribution maps generated in the study showed the effect of rainfall and temperature change impacts on millet yields.

Conclusions

This analysis of impact of rainfall and temperature on millet yields in the six states in North Central (Benue, Kogi, Kwara, Nasarawa, Niger, and Plateau including Federal Capital Territory (FCT), Abuja) Nigeria in order to generate geo-spatial variation maps for the study area using GIS techniques shows that the response of annual millet yield to climatic changes varies from one state to another with Kwara state having the highest average annual millet yield, while Kogi state has the lowest average annual millet yield. Thus, geo-spatial mapping has helped in the assessment of the vulnerability of millet production to climate change parameters. There is therefore the need to reduce the adverse impacts of erratic rainfall and increasing temperature. These findings may guide us better in adopting smart agriculture to cope with the negative effects of climate change to increase the productivity of crop and to provide a solution to the challenges of food scarcity. Geo-spatial analysis techniques may be applied for a realistic assessment of the impact of rainfall and temperature changes on millet production in other parts of the world, where it is cultivated. There is also the need for further work on identifying varieties that are more resilient to increased temperature during growth period.

References

Adejuwon JO (2004) Crop yield response to climate variability in the Sudano-Sahelian ecological zones of Nigeria in southwestern Nigeria. In: AIACC report of workshop for Africa and Indian Ocean Island, Dakar, Senegal, pp 15–16

Adejuwon JO, Odekunle TO (2004) Variability and intensity of the "little dry season" in south-western Nigeria. In: AIACC report of workshop for Africa and Indian Ocean Island, Dakar, pp 8–9

Anyanwu AC, Anyanwu BO, Anyanwu VA (2003) A text book for agricultural science for schools and colleges. Afp Africana Feb Publishers Ltd, Onitsha

FAO (2007) Annual publication. Food and Agricultural Organization, Rome

Ikwelle MC (1998) Pearl millet in Nigerian agriculture. In: Pearl millet in Nigerian agriculture: production, utilization and research priorities. In: Proceedings of the pre-season national co-ordination and planning meeting of the nationally co-ordinated research programme on pearl millet, Maiduguri, pp 1–8, 21–24 April 1997 (Emechebe AM, Ikwelle MC, Ajayi O, Aminu Kano M, Anaso AB ed)

Manyong VM, Ikpi A, Olayemi JK, Yusuf SA, Omonona BT, Okoruwa V, Idachaba FS (2005) Agriculture in Nigeria: identifying opportunities for increased commercialization and investment. IITA, Ibadan, pp 159

National Bureau of Statistics (NBS, Nigeria) (2008) Nigeria Annual Abstract of Statistics 2008. Available: http://ghdx.healthdata.org/record/nigeria-annual-abstract-statistics-2008

Nortes P, Gonzalez-Real MM, Egea G, Baille A (2009) Seasonal effects of deficit irrigation on leaf photosynthesis traits of fruiting and non-fruiting shoots in almond trees. Tree Physiol 29:375–388

NPC (2006) National Population Commission: 2007 Census Results. Federal Government of Nigeria

Ong CK, Monteith JL (1994) Response of pearl millet to light and temperature. Paper presented at International Crops Research Institute for the Semi-Arid Tropics (ICRISAT): Agrometeorology of Sorghum and Millet in the Semi-Arid Tropics: ICRISAT Center, India, pp 129–142, Patancheru, November 15–20

Schlenker W, Lobell D (2010) Robust negative impacts of climate change on African agriculture. Environ Res Lett 5:014010

Thornton PK, Jones PG, Alagarswamy A, Andresen J (2009) Spatial variation of crop yield responses to climate change in East. Africa Glob Environ Change 19:54–65

Tim H (2005) The impact of climatic variability over the period 1961 1990 on the soil water balance of upland soils in the North East Arid Zone of Nigeria, U.K. Met Office (2000): climate Change: an update of recent research from the Hadley Centre

UNFCCC (2007) Vulnerability and adaptation to climate change in small island developing States – Background paper for the expert meeting on adaptation for small island developing States. UNFCCC Secretariat. Bonn. http://unfccc.int/files/adaptation/adverse_effects_and_response_measures_art_48/application/pdf/200702_sids_adaptation_bg.pdf

Wheeler TR, Craufurd PQ, Ellis RH, Porter JR, Vara Prasad PV (2000) Temperature variability and the yield of annual crops. Agric Ecosyst Environ 82:159–167

Woodward G, Perkins DM, Brown L (2010) Climate change in freshwater ecosystems: impacts across multiple levels of organisation. Philos Trans R Soc Lond B 365:2093–2106

Climate Change Vulnerability Among Pastoralists and Farmers in Ethiopia

66

Desalegn Yayeh Ayal, Walter Leal Filho, Muluneh Woldetisadik, Solomon Desta, and Chunlan Li

Contents

D. Y. Ayal
Center for Food Security Studies, College of Development Studies, Addis Ababa University, Addis Ababa, Ethiopia
e-mail: desalula@gmail.com

W. Leal Filho (✉)
International Climate Change Information and Research Programme (ICCIRP), Faculty of Life Sciences, Hamburg University of Applied Sciences, Hamburg, Germany

Research and Transfer Centre Sustainable Development and Climate Change Management, Hamburg University of Applied Sciences, Hamburg, Germany
e-mail: walter.leal2@haw-hamburg.de; walter.leal@haw-hamburg.de; info@iccip.net

M. Woldetisadik
Addis Ababa University, Addis Ababa, Ethiopia
e-mail: mabshare@gmail.com

S. Desta
MARIL (Managing Risk for Improved Livelihood) Consulting Firm, Addis Ababa, Ethiopia
e-mail: solomon.desta82@gmail.com

C. Li
School of Geographic Sciences, East China Normal University, Shanghai, China
e-mail: 15598022233@163.com

W. Leal Filho (ed.), *Handbook of Climate Change Resilience*,
https://doi.org/10.1007/978-3-319-93336-8_190

Abstract

This paper has examined the vulnerability of pastoralists and farmers using a short-term scenario (2010–2039), employing a set of 22 exposure, sensitivity, and adaptive capacity vulnerability indicators. A combination of participatory, qualitative, and quantitative methods was used for primary and secondary data collection, so as to explore the pastoralists' and farmers' vulnerability. Assessment of pastoralists' and farmers' vulnerability to climate variability revealed the importance of comprehending exposure, sensitivity, and adaptive capacity indices. Differences in the trend of rainfall, temperature, and drought frequency resulted in varying adaptive responses so that the nature and magnitude of vulnerability differed according to the adaptive capacity of pastoralists and farmers to crop and livestock production hazards and agroecology differences. Given projected climate change, it is recommended that in both lowland and highland areas, future adaptation strategies should capitalize on building the adaptive capacity of pastoralists and farmers to excess rainfall, flooding, and severe drought.

Keywords

Capacity · Climate · Exposure · Smallholder · Sensitivity · Vulnerability

Introduction

Agriculture is inherently sensitive and vulnerable to climate variability and change (Aleksandrova et al. 2016; IPCC 2007. It is also the largest non-carbon dioxide (CO_2) greenhouse gas contributor (Fellmann et al. 2018). Climate models suggested that future climate variability and extremes will particularly undermine the performance of the agriculture sector (IPCC 2007, 2014). The impact could manifest in reduced crop production, rangeland degradation, water depletion, shortage of pasture, and spread of pest and diseases. Developing countries that heavily depend on rainfed and labor-intensive agriculture are the most affected (Heltberg et al. 2009; IPCC 2007; Bewket 2007; Ziervogel et al. 2006). Therefore, poor pastoralists and farmers are particularly vulnerable to current and future climate variability and change (Leary et al. 2008; IPCC 2007; Sivakumar et al. 2005).

The concept of vulnerability has been used differently in different disciplines, institutions, and scientific communities (Brauch et al. 2011). According to the IPCC (2001, 2007), vulnerability to climate change and variability have three components, (a) exposure, (b) sensitivity, and (c) adaptive capacity (TERI 2003), which are influenced by a range of biophysical and socioeconomic factors (IPCC 2007, 2014). In this chapter, exposure refers to the types and magnitude of climate extremes directly affecting farmers and pastoralists, while the frequency of such calamities determines their sensitivity. Sensitivity refers to the human-environment conditions that worsen climate change hazards such as population density, agricultural practices, sources of water, runoff erosion "risks," and livestock disease and pest and diseases. Sensitivity to climate variability and change determines the system's state of damage or benefits from climate-related stimuli. Vulnerability relates to how the rate and magnitude of climate extremes endanger the adaptive capacity of a given agricultural system (IPCC 2001).

Adaptive capacity relates to the ability of farming communities to mitigate effects of exposure to climate extremes (Leary et al. 2008) which is a function of local ecological potential, culture, and institutional realities (Masozera et al. 2007; Smit et al. 2000). Hence, the scope of adaptive capacity is influenced by objective and subjective adaptive capacity of individuals or communities at large. As Smit et al. (1999) and Downing et al. (1997) argued, an agricultural system will not be seriously damaged so long as climate extremes are not beyond the system's coping range. However, people may not know the scope of their subjective and objective adaptive capacity until a point when this capacity has been exceeded (Grothmanna and Patt 2005; Smit et al. 2000). Communities and individuals' subjective adaptive capacity (e.g., risk sharing experience, culture, norms, and believes) and objective capacity (e.g., basic infrastructures, early warning, crop and livestock asset, financial asset, saving and credit facilities and services) are a key factor to manage climate change-related risks (Ayal et al. 2017a, b). Wrong perceptions of objective and subjective adaptive capacities lead to underestimation and/or overestimation of risks (Ayal et al. 2017a, b; Grothmanna and Patt 2005). Therefore, adaptive capacity range in the chapter was investigated by assessing environmental, socioeconomic, and cultural realities in the research sites.

Table 1 illustrates that vulnerability is the function of physical, sociocultural, and political factors (Rosenzweig and Hillel 2008; IPCC 2001, 2007; Grothmanna

Table 1 Some of farming community climate variability adaptive capacity influencing factors

Items	Implications
Substantial decadal and interannual temperature and rainfall variability	Significant variations on crop yields and livestock production and productivity
Lack for farmer-level experience sharing	Limitations in the identification of adaptive approaches
Limited access to irrigation, agricultural extension services, early warning system, farm technologies, basic infrastructures, e.g., market, health center, veterinary services and facilities, road, etc.	Greater vulnerability to climate variability and extremes events, e.g., drought, flood Unable to adjust production decisions to variable climatic conditions
Land degradation due to overgrazing, over-cultivation, deforestation, etc.	Increased risk of damages by droughts, floods, or erosion
Exclusive dependency on agriculture	No diversification within the agricultural and agri-food sector as well as in other sectors of the economy Increased vulnerability to climate variability and extreme events
Weak sociocultural institutions, lack of climate change adaptation good governance	Aggravate climate variability adverse impact

Sources: Authors' constructed from field data and literature and Patt 2005). IPCC (2001) indicated that agriculture, water, and health sectors of Ethiopia are most vulnerable to climate variability and change. Ethiopian various sectors' vulnerability to climate variability and change has been documented (Bewket et al. 2015; Gebrehiwot and Veen 2013; Adem 2011; Füssel 2009; Amsalu and Adem 2009; Deressa et al. 2008; IPCC 2001; Devereux 2000). However, majority of the existing literature focused on homogenous biophysical, livelihood, and sociocultural settings. But despite the importance of a better understanding of the connections between climate and agriculture, the fact is that, in Ethiopia, agroecological-based comparison of pastoralists' and farmers' to climate variability has not yet been studied at a great level of detail. The main aim here is to examine the nature and magnitude of the highland farmers' and lowland pastoralists' vulnerability to climate variability, hence providing better understandings of the elements, which influence it.

Study Area and Research Design and Methodology

Study Area

The lowland study site is found in southern Ethiopia adjacent woredas of Yabello and Arero between 40 35′–50 5′ N and 380 15′–380 40′ E, covering an area of 900 km^2. The land use classification of the area shows that potentially cultivable land, cultivated land, covered by bush and forest, rangeland, range with encroachment, and others account for about 11.2%, 2.25%, 7.15%, 59.2%, 20%, and 0.5%, respectively. Population density is low; the population size of the two woredas is estimated at 150,493 and about 94% of those inhabits in rural areas, compared to only 6% urban dwellers which does not seem to offer ideal market condition (CSA 2008).

The highland study site is located in northwest Ethiopia, in the areas of Enemay and Gozamin woredas between 100 45′–110 15′ and 370 00′–380 30′. The total population of the Gozamin and Enemay woredas is 299,175 with an almost even male-female ratio, and less than 3% of the population lives in urban areas. The average population density is 168 people per square kilometer (Ayal and Muluneh 2014). High population pressure has limited the size of landholding which, through expanding, seems to have reached the limits of further expansion (Tafesse 2004).

Research Design and Methodology

Research Design

The researchers focused on the nature and magnitude of pastoralists' and farmers' vulnerability. Research is more reliable and valid (Johnson and Onwuegbuzie 2004) and best explainable and exploring when mixed methods are used. Hence, a mixed research design with concurrent triangulation approach was applied in this chapter (Creswell 2008). The quantitative, qualitative, and participatory methods were used to collect and correlate the relevant information on climate variability and vulnerability, using a set of local specific indicators. Local-level climate vulnerability indicators enable to understand the nature and magnitude of local climate change vulnerability (Ahumada-Cervantes et al. 2017).

Research Methodology

Sampling Techniques

The people in different regions have different culture, have access to livelihood resources, vary in their degree of vulnerability to various natural and human-made calamities, and differ in ways and capabilities to cope with stresses (Adger et al. 2003). To examine the magnitude and identify the nature of pastoralists' and farmers' vulnerability, the chapter adopted concurrent design using multistage sample for the quantitative and qualitative components of the study. Concurrent multistage sampling design allows a researcher to include different sets of sample participants such as pastoralists and farmers and key informants (development agents, experts, and community members). The Yabello and Arero *woredas* (districts) of Borana zone and Gozamin and Enemay woredas of East Gojjam and then one rural *kebele* (the smallest administration units) in each lowland woreda and five *kebeles* in each highland woreda (totally twelve) were selected purposefully due to their physical accessibility and manifestation of climate variability impacts on pastoralists and farmers settings (Table 2).

Accordingly 1 expert in each Amhara region and Oromia region, Agriculture and Rural Development Bureaus, Environmental Protection Bureaus, regional agricultural research institutes, and Regional Disaster Prevention and Preparedness, 1 expert in each woreda, and 1 development agent (DA) in each kebele, as well as 8 community key informants and 47 focus crop discussion (FGD) participant, were selected purposively. Furthermore, stratified random sampling techniques on the basis of agroecological zones, sex, and wealth status were used to select 300 sample households. Traditional wealth ranking method was applied to select proportional male and female household participants at village level. Site, FGD, and key informant selection process were conducted in consultation with the DAs and community key informants.

Sources and Data Collection Methods

Data on exposure, sensitivity, and adaptive capacity indicators were gathered from pastoralists, farmers, and other stakeholders using a questionnaire survey, Participatory Rural Appraisal (PRA), National Meteorological Services Agency (NMSA), Central Statistical Agency (CSA), General Circulation Model (GCM), and report document from the respective woredas. To provide a broad overview of the issues, 22 natural, biophysical, and socioeconomic vulnerability indicators were selected and used in this study (see Table 3). Qualitative and quantitative data collection and analysis methods were applied to offset the inherent limitation of each method and hence increase the validity and reliability of the findings (Creswell 2008).

Using a questionnaire survey, data on household socioeconomic; culture; demographic; accesses to rural infrastructures, such as veterinary, school, health center, and market; and climate-related risks were collected from households in the sampled

Table 2 Summary for research sites and total survey and PRA participants

Country	Ethiopia											
Region	Oromia		Amhara									
Zone	Borana		East Gojjam									
Woreda	Yabello	Arero	Gozamin					Enemay				
Name of RKAUs	Didahara	Alona	Libanos	Denba	Yebona Erjina	Wonka	Enerata	Yeser Eysus	Mahibre Birhan	Yetenbinaweyinam	Sekela	Dema
Total no. of survey participants												
	60	53	20	20	20	20	20	17	19	17	16	18
Total no. of focus group participants												
	12	11	12					12				
Total no. of key informant expert and DA informants												
	8		16									
Total no. of community key informants												
	2	2	2					2				

Table 3 List of identified indicators and their relationship with climate vulnerability

Vulnerability components	Indicators	Description of indicators	Relationship with vulnerability to climate change
Exposure	Extreme climate events	Frequency of drought	Increasing
	Change in climate variables from the base year	Change in maximum and minimum temperature, change in rainfall	Increasing
Sensitivity	Population density	Total kebeles' population per km^2	Increasing
	Irrigation practices	Percentage of household utilizing irrigation	Decreasing
	Agricultural diversification	Percentage of household practicing mixed farming system	Decreasing
	Access to diverse water sources	Percentage of household access and utilizing surface and groundwater for domestic and livestock use	Decreasing
	Crop diseases and pest damage	Amount of crop damage due to disease and pest	Increasing
	Rain water erosion risk	The proportion of farmland affected by rain water erosion The amount of crop lost by flooding Duration of grazing land covered by flood	Increasing
	Livestock diseases	Damage on livestock population caused by disease	Increasing
Adaptive capacity	Literacy rate	Proportion of households who can read and write and attendance formal and informal education	Decreasing
	Farm asset	Total farm income from livestock and crop production	Decreasing
	Crop production	Amount of crop produced in a harvesting year	Decreasing
	Access to credit and saving service	Proportion of household beneficiaries from credit association	Decreasing
	Improved farm technology	Proportion of households utilizing early maturing and drought-resistant crop varieties and improved breed and drought-resistant livestock varieties	Decreasing
	Livestock asset	Number of livestock owned	Decreasing
	Early warning information	Proportion of households utilizing climate forecasting information	Decreasing
	Access to health center	Average distance to health center	Decreasing
	Access to veterinary center	Average distance to veterinary center	Decreasing
	Access to market center	Average distance to market center	Decreasing

sites. The researchers also employed participatory methods, which allow wider community participation for the collection of more accurate information (Chambers 1994). Data was gathered from male and female community key informants and experts. The PRA was implemented using three major tools. Firstly, transect walk: This method was employed in Didahara and Libanos on distance covering 6 km in each site. Accordingly, information regarding farming practices, type and state of grazing land, livestock body condition, type of crop widely cultivated, and biophysical features of the area were collected. Secondly, the researcher carried out FDG to get in-depth information and insight about the socioeconomic, psychological, and biophysical effect of climate variability and their adaptive response. In FGD, a trend line was employed to collect data on rainfall and temperature trends and variations, drought occurrence frequency and severity, as well as variations of agricultural production in the past two decades. Similarly key informant interview was used to collect data on the events and trends of climate variability and change, its adverse impact on farmers' and pastoralists' livelihood and availability, and status of basic facilities and services.

HadGEM2-CC (CMIP5) delta scenarios under Representative Concentration Pathway (RCP) 4.5 and 8.5 concentration pathways were generated based on historical baseline daily weather data, with each day's weather variables perturbed using the changes in climate model outputs for future time periods versus those same model outputs for the historical period. Monthly changes in mean maximum temperature, minimum temperature, and precipitation were calculated by comparing future 30-year climate periods (2010–2039) to the baseline climate period (1980–2009) from the same GCM. The GCM data were obtained from Program for Climate Model Diagnosis and Intercomparison (PCMDI) portal (http://cmip-pcmdi.llnl.gov/cmip5/data_portal.html). CSA census reports and respective agriculture and rural development office survey reports on population density, crop, and livestock were also sources of data to the current chapter.

Data Analysis Techniques

An integrated vulnerability assessment approach of proxy biophysical and the socioeconomic indicators was employed in the 12 selected highland and lowland representative kebeles of Ethiopia. As suggested by Weis et al. (2016), initial indicators identified from the existing literature were commented by experts and local community key informants, and feedbacks were incorporated. The exposure index was computed using changes in rainfall and temperature based on daily data obtained from National Meteorological Services Agency (NMSA). Agnew and Chappel (1999) and McKee et al. (1993) noted that Standardized Precipitation Index (SPI) is the most widely used index for quantifying the frequency of drought events. Drought frequency and severity was calculated first using 30-year stations rainfall data by employing Standardized Precipitation Index (SPI) model (Agnew and Chappel 1999), and then data were interpolated according to Inverse Distance Weighted (IDW) interpolation techniques in geographic information system (GIS) environment. Then the data were classified to portray the spatial trends of drought frequency. The projection of climate change was made using gridded historical rainfall and temperature data as base period which was then ensemble with two downloaded Coupled Model Intercomparison Project 5 (CIMP5) GCM (HadGM2 CC and HadGM2 ES) climate data for 2050s (2040–2069) under RCP 4.5 scenario. The result generated a picture that quantifies the possible changes in rainfall and temperature.

A vulnerability index was created using exposure, sensitivity, and adaptive capacity indicators. The vulnerability index (VI) was computed using the formula below, Eqs. 1 and 2, by considering functional relationship between the indicators and vulnerability to climate change (Vincent 2004). This was done so as to obtain figures free from units and standardize their values. Indicators were normalized into values that range between 0 and 1 where 0 denotes least or no vulnerability and 1 denotes high vulnerability. Then, a standard deviation-based weighting technique was used to assign weights to all indicators in order to ensure that large variation in any one of the indicators (see Table 3) would not largely dominate the contribution of the rest of the indicators (Iyengar and Sudarshan 1982). This unequal weighting method is statistically sound and well suited to develop a composite index of vulnerability to climate change. Hence, the following formula was used where M stands for regions/districts and K for indicators of vulnerability and xij, i = 1, 2...M; J = 1, 2... are the normalized scores. The level or stage of development of ith zone, yi, is assumed to be a linear sum of xij (see Eq. 3).

$$\mathrm{VI} = \frac{\mathrm{AcV} - \mathrm{MinV}}{\mathrm{MaxV} - \mathrm{MinV}} \text{if the indicator is hypothesized to increase vulnerability} \tag{1}$$

$$\mathrm{VI} = \frac{\mathrm{MaxV} - \mathrm{AcV}}{\mathrm{MaxV} - \mathrm{MinV}} \text{if the indicator is hypothesized to decrease vulnerability} \tag{2}$$

where VI is the vulnerability index, AcV is the actual value, MinV is the minimum value, and MaxV is the maximum value.

$$y = \sum_{j=1}^{M/k} w\,j\,xji \tag{3}$$

To measure the level of vulnerability, index map was produced using the weighted sum of all indicators in the three sub-indices: exposure, sensitivity, and adaptive capacity in GIS environment. Finally an aggregated vulnerability index was developed from the three socioeconomic and biophysical indicators in Table 3. On the basis of index, the selected *kebeles* were classified into high, medium, and low level of vulnerability to climate variability risk. Medium level of vulnerability was defined as an index within one standard deviation unit of the whole kebele index mean (AVG − STD) < VI < (AVG + STDV), while high (VI ≥ (AVG + STDV)) and low (VI ≤ (AVG + STDV)) levels were greater than and less than or equal to one standard deviation unit above or below the whole *kebele* index mean, respectively.

Moreover, agroecological base vulnerability comparison was made using pentagon based on the vulnerability index (VI) value of each sub-index.

The qualitative data collected from interviews, individual household and focus group discussions, and observations were analyzed thematically. Paraphrasing, identification, and characterization of the recurring themes were performed in the analytical procedure. Thus, the qualitative information was used to substantiate the index results.

Results and Discussion

Exposure Index

The preliminary vulnerability sub-index result revealed that in the lowland agro-ecology, Alona and Didahara have relatively medium level of exposure to the risk of drought and future change in rainfall and temperatures (Fig. 1).

In the highland Yeser Eysus, Mahibre Birhan, Yetenbinaweyinam, Sekela and Dema *kebeles* are relatively highly exposed to the risk of drought and future temperatures and rainfall change (Fig. 2). On the other hand, Libanos, Denba, Yebona Erjina, Wonka, and Enerata *Kebeles* are relatively lower exposed. Therefore, magnitude of effect of change in rainfall and temperature in the highland *kebeles* will range from high to low. The result helps in indicating the future likely occurrence of climate induced risks on agriculture in lowland and highland study sites.

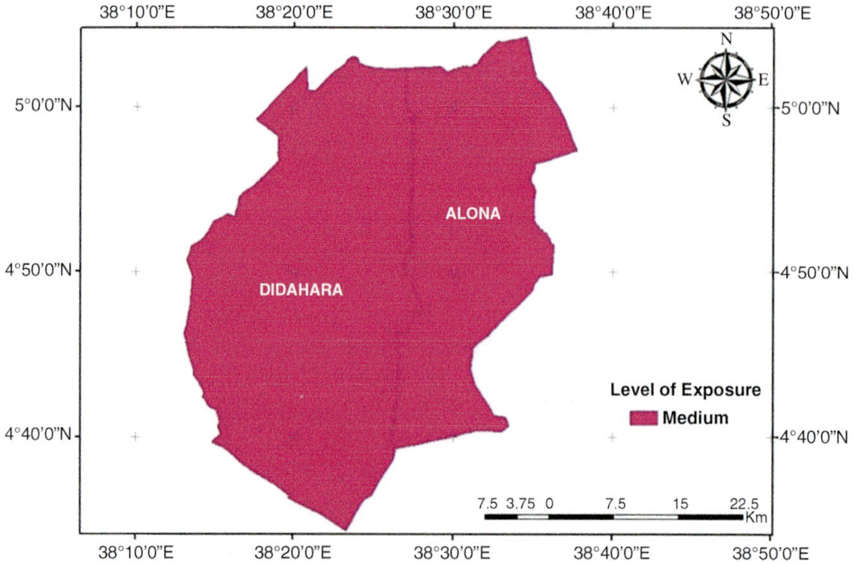

Fig. 1 Vulnerability sub-indices map: Exposure index of lowland kebeles. (Source: Authors' constructed from field survey, 2011 and 2012, NMSA, CSA, GCM model)

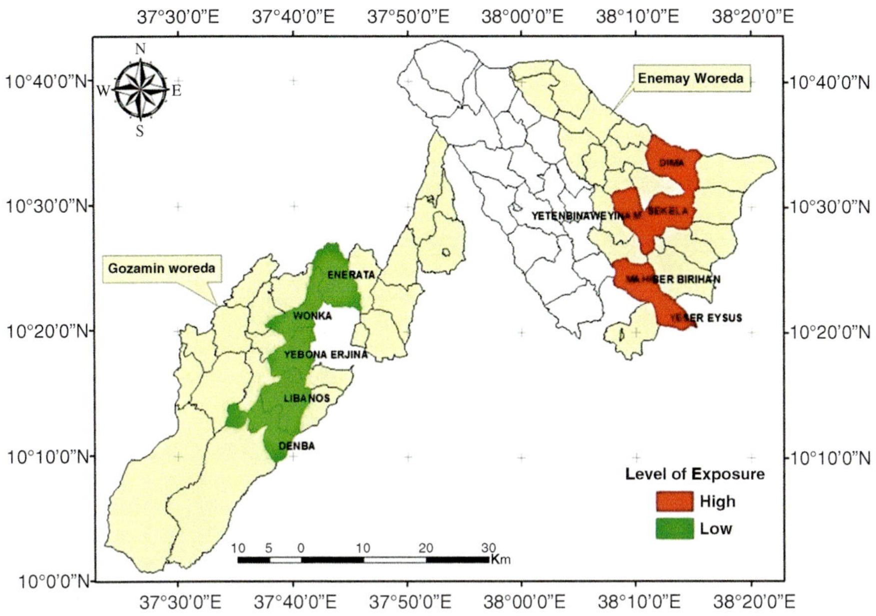

Fig. 2 Vulnerability sub-indices map: Exposure index of selected kebeles in highland. (Source: Authors' constructed from field survey, 2011 and 2012, NMSA, CSA, GCM model)

Sensitivity Index

Pastoralists and farmers heavily rely on subsistence rainfed agriculture and a degraded environment using traditional technology. Sensitivity analysis revealed that Enerata, Denba, and Mahibre Birhan highland *kebeles* are relatively highly sensitive to the adverse effects of climate change due to high human-environment interactions (Fig. 3). This is mainly caused by the combined effects of high population density, erosion, and crop disease and pest damage occurrences. The least sensitive highland *kebeles* are Wonka, Libanos, and Sekela.

In the lowland region, Alona kebele will be relatively highly sensitive to the adverse effects of climate variability and change (see Fig. 4) due to high crop disease and pest damage which might be directly related to the occurrence of high temperatures. Besides the Alona pastoralists' livelihood is dominated by livestock husbandry rather than mixed crop and livestock production. Didahara has medium level of sensitivity. In both lowland *kebeles*, the livelihood of pastoralists seems sensitive to the absence of alternative surface and underground water sources for agriculture and domestic purpose. The problem would affect the vitality of livestock production by decreasing the market demand of livestock and livestock products. In line with our finding, Kathleen et al. (2001) reported that water stress and other climate change manifestations threatened East African pastoralists and elsewhere in the world (Aleksandrova et al. 2016).

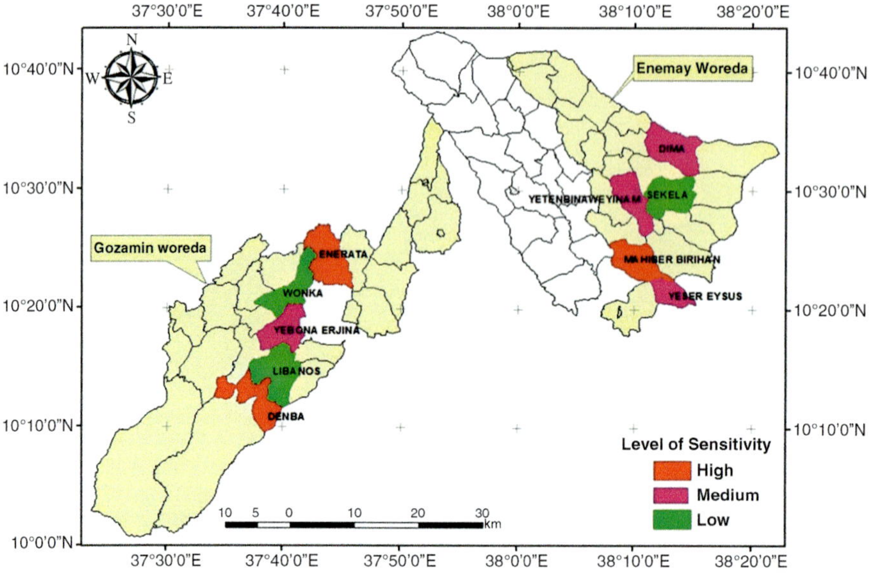

Fig. 3 Vulnerability sub-indices map: Sensitivity index of selected kebeles in highland. (Source: Authors' constructed from field survey, 2011 and 2012, NMSA, CSA, GCM model)

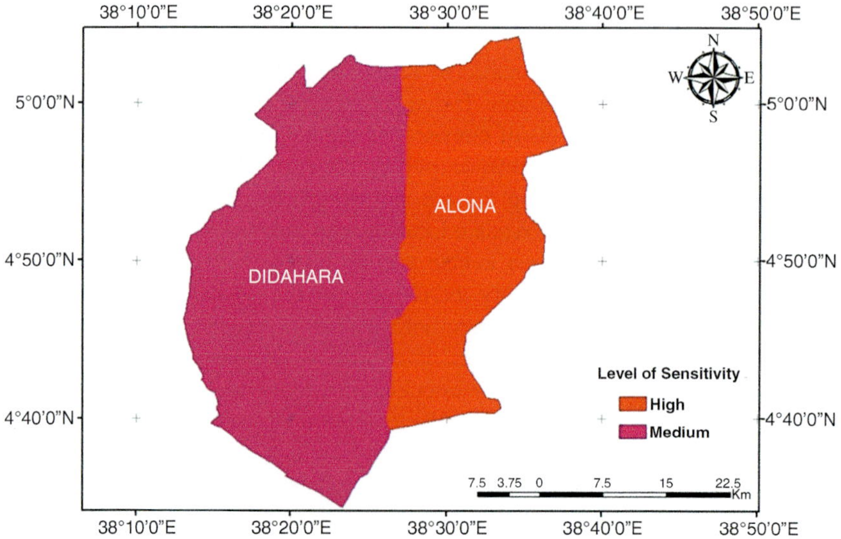

Fig. 4 Vulnerability sub-indices map: Sensitivity index of selected kebeles in lowland. (Source: Authors' constructed from field survey, 2011 and 2012, NMSA, CSA, GCM model)

Previous studies underlined that the less diversified a livelihood is, the more sensitive it would be to climate-related risks (Devereux 2000). Drastic population growth which could be translated to a shortage of farmland and grazing land and hence declining land productivity is another sensitivity indicator for the Ethiopian pastoralists and farmers to climate-related problem. Moreover, rapid population growth increased water demand and aggravate water scarcity. As confirmed by IPCC (2001), the Ethiopian agricultural rainfed system is more sensitive to climate-related risks. FGD and key informants in the highland anticipated that their livelihood sensitivity will be increased due to increased landlessness, declining soil fertility due to over cultivation and the erosive nature of rainfall, ever-increasing agricultural input price, prevalence of livestock diseases, and invasion of crops with pest and diseases. Their livelihoods are further harmed by accelerated loss of soil fertility, deforestation, and poor access of modern agricultural inputs. Similarly FGD and key informants in the lowland also explained the likely increment of their livelihood sensitivity due to rangeland degradation through population pressure, bush encroachment, and depletion of well water, prevalence of pest and disease, and decreasing of livestock diseases resistance.

Adaptive Capacity Index

Adaptive capacity measures the ability of a given system to adjust to climate-related shocks and risks, to moderate potential damages, to take advantage of opportunities, or to cope with the consequences (IPCC 2001, 2014). The result of adaptive capacity index shows that Mahibre Birhan is the only *kebele* in the highland that will have relatively higher adaptive capacity to avert the adverse effects of climate variability (Fig. 5). Dema and Denba *kebeles* will have the lowest adaptive capacity, while the rest will have medium-level adaptive capacity.

Although pastoralists in Didahara have an inferior physical asset base, such as in matters of livestock and crop production, Didahara has a relatively high adaptive capacity, while Alona will has low adaptive capacity (Fig. 6). The relatively higher adaptive capacity of Didahara is attributable to its physical access to basic infrastructure and credit and saving facilities. Pastoralists risk-sharing experience has also contributed for higher adaptive capacity in Didahara. Hence, the nature of adaptive capacity indicators in Didahara stems from social capital and its services. Generally, the result implies that differential socioeconomic and infrastructural asset distribution is responsible for the unequal adaptive capacity of some *kebeles* within the same agroecology to respond to the climate change risks.

Agroecological Comparative Exposure, Sensitivity, and Adaptive Capacity

The agroecological exposure, sensitivity, and adaptive capacity index result in Fig. 7 shows that in 2040 the highland sites will be relatively highly exposed than the lowland sites to the direct effect of climate variability and change. Highland and

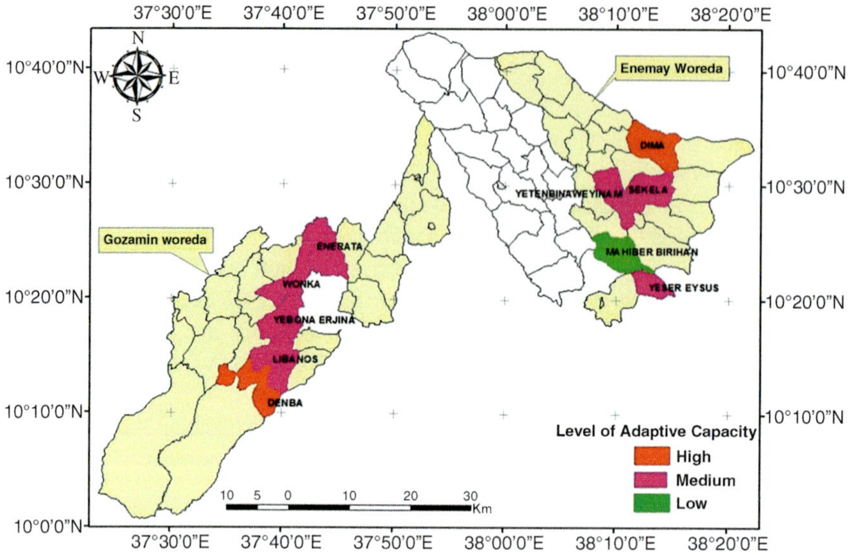

Fig. 5 Vulnerability sub-indices map: Adaptive capacity of selected kebeles in highland. (Source: Authors' constructed from field survey, 2011 and 2012, NMSA, CSA, GCM model)

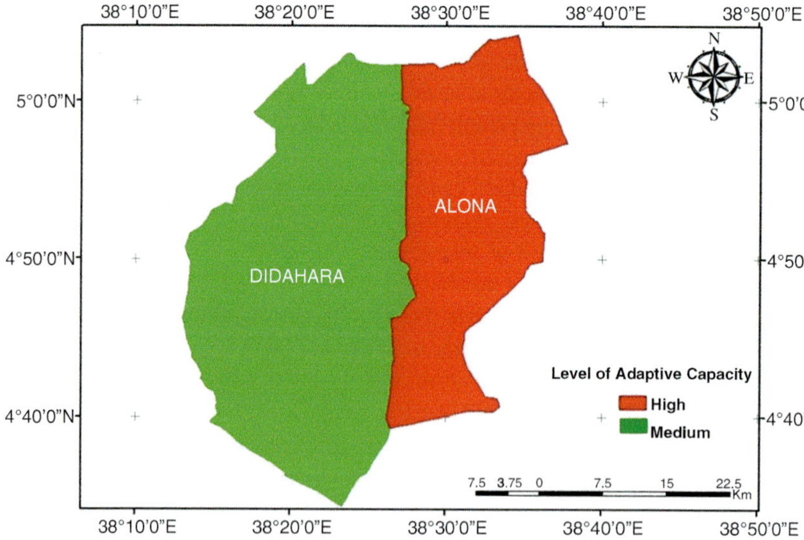

Fig. 6 Vulnerability sub-indices map: Adaptive capacity index of selected kebeles in lowland. (Source: Authors' constructed from field survey, 2011 and 2012, NMSA, CSA, GCM model)

lowland regions will be equally exposed to frequent drought occurrences. Concurrently, highland and lowland regions will be exposed to ever-increasing temperature at different magnitudes. Mean temperature will increase by 0.03 and 0.33 in the

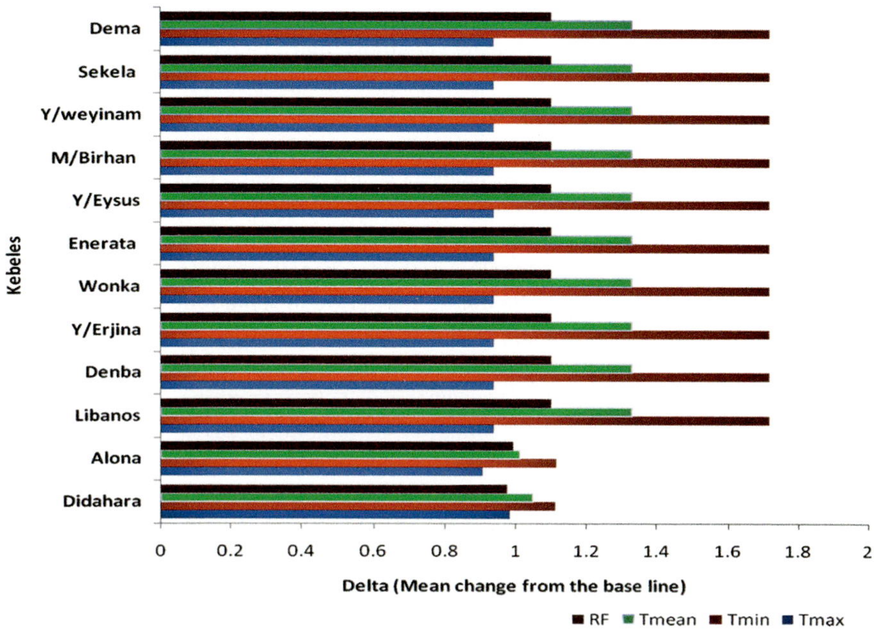

Fig. 7 Mean change (delta) of rainfall and temperature from the baseline. (Source: Authors' constructed from field survey, 2011 and 2012, NMSA, CSA, GCM model)

lowland and highland, respectively. More specifically the lowlands will be relatively highly exposed to changes in connection with the maximum temperature when compared with the highlands, whereas the highland region will be relatively highly exposed to minimum temperature change than the lowland sites (see Fig. 7).

Mean temperature historical records of the lowland and highland increased by 0.26 °C and 0.31 °C per decade, respectively. In comparison, for the last three decades, mean temperature increment was faster in the lowland than the highland. The result is consistent with the assertion that temperature is increasing in all regions and more pronouncedly in the lowland (Adem 2011).

The near-term prediction shows that in the highland rainfall will increase by 10%, while rainfall in the lowland will decrease by 2% (see Fig. 7). Hence, highlanders and lowlanders will face more excessive rainfall and rainfall deficiency, respectively. Therefore, the rainfall change will not be hospitable for highland and lowland regions since less change suggests the continuity of the existing imperfect rainfall condition. The predicated change in rainfall in the highland could hasten flooding and attendant crop pest and disease, livestock disease, crop failure, and shortage of pasture. Similarly the rainfall reduction in the lowland could aggravate shortage of water and related biological and epidemiological challenges which the region is already experiencing (Ayal and Muluneh 2014).

The result appears in agreement with a projection that future climate variability will affect the crop sector (Mendelsohn et al. 2007), and it is acute, especially since the highland would experience erratic rainfall and increase in minimum and maximum temperatures. Therefore, climate variability and change are more likely to adversely affect the highlands that are currently the main crop production area which supports the livelihood of the bulk of the population and reduces the contribution of the crop sector to Ethiopia GDP, and its performance to absorb labor force would decrease considerably. Instrumental analysis of the last three decades also reveals that the lowland sites were highly exposed to rainfall variability than the highland. However, looking at pentad and monthly levels, sometimes the highland region experienced relatively higher rainfall variability than the lowland region. In the lowland and highland research sites during the period of small rainy season, rainfall variability was very high (Ayal and Muluneh 2014).

It could be assumed that pastoralists and farmers' murky climate prediction enhances their subjective adaptive capacity so that they would be disposed to take measure and adopt technologies that minimize vulnerability. FGD and key informants in the lowland and highland anticipated that temperature extremes such as the number of hot days and nights will increase. In the future temperature will increase and rainfall amount will decrease. They underline that water stress for agriculture could increase due to increased evapotranspiration, high runoff, decreasing rainfall amount, and increasing water demand as the result of population growth. Key informants in the lowland region anticipated the occurrence of more frequent severe droughts in the future. In addition highlanders also predicted a decline of the number of rainy days, increase of rainfall intensity, and abandonment of the small rainy season. They also expected more events of erratic rainfall distribution, flood, and frost.

Figure 8 illustrates that the highland region will be relatively highly sensitive to climate change than the lowland region. However, the nature and magnitude of pastoralists and farmers' livelihood sensitivity to climate variability vary across agroecological regions. The lowland region will be relatively highly sensitive to climate variability due to the absence of irrigational practices, lack of mixed farming, absence of alternative surface and groundwater sources for livestock and domestic purpose, and prevalence of livestock disease compared with the highland region. On the other hand, the highland regions are relatively highly sensitive to climate variability compared with the lowland due to high population density, crop damage by pest and diseases, and high runoff erosion damage. On the whole however, except demographic pressure in the lowland, lowland and highland research sites are sensitive to climate variability owing to poor irrigation, shortage of water, and erosion, problems that would only increase with the future change in rainfall and temperature (see Fig. 7).

Pastoralists and farmers anticipated that crop and livestock production will be challenged by climate variability. Their anticipation about future climate variability adverse effect is confirmed by scholars' predictions, i.e., the increase of temperature and water stress and its negative effect on agriculture in Africa will shorten the grain-filling stage (Hatfield et al. 2011; Gbetibouo and Ringler 2009; Mendelsohn and Tiwari 2000). By posing physiological stress on the vegetative and reproductive stages of all crops, rising temperature causes a drop in yield amounts (Hatfield et al. 2011). According to Kucharik and Serbin (2008) where precipitation is constant,

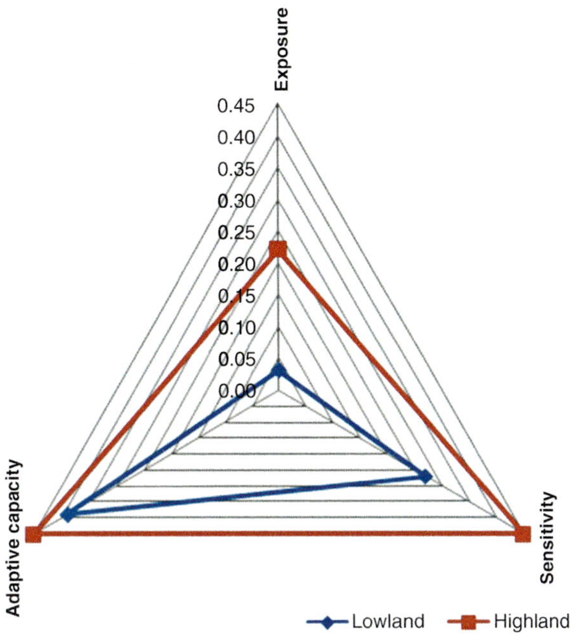

Fig. 8 The magnitude of pastoralists and farmers' exposure, sensitivity, and adaptive capacity to climate change in the lowland and highland. (Source: Authors' constructed from field survey, 2011 and 2012, NMSA, CSA, GCM model)

every increase in the degree of temperature is tantamount to cause yield decline, which varies from crop to crop, 13% for corn and 16% for soybean. Others argue that due to slow temperature increases, yield amount could fall by about 30–46%, while a 63–82% yield decline is expected in case of rapid temperature increment (Schlenker and Roberts 2009). It is noteworthy that maize is one of the most cultivated crops in the study sites whose yield is predicted to fall due to climate variability (Schlenker and Lobell 2010; Lobell and Field 2007; Easterling et al. 2000). Therefore, pastoralists' and farmers' exposure even for a single incident could make worse the sensitivity of agricultural production.

Therefore, pastoralists and farmers have no problem of subjective adaptive capacity (risk perception and motivation); the main adaptation challenge would be enabling environment, i.e., objective adaptive capacity, and physical environment. Where the use of modern agricultural inputs and provision of financial, marketing, health, information services is poor, it is doubtful to expect that people in the study areas would be successful to adapt the future climate variability.

Figure 8 also shows that farmers will have relatively higher adaptive capacity than pastoralists to climate change risks. Explicitly except the advantage in risk-sharing experience in bad times and livestock asset, pastoralists have relatively lower adaptive capacity than farmers in other indicators. Remarkable difference is registered particularly in areas of literacy, crop productivity, improved farm technology, modern early warning information systems, and basic infrastructures (such as access to health, veterinary, and market services). The availability and physical distance of basic infrastructure could determine pastoralists' resilience to climate-related shocks (ILRI 2011). In the highland farmers, risk-sharing experience in the time of climate and non-climate-related shocks is relatively lower contrary to the situation in the lowland. Hence, it seems clear that social bonds through collective action, which could contribute to a quick recovery, are weak among highlanders.

Although the magnitude of lowlanders' exposure and sensitivity to climate change is relatively lower than highlanders, they are more vulnerable to climate variability than the latter. The nature and magnitude of their vulnerability emanate from lower adaptive capacity. Pastoralists' and farmers' vulnerability to climate risk within and across agroecology is determined by the nature and magnitude of exposure, sensitivity, and adaptive capacity indices. Therefore, the function of the three vulnerability dimensions would make pastoralists more vulnerable to climate variability-related risks than their counterpart. As Gbetibouo and Ringler (2009) noted, regions with a relatively high exposure and sensitivity to climate variability could absorb shocks if they have high adaptive capacity. At kebele level, Mahibre Birhan has relatively high level of sensitivity and exposure for climate vulnerability indicators, but it has high adaptive capacity which makes the kebele less vulnerable for climate-related shocks. Therefore, what matters to the magnitude of vulnerability is the cumulative effect of potential impacts and adaptive capacity. The current research finding is not in agreement to Tesso et al. (2012) who argued that pastoralists in the lowland are relatively less vulnerable than highland farmers to climate-related shocks claiming that they are familiar with the problems of drought, disease outbreaks, and alien weeds which they can manage better than highlanders who have little experience with similar problems. But this is not the case based on the findings of this chapter.

FGD and key informants underlined that their livestock asset base is decreasing and will continue so in the future due to rising fodder prices and declining productivity of grazing land. The veterinary services and facilities are poorly equipped with medical supplies. The extension service is biased toward the crop sector. Key informants anticipated that with the existing livestock management system, the potential impact of climate variability will increase in the future. The crop sector also suffers from rising cost of agricultural input (e.g., fertilizer, selected seeds, pesticides, and herbicides), decreasing soil fertility, and increased damage by climate extremes. Hence, the net farm income from the crop and livestock sector will decrease because of climate stressors. Participants in the lowland and highland research sites reported that improved farm technologies such as water harvesting technology are unaffordable. Consistent with our finding, Fellmann et al. (2018) also indicated that climate change affects agricultural input prices. Therefore, the poor pastoralists and farmers who have no access to extension, credit, and savings services will be more vulnerable.

Overall Vulnerability Index

The overall vulnerability index map which is a composite of the three sub-indices maps (exposure, sensitivity, and adaptive capacity) revealed that Denba and Dema

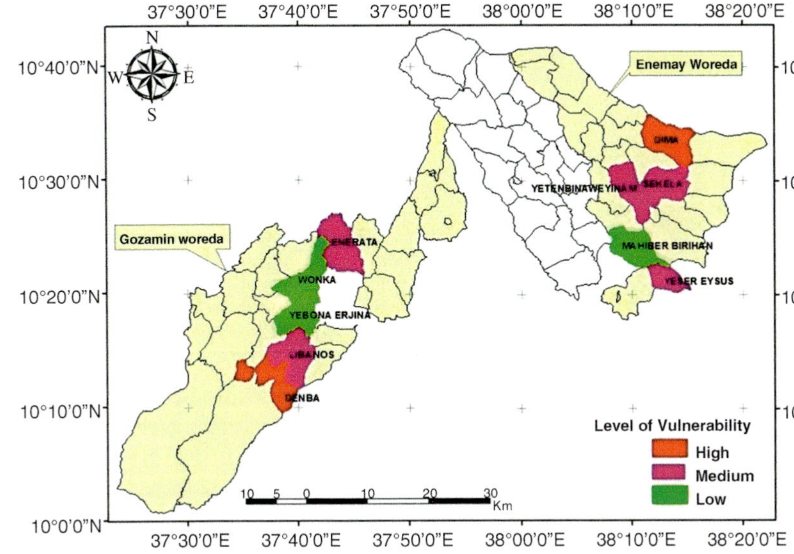

Fig. 9 Aggregate vulnerability index map of selected kebeles in highland

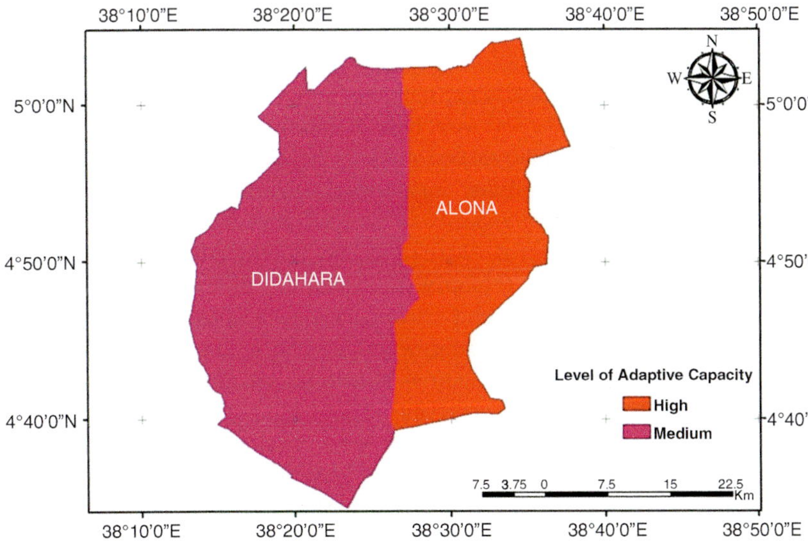

Fig. 10 Aggregate vulnerability index map of selected kebeles in lowland sites

from the highland are relatively highly vulnerable to the impact of climate change risk (Fig. 8), while Alona kebele is similarly highly vulnerable in the lowland sites (Figs. 9 and 10). The rest of the *kebeles* are under medium and low level of vulnerability to the risk of climate variability in the lowland and highland research kebeles. Compared to other *kebeles* in highland and lowland, Denba, Dema, and Alona *kebeles* should be provided with basic infrastructures and modern early warning information to improve their adaptive capacity and adaptation measures toward the adverse impacts of climate change.

Conclusions

The exposure and sensitivity of the chapter areas to climate variability and its related risks revealed that vulnerability is not necessarily determined by agroecology alone. Rather, whereas the agroecology models used are vulnerable to climate variability because of low technology, environmental degradation, and poor infrastructure, the magnitude of vulnerability is also influenced by other factors, for instance, increases in temperature, crop diseases, and population pressures. These appeared to be higher in frequency in the highland than in the lowland pastoral areas. On the other hand, poor irrigation, poor agricultural diversity, and rampant livestock diseases made the lowland more vulnerable to climate variability. However, within the same agroecology, areas could have different vulnerability levels to climatic variability. This has been noticed in the highland and lowland agroecology study site kebeles. By the same token, adaptive capacity varied – and will be likely to vary – within and between agroecologies because of differences in literacy level, technology adoption, and access to infrastructure, credit, market, and information and climate change adaptation governance.

The predominant rainfed subsistence nature of agricultural practice, coupled with high population growth and the low asset base of pastoralists and farmers, will make them more vulnerable to climate-related risks in the near term (2010–2039). This shows the importance of improving institutional, governance, and sociocultural factors to enhance the adaptive capacity of pastoralist and farmers' communities. As indicated by Aleksandrova et al. (2016) and Gbetibouo and Ringler (2009), in the highland and lowland context of Ethiopia, the availability of resources, feasible land and water management policies, and climate change adaptation governance are decisive factors for pastoralists and farmers, so as to reduce their vulnerability. Therefore, extension services and facilities should be equally expanded to remote areas and designed in a way that also serves the poor and female-headed households in the lowland and highland regions. Moreover, environmentally friendly and affordable adaptation technologies which may help people to respond to shocks should be adopted. Finally, awareness creation programs should target the promotion of livelihood diversification, as well as more sustainable land and water management, hence increasing the adaptive capacity of pastoralists and farmers.

Cross-References

▶ Building Community Capacity in Fragile Environments: Case Study of the Mara Serengeti Ecosystem

- Climate Change, Vulnerability, and Adaption Under the Small Farming Households of Konso Community, Southern Ethiopia
- Drivers of Deforestation and Land-Use Change in Southwest Nigeria
- Effect of Climate Variability on Crop Income in Central Highlands of Ethiopia
- Sustainability of Climate Change Adaptation Measures in Rivers State, South-South, Nigeria

References

Adem A (2011) Climate change and rural livelihoods in northern Ethiopia. Impacts, local adaptation strategies and implications for institutional interventions. Forum for Social Studies, Addis Ababa

Adger W, Huq S, Brown K, Conway D, Hulme M (2003) Adaptation to climate change in the developing world. Prog Dev Stud 3(3):179–195

Agnew C, Chappel A (1999) Drought in the Sahel. GeoJournal 48(4):299–311

Ahumada-Cervantes R, Velázquez-Angulo G, Rodríguez-Gallegos HB, Flores-Tavizón E, Félix-Gastélum R, Romero-González J, Granados-Olivas A (2017) An indicator tool for assessing local vulnerability to climate change in the Mexican agricultural sector. J Mitig Adapt Strateg Glob Chang 22(1):137–152. https://doi.org/10.1007/s11027-015-9670-z

Aleksandrova M, Gain AK, Giupponi C (2016) Assessing agricultural systems vulnerability to climate change to inform adaptation planning: an application in Khorezm, Uzbekistan. J Mitig Adapt Strateg Glob Chang 21(8):1263–1287. https://doi.org/10.1007/s11027-015-9655-y

Amsalu A, Adem A (2009) Assessment of climate change-induced hazards, impact and responses in the southern lowlands of Ethiopia. FSS research report no 4. Forum for Social Studies, Addis Ababa

Ayal D, Muluneh A (2014) Smallholder farmers' vulnerability to climate variability in the highland and lowland of Ethiopia: implications to adaptation strategies. Doctoral Thesis, University of South Africa, Geography Department

Ayal D, Woldetisadik Abshare M, Desta S, Filho WL (2017a) Pastoralists and farmers coping and adaptation strategies to climate variability and their perceived success in Ethiopia. In: Leal Filho W, Belay S, Kalangu J, Menas W, Munishi P, Musiyiwa K (eds) Climate change adaptation in Africa: fostering African resilience and capacity to adapt, vol 1. Springer, Cham, pp 457–476. https://doi.org/10.1007/978-3-319-49520-0

Ayal D, Woldetisadik M, Kassa T, Tilahun G, Filho WL (2017b) Climate variability, the proliferation and expansion of major livestock diseases in East Gojjam, Northwestern Ethiopia. Int J Glob Warming 12(3–4):513–531. https://doi.org/10.1504/IJGW.2017.084787

Bewket W (2007) Rainfall variability and agricultural vulnerability in Amhara region, Ethiopia. Ethiop J Dev Res 29:1–16

Bewket W, Radeny M, Mungai C (2015) Agricultural adaptation and institutional responses to climate change vulnerability in Ethiopia. CCAFS working paper no 106. CGIAR Research Program on Climate Change, Agriculture and Food Security (CCAFS), Copenhagen. Accessed on 02 Oct 2017 from www.ccafs.cgiar.org

Brauch HG, Spring UO, Mesjasz C, Grin J, Kameri-Mbote P, Chourou B, Dunay P, Birkmann J (eds) (2011) Coping with global environmental change, disasters and security. Hexagon series on human and environmental security and peace, vol 5. Springer, Cham, https://doi.org/10.1007/978-3-642-17776-7_2

Chambers R (1994) The origins and practice of participatory rural appraisal. World Dev 22(7):953–969

Creswell JW (2008) Research design: qualitative, quantitative, and mixed methods approaches, 3rd edn. Sage, Thousand Oaks. ISBN-10: 1452226105

CSA (Central Statistical Agency) (2008) Summary and statistical report of the 2007 population and housing census. The Federal Republic of Ethiopia, CSA, Addis Ababa

Deressa T, Hassan R, Ringler C (2008) Measuring Ethiopian farmers' vulnerability to climate change across regional states. IFPRI Publications, Washington DC, discussion paper no 806

Devereux S (2000) Food insecurity in Ethiopia. A discussion paper for DFED. IDS, Brighton

Downing T, Ringius L, Hulme M, Waughary D (1997) Adaptation to climate change in Africa. J Mitig Adapt Strateg Glob Chang 2(1):19–44

Easterling D, Evans J, Groisman P, Karl T, Kunkel K, Ambenje P (2000) Observed variability and trends in extremes climate events: a brief review. Bull Am Meteorol Soc 81(3):417–425. Accessed on 02 Oct 2017 from https://doi.org/10.1175/1520-0477(2000)081<0417:OVATIE>2.3.CO;2

Fellmann T, Witzke P, Weiss F, Van Doorslaer B, Drabik D, Huck I, Salputra G, Jansson T, Leip A (2018) Major challenges of integrating agriculture into climate change mitigation policy frameworks. J Mitig Adapt Strateg Glob Chang 23:451–468. https://doi.org/10.1007/s11027-017-9743-2

Füssel H-M (2009) Review and quantitative analysis of indices of climate change exposure, adaptive capacity, sensitivity, and impacts. Background note to the World Development Report 2010. World Bank, Washington, DC. Accessed on 11 March 2018 from http://siteresources.worldbank.org/INTWDR2010/Resources/5287678-1255547194560/WDR2010_BG_Note_Fussel.pdf

Gbetibouo G, Ringler C (2009) Mapping South African farming sector vulnerability to climate change and variability: a sub-national assessment. IFPRI discussion paper 00885. Washington, DC

Gebrehiwot T, van der Veen A (2013) Climate change vulnerability in Ethiopia: disaggregation of Tigray Region. J East Afr Stud 7(4):607–629. https://doi.org/10.1080/17531055.2013.817162

Grothmanna T, Patt A (2005) Adaptive capacity and human cognition: the process of individual adaptation to climate change. Glob Environ Chang A 15(3):199–213. Accessed on 05 Jan 2018 from http://www.elsevier.com/locate/gloenvcha

Hatfield J, Boote K, Kimball B, Ziska L, Izaurralde R, Ort D, Thomson MA, Wolfe D (2011) Climate impacts on agriculture: implications for crop production. J Agron 103(2):351–370. https://doi.org/10.2134/agronj2010.0303

Heltberg R, Siegel B, Jorgensen L (2009) Addressing human vulnerability to climate change: toward a 'no-regrets' approach. Glob Environ Chang 19(1):89–99. https://doi.org/10.1016/j.gloenvcha.2008.11.003

ILRI (International Livestock Research Institute) (2011) Dialogue on Ethiopian agricultural development: report of a conference, Addis Ababa, Ethiopia, 12 November 2009. ILRI, Nairobi

IPCC (2001) Climate change 2001: impacts, adaptation, and vulnerability. In: McCarthy J, Canziani O, Leary N, Dokken D, White K (eds) Contribution of Working Group II to the Third Assessment Report of the Intergovernmental Panel on Climate Change. Cambridge University Press, Cambridge, UK

IPCC (2007) Climate change 2007: impacts, adaptation and vulnerability. Contribution of Working Group II to the 4th Assessment Report of the IPCC. Cambridge University Press, Cambridge

IPCC (2014) Climate change (2014) synthesis report. Contribution of Working Groups I, II and III to the Fifth Assessment Report of the Intergovernmental Panel on Climate Change. IPCC, Geneva

Iyengar N, Sudarshan P (1982) A method of classifying regions from multivariate data. Econ Polit Wkly 17:2047. Special Article, 2048-52

Johnson R, Onwuegbuzie A (2004) Mixed methods research: a research paradigm whose time has come. Educ Res 33(7):14–26

Kathleen AG, Philip KT, Randall BB, Jennifer S (2001) Impacts of climate variability on East African pastoralists: linking social science and remote sensing. Clim Res 19:161–172. https://doi.org/10.3354/cr019161

Kucharik CJ, Serbin SP (2008) Impacts of recent climate change on Wisconsin corn and soybean yield trends. Environ Res Lett 3:034003

Leary N, Conde C, Kulkarni J, Nyong A, Pulhin J (eds) (2008) Climate change and vulnerability. Earthscan, London/Sterling

Lobell D, Field C (2007) Global scale climate–crop yield relationships and the impacts of recent warming. Environ Res Lett 2:014002

Masozera M, Bailey M, Kerchner C (2007) Distribution of impacts of natural disasters across income groups: a case study of New Orleans. Ecol Econ 63(2–3):299–306. https://doi.org/10.1016/j.ecolecon.2006.06.013

McKee TB, Doesken NJ, Kleist J (1993) The relationship of drought frequency and duration to time scale. In: Proceedings of the eighth conference on applied climatology, Anaheim, California, 17–22 January 1993. American Meteorological Society, Boston, pp 179–184

Mendelsohn R, Tiwari D (2000) Two essays on climate change and agriculture: a developing country perspective. FAO economic and social development paper 145. FAO, Rome

Mendelsohn R, Basist A, Kurukulasuriya P, Dinar A (2007) Climate and rural income. Clim Chang 81:101–118

Rosenzweig C, Hillel D (2008) Climate variability and the global harvest: impacts of El Nino and other oscillations on agro ecosystems. Oxford University Press, New York, pp 1–280. ISBN: 9780195137637

Schlenker W, Lobell D (2010) Robust negative impacts of climate change on African agriculture. Environ Res Lett 5:1–8

Schlenker W, Roberts M (2009) Nonlinear temperature effects indicate severe damages to U.S. crop yields under climate change. Proc Natl Acad Sci USA 106(37):15594–15598

Sivakumar M, Das H, Brunini O (2005) Impacts of present and future climate variability and change on agriculture and forestry in the arid and semi-arid tropics. Climate Change 70(1–2):31–72. https://doi.org/10.1007/s10584-005-5937-9

Smit B, Burton I, Klein RJT (1999) The science of adaptation: a framework for assessment. Mitig Adapt Strateg Glob Chang 4:2

Smit B, Burton B, Klein R, Wandel J (2000) An anatomy of adaptation to climate change and variability. Climate Change 45(1):223–251

Tafesse T (2004) The migration, environment, conflict nexus: a case study from Eastern Wollega. Forum for Social Studies, Addis Ababa, studies on poverty no 4

TERI (The Energy Research Institute) (2003) Coping with global change: vulnerability and adaptation in Indian agriculture. The Energy and Resource Institute, New Delhi

Tesso G, Emana B, Ketema M (2012) Analysis of vulnerability and resilience to climate change induced shocks in North Shewa, Ethiopia. J Agric Sci 3(6):871–888. Accessed on 05 Jan 2018 from https://doi.org/10.4236/as.2012.36106

Vincent K (2004) Creating an index of social vulnerability to climate change for Africa. Working paper 56. Tydall Centre for Climate change Research, University of East Anglia, Norwich

Weis SWM, Agostini VN, Roth LM, Gilmer B, Schill SR, Knowles JE, Blyther R (2016) Assessing vulnerability: an integrated approach for mapping adaptive capacity, sensitivity, and exposure. Clim Chang 136:615–629. https://doi.org/10.1007/s10584-016-1642-0

Ziervogel G, Nyong A, Osman B, Conde C, Cortes S, Downing T (2006) Climate variability and change: implications for household food security. AICC Working Papers, AICC, Washington DC, no 20

Printed by Printforce, the Netherlands